T0189875

Lecture Notes in Computer Science   11731

Founding Editors

Gerhard Goos
Karlsruhe Institute of Technology, Karlsruhe, Germany
Juris Hartmanis
Cornell University, Ithaca, NY, USA

Editorial Board Members

Elisa Bertino
Purdue University, West Lafayette, IN, USA
Wen Gao
Peking University, Beijing, China
Bernhard Steffen
TU Dortmund University, Dortmund, Germany
Gerhard Woeginger
RWTH Aachen, Aachen, Germany
Moti Yung
Columbia University, New York, NY, USA

More information about this series at http://www.springer.com/series/7407

Igor V. Tetko · Věra Kůrková ·
Pavel Karpov · Fabian Theis (Eds.)

Artificial Neural Networks and Machine Learning – ICANN 2019

Workshop and Special Sessions

28th International Conference on Artificial Neural Networks
Munich, Germany, September 17–19, 2019
Proceedings

Springer

Editors
Igor V. Tetko (iD)
Helmholtz Zentrum München - Deutsches
Forschungszentrum für Gesundheit
und Umwelt (GmbH)
Neuherberg, Germany

Pavel Karpov (iD)
Helmholtz Zentrum München - Deutsches
Forschungszentrum für Gesundheit
und Umwelt (GmbH)
Neuherberg, Germany

Věra Kůrková (iD)
Institute of Computer Science
Czech Academy of Sciences
Prague 8, Czech Republic

Fabian Theis (iD)
Helmholtz Zentrum München - Deutsches
Forschungszentrum für Gesundheit
und Umwelt (GmbH)
Neuherberg, Germany

ISSN 0302-9743   ISSN 1611-3349   (electronic)
Lecture Notes in Computer Science
ISBN 978-3-030-30492-8   ISBN 978-3-030-30493-5   (eBook)
https://doi.org/10.1007/978-3-030-30493-5

LNCS Sublibrary: SL1 – Theoretical Computer Science and General Issues

© Springer Nature Switzerland AG 2019
16 chapters are licensed under the terms of the Creative Commons Attribution 4.0 International License
(http://creativecommons.org/licenses/by/4.0/). For further details see licence information in the chapters.
This work is subject to copyright. All rights are reserved by the Publisher, whether the whole or part of the
material is concerned, specifically the rights of translation, reprinting, reuse of illustrations, recitation,
broadcasting, reproduction on microfilms or in any other physical way, and transmission or information
storage and retrieval, electronic adaptation, computer software, or by similar or dissimilar methodology now
known or hereafter developed.
The use of general descriptive names, registered names, trademarks, service marks, etc. in this publication
does not imply, even in the absence of a specific statement, that such names are exempt from the relevant
protective laws and regulations and therefore free for general use.
The publisher, the authors and the editors are safe to assume that the advice and information in this book are
believed to be true and accurate at the date of publication. Neither the publisher nor the authors or the editors
give a warranty, expressed or implied, with respect to the material contained herein or for any errors or
omissions that may have been made. The publisher remains neutral with regard to jurisdictional claims in
published maps and institutional affiliations.

This Springer imprint is published by the registered company Springer Nature Switzerland AG
The registered company address is: Gewerbestrasse 11, 6330 Cham, Switzerland

Preface

The fast development of machine learning methods is influencing all aspects of our life and reaching new horizons of what we have previously considered being Artificial Intelligence (AI). Examples include autonomous car driving, virtual assistants, automated customer support, clinical decision support, healthcare data analytics, financial forecast, and smart devices in the home, to name a few, which contribute to the dramatic improvement in the quality of our lives. These developments, however, also bring risks for significant hazards, which were not imaginable previously, e.g., falsification of voice, videos, or even manipulation of people's opinions during elections. Many such developments become possible due to the appearance of large volumes of data ("Big Data"). These proceedings include the theory and applications of algorithms behind these developments, many of which were inspired by the functioning of the brain.

The International Conference on Artificial Neural Networks (ICANN) is the annual flagship conference of the European Neural Network Society (ENNS). The 28th International Conference on Artificial Neural Networks (ICANN 2019) was co-organized with the final conference of the Marie Skłodowska-Curie Innovative Training Network European Industrial Doctorate "Big Data in Chemistry" (http://bigchem.eu) project coordinated by Helmholtz Zentrum München (GmbH) to promote the use of machine learning in Chemistry. The conference featured the main tracks "Brain-Inspired Computing" and "Machine Learning Research." Within the conference the First International Workshop on Reservoir Computing as well as five special sessions were organized, namely:

Artificial Intelligence in Medicine
Informed and Explainable Methods for Machine Learning
Deep Learning in Image Reconstruction
Machine Learning with Graphs: Algorithms and Applications
BIGCHEM: Big Data and AI in chemistry

A Challenge for Automatic Dog Age Estimation (DogAge) also took place as part of the conference. The conference covered all main research fields dealing with neural networks. ICANN 2019 was held during September 17–19, 2019, at Klinikum rechts der Isar der Technische Universität München, Munich, Germany.

Following a long-standing tradition, the proceedings of the conference were published as Springer volumes belonging to the *Lecture Notes in Computer Science* series. The conference had a historical record of 494 article submissions. The papers went through a two-step peer-review process by at least two and in majority of cases by three or four independent referees. In total, 503 Program Committee (PC) members and reviewers participated in this process. The majority of PC members had Doctoral degrees (88%) and 52% of them were also Professors. These reviewers were assigned 46 articles. The others were PhD students in the last years of their studies, who reviewed one to two articles each. In total, for the 323 accepted articles, 975 and 985 reports were submitted for the first and the second revision sessions. Thus, on average, each accepted article received 6.1 reports. A list of reviewers/PC Members, who agreed to publish their names, are included in these proceedings.

Based on the reviewers' comments, 202 articles were accepted and more than 100 articles were rejected after the first review. The remaining articles received an undecided status. The authors of the accepted articles as well as of those with undecided status were requested to address the reviewers' comments within two weeks. On the basis of second reviewers' feedback, another 121 articles were accepted and the authors were requested to include reviewers' remarks into the final upload. Based on these evaluations, diversity of topics, as well as recommendations of reviewers, special session organizers, and PC Chairs, 120 articles were selected for oral presentations. Out of the total number of 323 accepted articles (65% of initially submitted), 46 manuscripts were short articles with a length of five pages each, while the others were full articles with an average length of 13 pages.

The accepted papers of the 28th ICANN conference were published as five volumes:

Volume I Theoretical Neural Computation
Volume II Deep Learning
Volume III Image Processing
Volume IV Text and Time series analysis
Volume V Workshop and Special Sessions

The authors of accepted articles came from 50 different countries. While the majority of the articles were from academic researchers, the conference also attracted contributions from manifold industries including automobile (Volkswagen, BMW, Honda, Toyota), multinational conglomerates (Hitachi, Mitsubishi), electronics (Philips), electrical systems (Thales), mobile (Samsung, Huawei, Nokia, Orange), software (Microsoft), multinational (Amazon) and global travel technology (Expedia), information (IBM), large (AstraZeneca, Boehringer Ingelheim) and medium (Idorsia Pharmaceuticals Ltd.) pharma companies, fragrance and flavor (Firmenich), architectural (Shimizu), weather forecast (Beijing Giant Weather Co.), robotics (UBTECH Robotics Corp., SoftBank Robotics Group Corp.), contract research organization (Lead Discovery Center GmbH), private credit bureau (Schufa), as well as multiple startups. This wide involvement of companies reflects the increasing use of artificial neural networks by the industry. Five keynote speakers were invited to give lectures on the timely aspects of intelligent robot design (gentle robots), nonlinear dynamical analysis of brain activity, deep learning in biology and biomedicine, explainable AI, artificial curiosity, and meta-learning machines.

These proceedings provide a comprehensive and up-to-date coverage of the dynamically developing field of Artificial Neural Networks. They are of major interest both for theoreticians as well as for applied scientists who are looking for new innovative approaches to solve their practical problems. We sincerely thank the Program and Steering Committee and the reviewers for their invaluable work.

September 2019

Igor V. Tetko
Fabian Theis
Pavel Karpov
Věra Kůrková

Organization

General Chairs

Igor V. Tetko	Helmholtz Zentrum München (GmbH), Germany
Fabian Theis	Helmholtz Zentrum München (GmbH), Germany

Honorary Chair

Věra Kůrková (ENNS President)	Czech Academy of Sciences, Czech Republic

Publication Chair

Pavel Karpov	Helmholtz Zentrum München (GmbH), Germany

Local Organizing Committee Chairs

Monica Campillos	Helmholtz Zentrum München (GmbH), Germany
Alessandra Lintas	University of Lausanne, Switzerland

Communication Chair

Paolo Masulli	Technical University of Denmark, Denmark

Steering Committee

Erkki Oja	Aalto University, Finland
Wlodzislaw Duch	Nicolaus Copernicus University, Poland
Alessandro Villa	University of Lausanne, Switzerland
Cesare Alippi	Politecnico di Milano, Italy, and Università della Svizzera italiana, Switzerland
Jérémie Cabessa	Université Paris 2 Panthéon-Assas, France
Maxim Fedorov	Skoltech, Russia
Barbara Hammer	Bielefeld University, Germany
Lazaros Iliadis	Democritus University of Thrace, Greece
Petia Koprinkova-Hristova	Bulgarian Academy of Sciences, Bulgaria
Antonis Papaleonidas	Democritus University of Thrace, Greece
Jaakko Peltonen	University of Tampere, Finland
Antonio Javier Pons Rivero	Universitat Politècnica de Catalunya, Spain
Yifat Prut	The Hebrew University Jerusalem, Israel
Paul F. M. J. Verschure	Catalan Institute of Advanced Studies, Spain
Francisco Zamora-Martínez	Veridas Digital Authentication Solutions SL, Spain

Program Committee

Nesreen Ahmed	Intel Labs, USA
Narges Ahmidi	Helmholtz Zentrum München (GmbH), Germany
Tetiana Aksenova	Commissariat à l'énergie atomique et aux énergies alternatives, France
Elie Aljalbout	Technical University Munich, Germany
Piotr Antonik	CentraleSupélec, France
Juan Manuel Moreno-Arostegui	Universitat Politècnica de Catalunya, Spain
Michael Aupetit	Qatar Computing Research Institute, Qatar
Cristian Axenie	Huawei German Research Center Munich, Germany
Davide Bacciu	University of Pisa, Italy
Noa Barbiro	Booking.com, Israel
Igor Baskin	Moscow State University, Russia
Christian Bauckhage	Fraunhofer IAIS, Germany
Costas Bekas	IBM Research, Switzerland
Barry Bentley	The Open University, UK
Daniel Berrar	Tokyo Institute of Technology, Japan
Soma Bhattacharya	Expedia, USA
Monica Bianchini	Università degli Studi di Siena, Italy
François Blayo	NeoInstinct, Switzerland
Sander Bohte	Centrum Wiskunde & Informatica, The Netherlands
András P. Borosy	QualySense AG, Switzerland
Giosuè Lo Bosco	Università di Palermo, Italy
Farah Bouakrif	University of Jijel, Algeria
Larbi Boubchir	University Paris 8, France
Maria Paula Brito	University of Porto, Portugal
Evgeny Burnaev	Skoltech, Russia
Mikhail Burtsev	Moscow Institute of Physics and Technology, Russia
Jérémie Cabessa	Université Panthéon Assas (Paris II), France
Francisco de Assis Tenório de Carvalho	Universidade Federal de Pernambuco, Brazil
Wolfgang Graf zu Castell-Ruedenhausen	Helmholtz Zentrum München (GmbH), Germany
Stephan Chalup	University of Newcastle, Australia
Hongming Chen	AstraZeneca, Sweden
Artem Cherkasov	University of British Columbia, Canada
Sylvain Chevallier	Université de Versailles, France
Vladimir Chupakhin	Janssen Pharmaceutical Companies, USA
Djork-Arné Clevert	Bayer, Germany
Paulo Cortez	University of Minho, Portugal
Gennady Cymbalyuk	Georgia State University, USA
Maximilien Danisch	Pierre and Marie Curie University, France
Tirtharaj Dash	Birla Institute of Technology and Science Pilani, India
Tyler Derr	Michigan State University, USA

Sergey Dolenko	Moscow State University, Russia
Shirin Dora	University of Amsterdam, The Netherlands
Werner Dubitzky	Helmholtz Zentrum München (GmbH), Germany
Wlodzislaw Duch	Nicolaus Copernicus University, Poland
Ujjal Kr Dutta	Indian Institute of Technology Madras, India
Mohamed El-Sharkawy	Purdue School of Engineering and Technology, USA
Mohamed Elati	Université de Lille, France
Reda Elbasiony	Tanta University, Egypt
Mark Embrechts	Rensselaer Polytechnic Institute, USA
Sebastian Engelke	University of Geneva, Switzerland
Ola Engkvist	AstraZeneca, Sweden
Manfred Eppe	University of Hamburg, Germany
Peter Erdi	Kalamazoo College, USA
Peter Ertl	Novartis Institutes for BioMedical Research, Switzerland
Igor Farkaš	Comenius University in Bratislava, Slovakia
Maxim Fedorov	Skoltech, Russia
Maurizio Fiasché	F-engineering Consulting, Italy
Marco Frasca	University of Milan, Italy
Benoît Frénay	Université de Namur, Belgium
Claudio Gallicchio	Università di Pisa, Italy
Udayan Ganguly	Indian Institute of Technology at Bombay, India
Tiantian Gao	Stony Brook University, USA
Juantomás García	Sngular, Spain
José García-Rodríguez	University of Alicante, Spain
Erol Gelenbe	Institute of Theoretical and Applied Informatics, Poland
Petia Georgieva	University of Aveiro, Portugal
Sajjad Gharaghani	University of Tehran, Iran
Evgin Goceri	Akdeniz University, Turkey
Alexander Gorban	University of Leicester, UK
Marco Gori	Università degli Studi di Siena, Italy
Denise Gorse	University College London, UK
Lyudmila Grigoryeva	University of Konstanz, Germany
Xiaodong Gu	Fudan University, China
Michael Guckert	Technische Hochschule Mittelhessen, Germany
Benjamin Guedj	Inria, France, and UCL, UK
Tatiana Valentine Guy	Institute of Information Theory and Automation, Czech Republic
Fabian Hadiji	Goedle.io, Germany
Abir Hadriche	University of Sfax, Tunisia
Barbara Hammer	Bielefeld University, Germany
Stefan Haufe	ERC Research Group Leader at Charité, Germany
Dominik Heider	Philipps-University of Marburg, Germany
Matthias Heinig	Helmholtz Zentrum München (GmbH), Germany
Christoph Henkelmann	DIVISIO GmbH, Germany

Jean Benoit Héroux	IBM Research, Japan
Christian Hidber	bSquare AG, Switzerland
Martin Holeňa	Institute of Computer Science, Czech Republic
Adrian Horzyk	AGH University of Science and Technology, Poland
Jian Hou	Bohai University, China
Lynn Houthuys	Thomas More, Belgium
Brian Hyland	University of Otago, New Zealand
Nicolangelo Iannella	University of Oslo, Norway
Lazaros Iliadis	Democritus University of Thrace, Greece
Francesco Iorio	Wellcome Trust Sanger Institute, UK
Olexandr Isayev	University of North Carolina at Chapel Hill, USA
Keiichi Ito	Helmholtz Zentrum München (GmbH), Germany
Nils Jansen	Radboud University Nijmegen, The Netherlands
Noman Javed	Université d'Orléans, France
Wenbin Jiang	Huazhong University of Science and Technology, China
Jan Kalina	Institute of Computer Science, Czech Republic
Argyris Kalogeratos	Université Paris-Saclay, France
Michael Kamp	Fraunhofer IAIS, Germany
Dmitry Karlov	Skoltech, Russia
Pavel Karpov	Helmholtz Zentrum München (GmbH), Germany
John Kelleher	Technological University Dublin, Ireland
Adil Mehmood Khan	Innopolis, Russia
Rainer Kiko	GEOMAR Helmholtz-Zentrum für Ozeanforschung, Germany
Christina Klüver	Universität Duisburg-Essen, Germany
Taisuke Kobayashi	Nara Institute of Science and Technology, Japan
Ekaterina Komendantskaya	University of Dundee, UK
Petia Koprinkova-Hristova	Bulgarian Academy of Sciences, Bulgaria
Irena Koprinska	University of Sydney, Australia
Constantine Kotropoulos	Aristotle University of Thessaloniki, Greece
Ilias Kotsireas	Wilfrid Laurier University, Canada
Athanasios Koutras	University of Peloponnese, Greece
Piotr Kowalski	AGH University of Science and Technology, Poland
Valentin Kozlov	Karlsruher Institut für Technologie, Germany
Dean J. Krusienski	Virginia Commonwealth University, USA
Adam Krzyzak	Concordia University, Canada
Hanna Kujawska	University of Bergen, Norway
Věra Kůrková	Institute of Computer Science, Czech Republic
Sumit Kushwaha	Kamla Nehru Institute of Technology, India
Anna Ladi	Fraunhofer IAIS, Germany
Ward Van Laer	Ixor, Belgium
Oliver Lange	Google Inc., USA
Jiyi Li	University of Yamanashi, Japan
Lei Li	Beijing University of Posts and Telecommunications, China

Spiros Likothanassis	University of Patras, Greece
Christian Limberg	Universität Bielefeld, Germany
Alessandra Lintas	University of Lausanne, Switzerland
Viktor Liviniuk	MIT, USA, and Skoltech, Russia
Doina Logofatu	Frankfurt University of Applied Sciences, Germany
Vincenzo Lomonaco	Università di Bologna, Italy
Sock Ching Low	Institute for Bioengineering of Catalonia, Spain
Abhijit Mahalunkar	Technological University Dublin, Ireland
Mufti Mahmud	Nottingham Trent University, UK
Alexander Makarenko	National Technical University of Ukraine - Kiev Polytechnic Institute, Ukraine
Kleanthis Malialis	University of Cyprus, Cyprus
Fragkiskos Malliaros	University of Paris-Saclay, France
Gilles Marcou	University of Strasbourg, France
Urszula Markowska-Kaczmar	Wroclaw University of Technology, Poland
Carsten Marr	Helmholtz Zentrum München (GmbH), Germany
Giuseppe Marra	University of Firenze, Italy
Paolo Masulli	Technical University of Denmark, Denmark
Siamak Mehrkanoon	Maastricht University, The Netherlands
Stefano Mclacci	Università degli Studi di Siena, Italy
Michael Menden	Helmholtz Zentrum München (GmbH), Germany
Sebastian Mika	Comtravo, Germany
Nikolaos Mitianoudis	Democritus University of Thrace, Greece
Valeri Mladenov	Technical University of Sofia, Bulgaria
Hebatallah Mohamed	Università degli Studi Roma, Italy
Figlu Mohanty	International Institute of Information Technology at Bhubaneswar, India
Francesco Carlo Morabito	University of Reggio Calabria, Italy
Jerzy Mościński	Silesian University of Technology, Poland
Henning Müller	University of Applied Sciences Western Switzerland, Switzerland
Maria-Viorela Muntean	University of Alba-Iulia, Romania
Phivos Mylonas	Ionian University, Greece
Shinichi Nakajima	Technische Universität Berlin, Germany
Kohei Nakajima	University of Tokyo, Japan
Chi Nhan Nguyen	Itemis, Germany
Florian Nigsch	Novartis Institutes for BioMedical Research, Switzerland
Giannis Nikolentzos	École Polytechnique, France
Ikuko Nishikawa	Ritsumeikan University, Japan
Harri Niska	University of Eastern Finland
Hasna Njah	ISIM-Sfax, Tunisia
Dimitri Nowicki	Institute of Cybernetics of NASU, Ukraine
Alessandro Di Nuovo	Sheffield Hallam University, UK
Stefan Oehmcke	University of Copenhagen, Denmark

Erkki Oja	Aalto University, Finland
Luca Oneto	Università di Pisa, Italy
Silvia Ortin	Institute of Neurosciences (IN) Alicante, Spain
Ivan Oseledets	Skoltech, Russia
Dmitry Osolodkin	Chumakov FSC R&D IBP RAS, Russia
Sebastian Otte	University of Tübingen, Germany
Latifa Oukhellou	The French Institute of Science and Technology for Transport, France
Vladimir Palyulin	Moscow State University, Russia
George Panagopoulos	École Polytechnique, France
Massimo Panella	Università degli Studi di Roma La Sapienza, Italy
Antonis Papaleonidas	Democritus University of Thrace, Greece
Evangelos Papalexakis	University of California Riverside, USA
Daniel Paurat	Fraunhofer IAIS, Germany
Jaakko Peltonen	Tampere University, Finland
Tingying Peng	Technische Universität München, Germany
Alberto Guillén Perales	Universidad de Granada, Spain
Carlos Garcia Perez	Helmholtz Zentrum München (GmbH), Germany
Isabelle Perseil	INSERM, France
Vincenzo Piuri	University of Milan, Italy
Kathrin Plankensteiner	Fachhochschule Vorarlberg, Austria
Isabella Pozzi	Centrum Wiskunde & Informatica, The Netherlands
Mike Preuss	Leiden University, The Netherlands
Yifat Prut	The Hebrew University of Jerusalem, Israel
Eugene Radchenko	Moscow State University, Russia
Rajkumar Ramamurthy	Fraunhofer IAIS, Germany
Srikanth Ramaswamy	Swiss Federal Institute of Technology (EPFL), Switzerland
Beatriz Remeseiro	Universidad de Oviedo, Spain
Xingzhang Ren	Alibaba Group, China
Jean-Louis Reymond	University of Bern, Switzerland
Cristian Rodriguez Rivero	University of California, USA
Antonio Javier Pons Rivero	Universitat Politècnica de Catalunya, Spain
Andrea Emilio Rizzoli	IDSIA, SUPSI, Switzerland
Florian Röhrbein	Technical University Munich, Germany
Ryan Rossi	PARC - a Xerox Company, USA
Manuel Roveri	Politecnico di Milano, Italy
Vladimir Rybakov	WaveAccess, Russia
Maryam Sabzevari	Aalto University School of Science and Technology, Finland
Julio Saez-Rodriguez	Medizinische Fakultät Heidelberg, Germany
Yulia Sandamirskaya	NEUROTECH: Neuromorphic Computer Technology, Switzerland
Carlo Sansone	University of Naples Federico II, Italy
Sreela Sasi	Gannon University, USA
Burak Satar	Uludag University, Turkey

Axel Sauer	Munich School of Robotics and Machine Intelligence, Germany
Konstantin Savenkov	Intento, Inc., USA
Hanno Scharr	Forschungszentrum Jülich, Germany
Tjeerd olde Scheper	Oxford Brookes University, UK
Rafal Scherer	Czestochowa University of Technology, Poland
Maria Secrier	University College London, UK
Thomas Seidl	Ludwig-Maximilians-Universität München, Germany
Rafet Sifa	Fraunhofer IAIS, Germany
Pekka Siirtola	University of Oulu, Finland
Prashant Singh	Uppsala University, Sweden
Patrick van der Smagt	Volkswagen AG, Germany
Maximilian Soelch	Volkswagen Machine Learning Research Lab, Germany
Miguel Cornelles Soriano	Campus Universitat de les Illes Balears, Spain
Miguel Angelo Abreu Sousa	Institute of Education Science and Technology, Brazil
Michael Stiber	University of Washington Bothell, USA
Alessandro Sperduti	Università degli Studi di Padova, Italy
Ruxandra Stoean	University of Craiova, Romania
Nicola Strisciuglio	University of Groningen, The Netherlands
Irene Sturm	Deutsche Bahn AG, Germany
Jérémie Sublime	ISEP, France
Martin Swain	Aberystwyth University, UK
Zoltan Szabo	Ecole Polytechnique, France
Kazuhiko Takahashi	Doshisha University, Japan
Fabian Theis	Helmholtz Zentrum München (GmbH), Germany
Philippe Thomas	Université de Lorraine, France
Matteo Tiezzi	University of Siena, Italy
Ruben Tikidji-Hamburyan	Louisiana State University, USA
Yancho Todorov	VTT, Finland
Andrei Tolstikov	Merck Group, Germany
Matthias Treder	Cardiff University, UK
Anton Tsitsulin	Rheinische Friedrich-Wilhelms-Universität Bonn, Germany
Yury Tsoy	Solidware Co. Ltd., South Korea
Antoni Valencia	Independent Consultant, Spain
Carlos Magno Valle	Technical University Munich, Germany
Marley Vellasco	Pontifícia Universidade Católica do Rio de Janeiro, Brazil
Sagar Verma	Université Paris-Saclay, France
Paul Verschure	Institute for Bioengineering of Catalonia, Spain
Varvara Vetrova	University of Canterbury, New Zealand
Ricardo Vigário	University Nova's School of Science and Technology, Portugal
Alessandro Villa	University of Lausanne, Switzerland
Bruno Villoutreix	Molecular informatics for Health, France

Paolo Viviani	Università degli Studi di Torino, Italy
George Vouros	University of Piraeus, Greece
Christian Wallraven	Korea University, South Korea
Tinghuai Wang	Nokia, Finland
Yu Wang	Leibniz Supercomputing Centre (LRZ), Germany
Roseli S. Wedemann	Universidade do Estado do Rio de Janeiro, Brazil
Thomas Wennekers	University of Plymouth, UK
Stefan Wermter	University of Hamburg, Germany
Heiko Wersing	Honda Research Institute and Bielefeld University, Germany
Tadeusz Wieczorek	Silesian University of Technology, Poland
Christoph Windheuser	ThoughtWorks Inc., Germany
Borys Wróbel	Adam Mickiewicz University in Poznan, Poland
Jianhong Wu	York University, Canada
Xia Xiao	University of Connecticut, USA
Takaharu Yaguchi	Kobe University, Japan
Seul-Ki Yeom	Technische Universität Berlin, Germany
Hujun Yin	University of Manchester, UK
Junichiro Yoshimoto	Nara Institute of Science and Technology, Japan
Qiang Yu	Tianjin University, China
Shigang Yue	University of Lincoln, UK
Wlodek Zadrozny	University of North Carolina Charlotte, USA
Danuta Zakrzewska	Technical University of Lodz, Poland
Francisco Zamora-Martínez	Veridas Digital Authentication Solutions SL, Spain
Gerson Zaverucha	Federal University of Rio de Janeiro, Brazil
Junge Zhang	Institute of Automation, China
Zhongnan Zhang	Xiamen University, China
Pengsheng Zheng	Daimler AG, Germany
Samson Zhou	Indiana University, USA
Riccardo Zucca	Institute for Bioengineering of Catalonia, Spain
Dietlind Zühlke	Horn & Company Data Analytics GmbH, Germany

Keynote Talks

Recurrent Patterns of Brain Activity Associated with Cognitive Tasks and Attractor Dynamics (John Taylor Memorial Lecture)

Alessandro E. P. Villa

NeuroHeuristic Research Group, University of Lausanne,
Quartier UNIL-Chamberonne, 1015 Lausanne, Switzerland
alessandro.villa@unil.ch
http://www.neuroheuristic.org

The simultaneous recording of the time series formed by the sequences of neuronal discharges reveals important features of the dynamics of information processing in the brain. Experimental evidence of firing sequences with a precision of a few milliseconds have been observed in the brain of behaving animals. We review some critical findings showing that this activity is likely to be associated with higher order neural (mental) processes, such as predictive guesses of a coming stimulus in a complex sensorimotor discrimination task, in primates as well as in rats. We discuss some models of evolvable neural networks and their nonlinear deterministic dynamics and how such complex spatiotemporal patterns of firing may emerge. The attractors of such networks correspond precisely to the cycles in the graphs of their corresponding automata, and can thus be computed explicitly and exhaustively. We investigate further the effects of network topology on the dynamical activity of hierarchically organized networks of simulated spiking neurons. We describe how the activation and the biologically-inspired processes of plasticity on the network shape its topology using invariants based on algebro-topological constructions. General features of a brain theory based on these results is presented for discussion.

Unsupervised Learning: Passive and Active

Jürgen Schmidhuber

Co-founder and Chief Scientist, NNAISENSE, Scientific Director,
Swiss AI Lab IDSIA and Professor of AI, USI & SUPSI, Lugano, Switzerland

I'll start with a concept of 1990 that has become popular: unsupervised learning without a teacher through two adversarial neural networks (NNs) that duel in a mini-max game, where one NN minimizes the objective function maximized by the other. The first NN generates data through its output actions while the second NN predicts the data. The second NN minimizes its error, thus becoming a better predictor. But it is a zero sum game: the first NN tries to find actions that maximize the error of the second NN. The system exhibits what I called "artificial curiosity" because the first NN is motivated to invent actions that yield data that the second NN still finds surprising, until the data becomes familiar and eventually boring. A similar adversarial zero sum game was used for another unsupervised method called "predictability minimization," where two NNs fight each other to discover a disentangled code of the incoming data (since 1991), remarkably similar to codes found in biological brains. I'll also discuss passive unsupervised learning through predictive coding of an agent's observation stream (since 1991) to overcome the fundamental deep learning problem through data compression. I'll offer thoughts as to why most current commercial applications don't use unsupervised learning, and whether that will change in the future.

Machine Learning and AI for the Sciences— Towards Understanding

Klaus-Robert Müller

Machine Learning Group, Technical University of Berlin, Germany

In recent years machine learning (ML) and Artificial Intelligence (AI) methods have begun to play a more and more enabling role in the sciences and in industry. In particular, the advent of large and/or complex data corpora has given rise to new technological challenges and possibilities.

The talk will connect two topics (1) explainable AI (XAI) and (2) ML applications in sciences (e.g. Medicine and Quantum Chemistry) for gaining new insight. Specifically I will first introduce XAI methods (such as LRP) that are now readily available and allow for an understanding of the inner workings of nonlinear ML methods ranging from kernel methods to deep learning methods including LSTMs. In particular XAI allows unmasking clever Hans predictors. Then, ML for Quantum Chemistry is discussed, showing that ML methods can lead to highly useful predictors of quantum mechanical properties of molecules (and materials) reaching quantum chemical accuracies both across chemical compound space and in molecular dynamics simulations. Notably, these ML models do not only speed up computation by several orders of magnitude but can give rise to novel chemical insight. Finally, I will analyze morphological and molecular data for cancer diagnosis, also here highly interesting novel insights can be obtained.

Note that while XAI is used for gaining a better understanding in the sciences, the introduced XAI techniques are readily useful in other application domains and industry as well.

Large-Scale Lineage and Latent-Space Learning in Single-Cell Genomic

Fabian Theis

Institute of Computational Biology, Helmholtz Zentrum München (GmbH),
Germany
http://comp.bio

Accurately modeling single cell state changes e.g. during differentiation or in response to perturbations is a central goal of computational biology. Single-cell technologies now give us easy and large-scale access to state observations on the transcriptomic and more recently also epigenomic level, separately for each single cell. In particular they allow resolving potential heterogeneities due to asynchronicity of differentiating or responding cells, and profiles across multiple conditions such as time points and replicates are being generated.

Typical questions asked to such data are how cells develop over time and after perturbation such as disease. The statistical tools to address these questions are techniques from pseudo-temporal ordering and lineage estimation, or more broadly latent space learning. In this talk I will give a short review of such approaches, in particular focusing on recent extensions towards large-scale data integration using single-cell graph mapping or neural networks, and finish with a perspective towards learning perturbations using variational autoencoders.

The Gentle Robot

Sami Haddadin

Technical University of Munich, Germany

Enabling robots for interaction with humans and unknown environments has been one of the primary goals of robotics research over decades. I will outline how human-centered robot design, nonlinear soft-robotics control inspired by human neuromechanics and physics grounded learning algorithms will let robots become a commodity in our near-future society. In particular, compliant and energy-controlled ultra-lightweight systems capable of complex collision handling enable high-performance human assistance over a wide variety of application domains. Together with novel methods for dynamics and skill learning, flexible and easy-to-use robotic power tools and systems can be designed. Recently, our work has led to the first next generation robot Franka Emika that has recently become commercially available. The system is able to safely interact with humans, execute and even learn sensitive manipulation skills, is affordable and designed as a distributed interconnected system.

Contents

**Special Session: Informed and Explainable Methods
for Machine Learning**

Special Session: Deep Learning in Image Reconstruction

Special Session: Machine Learning with Graphs: Algorithms and Applications

Special Session: BIGCHEM: Big Data and AI in Chemistry

Abstracts of the BIGCHEM Session

Workshop: Reservoir
Computing - Methodology

The First International Workshop on Reservoir Computing (RC 2019)

Reservoir Computing (RC) is a leading-edge paradigm for the design of efficiently trainable recurrent neural network models. The approach has rapidly become popular among practitioners due to its simplicity of software implementation, great effectiveness in applications, and striking efficiency. From a more theoretic-oriented perspective, RC is of increasing interest in the neural networks research community, allowing a greater depth of study into intrinsic characterizations and smart initialization of dynamical neural models, at the same time improving our understanding of the processes behind fruitful exploitation of dynamical systems in machine learning. In recent years, studies on neuromorphic implementations of RC, based e.g. on photonic hardware, promise to open the way to breakthrough advancements enabling ultra-fast learning in the temporal domain.

The First International Workshop on Reservoir Computing (RC 2019) intended to bring together researchers and practitioners to discuss the state-of-the-art and open challenges for the field of RC, in all its declinations. The workshop provided an open forum for researchers to meet and present recent contributions and ideas in a highly interdisciplinary environment.

This first edition of the workshop was greeted with enthusiasm by the community. After an accurate and thorough peer-review process we selected 22 contributions (10 full papers and 12 short papers) for presentation at the workshop. These provided an overview of the main streams of current research in the RC field, covering, without being limited to, theoretical analysis of reservoirs, neuromorphic computing and non-conventional hardware implementations of RC, optimization of reservoir configurations, emerging developments of RC, including conceptors and deep reservoirs, as well as novel application fields for the RC paradigm. We were also delighted to host within the workshop program two invited keynote presentations by pioneers of RC: Prof. Wolfgang Maass (Technische Universität Graz, Austria) with a talk on "Bringing Reservoirs Up to Speed for Deep Learning," and Prof. Peter Tiňo (The University of Birmingham, UK) with a talk on "Reservoirs as Temporal Filters and Feature Mappings."

We sincerely thank all the members of the Program Committee and the reviewers for their invaluable work, and the organizers of the ICANN conference for hosting and supporting this workshop.

Organization

This workshop was organized as part of the activities of the IEEE Task Force on Reservoir Computing (TFRC).

Organizers and Program Chairs

Claudio Gallicchio	University of Pisa, Italy
Alessio Micheli	University of Pisa, Italy
Simone Scardapane	La Sapienza University of Rome, Italy
Miguel C. Soriano	University of the Balearic Islands, Spain

Program Committee and Reviewers

Piotr Antonik	CentraleSupélec, France
Apostolos Argyris	IFISC (CSIC-UIB), Spain
Julián Bueno	University of Strathclyde, UK
John Butcher	Keele University, UK
Luca Cerina	Politecnico di Milano, Italy
Danilo Comminiello	La Sapienza University of Rome, Italy
Stefano Dettori	Scuola Superiore Sant'Anna Pisa, Italy
Igor Farkaš	Comenius University in Bratislava, Slovakia
Giuseppe Franco	University of Pisa, Italy
Doreen Jirak	University of Hamburg, Germany
Fatemeh Hadaeghi	University Medical Center Hamburg-Eppendorf, Germany
Jean Benoit Héroux	IBM Research Tokyo, Japan
Xavier Hinaut	Inria Bordeaux, France
Masanobu Inubushi	Osaka University, Japan
Andrew Katumba	Gent University, Belgium
Petia Koprinkova-Hristova	Bulgarian Academy of Sciences, Bulgaria
Lorenzo Livi	University of Manitoba, Canada
Silvia Ortín	Instituto de Neurociencias de Alicante, Spain
Massimo Pannella	La Sapienza University of Rome, Italy
Luca Pedrelli	University of Pisa, Italy
Moritz Pflüger	IFISC (CSIC-UIB), Spain
Xavier Porte Parera	FEMTO-ST Institute, UBFC, France
Michele Scarpiniti	Sapienza University of Rome, Italy
Marek Šuppa	Comenius University in Bratislava, Slovakia
Guy Van der Sande	Vrije Universiteit Brussel, Belgium
Gouhei Tanaka	The University of Tokyo, Japan
Steven Van Vaerenbergh	University of Cantabria, Spain
Thomas Van Vaerenbergh	Hewlett Packard Labs, USA

Abstracts of Talks
of Invited Speakers

Bringing Reservoirs up to Speed for Deep Learning

Wolfgang Maass

Technische Universität Graz, Austria

Reservoirs are good at temporal processing and easy to train. They are suitable for temporal processing because they employ a recurrent neural network or some other type of dynamical system. They are easy to train, since they only adapt linear readouts from this dynamical system for a specific task, rather than tackling the more difficult problem of training the recurrent network itself. They also appear to be more plausible as models for brain processing than many models that arise in deep learning.

In deep learning and modern learning-driven AI one rarely uses reservoirs for tasks that have a temporal dimension because their performance is sub-optimal. Instead, one relies on recurrent network of LSTM (Long-Short Term Memory) units that are trained by backpropagation through time (BPTT). In addition, one applies Learning-to-Learn (L2L) methods in order to enable networks to learn from few examples.

I will discuss recent progress in narrowing the performance gap between reservoir computing and deep learning for the case of spike-based reservoirs.

References

1. Subramoney, A., Scherr, F., Maass, W.: Reservoirs learn to learn. To appear in Reservoir Computing: Theory and Physical Implementations, Nakajima, K., Fischer, I., (eds.) Springer, to appear, draft on arxiv (2019)
2. Bellec, G., Salaj, D., Subramoney, A., Legenstein, R., Maass, W.: Long short-term memory and learning-to-learn in networks of spiking neurons. NeurIPS 2018
3. Bellec, G., Scherr, F., Hajek, E., Salaj, D., Legenstein, R., Maass, W.: Biologically inspired alternatives to backpropagation through time for learning in recurrent neural nets (2019). arxiv.org/abs/1901.09049

Reservoirs as Temporal Filters
and Feature Mappings

Peter Tiňo

The University of Birmingham, UK

Parametrized state space models in the form of recurrent networks are often used in machine learning to learn from data streams exhibiting temporal dependencies. To break the black box nature of such models it is important to understand the dynamical features of the input driving time series that are formed in the state space.

I will talk about a framework for rigorous analysis of such state representations in vanishing memory state space models, such as echo state networks (ESN). In particular, we will view the state space as a temporal feature space and the readout mapping from the state space as a kernel machine operating in that feature space. This viewpoint leads several rather surprising results linking the structure of the reservoir coupling matrix with properties of the dynamic feature space.

A Reservoir Computing Framework for Continuous Gesture Recognition

Stephan Tietz[1,2]($\boxtimes$), Doreen Jirak[1]($\boxtimes$), and Stefan Wermter[1]

[1] Department of Computer Science, University of Hamburg,
Vogt-Kölln-Str. 30, 22527 Hamburg, Germany
{jirak,wermter}@informatik.uni-hamburg.de
[2] Department of Computer Science, Technical University Berlin,
Straße des 17. Juni 135, 10623 Berlin, Germany
stietz@win.tu-berlin.de

Abstract. We present a novel gesture recognition system for the application of continuous gestures in mobile devices. We explain how meaningful gesture data can be extracted from the inertial measurement unit of a mobile phone and introduce a segmentation scheme to distinguish between different gesture classes. The continuous sequences are fed into an Echo State Network, which learns sequential data fast and with good performance. We evaluated our system on crucial network parameters and on our established metric to compute the number of successfully recognized gestures and the number of misclassifications. On a total of ten gesture classes, our framework achieved an average accuracy of 85%.

Keywords: Echo State Networks · Continuous gesture recognition

1 Introduction

Mobile phones rapidly advanced from a portable telephone to a device with a multitude of technological features. While most smartphones are controlled through a touchscreen or by voice, another feature that can be easily exploited is the inbuilt Inertial Measurement Unit (IMU). It enables using a phone as a direct gesture input device, similar to what has been explored by the gesture recognition community adopting the Wii Remote Controller around a decade ago.

This third, gesture based interaction channel would allow to better operate the phone when only one hand is available to the user or when the environment is too noisy to deliver speech commands. Human-robot interaction (HRI) research shows that using IMU sensor data to control a robot with gestures (e.g. showing directions in a navigation task) can be an interesting approach in contrast to vision-based approaches, which rely more on controlled environmental conditions and are thus more challenging to realize in real world applications.

Our present study introduces a framework for continuous gesture recognition exploiting the IMU sensor availability in standard mobile phones and is further

© Springer Nature Switzerland AG 2019
I. V. Tetko et al. (Eds.): ICANN 2019, LNCS 11731, pp. 7–18, 2019.
https://doi.org/10.1007/978-3-030-30493-5_1

motivated to foster research on our chosen model for classifying gestures: Echo State Networks (ESN) [6]. ESNs implement a specific form of recurrent neural networks (RNN) that deviate from the classic RNN training scheme. The important distinction to the usual gradient-descent methods lies in a functional separation between a neuronal 'reservoir' providing a high-dimensional processing of the input data and a read out component, in its most basic form simply a linear model.

We want to work on the time stream directly which necessitates a model that can deal with temporal data, i.e. inputs of different length. While Hidden Markov Models (HMMs) would be a more classical choice and Long Short-Term Memory Networks (LSTMs) would be more powerful, we chose ESNs as they provide a good trade off between flexibility and ease of training.

ESNs together with other models that follow a similar computational scheme are summarized under the term Reservoir Computing (RC) [12]. They have been primarily studied in the language and robotic domain including navigation [1,2,5,11] while the research on gestures is rather sparse [3,8]. This is surprising as all fields share important properties like long-range dependencies and high performance variances across subjects (e.g. the speakers' speed, gesture movement speed).

In the present paper, we address the task of continuous gesture recognition implemented within an ESN-based framework. We highlight the specific requirements for the segmentation and classification of gestures embedded in a continuous data stream recorded from movements using a smartphone. Our study shows that ESNs are a potentially interesting computational method for this task and give further directions to extend our framework for future applications in the gesture recognition domain.

2   Data Recordings and Methods

We defined a gesture protocol to standardize our data recordings and developed a 'SensorTracker App' to enable experimental reproducibility. The data and the code including the ESN are publicly available[1]. We collected IMU sensor data from a Samsung Galaxy S5 neo smartphone (Android OS). Figure 1 displays the ten gestures we used in our experiments (following [9]) recorded from a total of five subjects. The subjects were seated at a table and instructed to hold

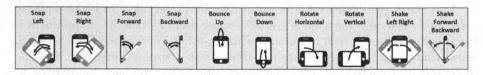

Fig. 1. Gestures performed with a smartphone used in our experiments. Arrows show the direction. Temporary positions are shown in grey.

---

[1] https://github.com/swtietz/UHH-IMU-gestures.

the phone in their dominant hand. Each gesture performance started in the so-called 'portrait' mode (i.e. the user faces the display) with elbows at around 90°. A supervisor responsible to set the time markers (i.e. gesture start and end phase) was seated opposite to the participant but gave no instructions on how to perform the gesture specifically. Each gesture type was performed ten times with set time breaks in between to avoid fatigue and task habituation. The access to the sensor values including preprocessing steps to gain reasonable input values are described in the reference work [9]. Each time step is described by nine values representing the 3D acceleration, 3D rotation velocity (local coordinate system) and 3D orientation (world coordinate) system. Figure 2 shows an exemplary input stream for the *snap right* gesture and the markers obtained from the supervisor (blue) and our average computation (cf. Eq. 1) (green). In total, we obtained 500 gesture samples, from which we created activity-based ground truth labels:

$$target(t) = \begin{cases} 1, & 1/l * \sum_{i=t-l}^{t} \sum_{s}^{signals} |input(i,s)| \geq \theta \\ 0, & else \end{cases} \tag{1}$$

where t is the current time step, l the length of a sliding window and θ denotes a threshold. We used an input-driven ESN, so we analyzed our data before feeding it into the network regarding temporal variations and input changes, as this can have an influence on the parametrization. Figure 3 demonstrates the experimental variances across gestures and subjects. Generally, almost all gestures show only small deviations between subjects (small boxes) except outlier for *snap right* and *snap forward* and performances range in the interval of 10 to 40 time steps (≈ 0.3 to 1.3 s). In contrast, the *shaking updown* and *shaking left-right* gestures took around 50–70 time steps (≈ 1.6 to 2.3 s) and showed highest variability regarding the subjects performances and between the two gestures.

Also, we consider gestures that differ in the required rotation (rot) or acceleration (acc). We call the performance strength the 'energy':

$$Energy(t) = E_{acc}(t) + E_{rot}(t) \tag{2}$$

$$E_{acc}(t) = \sqrt{acc_x(t)^2 + acc_y(t)^2 + acc_z(t)^2} \tag{3}$$

$$E_{rot}(t) = \sqrt{rot_x(t)^2 + rot_y(t)^2 + rot_z(t)^2} \tag{4}$$

The gesture energy is summed over time t and the value is normalized by dividing its sample length to avoid a bias towards longer gesture samples. Notably, the absolute rotation value is constantly smaller than the acceleration due to the different scales (rad/s vs. m/s^2). This fact has to be taken into account for the input scaling before feeding the data into the reservoir.

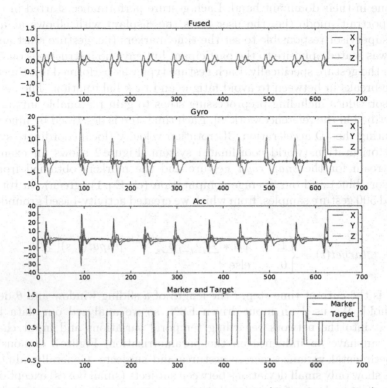

Fig. 2. Sensor data for the *snap right* gesture. (Color figure online)

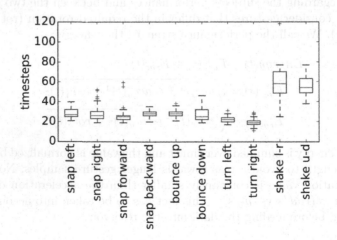

Fig. 3. Gesture variances within gestures and across the different gesture types. Black crosses denote outliers. (Color figure online)

2.1 Echo State Networks

Echo State Networks (ESN) [6] have proven successful in prediction benchmarks including chaotic time series. Here, we follow the standard notation for leaky-integrator ESN (LI-ESN) [7] and describe our scheme to search for optimized network parameters.

An ESN is functionally separated into a reservoir of recurrently connected neurons, whose activation, or simply states, x are updated as:

$$\tilde{x}(t+1) = f(u(t+1)W^{in} + x(t)W^{res} + y(t)W^{fb} + \nu(t)) \tag{5}$$

$$x(t+1) = (1-\alpha)x(t) + \alpha\tilde{x}(t+1) \tag{6}$$

where f is the activation function (here $tanh$) and u is the input. The layer-wise connectivity matrices W^* for the input, reservoir, and feedback remain fixed while training the network. We initialize W^{in} sparsely with only 10% of all weights set. W^{res} is initialized fully connected with gaussian weights and then rescaled to the desired spectral radius. As we are using the ESN for supervised learning, we set the feedback matrix W^{fb} and the noise term ν to 0.

The neural activation are collected into a state matrix X and then fitted using a linear model. Given the output Y (supervised learning):

$$Y = g(W^{out}X) \tag{7}$$

with linear output function g, standard ESN training contracts to a regression on the output weights W^{out}. Using common matrix notation for regression it can be computed:

$$W^{out} = YX^T(XX^T + \lambda\mathbf{I})^{-1} \tag{8}$$

where λ is a penalty term applied in ridge regression and $\mathbf{I}$ is the identity matrix.

Our input vectors contain nine sensor values and the output layer has ten neurons representing the gesture classes. The fact that temporal patterns excite the reservoir differently can be represented in the readout [10]. Figure 4 shows three possible shapes for the target signal. First, a rectangular shape whenever a signal is detected, which we obtain when applying our threshold scheme explained below. Second, a Gaussian shape providing a model for the low activity in the beginning of every gesture (start and end) and greatest activity during the gesture performance (stroke). The last option creates a time window that focuses only on the last time steps assuming that most information is gathered after sufficient reservoir excitement. In our experiments, we obtained the best results when using the first option (Fig. 4a).

2.2 Training and Parameter Validation

The data is split into a $n-1$ persons training set, where it is further split into segments containing only one target sample, shuffled randomly and then concatenated to four different training sets. The remaining person is used as the test set. We performed a grid search to obtain a reasonable network parametrization.

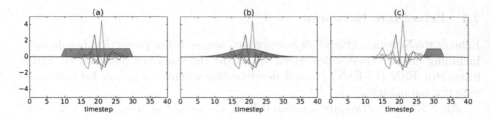

Fig. 4. Sketch of three different shapes for the target signal whenever activity occurs. (a) A rectangular shape (b) A Gaussian shape (c) A time window considering only the final activity.

For each combination of model parameters, an ESN is evaluated via leave-one-out cross validation. As this process is an exhaustive search over the parameter space, Table 1 shows the final collection of values, where we show in bold the chosen values giving the lowest validation error and which are used for the gesture experiments.

Table 1. Reservoir Parameters: Here ρ stands for spectral radius, α for the leak rate, κ for input connectivity and λ for the regularization coefficient from ridge regression.

Parameter	Domain
Reservoir size	400
Input scaling	1, 5, 9, **13**
ρ	0.1, 0.4, 0.7, **1.0**, 1.3
α	0.1, **0.3**, 0.5, 0.7, 0.9
κ	0.1
λ	**0.01**, 0.1, 1, 10

In summary, we observe from our validations on network parameters, that obeying the reciprocal relationship between connectivity and input scaling as well as keeping the processing power in the reservoir by a low leakage rate α and a (relatively) high spectral radius ρ is beneficial for our task.

3   Classification Issues in a Continuous Data Stream

In many gesture recognition tasks the network is presented with a short segment that only contains one isolated sample of a gesture with predetermined start-stroke-end scheme.

In strong contrast to this approach we train our network on a continuous time stream containing multiple, different gestures. The network does not only need to learn which gesture it sees but also when each gesture starts and ends.

This is a much more realistic, but also more challenging, setup than the classical classification task.

We now explain how we detect and distinguish between active segments and nonactive gesture segments and their corresponding labels. This is an important process to ensure proper network evaluation.

3.1 The 'No-Gesture' Class

As we work on a continuous time stream, the network outputs label information at every time step, as opposed to just a single label after the gesture has been presented. Therefore we need a way to deal with segments in between gestures that do not contain any information.

The simplest approach would be to use a constant cutoff and only label a time step if any neurons activity is above the threshold. A more flexible, neural way would be to add another neuron to the output layer that functions as a 'No-Gesture' class. The target for this neuron is to be active whenever the smart phone is resting and to be inactive while a gesture is being performed, i.e. it is one whenever all other targets are zero.

However, our experiments reveal that the activation of this signal can lead to unstable behavior, consequently producing misclassifications, as demonstrated in Fig. 5. While for some active segments the threshold signal decreases, e.g. at time step 4800, and a *turn left* gesture is classified, at around time steps 5050–5150 the threshold oscillates, distracting the classifier.

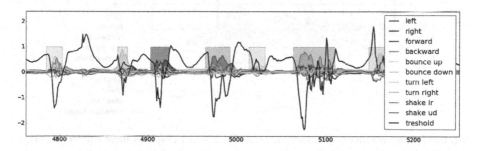

Fig. 5. Example of a threshold (black line) learnt to discriminate between an active and a non-active gesture segment. The colored areas show the actual gesture hypotheses. Its activity increases when no gestures are classified and decreases otherwise. The signal also tends to fluctuate with some negative impact on the classification performance of our system, although overall, the approach is robust. (Color figure online)

We run an exemplary experiment on a random subset of our dataset comparing the two approaches, the additional no gesture output-neuron versus a fixed cutoff value. We obtained an average F1-score of 0.874 for the neural approach and an average F1-score of 0.949 for a fixed threshold value of 0.5. Consequently we chose the constant threshold method for further evaluations.

3.2   Gesture Segmentation Process

In a continuous gesture stream, the determination of start and end as well as the distinction between a gesture performed and 'No-Gesture' are specific challenges of gesture recognition systems. As we are using supervised learning, a first approach to tackle this issue would be to segment gestures when the predicted label changes. This is problematic due to the so-called subgesture problem. A subgesture is when a specific gesture can either have a standalone character with its own meaning or be an integral part of another gesture. Figure 6 demonstrates how subgestures can influence the classification. We differentiate between the 'ArgMax' approach, which at every time step simply hypothesizes a gesture based on the neuron with highest activity and the actual desired segmentation. At the beginning of this example the 'ArgMax' approach classifies a *snap forward* gestures but in effect the gesture is a *shake up-down*. The former gesture is a subgesture of the latter, and as such every time a *shake* gesture is performed it is highly likely that misclassifications occur which reduces the overall gesture system performance.

To avoid this issue we start summing up the activity of every output neuron over time, as soon as it hits a certain threshold. When activity of all neurons falls below the threshold we label the gesture based on the neuron with highest total activation since the threshold has been exceeded.

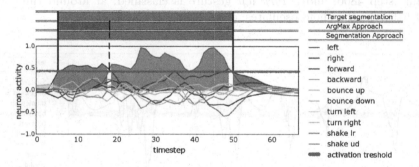

Fig. 6. Gesture segmentation showing the effect when gestures have the same subgesture. Here, the actual gesture is *shake up-down* but the first movements are detected as *snap forward*.

With the resultant segmentation we proceed to identify a mapping between the predicted outcome of the classification and the true labels. This mapping is important to provide a meaningful evaluation scheme on correctly but also incorrectly classified gestures. Notably, the prediction and target segments do not need to have matching segmentation borders. Further they do not need to have the same number of segments. Our rules for the matching between the (segmented) gesture prediction and the actual correct gesture are as follows:

- A prediction and a target segment are mapped together when they are overlapping in at least one time step. No care is given whether the length of the predicted segment matches the original signal length.
- Each predicted gesture segment is mapped exactly once.
- There is only one true positive (TP) which counts as the target gesture segments; otherwise, the segment is counted as false positives (FP) or wrong gestures (WG).
- If a predicted segment does not overlap with a target class signal, it will be mapped and labeled as WG. It will be mapped to 'No-Gesture' and scored as FP if neither the original class nor any other class can be mapped.
- If a target segment remains without any mapping, it will be marked as a 'No-Gesture' segment and labeled as false negative (FN).
- For 'No-Gesture' segments an unlimited amount of matches is allowed.

The set of rules allows to label all segments in a sequence while preserving the temporal dependencies. An example of the rule application is demonstrated in Fig. 7. Our approach has a disadvantage when considering the definition of true positives (TP), i.e. when the actual gesture was correctly classified - it assumes a one-to-one correspondence from the target signal to the classifier signal. However, a target signal can be recognized a second time by another prediction signal, which will result in a double entry of a target signal in its corresponding row of the confusion matrix. Is this the case, the row sums define an upper bound for the positive entries per class but do not tell the absolute value anymore. A solution would be to normalize over the row sums, however, we prefer to evaluate the number of misclassifications rather than argue over normalized values.

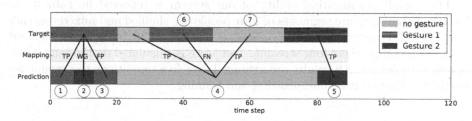

Fig. 7. Example of the mapping between predicted and actual labels when applying our rules.

4   Evaluation

We evaluated our system using the F1-score obtained from the segmentation and mapping procedure. Additionally, we evaluated our system on an individual basis to show how *inter-subject variability* across subjects can influence the final classification. Finally, we provide an overall evaluation of our system. The focus of this paper is not on achieving the highest possible performance but rather on showing a novel evaluation approach without the need for segmenting the signal before classification. We therefore refrain from reporting comparison scores.

Table 2 summarizes the results from a leave-one-out test and evaluation on each participant (names are anonymised). The varying F1-scores underpin the performance variations introduced in our experiments, however, the evaluation on gesture misclassifications gives us more valuable information. Therefore, we also provide a confusion matrix for an average subject (Ni) in Fig. 8.

In accordance with the previously mentioned subgesture problem misclassifications mostly originate from confusions of the *shake up-down* ('shake ud') gesture with the forward and backward gesture and respectively of the *shake left-right* ('shake lr') gesture with a *left* or *right* gesture.

Table 2. Subject evaluation: network is trained in a hold one out fashion on data of 4 people, then evaluated on a fifth, unseen person. Scores are reported for every fold. Train error is calculated as $1 - F1\ score$

Test set	Train	Train error	F1 test score
J	{L, S, Ni, Na}	0.088	0.624
Ni	{Na, J, L, S}	0.096	0.791
S	{Ni, Na, J, L}	0.076	0.809
Na	{J, L, S, Ni}	0.074	0.831
L	{S, Ni, Na, J}	0.093	0.926

The overall classification ability of our system is reported in Table 3. As described above gesture segments of four people are shuffled and mixed together into four different time streams. Validation score is calculated by leave one out cross validation on these four streams. Test score is calculated on the remaining, unseen fifth person. All results are averaged over five repetitions to account for stochastic effects during initialization and training.

Table 3. F1 scores and accuracy for final experiment. Reported values are averaged over five repetitions, standard deviation shown in brackets.

Validation		Test	
F1 score	Accuracy	F1 score	Accuracy
0.923 (0.00)	0.932 (0.00)	0.734 (0.03)	0.846 (0.02)

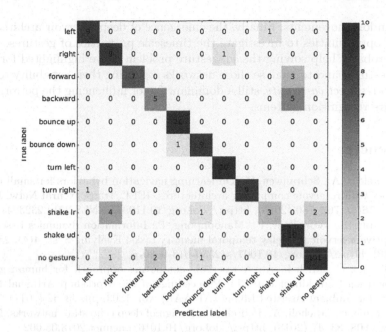

Fig. 8. Confusion matrix showing an average performance among the five subjects.

5   Discussion and Future Work

We have demonstrated a successful ESN-based framework for the task of continuous gesture recognition. We introduced a scheme for the meaningful segmentation of gestures in a stream of sensor values and their mapping to labels to establish a reasonable evaluation scheme. This way, we laid the foundation for future IMU data collections. We presented the parametrization of an ESN for optimal classification and evaluated our system on individual subjects. Although we highlighted the *inter-subject variability* as an influential factor for classification, our system achieved overall an accuracy of 85% with notably few standard deviation.

From our experiments, we propose some ideas for future applications which would improve our system. First, we are aware that our dataset is rather small compared to common gesture benchmark datasets used in machine learning. Although IMU data is used for different applications, no other, probably bigger, dataset is publicly available. Therefore, we provide our IMU dataset along with our implementation to encourage further explorations on the data including our segmentation scheme as well as analysis on the ESN model. Regarding the ESN we think of an additional online learning phase for newly introduced people to increase the system adaptability. We also recommend revising our segmentation process to be independent on the zero activity condition applied so far. For future application, the *No-Gesture* condition should be removed by a mechanism which can better distinguish between a meaningful gesture and some random movements. Finally, the emergence of deep reservoir architectures [4] offer opportunities to investigate the timescale properties of gestures, which would probably help solving the subgesture problem as we highlighted for some of our gestures, and to analyse those networks regarding their capability to cope with *inter-subject variability*, still a dominant factor influencing the performance in gesture recognition systems.

References

1. Antonelo, E.A., Schrauwen, B.: On learning navigation behaviors for small mobile robots with reservoir computing architectures. IEEE Trans. Neural Netw. Learn. Syst. **26**(4), 763–780 (2015). https://doi.org/10.1109/TNNLS.2014.2323247
2. Dasgupta, S., Wörgötter, F., Manoonpong, P.: Information dynamics based self-adaptive reservoir for delay temporal memory tasks. Evolving Syst. **4**(4), 235–249 (2013). https://doi.org/10.1007/s12530-013-9080-y
3. Gallicchio, C., Micheli, A.: A reservoir computing approach for human gesture recognition from kinect data. In: Proceedings of the Workshop Artificial Intelligence for Ambient Assisted Living (AI*AAL), vol. 1803, pp. 33–42 (2017)
4. Gallicchio, C., Micheli, A., Pedrelli, L.: Design of deep echo state networks. Neural Netw. **108**, 33–47 (2018). https://doi.org/10.1016/j.neunet.2018.08.002
5. Hinaut, X., Petit, M., Pointeau, G., Dominey, P.F.: Exploring the acquisition and production of grammatical constructions through human-robot interaction with echo state networks. Front. Neurorobot. **8**, 16 (2014). https://doi.org/10.3389/fnbot.2014.00016
6. Jaeger, H.: Tutorial on training recurrent neural networks, covering BPPT, RTRL, EKF and the "echo state network" approach. GMD-Forschungszentrum Informationstechnik (2002)
7. Jaeger, H., Lukoševičius, M., Popovici, D., Siewert, U.: Optimization and applications of echo state networks with leaky- integrator neurons. Neural Netw. **20**(3), 335–352 (2007). Echo State Networks and Liquid State Machines: https://doi.org/10.1016/j.neunet.2007.04.016
8. Jirak, D., Barros, P., Wermter, S.: Dynamic gesture recognition using echo state networks. In: Proceedings of the European Symposium of Artificial Neural Networks and Machine Learning, pp. 475–480 (2015)
9. Lee, M.C., Cho, S.B.: Mobile gesture recognition using hierarchical recurrent neural network with bidirectional long short-term memory. In: Proceedings of UBICOMM, vol. 6, pp. 138–141 (2012)
10. Lukoševičius, M., Jaeger, H.: Reservoir computing approaches to recurrent neural network training. Comput. Sci. Rev. **3**(3), 127–149 (2009). https://doi.org/10.1016/j.cosrev.2009.03.005
11. Tong, M.H., Bickett, A.D., Christiansen, E.M., Cottrell, G.W.: Learning grammatical structure with echo state networks. Neural Netw. **20**(3), 424–432 (2007). Echo State Networks and Liquid State Machines: https://doi.org/10.1016/j.neunet.2007.04.013
12. Verstraeten, D., Schrauwen, B., D'Haene, M., Stroobandt, D.: An experimental unification of reservoir computing methods. Neural Netw. **20**(3), 391–403 (2007). Echo State Networks and Liquid State Machines: https://doi.org/10.1016/j.neunet.2007.04.003

Using Conceptors to Transfer Between Long-Term and Short-Term Memory

Anthony Strock[1,2,3(✉)], Nicolas Rougier[1,2,3], and Xavier Hinaut[1,2,3]

[1] Inria Bordeaux Sud-Ouest, Bordeaux, France
Anthony.Strock@inria.fr
[2] LaBRI, Université de Bordeaux, CNRS, UMR 5800, Talence, France
[3] IMN, Université de Bordeaux, CNRS, UMR 5293, Bordeaux, France

Abstract. We introduce a model of working memory combining short-term and long-term components. For the long-term component, we used Conceptors in order to store constant temporal patterns. For the short-term component, we used the Gated-Reservoir model: a reservoir trained to hold a triggered information from an input stream and maintain it in a readout unit. We combined both components in order to obtain a model in which information can go from long-term memory to short-term memory and vice-versa.

1 Introduction

Jaeger recently showed how to store and retrieve several temporal patterns using an extension of reservoirs [1,2]. The key idea is to project, using Conceptors, the network activity into a lower dimensional space specific to the patterns. These Conceptors can be considered as a long-term memory (of the temporal patterns). In the meantime, we have shown [3] how a reservoir, using a gating signal, can faithfully memorize an information for a short delay (i.e. working memory) from a stream of random inputs. In this model, the maintenance of information in readout unit(s) is remarkably precise for rather long periods of time. However, for very long periods of time, the precision is lost because of a slow drift in the dynamics. In the present work, we introduce a preliminary attempt to link short-term and long-term memories by combining these two approaches: (1) a gated reservoir maintaining short-term information and (2) several conceptors maintaining long-term information.

2 Methods

We consider a reservoir of 1000 neurons that has been trained to solve a gating task described in [3]. In this task the model receives an input continuously varying over time (the value) and another input being either 0 or 1 (the trigger). To succeed the task, the output has to be updated to the value received when the trigger is active but has to stay still otherwise (similarly to a line attractor). In

© Springer Nature Switzerland AG 2019
I. V. Tetko et al. (Eds.): ICANN 2019, LNCS 11731, pp. 19–23, 2019.
https://doi.org/10.1007/978-3-030-30493-5_2

other words, the trigger acts as a gate that controls the entry of the value in the memory (the output). The overall dynamics of the network we consider is described by the following equations:

$$x[n] = C \tanh \left(W_{in} u[n] + W(x[n-1] + \xi) + W_{fb}(y[n-1]) \right) \text{ and } y[n] = W_{out} x[n] \qquad (1)$$

where $u[n]$, $x[n]$ and $y[n]$ are respectively the input, the reservoir and the output at time n. W, W_{in}, W_{fb}, W_{out} and C are respectively the recurrent, the input, the feedback, the output and the conceptor weight matrices and ξ is a uniform white noise term added to reservoir units. W, W_{in}, W_{fb} are uniformly sampled between -1 and 1. Only W is modified to have sparsity level equal to 0.5 and a spectral radius of 0.1. When W_{out} is computed to solve the gating task, the conceptor C is considered to be fixed and equal to the identity matrix ($C = I$). In normal mode, the conceptor C is equal to a conceptor C_m that is generated and associated to a constant value m. In order to compute this conceptor C_m, we impose a trigger ($T = 1$) as well as the input value ($V = m$) at the first time step, such that the reservoir has to maintain this value for 100 time steps. During these 100 time steps, we use the identity matrix in place of the conceptor. The conceptor C_m is then computed according to $C_m = X X^T \left(X X^T + \frac{I}{a} \right)^{-1}$, where X corresponds to the concatenation of all the 100 reservoir states after the trigger, each row corresponding to a time step, I the identity matrix and a the aperture. In all the experiments the aperture has been fixed to $a = 10$. For the conceptors pre-computed in Figs. 1 and 2, the reservoir have been initialised with its last training state.

3 Results

Transfer Between Long-Term and Short-Term Memory: Figure 1 displays the two core ideas of our approach: **(1)** How to transfer short-term to long-term memory and **(2)** How to retrieve (in short-term memory) an information stored in long-term memory. **(1)** The long-term memory we consider is the conceptor C_m associated to the value m maintained in short-term memory. We compute C_m using the 100 first time steps after a trigger without conceptors ($C = I$), after what we update C with C_m. On Fig. 1B we can see that it doesn't seem to cause any interference in the short-term memory. However now the memory lies both in the conceptor C (long-term) and in the output unit y (short-term). A more extensive and quantitative analysis could study whether applying C_m would be a way to stabilize the short-term memory. **(2)** Now, the long-term memory we consider are only conceptors C_m^D associated to discrete values between -1 and 1 (11 values uniformly spread between -1 and 1). Similarly as before, after a trigger we compute a new conceptor C_m^D using the 100 first time steps and without conceptors ($C = I$). Then, we search for the closest conceptor C_m^i among the conceptors with discrete values C_m^D using a distance between conceptors and we update C with this conceptor. On Fig. 1C, we see the following behavior: after a trigger, the value is correctly updated in short-term memory and remains stable until C is updated (after 100 time steps) and then the output jumps to the closest discrete representation of the memory.

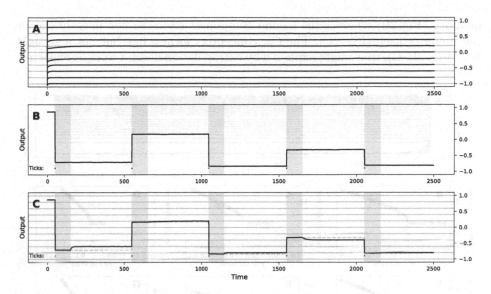

Fig. 1. Approximation with conceptors or discrete conceptors. Black: Evolution of the short-term memory (i.e. the readout y). Gray lines: the discrete value considered. Light gray areas: time when conceptors are computed for the current value stored in short-term memory. **A.** Discrete conceptors are applied **B-C.** Exact conceptors are computed using 100 time steps after a trigger while $C = I$. **B.** The exact conceptor is applied to the following time steps. **C.** The closer conceptor among the discretized conceptor is applied for the following time steps. Dashed lines represents the memory that should have been kept if not discretized.

Generalizing Long-Term Memory: On Fig. 2, we show two main ideas: **(1)** how a linear interpolation between two conceptors can allow to generalize to gating of other values, and **(2)** a representation of the space in which lies the conceptors and their link to the memory they encode. **(1)** Interpolation and extrapolation C of conceptor $C_{0.1}$ and conceptor $C_{1.0}$ has been computed as $C = \lambda C_{1.0} + (1 - \lambda) C_{0.1}$ with 31 λ values uniformly spread between -1 and 2. Even though the interpolated ($\lambda \in [0, 1]$) conceptors obtained are not exactly equivalent to C_m conceptors obtained in Fig. 1, they seem to also correspond to a retrieved long-term memory value to be maintained. The mapping between λ and the value is non-linearly encoded. For right-extrapolation ($\lambda \in [1, 2]$) the conceptor seems to be linked to a noisy version of a C_m conceptor. A value seems still to be retrieved from long-term memory and maintained in short-term memory: the output activity is not constant, but its moving average is constant. For left-extrapolation ($\lambda \in [-1, 0]$), the conceptor obtained seems not to encode any more information than the conceptor C_0: all the output activities collapse to zero. **(2)** Principal component analysis have been performed using 201 pre-computed conceptors associated to values uniformly spread between -1 and 1. The first three components explain already around 85% of the variance. The first component seems to non-linearly encode the absolute value of the memory (Fig. 2B) whereas the second component seems to non-linearly encode the memory itself (Fig. 2C). The C_m conceptors form a line but not a straight line (Fig. 2E–G), that might explain why extrapolation doesn't work as expected.

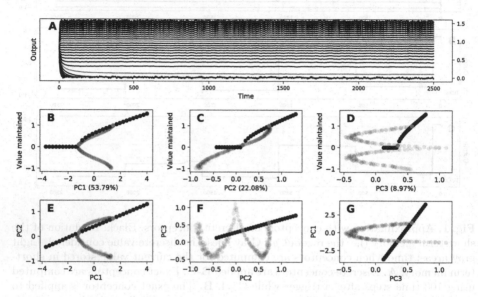

Fig. 2. Generalisation of C_m conceptors. Red: two learned conceptors $C_{0.1}$ and $C_{1.0}$. Black: Linear interpolation between conceptor $C_{0.1}$ and conceptor $C_{1.0}$. Gray: learned C_m for 201 value of m uniformly spread between -1 and 1. **A** Evolution of the short-term memory against time for the different conceptors. **B–D** Link between principal components of the constant conceptors and the memory it is encoding. For the interpolated conceptors, the memory is considered as the mean in the last 1000 time steps. **E–G** Representation of the conceptors in the three principal components of the C_m conceptors. (Color figure online)

4   Discussion

This preliminary study introduces the basis for establishing a link between long-term and short-term memory. Future work will concentrate on removing the engineered steps, namely, the offline computation and selection of the closest conceptor. This could be realized using the autoconceptors introduced in [2]. Moreover, this study raised concerns concerning the best way to interpolate and extrapolate conceptors since, as we have shown, a mere linear combination might not represent the best solution.

References

1. Jaeger, H.: Controlling recurrent neural networks by conceptors. arXiv preprint arXiv:1403.3369 (2014)
2. Jaeger, H.: Using conceptors to manage neural long-term memories for temporal patterns. J. Mach. Learn. Res. **18**(13), 1–43 (2017)
3. Strock, A., Hinaut, X., Rougier, N.P.: A robust model of gated working memory. biorXiv (2019). https://doi.org/10.1101/589564

Bistable Perception in Conceptor Networks

Felix Meyer zu Driehausen[1]([⊠])(iD), Rüdiger Busche[1,2](iD), Johannes Leugering[1], and Gordon Pipa[1](iD)

[1] Institute of Cognitive Science, Osnabrück University,
Wachsbleiche 27, 49076 Osnabrück, Germany
{fmeyerzudrie,rbusche,jleugeri,gpipa}@uni-osnabrueck.de
[2] AIM Agile IT Management GmbH,
Habichtshorststraße 1, 30655 Hannover, Germany

Abstract. Bistable perception describes the phenomenon of perception alternating between stable states when a subject is presented two incompatible stimuli. Besides intensive research in the last century many open questions remain. As a phenomenon occurring across different perceptual domains, understanding bistable perception can help to reveal properties of information processing in the human brain. It becomes apparent that bistable perception involves multiple distributed processes and several layers in the hierarchy of sensory processing. This observation directs research towards general models of perceptual inference and to the question whether these models can account for the spontaneous subjective changes in percepts that subjects experience when shown rivalling stimuli. We implemented a recurrent generative model based on hierarchical conceptors to investigate its behaviour when fed an ambiguous signal as input. With this model we can show that (1) it is possible to obtain precise predictions about the properties of bistable perception using a general model for perceptual inference, (2) hierarchical processes allow for reduction in prediction error, (3) random switches in the percept of the network are due to noise in the input and (4) dominance times exhibit a gamma distribution of stimulus dominance times compatible with experimental findings in psychophysics. Code for the experiments is available at https://github.com/felixmzd/Conceptors.

Keywords: Bistable perception · Predictive coding · Conceptors

1 Introduction

Bistable perception arises when a sensual modality is presented with a stimulus that is too ambiguous to be resolved by a unique interpretation. While there are many examples of visual stimuli that evoke bistable perception such as the Neckar cube [14] or binocular rivalry [15], the phenomenon has also been observed for other sensual modalities such as olfaction [16], audition [7] and the tactile sense [5].

© Springer Nature Switzerland AG 2019
I. V. Tetko et al. (Eds.): ICANN 2019, LNCS 11731, pp. 24–34, 2019.
https://doi.org/10.1007/978-3-030-30493-5_3

As such a general phenomenon, bistable perception seems to be a direct result of properties inherent in the information processing of the human brain.

The common characteristic of bistable perception across sensual modalities are spontaneous alternations in percept between the interpretations of the stimulus while the presented stimulus itself remains constant. The exact timepoints of change of the percept can not be predicted and are apparently random. However, the distribution of dominance time durations was shown to be relatively constant across examples of bistable stimuli and resembles a gamma like or right-skewed normal distribution [3].

Several models have been proposed that attempt to account for the established results on the timing of dominance intervals. Some also take more recent evidence on the distributed neural processing of rivalling stimuli into account [1]. A model for the condition of binocular rivalry by Freeman consists of four parallel visual channels, two driven by the left eye and two by the right. Therein, the succession of cortical levels is represented by several consecutive processing stages for each channel [8]. Dayan describes a model wherein the alternation between the percepts can be generated by competition between top-down cortical explanations for the inputs instead of direct competition between the inputs [6]. In a similar spirit, Hohwy et al. offer an explanation of the binocular rivalry condition in terms of predictive coding [10].

While computational models accurately predict the properties of bistable perception the often lack applicability to other perceptual processes [4]. Here, we present a model based on the hierarchical random feature conceptor architecture proposed by Jaeger [12]. Hierarchical random feature conceptors have successfully been applied to denoising tasks, which presents a core function of general perception. Conceptors in general have been proposed as a solution to the neuro-symbolic integration problem by implementing a filter mechanism on the hidden state dynamics of echo state networks [11]. A conceptor is inserted in the state update rule of the echo state network. It suppresses activity in atypical directions in the network dynamics while activity in typical directions remains unaffected. Typicality of the directions can in this setting be determined by observed activity during training under the same input pattern.

2   Experimental Setup

In order to simulate the condition of bistable perception, we utilized the hierarchical random feature conceptor network, as it was presented in "Controlling recurrent neural networks by conceptors", Chap. 3.16, by Jaeger in 2014. The network consists of three identical echo state networks arranged in a hierarchy of three levels. This echo state network was chosen to consist of 100 neurons with a feature space consisting of 700 neurons.

2.1   Learning of Prototype Patterns

In preparation to the experimental condition, the echo state network is presented with the clean signals of two sine waves with periods 13.190045 and 4.8342522 sampled at integer t, henceforth referred to as sine 1 and sine 2. After the prototype conceptors were learned for the clean signals, a bistable signal composed out of a superposition of these signals and noise is presented to the network. For each of the sine waves, the system is run through three periods:

1. For a washout period of 200 timesteps, during which the networks response starts to be correlated with the driver, no network responses are collected. Then the system is run in the conceptor adaptation mode for 2000 timesteps, wherein the prototype conceptor for that pattern is learned. Finally the system is run for 600 timesteps with the adapted conceptor in the network state update loop, and the network's response is collected.
2. In the following, two learning steps are performed.
   (a) The output weights W_{out} are computed by ridge regression with all collected reservoir states as arguments and the corresponding prototype patterns as targets. The normalized root mean squared deviation (NRMSD) between the output of the system, utilizing the calculated output weights, and the prototype pattern is computed.
   (b) In the second learning step, the loading, an input simulation matrix D is obtained. This is done by ridge regression, with the objective to reproduce the same network activations as they were elicited by the driver, but in absence of the driver.
3. Subsequently, the success of the learning steps was tested by a recall period. For every pattern the trained system was run under the respective conceptor for 200 washout steps. This allows for the adaptation of the network dynamics to the control of the current conceptor. Afterwards the output of the system was collected for 200 timesteps and compared to the original prototype pattern.

This describes the setup of one module of the random feature conceptor architecture with two sine waves of different periods learned. In the hierarchical random feature conceptor system, three layers of this architecture are bidirectionally connected. Weight matrices are learned beforehand and are then shared between layers. Conception weights and inputs evolve independently for each layer.

Bottom-up and top-down processing is mediated by trust variables that are adapted based on discrepancies between predicted and actual input in each layer. The input to each layer is a mixture of predicted input and the signal from the lower layer, mediated by the trust variable.

There exists a bottom-up as well as a top-down flow of information. The output of the lower layer is fed to the higher layer, constituting the bottom-up flow. The top-down flow is the influence of conception weights of a higher module on a lower one. Both are mediated by the trust variables.

The top-down pathway influences the conception weights in each layer of the hierarchy. The top-layer hypothesis is passed downwards in the hierarchy. In each of the lower levels 2 and 1 an autoconceptor adaptation process is taking place, yielding layer internal conception weight vectors. These are linearly mixed with the conception weight vector from the next higher layer, using the trust variables as mixing coefficients. The topmost layer is a special case, as there is no layer above which can have any influence on its conceptor. By design its conceptor is constrained to be a disjunction of prototype conceptors of the two sine waves.

The bottom-up pathway influences the input to the higher levels 2 and 3. These levels have a self generated input simulation signal. Additionally to this, they receive the output from the next lower layer. Again, the trust variables determine how much influence the bottom-up pathway has against the self generated input simulation signal by serving as mixing coefficients.

2.2 Experiments with the Bistable Stimulus

In addition to the hierarchical random feature architecture we introduced a feedback loop from the top level hypothesis to the input of the system. This feedback loop suppresses those parts of the input signal that can be explained or predicted under the current hypothesis of the system.

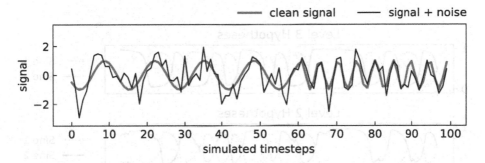

Fig. 1. A sample of the effective, ambiguous input, with influence from the feedback loop. Up to timepoint 62 the signal consists of sine wave 1 and noise, thereafter of sine wave 2 plus noise. The hypothesis that sine wave 2 is the source in the signal was winning until timestep 62. Therefore the signal of sine wave 1, which is not predictable under this hypothesis, remained in the input signal. From timestep 62 on the same reasoning holds, with hypothesis 1 being the winning hypothesis and sine wave 2 remaining as unpredicted residuum in the input signal.

The input to the hierarchical architecture at the lowest level is the sum of the two irrational sine patterns and normally distributed noise, with the signal to noise ratio of 1 with respect to the clean sine wave input. The noise was found to be necessary to push the system into an oscillating regime [2]. When the system settles on a hypothesis on the highest level of the hierarchy, the part of the input signal that can be explained by that hypothesis is subtracted from the input to the lowest level of the hierarchy. Importantly, we defined the winning hypothesis by the procedure of 'the winner takes it all'. This affects the input drastically as the complete clear signal that belongs to the winning hypothesis is subtracted from the input. Thereby the effective input to the system is usually a composition of noise and one signal source. A sample of this effective input is shown in Fig. 1.

The system is run for 50.000 timesteps. Over the course of this simulation the hypothesis of the system about the source of the driver is collected on all three levels of the hierarchy. Moreover the dynamics of the trust variables that operate between the levels are saved. Experiments were recorded using the `Sacred` library [9] to ensure reproducibility.

3   Results

Initial learning of the prototype conceptors was performed successfully. The NRMSD for computing the output weights was 0.0027. The NRMSD per neuron between the input driven network response and the network response elicited by D was 0.0005 on average per neuron. Both learned sine wave patterns were recalled. After the correction of an inevitable phase shift, the NRMSD for the sine 1 was 0.025 and the NRMSD for sine 2 was 0.059.

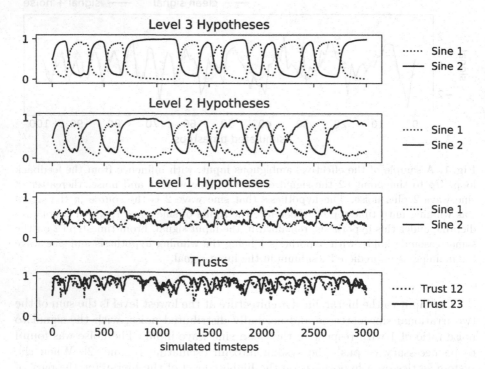

Fig. 2. Developments of hypotheses and trusts in the network. Displayed are the first 3000 simulated timesteps. The three topmost plots show the evolution the of hypothesis vectors for the three levels of the hierarchy. The bottom-most plot shows the trust variables operating at the intersection of the levels of the hierarchy. For details see the Results section.

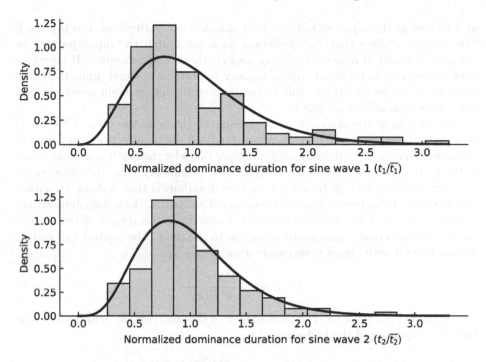

Fig. 3. Distribution of dominance times, separately for each sine wave. Both histograms were fit to a gamma distribution function (black line). The distribution of dominance times in the simulation is similar to data acquired from experiments in humans, when they were viewing rivalling stimuli.

Figure 2 shows the results of then presenting the combined signal for the first 3000 out of the total of 50000 simulated timesteps. A few observations can be made: On level 1 the hypotheses are not yet really differentiated with relatively long periods where both hypotheses are almost equally likely. On level 2 this differentiation is far better, surpassed only by a little in layer 3. Moreover, a small delay in the processing of the system can be observed. Comparing level 2 and level 3 hypotheses, it can be seen that level 3 reacts similar but has a delay on the order of 100 to 300 timesteps with regard to level 2. Between level 1 and level 2 this is less obvious, but can also be observed. It is also far more difficult to see, because on level 1 the structure of the hypothesis peaks is still very different compared to the higher levels. Most importantly an oscillation between the hypotheses can be observed on all levels. The top level hypothesis vector can be interpreted as the perception of the system, switching from one sine wave to the other, back to the first one, and so on. This resembles the perception human observers have when they are viewing rivalling stimuli. The bottommost plot displays the trust variables that operate between the levels. They both stay at a high level during the stimulation, indicating that the system is confident to generate the correct pattern most of the time. Especially for the trust variable between level 1 and 2 several small dips can be observed. These can correspond to a switch in the input signal due to a change in hypothesis on the top level. The system realizes that its prediction does not match the input pattern as much as it would, if it were to change its hypothesis and conceptor. It therefore operates shortly in an input driven manner to find the optimal input matching hypothesis and settles again, only to be tempted to change again as soon as the new hypothesis affects its input.

We calculated the distribution of dominance times on the data of the third level hypothesis vector. We measured normalized dominance times in terms of simulated time steps t, dividing all dominance times by the mean dominance time $\bar{t}$. We in particular calculated the normalized dominance time distribution for each sine wave separately, to not get a mixed distribution that is skewed to either side because the patterns have different signal strengths. Both distributions are plotted in Fig. 3. The distributions show similarity to the results of Levelt [13]. As Levelt did, we fit a gamma distribution to the data. The gamma pdf can be parameterized with shape k and scale θ as

$$f(t; k, \theta) = \frac{t^{k-1}e^{-\frac{t}{\theta}}}{\theta^k \Gamma(k)}$$

We estimated the parameters k and scale θ of the distribution, which yields the following equation of the fit for sine 1

$$f(t; 3.179, 0.315) = \frac{t^{3.179-1}e^{-\frac{t}{0.315}}}{0.315^{3.179}\Gamma(3.179)}$$

and for sine 2 respectively

$$f(t; 7.237, 0.138) = \frac{t^{7.237-1}e^{-\frac{t}{0.138}}}{0.138^{7.237}\Gamma(7.237)}$$

3.1 Stability

We find the phenomenon of bistable perception in our architecture to occur across different input patterns. We tested our architecture with different combination of sine waves and a combination of random periodic patterns and sine waves. We found the hypotheses at the highest level to consistently switch between the two presented patterns for all tested pattern combinations. We also found the distributions of dominance times to consistently form gamma-like right-skewed distributions, that resemble the distributions observed in psychophysics experiments.

However, the exhibition of the phenomenon is dependant on the right interplay between input pattern, aperture and initialization.

4 Discussion

We used the hierarchical random feature conceptor as a model for bistable perception, thereby providing an application of the conceptor architecture in a cognitive modeling task.

4.1   Comparing Dominance Times to Levelt's Work

The distribution of dominance times that we obtained from the simulation is remarkably similar to the dominance time distribution of Levelt's work in the 60 s, which is shown in Fig. 4. Across different combinations of input pattern, we especially find that normalized dominance times concentrate in the range between zero and three. Also, we consistently observe a right skewed distribution of dominance time durations, which is a well established result in research on bistable perception across modalities. These results encourage further investigations in how far the hierarchical random feature conceptor architecture is a suitable model for general human perception.

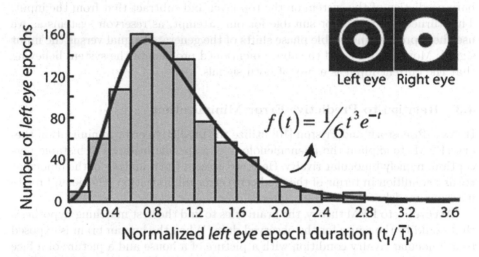

$$f(t) = \tfrac{1}{6} t^3 e^{-t}$$

Fig. 4. Distribution of dominance times for binocular rivalry as reported by Levelt [13]. This figure is reproduced from Brascamp et al. [4].

4.2   Bistable Perception

In our simulation the system has learned two prototype patterns. These two patterns are "the world" for the system. Besides the driving input itself, its internal representation of the prototype patterns is the only information it has access to during runtime. As the system is also not adapting or learning any new patterns during the course of the simulation, the only hypothesis it can make up involve the two prototype patterns. The simulation of the bistable perception shows that the system adapts its hypothesis about the current input in accordance to the input. On the level of the hypotheses it shows an alternating behaviour, just as it is the key observation in bistable perception in humans. Insofar we have a working example of a challenging situation for a perceptual system. The system has only the option to make up hypotheses from the two prototype pattern it knows. It can, however, settle on a mixture of these, maintaining for example the hypothesis that a mixture of the prototype patterns causes the current sensory input. This is in fact the case, if the system is run without the effect of the feedback loop. In many situations this is highly desirable and it is a research project in its own right in how far conceptor combinations really are able to combine concepts. Nevertheless in the special case of humans viewing rivalling stimuli, the hypothesis of a mixture of both stimuli is a priori highly unlikely. For the concrete example of binocular rivalry, face-house compounds do usually not appear in the world. We reflect this low prior probability for the compound hypothesis by subtracting the winning hypothesis from the input. This design choice is supported by a strong effect of the prediction of the system on the actual perception. The predicted signal is completely explained and therefore can be subtracted from the input signal. In the original approach we tried to take the bare prediction of the system on the top layer and subtract that from the input. This turned out to be not suitable for our attempt, as reservoir systems as we use them produce inevitable phase shifts of the generated signal versus the input signal. Moreover we faced the above mentioned problem of the system believing that the current input is a mix of both signals.

4.3   Relation to Predictive Error Minimization

Hohwy, Roepstorff and Friston [10] utilize the predictive error minimization theory (PEM) to explain the phenomenology of a specific instance of bistable perception, namely binocular rivalry. Here, we present their analysis of the binocular rivalry condition in terms of the predictive error minimization scheme and relate it to our model.

According to PEM theory, the brain tries to find the best matching hypothesis that could be the cause for the observed data. When the human brain is exposed to a binocular rivalry condition with a picture of a house and a picture of a face presented to separate eyes at the same time the brain of a subject might settle on the hypothesis that a house caused the visual stimulation. Under this hypothesis, the brain, as a hierarchical generative model, would predict some features of a house which will match with parts of the sensory data. The sensory drive that is generated by the face would remain as a residuum and as a prediction error that is not accounted for by the prediction of the brain. This error is on about the same order of magnitude as the explained data, namely the part of the stimulus that belongs to the house. Due to this balance of information content between both parts of the stimulus and due to noise, the hypothesis that the face generated the sensory drive would overtake. This oscillation describes the alternation between different percepts that is observed when humans view rivalling stimuli.

The hierarchical random feature architecture tries to minimize prediction error by selecting the best hypothesis in order to predict the incoming sensory data. The residuum of the incoming data which can not be explained is called the prediction error. In contrast to PEM theory, our proposed architecture does not signal the prediction error upwards in the hierarchy, but a denoised version of the sensory input. 'Denoised' means in this context that parts of the signal which are not predictable under the current hypothesis are regarded as noise and are suppressed. This, in fact, leads to less prediction error on higher layers of the hierarchy, as the prediction error is suppressed by each layer. This mechanism is therefore actually minimizing prediction error, but in a slightly different fashion than the usually in predictive coding proposed upwards signalling of the residual signals or prediction errors. Minimizing prediction error just by suppressing all signals that can not be predicted on its own does not seem very useful. But this process is aided by a general assessment of fit of all prototype patterns to the input signal. This is inherent in the conceptor mechanism. Therefore the mechanism for prediction error minimization is different in the hierarchical random feature conceptor as compared to the usual notion in predictive coding. This issue is still in debate, also for the predictive coding research community, as we are not aware of any clear cut evidence in favour of and against other possible realizations of error signalling.

5   Conclusion

We implemented a recurrent generative model based on hierarchical conceptors to investigate its behaviour with regards to bistable perception. We were able to show that the network exhibits random switches in its percepts. The distribution of dominance durations furthermore resemble well established findings from psychophysical experiments on bistable perception in humans. Moreover, the hierarchical organization and the message passing between the levels of the hierarchy allows for noise suppression and prediction error minimization. Therefore, we were able to construct an accurate model for bistable perception that is based on a model for general perception and is applicable to other tasks. Overall, we conclude that the hierarchical random feature conceptor architecture is a promising model for general human perception. Further work has to be done in order to investigate whether the architecture can account for more perceptual phenomena.

References

1. Alais, D., Blake, R.: Binocular Rivalry. MIT Press, Cambridge (2005)
2. Brascamp, J.W., Van Ee, R., Noest, A.J., Jacobs, R.H., van den Berg, A.V.: The time course of binocular rivalry reveals a fundamental role of noise. J. Vis. **6**(11), 8 (2006)
3. Brascamp, J.W., Van Ee, R., Pestman, W.R., Van Den Berg, A.V.: Distributions of alternation rates in various forms of bistable perception. J. Vis. **5**(4), 1 (2005)
4. Brascamp, J., Klink, P., Levelt, W.J.: The 'laws' of binocular rivalry: 50 years of levelt's propositions. Vis. Res. **109**, 20–37 (2015)
5. Carter, O., Konkle, T., Wang, Q., Hayward, V., Moore, C.: Tactile rivalry demonstrated with an ambiguous apparent-motion quartet. Curr. Biol. **18**(14), 1050–1054 (2008)
6. Dayan, P.: A hierarchical model of binocular rivalry. Neural Comput. **10**(5), 1119–1135 (1998)
7. Deutsch, D.: An auditory illusion. Nature **251**(5473), 307 (1974)

8. Freeman, A.W.: A multi-stage model for binocular rivalry. J. Neurophysiol. **94**, 4412–4420 (2005)

9. Greff, K., Klein, A., Chovanec, M., Hutter, F., Schmidhuber, J.: The Sacred infrastructure for computational research. In: Proceedings of the 16th Python in Science Conference, pp. 49–56. SciPy (2017)

10. Hohwy, J., Roepstorff, A., Friston, K.: Predictive coding explains binocular rivalry: an epistemological review. Cognition **108**(3), 687–701 (2008)

11. Jaeger, H.: The "echo state" approach to analysing and training recurrent neural networks-with an erratum note. Bonn, Germany: German National Research Center for Information Technology GMD Technical report **148**, 34 (2001)

12. Jaeger, H.: Controlling recurrent neural networks by conceptors. arXiv preprint arXiv:1403.3369 (2014)

13. Levelt, W.J.: On binocular rivalry. Ph.D. thesis, Van Gorcum Assen (1965)

14. Necker, L.A.: LXI. observations on some remarkable optical phænomena seen in switzerland; and on an optical phænomenon which occurs on viewing a figure of a crystal or geometrical solid. The London, Edinburgh, and Dublin Philos. Mag. J. Sci. **1**(5), 329–337 (1832)

15. Wheatstone, C.: Contributions to the physiology of vision.-part the first. on some remarkable, and hitherto unobserved, phenomena of binocular vision. Philos. Trans. Roy. Soc. Lond. **128**, 371–394 (1838)

16. Zhou, W., Chen, D.: Binaral rivalry between the nostrils and in the cortex. Curr. Biol. **19**(18), 1561–1565 (2009)

Continual Learning Exploiting Structure of Fractal Reservoir Computing

Taisuke Kobayashi$^{(\boxtimes)}$ (iD) and Toshiki Sugino

Nara Institute of Science and Technology, Nara, Japan
{kobayashi,sugino.toshiki.so7}@is.naist.jp
https://kbys_t.gitlab.io/en/

Abstract. Neural network has a critical problem, called catastrophic forgetting, where memories for tasks already learned are easily overwritten with memories for a task additionally learned. This problem interferes with continual learning required for autonomous robots, which learn many tasks incrementally from daily activities. To mitigate the catastrophic forgetting, it is important for especially reservoir computing to clarify which neurons should be fired corresponding to each task, since only readout weights are updated according to the degree of firing of neurons. We therefore propose the way to design reservoir computing such that the firing neurons are clearly distinguished from others according to the task to be performed. As a key design feature, we employ fractal network, which has modularity and scalability, to be reservoir layer. In particular, its modularity is fully utilized by designing input layer. As a result, simulations of control tasks using reinforcement learning show that our design mitigates the catastrophic forgetting even when random actions from reinforcement learning prompt parameters to be overwritten. Furthermore, learning multiple tasks with a single network suggests that knowledge for the other tasks can facilitate to learn a new task, unlike the case using completely different networks.

Keywords: Continual learning · Reservoir computing · Fractal network

1 Introduction

In recent years, reinforcement learning (RL) [15] combined with neural network (NN) has been greatly succeeded in acquiring tasks that are difficult to be solved analytically [10,16]. These studies employ NN as a function approximator of value function, policy, etc., and learn its parameters (i.e., weights of connection between neurons) based on observed samples, thereby yielding the optimal policy, which maximizes the sum of reward given from environment. Even in the case without RL, NN, more specifically deep learning, has achieved image recognition higher accuracy than human by preparing a huge dataset [9].

Supported by JSPS KAKENHI, Grant-in-Aid for Young Scientists (B), Grant Number 17K12759.

© Springer Nature Switzerland AG 2019
I. V. Tetko et al. (Eds.): ICANN 2019, LNCS 11731, pp. 35–47, 2019.
https://doi.org/10.1007/978-3-030-30493-5_4

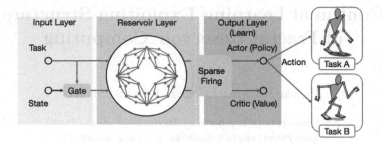

Fig. 1. Conceptual structure of reservoir computing: the fractal reservoir network is fully utilized to mitigate the catastrophic forgetting by adding the activation function sparsely firing to output non-zero value only for large input, the task inputs, and the gates for them.

However, NN has a fatal problem, called catastrophic forgetting [3,11], where memories about the task previously learned is overwritten when a new task is learned since all the parameters of NN is specialized for the new task by back-propagation. This problem prevents agents like autonomous robots from incrementally learning various tasks (i.e., continual or lifelong learning [2,5,6,18]). In particular, its adverse effects in RL are remarkable because NN is optimized through exploration behaviors, which cause frequent updates of parameters.

A main idea to mitigate the catastrophic forgetting is to keep the parameters with high contribution to corresponding tasks. To achieve this idea, two approaches are mainly raised: (i) regularization toward the parameters optimal for previous tasks while adjusting its magnitude according to the importance of parameters for the previous tasks [5,6], and (ii) modularization of the network structure for tasks as much as possible [2,18]. Previous work for (ii) assumed that the number of tasks is known, and therefore, adding a new task involves expensive computational cost due to reconstruction of the network structure.

To modularize the network structure in a feasible way, a new network is designed to mitigate the catastrophic forgetting exploiting a model that learns only output (called readout) layer of recurrent NN (RNN), named reservoir computing (RC) [4]. Specifically, the dynamical network in RC, named reservoir layer, is firstly designed by a fractal complex network [14] (see Fig. 1), which has fractality giving modularity and scalability. To exploit the features, RC is further designed with the following three way: (i) designing a new activation function that makes neuron fire sparsely, in other words, output non-zero value only for large input, to clarify the neurons used for tasks; (ii) applying bias inputs for tasks, named task inputs, into corresponding modules to clarify the modules used for tasks; and (iii) installing gates for state inputs opened by the corresponding task inputs to facilitate modularization.

The proposed RC is employed as the function approximator in RL to acquire multiple policies for different tasks. In simulations, it is verified that the proposed RC can perform the tasks after learning the other tasks, that is, the catastrophic forgetting can be mitigated, although the general RC (without the proposed design) forgets a part of the learned tasks. In addition, we suggest that, thanks to learning tasks using a single network, knowledge for the previous tasks can facilitate to learn a new task, unlike the case using completely different networks.

2   Preliminaries

2.1   Reinforcement Learning: RL

RL makes an agent learn optimal policy, π^*, from numerous interactions with environment, which are represented by state, $s \in \mathcal{S}$, action, $a \in \mathcal{A}$, and reward, $r \in \mathbb{R}$. Here, π^* refers to a policy that obtains maximum return from current time t, $\sum_{k=0}^{\infty} \gamma^k r_{t+k+1}$, where $\gamma \in [0,1)$ is a discount factor. The advantage of RL is the capability to learn π^* without specific supervision, whereas overwriting parameters in various ways with exploratory actions is likely to cause the catastrophic forgetting.

Procedure of RL is defined as Markov decision process (MDP) as below. That is, the agent firstly observes the initial state s_0 sampled from arbitrary probability distribution. At each time, the policy π choices the action $a_t \sim \pi(a \mid s_t)$ stochastically, which works on the environment. Through interaction, a new state s_{t+1} and and a reward $r_t := r(s_t, a_t, s_{t+1})$ are gained depending on the transition probability, $p(s_{t+1} \mid s_t, a_t)$.

Now, let us briefly introduce actor-critic algorithm [8,12], which is employed in this paper. It has two learners: an actor for optimizing the policy and; a critic for evaluating the policy. The actor learns the policy to transition to a state where the value function V, which is learned in the critic, increases. To this end, a temporal difference (TD) error, δ_t, is given as follows:

$$\delta_t = r_t + \gamma V_{t+1} - V_t \tag{1}$$

V is simply updated by δ_t as its gradient. In addition, this paper employs True Online TD(λ) [17] to accelerate its learning.

The actor learns its policy by using REINFORCE [19]. Specifically, the gradient of the policy (its parameters more specifically), g, is given as follows:

$$g_t = \delta_t \nabla_\pi \log \pi(a_t \mid s_t) \tag{2}$$

Improving the policy by this gradient with learning rate α is expected to acquire a locally optimal policy.

To this point, the value function V and the policy π are described as general forms, but their accurate functions for the environment (i.e., the given task) are actually black-box. Function approximation like deep learning is therefore required to learn them approximately. In addition, the policy firstly requires to assume a specific stochastic model to learn its parameters with tractable cost. According to ref. [7], this paper employs multivariate diagonal student-t distribution, which regards the exploration behaviors of animals (i.e., Lévy flight

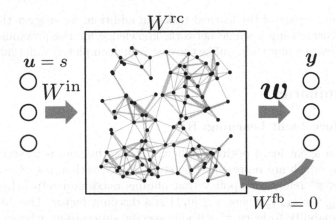

Fig. 2. Structure of reservoir computing [4]: it is RNN with fixed parameters except readout weights w; for simplicity, a feedback term is omitted in this paper.

or Lévy walk [1]). When action space $\mathcal{A}$ is d-dimensional, the policy $\pi(a \mid s)$ is given as follows:

$$\pi(a \mid s) \sim \mathcal{T}(a \mid \boldsymbol{\mu}(s), \mathrm{diag}(\boldsymbol{\sigma}^2(s)), \nu(s)) \tag{3}$$

$$\mathcal{T}(\boldsymbol{x} \mid \boldsymbol{\mu}, \Sigma, \nu) = \frac{\Gamma(\frac{d+\nu}{2})}{\Gamma(\frac{\nu}{2})(\nu\pi)^{\frac{d}{2}} |\Sigma|^{\frac{1}{2}}} \left[1 + \frac{1}{v}(\boldsymbol{x} - \boldsymbol{\mu})^\top \Sigma^{-1}(\boldsymbol{x} - \boldsymbol{\mu}) \right]^{-\frac{d+\nu}{2}} \tag{4}$$

where $\boldsymbol{\mu}$, $\boldsymbol{\sigma}$, and ν are the parameters to be learned using the function approximation. $\Gamma(\cdot)$ denotes the gamma function.

2.2 Reservoir Computing: RC

To approximate and learn V, $\boldsymbol{\mu}$, $\boldsymbol{\sigma}$, and ν, RC [4] is employed. RC is a model of cerebellum in biology, and as a linear regression model mathematically, which is easy to learn with small samples unlike deep learning requiring backpropagation. That is why the damage by the catastrophic forgetting would be small, and furthermore, it is easy to investigate the effects of network design on the catastrophic forgetting.

The simplest dynamics of RC with N neurons' internal states $\boldsymbol{x}$ (see Fig. 2) is described as follows:

$$\boldsymbol{x}_t = f(W^{\mathrm{rc}} \boldsymbol{x}_{t-1} + W^{\mathrm{in}} s_t) \tag{5}$$

$$\boldsymbol{y}_t = g(\boldsymbol{w}^\top [\boldsymbol{x}_t^\top, s_t^\top]^\top) \tag{6}$$

where $f(\cdot)$ denotes the activation function, which is hyperbolic tangent function in general, whereas a new one is proposed in this paper. W^{rc} and W^{in} are fixed weights, which are randomly given in general, whereas they are appropriately designed in this paper. $g(\cdot)$ guarantees the domain of the approximated function

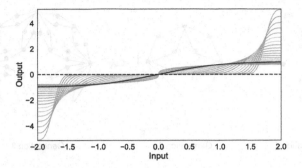

Fig. 3. Proposed activation function according to p: as p increases, the dead zone where the output is almost 0 becomes wider; that is, when $p < 1/3$ or $p > 1/3$, this converges to sign or step function.

(e.g., exponential function for σ to ensure positive real value). W^{rc} is resized to guarantee that its spectral radius ρ, i.e., $\max |\lambda|$ where λ eigenvalues of W^{rc}, is smaller than 1 for stability. To improve outputs y, w is updated by applying a gradient method with learning rate α.

2.3   Elastic Weight Consolidation: EWC

If the important parameters for the old tasks are discriminated, letting their values be invariant would avoid to overwrite them. The methods, represented by elastic weight consolidation (EWC) [5], regularize the parameters toward the optimal values for the old tasks, w^*, by a following additional gradient.

$$\therefore g_{\mathrm{EWC}} = \lambda_{\mathrm{EWC}} F \odot (w - w^*) \tag{7}$$

where λ_{EWC} is the magnitude of regularization and F is the importance of the parameters, which have been defined as the diagonal of Fisher information matrix in the paper of EWC. Specifically, w^* and F correspond to the mean and the precision of diagonal multivariate Gaussian distribution of w, respectively. Note that several types of relatives have been proposed to approximates approximate w^* and/or F [6,20] Due to a non-verification target, this paper employs a moving average to estimate w^* and F for simplicity.

3   Proposal

3.1   Sparse Activation Function

In RC (see Eq. (5)), the activation function for implying firing rate of neuron is usually given as hyperbolic tangent function, which is basically non-zero. In that case, w is always updated by the gradient for optimization due to its linearity for x. Such frequent update would overwrite even the parameters important

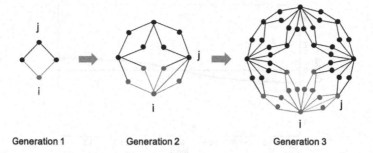

Fig. 4. Example of the generation process of (u, v)-flower: $(2, 2)$-flower up to the third generation is illustrated with module indicated in red; this structure is represented as adjacency matrix A. (Color figure online)

for the previous tasks, thereby causing the catastrophic forgetting. Alternative activation function is therefore required to hold the important parameters.

Focusing on the fact that RC does not use backpropagation, this paper proposes a more non-linear and sparsely-firing activation function. Specifically, the following equation is defined as the activation function $f(\cdot) = \mathrm{ls}(\cdot)$ in Eq. (5).

$$\mathrm{ls}(x; p) = \begin{cases} A(p)\mathrm{sgn}(x)\left\{\frac{1}{2}|x| - \frac{1}{2\pi}\sin(\pi|x|)\right\}^p & |x| \le 2 \\ A(p)\mathrm{sgn}(x) & |x| > 2 \end{cases} \qquad (8)$$

where p is the hyperparameter to determine the sparsity of firing (i.e., the width of the dead zone where the output is almost 0). $A(p)$ is given in accordance with p to standardize the integral value within $x \in [-2, 2]$.

The effect of p is illustrated in Fig. 3. When p is large, the input for firing (i.e., getting the non-zero output) and the maximum output increase; otherwise, it converges to sign (constant) function. Note that the proposed function with $p = 1/3$ is almost consistent with the hyperbolic tangent function. In summary, the neurons with large p would become task-specific due to the sparsity of firing, and the neurons with small p would be for common tasks for all the tasks.

3.2 Fractal Network for Reservoir Layer

In general RC [4], the weights for the input and reservoir layers, W^{rc} and W^{in}, are given at random as constants. Such randomness is useful when learning a single task (or multiple tasks simultaneously) since it generates statistically diverse neurons. In the context of continual learning, however, these network structure can be designed to separate the regions of the network occupied for the respective tasks more efficiently. Our contribution in this paper is here: the design of W^{rc} and W^{in} suitable for the continual learning.

Here, W^{rc} for the dynamical network in RC is designed. To separate the regions of the network occupied for the respective tasks, modularity is obviously indispensable, and scalability is also important to incrementally learn the multiple tasks. As a network that meets these conditions, (u, v)-flower network, a kind of hierarchically structured fractal network [14], is employed. Note that $u = 1$ has no modularity, so (u, v)-flower network is given under the condition: $v \geq u \geq 2$. It has $u + v$ hubs with the maximum degree, which are on the center of respective main modules. In other words, submodules can easily be increased while the previous main modules are stored when its generation progresses.

The generation algorithm for (u, v)-flower is described below, and its example is illustrated in Fig. 4.

1. define a cycle C_1 composed of two completely different paths with u and v edges, respectively, as the first generation graph;
2. replace all the edges in the current generation with C_1 while retaining existing nodes in the current generation;
3. and repeat n-th generation.

After this procedure, the adjacency matrix A, which is binary representation of the presence or absence of connections between neurons, is obtained. That is, W^{rc} is given combined with A and a random matrix R.

$$W^{\mathrm{rc}} = A \circ R$$

$$\text{s.t. } \rho < 1 \tag{9}$$

where $\circ$ denotes Hadamard product.

3.3   Design of Input Layers

Next, we focus on the input layer, W^{in} and inputs themselves, to fully exploit the properties of (u, v)-flower. Specifically, two designs are implemented: task inputs as biases and their weights to clarify the modules to be used for the commanded tasks; and gates for state inputs (state s) to facilitate modularization.

Task Inputs to Modules. First of all, we assume that the task that the agent should do can be commanded as one-hot vector or probabilities of all the tasks that the agent has, b. In that case, b is added to inputs to RC instead of a bias, $[s^\top, b^\top]^\top$. The submatrix of W^{in} is only for b, so-called bias weight W^{bs}.

Here, W^{bs} is designed to activate the corresponding modules by b. As a naive idea, when giving W_i^{bs} of i-th neuron according to the distance to the hubs for the respective tasks, they are easily fired for the respective tasks, and the neurons around the boundaries between the modules have the role of representing common knowledge for all the tasks. Therefore, we design W^{bs} according to the following procedure:

1. choose hubs for all the tasks to be learned, h from all the neurons n;
2. set a hyperparameter of KCS kernel $k(\cdot, \cdot)$ [13], which is modified Gaussian kernel to consider upper and lower bounds of the distance between two inputs, so that the bounds are on the intermediate points with adjacent hubs;

3. set W^{bs} of h according to the distance between them and n with uniformly random plus-minus sign.

$$W^{\mathrm{bs}} = \pm K(n, h) \tag{10}$$

where $K(\cdot, \cdot)$ denotes Gram matrix for $k(\cdot, \cdot)$.

Even when the generation of (u, v)-flower is increment to add new tasks, it is expected that the distance between the adjacent hubs would not be smaller than before generation increment. Namely, this design does not lose scalability.

Gates for State Inputs. The task inputs are effective to clarify which modules are activated for the respective tasks. In the tasks with large state space, however, it can be expected that state inputs would make it easier for any neurons to fire even without the task inputs. To facilitate modularization, we should properly suppress the dominant influence of the state inputs.

To this end, a gate matrix, $G(W^{\mathrm{bs}}, b)$, is additionally implemented to smoothly cut the state inputs to the neurons close to the hubs (the center of the modules) unactivated. $G(W^{\mathrm{bs}}, b)$ is simply designed as follows:

$$G_{ij} = \begin{cases} 0 & i \neq j \\ \mathrm{ls}(2x_i; 1) & i = j \end{cases} \tag{11}$$

$$x_i = \left| \frac{W_i^{\mathrm{bs}} b}{\max(1, \|W_i^{\mathrm{bs}}\|) \max(1, \|b\|)} \right| + \left(1 - \sqrt{\min(1, \|W_i^{\mathrm{bs}}\|)} \right)$$

where W_i^{bs} is the vector for i-th neuron's bias weight. This gate is open regardless of b if W_i^{bs} is small; otherwise, it is 1 or 0 if the corresponding b is (not) given.

Redefinition of Reservoir Dynamics. The dynamics of RC, in particular, the term for the state inputs in Eq. (5), should be redefined according to the above designs. Let W^{in} be given from uniform random distribution $[-1, 1]$. In that case, a new dynamics is therefore given as follows:

$$x_t = \mathrm{ls}\left(W^{\mathrm{rc}} x_{t-1} + \begin{bmatrix} GW^{\mathrm{in}} & W^{\mathrm{bs}} \end{bmatrix} \begin{bmatrix} s_t \\ b_t \end{bmatrix} ; p \right) \tag{12}$$

$$y_t = g(w^\top [x_t^\top, s_t^\top, b_t^\top]^\top) \tag{13}$$

where the hyperparameters for the respective neurons p are also randomly given in this paper. Note that b can freely be given at each time, although it is constant during episode in this paper.

4   Simulations

4.1   Conditions

The performance of the proposed design is verified in two kinds of RL simulations: (a) Acrobot and (b) BallArm, as shown in Fig. 5 (the same environments

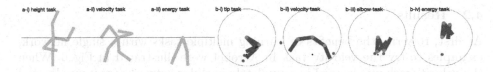

a-i) height task   a-ii) velocity task   a-iii) energy task   b-i) tip task   b-ii) velocity task   b-iii) elbow task   b-iv) energy task

Fig. 5. Snapshots of simulations: in Acrobot, an agent outputs torque for second joint and gets joint angles and angular velocities; in BallArm, an agent outputs torques for both joints and gets joint angles, its angular velocities, ball position, and its velocities.

Table 1. Parameters to be used

Number of neurons N	Discount factor γ	Learning rate α	Spectral radius ρ	Gain for EWC λ_{EWC}
1038	0.99	1e$-2/N$	0.5	1e-10

used in ref. [6]). Acrobot has three tasks: (i) swinging up the tip of a two-link arm and keeping it stand, named height task; (ii) maximizing the angular veloc-ity of a root joint, named velocity task; and (iii) stopping motion not to waste energy, named energy task. BallArm has four tasks: (i) bringing the tip of a two-link arm to a ball, named tip task; (ii) hitting the ball to maximize its speed, named velocity task; (iii) bringing the elbow of the two-link arm to the ball, named elbow task; and (iv) stopping motion as well as the case of Acrobot, named energy task. Note that the catastrophic forgetting was confirmed in these environments without any approaches to mitigate it.

As a learning procedure, the target task to be learned, i.e., b with one-hot encoding, is switched every 300 episodes by turns. After first rounding, the second round is used for evaluation. A score is a weighted mean of returns normalized by the maximum one with inverse of episodes as weights, which are mainly affected by up to first 50 episodes. This procedure is conducted 20 times.

We basically compare two networks. One is designed as a baseline (named random network) that the reservoir layer W^{rc}, even its connections, is randomly given, and as for the bias weight matrix W^{bs}, each neuron is assigned to only one task with a random value. Another is designed as a proposal (named fractal network) that the reservoir layer W^{rc} is given as 4-th generation of $(u = 3, v = 3)$-flower with random values, and W^{bs} is given by the proposed procedure in the Sect. 3.3 while choosing the hubs so that each one is the most apart from each other. By the comparison of the scores between them, we can verify the benefits of modularity of the fractal network structure.

The other configurations of them are naturally the same as each other. Specif-ically, p in Eq. (8) for each neuron is randomly given from $[1/3, 50]$. The gate matrix is given by Eq. (11). EWC [5] is employed in both networks as an auxil-iary for mitigating the catastrophic forgetting. The parameters for learning (see Table 1) are decided empirically.

4.2   Results

At first, to verify the benefits of learning multiple tasks with a single network, examples to try the velocity task in Acrobot were illustrated in Fig. 6. When the velocity task was learned by completely modular networks (or it was learned firstly before the height task), the agent could not find the global optimum, i.e., rotating the base link, since it did not know how to swing up the base link. In contrast, after learning the height task with the single network, the agent could find the global optimum thanks to the knowledge of swinging up the base link acquired during the height task. That is, we suggested that the common skills among tasks can be utilized implicitly because the modules for all the tasks are not completely separated, although the catastrophic forgetting would occur in the single network.

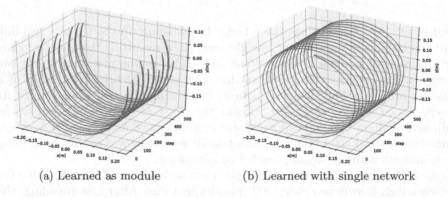

(a) Learned as module   (b) Learned with single network

Fig. 6. Loci of tip of base link in Acrobot: (a) when all the tasks are learned by the respective networks separately (or the velocity task is learned firstly), the agent failed to swing up the base link, in other words, the agent trapped into a local optimum; (b) when all the tasks are learned by the single networks, and the velocity task is learned after the height task, the skill for swinging up the base link acquired when the height task helped the agent to acquire the global optimum.

Next, learning curves were depicted in Fig. 7. The left halves of these curves denoted the first-round learning of the given tasks and the right halves were for the second-round learning. That is, it can be said that the catastrophic forgetting is mitigated as the right halves are horizontal with high values. Thanks to EWC, both networks succeeded in keeping the parameters important for the previous tasks to be mostly constant, while their performances were different.

As can be seen in the left halves of the learning curves, the proposed fractal network was able to obtain high values even at the first-round learning. Accordingly, the right halves (i.e., at the second-round learning) were basically higher than the baseline. This is because the dynamics of the fractal network is well arranged, so firing the neurons that does not correspond to the commanded task is hardly observed, unlike the dynamics of the random network, where the

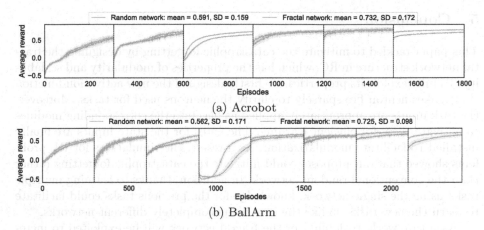

(a) Acrobot

(b) BallArm

Fig. 7. Learning curves of simulations with multiple tasks: the commanded task *b* was changed before and after dashed lines; the proposal with the fractal network outperformed the baseline with the random network in both environments, although it could not hold the essential parameters for the multiple tasks completely.

neurons assigned to the tasks would scattered while causing many unnecessary firings. Such a few firing unrelated to the commanded task resulted in minimizing the number of parameters should be stored by EWC and adding new tasks. That is, our proposal is easy to learn new tasks incrementally and while mitigating the catastrophic forgetting (Fig. 8).

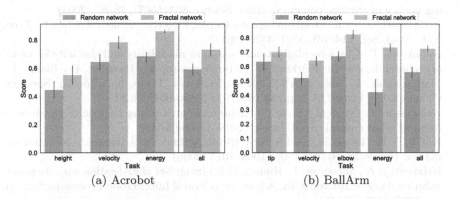

(a) Acrobot

(b) BallArm

Fig. 8. Scores in terms of mitigating the catastrophic forgetting: black line segments on bars denote 95% confidence interval; the proposed fractal network outperformed the conventional random network on all the tasks; in particular, significant difference in BallArm was certainly confirmed.

5   Conclusion

This paper tackled to mitigate the catastrophic forgetting by designing the fractal network structure in RC, which has the properties of modularity and scalability. To fully exploit its properties, we firstly designed the new activation function that makes neuron fire sparsely to clarify the neurons used for tasks. Moreover, the task inputs are appropriately applied as bias into the corresponding modules to activate them for efficient learning. The gates for the state inputs are finally installed to facilitate modularization. As a result, the simulations for RL problems showed that our proposal could mitigate the catastrophic forgetting rather than the conventional random network. In addition, thanks to learning multiple tasks using the single network, knowledge for the previous tasks could facilitate to learn the new tasks, unlike the case using completely different networks.

As future work, scalability of the fractal network will be exploited to increment the number of tasks to be learned in real autonomous robots.

References

1. Bartumeus, F., da Luz, M.E., Viswanathan, G., Catalan, J.: Animal search strategies: a quantitative random-walk analysis. Ecology **86**(11), 3078–3087 (2005)
2. Ellefsen, K.O., Mouret, J.B., Clune, J.: Neural modularity helps organisms evolve to learn new skills without forgetting old skills. PLoS Comput. Biol. **11**(4), e1004128 (2015)
3. French, R.M.: Catastrophic forgetting in connectionist networks. Trends Cogn. Sci. **3**(4), 128–135 (1999)
4. Jaeger, H., Haas, H.: Harnessing nonlinearity: predicting chaotic systems and saving energy in wireless communication. Science **304**(5667), 78–80 (2004)
5. Kirkpatrick, J., et al.: Overcoming catastrophic forgetting in neural networks. Proc. Nat. Acad. Sci. **114**(13), 3521–3526 (2017)
6. Kobayashi, T.: Check regularization: combining modularity and elasticity for memory consolidation. In: Kůrková, V., Manolopoulos, Y., Hammer, B., Iliadis, L., Maglogiannis, I. (eds.) ICANN 2018. LNCS, vol. 11140, pp. 315–325. Springer, Cham (2018). https://doi.org/10.1007/978-3-030-01421-6_31
7. Kobayashi, T.: Student-t policy in reinforcement learning to acquire global optimum of robot control. Appl. Intell. (2019, Online first)
8. Konda, V.R., Tsitsiklis, J.N.: Actor-critic algorithms. In: Advances in Neural Information Processing Systems, pp. 1008–1014 (2000)
9. Krizhevsky, A., Sutskever, I., Hinton, G.E.: ImageNet classification with deep convolutional neural networks. In: Advances in Neural Information Processing System, pp. 1097–1105 (2012)
10. Luo, J., Edmunds, R., Rice, F., Agogino, A.M.: Tensegrity robot locomotion under limited sensory inputs via deep reinforcement learning. In: IEEE International Conference on Robotics and Automation, pp. 6260–6267. IEEE (2018)
11. McCloskey, M., Cohen, N.J.: Catastrophic interference in connectionist networks: the sequential learning problem. In: Psychology of Learning and Motivation, vol. 24, pp. 109–165. Elsevier, Amsterdam (1989)
12. Peters, J., Schaal, S.: Natural actor-critic. Neurocomputing **71**(7–9), 1180–1190 (2008)

13. Remaki, L., Cheriet, M.: KCS-new kernel family with compact support in scale space: formulation and impact. IEEE Trans. Image Process. **9**(6), 970–981 (2000)
14. Rozenfeld, H.D., Havlin, S., Ben-Avraham, D.: Fractal and transfractal recursive scale-free nets. New J. Phys. **9**(6), 175 (2007)
15. Sutton, R.S., Barto, A.G.: Reinforcement Learning: An Introduction. MIT Press, Cambridge (1998)
16. Tsurumine, Y., Cui, Y., Uchibe, E., Matsubara, T.: Deep reinforcement learning with smooth policy update: application to robotic cloth manipulation. Robot. Auton. Syst. **112**, 72–83 (2019)
17. Van Seijen, H., Mahmood, A.R., Pilarski, P.M., Machado, M.C., Sutton, R.S.: True online temporal-difference learning. J. Mach. Learn. Res. **17**(145), 1–40 (2016)
18. Velez, R., Clune, J.: Diffusion-based neuromodulation can eliminate catastrophic forgetting in simple neural networks. PLoS ONE **12**(11), e0187736 (2017)
19. Williams, R.J.: Simple statistical gradient-following algorithms for connectionist reinforcement learning. Mach. Learn. **8**(3–4), 229–256 (1992)
20. Zenke, F., Poole, B., Ganguli, S.: Continual learning through synaptic intelligence. In: International Conference on Machine Learning, pp. 3987–3995 (2017)

Continuous Blood Pressure Estimation Through Optimized Echo State Networks

Giuseppe Franco[2](✉), Luca Cerina[1], Claudio Gallicchio[2], Alessio Micheli[2],
and Marco Domenico Santambrogio[1]

[1] Dipartimento di Elettronica, Informazione e Bioingegneria,
Politecnico di Milano, Milan, Italy
{luca.cerina,marco.santambrogio}@polimi.it
[2] Department of Computer Science, University of Pisa, Pisa, Italy
g.franco4@studenti.unipi.it, {gallicch,micheli}@di.unipi.it

Abstract. Technology is pushing cardiology towards non-invasive recording of continuous blood pressure, with methods that do not require the insertion of a pressure transducer in the Aorta. Although novel analyses based on the Electrocardiogram (ECG) and Photoplethysmography (PPG) provided an elegant model of the interaction between the heart and blood vessels necessary estimate systolic/diastolic points, these methods lack long-term stability and require intermittent re-calibrations. On the other hand, time-series centric algorithms are potentially able to map this coupling in a continuous way. Recurrent Neural Networks (RNN) in particular shown high accuracy and stability. We propose here a system that automatically optimizes, among other hyper-parameters, the input scaling of ECG and PPG signals individually, exploiting an efficient Bayesian Optimization process, on a Echo State Networks (ESN), to quantify continuous blood pressure. Compared to other RNN architectures, the ESN provide a faster training time and lower computational constraints that may allow the deployment on an embedded monitoring device. Preliminary results showed an accuracy of 80% and 98% in terms of Normalized Root Mean Squared Error and Median Symmetric Accuracy, respectively. Considering peaks' estimation, the system achieves grade A (British Hypertension Society) for both Systolic and Diastolic points, making it comparable with clinical recommended devices.

Keywords: Echo State Network · Bayesian Optimization · Continuous Blood Pressure Estimation

1 Introduction

Among all medical parameters, one of the most valuable is certainly the Blood Pressure (BP): it provides insights on overall hemodynamic health and it can be helpful to prognosticate many preventable illnesses, such as brain strokes, coronary and kidney diseases. Although BP is inherently a continuous signal, the points of utmost clinical interest are the systolic peak and the diastolic valley,

© Springer Nature Switzerland AG 2019
I. V. Tetko et al. (Eds.): ICANN 2019, LNCS 11731, pp. 48–61, 2019.
https://doi.org/10.1007/978-3-030-30493-5_5

i.e. its maximum and minimum, respectively. These points are usually measured using a sphygmomanometer, which collapses vessels through an inflatable cuff and determines BP by measuring heartbeat oscillations. However, the analysis of the continuous signal could provide many interesting insights, such as the autonomic control of baroreflex under general anesthesia [7], or it might allow to detect dangerous hypotensive events in ambulatory settings [11]. This is usually done inserting a catheter inside an artery, a really invasive procedure, or using devices such as FinaPres [21], which need intermittent calibrations with regular cuffs and are uncomfortable to wear for more than a few minutes. Another indirect way to measure BP is to monitor Electrocardiogram (ECG) and Photoplethysmography (PPG) simultaneously. These signals are generally easier to obtain with non-expensive (possibly wearable) devices, and stay accurate in the long-term. However, although their relationship with BP is known since a long time, it is very difficult to define the exact mapping that links the three signals. In fact, it depends on individual physiological parameters (e.g. arterial thickness and blood density) which are quite difficult to obtain for each user, and are heavily time-dependent.

As we will see in Sect. 2, many alternatives have been proposed to overcome this obstacle and measure BP non-invasively, with mixed results. Among these, solutions based on Recurrent Neural Networks (RNN), and in particular on Long-Short Term Memory (LSTM), are capable of achieving among the best results in the field. Although promising, these can not be considered a final solution to the problem, given the complexity of the structure used, and, in general, the computational load of the LSTM, which prevents the ability to move the computation on wearable devices, that would allow an on-site and non-invasive monitoring of the blood pressure. Moreover, all the systems developed are limited to the prediction of the Systolic and Diastolic Peaks, without the advantages given by a continuous estimation of the BP.

In this scenario, Reservoir Computing (RC) [18,26], and Echo State Networks (ESNs) [12,13] in particular, represent a valuable alternative to express this relationship through the dynamical richness of reservoir states. The intrinsic ability of the ESN to simulate non-linear dynamic systems is paired by their efficiency and low computational complexity. This could be the key to find and approximate the hidden relationship between PPG-ECG and BP, allowing also to embed the computation on devices with limited resources [2]. Nonetheless, the full capabilities of the ESN approach is still being explored: its extensive usage in real-world problems often encounters practical bottlenecks. One of the main reasons is related to the necessity to fine tune the hyper-parameters that characterize the network, which in a randomized context have a decisive influence on the performance, according to the specific case study or learning task-at-hand.

In order to fully exploit the potential of ESN and reducing the effort required for their configuration, we propose EchoBay, a library that automatizes the process for creating, configuring, and testing an ESN. Moreover, exploiting the Bayesian Optimization (BO) approach, it allows finding the best configuration of the network for each specific case study, reducing exponentially the time required to have an optimally configured ESN.

To summarize, the main contribution of this paper is the development of an ESN-tool that allows a reliable continuous estimation of BP from ECG and PPG time-series, at the same time easing and speeding-up the overall set-up process for the network.

The rest of the paper is organized as follow: in Sect. 2 is presented an overview regarding the developed solution for BP monitoring. ESN are introduced in Sect. 3. The EchoBay library and its advantages are outlined in Sect. 4, while Sect. 5 presents the experimental setting adopted for this work, and the results obtained. Finally, the conclusions are presented in Sect. 6.

2 Related Work

We can group the related literature regarding blood pressure estimation into hierarchies: continuous estimation vs. systolic/diastolic level, and feature-based vs. signal-based. With regards to the signal sources employed, part of the literature focuses on PPG as the less invasive signal available. The authors in [1] tried to infer the PPG-to-BP dynamical system as a continuous ARX model. Conversely, in [6] and [15] the PPG was coupled with morphological features and 2nd order derivative.

The most studied coupling, however, is given by the ECG, which represents the electrical activity of the *source* of blood pressure, and the PPG, which is its direct expression after the pressure wave traveled through the blood vessels. The Pulse Transit Time (PTT) commonly expresses this relationship as the time interval that measures how fast the pulse propagates from the heart to the peripheral site. Recently [27] mapped BP and PTT relationship with a non-linear model, obtaining slightly better results compared to simpler linear or exponential models. To improve the model, both [5] and [23] introduced the Photoplethysmographic Intensity Ratio (PIR) indicator, the first as a factor of direct proportionality, the latter in a logarithmic model. Nevertheless, the PTT method remains controversial, due to a lower accuracy observed in real-life situations and the need of intermittent re-calibrations with cuff methods.

As in other time-series based fields, RNN are outdating feature-based methods in favour of more *holistic* models. For example, a LSTM network with residual inputs and bidirectional structure is implemented in [24] with optimal results in the estimation of systolic/diastolic points and long-term resilience. Similarly, the authors in [16] pre-processed the Arterial Blood Pressure (ABP) signal to remove unnecessary information and fed ECG and PPG to the LSTM network. Lastly, the authors in [25] exploits a complex hierarchical model consisting of both fully-connected Feed-Forward Neural Networks, to extract signal features, and LSTM network to build temporal relationships with promising results.

Although RNN are capable of achieving the best results in terms of accuracy, the current solutions are limited by their complexity and strict requirements in terms of computational power required by the architecture where they are implemented. In this scenario, an ESN solution could be able to grant the great performance typical of a RNN, while providing the efficiency and lightweight computation proper of the RC paradigm.

3 Echo State Networks

The main principle of ESN [12] is to have a large untrained recurrent reservoir that performs a non-linear encoding of the time-varying input into a high dimensional stat representation. Then, a feed-forward readout component uses the reservoir states to compute the output. The operation performed by the ESN is characterized by the following three matrices:

- **Win:** an $Nr \times Nu$ input weight matrix;
- **Wr:** an $Nr \times Nr$ recurrent reservoir weight matrix;
- **Wout:** an $No \times Nr$ readout weight matrix.

Here Nu denotes the dimensionality of the input signal, Nr represents the number of units in the reservoir, while No is the dimensionality of the output signal. Resorting to the case of leaky integrate recurrent units [13], the state update equation that rules the reservoir computation, and the output function computed by the readout tool, are described by the following equations[1]:

$$\begin{cases} \mathbf{x}(t) = (1 - \alpha)\,\mathbf{x}(t-1) + \alpha\,f(\mathbf{Wr}\,\mathbf{x}(t-1) + \mathbf{Win}\,\mathbf{u}(t)) \\ \mathbf{y}(t) = \mathbf{Wout}\,\mathbf{x}(t), \end{cases} \tag{1}$$

where, at for each time step t, $\mathbf{u}(t) \in \mathbb{R}^{Nu}$ represents the input signal, $\mathbf{x}(t) \in \mathbb{R}^{Nr}$ is the reservoir state, and $\mathbf{y}(t) \in \mathbb{R}^{No}$ is the output signal. Moreover, $f(\cdot)$ denotes the activation function for the reservoir units, and in our case corresponds to *tanh*. The term $\alpha \in (0, 1]$ represents the *leaky factor*, which controls the dynamic of the reservoir; in particular, a low value of alpha means that the reservoir states will vary slowly and thus could be better suited to handle signals sampled at very high frequencies. To setup initial conditions for the reservoir dynamics described in the first line of Eq. 1, the state is typically initialized to a null vector, i.e. $\mathbf{x}(0) = \mathbf{0}$. Besides, to account for initial transients of the dynamics, a first set of washout states are typically discarded for training purposes.

The parameters of the reservoir component, i.e. the weight values in **Win** and **Wr**, are left untrained after initialization based on asymptotic stability constrained expressed by the Echo State Property (ESP) [12]. This step is typically and accomplished by controlling algebraic properties of the recurrent weight matrix. A common practice in RC applications is to control the value of the effective spectral radius, hereafter denoted by ρ, and representing the largest among the moduli of the eigenvalues of matrix $(1-\alpha)\,\mathbf{I} + \alpha\,\mathbf{Wr}$. Accordingly, the elements in **Wr** are initialized randomly from a uniform distribution in $[-1, 1]$, and then are re-scaled to achieve a desired value of ρ, typically exploring values

---

[1] Bias terms are omitted here for the ease of notation.

not exceeding 1 (in line with the necessary condition for the ESP). However, values of ρ slightly exceeding 1 are sometimes considered, possibly leading to good reservoir initializations (see e.g. [8,26]). The elements in **Win** are commonly randomly initialized from a uniform distribution in $[-\omega^{in}, \omega^{in}]$, where ω^{in} is an input-scaling parameter.

The only part of the network that is trained is the readout layer. This means that only the elements in the output weight matrix **Wout** need to be adjusted to fit a training set, a process that is commonly performed in closed form by means of pseudo-inversion or ridge regression [18]. In the latter case, a further hyper-parameter is required, the readout regularization coefficient (λ).

Overall, given the randomization aspect associated especially with the reservoir construction, a successful application of the ESN approach entails the careful control of several hyper-parameters [17]. In this regard, the most relevant ones to consider include: the number of reservoir units Nr, the spectral radius ρ, the input scaling ω^{in}, the leaky factor α and the readout regularization λ. The appropriate selection of these hyper-parameters is fundamental since they need to match the intrinsic dynamic characteristics of the learning task-at-hand, thus related to both input and target signals. The usual practice in literature is to perform a grid search on a sub-selection of the hyper-parameter space and choose the one that (across multiple trials) achieves the best performance. This procedure, although potentially exhaustive, is very time consuming, and requires certain expertise from the user-side, especially in the selection of what parameters to optimize and what values to test.

For the sake of completeness, we finally mention that other minor hyper-parameters (typically left apart from optimization) are related to the density of the reservoir, which is usually very sparse in order to speed-up the computation and the length of initial transient to discard (which is commonly in the order of hundreds to thousand time steps).

4 EchoBay: Automatic Hyper-parameter Configuration for ESN

In order to overcome the shortcomings associated with the grid search approach, in our work the BO approach [3] is used. Instead of testing blindly every possible combination of hyper-parameters, the algorithm exploits the knowledge gained by each observation to learn which are the most promising areas to explore, and which one to avoid from further testing. In this way it is possible to perform a smart sampling of the hyper-parameter space, thus reducing dramatically the number of trials required to find the optimal configuration. Another great advantage is given by the possibility to select continuous ranges of value where to perform the search, instead of particular points, which remove possible biases associated to how dense the hyper-parameter space is mapped. More details about the relationship between BO and ESN can be found in [28].

Overall, the BO approach is useful since it allows to reduce the amount of expertise related to the ESN hyper-parameters and their influence on one another. Starting from this concept, we wanted to reduce to the minimum the effort on the user-side when approaching a new problem that can be solved with ESN. In fact, although there are many different libraries for performing BO (such as www.pypi.org/project/bayesian-optimization/ in python and www.mathworks.com/help/stats/bayesian-optimization-algorithm.html in Matlab), the user has still to write the code related to the creation of the particular ESN model and integrate it with the particular BO framework chosen. This may represent a difficult challenge, especially for newcomers in the machine learning field.

For these reasons, we designed EchoBay [4], a library for automatic optimization of ESN, exploiting the BO algorithm. The framework allows to cut down the time required for the selection of the optimal configuration, given the smart sampling of the hyper-parameter space, and it eases the processing of creating and testing an ESN. In fact, from the user-side, it is required only to provide the dataset (divided in Training, Validation, and Test Set), and a configuration YAML file, which specifies the location of the dataset, and which hyper-parameters to optimize (and the respective range of search) or to keep fixed.

At the end of the optimization process, the system outputs the optimal hyper-parameter configuration, together with the accuracy obtained on the Test Set, and different realizations of the matrices created with that configuration. Another benefit given by the BO approach is that there is no more an exponential relationship between the number of hyper-parameters to optimize and all their possible combinations that need to be tested, as with grid search.

For this reason, and starting from the considerations made in [17], in this case study we decided to evaluate the effect of optimizing a different ω^{in} for each dimension of the input signal. In particular, the matrix **Win** was split along its columns and initialized it with the following setup:

$$\mathbf{Col_i} = \omega_i^{in} \, rand(Nr), \qquad\qquad i = ECG, PPG, Bias$$
$$\mathbf{Win} = [Col_{ECG}, Col_{PPG}, Col_{Bias}]$$

(2)

where $rand(N)$ is a function that generates N random values uniformly taken in the range $[-1, 1]$, and ω_i^{in} represent the value of scale associated to the ith input dimension.

5   Experimental Analysis

This section presents the dataset used for this paper (Sect. 5.2), the experimental settings for our tests (Sect. 5.3), and an overview of the results obtained (Sect. 5.4).

5.1   Relationship Between the Observed Physiological Signals

As we stated before in Sect. 2, the electrocardiogram, the photoplethysmogram, and the blood pressure wave are tied together as they have complementary roles in the representation of the heart activity. In Fig. 1 we can observe the ECG, which is characterized by sudden variations due to the high speed of the electrical wave in the heart. Subsequently, the contraction of the heart propagates a pressure wave which travels through the vessels. The arterial blood pressure is measured using a pressure transducer; some subjects exhibit also a secondary flexion, called dicrotic notch, which measures the transient increase of pressure after the closure of the aortic valve. Lastly, the pressure wave reaches the arterioles and it could be measured non-invasively through the reflectance of oxygenated hemoglobin in the blood vessels. Compared to the arterial BP, the PPG signal is more shallow, since arteria acts as a low-pass filter due to their elasticity, therefore, it is quite difficult to determine the original arterial BP signal using only the PPG.

It could be observed in Fig. 1 that the systolic point (red dot) moves forward in time in the different signals. This delay is dependent by the time necessary to the pressure wave to run through the blood vessels. This dynamic temporal relation, coupled with unknown mechanic properties of the blood vessels, represent one of the most critical points in the determination of the original arterial blood pressure signal.

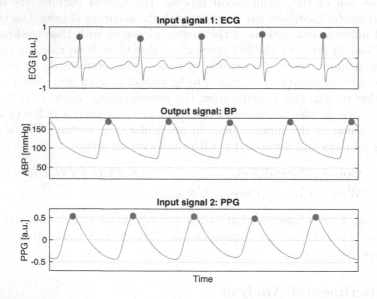

Fig. 1. An example of the input signals, ECG and PPG, and the arterial blood pressure (ABP) *output* signal. The red dot represent the systolic peak in each signal. (Color figure online)

5.2   Dataset and Pre-processing

The dataset used in our experiments is a pre-processed version of the MIMIC II dataset [10], made available by [14], where part of the noise sources and artifacts were removed. The dataset contains the three signals of interest, PPG, ECG, and BP, sampled at 125 Hz and synchronized together. The dataset is divided into five parts, each one composed by 3000 fragments of the signal of varying length (from 1000 to over 60000 samples). From a first analysis, it was clear that the dataset required extensive pre-processing and for this reason, we were forced to focus just on the analysis of the first part of the dataset. Many of the signals were very noisy and impossible to use for this analysis. On the other hand, looking at the shape and range of the signals, it was clear that multiple consecutive fragments belonged to the same subject, and then were divided due to noise in-between. For our analysis, we re-grouped adjacent fragments that belonged to the same subject, ending up with 248 unique subjects. Then the following steps were carried out for each subject individually: The input signals (ECG, PPG) were first filtered using a wavelet transform filter [22] with tunable Q-factor. The filter allows to expand the signal in multiple wavelet sub-bands and then precisely remove noise factors such as baseline wanders or random spikes. The filtered signals were then rescaled in the range $[-1, 1]$, while the output signal (BP) has been left in its original scale. The tracks were divided in sequences whose length would correspond to a 30 s acquisition. We set a minimum and a maximum number of sequences for each subject, respectively 6 and 18. Starting from this division, 55% of sequences were randomly picked for the training set, 20% for validation, and the remaining 25% for the test set. At the end of each sequence, the internal state of the ESN is reset and a washout period is added.

The continuous prediction of BP can be modeled as regression task on time-series, where for each time step the inputs of the system are the ECG and the PPG signal, and the output is the predicted BP value.

5.3 Experimental Settings

In our work we focused on the optimization of the major ESN hyper-parameters exploring the ranges reported below:

- Nr: range $[50, 400]$;
- ρ: range $[0.5, 1.1]$;
- ω_i^{in}: range $[0.1, 1]$, i = ECG, PPG, Bias;
- α: range $[0.1, 1]$;
- λ: range $[0, 0.1]$;

Other minor hyper-parameters remained fixed, such as the number of washout samples after each reset (1000 samples) and the density of the reservoir matrix **Wr** (10%).

5.4 Results

For our experiments, two different optimization processes have been carried out, using two objective functions. In particular, the two Error Functions $E(\mathbf{y}, \mathbf{z})$

used are the Normalized Root Mean Squared Errors (NRMSE) and the Median Symmetric Accuracy (MSA). The NRMSE is defined as:

$$NRMSE = \sqrt{\left(\frac{1}{N}\sum_{t=1}^{N}(y_t - z_t)^2\right)/\left(var(\mathbf{z})\right)}, \quad (3)$$

where y_t and z_t are respectively the tth predicted and actual values. The measure is normalized with respect to the variance of the actual values.

The MSA is defined as follows:

$$MSA = (exp(M(|Q_t|)) - 1),$$
$$Q_t = ln(\frac{y_t}{z_t}), \quad (4)$$

where the function $M(\cdot)$ represent the median operator. The MSA measure is a particular metric that can be used when $\mathbf{y}$ is non-negative, and represents a more robust version of the Mean Absolute Percentage Error (MAPE) estimator [19]: compared to the MAPE and NRMSE metrics, the MSA function is preferable for Bayesian Optimization as it can be bounded to a percentage error ignoring the magnitude of data observed, it is symmetric against over/under-prediction, and it's resistant to outliers. Since our objective is to maximize the Fitness Function $F(\mathbf{y}, \mathbf{z})$, we define it as $F(\mathbf{y}, \mathbf{z}) = 1 - E(\mathbf{y}, \mathbf{z})$.

During the optimization phase, for each configuration of hyper-parameters, five different ESN realizations have been tested (with different random seeds for initialization), and then the average fitness among them is used as representative for the particular configuration.

Continuous Blood Pressure Estimation. In order to assess whether the optimization of multiple ω^{in} is beneficial for our specific case study, following the paradigm described in Sect. 5.3, we conducted two optimization processes for each error function: **Multi** ω^{in}, where we optimized multiple ω^{in}, one for each input; **Single** ω^{in}, where a single ω^{in} was optimized for all inputs. The fitness values of each subject are the average of guesses on the test set, which are then averaged to described the population results. In Table 1 are presented the results obtained with the two error functions. As it is clear from the table, the fitness values in terms of MSA are considerably higher compared to those obtained with NRMSE. The main reason behind this regards the fact that the errors in the predicted BP are mainly due to noise in the input signals that cause sudden and brief spikes in the predicted output, whereas the rest of the prediction values are very accurate, as it is possible to see in Fig. 2, where an extract of the two input signals and the true and predicted values of BP are presented. The median operation in the MSA allows to ignore these outliers, whilst penalizing more other errors. Focusing on the comparison between the two different optimization strategies for the input scaling, the *p-value* related to the NRMSE shows a statistical significant increase of performance (p = 0.058, with right-tail Wilcoxon Signed Rank Test) when using the **Multi** ω^{in} strategy. The increase

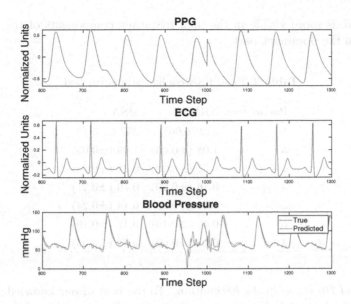

Fig. 2. An example of the two input signals, and the output signal, together with the prediction computed by ESN model

in performance is present, although less evident, with MSA with respect to when using NRMSE as the error function, due to the considerations made above.

Table 1. Performance, averaged across subjects, with single and multiple ω^{in}.

Error function	Multi ω^{in}	Single ω^{in}
NRMSE	79.85 ($\pm$9.25)	78.60 ($\pm$9.86)
MSA	97.96 ($\pm$1.04)	97.83 ($\pm$1.17)

In Table 2 are reported the values of the optimal configuration of hyper-parameters reached with the **Multi** ω^{in} optimization for both error functions, averaged across subjects. It is worth noticing that the parameter ρ tends to assume values greater than 1 and close to the upper bound of the explored range, with little to no variance. The same tendency is reflected with a bit more variance for the reservoir size **Nr**, showing the tendency of the system to maximize its *memory*. Considering the rest of hyper-parameters, the two optimization process reaches more or less the same configurations, which can be considered a sign of the *stability* of the EchoBay approach. Regarding the two final values of ω^{in}, we can see that ω^{in}_{PPG} is slightly larger than ω^{in}_{ECG}. This is a sign of the relation between the PPG and BP signals, which are both an expression of the same physical phenomenon. Complementary, the heart's activity that determines the

ECG signal, is more visible in the high frequency components of the BP wave rather than the dominant part.

Table 2. Final ESN configurations reached.

Parameters	NRMSE	MSA
Nr	342 ($\pm$66)	357 ($\pm$48)
ρ	1.08 ($\pm$0.06)	1.09 ($\pm$0.03)
ω_{Bias}^{in}	0.57 ($\pm$0.28)	0.71 ($\pm$0.24)
ω_{PPG}^{in}	0.76 ($\pm$0.20)	0.76 ($\pm$0.21)
ω_{ECG}^{in}	0.45 ($\pm$0.23)	0.48 ($\pm$0.24)
α	0.46 ($\pm$0.19)	0.47 ($\pm$0.19)
λ	0.03 ($\pm$0.03)	0.03 ($\pm$0.03)

Systolic and Diastolic Peaks Estimation. To the best of our knowledge, all the RNN-based solutions in literature are limited to the estimation of the Diastolic and Systolic peaks. In order to evaluate how our work compares with the state of the art, we extracted those two quantities as well. In particular, we used the already computed continuous BP estimations obtained using NRMSE as the error function, with the **Multi** ω^{in} strategy, without performing a new learning task. Among the five guesses obtained on the test set, we chose to consider the worst case scenario, thus using the estimation with the largest NRMSE.

After identifying the Systolic and Diastolic peaks, the error was computed in terms of Root Mean Squared Errors (RMSE), since it is one of the most common error measure adopted in literature. The RMSE has a similar structure to the NRMSE, without the normalization with respect to the variance of the true signal.

In our case, the error achieved for Diastolic and Systolic peaks estimation in terms of RMSE are 4.99 and 7.59 respectively, compared to 1.60 (Diastolic) and 2.75 (Systolic) achieved by [16], and 0.52 (Diastolic) and 0.93 (Systolic) achieved by [25]. In both cases an LSTM architecture is used, in the second case paired also with a Feed-Forward neural network, meaning that these proposed solutions are more computationally complex and almost impossible to be implemented on an embedded device. Conversely, the ESN active units have a simple structure and their computational time is minimal, compared to the LSTM case, where each active unit has several gates and weights that need to be evaluated. In fact, as reported in [9], ESN are faster than LSTM with an equivalent number of free parameters, while being able to have a higher number of recurrent units. This factor, together with the intrinsic sparsity of the reservoir, makes ESN the best candidate for computational efficient implementations.

Since our system should also work in a real-case scenario, we compared our results to the standard set by the British Hypertension Society (BHS) [20], which grades BP measurement systems based on the cumulative percentage error (in terms of absolute error) under three thresholds, i.e. at 5, 10, and 15 mmHg (see Table 3).

In order for a device to be *recommended*, it must achieve B or better. According to Table 4, our system is able to reach grade A for both Systolic and Diastolic peaks prediction thus meaning that our proposed solution fulfills the requirements of the BHS for clinical recommended devices.

Table 3. BHS standard.

Grade	5 mmHg	10 mmHg	15 mmHg
A	60%	85%	95%
B	50%	75%	90%
C	40%	65%	85%

Table 4. Comparison with BHS standard.

Error (MAE)	5 mmHg	10 mmHg	15 mmHg	Grade
Systolic peak	77%	91%	95%	A
Diastolic peak	85%	95%	97%	A

6 Conclusion

In this paper we presented a reliable, ESN-based architecture for a continuous, non-invasive estimation of the BP, starting from ECG and PPG signals. The system proposed is based on EchoBay, a library that eases the configuration and optimization process of ESN. In particular, the library exploits the BO approach in order to perform a smart and automatic selection of the ESN hyperparameters, suited for the specific learning task at-hand. In our particular case study, the solution proposed is able to provide a reliable, continuous estimation of BP, capable to achieve an accuracy of 98% in terms of MSA. Relevantly, the developed model reaches grade A in both Systolic and Diastolic peaks estimation, according to the BHS standard, hence providing a valuable result for real-world clinical applications. Moreover, the intrinsic efficiency of ESN allowed to create a lightweight system from a computational point of view, especially compared to other RNN solutions proposed in the literature. This last point could open up the possibility to create a wearable device that performs on-site recording of ECG and PPG, while providing also an accurate BP estimation. Another point that should be investigated regards the possibility, given an appropriate dataset with specific constraints in the acquisition procedure, to create a general, inter-subject model that would not require an individualized training.

References

1. Acciaroli, G., Facchinetti, A., Pillonetto, G., Sparacino, G.: Non-invasive continuous-time blood pressure estimation from a single channel PPG signal using regularized ARX models. In: 40th International Conference of the IEEE Engineering in Medicine and Biology Society (EMBC), pp. 3630–3633. IEEE (2018). https://doi.org/10.1109/EMBC.2018.8512944
2. Bacciu, D., Barsocchi, P., Chessa, S., Gallicchio, C., Micheli, A.: An experimental characterization of reservoir computing in ambient assisted living applications. Neural Comput. Appl. **24**(6), 1451–1464 (2014). https://doi.org/10.1007/s00521-013-1364-4
3. Brochu, E., Cora, V.M., De Freitas, N.: A tutorial on Bayesian optimization of expensive cost functions, with application to active user modeling and hierarchical reinforcement learning. arXiv preprint arXiv:1012.2599 (2010)
4. Cerina, L., Franco, G., Santambrogio, M.D.: Lightweight autonomous Bayesian optimization of echo-state networks. In: Proceedings of ESANN 2019. i6doc (2019)
5. Ding, X., Yan, B.P., Zhang, Y.T., Liu, J., Zhao, N., Tsang, H.K.: Pulse transit time based continuous cuffless blood pressure estimation: a new extension and a comprehensive evaluation. Sci. Rep. **7**(1), 11554 (2017)
6. Duan, K., Qian, Z., Atef, M., Wang, G.: A feature exploration methodology for learning based cuffless blood pressure measurement using photoplethysmography. In: 38th International Conference of the IEEE Engineering in Medicine and Biology Society (EMBC), pp. 6385–6388. IEEE (2016). https://doi.org/10.1109/EMBC.2016.7592189
7. Faes, L., Bari, V., Ranucci, M., Porta, A.: Multiscale decomposition of cardiovascular and cardiorespiratory information transfer under general anesthesia. In: 40th International Conference of the IEEE Engineering in Medicine and Biology Society (EMBC), pp. 4607–4610. IEEE (2018). https://doi.org/10.1109/EMBC.2018.8513191
8. Gallicchio, C.: Chasing the echo state property. In: Proceedings of ESANN 2019. i6doc (2019)
9. Gallicchio, C., Micheli, A., Pedrelli, L.: Comparison between DeepESNs and gated RNNs on multivariate time-series prediction. arXiv preprint arXiv:1812.11527 (2018)
10. Goldberger, A.L., et al.: PhysioBank, PhysioToolkit, and PhysioNet: components of a new research resource for complex physiologic signals. Circulation **101**(23), e215–e220 (2000). https://doi.org/10.1161/01.CIR.101.23.e215
11. Ilies, C., et al.: Detection of hypotension during caesarean section with continuous non-invasive arterial pressure device or intermittent oscillometric arterial pressure measurement. Br. J. Anaesth. **109**(3), 413–419 (2012). https://doi.org/10.1093/bja/aes224
12. Jaeger, H., Haas, H.: Harnessing nonlinearity: predicting chaotic systems and saving energy in wireless communication. Science **304**(5667), 78–80 (2004). https://doi.org/10.1126/science.1091277
13. Jaeger, H., Lukoševičius, M., Popovici, D., Siewert, U.: Optimization and applications of echo state networks with leaky-integrator neurons. Neural Netw. **20**(3), 335–352 (2007). https://doi.org/10.1016/j.neunet.2007.04.016
14. Kachuee, M., Kiani, M.M., Mohammadzade, H., Shabany, M.: Cuff-less high-accuracy calibration-free blood pressure estimation using pulse transit time. In: 2015 IEEE International Symposium on Circuits and Systems (ISCAS), pp. 1006–1009. IEEE (2015). https://doi.org/10.1109/ISCAS.2015.7168806

15. Liu, M., Po, L.M., Fu, H.: Cuffless blood pressure estimation based on photo-plethysmography signal and its second derivative. Int. J. Comput. Theory Eng. **9**(3), 202 (2017). https://doi.org/10.7763/IJCTE.2017.V9.1138

16. Lo, F.P.W., Li, C.X.T., Wang, J., Cheng, J., Meng, M.Q.H.: Continuous systolic and diastolic blood pressure estimation utilizing long short-term memory network. In: 39th International Conference of the IEEE Engineering in Medicine and Biology Society (EMBC), pp. 1853–1856. IEEE (2017). https://doi.org/10.1109/EMBC.2017.8037207

17. Lukoševičius, M.: A practical guide to applying echo state networks. In: Montavon, G., Orr, G.B., Müller, K.-R. (eds.) Neural Networks: Tricks of the Trade. LNCS, vol. 7700, pp. 659–686. Springer, Heidelberg (2012). https://doi.org/10.1007/978-3-642-35289-8_36

18. Lukoševičius, M., Jaeger, H.: Reservoir computing approaches to recurrent neural network training. Comput. Sci. Rev. **3**(3), 127–149 (2009). https://doi.org/10.1016/j.cosrev.2009.03.005

19. Morley, S.K., Brito, T.V., Welling, D.T.: Measures of model performance based on the log accuracy ratio. Space Weather **16**(1), 69–88 (2018). https://doi.org/10.1002/2017SW001669

20. O'brien, E., Waeber, B., Parati, G., Staessen, J., Myers, M.G.: Blood pressure measuring devices: recommendations of the European society of hypertension. BMJ **322**(7285), 531–536 (2001). https://doi.org/10.1136/bmj.322.7285.531

21. Polito, M.D., Farinatti, P.T., Lira, V.A., Nobrega, A.C.: Blood pressure assessment during resistance exercise: comparison between auscultation and finapres. Blood Press. Monit. **12**(2), 81–86 (2007). https://doi.org/10.1097/MBP.0b013e32809ef9f1

22. Selesnick, I.W.: Wavelet transform with tunable Q-factor. IEEE Trans. Sig. Process. **59**(8), 3560–3575 (2011). https://doi.org/10.1109/TSP.2011.2143711

23. Sharifi, I., Goudarzi, S., Khodabakhshi, M.B.: A novel dynamical approach in continuous cuffless blood pressure estimation based on ECG and PPG signals. Artif. Intell. Med. (2018). https://doi.org/10.1016/j.artmed.2018.12.005

24. Su, P., Ding, X., Zhang, Y., Liu, J., Miao, F., Zhao, N.: Long-term blood pressure prediction with deep recurrent neural networks. In: 2018 IEEE EMBS International Conference on Biomedical Health Informatics (BHI), pp. 323–328, March 2018. https://doi.org/10.1109/BHI.2018.8333434

25. Tanveer, M., Hasan, M., et al.: Cuffless blood pressure estimation from electro-cardiogram and photoplethysmogram using waveform based ANN-LSTM network. arXiv preprint arXiv:1811.02214 (2018)

26. Verstraeten, D., Schrauwen, B., d'Haene, M., Stroobandt, D.: An experimental unification of reservoir computing methods. Neural Netw. **20**(3), 391–403 (2007). https://doi.org/10.1016/j.neunet.2007.04.003

27. Wibmer, T., et al.: Pulse transit time and blood pressure during cardiopulmonary exercise tests. Physiol. Res. **63**(3) (2014). https://doi.org/10.1038/s41598-017-11507-3

28. Yperman, J., Becker, T.: Bayesian optimization of hyper-parameters in reservoir computing. arXiv preprint arXiv:1611.05193 (2016)

Reservoir Topology in Deep Echo State Networks

Claudio Gallicchio$^{(\boxtimes)}$ and Alessio Micheli

Department of Computer Science, University of Pisa,
Largo B. Pontecorvo, 3, 56127 Pisa, Italy
{gallicch,micheli}@di.unipi.it

Abstract. Deep Echo State Networks (DeepESNs) recently extended
the applicability of Reservoir Computing (RC) methods towards the field
of deep learning. In this paper we study the impact of constrained reservoir topologies in the architectural design of deep reservoirs, through
numerical experiments on several RC benchmarks. The major outcome
of our investigation is to show the remarkable effect, in terms of predictive performance gain, achieved by the synergy between a deep reservoir
construction and a structured organization of the recurrent units in each
layer. Our results also indicate that a particularly advantageous architectural setting is obtained in correspondence of DeepESNs where reservoir
units are structured according to a permutation recurrent matrix.

Keywords: Deep Echo State Networks · Deep Reservoir Computing ·
Reservoir topology

1 Introduction

Reservoir Computing (RC) [21,28] delineates a class of Recurrent Neural Network (RNN) models based on the idea of separating the non-linear dynamical
component of the network, i.e. the recurrent hidden reservoir layer, from the
feed-forward linear readout layer. The reservoir is initialized randomly under
stability constraints and then is left untrained, leaving the burden of training
to fall only on the readout part of the architecture, hence resulting in a strikingly efficient approach to RNN design. In this context, the Echo State Network
(ESN) model [16,18] is a popular realization of the RC paradigm based on implementing the reservoir in terms of a discrete-time non-linear dynamical system.
Being featured by untrained dynamics, ESNs represent an important tool to
understand and characterize the operation and potentialities of recurrent neural
models. Shaping the reservoir architecture in order to achieve desired properties
and optimized performance in applications, even in the absence of training of
the recurrent connections, is one of the key goals of RC research [9].

In this paper we bring together two major trends in the area of ESN architectural studies. The first one focuses on the pattern of connectivity among the

© Springer Nature Switzerland AG 2019
I. V. Tetko et al. (Eds.): ICANN 2019, LNCS 11731, pp. 62–75, 2019.
https://doi.org/10.1007/978-3-030-30493-5_6

recurrent units. In this case, the aim is to constrain the random reservoir initialization process towards topologies that determine specific algebraic properties of the resulting recurrent weight matrices. A relevant class of reservoir variants in this regard is given by ESNs with orthogonal recurrent matrices [15,30], which were shown to lead to improved performance with respect to random reservoirs both in terms of memorization skills and in terms of predictive performance on non-linear tasks. In particular, reservoirs whose structure is based on permutation matrices represent particularly appealing instances of orthogonal ESNs [15,26], entailing a simple and very sparse pattern of connectivity among the recurrent units. Other relevant architectural variants are given by reservoirs structured according to a ring topology or to form a chain of units [24,26]. The second major line of research that we consider regards the construction of hierarchically structured reservoir models. While initial studies in this context focused on composing multiple ESN modules to form ad-hoc architectures [19,27], recent works started analyzing the effects of stacking multiple untrained reservoir layers with the introduction of the DeepESN model [8]. On the one hand, the analysis of DeepESN dynamics contributes to uncover the intrinsic computational properties of deep neural networks in the temporal domain [8,13]. On the other hand, a proper architectural design of deep reservoirs might have a huge impact in real-world applications [12], enabling effective multiple time-scales processing and at the same time preserving the training efficiency typical of RC models.

In this paper we analyze the impact on the predictive performance given by a constrained reservoir topology in DeepESNs. Specifically, we consider deep architectures in which the individual reservoir layers are implemented based on permutation matrices, as well as on ring and on chain topologies. Our study is conducted in comparison to shallow ESN counterparts through numerical experiments on several benchmarks in the RC area.

The rest of this paper is structured as follows. The DeepESN model is introduced in Sect. 2, while the investigated reservoir topologies are described in Sect. 3. The experimental analysis is reported in Sect. 4. Finally, Sect. 5 draws conclusions and delineates future research directions.

2   Deep Echo State Networks

A DeepESN is an RC model in which the reservoir part is organized into a stacked composition of multiple untrained recurrent hidden layers. The external input is propagated only to the first reservoir layer, while each successive level in the deep architecture is fed by the output of the previous one, as graphically illustrated in Fig. 1.

To fix our notation, we use L to indicate the number of layers in the deep reservoir, while N_U and N_Y respectively denote the sizes of input and output spaces. For the sake of simplicity in the presentation of the DeepESN model, here we make the assumption that all the reservoir layers are featured by the same number of units, indicated by N_R. The operation of each reservoir layer can be described in terms of a discrete-time non-linear dynamical system, whose

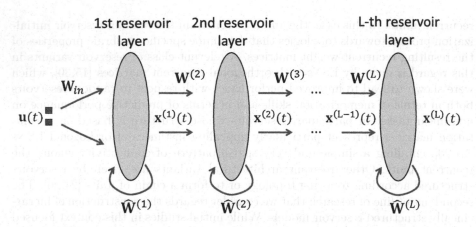

Fig. 1. Hierarchical reservoir architecture in a DeepESN.

state update equation is given in the form of an iterated mapping. In particular, at time-step t, the state of the first layer, i.e. $\mathbf{x}^{(1)}(t) \in \mathbb{R}^{N_R}$, is computed as follows:

$$\mathbf{x}^{(1)}(t) = \tanh(\mathbf{W}_{in}\mathbf{u}(t) + \hat{\mathbf{W}}^{(1)}\mathbf{x}^{(1)}(t-1)), \tag{1}$$

while the state of each successive layer $l > 1$, i.e. $\mathbf{x}^{(l)}(t) \in \mathbb{R}^{N_R}$, is given by:

$$\mathbf{x}^{(l)}(t) = \tanh(\mathbf{W}^{(l)}\mathbf{x}^{(l-1)}(t) + \hat{\mathbf{W}}^{(l)}\mathbf{x}^{(l)}(t-1)). \tag{2}$$

Here, tanh indicates the element-wise application of the hyperbolic tangent non-linearity, $\mathbf{u}(t) \in \mathbb{R}^{N_U}$ represents the external input at time-step t, while $\mathbf{W}_{in}$, $\mathbf{W}^{(l)}$ and $\hat{\mathbf{W}}^{(l)}$ respectively denote the input weight matrix (that modulates the external input stimulation to the first layer), the inter-layer weight matrix for layer l (that modulates the strength of the connections from layer $l-1$ to layer l), and the recurrent reservoir weight matrix for layer l. In both the above Eqs. 1 and 2 we omitted the bias terms for the ease of notation. The interested reader can find in [8] a more detailed description of the deep reservoir equations, framed in the more general context of leaky integrator reservoir units. In order to set up initial conditions for the state update Eqs. 1 and 2, at time-step 0 all reservoir layers are set to a null state, i.e. $\mathbf{x}^{(l)}(0) = \mathbf{0}$ for all $l = 1, \ldots, L$. Given this framework, it is worth noticing that a standard shallow ESN model can be seen as a special case of DeepESN in which a single reservoir layer is considered, i.e. $L = 1$.

As in standard RC approaches, the parameters of the entire reservoir component, i.e. the elements in all the weight matrices in Eqs. 1 and 2, are left untrained after initialization subject to stability constraints. These are required to avoid the system dynamics to fall into unstable regimes, which would make them unsuitable for robust processing of time-series data. In the context of ESNs, the analysis of asymptotic stability is usually described in terms of the Echo State Property (ESP) [16, 21], providing simple algebraic conditions for the initialization of reservoir weight matrices that have been recently extended to cope with the case of deep reservoirs in [6]. Under a practical view-point, the analysis in [6] suggests to carefully control the spectral radius of all the reservoir weight matrices in the deep reservoir. In this paper, we use $\rho^{(l)}$ to denote the spectral radius in layer l, i.e. the largest among the absolute values of the eigenvalues of $\hat{\mathbf{W}}^{(l)}$. A simple initialization procedure for the reservoir of a DeepESN then consists in choosing the elements in $\hat{\mathbf{W}}^{(l)}$ randomly from a uniform distribution on $[-1, 1]$, subsequently re-scaling them to achieve desired values of $\rho^{(l)}$, typically not above unity. Similarly, the elements in $\mathbf{W}_{in}$ and those in $\mathbf{W}^{(l)}$ (for $l > 1$) are initialized randomly from a uniform distribution on $[-1, 1]$, and then are re-scaled to control the input scaling hyper-parameter $\omega_{in} = \|\mathbf{W}_{in}\|_2$, and the set of inter-layer scaling hyper-parameters $\omega_{il}^{(l)} = \|\mathbf{W}^{(l)}\|_2$.

The output of the DeepESN is computed by a simple readout tool, which linearly combines the reservoir representations developed in all the layers of the deep architecture. In formulas, the output at time-step t, denoted as $\mathbf{y}(t) \in \mathbb{R}^{N_Y}$, is computed by the following equation:

$$\mathbf{y}(t) = \mathbf{W}_{out} [\mathbf{x}^{(1)}(t); \ldots; \mathbf{x}^{(L)}(t)], \tag{3}$$

where $\mathbf{W}_{out}$ is the output weight matrix, and $[\mathbf{x}^{(1)}(t); \ldots; \mathbf{x}^{(L)}(t)]$ represents the global deep reservoir state at time-step t, expressed as the concatenation of all the states in the architecture. The elements in $\mathbf{W}_{out}$ represent the only learnable weights of the DeepESN, and are typically adjusted to fit a training set by exploiting non-iterative training algorithms as in the case of standard RC models [21]. Notice that, although different patterns of reservoir-to-readout connectivity are possible [23], the one employed here, where all reservoir layers are used to feed the readout, has the advantage to allow the training algorithms to modulate and exploit differently the variety of representations provided by the different levels in the deep architecture.

A more comprehensive description of the characteristics and advantages of the DeepESN approach can be found in [7], while a constantly updated overview on the advancements achieved in this research field is given in [10]. To date, software implementations of the DeepESN model are made publicly available as libraries for Python[1], Matlab[2] and Octave[3].

3   Reservoir Topology

We consider DeepESN architectural variants where the recurrent weight matrix in each layer l, i.e. $\hat{\mathbf{W}}^{(l)}$, is characterized by a specific structure, according to the topologies described in the following. The resulting patterns of reservoir connectivity are graphically exemplified in Fig. 2.

Sparse: Each reservoir unit is randomly connected to a subset of the others, determining a *sparse* recurrent matrix $\hat{\mathbf{W}}^{(l)}$ (see Fig. 2(a)). This corresponds

---

[1] https://github.com/lucapedrelli/DeepESN.

[2] https://it.mathworks.com/matlabcentral/fileexchange/69402-deepesn.

[3] https://github.com/gallicch/DeepESN_octave.

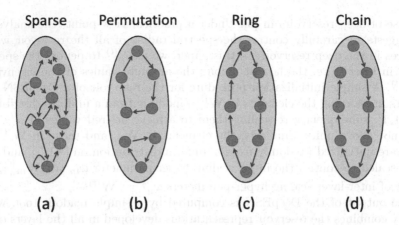

Fig. 2. Reservoir topologies of DeepESN layers.

to a common setting used in RC practice and serves here as baseline for our analysis.

Permutation: The structure of the recurrence matrix $\hat{\mathbf{W}}^{(l)}$ is given by a *permutation* matrix $\mathbf{P}$, i.e. we have:

$$\hat{\mathbf{W}}^{(l)} = \lambda \mathbf{P}, \tag{4}$$

where $\mathbf{P}$ is obtained by randomly permuting the columns of the identity matrix, and λ is a multiplicative constant that specifies the value of the non-zero recurrent weights. In this case, the spectral radius of $\hat{\mathbf{W}}^{(l)}$ is determined by the value of λ, i.e. $\rho^{(l)} = \lambda$. The permutation topology implies that each row and each column of the recurrence matrix have exactly one non-zero element, resulting into a reservoir architecture that presents a variable number of disjoint cyclic structures, as graphically exemplified in Fig. 2(b). The levels in the deep reservoir architecture are allowed to employ different permutations, i.e. the number of cycles in each reservoir layer can be different.

In the context of shallow ESNs, this kind of topology has been empirically studied in [2], where it was shown to achieve good memorization skills at the same time improving the performance of randomly initialized reservoirs in tasks involving non-linear mappings. Interestingly, the permutation topology has been investigated in [15] as a way to implement orthogonal reservoir matrix structures, under the name of Critical ESNs.

Ring: The reservoir units are organized to form a single *ring*, as shown in Fig. 2(c). Accordingly, the recurrent weight matrix $\hat{\mathbf{W}}^{(l)}$ is expressed as:

$$\hat{\mathbf{W}}^{(l)} = \lambda \begin{bmatrix} 0 & 0 & \dots & 1 \\ 1 & 0 & \dots & 0 \\ \vdots & \ddots & \ddots & \vdots \\ 0 & \dots & 1 & 0 \end{bmatrix}, \tag{5}$$

where λ is the value of non-zero recurrent weights, and determines the spectral radius of $\hat{\mathbf{W}}^{(l)}$, i.e. $\rho^{(l)} = \lambda$. The ring topology can be easily seen as a special case of the permutation topology, where the pattern of reservoir connectivity is ruled by the specific permutation matrix in Eq. 5, and the reservoir units are all part of the same cyclic structure.

Reservoirs following this architectural organization have been subject of several studies in literature on shallow RC. Notable instances in this regard are given by the work in [26], in which the ring topology is studied in the context of orthogonal reservoir structures, and by the work in [24], where the study is carried out under the perspective of architectural design simplification for minimum complexity ESN construction. One interesting outcome of previous analysis on the ring topology is that, compared to randomly initialized reservoirs, it shows superior memory capacity that, at least in the linear case, approaches the optimal value [24]. While this optimal memory characterization has been extensively analyzed in literature for the more general class of orthogonal recurrent weight matrices (see e.g. [3,17,30]), the ring topology presents the advantage of a strikingly simple (and sparse) dynamical network construction.

Chain: The recurrent units are arranged in a pipeline, where each unit - except for the first one - receives in input the activation of the previous one, forming a *chain* as in the example in Fig. 2(d). The only non-zero elements in $\hat{\mathbf{W}}^{(l)}$ are located in the lower sub-diagonal, i.e. we have:

$$\hat{\mathbf{W}}^{(l)} = \lambda \begin{bmatrix} 0 & 0 & \dots & 0 \\ 1 & 0 & \dots & 0 \\ \vdots & \ddots & \ddots & \vdots \\ 0 & \dots & 1 & 0 \end{bmatrix}, \tag{6}$$

where as in previous cases λ identifies the value of non-zero weights. Although in this case $\hat{\mathbf{W}}^{(l)}$ is nilpotent and hence its spectral radius is always 0, we still operate on λ to control the magnitude of recurrent weights. As such, with a slight abuse of notation, also in this case we set $\rho^{(l)} = \lambda$. Overall, the chain topology results in a particularly simple design strategy that, from the architectural perspective, applies a further simplification to the ring topology by removing one of the connections between the internal units.

Literature works on shallow ESN models pointed out the merits of reservoir organizations based on a chain topology (also called delay-line reservoirs), as a very simple approach to the architectural design of the network, resulting in a model that is easier to analyze [30] and that leads to comparable or even better performance than standard ESNs [24,26].

4 Experiments

In this section we illustrate the experimental analysis conducted in this paper. Specifically, in Sect. 4.1 we detail the datasets considered and the experimental settings adopted in our work, whereas in Sect. 4.2 we report and discuss the achieved numerical results.

4.1 Datasets and Experimental Settings

In our experiments, we considered benchmark datasets featured by univariate time-series (i.e., $N_U = N_Y = 1$).

The first dataset is obtained from a non-linear auto-regressive moving average system of the 10-th order (NARMA10). At each time-step, the input $u(t)$ comes from a uniform distribution over $[0, 0.5]$, whereas the corresponding target output $y_{tg}(t)$ is given by the following relation:

$$y_{tg}(t) = 0.3\,y_{tg}(t-1) + 0.05\,y_{tg}(t-1) \sum_{i=1}^{10} y_{tg}(t-i) + 1.5\,u(t-10)\,u(t-1) + 0.1. \quad (7)$$

The second dataset that we considered is the Santa Fe Laser time-series [29], where the input values $u(t)$ are sampled intensities from a far-infrared laser in chaotic regime, re-scaled by a factor of 0.01. We used the Laser dataset to define a next-step prediction task, where $y_{tg}(t) = u(t + 1)$ for each time-step t.

The last two datasets are instances of the Mackey-Glass (MG) [4,22] time-series, obtained by discretizing the following non-linear differential equation:

$$\frac{\delta u(t)}{\delta t} = \frac{0.2\,u(t - \tau)}{1 + u(t - \tau)^{10}} - 0.1\,u(t), \quad (8)$$

where τ is a parameter of the system influencing its dynamical behavior. We generated two MG time-series using $\tau = 17$ (MG17) and $\tau = 30$ (MG30), representing cases with increasingly complex chaotic behavior. In both cases, the elements of the time-series where shifted by -1 and then passed through the tanh squashing function as in [16,18]. The two MG time-series allowed us to set up two next-step prediction tasks, where $y_{tg}(t) = u(t + 1)$ for each time-step t.

For NARMA10, MG17 and MG30 we generated datasets with 10000 time-steps, while the Laser dataset contained a number of 10092 samples. In all the cases, the available data was split into a training set, comprising the first 5000 time-steps, and a test set, comprising the remaining samples. The first 100 time-steps were used as transient to wash out the initial conditions. The performance of the considered RC models was evaluated in terms of mean squared error (MSE) in all the tasks.

In our experiments, we considered DeepESNs with a total number of 500 recurrent reservoir units, distributed evenly across the layers of the deep architecture, varying the number of layers L from 2 to 5[4]. All the reservoir layers in the deep architecture shared the same values for the scaling hyper-parameters ρ and ω_{il}, i.e. $\rho = \rho^{(1)} = \ldots = \rho^{(L)}$ and $\omega_{il} = \omega_{il}^{(2)} = \ldots = \omega_{il}^{(L)}$. To account for sparsity, each reservoir unit was randomly connected to 5 units in the previous layer and to 5 units in the same layer. Of course, when considering permutation, ring and chain reservoir topologies, the connectivity of the reservoir units in each layer followed the corresponding specific structure described in Sect. 3.

---

[4] With the only exception of the case $L = 3$, where the first two layers contained 167 units and the last one contained 166 units.

In all the cases, we used fully-connected input weight matrices. For every task and choice of the reservoir topology, the DeepESN hyper-parametrization was chosen by model selection on a validation set comprising the last 1000 time-steps of the training split. To this end, we performed a random search with 50 networks configurations for each number of layers, sampling the value of ρ from a uniform distribution in $(0.1, 1)$, and the values of ω_{in} and ω_{il} from uniform distributions in $(0.1, 2)$. The achieved results were averaged on 10 network guesses for each hyper-parametrization explored, and readout training was performed by using pseudo-inversion. Finally, our experimental analysis was conducted in comparison with shallow ESN setups, considering the same reservoir topologies investigated in the DeepESN case, and using the same experimental setting described above, with the only crucial exception that all the available reservoir units were organized into a single layer (i.e., $L = 1$). Also note that, to provide a fair comparative analysis, the shallow reservoir configuration is not accounted in our experiments with DeepESNs (i.e., for DeepESNs we always consider $L > 1$).

4.2   Results

The test MSE values obtained by DeepESNs in correspondence of all the considered types of layer-wise reservoir topology are reported in Table 1. For the sake of comparison, the same table shows the results achieved by shallow ESNs under the same architectural conditions examined in the deep case. In all the cases, the sparse reservoir topology is considered as a baseline setup for our analysis. For completeness, in Appendix A we report the hyper-parametrization values selected on the validation set for all the considered architectural settings.

The performance values reported in Table 1 allow us to draw several lines of observations. First of all, our results confirm the goodness of the considered reservoir architectural variants already in the shallow setup, showing improved performance (i.e., a smaller MSE) with respect to the sparse baseline in all the cases (with the sole exception of permutation shallow reservoirs on Laser). Second, we observe that the performance of DeepESN with constrained topology (i.e. permutation, ring and chain) enhances that one of sparse DeepESN in all the considered tasks (with the only exception of deep reservoirs with chain architecture on Laser). Moreover, we can see that DeepESN improves the results of shallow ESN in all the tasks and for all the choices of reservoir topology, both in the constrained architectural cases and for the base sparse reservoir setup. Taken together, results in Table 1 clearly indicate the performance advantage arising from the synergy between deep organization and constrained topology as factors of architectural design of reservoirs. Giving a structure to the architecture of reservoirs both at a coarser level, i.e. organizing the recurrent units into layers, and at a finer level, i.e. organizing individual layers' units into cyclic or chain structures, amplifies the benefits brought by the two factors individually. Finally, we notice that the best performing architecture in our experiments is the DeepESN with permutation reservoir topology, which obtained the smallest

Table 1. Test MSE (and std) achieved by shallow ESN and DeepESN settings for different choices of the reservoir topology. The last column reports the number of layers selected for DeepESN. Best results for each task are underlined.

Topology	ESN	DeepESN
NARMA10		
Sparse	$1.658\ 10^{-4}\ (3.367\ 10^{-5})$	$1.647\ 10^{-4}\ (3.415\ 10^{-5})$
Permutation	$1.354\ 10^{-4}\ (1.589\ 10^{-5})$	$\underline{1.243\ 10^{-4}}\ (1.464\ 10^{-5})$
Ring	$1.494\ 10^{-4}\ (1.547\ 10^{-5})$	$1.482\ 10^{-4}\ (1.713\ 10^{-5})$
Chain	$1.571\ 10^{-4}\ (1.780\ 10^{-5})$	$1.569\ 10^{-4}\ (2.594\ 10^{-5})$
Laser		
Sparse	$1.226\ 10^{-3}\ (1.037\ 10^{-4})$	$8.228\ 10^{-4}\ (2.309\ 10^{-4})$
Permutation	$1.312\ 10^{-3}\ (1.385\ 10^{-4})$	$\underline{6.633\ 10^{-4}}\ (8.861\ 10^{-5})$
Ring	$1.161\ 10^{-3}\ (7.541\ 10^{-5})$	$7.640\ 10^{-4}\ (4.331\ 10^{-5})$
Chain	$9.496\ 10^{-4}\ (1.183\ 10^{-4})$	$8.555\ 10^{-4}\ (8.302\ 10^{-5})$
MG17		
Sparse	$3.739\ 10^{-9}\ (1.387\ 10^{-9})$	$2.328\ 10^{-9}\ (8.299\ 10^{-10})$
Permutation	$3.093\ 10^{-9}\ (3.241\ 10^{-10})$	$\underline{4.576\ 10^{-10}}\ (6.280\ 10^{-10})$
Ring	$1.585\ 10^{-9}\ (2.989\ 10^{-10})$	$5.043\ 10^{-10}\ (3.891\ 10^{-10})$
Chain	$1.950\ 10^{-9}\ (3.745\ 10^{-10})$	$4.913\ 10^{-10}\ (2.535\ 10^{-10})$
MG30		
Sparse	$1.476\ 10^{-8}\ (1.781\ 10^{-9})$	$1.172\ 10^{-8}\ (1.406\ 10^{-9})$
Permutation	$1.027\ 10^{-8}\ (5.412\ 10^{-10})$	$\underline{8.618\ 10^{-9}}\ (1.457\ 10^{-9})$
Ring	$1.197\ 10^{-8}\ (1.549\ 10^{-9})$	$1.078\ 10^{-8}\ (2.066\ 10^{-9})$
Chain	$1.086\ 10^{-8}\ (9.519\ 10^{-10})$	$9.096\ 10^{-9}\ (1.803\ 10^{-9})$

errors on all the tasks[5], and is put forward here as a particularly effective (yet sparse and efficient) approach to the architectural design of reservoir models. We leave to further studies the analysis of the dynamical properties that make Deep-ESN constructions based on permutation matrices so effective in applications, while here we limit ourselves to intuitive yet insightful considerations that might explain the observed results. On the one hand, recent literature works (e.g., [5,8]) provided empirical evidence of the fact that higher layers in deep reservoirs tend to develop progressively more abstract temporal representations of the driving input, and are naturally featured by longer memory. On the other hand, reservoir architectures based on permutation topology present multiple ring sub-structures that (at least in the linear approximation) are possibly featured by maximized memory. The resulting reservoir provides a variety of memories (that can be

---

[5] Performance differences between DeepESN with permutation topology and all the other architectures are confirmed by Wilcoxon rank-sum test performed at 1% significance level on all the tasks (with the only exceptions of the comparisons with ESN using chain topology on Laser, and ESN using permutation topology on MG30).

easily controlled by scaling the strength of the recurrent connections), and the developed state representations are enriched [15]. Our results show that Deep-ESNs with permutation reservoir topology are able to effectively exploit both the advantages of deep recurrent architectures and multiple ring sub-structures.

5 Conclusions

In this paper we have investigated the role of reservoir topology in the architectural design of DeepESNs. Specifically, we focused on analyzing the effects of constraining the recurrent weight matrix of each layer according to permutation, ring and chain topologies. Numerical results on several RC benchmarks pointed out a striking beneficial effect arising from the combination of a deep reservoir construction with a structured organization of the recurrent units in each layer. Our results indicate that DeepESN with reservoir units arranged to obey a permutation scheme (i.e., forming multiple rings) provides a particularly advantageous design strategy for reservoirs, leading to the best performance in all the explored tasks.

While already giving interesting empirical evidences on the potentialities of deep RC architectures, the study presented in this paper opens the way to several directions for further research. First of all, the experimental analysis described here suggests that the use of simplified deep RC models has a great potential that can be exploited massively in real-world applications. Leveraging the parsimonious design approach resulting from structured sparsity of reservoir units, the class of deep neural models studied in this work seems an ideal candidate e.g. for embedding advanced learning capabilities on resource-constrained computing devices. On the methodological side, a natural extension of the work in this paper is to analyze the effect of a broader pool of reservoir architectural variants, including e.g. small-world [20], cycles with regular jumps [25] and concentric [1] reservoirs. Moreover, future research could pursue even further the simplification of architectural construction in deep RC models, reducing the impact of randomness in the network initialization in the same vein as the works on minimum complexity ESNs [24,25]. Simplifying the reservoir structure locally to each layer can also be exploited from a more theoretically-oriented perspective, easing the mathematical analysis of dynamical properties naturally emerging in deep RNNs. In this concern, it is certainly interesting to extend fundamental mathematical results, e.g. pertaining to short-term memory capacity [17,24,30], or to approximation properties [14] of shallow reservoirs to the case of DeepESN. In addition to this, we believe that the role of orthogonality in deep reservoirs, studied in this paper in relation to the individual layers of the architecture, is an intriguing concept that deserves to be investigated also at the level of global (instead of local) DeepESN dynamics. Finally, the advantages of constrained DeepESN architectures delineated in this paper can be extended to larger classes of models, including e.g. deep RC for complex data structures [11], as well as fully trained deep RNNs.

A   Selected Hyper-parameters

Table 2 reports the DeepESN hyper-parameters selected by model selection for the experiments reported in Sect. 4. The reported values are the following: spectral radius ρ, input scaling ω_{in}, inter-layer scaling ω_{il}, and number of layers L. We recall from Sect. 4.1 that the values of ρ and ω_{il} are shared by all the layers. The selected hyper-parametrization for (shallow) ESN, are given in Table 3, where we report the chosen values of ρ and ω_{in}. We also recall from Sect. 4.1 that the total number of reservoir units is set to 500 for both DeepESN and ESN. While in the latter case all the 500 units form a single recurrent layer, in the former they are evenly distributed across the layers in the deep reservoir.

Table 2. Selected DeepESN hyper-parameters: spectral radius ρ, input scaling ω_{in}, inter-layer scaling ω_{il}, and number of layers L.

Topology	ρ	ω_{in}	ω_{il}	L
NARMA10 - DeepESN				
Sparse	0.9578	1.6466	1.0125	2
Permutation	0.8071	0.5883	1.0448	2
Ring	0.8190	1.1392	0.6026	2
Chain	0.8644	1.1138	1.5534	2
Laser - DeepESN				
Sparse	0.6866	1.7247	1.6930	5
Permutation	0.4525	1.7606	1.8147	5
Ring	0.4860	1.8217	1.5893	4
Chain	0.4755	1.8255	1.5886	4
MG17 - DeepESN				
Sparse	0.9972	1.9762	0.5559	2
Permutation	0.9364	1.9384	1.3763	5
Ring	0.9658	1.1014	1.9231	5
Chain	0.9904	1.7046	1.9654	3
MG30 - DeepESN				
Sparse	0.9756	1.5653	1.7706	2
Permutation	0.9690	1.9918	1.0660	3
Ring	0.9357	1.7230	1.6304	5
Chain	0.9884	1.7204	1.1962	3

Interestingly, from Table 2 we can observe that constrained reservoir topologies in DeepESNs generally tend to show smaller values of the spectral radius and a deeper architecture than basic (i.e., sparse) reservoir settings. Comparing

Table 3. Selected ESN hyper-parameters: spectral radius ρ and input scaling ω_{in}.

Topology	ρ	ω_{in}
NARMA10 - ESN		
Sparse	0.9251	1.4523
Permutation	0.8284	0.6058
Ring	0.8324	1.1713
Chain	0.8037	1.1165
Laser - ESN		
Sparse	0.5094	1.8544
Permutation	0.4594	1.9509
Ring	0.4916	1.8586
Chain	0.6964	1.9850
MG17 - ESN		
Sparse	0.9406	1.5544
Permutation	0.9615	0.9063
Ring	0.9784	1.6432
Chain	0.9437	1.4265
MG30 - ESN		
Sparse	0.9556	1.2408
Permutation	0.9196	1.9682
Ring	0.9468	1.1950
Chain	0.9891	1.8835

Tables 2 and 3 we also note that the values of spectral radius and input scaling selected for DeepESN and ESN correspond quite well in all the analyzed reservoir settings.

References

1. Bacciu, D., Bongiorno, A.: Concentric ESN: assessing the effect of modularity in cycle reservoirs. In: 2018 International Joint Conference on Neural Networks (IJCNN), pp. 1–8. IEEE (2018)
2. Boedecker, J., Obst, O., Mayer, N.M., Asada, M.: Studies on reservoir initialization and dynamics shaping in echo state networks. In: Proceedings of the 17th European Symposium on Artificial Neural Networks (ESANN), pp. 227–232. d-side publi. (2009)
3. Farkaš, I., Bosák, R., Gergeľ, P.: Computational analysis of memory capacity in echo state networks. Neural Netw. **83**, 109–120 (2016). https://doi.org/10.1016/j.neunet.2016.07.012
4. Farmer, J.D.: Chaotic attractors of an infinite-dimensional dynamical system. Physica D **4**(3), 366–393 (1982). https://doi.org/10.1016/0167-2789(82)90042-2

5. Gallicchio, C.: Short-term memory of deep RNN. In: Proceedings of the 26th European Symposium on Artificial Neural Networks (ESANN), pp. 633–638 (2018)
6. Gallicchio, C., Micheli, A.: Echo state property of deep reservoir computing networks. Cogn. Comput. **9**(3), 337–350 (2017). https://doi.org/10.1007/s12559-017-9461-9
7. Gallicchio, C., Micheli, A.: Why layering in RNN? A DeepESN survey. In: Proceedings of the 2018 International Joint Conference on Neural Networks (IJCNN), pp. 1–8. IEEE (2018)
8. Gallicchio, C., Micheli, A., Pedrelli, L.: Deep reservoir computing: a critical experimental analysis. Neurocomputing **268**, 87–99 (2017). https://doi.org/10.1016/j.neucom.2016.12.089
9. Gallicchio, C., Micheli, A., Tiňo, P.: Randomized recurrent neural networks. In: 26th European Symposium on Artificial Neural Networks, Computational Intelligence and Machine Learning (ESANN 2018), pp. 415–424. i6doc.com publication (2018)
10. Gallicchio, C., Micheli, A.: Deep echo state network (DeepESN): a brief survey. arXiv preprint arXiv:1712.04323 (2017)
11. Gallicchio, C., Micheli, A.: Deep reservoir neural networks for trees. Inf. Sci. **480**, 174–193 (2019). https://doi.org/10.1016/j.ins.2018.12.052
12. Gallicchio, C., Micheli, A., Pedrelli, L.: Design of deep echo state networks. Neural Netw. **108**, 33–47 (2018). https://doi.org/10.1016/j.neunet.2018.08.002
13. Gallicchio, C., Micheli, A., Silvestri, L.: Local lyapunov exponents of deep echo state networks. Neurocomputing **298**, 34–45 (2018). https://doi.org/10.1016/j.neucom.2017.11.073
14. Grigoryeva, L., Ortega, J.P.: Echo state networks are universal. Neural Netw. **108**, 495–508 (2018). https://doi.org/10.1016/j.neunet.2018.08.025
15. Hajnal, M.A., Lőrincz, A.: Critical echo state networks. In: Kollias, S.D., Stafylopatis, A., Duch, W., Oja, E. (eds.) ICANN 2006. LNCS, vol. 4131, pp. 658–667. Springer, Heidelberg (2006). https://doi.org/10.1007/11840817_69
16. Jaeger, H.: The "echo state" approach to analysing and training recurrent neural networks - with an erratum note. Technical report, GMD - German National Research Institute for Computer Science (2001)
17. Jaeger, H.: Short term memory in echo state networks. Technical report, German National Research Center for Information Technology (2001)
18. Jaeger, H., Haas, H.: Harnessing nonlinearity: predicting chaotic systems and saving energy in wireless communication. Science **304**(5667), 78–80 (2004). https://doi.org/10.1126/science.1091277
19. Jaeger, H.: Discovering multiscale dynamical features with hierarchical echo state networks. Technical report, Jacobs University Bremen (2007)
20. Kawai, Y., Park, J., Asada, M.: A small-world topology enhances the echo state property and signal propagation in reservoir computing. Neural Netw. (2019). https://doi.org/10.1016/j.neunet.2019.01.002
21. Lukoševičius, M., Jaeger, H.: Reservoir computing approaches to recurrent neural network training. Comput. Sci. Rev. **3**(3), 127–149 (2009). https://doi.org/10.1016/j.cosrev.2009.03.005
22. Mackey, M.C., Glass, L.: Oscillation and chaos in physiological control systems. Science **197**(4300), 287–289 (1977). https://doi.org/10.1126/science.267326
23. Pascanu, R., Gulcehre, C., Cho, K., Bengio, Y.: How to construct deep recurrent neural networks. arXiv preprint arXiv:1312.6026v5 (2014)
24. Rodan, A., Tino, P.: Minimum complexity echo state network. IEEE Trans. Neural Networks **22**(1), 131–144 (2011). https://doi.org/10.1109/TNN.2010.2089641

25. Rodan, A., Tiňo, P.: Simple deterministically constructed cycle reservoirs with regular jumps. Neural Comput. **24**(7), 1822–1852 (2012). https://doi.org/10.1162/NECO_a_00297

26. Strauss, T., Wustlich, W., Labahn, R.: Design strategies for weight matrices of echo state networks. Neural Comput. **24**(12), 3246–3276 (2012). https://doi.org/10.1162/NECO_a_00374

27. Triefenbach, F., Jalalvand, A., Schrauwen, B., Martens, J.P.: Phoneme recognition with large hierarchical reservoirs. In: Advances in Neural Information Processing Systems, pp. 2307–2315 (2010)

28. Verstraeten, D., Schrauwen, B., d'Haene, M., Stroobandt, D.: An experimental unification of reservoir computing methods. Neural Netw. **20**(3), 391–403 (2007). https://doi.org/10.1016/j.neunet.2007.04.003

29. Weigend, A.S.: Time Series Prediction: Forecasting the Future and Understanding the Past. Routledge, Abingdon (2018)

30. White, O.L., Lee, D.D., Sompolinsky, H.: Short-term memory in orthogonal neural networks. Phys. Rev. Lett. **92**(14), 148102 (2004). https://doi.org/10.1103/PhysRevLett.92.148102

Multiple Pattern Generations and Chaotic Itinerant Dynamics in Reservoir Computing

Hiromichi Suetani[1,2](✉) (iD)

[1] Oita University, 700 Dannoharu, Oita, Oita 870-1192, Japan
suetani@oita-u.ac.jp
[2] RIKEN Center for Brain Science,
2-1 Hirosawa, Wako, Saitama 351-0198, Japan

Abstract. It is an open question how spontaneous or "existing" activity plays a constructive role in information processing of the central nervous systems in the brain. In this study, using the FORCE learning framework, we investigate the problem of multiple temporal pattern generations by a single recurrent neural network (RNN) pushed by appropriate combinations of input pulses. We show that weak chaos meaning that the maximal Lyapunov exponent is small but strictly positive (this is *not* the so-called "edge of chaos") is required as a generic property of the RNN for giving optimal performances. Furthermore, such a network exhibits intermittent switching processes among several quasi-stable patterns when there is no external inputs, which reminds us a prominent behavior of the high-dimensional nonlinear dynamical systems called *chaotic itinerancy*. We also characterize this itinerant dynamics in terms of the fluctuations of the finite-time Lyapunov exponents and the eigenvalue spectra of the recurrent weight matrices after learning.

Keywords: Spontaneous activity · Chaotic itinerancy · FORCE learning

1 Introduction

In his seminal book [1], Hebb conducted that the brain is continuously active through the electrophysiological evidence at that time and considered how such an "existent activity" in the central nervous system builds the psychological behaviors such as attention, set and expectancy. Although the importance of spontaneous activity of the brain has been paid less attention and much of interests has mainly focused on the input-output functional relationships until recently, intensive developments of methodological tools reveals that spontaneous activity in the brain is not merely random but rather is more structured with

Supported by MEXT KAKENHI Grant Numbers 16H0167 and 18H05136, and by JSPS KAKENHI Grant Number 19H041183.

© Springer Nature Switzerland AG 2019
I. V. Tetko et al. (Eds.): ICANN 2019, LNCS 11731, pp. 76–81, 2019.
https://doi.org/10.1007/978-3-030-30493-5_7

coherent patterns [5]. In practice, there are several studies showing that the spatiotemporal patterns of spontaneous activity are closely related to those during stimulus-driven or task-related activities. For example, Kenet et al. found that spontaneous activity in the primary visual cortex of cat consists of a sequence of dynamically switching among cortical states corresponding to orientation preference maps [4].

In this study, we discuss a constructive role of spontaneous activity in the information processing of the brain from the viewpoint of of the reservoir computing (RC) paradigm [2,6]. Especially, using FORCE-learning framework proposed by Sussillo and Abbott [8], we consider the problem of multiple temporal pattern generations via a single recurrent neural network (RNN) that are selectable by the combination of pushing pulses through multiple channels. We show that weak chaos meaning that the maximal Lyapunov exponent is small but strictly positive (this is *not* the so-called "edge of chaos") is required as a generic property of the RNN for giving optimal performances. Furthermore, such a network exhibits intermittent switching processes among several quasi-stable patterns when there is no external inputs, which reminds us a prominent behavior of the high-dimensional nonlinear dynamical systems called *chaotic itinerancy* [3].

2   Model

In this paper, we consider the following system [7,8] (Fig. 1(a)) consisting of a continuous-time RNN (reservoir) with M input neurons, N internal neurons, and L readout neurons defined as

$$\tau \dot{x}_n(t) = -x_n(t) + g \sum_{n'=1}^{N} w_{nn'}^{\text{rec}} \tanh(x_{n'}(t))$$

$$+ \sum_{m=1}^{M} w_{nm}^{\text{in}} u_m(t) + \sum_{l=1}^{L} w_{nl}^{\text{fb}} z_l(t) + b_n, \quad n = 1, ..., N. \tag{1}$$

Here, x_n is the membrane potential of the n-th neuron in the reservoir, u_m is input of the m-th neuron applied to the reservoir, and $z_l(t) = \sum_{n=1}^{N} w_{ln}^{\text{out}} \tanh(x_n(t))$, $l = 1, ..., L$ is the l-th readout neuron as a weighted sum of the firing rates in the reservoir. The parameter g is the gain control parameter on the recurrent connections within the reservoir, which can induce the chaotic dynamics depending on the value, and τ is the parameter determining the characteristic time-scale of dynamics. Weights are as follows: $\boldsymbol{W}^{\text{rec}} = (w_{nn'}^{\text{rec}})$ for connections between neurons in the reservoir, $\boldsymbol{W}^{\text{in}} = (w_{nm}^{\text{in}})$ for those between the input and the reservoir neurons, $\boldsymbol{W}^{\text{fb}} = (w_{nl}^{\text{fb}})$ for a feedback from the output to the reservoir neurons, and a $\boldsymbol{W}^{\text{out}} = (w_{ln}^{\text{out}})$ for those between the reservoir and the readout neurons. The term b_n is also introduced as a small bias of the n-th neuron that is given as Gaussian random variable.

In the framework of RC, the matrices $\boldsymbol{W}^{\text{rec}}$, $\boldsymbol{W}^{\text{in}}$, and $\boldsymbol{W}^{\text{fb}}$ are fixed and do not change in training. Only the matrix $\boldsymbol{W}^{\text{out}}$ is updated in an online way using, e.g., the recursive least squares (RLS) method in order to minimize the root-mean-squared error between the output $z(t)$ and the teacher signal $z^{\text{target}}(t)$. In this paper, the matrix W^{rec} is sparse, i.e., the probability whether there exists the connection between a pair of two neurons in the reservoir or not is set to $p = 0.1$, and its intensity obeys Gaussian distribution as $w_{nn'}^{\text{rec}} \sim (g/\sqrt{N})\mathcal{N}(0,1)$ if there is a connection between the n and the n'-th neurons. Weights in the matrices W^{in} and W^{fb} obey the uniform distributions.

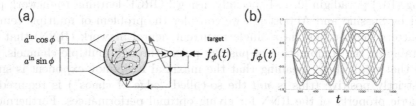

Fig. 1. (a) A schematic figure of FORCE learning. (b) Target patterns used in this study

3   Results

As a demonstration, we trained a RNN to produce a set of 16 different patterns shown in Fig. 1(b). These patterns are characterized by the value of a single parameter ϕ. In the training phase, first the RNN is pushed by pulses through two channels, whose amplitude is changed according to the value of ϕ associated with each of target patterns shown in Fig. 1(a), then the weights in the matrix W^{out} are updated by applying FORCE learning as explained in the previous section. We repeat this process for a number of times by changing the initial conditions. Figure 2 shows the outputs $z(t)$ of the system after training for three different values of g. We see that the performance depends on the value of the parameter g. When $g = 0.9$, in the case where the dynamics of the reservoir converges a fixed point, all of responses to input pulses shrinks and differs from the correct targets. On the other hand, when $g = 1.5$, in the case where the dynamics of the reservoir exhibits developed chaos, all of responses oscillates violently, the performance is still worse. However, when $g = 1.2$, responses to input pulses mimic the target patterns, which gives the optimal performance. It is noted that when $g = 1.2$, the reservoir exhibits chaotic dynamics in the sense that the maximal Lyapunov exponent is small but strictly positive, which is *not* the case of edge of chaos.

Next, we show that what kinds of spontaneous activity, i.e., any input is not applied to the reservoir, are observed after training. Figure 3 shows the time series of the readout neuron $z(t)$ without input pulses after training for $g = 1.2$. We observe that there are several quasi-stable states and the dynamics is wondering

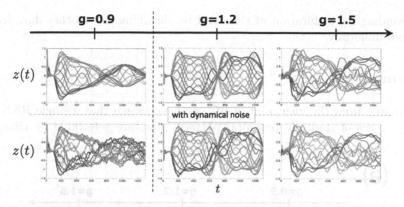

Fig. 2. Responses from the reservoir to pushing pulses after FORCE learning. In lower panels, the dynamical noise with Gaussian distribution is added to the reservoir.

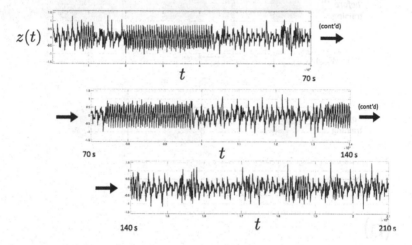

Fig. 3. Spontaneous activity of the readout neuron after FORCE learning.

among these states. This behavior reminds us a prominent behavior of the high-dimensional nonlinear dynamical systems, called *chaotic itinerancy* [3].

Finally, we investigate the mechanism causing such an itinerant behavior. To this end, we investigate that spectra of the eigenvalues of the network connectivities, i.e., $\boldsymbol{W}^{\mathrm{rec}} + \boldsymbol{W}^{\mathrm{fb}} \cdot \boldsymbol{W}^{\mathrm{out}}$ before and after training (in the case of pre-training, the eigenvalues are equal to those of $\boldsymbol{W}^{\mathrm{rec}}$ because $\boldsymbol{W}^{\mathrm{out}}$ is set to zero at the beginning), and the histograms of the finite-time (maximal) Lyapunov exponents (FTLEs) for three different values of g. Results are shown in Fig. 4. When the network yields the optimal performance, the corresponding eigenspectrum has a pair of isolated eigenvalues whose real part is negative as marked in two red circles in Fig. 4(a), which enables the network to be locally stable. Furthermore, as shown in Fig. 4(b), the fluctuation of FTLEs becomes large for the optimal performance case, which means that the trajectory of the system continues to be expanding for a duration of time and be shrinking for another duration of time non-uniformly.

4 Summary

In summary, from the viewpoint of RC, we have shown that a single RNN can produce a set of multiple temporal patterns that are selectable by changing

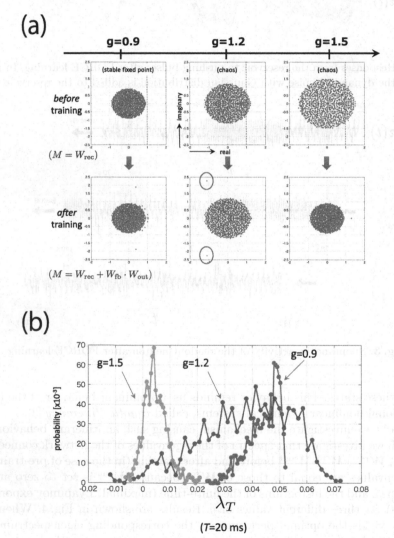

Fig. 4. (a) Eigenvalue spectra of the network connectivity before (upper panels) and after (lower panels) training for three different values of g. (b) Histograms of the finite-time maximal Lyapunov exponents for three different values of g.

the amplitude of input pulses. But, the network is required to be chaotic for optimal performance. By choosing the values of parameters appropriately, the network take advantage of a rich variety of trajectories originated from chaos as well as the harnessing sensitive dependence on initial conditions. Although our present study is just obtained from a mathematical toy model, it gives a clue for answering why spontaneous activity is in need of the information processing in the brain [1].

References

1. Hebb, D.: The Organization of Behavior: A Neuropsychological Theory. Wiley, Hoboken (1949)
2. Jäger, H.: The echo state approach to analysing and training recurrent neural networks. GMD Technical report **148**, 13, German National Research Center for Information Technology, Bonn, Germany (2001)
3. Kaneko, K., Tsuda, I.: Chaotic itinerancy. Chaos **13**, 926 (2003). https://doi.org/10.1063/1.1607783
4. Kenet, T., Bibitchkov, D., Tsodyks, M., Grinvald, A., Arieli, A.: Spontaneously emerging cortical representations of visual attributes. Nature **425**, 954 (2003). https://doi.org/10.1038/nature02078
5. Luczak, A., Barthó, P., Harris, K.D.: Spontaneous events outline the realm of possible sensory responses in neocortical population. Neuron **62**, 413 (2009). https://doi.org/10.1016/j.neuron.2009.03.014
6. Maas, W., Natschläger, T., Markram, H.: Real-time computing without stable states: a new framework for neural computation based on perturbations. Neural Comput. **14**, 2351 (2002). https://doi.org/10.1162/089976602760407955
7. Suetani, H.: Weak sensitivity to initial conditions for generating temporal patterns in recurrent neural networks: a reservoir computing approach. In: Sanayei, A., Rössler, O., Zelinka, I. (eds.) Interdisciplinary Symposium on Complex Systems (ISCS), vol. 13, pp. 47–55. Springer, Cham (2015). https://doi.org/10.1007/978-3-319-10759-2_6
8. Sussillo, D., Abbott, L.: Generating coherent patterns of activity from chaotic neural networks. Neuron **63**, 544 (2009). https://doi.org/10.1016/j.neuron.2009.07.018

Echo State Network with Adversarial Training

Takanori Akiyama[1]($\boxtimes$) and Gouhei Tanaka[1,2]

[1] Graduate School of Information Science and Technology,
The University of Tokyo, Tokyo 113-8656, Japan
{taka_akiyama,gouhei}@sat.t.u-tokyo.ac.jp
[2] Graduate School of Engineering,
The University of Tokyo, Tokyo 113-8656, Japan

Abstract. Reservoir Computing (RC) is a high-speed machine learning framework for temporal data processing. Especially, the Echo State Network (ESN), which is one of the RC models, has been successfully applied to many temporal tasks. However, its prediction ability depends heavily on hyperparameter values. In this work, we propose a new ESN training method inspired by Generative Adversarial Networks (GANs). Our method intends to minimize the difference between the distribution of teacher data and that of generated samples, and therefore we can generate samples that reflect the dynamics in the teacher data. We apply a feedforward neural network as a discriminator so that we don't need to use backpropagation through time in training. We justify the effectiveness of the proposed method in time series prediction tasks.

Keywords: Echo State Network · Recurrent Neural Network · Generative Adversarial Network · Nonlinear time series prediction

1 Introduction

Reservoir Computing (RC) has been widely researched as a fast machine learning method. RC is the Recurrent Neural Network (RNN) based framework that only trains the linear readout layer and fixes parameters in other layers before training. RC shows excellent performance in various benchmark tasks despite its simple training algorithm. Moreover, another major advantage is that RC is suitable for hardware implementation with a wide variety of physical systems [8].

Especially, Echo State Networks (ESNs) were initially proposed by Jaeger [2] and it was shown that they were useful in nonlinear time series prediction [4]. The underlying principle is that a well-trained ESN is able to reproduce the attractor of given nonlinear dynamical systems. However, the performance of the ESNs is significantly sensitive to the settings of hyperparameters such as the input scaling and the spectral radius of the reservoir connection weight matrix. In the case of physical implementation, it is hard to change and adjust such

© Springer Nature Switzerland AG 2019
I. V. Tetko et al. (Eds.): ICANN 2019, LNCS 11731, pp. 82–88, 2019.
https://doi.org/10.1007/978-3-030-30493-5_8

hyperparameters. Therefore, a reduction of the hyperparameter sensitivity only by changing a training method is regarded as an important research topic.

In this work, we incorporate the concept of Generative Adversarial Networks (GANs) [1] into ESNs to solve the abovementioned problem. A GAN consists of two networks, a discriminator to distinguish between teacher data and generated samples and a generator to deceive the discriminator. Original GANs can minimize the Jensen-Shannon divergence between the real data distribution and the generated one instead of minimizing the squared error. In our method, a generator is an ESN and a Deep Neural Network (DNN) based discriminator distinguishes between the real time series data and those generated by the ESN. Then we use the weighted sum of the conventional squared error and the adversarial loss. By introducing the adversarial loss, it is expected that the ESN can generate samples which reflect the dynamics underlying the given data better than the conventional ESN training based on the least squared error.

There are three major advantages in the proposed method. First, the prediction accuracy can be improved even when the settings of hyperparameters in the ESN are inappropriate. Only by introducing the adversarial loss in the training step, we can construct a high-quality predictor with 'bad' reservoirs. Second, the computational cost for training in our method is much smaller than that of RNNs. We use simple feedforward neural networks as a discriminator, instead of temporal neural networks like RNNs, to avoid using computationally expensive backpropagation through time (BPTT) [9] in training. Simultaneously, we introduce the concept of time-delay embeddings to construct the input to a discriminator. Therefore, we can consider time-dependent features in a discriminator without BPTT. Third, trained parameters in the ESN are only those in the readout layer, and therefore our method can be applied to other types of reservoir models and physical reservoirs.

We demonstrate the effectiveness of our method for benchmark nonlinear time series prediction tasks. Especially, when the settings of hyperparameters are not so appropriate, our training method outperforms the conventional one.

2   Methods

2.1   Echo State Network

The ESN uses an RNN-based reservoir composed of discrete-time artificial neurons. The time evolution of the reservoir state vector is described as follows:

$$r(t + \Delta t) = \tanh\left[Ar(t) + W_{in}u(t)\right], \ \hat{y}(t) = W_{out}r(t), \tag{1}$$

where $r(t)$ denotes the state vector of reservoir units, W_{in} is the weight matrix of the input layer, and A is the reservoir connection weight matrix. The readout layer gives the linear combination of the states of reservoir neurons with W_{out}, which denotes the weight matrix of the readout layer.

In the training period $-T \leq t < 0$, the readout weight is adjusted so as to minimize the squared error between the predicted value $\hat{y}(t)$ and the teacher output $y(t)$ as follows:

$$\hat{W}_{out} = \underset{W_{out}}{\operatorname{argmin}} \sum_{-T \leq t < 0} \|W_{out} r(t) - y(t)\|^2 + \beta \|W_{out}\|^2, \tag{2}$$

where $\beta > 0$ is the Tikhonov regularization parameter to avoid overfitting.

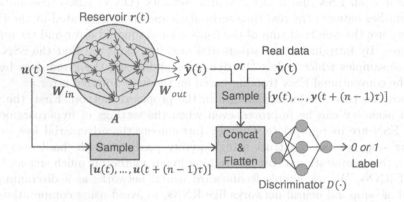

Fig. 1. Architecture of the ESN with adversarial training

2.2   Echo State Network with Adversarial Training

In our method, adversarial training [6] is used to optimize the readout weight in the ESN. Figure 1 shows the architecture of the proposed model. The discriminator D is trained such that the output represents the probability that the input is drawn from teacher data. $u(t)$ and $\hat{y}(t)$ are embedded into the same time-delay coordinates, and then the concatenation of them are fed to the discriminator as

$$D\left(\begin{bmatrix} u(t), \cdots, u(t + (n-1)\tau) \\ \hat{y}(t), \cdots, \hat{y}(t + (n-1)\tau) \end{bmatrix} \right) := D(u, \hat{y}, t), \tag{3}$$

where n and τ represent the dimension and time delay of the time-delay coordinates, respectively.

We define the discriminator loss as follows:

$$L_D = \mathbb{E}\left[-\log\left(1 - D(u, \hat{y}, t)\right)\right] + \mathbb{E}\left[-\log\left(D(u, y, t)\right)\right]. \tag{4}$$

Then we define the generator loss as follows:

$$L_G = w_D L_G^{ADV} + (1 - w_D) \frac{E_{L_{ADV}}}{E_{L_{SE}}} L_G^{SE}, \tag{5}$$

$$L_G^{ADV} = \mathbb{E}\left[-\log\left(D(u, \hat{y}, t)\right)\right], \quad L_G^{SE} = \sum_{-T \leq t < 0} \|\hat{y}(t) - y(t)\|^2, \tag{6}$$

where $E_{L_{ADV}}$ and $E_{L_{SE}}$ denote the expectation values of L_G^{ADV} and L_G^{SE}, respectively. w_D is the weight of the adversarial loss in the generator loss. The discriminator loss L_D and the generator loss L_G are minimized alternately. Note that only $\boldsymbol{W}_{out}$ is optimized in the training of the generator. The procedure is formally presented in Algorithm 1.

3   Results

We demonstrate the effectiveness of the proposed method in prediction tasks with two benchmark time series, NARMA10 and the Lorenz system. In these two experiments, we set the reservoir size at 100 and the scarcity of the connection weight matrix at 0.95. Before adversarial training, we pretrained the output weight in the ESN using Tikhonov regularization with $\beta = 10^{-4}$. The architecture of the discriminator model is a feedforward network that consists of four hidden layers of 32 ReLU units and we set $n = 20$.

Algorithm 1. Stochastic gradient descent adversarial training of the ESN. We used $k_G = 2$ in our experiments.

Require: n, τ: the parameter of time-delay embeddings, w_D: the weight of the adversarial loss, m: the batch size, k_G: the number of generator iterations per the discriminator iteration.
1: Pretrain the readout weight $\boldsymbol{W}_{out}$ using conventional least squares regression.
2: **for** number of training iterations **do**
3:     Sample a batch $t^{(1)}, \cdots, t^{(m)}$ from $-T \leq t < 0$.
4:     Update the discriminator with the stochastic gradient method:

$$\nabla_{\theta_D} \frac{1}{m} \sum_{i=1}^{m} \left[-\log(1 - D(\boldsymbol{u}, \hat{\boldsymbol{y}}, t^{(i)})) - \log D(\boldsymbol{u}, \boldsymbol{y}, t^{(i)}) \right]. \qquad (7)$$

5:     Calculate $E_{L_{ADV}}$ and $E_{L_{SE}}$.
6:     **for** k_G steps **do**
7:         Sample a batch $t^{(1)}, \cdots, t^{(m)}$ from $-T \leq t < 0$.
8:         Update $\boldsymbol{W}_{out}$ with the stochastic gradient descent method:

$$\nabla_{\boldsymbol{W}_{out}} \left(\frac{\omega_D}{m} \sum_{i=1}^{m} \left[-\log D(\boldsymbol{u}, \hat{\boldsymbol{y}}, t^{(i)}) \right] + (1 - \omega_D) \frac{E_{L_{ADV}}}{E_{L_{SE}}} L_G^{SE} \right). \qquad (8)$$

9:     **end for**
10: **end for**

3.1   NARMA10

The NARMA10 task [3] is the identification of the order-10 discrete-time nonlinear dynamical system. We used $-900 \leq t < 0$ for training and $0 \leq t < 100$ for testing. In this task, we set the input scaling of the ESN at 1.0. Time delay in embeddings is $\tau = 1$. We conducted experiments for two different cases, where the spectral radius is good ($\rho = 0.8$) and bad ($\rho = 0.4$).

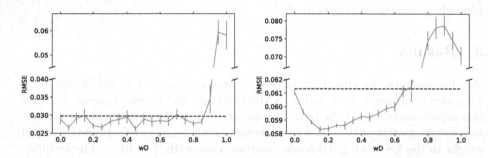

Fig. 2. RMSEs for different values of w_D in the NARMA10 task. The spectral radius of A is set at 0.8 (left) and 0.4 (right). Block dotted lines are RMSEs in the ESN with the least squared method. The error bars represent normalized errors.

Figure 2 shows the root mean squared error (RMSE) for the NARMA10 task, plotted against the value of w_D. In the case with $\rho = 0.8$, we can see that the prediction performance is improved by the introduction of the adversarial loss in some settings. In addition, in the case with $\rho = 0.4$, the prediction accuracy for $0.05 \leq w_D \leq 0.6$ is lower than the case when we use only the squared error. From this result, we can conclude that the adversarial loss in the ESN training improves the prediction accuracy, especially when the settings of the hyperparameter in the ESN are not so good.

3.2   Lorenz Systems

The Lorenz system [5] is a continuous-time nonlinear dynamical system which shows chaotic bahavior and is described by the following differential equations:

$$\frac{dx}{dt} = 10(y - x),$$
$$\frac{dy}{dt} = x(28 - y) - y,$$
$$\frac{dz}{dt} = xy - \frac{8}{3}z. \tag{9}$$

In this experiment, we predict the first variable $x(t)$ to evaluate the performance of the proposed model. We used $-100 \leq t < 0$ for training and $0 \leq t < 25$ for testing and set $\Delta t = 0.02$. The input scaling of the ESN is 0.1 and time delay τ is 0.08. We conducted an experiment for a bad parameter setting where the spectral radius $\rho = 0.4$.

Figure 3 shows the RMSE in this task, plotted against the value of w_D. The proposed method improves the prediction performance compared with the ESN

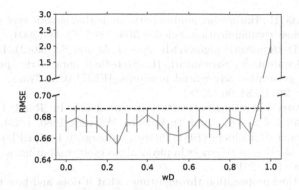

Fig. 3. RMSEs for different values of w_D in the prediction of the first variable of the Lorenz system. Block dotted lines are RMSEs in the ESN with the least squared method. The error bars represent normalized errors.

with the conventional training in a wide range of w_D (the optimal setting of w_D appears 0.25). Our proposed method uses the concept of time-delay embeddings, and thus we can conclude that the generator can reflect the overall dynamics even when we can observe only one variable on the basis of Takens Embedding Theorem [7].

4   Conclusion

In this work, we proposed a new ESN training method using adversarial training where the loss function is described as the weighted sum of the conventional squared error and the adversarial loss. Then we demonstrated that the proposed method can improve the prediction accuracy in nonlinear time series prediction tasks. In future work, we will test another model as a discriminator and check the effectiveness for other tasks.

Acknowledgment. This work was partially supported by JSPS KAKENHI Grant Number JP16K00326 (GT) and based on results obtained from a project subsidized by the New Energy and Industrial Technology Development Organization (NEDO).

References

1. Goodfellow, I., et al.: Generative adversarial nets. Adv. Neural Inf. Process. Syst. **27**, 2672–2680 (2014)
2. Jaeger, H.: The "echo state" approach to analysing and training recurrent neural networks. GMD Report 148, GMD - German National Research Institute for Computer Science (2001)
3. Jaeger, H.: Adaptive nonlinear system identification with echo state networks. In: Proceedings of the 15th International Conference on Neural Information Processing Systems, pp. 609–616 (2002)

4. Jaeger, H., Haas, H.: Harnessing nonlinearity: predicting chaotic systems and saving energy in wireless communication. Science **304**(5667), 78–80 (2004)
5. Lorenz, E.N.: Deterministic nonperiodic flow. J. Atmos. Sci. **20**(2), 130–141 (1963)
6. Saito, Y., Takamichi, S., Saruwatari, H.: Statistical parametric speech synthesis incorporating generative adversarial networks. IEEE/ACM Trans. Audio Speech Lang. Process. **26**(1), 84–96 (2018)
7. Takens, F.: Detecting strange attractors in turbulence. In: Rand, D., Young, L.-S. (eds.) Dynamical Systems and Turbulence, Warwick 1980. LNM, vol. 898, pp. 366–381. Springer, Heidelberg (1981). https://doi.org/10.1007/BFb0091924
8. Tanaka, G., et al.: Recent advances in physical reservoir computing: a review. Neural Netw. **115**, 100–123 (2019)
9. Werbos, P.J.: Backpropagation through time: what it does and how to do it. Proc. IEEE **78**(10), 1550–1560 (1990)

Hyper-spherical Reservoirs for Echo State Networks

Pietro Verzelli[1]([✉]), Cesare Alippi[1,2], and Lorenzo Livi[3,4]

[1] Faculty of Informatics, Università della Svizzera Italiana,
69000 Lugano, Switzerland
pietro.verzelli@usi.ch
[2] Department of Electronics, Information and Bioengineering,
Politecnico di Milano, 20133 Milan, Italy
[3] Departments of Computer Science and Mathematics,
University of Manitoba, Winnipeg, MB R3T 2N2, Canada
[4] Department of Computer Science, College of Engineering,
Mathematics and Physical Sciences, University of Exeter,
Exeter EX4 4QF, UK

Abstract. In this paper, we propose a model of ESNs that eliminates critical dependence on hyper-parameters, resulting in networks that provably cannot enter a chaotic regime and, at the same time, denotes nonlinear behaviour in phase space characterised by a large memory of past inputs, comparable to the one of linear networks. Our contribution is supported by experiments corroborating our theoretical findings, showing that the proposed model displays dynamics that are rich-enough to approximate many common nonlinear systems used for benchmarking.

Keywords: Echo State Networks · Edge of criticality ·
Memory–nonlinearity tradeoff

Although the use of *Recurrent Neural Networks (RNNs)* in machine learning is boosting, also as effective building blocks for deep learning architectures, a comprehensive understanding of their working principles is still missing [4,26]. Of particular relevance are Echo State Networks (ESNs), introduced by Jaeger [13] and independently by Maass *et al.* [16] under the name of Liquid State Machine (LSM), which emerge from RNNs due to their training simplicity. The basic idea behind ESNs is to create a randomly connected recurrent network, called reservoir, and feed it with a signal so that the network will encode the underlying dynamics in its internal states. The desired – task dependent – output is then generated by a readout layer (usually linear) trained to match the states with the desired outputs. Despite the simplified training protocol, ESNs are universal function approximators [10] and have shown to be effective in many relevant tasks [2,3,7,19–22].

These networks are known to be sensitive to the setting of hyper-parameters like the Spectral Radius (SR), the input scaling and the sparseness degree [13],

© Springer Nature Switzerland AG 2019
I. V. Tetko et al. (Eds.): ICANN 2019, LNCS 11731, pp. 89–93, 2019.
https://doi.org/10.1007/978-3-030-30493-5_9

which critically affect their behaviour and, hence, the performance at task. Fine tuning of hyper-parameters requires cross-validation or ad-hoc criteria for selecting the best-performing configuration. Experimental evidence and some results from the theory show that ESNs performance is usually maximised in correspondence of a very narrow region in hyper-parameter space called Edge of Chaos (EoC) [1,6,14,15,23–25,30]. However, we comment that beyond such a region ESNs behave chaotically, resulting in useless and unreliable computations. At the same time, it is everything but trivial configuring the hyperparameters to lie on the EoC still granting a non-chaotic behaviour. A very important property for ESNs is the Echo State Property (ESP), which basically asserts that their behaviour should depend on the signal driving the network only, regardless of its initial conditions [32]. Despite being at the foundation of theoretical results [10], the ESP in its original formulation raises some issues, mainly because it does not account for multi-stability and is not tightly linked with properties of the specific input signal driving the network [17,31,32].

In this context, the analysis of the memory capacity (as measured by the ability of the network to reconstruct or remember past inputs) of input-driven systems plays a fundamental role in the study of ESNs [8,9,12,27]. In particular, it is known that ESNs are characterized by a memory–nonlinearity trade-off [5,11,28], in the sense that introducing nonlinear dynamics in the network degrades memory capacity. Moreover, it has been recently shown that optimizing memory capacity does not necessarily lead to networks with higher prediction performance [18].

In a recent paper [29], we proposed an ESN model that eliminates critical dependence on hyper-parameters, resulting in models that cannot enter a chaotic regime. In addition to this major outcome, we showed that such networks denote nonlinear behaviour in phase space characterised by a large memory of past inputs (see Fig. 1): the proposed model generates dynamics that are rich-enough to approximate nonlinear systems typically used as benchmarks. Our contribution was based on a nonlinear activation function that normalizes neuron activations on a hyper-sphere. We showed that the spectral radius of the reservoir, which is the most important hyper-parameter for controlling the ESN behaviour, plays a marginal role in influencing the stability of the proposed model, although it has an impact on the capability of the network to memorize past inputs. Our theoretical analysis demonstrates that this property derives from the impossibility for the system to display a chaotic behaviour: in fact, the maximum Lyapunov exponent is always null. An interpretation of this very important outcome is that the network always operates on the EoC, regardless of the setting chosen for its hyper-parameters.

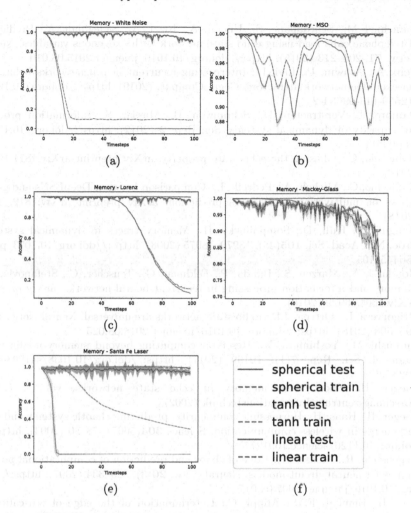

Fig. 1. Results of the experiments on memory for different benchmarks. Panel (a) displays the white noise memorization task, (b) the MSO, (c) the x-coordinate of the Lorenz system, (d) the Mackey-Glass series and (e) the Santa Fe laser dataset. As described in the legend (f), different line types account for results obtained on training and test data. The shaded areas represent the standard deviations, computed using 20 different realization for each point.

References

1. Bertschinger, N., Natschläger, T.: Real-time computation at the edge of chaos in recurrent neural networks. Neural Comput. **16**(7), 1413–1436 (2004). https://doi.org/10.1162/089976604323057443
2. Bianchi, F.M., Scardapane, S., Løkse, S., Jenssen, R.: Reservoir computing approaches for representation and classification of multivariate time series. arXiv preprint arXiv:1803.07870 (2018)

3. Bianchi, F.M., Scardapane, S., Uncini, A., Rizzi, A., Sadeghian, A.: Prediction of telephone calls load using echo state network with exogenous variables. Neural Netw. **71**, 204–213 (2015). https://doi.org/10.1016/j.neunet.2015.08.010
4. Ceni, A., Ashwin, P., Livi, L.: Interpreting recurrent neural networks behaviour via excitable network attractors. Cogn. Comput. (2019). https://doi.org/10.1007/s12559-019-09634-2
5. Dambre, J., Verstraeten, D., Schrauwen, B., Massar, S.: Information processing capacity of dynamical systems. Sci. Rep. **2** (2012). https://doi.org/10.1038/srep00514
6. Gallicchio, C.: Chasing the echo state property. arXiv preprint arXiv:1811.10892 (2018)
7. Gallicchio, C., Micheli, A., Pedrelli, L.: Comparison between DeepESNs and gated RNNs on multivariate time-series prediction. arXiv preprint arXiv:1812.11527 (2018)
8. Ganguli, S., Huh, D., Sompolinsky, H.: Memory traces in dynamical systems. Proc. Nat. Acad. Sci. **105**(48), 18970–18975 (2008). https://doi.org/10.1073/pnas.0804451105
9. Goudarzi, A., Marzen, S., Banda, P., Feldman, G., Teuscher, C., Stefanovic, D.: Memory and information processing in recurrent neural networks. arXiv preprint arXiv:1604.06929 (2016)
10. Grigoryeva, L., Ortega, J.P.: Echo state networks are universal. Neural Netw. **108**, 495–508 (2018). https://doi.org/10.1016/j.neunet.2018.08.025
11. Inubushi, M., Yoshimura, K.: Reservoir computing beyond memory-nonlinearity trade-off. Sci. Rep. **7**(1), 10199 (2017). https://doi.org/10.1038/s41598-017-10257-6
12. Jaeger, H.: Short term memory in echo state networks, vol. 5. GMD-Forschungszentrum Informationstechnik (2002)
13. Jaeger, H., Haas, H.: Harnessing nonlinearity: predicting chaotic systems and saving energy in wireless communication. Science **304**(5667), 78–80 (2004). https://doi.org/10.1126/science.1091277
14. Legenstein, R., Maass, W.: Edge of chaos and prediction of computational performance for neural circuit models. Neural Netw. **20**(3), 323–334 (2007). https://doi.org/10.1016/j.neunet.2007.04.017
15. Livi, L., Bianchi, F.M., Alippi, C.: Determination of the edge of criticality in echo state networks through Fisher information maximization. IEEE Trans. Neural Netw. Learn. Syst. **29**(3), 706–717 (2018). https://doi.org/10.1109/TNNLS.2016.2644268
16. Maass, W., Natschläger, T., Markram, H.: Real-time computing without stable states: a new framework for neural computation based on perturbations. Neural Comput. **14**(11), 2531–2560 (2002). https://doi.org/10.1162/089976602760407955
17. Manjunath, G., Jaeger, H.: Echo state property linked to an input: exploring a fundamental characteristic of recurrent neural networks. Neural Comput. **25**(3), 671–696 (2013). https://doi.org/10.1162/NECO_a_00411
18. Marzen, S.: Difference between memory and prediction in linear recurrent networks. Phys. Rev. E **96**(3), 032308 (2017). https://doi.org/10.1103/PhysRevE.96.032308
19. Palumbo, F., Gallicchio, C., Pucci, R., Micheli, A.: Human activity recognition using multisensor data fusion based on reservoir computing. J. Ambient Intell. Smart Environ. **8**(2), 87–107 (2016)
20. Pathak, J., Hunt, B., Girvan, M., Lu, Z., Ott, E.: Model-free prediction of large spatiotemporally chaotic systems from data: a reservoir computing approach. Phys. Rev. Lett. **120**(2), 024102 (2018)

21. Pathak, J., Lu, Z., Hunt, B.R., Girvan, M., Ott, E.: Using machine learning to replicate chaotic attractors and calculate Lyapunov exponents from data. Chaos: Interdisc. J. Nonlinear Sci. **27**(12), 121102 (2017). https://doi.org/10.1063/1.5010300
22. Pathak, J., et al.: Hybrid forecasting of chaotic processes: using machine learning in conjunction with a knowledge-based model. Chaos: Interdisc. J. Nonlinear Sci. **28**(4), 041101 (2018). https://doi.org/10.1063/1.5028373
23. Rajan, K., Abbott, L.F., Sompolinsky, H.: Stimulus-dependent suppression of chaos in recurrent neural networks. Phys. Rev. E **82**(1), 011903 (2010). https://doi.org/10.1103/PhysRevE.82.011903
24. Rivkind, A., Barak, O.: Local dynamics in trained recurrent neural networks. Phys. Rev. Lett. **118**, 258101 (2017). https://doi.org/10.1103/PhysRevLett.118.258101
25. Sompolinsky, H., Crisanti, A., Sommers, H.J.: Chaos in random neural networks. Phys. Rev. Lett. **61**(3), 259 (1988). https://doi.org/10.1103/PhysRevLett.61.259
26. Sussillo, D., Barak, O.: Opening the black box: low-dimensional dynamics in high-dimensional recurrent neural networks. Neural Comput. **25**(3), 626–649 (2013). https://doi.org/10.1162/NECO_a_00409
27. Tiňo, P., Rodan, A.: Short term memory in input-driven linear dynamical systems. Neurocomputing **112**, 58–63 (2013). https://doi.org/10.1016/j.neucom.2012.12.041
28. Verstraeten, D., Dambre, J., Dutoit, X., Schrauwen, B.: Memory versus nonlinearity in reservoirs. In: IEEE International Joint Conference on Neural Networks, pp. 1–8. IEEE, Barcelona (2010)
29. Verzelli, P., Alippi, C., Livi, L.: Echo state networks with self-normalizing activations on the hyper-sphere. arXiv preprint arXiv:1903.11691 (2019)
30. Verzelli, P., Livi, L., Alippi, C.: A characterization of the edge of criticality in binary echo state networks. In: 2018 IEEE 28th International Workshop on Machine Learning for Signal Processing (MLSP), pp. 1–6. IEEE (2018). https://doi.org/10.1109/MLSP.2018.8516959
31. Wainrib, G., Galtier, M.N.: A local echo state property through the largest Lyapunov exponent. Neural Netw. **76**, 39–45 (2016). https://doi.org/10.1016/j.neunet.2015.12.013
32. Yildiz, I.B., Jaeger, H., Kiebel, S.J.: Re-visiting the echo state property. Neural Netw. **35**, 1–9 (2012). https://doi.org/10.1016/j.neunet.2012.07.005

Echo State vs. LSTM Networks for Word Sense Disambiguation

Alexander Popov⬤, Petia Koprinkova-Hristova(✉)⬤, Kiril Simov⬤,
and Petya Osenova⬤

IICT, Bulgarian Academy of Sciences, Sofia, Bulgaria
{alex.popov,kivs,petya}@bultreebank.org, pkoprinkova@bas.bg

Abstract. Inspired by bidirectional long short-term memory (LSTM) recurrent neural network (RNN) architectures, commonly applied in natural language processing (NLP) tasks, we have investigated an alternative bidirectional RNN structure consisting of two Echo state networks (ESN). Like the widely applied BiLSTM architectures, the BiESN structure accumulates information from both the left and right contexts of target word, thus accounting for all available information within the text. The main advantages of BiESN over BiLSTM are the smaller number of trainable parameters and a simpler training algorithm. The two modelling approaches have been compared on the word sense disambiguation task (WSD) in NLP. The accuracy of several BiESN architectures is compared with that of similar BiLSTM models trained and evaluated on the same data sets.

Keywords: Recurrent neural networks · Echo state network ·
Long short-term memory · Natural language processing ·
Word sense disambiguation

1 Introduction

An important recent development in natural language processing (NLP) has been the adoption of recurrent neural networks (RNNs) as a viable tool for language modeling. Word sense disambiguation (WSD) is an NLP task aimed at assigning proper categories of meaning to words that are ambiguous (i.e. they can have several related or unrelated meanings depending on the context). For instance, the word "speaker" can refer to a person who speaks ("The next speaker will talk about new scientific achievements.") or a device reproducing speech ("The speaker sound was of very poor quality."). In order to do WSD we need to account for all words (the context) not only before but also after the target word ("speaker" in our example). In some cases the task might require examining even the context preceding a sentence boundary, and on rare occasions it might even depend on looking forward into and beyond the current sentence. Thus the ability of RNNs to keep a memory trace of past "events" at theoretically

© Springer Nature Switzerland AG 2019
I. V. Tetko et al. (Eds.): ICANN 2019, LNCS 11731, pp. 94–109, 2019.
https://doi.org/10.1007/978-3-030-30493-5_10

arbitrary distances from the present step is an obvious advantage over algorithms that collect information from a fixed window around the target word.

For a long time RNNs were considered difficult to train, as their memory capabilities are often thwarted in practice by the *exploding/vanishing gradients problem*. While the exploding gradients can be capped, a more elaborate solution was needed to combat the vanishing part. As an attempt to solve these problems *long short-term memory* cells [12] were designed to selectively forget information about old states and pay attention to new inputs (a good introduction to LSTMs can be found in [8]; another similar and newer development are *gated recurrent units* [2]). A further enhancement to such an architecture is making it *bidirectional* [9], i.e. the input sequence is fed into two recurrent context layers – one running from the beginning to the end of the sequence and the other one running in reverse. Bidirectional LSTMs (BiLSTMs) have been successfully applied to a number of sequence-to-sequence tasks in NLP [13, 31–33].

Although LSTM architectures cope well with the *exploding/vanishing gradients problem*, their training via the gradient descent algorithm is a computationally demanding task, especially in the case of very deep networks. Aimed at the development of fast trainable RNNs, an alternative approach was proposed in 2002 independently by [18] under the name *liquid state machines* (LSM) and by [14] under the name *echo state networks* (ESN). Nowadays these are collectively referred to as *reservoir computing* (RC) [17] approaches. The idea consists of generating a random and sparsely connected recurrent reservoir of neurons whose mutual connection weights are not subject to training and a linear readout layer that can be tuned in one shot (presenting each training sample only once and solving the least squares problem). Since their emergence, RC approaches have been widely used for the modeling of a variety of dynamical systems [1]. Gallicchio et al. [7] compared deep ESN architectures with popular deep gated RNNs (like LSTM and GRU architectures) for time series prediction and demonstrated that in most benchmark problems deep ESNs outperform the fully trained RNNs.

Application of ESNs in NLP have started to appear only recently, so there are only a few works in this area. The possibility of language modelling via ESN was investigated in [5, 11]. In [29, 30], ESNs are applied for semantic role labeling in a multimodal robotic architecture. Other NLP applications are in the area of speech processing—[26, 27], and in language modeling—[28]. To the best of our knowledge, there are yet no examples of ESNs being used for WSD. Our preliminary attempt to solve the WSD task using a single ESN reservoir [16] has shown that although the training and testing errors on predicted sense embedding vectors are quite small, the vector representations of the possible senses per word are very close to one another in the embedding space, resulting in low accuracy scores.

That is why, in an attempt to increase WSD accuracy, in present work we have adopted a bidirectional reservoir approach proposed initially in [6, 24]. Similar to BiLSTMs, bidirectional reservoir structures (BiESN) were applied for feature extraction from time series – from forward and backward contexts (left and right ESN reservoir state vectors); the features are further used for classification purposes. In our model the readout is a linear combination of the current input and both reservoir states, and the training of the output connections is done via the *recursive least squares* (RLS) algorithm. Here we also compare BiESN models with similar BiLSTM models [22] evaluated on the same task of WSD.

The structure of the paper is as follows: the next section introduces ESN and BiESN architectures and compares them with similar LSTM models; Sect. 3 briefly describes the available research on BiLSTMs for WSD; the results obtained on WSD with BiESNs are presented next and compared with the accuracy of BiLSTM architectures; a section dedicated to conclusions and future work closes the paper.

2   Echo State Networks vs LSTM Architectures

2.1   ESN Basics

The structure of an ESN is shown on Fig. 1. It incorporates a dynamic reservoir of neurons with a sigmoid activation function (usually the hyperbolic tangent) and randomly generated recurrent connections W^{res}. The reservoir state $R(k)$ depends both on its previous state $R(k-1)$ and on the current input $in(k)$—Eq. 1. The network output is generated as a linear combination of the reservoir and input states—Eq. 2:

$$R(k) = (1-a)R(k-1) + a\tanh(W^{in}in(k) + W^{res}R(k-1)) \qquad (1)$$

$$out(k) = W^{out}[in(k), R(k)] \qquad (2)$$

Here W^{in} and W^{res} are $n_{in} \times n_R$ and $n_R \times n_R$ matrices that are randomly generated and are not trainable; n_{out}, n_{in} and n_R are the sizes of the corresponding vectors *out*, *in* and *R*; the parameter a, called leaking rate, influences the reservoir short-term memory and in many applications is omitted, i.e. a is set to 1; W^{out} is a $n_{out} \times (n_{in} + n_R)$ trainable matrix.

According to recipes in [14], the reservoir connection matrix W^{res} should be generated so as to guarantee the "echo state property" of the ESN, i.e. the changes in the input vector must be reflected like an "echo" in the output vector (meaning that the response effect should vanish gradually over time). This is achieved by proper normalization of the matrix W^{res} so that its spectral radius becomes smaller than 1 (although for some applications it has been shown that a spectral radius above 1 could also work). The output weights can be tuned by the least squares (LS) approach (or ridge regression) in one shot after the single presentation of all input/output training samples, or iteratively using its recursive version (RLS) – presenting training input/output samples one by one [14].

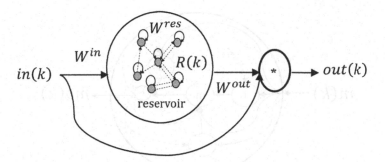

Fig. 1. Echo state network structure.

2.2   Echo State vs. Long Short-Term Memory Networks

A similar recurrent structure widely used in the area of NLP is the long short-term memory (LSTM) network [10], shown in Fig. 2. The LSTM cell model is described by the following set of equations:

$$m(k) = f(k) \circ m(k-1) + i(k) \circ \tanh(W_m^{in}in(k) + W_m^{res}mo(k-1) + b_m) \quad (3)$$

$$mo(k) = o(k) \circ \sigma(m(k)) \quad (4)$$

$$out(k) = W_m^{out}mo(k) \quad (5)$$

Here $\circ$ denotes Hadamard product; $m(k)$ is an internal state (memory) vector, $mo(k)$ is a memory units output vector and $out(k)$ is readout from the structure (following the analogy with ESNs) at the current time instant k; $f(k)$, $i(k)$ and $o(k)$ denote the states of the forget, input and output gates respectively. The dynamics of the gates is governed by the following recurrent equations:

$$i(k) = \sigma(W_i^{in}in(k) + W_i^{res}mo(k-1) + b_i) \quad (6)$$

$$f(k) = \sigma(W_f^{in}in(k) + W_f^{res}mo(k-1) + b_f) \quad (7)$$

$$o(k) = \sigma(W_o^{in}in(k) + W_o^{res}mo(k-1) + b_o) \quad (8)$$

Here σ denotes the sigmoid function. In order to maintain the analogy with ESNs, the recurrent connection weights are denoted by the superscript res, while the input connection weights – by the superscript in. The dimension of the weight matrices depends on the number of memory units n_m and the input vector size n_{in} as follows: W_*^{in}, where $*$ is for m, i, f or o respectively, are matrices of size $n_{in} \times n_m$; W_*^{res} are matrices of size $n_m \times n_m$; b_* are vectors of size n_m.

Looking at Figs. 1 and 2 and the corresponding Eqs. 3–8 and 1–2, the similarity is obvious. Both accumulate temporal information – in the reservoir (Eq. 1) or in the memory (Eq. 3) of the recurrent structures. The only difference is that ESNs have no bias term (although it could be added) and that its parameters analogous to the input, forget and output gates (according to Eq. 1 they are a, $(1 - a)$ and 1, respectively) are constant, while the LSTM gates are dynamically changing (Eqs. 6–8, respectively). In contrast to ESNs, whose trainable

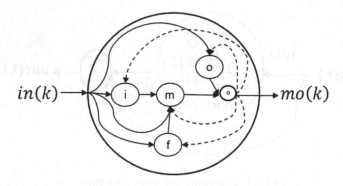

Fig. 2. LSTM cell structure. Dashed lines represent recurrent connections, while solid lines – feedforward connections.

parameters are only in the reservoir readout matrix, all weight matrices from Eqs. 3, 5, as well as 6–8, have to be trained. Thus the total number of trainable parameters of LSTMs and ESNs having the same number of internal units ($n_m = n_R = n_{units}$), as well as the same input and output vector sizes are:

$$n_{tr.params}^{LSTM} = 4n_{units}(n_{in} + n_{units} + 1) + n_{out}n_{units} \tag{9}$$

$$n_{tr.params}^{ESN} = n_{out}n_{in} + n_{out}n_{units} \tag{10}$$

An additional advantage of ESNs is that the matrix containing the recurrent connection weights could be sparse, so that the memory necessary to store its parameters decreases even further. Another difference is found in the training procedure: LSTMs are trained via backpropagation through the time states. The gradient descent algorithm used for the training needs at least several epochs (iterations over all training data) to settle into a possible local error minimum and a lot of memory to keep all intermediate state variables necessary for gradient calculation. ESNs, on the other hand, are trained in one epoch by solving ridge regression equations or using an RLS algorithm, thus finding optimal parameters in one shot. Hence the main advantages of ESNs are the smaller number of parameters and the simpler training algorithm, which can significantly decrease the necessary computational resources. All of this is achieved at the expense of simplifications (constant gates) and the random choice of ESN non-trainable parameters (reservoir connection matrix, its sparsity and leakage rate) that could make its application trickier.

2.3   Bidirectional ESN Architecture

In order to compare ESN performance with that of widely applied LSTM network structures, here we adopt a bidirectional reservoir structure (BiESN) similar to the one proposed in [6] and shown in Fig. 3. The information for the left and right contexts is accumulated in two independent reservoirs – ESN_L and ESN_R.

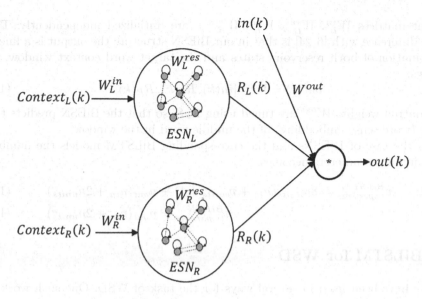

Fig. 3. Bidirectional structure composed of two ESNs for the left and right contexts.

First, for the k-th word in a series of tokens, containing n words in total, we scan the text from left to right (Eq. 11) and from right to left (Eq. 12):

$$Context_L(k) = [in(0), in(1), ...in(k)] \tag{11}$$

$$Context_R(k) = [in(n), in(n-1), ...in(k)] \tag{12}$$

in order to accumulate the left and right contexts respectively. Here $in(t)$ denotes the vector concatenation of the embeddings of the words within the window centered on the t-th word. In the case of a fixed-size context window applied to separate sentences it is necessary to have zero padding at the beginning and end of sentences. However, since the reservoirs themselves are able to accumulate information from context series with indefinite lengths, in our application we scan training/testing text documents from beginning to end (left context) and vice versa (right context). This should allow them to build up a richer contextual representation beyond sentence boundaries. In this sense the architecture behaves differently from the BiLSTM. LSTMs typically read only one sentence at a time, since they are much more difficult to train over such long dependencies.

The two separate reservoir states calculated upon the presentation of the left and right context sequences are obtained as follows:

$$R_L(k) = ESN_L(Context_L(k)) \tag{13}$$

$$R_R(k) = ESN_R(Context_R(k)) \tag{14}$$

Here $ESN_*(Context_*(k))$ stands for the iterative calculation of Eq. (1), as applied to input sequences (11) and (12) respectively. Notice that the parameters for the two ESNs – leaking rates a_L and a_R and the input and reservoir weight matrices W_L^{in}, W_L^{res}, W_R^{in}, W_R^{res} – are initialized independently. The main difference with [6, 24] is that in our BiESN structure the output is a linear combination of both reservoirs states and the target word context window, as follows:

$$out(k) = W^{out}[in(k), R_L(k), R_R(k)] \tag{15}$$

The output weights W^{out} are tuned using RLS, so that the BiESN predicts the sense (word sense embedding) of the middle word in the window.

In the case of BiESNs and the corresponding BiLSTM models the number of trainable parameters becomes:

$$n_{tr.params}^{BiLSTM} = 8n_{units}(n_{in} + n_{units} + 1) + n_{out}(n_{in} + 2n_{units}) \tag{16}$$

$$n_{tr.params}^{BiESN} = n_{out}(n_{in} + 2n_{units}) \tag{17}$$

3   BiLSTM for WSD

RNNs have been used in several ways for the task of WSD. One such work is [15], which uses a BiLSTM to solve a *lexical sample task* – that is, the model is disambiguating one word per sentence only. The model is on par (or slightly better) with state-of-the-art systems, but uses no other features apart from input word embeddings. Popov [22] presents two BiLSTM architectures for solving the *all-words task* (i.e. disambiguating all open-class words in a context). Figure 4 is a combined representation of the two architectures that share the same embedding and BiLSTM layers design; the main difference is in the output layers.

The recurrent BiLSTM layer consists of two LSTM cells processing the incoming embeddings in forward and reverse order; the outputs of the forward and backward LSTM cells are then concatenated and fed into the linear output layer to be re-sized according to the training data dimension (the size of the **vocabulary** n_v for *Architecture A* and the size of the *embedding vectors* n_{emb} for *Architecture B* respectively). *Architecture A* has an additional **softmax** layer calculating the probability distribution of the output layer over the vocabulary of synonym sets (**synsets**).

The training data for *Architecture A* consists of one-hot vectors **labels**. The training data for *Architecture B* consists of real-valued *embedding vectors* of word senses obtained from the gold labels (for a description of how these vectors are generated, see [25] and [22]). The embedding model used for the word senses is the same as the model used for the input tokens, as this should facilitate training.

Thus the loss functions subject to minimization during training via the gradient descent procedure are: **cross entropy** in the case of *Architecture A* and **mean square error** between predicted and target word synset embeddings in the case of *Arcitechture B*. In the case of *Arcitechture B* the classification itself is done by choosing the closest possible synset label, as measured in terms of cosine similarity between synset embeddings. Thus, *Architecture A* is a classifier which learns to distinguish directly between synset categories (hence the use of cross entropy – the network is trained to predict probability distributions), whereas *Architecture B* learns to embed the context of usage of each open-class word – in terms of the semantico-syntactic features in the embedding models (hence the mean square error loss function is used – it allows the network to learn to predict each dimension of meaning).

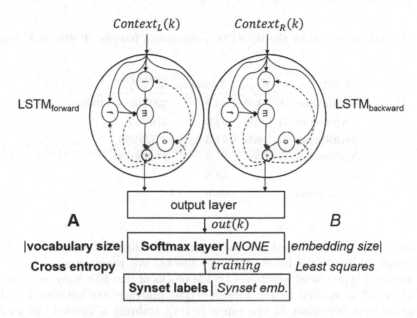

Fig. 4. Recurrent neural networks for WSD: where diverging, the two pathways (*Architectures A and B*) are marked by a separator (/); the left pathway (marked in **bold**) corresponds to elements from *Architecture A*; the right pathway (marked in *italic*) corresponds to elements from *Architecture B*. The *NONE* symbol signals that no corresponding element is present for the particular architecture.

Both architectures are trained on the SemCor data set [20], which is traditionally used for the WSD task (it is used for the ESN models below as well). For evaluation, the concatenation of a number of Senseval and SemEval data sets is used, as described in [22] and here called *WSDEval-ALL*; it contains 7254 word sense annotations in total. The Senseval-2 data set is used as a development set [3].

The word/synset embeddings are obtained via the Word2Vec package [19], using training data obtained from a *pseudo corpus* of artificial sentences generated from an enriched version of the WordNet knowledge graph [4] (for more details see [22]). The model here is slightly richer, as a Wikipedia dump is concatenated to the pseudo corpus prior to training, so that the model would reflect not just the explicit relational knowledge encoded in the WordNet knowledge graph but knowledge about natural language text as well. The same embeddings are used to train the BiESN models as well, as described in the next section.

It is important to note that for some of these recurrent architectures this embedding model is not optimal. Widely used models such as GloVe [21] have been shown to yield results which are closer to the state of the art (e.g. [22]). However, since some of the architectures presented here need access to both word and word sense (synset) embeddings, the model described above is used in all cases in the interest of making a fair comparison.

Table 1. WSD accuracy for the BiLSTM architectures (on the *WSDEval-ALL* data set).

$Model$	n_{units}	$Accuracy$	$n_{tr.params}^{BiLSTM}$
Architecture A	400	66.2	$320\,800 + 1100n_v$
Architecture B	100	62.6	$470\,800$
Architecture B	400	61.3	$2\,573\,200$
Architecture B	1000	61.9	$11\,098\,000$
MFS		64.8	
IMS-s+emb		69.6	

Table 1 shows the recorded accuracy scores. The BiLSTM architectures both use a single bidirectional layer of LSTM blocks. We report a result with an *Architecture A* type model which has LSTM blocks of size 400; input and output dropout of 0.2 is applied to the blocks, whose matrices are initialized with a random uniform initializer in the range $[-1; 1]$; training is carried out with a SGD optimizer with learning rate 0.2, on batches of 100 training sentences [22]. As for *Architecture B* type models, we report results with LSTM blocks of sizes 100, 400 and 1000, so as to give a fuller sense of how those compare with their reformulation as ESN networks (the rest of the hyperparameters are the same as with the *A*-model, only the dropout parameter is set to a different value – 0). Additional improvements to the RNNs are possible, such as adding attention mechanisms and training on auxiliary tasks (see [23]), but since our purpose is to provide a relatively fair comparison between the ESN and LSTM mechanisms, we refrain from exploring them. The results from the *IMS-s+emb* system are used here as representative of the state of the art, as reported in [22]; the model is a trained SVM using tailor-made features, which include word embeddings. The most frequent sense (MFS) baseline, which is a very strong one and is also described in [22], is provided as well.

4   BiESNs for WSD

In this section we evaluate the BiESN architecture with linear readout (described previously). For the sake of comparison, we have also trained BiESN models with an additional softmax layer at the output.

4.1 BiESN with Linear Readout

Since the reservoir, being a recurrent structure, has its own internal memory, we set the window size to one word. The left and right contexts are obtained by starting from the beginning and the end of each text in the training corpus. Hyper-parameters of ESN reservoirs were not subject of optimization in present work. Based on our previous experience on WSD with one reservoir [16], we set hyper-parameters of both reservoirs as follows: random input weight matrices W^{in} in range $[-0.5; 0.5]$; recurrent connection matrices W^{res} with 50% sparsity were initialized randomly and were scaled to achieve spectral radius at the edge of stability (1.25), the leakage rate a was set to 1. The following Table 2 presents accuracy scores for trained BiESN models that have different reservoir sizes and, correspondingly, varying numbers of trainable parameters.

Table 2. Accuracy of WSD by BiESN models with linear readout.

n_R	$Accuracy$	$n_{tr.params}^{BiESN}$
100	61.712	150 000
1000	63.201	690 000
3000	64.387	1 890 000
5000	65.049	3 090 000

Comparison with the reported in Table 1 accuracy for BiLSTM *Architecture B* from Fig. 4 (which is more akin to the BiESN structure) shows that the BiESN architecture outperforms BiLSTM significantly. The total number of trainable parameters of the BiLSTM *Architecture B* that achieved highest accuracy of 62.6% ($n_{units} = 100$) was 470800; this corresponds to a BiESN structure with much more (about 770) units. However, further increase of the BiLSTM size decreased WSD accuracy. In contrast, a BiESN with only 100 units and 150000 trainable parameters achieved accuracy of 61.7%, which corresponds to accuracy of 61.9% for the BiLSTM *Architecture B* with 1000 units and about 74 times bigger number (11098000) of trainable parameters. Further increase of BiESN model size led to increased accuracy so that even a BiESN with 1000 units outperforms a BiLSTM with the same number of units (and many more trainable parameters), achieving 63.2% accuracy. Moreover, our best BiESN model achieved accuracy of more than 65% (for $n_R = 5000$). The BiESN architecture thus offers a simpler and easily trainable alternative that is able to beat the difficult MFS baseline and edge towards the state of the art. Moreover, the same kinds of improvements applicable to LSTMs (more expressive embedding models, attention mechanisms, etc.) can be implemented for ESNs as well.

4.2 BiESNs with Softmax Output Layer

Next, we present a few experiments with a BiESN architecture trained to perform direct classification of word senses. It is to a great extent analogous with the BiLSTM *Architecture A* from Fig. 4. The concatenated vector of both reservoirs states and the input vector is processed by a layer of neurons with ReLU activation functions whose dimension corresponds to the size of the dictionary n_v. Then the gradient procedure minimizing the mean square error is used to train the weight matrices of both ReLU and softmax layers positioned after the reservoirs. Finally, the softmax probability distribution is computed and a cross entropy comparison is performed against the gold labels (one-hot vectors) like in the case of BiLSTM *Architecture A*. Table 3 displays the results for the softmax version of the BiESN.

Table 3. Accuracy of WSD with BiESN + ReLU + softmax.

n_R	$Accuracy$	$n_{tr.params}^{BiESN}$
300	64.856	$900n_v$
1000	65.008	$2300n_v$
3000	64.994	$6300n_v$

We are yet to comprehensively optimize the hyperparameters of the network, but even a preliminary comparison shows that BiESNs are able to achieve results similar to those associated with vanilla BiLSTMs. In all cases the accuracy is around 65%, i.e. slightly below the results obtained with BiESNs trained via RLS method on the linear readout (see Table 2). We have to mention that in our experiments with BiLSTM models training is done only in one iteration over all data with constant learning rate (1.0). Therefore using smaller or variable learning rates in combination with multiple iterations could improve WSD accuracy further at the expense of increased training time.

4.3   Linguistic Error Analysis

The error analysis was performed by comparing the outputs of the best-performing BiESN network with the gold annotations. A table with all error cases was extracted, aligning the gold synset and gold synset gloss (i.e. definition) per error with the selected synset and its gloss. In each case the cosine similarity between the gold and selected synset embeddings was included. The totality of error cases were then ordered in descending order, starting from the most similar pairs and proceeding to the least similar ones.

In the range of pair cosine similarity between 0.97 and 0.90 there are 392 examples. Inside this group examples of all content words have been found: nouns, verbs, adjectives, adverbs. We chose highly similar cases in order to inspect whether such small differences reflect genuine distinctions or are rather misleading.

These pairs containing highly similar but non-identical synsets can be divided into several categories. In the first case the non-match is due to the selection of another meaning of the lemma which is very close to the gold sense (in terms of the synset glosses). For example, the noun "week" has three meanings. The gold one is sense number three: "a period of seven consecutive days starting on Sunday". The selected one is the first sense: "any period of seven consecutive days; it rained for a week". Another example is the verb "say", which has 11 senses. The gold one is number 2: "report or maintain". The selected sense is number 1: "express in words". Interestingly enough, in another case the same verb is annotated in the gold corpus with sense number 8: "utter aloud". The sense selected by the system is again number 1: "express in words", which could indicate that it is in fact a generalization of the other two. Another example is the adjective "new". It has 10 senses. The gold one is number 3: "original and of a kind not seen before". The selected one is number 1: "not of long duration".

In this group we observe that the gold annotation provides some non-first sense of the lemma while the selected sense is always the first meaning. This observation can be used to put forward two hypotheses. The first one is that apparently the partition of meanings in WordNet includes some overlap between the different senses (or even subsumption in some cases). The other hypothesis is that the algorithm tends to select the first meaning whenever there is insufficient knowledge to distinguish between more fine-grained senses of the same lemma.

The second category of errors includes examples where the gold sense and the selected sense belong to different domains. For example, for the lemma "window" the gold sense is number 8 and belongs to the domain of computer science: "a rectangular part of a computer screen that contains a display different from the rest of the screen". The selected sense is again number 1 but it belongs to a totally different domain: "a framework of wood or metal that contains a glass windowpane and is built into a wall or roof to admit light or air". Another example is the noun "trial", which has 6 senses altogether. In the gold annotation it is used in its sense #3 in the law domain: "(law) the determination of a person's innocence or guilt by due process of law". In the test data however sense 1 is selected, which is more general: "the act of testing something".

Within this set of highly similar pairs one can find also mismatches between highly dissimilar senses, irrespective of any domain misalignments. For example, the verb "crawl" has 5 meanings. The gold one is sense 2: "feel as if crawling with insects". The selected one is, not unexpectedly, sense 1: "move slowly; in the case of people or animals with the body near the ground". Here the meanings are neither close, nor overlapping. As a subtype here we can consider also senses that differ not so much in their lexical meaning but rather in the participants involved. For example, the verb "kill", with 15 senses, is in one case annotated in the gold corpus with meaning number 9: "deprive of life". The selected sense number 1 is: "cause to die; put to death, usually intentionally or knowingly". In the former case the causing force is underspecified, it could be due to something like a disease. In the latter case it necessarily involves humans.

There are also some grey areas where the domain is not explicitly stated in the definition but it might be presupposed. For example the noun "pound", with 12 senses, is in the gold corpus given meaning number 2: "the basic unit of money in Great Britain and Northern Ireland; equal to 100 pence". The selected sense is number 1: "16 ounces avoirdupois".

From the point of view of linguistics, and more precisely—that of modeling lexical senses, our survey confirms previous observations—that in some cases the distinct senses per lemma overlap and become virtually indistinguishable for algorithms. Thus, a productive strategy could be merging them in more general senses. From the point of view of computational methods, more knowledge is necessary in order to make these distinctions, especially new training sets which provide good examples of the senses currently used for evaluation. Last but not least, resorting to the first sense (i.e. usually the most frequent one) when no other clues are available remains the preferred strategy for computational models.

5   Conclusions

The results reported in this paper outperform our previous work [16], thereby confirming the ability of the BiESN architecture to capture more context information. Moreover, the results obtained for WSD are comparable with the performance of BiLSTM architectures. One important takeaway is that since the computational time and resources necessary for the training of reservoir architectures are considerably less in comparison with the gradient descent training of LSTM architectures, ESNs could emerge as a more attractive tool in the near future.

We have to point out that since in the present work we do not optimize the ESN hyperparameters (they were set on the basis of some preliminary observations), further work towards their optimal tuning could increase even more the accuracy of BiESN models. In addition to this, since recent reports on deep ESN architectures are also very promising, our next aim will be to apply them as well in the context of WSD. One potential direction for future work is to apply the same approach over different kinds of input constructions: not just over linear text, but also over graphs. In this way we will be able to explore the syntactic and discourse structures of text, in addition to its linear make-up.

Acknowledgments. This research has been funded by the Bulgarian National Science Fund grant number 02/12/2016—*Deep Models of Semantic Knowledge (DemoSem)* and was partially supported by the National Scientific Program "Information and Communication Technologies for a Single Digital Market in Science, Education and Security (ICTinSES)", financed by the Ministry of Education and Science. Alexander Popov was also partially supported by the Bulgarian Ministry of Education and Science under the National Research Programme "Young scientists and postdoctoral students" approved by DCM # 577/17.08.2018.

References

1. Butcher, J.B., Verstraeten, D., Schrauwen, B., Day, C.R., Haycock, P.W.: Reservoir computing and extreme learning machines for non-linear time-series data analysis. Neural Netw. **38**, 76–89 (2013). https://doi.org/10.1016/j.neunet.2012.11.011
2. Cho, K., van Merriënboer, B., Bahdanau, D., Bengio, Y.: On the properties of neural machine translation: encoder-decoder approaches. In: Proceedings of SSST-8, Eighth Workshop on Syntax, Semantics and Structure in Statistical Translation, pp. 103–111. Association for Computational Linguistics, Doha, October 2014. https://doi.org/10.3115/v1/W14-4012
3. Edmonds, P., Cotton, S.: SENSEVAL-2: overview. In: The Proceedings of the Second International Workshop on Evaluating Word Sense Disambiguation Systems, SENSEVAL 2001, pp. 1–5. Association for Computational Linguistics, Stroudsburg (2001). http://dl.acm.org/citation.cfm?id=2387364.2387365
4. Fellbaum, C.: WordNet. In: Poli, R., Healy, M., Kameas, A. (eds.) Theory and Applications of Ontology: Computer Applications, pp. 231–243. Springer, Dordrecht (2010). https://doi.org/10.1007/978-90-481-8847-5_10
5. Frank, S.L., Čerňanský, M.P.: Generalization and systematicity in echo state networks. In: The Annual Meeting of the Cognitive Science Society, pp. 733–738 (2008)
6. Gallicchio, C., Micheli, A.: A reservoir computing approach for human gesture recognition from kinect data. In: Proceedings of the AI for Ambient Assisted Living (2016)
7. Gallicchio, C., Micheli, A., Pedrelli, L.: Comparison between DeepESNs and gated RNNs on multivariate time-series prediction. CoRR (2018). http://arxiv.org/abs/1812.11527
8. Graves, A.: Supervised Sequence Labelling with Recurrent Neural Networks. SCI, vol. 385. Springer, Heidelberg (2012). https://doi.org/10.1007/978-3-642-24797-2
9. Graves, A., Schmidhuber, J.: Framewise phoneme classification with bidirectional LSTM and other neural network architectures. Neural Netw. **18**(5 6), 602–610 (2005). https://doi.org/10.1016/j.neunet.2005.06.042
10. Greff, K., Srivastava, R.K., Koutník, J., Steunebrink, B.R., Schmidhuber, J.: LSTM: a search space odyssey. IEEE Trans. Neural Netw. Learn. Syst. **28**(10), 2222–2232 (2017). https://doi.org/10.1109/TNNLS.2016.2582924
11. Hinaut, X., Dominey, P.F.: Real-time parallel processing of grammatical structure in the fronto-striatal system: a recurrent network simulation study using reservoir computing. PLOS ONE **8**(2), 1–18 (2013). https://doi.org/10.1371/journal.pone.0052946
12. Hochreiter, S., Schmidhuber, J.: Long short-term memory. Neural Comput. **9**(8), 1735–1780 (1997). https://doi.org/10.1162/neco.1997.9.8.1735
13. Huang, Z., Xu, W., Yu, K.: Bidirectional LSTM-CRF models for sequence tagging. CoRR (2015). http://arxiv.org/abs/1508.01991
14. Jaeger, H.: Tutorial on training recurrent neural networks, covering BPPT, RTRL, EKF and the echo state network approach. GMD Report 159, German National Research Center for Information Technology (2002)
15. Kågebäck, M., Salomonsson, H.: Word sense disambiguation using a bidirectional LSTM. In: Proceedings of the 5th Workshop on Cognitive Aspects of the Lexicon (CogALex - V), pp. 51–56. The COLING 2016 Organizing Committee, Osaka, December 2016. https://www.aclweb.org/anthology/W16-5307
16. Koprinkova-Hristova, P., Popov, A., Simov, K., Osenova, P.: Echo state network for word sense disambiguation. In: Proceedings of the Artificial Intelligence: Methodology, Systems, and Applications - 18th International Conference, AIMSA 2018,

Varna, Bulgaria, 12–14 September 2018, pp. 73–82 (2018). https://doi.org/10. 1007/978-3-319-99344-7_7

17. Lukosevicius, M., Jaeger, H.: Reservoir computing approaches to recurrent neural network training. Comput. Sci. Rev. **3**, 127–149 (2009). https://doi.org/10.1016/ j.cosrev.2009.03.005

18. Maass, W., Natschläger, T., Markram, H.: Real-time computing without stable states: a new framework for neural computation based on perturbations. Neural Comput. **14**(11), 2531–2560 (2002). https://doi.org/10.1162/089976602760407955

19. Mikolov, T., Chen, K., Corrado, G.S., Dean, J.: Efficient estimation of word representations in vector space. CoRR (2013). https://arxiv.org/abs/1301.3781

20. Miller, G.A., Leacock, C., Tengi, R., Bunker, R.T.: A semantic concordance. In: Proceedings of the Workshop on Human Language Technology, HLT 1993, pp. 303–308. Association for Computational Linguistics, Stroudsburg (1993). https:// doi.org/10.3115/1075671.1075742

21. Pennington, J., Socher, R., Manning, C.: Glove: global vectors for word representation. In: Proceedings of the 2014 Conference on Empirical Methods in Natural Language Processing (EMNLP), pp. 1532–1543. Association for Computational Linguistics, Doha, October 2014. https://doi.org/10.3115/v1/D14-1162

22. Popov, A.: Word sense disambiguation with recurrent neural networks. In: Proceedings of the Student Research Workshop Associated with RANLP 2017, pp. 25–34. INCOMA Ltd., Varna, September 2017. https://doi.org/10.26615/issn.1314-9156. 2017_004

23. Raganato, A., Camacho-Collados, J., Navigli, R.: Word sense disambiguation: a unified evaluation framework and empirical comparison. In: Proceedings of the 15th Conference of the European Chapter of the Association for Computational Linguistics: Long Papers, vol. 1, pp. 99–110. Association for Computational Linguistics, Valencia (2017). https://www.aclweb.org/anthology/E17-1010

24. Rodan, A., Sheta, A.F., Faris, H.: Bidirectional reservoir network strained using SVM+ privileged information for manufacturing process modeling. Soft Comput. **21**(22), 6811–6824 (2017). https://doi.org/10.1007/s00500-016-2232-9

25. Simov, K., Osenova, P., Popov, A.: Comparison of word embeddings from different knowledge graphs. In: Gracia, J., Bond, F., McCrae, J.P., Buitelaar, P., Chiarcos, C., Hellmann, S. (eds.) LDK 2017. LNCS (LNAI), vol. 10318, pp. 213–221. Springer, Cham (2017). https://doi.org/10.1007/978-3-319-59888-8_19

26. Skowronski, M., Harris, J.: Minimum mean squared error time series classification using an echo state network prediction model. In: 2006 IEEE International Symposium on Circuits and Systems. IEEE (2006). https://doi.org/10.1109/ISCAS.2006. 1693294

27. Squartini, S., Cecchi, S., Rossini, M., Piazza, F.: Echo state networks for real-time audio applications. In: Liu, D., Fei, S., Hou, Z., Zhang, H., Sun, C. (eds.) ISNN 2007. LNCS, vol. 4493, pp. 731–740. Springer, Heidelberg (2007). https://doi.org/ 10.1007/978-3-540-72395-0_90

28. Tong, M.H., Bickett, A.D., Christiansen, E.M., Cottrell, G.W.: Learning grammatical structure with echo state networks. Neural Netw. **20**(3), 424–432 (2007). https://doi.org/10.1016/j.neunet.2007.04.013

29. Twiefel, J., Hinaut, X., Soares, M.B., Strahl, E., Wermter, S.: Using natural language feedback in a neuro-inspired integrated multimodal robotic architecture. In: 25th IEEE International Symposium on Robot and Human Interactive Communication, RO-MAN 2016, New York, NY, USA, 26–31 August 2016, pp. 52–57 (2016). https://doi.org/10.1109/ROMAN.2016.7745090

30. Twiefel, J., Hinaut, X., Wermter, S.: Semantic role labelling for robot instructions using echo state networks. In: 24th European Symposium on Artificial Neural Networks, ESANN 2016, Bruges, Belgium, 27–29 April 2016 (2016). http://www.elen.ucl.ac.be/Proceedings/esann/esannpdf/es2016-168.pdf
31. Wang, P., Qian, Y., Soong, F.K., He, L., Zhao, H.: Part-of-speech tagging with bidirectional long short-term memory recurrent neural network. CoRR (2015). http://arxiv.org/abs/1510.06168
32. Wang, P., Qian, Y., Soong, F.K., He, L., Zhao, H.: A unified tagging solution: bidirectional LSTM recurrent neural network with word embedding. CoRR (2015). http://arxiv.org/abs/1511.00215
33. Wang, W., Chang, B.: Graph-based dependency parsing with bidirectional LSTM. In: Proceedings of the 54th Annual Meeting of the Association for Computational Linguistics: Long Papers, vol. 1, pp. 2306–2315. Association for Computational Linguistics, Berlin, August 2016. https://doi.org/10.18653/v1/P16-1218

Echo State Networks for Named Entity Recognition

Rajkumar Ramamurthy[1,2,3]([📧]), Robin Stenzel[1,2,3], Rafet Sifa[1,2,3],
Anna Ladi[1,2,3], and Christian Bauckhage[1,2,3]

[1] Fraunhofer Center for Machine Learning, Sankt Augustin, Germany
[2] Fraunhofer IAIS, Sankt Augustin, Germany
{rajkumar.ramamurthy,marc.robin.stenzel,rafet.sifa,
anna.ladi,christian.bauckhage}@iais.fraunhofer.de
[3] University of Bonn, Bonn, Germany

Abstract. This paper explores a simple method for obtaining contextual word representations. Recently, it was shown that random sentence representations obtained from echo state networks (ESNs) were able to achieve near state-of-the-art results in several sequence classification tasks. We explore a similar direction while considering a sequence labeling task specifically named entity recognition (NER). The idea is to simply use reservoir states of an ESN as contextual word embeddings by passing pre-trained word-embeddings as its input. Experimental results show that our approach achieves competitive results in terms of accuracy and faster training times when compared to state-of-the-art methods. In addition, we provide an empirical evaluation of hyper-parameters that influence this performance.

Keywords: Echo state networks · Recurrent Neural Networks · Named entity recognition · Natural language processing

1 Introduction

Natural language processing (NLP) comprises a broad spectrum of tasks such as sentence labeling [2, 31], question answering [24], sequence-to-sequence learning [28], natural language interference [3] etc. One of the core components crucial to all these tasks is obtaining an appropriate representation of text. For instance, in a sequence tagging task, each word in a sentence is assigned a linguistic tag (eg. a named entity or a part of speech). In order to be processed by supervised machine learning models, each word is represented by a real-valued vector that encodes both context and semantics. Earlier systems considered word representations that are obtained using latent-semantic analysis methods such as Singular Value Decomposition [6] or GloVe [22] that uses word co-occurrences and context-window based word representations such as Word2Vec [19]. Although effective,

R. Ramamurthy and R. Stenzel—Contributed equally.

© Springer Nature Switzerland AG 2019
I. V. Tetko et al. (Eds.): ICANN 2019, LNCS 11731, pp. 110–120, 2019.
https://doi.org/10.1007/978-3-030-30493-5_11

these methods are not robust against morphological variations or misspelled words. Tackling this and extending on Word2Vec, FastText [13] represents each word with a bag of character n-grams. Following this trend, recent methods such as ELMo [23] and Flair [1] combine character-level language models and deep recurrent architectures to obtain contextual word representations and have become state-of-the-art methods.

In a sequence classification task, too, the task boils down to obtaining compact representations of sentences. While classic methods using bags-of-words (BOW) or term frequency-inverse document frequency (TF-IDF) are simple and effective, they ignore word ordering and suffer from high dimensionality. In recent years, several methods [14,27] have been developed to learn compositional operators that convert word representations to sentence representations using different neural architectures. One drawback of these approaches is that they are trained for a particular task and require supervised learning. However, it is desired to obtain task agnostic representations that can be shared across a wide range of tasks. An approach tailored to this is SkipThought [15] which is similar to Skip-gram architecture of Word2Vec but encode sentences directly. Similarly, InferSent [5] learns a generic representation which is trained on a standard natural language inference task yields state-of-the-art results on several tasks.

While the majority of representation learning methods nowadays are driven by choosing methods that vary the encoder architecture or the type of network (RNNs, CNNs), there has been an alternative line of work [32] that caught attention recently. These methods do not train any sentence encoders explicitly; rather they use pre-trained word embeddings as inputs to randomized recurrent neural networks such as bi-directional long short term memory (BiLSTMs) and echo state networks (ESNs). They showed that state-of-the-art sentence encoders do not significantly improve performance over random encoders.

Inspired by these works, we explore contextualized word representations using echo state networks. In particular, we propose to solve a sequence labeling task namely named entity recognition (NER). Current approaches train a bi-directional LSTM which takes pre-trained word embeddings as its inputs to train a contextual word representation for a NER classifier. Instead, we propose to use echo state networks which provide a random context for its input *without* training any of the recurrent connections. Therefore, our main contribution is evaluating echo state network-based random contextual encoder and look how close they can match the performance of trained contextual encoders.

Providing random context has been studied under the title "reservoir computing". Echo state networks which follow this paradigm, have been successfully applied to provide context in several sequential tasks; for neural cryptography to memorize sequences [25], to learn policies in reinforcement learning [26] and time series prediction [11,16]. Extending this to deep architectures, deep ESNs [9] have been applied to detect Parkinson's disease. However, its application in natural language processing is relatively an unexplored area, and is limited to learn grammatical structures [10,30] and systematicity in natural language [8]. Differing from these methods, our approach focuses on obtaining contextual word representations to be used in NER which, to the best of our knowledge, has not been studied with echo state networks.

2 Preliminaries

2.1 Named Entity Recognition

Named entity recognition (NER) is the task of detecting named entities in text and assigning them an appropriate label such as organization, location or person. For example: "[Jane Green]$_{PER}$ became the youngest director of the world leading sports equipment manufacturer [ActPro Equipment]$_{ORG}$". More generally NER can be typically considered as a sequence labeling problem, where words or characters in a sequence must be classified to one of the predefined classes. Detecting named entities can either be a standalone task (e.g for anonymizing sensitive entities in documents) or part of a text pre-processing pipeline.

Early NER systems relied heavily on domain-specific knowledge in the form of lexicons and simple hand-crafted rules and features; for instance, the morphology of the word, trigger context words and term frequency. Further development on feature-engineering systems, included the replacement of hand-crafted rules with supervised machine learning models. Additionally, more sophisticated features were used, such as part-of-speech tagging, word embedding, context features. For a detailed review of these earlier approaches, the reader is referred to [20].

One of the limitations of feature-engineering models is that they rely on a domain expert, either to hand-craft the rules or to define meaningful features for a specific application. In recent years, research has moved away from this paradigm, towards NN-based feature inferring systems, as showcased in a recent review [33]. Most of the recently developed NER systems consist of the same core, namely a bi-directional long short-term memory (LSTM). The LSTM takes a sequence of word embeddings as its input and returns a sequence of contextual word embeddings by encoding them into its sentence context. This contextual word embedding is obtained by concatenating the LSTM's hidden state of a word for both directions. The LSTM is often coupled with a Conditional Random Field (CRF), a probabilistic method that can predict labels of sequences by taking the neighboring labels into account [29].

2.2 Echo State Networks

Recurrent Neural Networks are a very powerful tool in NLP. However, both training and parameter tuning of an RNN can be cumbersome, due to the size and the connectivity of the network [7, 21]. This problem can be addressed by an alternative paradigm, reservoir computing. Reservoir computing is based on the notion of an interconnected randomly generated network that processes sequential data. This concept is the core of the Echo State Network (ESN), introduced by Jäger [12].

An ESN consists of a reservoir of recurrently and sparsely connected nodes, which are then connected to an output layer, responsible for transforming the response of the network to the appropriate target output. The basic benefit of ESNs is that only the output layer of the network needs to be trained, while the reservoir essentially serves as a context-aware, non-linear mapping of the input data to a high dimensional space.

For an ESN with n_{in} input nodes, n_{out} output nodes and n_{res} reservoir nodes, we can define three sets of weights: the input weights $\boldsymbol{W}^{in} \in \mathbb{R}^{n_{res} \times n_{in}}$ that connect the input nodes to the reservoir, the reservoir weights $\boldsymbol{W}^{res} \in \mathbb{R}^{n_{res} \times n_{res}}$ that interconnect the reservoir nodes and the output weights $\boldsymbol{W}^{out} \in \mathbb{R}^{n_{out} \times n_{res}}$ that connect the reservoir to the output nodes.

At time step t, let $\boldsymbol{x}_t \in \mathbb{R}^{n_{in}}$, $\boldsymbol{r}_t \in \mathbb{R}^{n_{res}}$ and $\boldsymbol{y}_t \in \mathbb{R}^{n_{out}}$ be the state of the input, reservoir and output neurons respectively. The update of the network is then as follows:

$$r_t = (1 - \alpha)r_{t-1} + \alpha f_{res}(\boldsymbol{W}^{res}r_{t-1} + \boldsymbol{W}^{in}\boldsymbol{x}_t) \tag{1}$$

$$\boldsymbol{y}_t = f_{out}(\boldsymbol{W}^{out}r_t), \tag{2}$$

where $f_{res}(.)$ and $f_{out}(.)$ are component-wise activation functions and $\alpha \in [0, 1]$ is the leaking rate. For $f_{res}(.)$ typically a sigmoidal function is chosen for the reservoir neurons, while for the output layer the selection is task dependent; usually, a linear or softmax function is used. The leaking rate has to be tuned for the given task, as it relates to the required memory of the network [17].

As discussed earlier, the generation of the reservoir and input weight matrices is random. However, it is controlled by certain parameters, such as the distribution from which the weights are chosen, the spectral radius (maximum eigenvalue of $\boldsymbol{W}^{res}$), or the input scaling. Similarly to the leaking rate, these parameters are task dependent and need to be carefully set, in order to ensure reasonable performance. For a detailed guideline on how such hyper-parameters can be set the reader is referred to the review in [17]. Training a network to perform a certain task, comes then to training the output weights, $\boldsymbol{W}^{out}$ which is explained in Sect. 3 with respect to named entity recognition.

3   Using ESN for Contextual Word Representations

We are concerned with obtaining contextual-word representations using echo state networks; In particular, we are given pre-trained word embeddings of sentences. To that end, let us consider a sentence s consisting of L tokens which can be regarded as a sequence of its pre-trained token embeddings ie. $s = (\boldsymbol{x_1}, \boldsymbol{x_2}, \ldots \boldsymbol{x_L})$. Given this input sequence and an echo state network with randomly initialized input and reservoir weights, we pass the given input sentence one token at a time and collect its reservoir states at every time step as $c = (\boldsymbol{r_1}, \boldsymbol{r_2}, \ldots \boldsymbol{r_L})$ by following Eq. (1).

By computing context in this manner, the representation of a token includes the previous context. Whereas for named entity recognition, it is observed in [5,32] that it is beneficial to have bi-directional context. To do this, we pass the sequence also in reverse order once and collect its corresponding reservoir states.

Therefore, contextual word-representation for each token is the concatenated reservoir states obtained from both directions ie. $c = (r_1, r_2, \dots r_L)$ where each r_i is $[\overrightarrow{r_i}, \overleftarrow{r_i}]$. Note, we use the same echo state network for both directions, unlike LSTM-based methods which use two separate hidden layers parsing in each direction. This is primarily because W_{res} is randomly initialized.

To solve a sequence labeling task such as NER, we pass this sequence c of contextual-word representations to the readout layer with weights W_{out} and its output can be collected in a sequence $y = (y_1, y_2, \dots y_L)$ following Eq. (2) with f_{out} as softmax activation function. Given a ground truth sequence $g = (g_1, g_2, \dots g_L)$ of NER tags where each g_i is a one-hot encoded vector of possible tags, then it is sufficient to optimize W_{out} by taking a gradient descent at the rate of η minimizing cross-entropy loss $\mathcal{L}$ between predicted sequence y and ground truth sequence g.

$$\mathcal{L}(y, g) = -\sum_{i=1}^{L} \sum_{j=1}^{C} g_i(j) \log(y_i(j)) \tag{3}$$

$$W_{out} = W_{out} - \eta \nabla_{W_{out}} \mathcal{L}(y, g) \tag{4}$$

As input and reservoir weight matrices W_{in} and W_{res} are kept constant throughout this training process, we can break this whole procedure naturally into (1) generating contextual-word representations for all sentences (2) fitting a readout layer. Step (1) can be done just once for each echo state network setting and contextual-representations can be stored. In Step (2), any classifier of our choice (shallow or deep) can be trained offline with a batch gradient descent.

4   Experimental Results

In our experiments, we evaluate the contextual-word representations computed by echo state networks on the task of Named Entity Recognition. Specifically, we test our approach on GermEval Dataset from 2014 [2] in which the task is to identify named entities in sentences. In total, there are 12 classes comprising of 4 main classes: PERson, LOCation, ORGanisation, OTHer and 2 subclasses for each of them: -deriv and -part. The subclasses complicate the task by introducing nested named entities and derived entities. One example of a nested entity is "University of Bonn" which is an ORG but at the same time contains a location entity "Bonn" (LOC-part). Similarly, the second subclass includes word derivation, for example, "das Bonner Theater" (translated as "the theater of Bonn") contains the word "Bonner" (LOC-deriv) derived from the entity "Bonn". The dataset has a predefined splitting into a training, a development and a test set with 24 000, 2 200 and 5 100 sentences respectively.

4.1   Hyper-parameter Tuning

As echo state networks have several hyper-parameters which must be set beforehand, we first tune them by cross-validating on the development set. To that end,

Table 1. Hyper-parameters tuning

Parameter	Value	Performance
Spectral radius	0.6	61.18
	0.7	**61.34**
	0.8	60.92
	0.9	60.60
	1.0	60.71
	1.1	60.50
	1.3	60.14
Leaking rate	0.3	60.92
	0.5	64.74
	0.6	65.11
	0.7	66.27
	0.75	**66.28**
	0.8	60.71
Input scaling	0.125	60.76
	0.25	**62.23**
	0.375	61.10
	0.5	60.92
	0.75	59.45

we use pre-trained word embeddings from FastText [13] as the inputs. For this purpose, we use the Flair implementation[1], which offers an easy to use framework for training and evaluating NER models along with access to different types of embeddings. To keep this tuning process tractable, as discussed before, we first generate contextual-word embeddings for different settings of ESN. Later, we fit a logistic regression model as our read out layer to train it to predict NER tags. For training this readout layer, we use Adadelta optimizer trained for 150 epochs with a learning rate of 0.1 and weight decay of 10^{-5}.

Spectral Radius: The spectral radius $\rho(W_{res})$ which is the maximal absolute eigenvalue of the reservoir weight matrix. As W_{res} is initialized randomly, it is essential to set the spectral radius to a value less than 1.0 to satisfy the echo state property [12]. However, it was shown empirically that sometimes even higher values satisfy this condition and deliver better performance [17]. We varied the spectral radius between 0.6 and 1.3 and measured the resulting F1 score. Table 1 shows the F1 scores for different spectral radii; the performance peaks around 0.7 and then decreases continuously apart from a small local maximum around the unit spectral radius.

---

[1] https://github.com/zalandoresearch/flair.

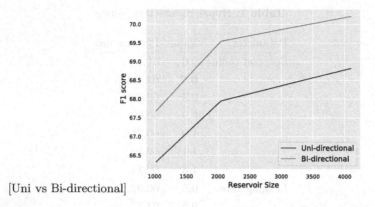

[Uni vs Bi-directional]

Fig. 1. Influence of reservoir size on performance; the larger the reservoir, the better the performance. The fact that the bi-directional variant performs better than the uni-directional suggests that task depends on context from either sides

Leaking Rate: Next, we examine the leaking rate α which determines the speed of reservoir updates for its inputs. A high leaking rate indicates that a lot of old information vanishes through the "leak" in favor of new inputs. Table 1 shows that the best values are somewhere found between 0.7 and 0.75. This indicates that for this task, a higher leaking rate is preferred indicating that a faster update is necessary. Optimal settings of both leaking rate and spectral radius suggest that it is sufficient to have a shorter temporal context.

Input Scaling: Another key parameter of echo state networks is the scaling of input weight matrix W_{in}. These weights can be controlled to influence the non-linearity of the reservoir responses. In our setting, we use $\tanh(\cdot)$ activations; for linear tasks, it is thus beneficial to have small weights around 0.0, where the activation is almost linear. On the other side, for a more complex task, it might be better to choose a high scaling in order to make use of the non-linearity of the activation function. However, it was shown in previous research [17] that large weights can make the ESN unstable. Also, there exists a trade-off between non-linear mapping and memory capacity of echo state networks, it was suggested to use Extreme Learning Machines (ELMs) to tackle that subject [4]. In Table 1, the effect of the input scaling is presented. The highest f1-score corresponds to input scaling of 0.25.

Reservoir Size: Next, we investigate the influence of reservoir size for both uni- and bi-directional ESNs. Since a bi-directional ESN produces an embedding of twice the reservoir size, we instead look at the final embedding sizes to obtain a fair comparison. Figure 1 shows that the bi-directional embedding leads to a continuously better f1-score of around 1.5% indicating the task benefits from having bi-directional temporal context.

Table 2. NER on word embeddings

Model	Embedding dimension	Run time	Performance
Logistic regression (LR)	300	33 min	52.75 ± 0.19
Bi-ESN + LR	4096	37 min	69.04 ± 0.22
Bi-ESN + NN	4096	47 min	70.54 ± 0.10
Bi-LSTM	4096	10 h 12 min	75.45 ± 0.10
Bi-LSTM	256	3 h 3 min	76.87 ± 0.21

4.2 Evaluation on Different Embeddings

Next, we investigate the performance of contextual-word representations obtained from ESN on the test set by considering two types of word embeddings: FastText and Stacked embedding of FastText (dimension of 300) and Flair (dimension of 4096) [1]. We also consider two variants of ESN, first with a logistic regression (Bi-ESN + LR) read out layer and second with a neural network (Bi-ESN + NN) with a hidden layer consisting of 1000 neurons and 0.5 dropout. In either case, we fix our size of reservoir to have 2048 neurons amounting to 4096 dimensional embedding (bi-directional). The optimal setting of other hyper-parameters is obtained from the analysis presented in the previous section. We compare our approach to several baseline models: (i) logistic regression (LR) which does not use any context (ii) bi-directional LSTM (Bi-LSTM) which learns contextual word-representations in an end-to-end fashion for NER task with hidden size 256 as chosen in [1]. For fair comparison, we also train a variant with hidden size that matches the size of ESN reservoir.

On FastText Embeddings: Table 2 summarizes the performance of all methods trained with FastText embeddings as inputs. It is evident that the contextual-word representations generated by ESN lead to a strong improvement over a logistic regression method. Comparing the two ESN variants, it is noticed that a readout with neural networks has ≈1.5% improvement over logistic regression read out layer. The most important result is that the difference between the LSTM and the ESN models amounts to a value of 6% only. This suggests that LSTM variants do not improve much over ESN methods but only incur longer training times of ≈3 h. On the other hand, ESN methods can be quickly trained in less than an hour ≈0.5 h. These result suggests that random contextual-word representations obtained from ESN are already competitive and can be used as a baseline while bench-marking LSTM-based NER models.

On Word + Flair Embeddings: Table 3 presents the same set of experiments as before but with the combination of Flair and FastText token embeddings as inputs. In these experiments, we choose the size of the reservoir as 4396 resulting in contextual-word embedding size of 8792 due to bi-direction. One important observation is that all methods have a considerable improvement over the results presented in Table 2 which shows that Flair embeddings are more powerful in NER task. As Flair embeddings already encode contextual information

Table 3. NER on Flair + word embeddings

Model	Embedding dimension	Run time	Performance
Logistic regression	4396	2 h 10 min	73.78 ± 0.18
Bi-ESN + LR	8792	3 h 40 min	76.74 ± 0.03
Bi-ESN + NN	8792	4 h 40 min	78.17 ± 0.19
Bi-LSTM	8792	75 h	81.24 ± 0.13
Bi-LSTM	256	4 h 35 min	83.52 ± 0.21

as opposed to FastText, one might expect no further improvement by applying a further contextual encoding using ESN or LSTM. Nevertheless, both the ESN and the LSTM increase the performance noticeably by 3 to 5% and 8 to 10% respectively. Comparing the LSTM with the ESN models, we observe that the performance gap is just around 5%. These findings concur with our previous analysis that ESNs are capable of achieving competitive performance as LSTMs while requiring only short period of training time.

5   Conclusion

In this paper, we explored a random contextual-word encoder using echo state networks. Although their input and recurrent connections are not adapted to the task, they are still capable of providing context which can be leveraged for named entity recognition. Experiments suggest that they can be trained quickly and achieve competitive accuracy when compared to state-of-the-art methods. As the performance between trained encoders such as LSTMs and our ESN-based random context encoders is not much, our method can be used as a baseline for evaluating other contextual-word learning methods. There are several aspects in echo state networks which can be further explored; these include sparsity of connections, the structure of reservoir (random graphs, scale-free networks). Also, there has been some research into ESNs with multiple layers [18] that could enhance the performance. As future work, we intend to pursue these ideas which would bring the performance gap further down.

References

1. Akbik, A., Blythe, D., Vollgraf, R.: Contextual string embeddings for sequence labeling. In: Proceedings of International Conference on Computational Linguistics (2018)
2. Benikova, D., Biemann, C., Kisselew, M., Pado, S.: Germeval 2014 Named Entity Recognition Shared Task (2014)
3. Bowman, S.R., Angeli, G., Potts, C., Manning, C.D.: A large annotated corpus for learning natural language inference. arXiv preprint arXiv:1508.05326 (2015)

4. Butcher, J.B., Verstraeten, D., Schrauwen, B., Day, C., Haycock, P.: Reservoir computing and extreme learning machines for non-linear time-series data analysis. Neural Netw. **38**, 76–89 (2013)
5. Conneau, A., Kiela, D., Schwenk, H., Barrault, L., Bordes, A.: Supervised learning of universal sentence representations from natural language inference data. arXiv preprint arXiv:1705.02364 (2017)
6. Deerwester, S., Dumais, S.T., Furnas, G.W., Landauer, T.K., Harshman, R.: Indexing by latent semantic analysis. J. Am. Soc. Inf. Sci. **41**(6), 391–407 (1990)
7. Doya, K.: Bifurcations in the learning of recurrent neural networks. In: Proceedings IEEE International Symposium on Circuits and Systems (1992)
8. Frank, S.L.: Learn more by training less: systematicity in sentence processing by recurrent networks. Connect. Sci. **18**(3), 287–302 (2006)
9. Gallicchio, C., Micheli, A., Pedrelli, L.: Deep echo state networks for diagnosis of Parkinson's disease. arXiv preprint arXiv:1802.06708 (2018)
10. Hinaut, X., Dominey, P.F.: On-line processing of grammatical structure using reservoir computing. In: Villa, A.E.P., Duch, W., Érdi, P., Masulli, F., Palm, G. (eds.) ICANN 2012. LNCS, vol. 7552, pp. 596–603. Springer, Heidelberg (2012). https://doi.org/10.1007/978-3-642-33269-2_75
11. Jaeger, H., Haas, H.: Harnessing nonlinearity: predicting chaotic systems and saving energy in wireless communication. Science **304**(5667), 78–80 (2004)
12. Jäger, H.: The "Echo State" approach to analysing and training recurrent neural networks. Technical report 148, GMD (2001)
13. Joulin, A., Grave, E., Bojanowski, P., Mikolov, T.: Bag of tricks for efficient text classification. arXiv preprint arXiv:1607.01759 (2016)
14. Kim, Y.: Convolutional neural networks for sentence classification. arXiv preprint arXiv:1408.5882 (2014)
15. Kiros, R., et al.: Skip-thought vectors. In: Proceedings of Advances in Neural Information Processing Systems, pp. 3294–3302 (2015)
16. Lin, X., Yang, Z., Song, Y.: Short-term stock price prediction based on echo state networks. Expert Syst. Appl. **36**(3), 7313–7317 (2009)
17. Lukoševičius, M.: A practical guide to applying echo state networks. In: Montavon, G., Orr, G.B., Müller, K.-R. (eds.) Neural Networks: Tricks of the Trade. LNCS, vol. 7700, pp. 659–686. Springer, Heidelberg (2012). https://doi.org/10.1007/978-3-642-35289-8_36
18. Ma, Q., Shen, L., Cottrell, G.W.: Deep-ESN: a multiple projection-encoding hierarchical reservoir computing framework. arXiv preprint arXiv:1711.05255 (2017)
19. Mikolov, T., Chen, K., Corrado, G., Dean, J.: Efficient estimation of word representations in vector space. arXiv preprint arXiv:1301.3781 (2013)
20. Nadeau, D., Sekine, S.: A survey of named entity recognition and classification. Lingvisticae Investigationes **30**(1), 3–26 (2007)
21. Pascanu, R., Mikolov, T., Bengio, Y.: On the difficulty of training recurrent neural networks. In: Proceedings of the ICML (2013)
22. Pennington, J., Socher, R., Manning, C.: Glove: global vectors for word representation. In: Proceedings of Conference on Empirical Methods in Natural Language Processing, pp. 1532–1543 (2014)
23. Peters, M.E., et al.: Deep contextualized word representations. arXiv preprint arXiv:1802.05365 (2018)
24. Rajpurkar, P., Zhang, J., Lopyrev, K., Liang, P.: SQuAD: 100,000+ questions for machine comprehension of text. arXiv preprint arXiv:1606.05250 (2016)

25. Ramamurthy, R., Bauckhage, C., Buza, K., Wrobel, S.: Using echo state networks for cryptography. In: Lintas, A., Rovetta, S., Verschure, P.F.M.J., Villa, A.E.P. (eds.) ICANN 2017. LNCS, vol. 10614, pp. 663–671. Springer, Cham (2017). https://doi.org/10.1007/978-3-319-68612-7_75

26. Ramamurthy, R., Bauckhage, C., Sifa, R., Wrobel, S.: Policy learning using SPSA. In: Kůrková, V., Manolopoulos, Y., Hammer, B., Iliadis, L., Maglogiannis, I. (eds.) ICANN 2018. LNCS, vol. 11141, pp. 3–12. Springer, Cham (2018). https://doi.org/10.1007/978-3-030-01424-7_1

27. Socher, R., et al.: Recursive deep models for semantic compositionality over a sentiment treebank. In: Proceedings of Conference on Empirical Methods in Natural Language Processing (2013)

28. Sutskever, I., Vinyals, O., Le, Q.V.: Sequence to sequence learning with neural networks. In: Proceedings of Advances in Neural Information Processing Systems, pp. 3104–3112 (2014)

29. Sutton, C., McCallum, A., et al.: An introduction to conditional random fields. Found. Trends Mach. Learn. 4(4), 267–373 (2012)

30. Tong, M.H., Bickett, A.D., Christiansen, E.M., Cottrell, G.W.: Learning grammatical structure with echo state networks. Neural Netw. 20(3), 424–432 (2007)

31. Toutanova, K., Klein, D., Manning, C.D., Singer, Y.: Feature-rich part-of-speech tagging with a cyclic dependency network. In: Proceedings of Conference on HLT-NAACL (2003)

32. Wieting, J., Kiela, D.: No training required: exploring random encoders for sentence classification. arXiv preprint arXiv:1901.10444 (2019)

33. Yadav, V., Bethard, S.: A survey on recent advances in named entity recognition from deep learning models. In: Proceedings of International Conference on Computational Linguistics (2018)

Efficient Cross-Validation of Echo State Networks

Mantas Lukoševičius[✉][iD] and Arnas Uselis

Kaunas University of Technology, Studentu st. 50—406, 51368 Kaunas, Lithuania
{mantas.lukosevicius,arnas.uselis}@ktu.edu

Abstract. Echo State Networks (ESNs) are known for their fast and precise one-shot learning of time series. But they often need good hyper-parameter tuning for best performance. For this good validation is key, but usually, a single validation split is used. In this rather practical contribution we suggest several schemes for cross-validating ESNs and introduce an efficient algorithm for implementing them. The component that dominates the time complexity of the already quite fast ESN training remains constant (does not scale up with k) in our proposed method of doing k-fold cross-validation. The component that does scale linearly with k starts dominating only in some not very common situations. Thus in many situations k-fold cross-validation of ESNs can be done for virtually the same time complexity as a simple single split validation. Space complexity can also remain the same. We also discuss when the proposed validation schemes for ESNs could be beneficial and empirically investigate them on several different real-world datasets.

Keywords: Echo State Networks · Reservoir computing ·
Recurrent neural networks · Cross-validation · Time complexity

1 Introduction

Echo State Network (ESN) [1–3] is a recurrent neural network training technique of reservoir computing type [4], known for its fast and precise one-shot learning of time series. But it often needs good hyper-parameter tuning to get the best performance. For this fast and representative validation is very important.

Validation aims to estimate how well the trained model will perform in testing. Typically ESNs are validated on a single data subset. This single training-validation split is just one shot at estimating the test performance. Doing several splits and averaging the results can make a better estimation. This is known as cross-validation ([5] is often cited as one of the earliest descriptions), as the same data can be used for training in one split and for validation in another and vice versa.

In this contribution we suggest several schemes for cross-validating ESNs and introduce an efficient algorithm for implementing them. We also test the validation schemes on five different real-world datasets.

© Springer Nature Switzerland AG 2019
I. V. Tetko et al. (Eds.): ICANN 2019, LNCS 11731, pp. 121–133, 2019.
https://doi.org/10.1007/978-3-030-30493-5_12

The goal of the experiments here is not to obtain the best possible performance on the datasets, but to compare different validation methods of ESNs on equal terms. The best performance here is often sacrificed for the simpler models and procedures. Therefore we use classical ESNs here, but the proposed validation schemes apply to any type of reservoirs with the same time and space complexity savings, as long as they have the same linear readouts.

We introduce our ESN model, training, and notation in Sect. 1.1, discuss different classes of tasks that might be important for validation in Sect. 2.1, discuss cross-validation nuances of time series in Sect. 2.2, suggest several cross-validation schemes for ESNs in Sect. 2.3, suggest several ways of producing the final trained model in Sect. 2.4, and introduce a time- and space-efficient algorithm for cross-validating ESNs in Sect. 2.5. We also report empirical experiments with different types of data in Sect. 3 and conclude with a discussion in Sect. 4.

1.1 Basic ESN Training

Here we introduce our ESN model and notation following [6].

The typical update equations of ESN are

$$\tilde{\mathbf{x}}(n) = \tanh\left(\mathbf{W}^{\text{in}}[1; \mathbf{u}(n)] + \mathbf{W}\mathbf{x}(n-1)\right), \tag{1}$$

$$\mathbf{x}(n) = (1 - \alpha)\mathbf{x}(n-1) + \alpha\tilde{\mathbf{x}}(n), \tag{2}$$

where $\mathbf{x}(n) \in \mathbb{R}^{N_x}$ is a vector of reservoir neuron activations and $\tilde{\mathbf{x}}(n) \in \mathbb{R}^{N_x}$ is its update, all at time step n, $\tanh(\cdot)$ is applied element-wise, $[\cdot; \cdot]$ stands for a vertical vector (or matrix) concatenation, $\mathbf{W}^{\text{in}} \in \mathbb{R}^{N_x \times (1+N_u)}$ and $\mathbf{W} \in \mathbb{R}^{N_x \times N_x}$ are the input and recurrent weight matrices respectively, and $\alpha \in (0, 1]$ is the leaking rate.

The linear readout layer is typically defined as

$$\mathbf{y}(n) = \mathbf{W}^{\text{out}}[1; \mathbf{u}(n); \mathbf{x}(n)], \tag{3}$$

where $\mathbf{y}(n) \in \mathbb{R}^{N_y}$ is the network output and $\mathbf{W}^{\text{out}} \in \mathbb{R}^{N_y \times N_r}$ the output weight matrix. We denote $N_r = 1 + N_u + N_x$ as the size of the "expanded" reservoir $[1; \mathbf{u}(n); \mathbf{x}(n)]$ for brevity.

Equation (3) can be written in a matrix notation as

$$\mathbf{Y} = \mathbf{W}^{\text{out}}\mathbf{X}, \tag{4}$$

where $\mathbf{Y} \in \mathbb{R}^{N_y \times T}$ are all $\mathbf{y}(n)$ and $\mathbf{X} \in \mathbb{R}^{N_r \times T}$ are all $[1; \mathbf{u}(n); \mathbf{x}(n)]$ produced by presenting the reservoir with $\mathbf{u}(n)$, both collected into respective matrices by concatenating the column-vectors horizontally over the training period $n = 1, \ldots, T$. We use here a single $\mathbf{X}$ instead of $[1; \mathbf{U}; \mathbf{X}]$ for notational brevity.

Finding the optimal weights $\mathbf{W}^{\text{out}}$ that minimize the squared error between $\mathbf{y}(n)$ and $\mathbf{y}^{\text{target}}(n)$ amounts to solving a system of linear equations

$$\mathbf{Y}^{\text{target}} = \mathbf{W}^{\text{out}}\mathbf{X}, \tag{5}$$

where $\mathbf{Y}^{\text{target}} \in \mathbb{R}^{N_y \times T}$ are all $\mathbf{y}^{\text{target}}(n)$, with respect to $\mathbf{W}^{\text{out}}$ in a least-square sense, i.e., a case of linear regression. In this context $\mathbf{X}$ can be called the *design matrix*. The system is typically overdetermined because $T \gg N_r$.

The most commonly used solution to (5) in this context is ridge regression:

$$\mathbf{W}^{\text{out}} = \mathbf{Y}^{\text{target}} \mathbf{X}^{\text{T}} \left(\mathbf{X}\mathbf{X}^{\text{T}} + \beta \mathbf{I}\right)^{-1}, \tag{6}$$

where β is a regularization coefficient and $\mathbf{I}$ is the identity matrix. It is advisable to set the first element of $\mathbf{I}$ to zero to exclude the bias connection from the regularization.

For more details on generating and training ESNs see [6].

2 Validation in Echo State Networks

In this chapter we discuss the validation options for ESNs, and propose an efficient algorithm for cross-validation.

2.1 Different Tasks

Some details of implementation and computational savings depend on what type of task we are learning. Let us distinguish three types of temporal machine learning tasks:

1. **Generative** tasks, where the computed output $\mathbf{y}(n)$ comes back as (part of) the input $\mathbf{u}(n + k)$. This is often pattern generation or multi-step time series prediction in a generative mode.[1]
2. **Output** tasks, where the computed output time series $\mathbf{y}(n)$ does not come back as part of input. This is often detection or recognition in time series, or deducing a signal from other contemporary signals.
3. **Classification** tasks, of separate (pre-cut) finite sequences, where a class $\mathbf{y}$ is assigned to each sequence $\mathbf{u}(n)$.

For the latter type of tasks we usually store only an averaged or a fixed number of states $\mathbf{x}(n)$ for every sequence in the state matrix $\mathbf{X}$. It is similar to a non-temporal classification task.

The experiments Sect. 3 is structured according to this distinction.

2.2 Cross-Validation in Time Series

k-fold cross-validation, arguably the most popular cross-validation type, is a standard technique in static (non-temporal) machine learning tasks where data points are independent of each other. Here the data are partitioned into k usually

---

[1] Note that this can alternatively be implemented with feedback connections $\mathbf{W}^{\text{fb}}$ from $\mathbf{y}(n - 1)$ to $\tilde{\mathbf{x}}(n)$ in (1).

equal folds, and k different train-validate splits of the data are done, where one (each time different) fold is used for validation and the rest for testing.

Temporal data, on the other hand, are time series or signals, often a single continuous one. They are position-dependent. Cross-validation in them is a bit less intuitive and popular.

A classical option for ESNs is the static split of the data into initialization, training, validation, and testing, in that order in time. The short initialization (also called transient) phase is used to get the state of the reservoir $\mathbf{x}(n)$ "in tune" with the input $\mathbf{u}(n)$ [6]. This sequence is finite, and often quite short, because ESNs possess the echo state property [7]. Initialization is only necessary before the first (training) phase, because the subsequent phases can take the last $\mathbf{x}(n)$ from the previous phase if data continue without gaps. In generative tasks the real (future) outputs $\mathbf{y}(n)$ are substituted with targets $\mathbf{y}^{\text{target}}(n)$ in inputs, known as "teacher forcing", to break the cyclic dependency.

Because the memory of ESN is preserved in its collected state, and the classical output that is learned is memory-less, we can do the instant switches between phases. Exploiting this same Markovian property, cross-validation with ESNs is rather straightforward. In other temporal models this can be more involved [8].

We see the following intuitive cases when using cross-validation on time series could be beneficial:

- When the data are scarce, cross-validation efficiently uses all the available data for both training and validation.
- Combining the models trained on different folds could be a form of (additional) regularization, improving stability.
- When the process generating the data are slightly non-stationary and it "wanders" around, cross-validation increases the chances that the testing interval is adequately covered by the model. However, if it "drifts" in one direction validating and tuning the hyper-parameters on the data interval adjacent to the testing one (i.e., the classic validation) might be the best option.

In particular, we do not expect cross-validation to be beneficial on stationary synthetic long time series, like chaotic attractors, as it does not matter which (and to some extent how much if the data are ample) sections are taken for training or validation.

2.3 Validation Schemes

Here we suggest several validation schemes for ESNs.

We firstly split the testing part off the end which is independent of the validation scheme and is left for testing it as illustrated in Fig. 1(b). The classical ESN validation scheme explained above is presented in (c). We will refer to it as **static validation** (SV). Here we also investigate alternative validation schemes where the data left from initialization and testing are used for training and validation differently and iteratively.

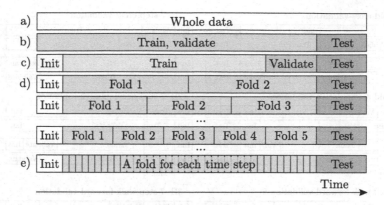

Fig. 1. Splitting the data: (a) all the available data; (b) splitting-off the testing set; (c) a static classical (SV) initialization, training, and validation split for ESNs; (d) splitting data into folds 2 and up for n-fold cross-validation; (e) the maximum amount of folds for leave-one-out cross-validation.

To investigate the k-fold cross-validation of ESNs we split the data into k-folds. The number of folds k can be varied from 2, as in Fig. 1(d), up to available data/time points ending up with leave-one-out cross-validation (e).

In addition to the classical SV split we investigate these validation schemes of using the data between the initialization and testing parts:

1. **k-fold cross-validation** (CV). The data are split into k equal folds. Training and validation are performed k times, each time taking a different single fold for validation and all the rest for training.
2. **k-fold accumulative validation** (AV). First, we split the "minimum" required amount for training only off the beginning, then we divide the rest of the data into k equal folds. Training and validation are performed k times, each time validating on a different fold, similarly to CV, but only training on all the data preceding the validation fold.
3. **k-fold walk forward validation** (FV) is similar to AV: the splitting is identical and validation is done on the same folds, but the training is each time done only on the same fixed "minimal" amount of data directly preceding the validation fold.

The three validation schemes are illustrated in the left column of Fig. 2. Each validation scheme has some rationale behind it.

In CV all the data are used for either training or validation in each split. Also, all the data have been used exactly $k - 1$ times for training and 1 time for validation. This has its benefits explained in Sect. 2.4. What is a bit unusual for temporal data here is that training data also come later in the sequence than the validation data. But this should not necessarily be considered a problem. In fact, for time series output and classification tasks the columns of $\mathbf{X}$ and $\mathbf{Y}^{\text{target}}$ could in principle be randomly shuffled, before applying k-fold cross-validation,

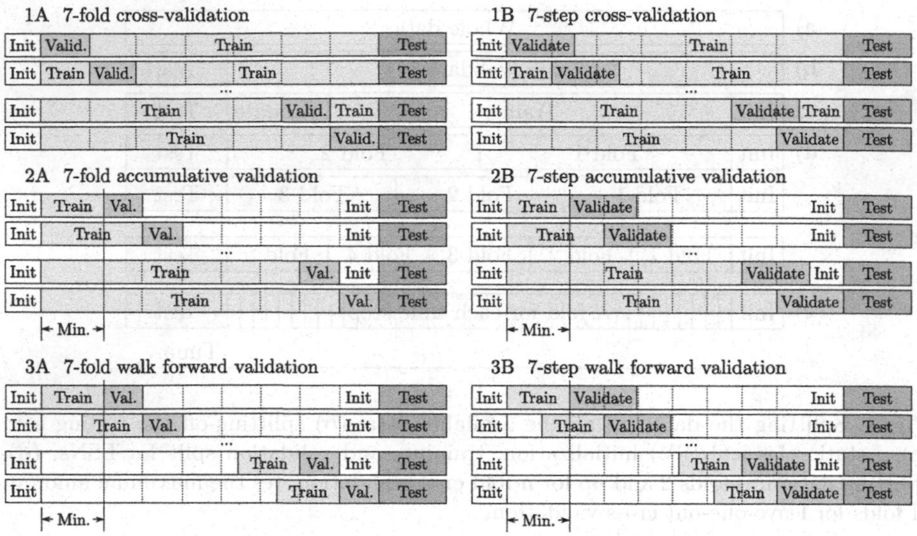

1A 7-fold cross-validation

Init	Valid.	Train		Test
Init	Train	Valid.	Train	Test

...

Init		Train	Valid.	Train	Test
Init		Train		Valid.	Test

1B 7-step cross-validation

Init	Validate	Train		Test
Init	Train	Validate	Train	Test

...

Init	Train	Validate	Train	Test
Init	Train		Validate	Test

2A 7-fold accumulative validation

Init	Train	Val.		Init	Test
Init	Train	Val.		Init	Test

...

Init	Train		Val.	Init	Test
Init	Train		Val.	Test

|← Min. →|

2B 7-step accumulative validation

Init	Train	Validate		Init	Test
Init	Train	Validate		Init	Test

...

Init	Train		Validate	Init	Test
Init	Train		Validate	Test

|← Min. →|

3A 7-fold walk forward validation

Init	Train	Val.		Init	Test
Init		Train	Val.	Init	Test

...

Init		Train	Val.	Init	Test
Init		Train	Val.	Test

|← Min. →|

3B 7-step walk forward validation

Init	Train	Validate		Init	Test
Init		Train	Validate	Init	Test

...

Init		Train	Validate	Init	Test
Init		Train	Validate	Test

|← Min. →|

Fig. 2. Different validation schemes used.

as is common in non-temporal tasks. CV is geared towards training the model once on a fixed well-representative dataset.

AV emulates the classical static training and validation SV k times by each time training on all the available data that come before the validation fold and then validating. This scenario would also happen when the model is repeatedly updated with newly available data. Here we do not allow our model to "peek into the future". The downsides are that models are not trained on the same amount of data and more models are trained on the beginning of the data than the ending.

FV is similar to AV but keeps the training data length constant like CV, which is more consistent when selecting good hyper-parameters. Training length can be set to match several folds, which is quite common when doing walk forward validation. Here the assumption of stationarity of generating process is weaker, models are trained and validated "locally" in time.

CV, AV, and FV all use progressively less data for training thus, in general, are progressively less advisable when data are scarce.

We have also investigated counterparts for the three validation schemes with validation "folds" set longer, independent of k. They are illustrated in the right column of Fig. 2. We call them "k-step" as opposed to "k-fold". We are inventing terminology here when we could not find (a consistent) one in the literature.

These "k-step" validation schemes are mostly relevant to realistically validate generative tasks, where errors tend to escalate over time when running in the generative mode. For other types of tasks, validation length is usually not important when we compute time-averaged errors. The validation length here can be set to match a testing length in a realistic use scenario. It can also be set to match the length of several steps, which would use data for training and validation more consistently.

2.4   Final Trained Model

Validation results are usually averaged over splits for every hyper-parameter set and the best set is selected. We investigate several ways of producing the final trained model with the best hyper-parameters for testing:

- **Retrain** ESN on all the training and validation data;
- **Average** $\mathbf{W}^{out}$s of ESNs that have been trained on the k splits;
- Select the ESN that validated **best** among the k.

Each method again has its own rationale. Retraining uses all the available data to an ESN in a straightforward way. However, this requires some additional computation and the hyper-parameters might no longer be optimal for the longer training sequence. Averaging $\mathbf{W}^{out}$s introduces additional regularization, which adds to the stability of outputs. This method might make less sense on AV, since models are trained on different amounts of data. The best-validated split, on the other hand, was likely trained on the hardest parts of the data (and validated on the easiest). A weighting scheme among splits could also be introduced when combining ESNs, as well as time step weighting as discussed in [6].

Regularization parameter is grid-searched for every split individually, as opposed to other hyper-parameters, that are searched in outside loops (see [6] for why this is efficient). For average and best ESNs with their best regularization for that particular fold can be used, whereas for retraining the regularization that is best on average (over the folds) is utilized. Regularization is most important in generative tasks.

For the classic static split SV we also have two options: to either retrain the model on the whole data or to use the validated model as it is.

2.5   Efficient Implementations

Running the ESN reservoir (1) is dominated by $\mathbf{Wx}$ which takes $\mathcal{O}(N_x^2)$ operations per time step with dense $\mathbf{W}$, or $\mathcal{O}(N_x^2 T)$ for all the data. This can be pushed down to $\mathcal{O}(N_x T)$ with sparse $\mathbf{W}$ [6] which is the same $\mathcal{O}(N_r T)$ required for collecting $\mathbf{X}$. For computing $\mathbf{W}^{out}$ (6), collecting $\mathbf{XX}^{\mathrm{T}}$ takes $\mathcal{O}(N_r^2 T)$ and the matrix inversion takes $\mathcal{O}(N_r^3)$ operations in practical implementations. Thus the whole training of ESN (dominated by collecting $\mathbf{XX}^{\mathrm{T}}$) is back to

$$\mathcal{O}(N_r^2 T). \tag{7}$$

The same (7) applies to training and validating ESN, as validation itself has the same complexity as simply running it. And doing a straightforward ESN k-fold cross-validation is

$$\mathcal{O}(k N_r^2 T). \tag{8}$$

Space complexity can be pushed from $\mathcal{O}(N_r T)$ for $\mathbf{X}$ down to $\mathcal{O}(N_r^2)$ when collecting $\mathbf{X}\mathbf{X}^T$ and $\mathbf{Y}^{target}\mathbf{X}^T$ on the fly, which also allows ESNs to be one-shot-trained on virtually infinite time sequences [6].

However, we do not need to rerun the ESN for every split. We can collect and store the matrices $\mathbf{X}\mathbf{X}^T$ and $\mathbf{Y}^{target}\mathbf{X}^T$ for the whole sequence once. Then in every split we only run the reservoir on the validation fold. Validation folds should be arranged consecutively like in Fig. 2.1A, so that after running one validation fold we can save the reservoir state $\mathbf{x}(n)$ for the next validation fold of the next split. We collect $\mathbf{X}\mathbf{X}^T$ and $\mathbf{Y}^{target}\mathbf{X}^T$ on the validation fold and subtract them from the global ones to compute $\mathbf{W}^{out}$ for the particular split. If we are doing output or classification task (see Sect. 2.1) and we can afford to store $\mathbf{X}$ of the validation fold in memory, we can reuse it to compute the validation output $\mathbf{y}(n)$ (3). If not, we need to rerun the validation fold one more time for this. *This way the ESN is rerun through the whole data only two or three times irrespective of k.*

Notice also, that the space complexity of such implementation remains $\mathcal{O}(N_r^2)$. We could alternatively also store $\mathbf{X}\mathbf{X}^T$ and $\mathbf{Y}^{target}\mathbf{X}^T$ for every fold and save one running through the data this way, by having space complexity $\mathcal{O}(kN_r^2)$.

The proposed method pushes down the time complexity of preparation of $\mathbf{X}\mathbf{X}^T$'s in k-means cross-validation from $\mathcal{O}(kN_r^2 T)$ which dominates in (8), to $\mathcal{O}(N_r^2 T)$. Adding the matrix inversions (6) which are now not necessarily dominated, the proposed more efficient implementation of ESN k-means cross-validation has time complexity

$$\mathcal{O}(N_r^2 T + kN_r^3). \tag{9}$$

Thus we get a k or T/N_r time complexity speedup in a more efficient implementation (9) compared to naive (8), depending on which multiplier is smaller.

When the data sample length T is many times larger than the ESN size N_r (a typical case and a one where optimization is most relevant) and thus k such that $k < T/N_r$, we can say that *the proposed efficient implementation permits doing ESN k-folds cross-validation with the same time complexity as a simple one-shot validation. The space complexity can also remain the same.*

We have outlined an efficient method for ESN k-folds cross-validation (CV), but it can easily be adapted to other types of validation schemes described in Sect. 2.3.

3   Experiments

Having established that different validation schemes for ESNs are possible and can be implemented quite efficiently, in this section we test them empirically on several different time series datasets.

As mentioned before, the goal here is not to obtain the best possible performance but to compare different validation methods of simple ESNs on equal terms.

3.1 Generative Mode

We evaluate the proposed validation methods by examining multiple univariate datasets of increasing sizes:

- **Labour:** "Monthly unemployment rate in US from 1948 to 1977" dataset[2];
- **Gasoline:** "US finished motor gasoline product supplied" dataset[3];
- **Sunspots:** "Monthly numbers of sunspots, as from the World Data Center" dataset[4];
- **Electricity:** "Half-hourly electricity demand in England" dataset [9].

Lengths of these datasets and testing, validation split parameters are presented in Table 1. "Min. ratio" here is the percentage of the whole data (excluding testing) used as the minimal training length in AV, or the whole training length in FV ("Min." in Fig. 2).

Table 1. Datasets and validation setup parameters.

Dataset	Samples T	Valid, test samples	Folds, steps k	Min. ratio
Labour	360	10	34	50%
Gasoline	1355	67	18	50%
Sunspots	3177	200	10	50%
Electricity	4033	200	18	50%

For k-fold CV, initial transient length and k are chosen in such a way, that the k folds would have the same size as testing. For k-step validation variants the overlapping validation block used also have the same length as the testing range.

We use a grid search to find the best ESN hyper-parameters. Reservoir size $N_x = 50$ was chosen, and candidates of leaking rate $\alpha \in \{0.1, 0.2, 0.3, ..., 1\}$, spectral radius $\rho \in \{0.1, 0.2, 0.3, ..., 1.5\}$ and regularization degree $\beta \in \{0, 10^{-9}, 10^{-8}, ..., 1\}$ were evaluated following [6].

The experiment results are presented in Table 2. We can see that in all the experiments either FV or AV find the hyper-parameters producing best generalizing models, it is never SV or CV. The relative underperformance of CV can probably be explained by the nature of the non-stationary of the temporal data. The generating processes likely have a one-directional "drift", thus validation

---

[2] Publicly available at https://data.bls.gov/timeseries/lns14000000.
[3] Publicly available at https://www.eia.gov/dnav/pet/hist/LeafHandler.ashx?n=PET &s=wgfupus2&f=W.
[4] Publicly available at http://www.sidc.be/silso/datafiles.

Table 2. Validation and testing NRMSEs on generative datasets

Method		Labour		Gasoline		Sunspots		Electricity	
Validation	Final	Valid	Test	Valid	Test	Valid	Test	Valid	Test
SV	As is	1.034	1.927	0.891	0.881	0.749	0.784	0.623	0.860
	Retrained		1.957		1.132		0.755		0.835
k-fold CV	Averaged	2.009	1.835	1.000	0.914	1.060	0.924	0.834	0.990
	Retrained		1.833		0.913		0.970		1.006
	Best		1.838		0.901		1.008		0.995
k-step AV	Averaged	1.927	4.469	1.040	0.867	0.703	0.835	0.812	0.829
	Retrained		1.171		0.962		0.742		1.006
	Best		4.546		**0.829**		0.855		0.820
k-step FV	Averaged	2.188	3.413	1.065	0.925	0.726	0.640	0.783	**0.733**
	Retrained		**0.681**		0.949		**0.612**		1.006
	Best		2.799		0.894		0.649		0.769

schemes FV and AV, that select models capable of predicting sequences directly following the training ones, win. They also win over SV, as this selection is validated over k splits instead of just one.

The main bottleneck with the Labour dataset is its scarcity: only 360 samples in total. We see that SV overfits the hyper-parameters on the single split; CV gets a better estimate; and AV and FV fail in averaged and best modes, as these use the scarce data inefficiently, but produce the very best overall results when retrained. When having more data the benefits of retraining are less evident.

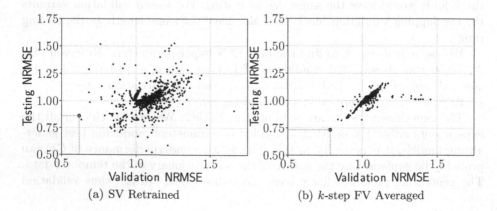

(a) SV Retrained   (b) k-step FV Averaged

Fig. 3. Results of grid search on Electricity dataset. Every point corresponds to one combination of hyper-parameters.

Validation vs. testing errors of two validation schemes on the Electricity dataset are illustrated in Fig. 3. We can see that while there exist outliers with very good testing performance in Fig. 3(a), they would not be picked up by the validation. In fact, k-step FV validation errors for these hyper-parameter sets were so bad, that they went off-scale in Fig. 3(b). This indicates that the lucky outliers were particular to the testing data and most likely would not do well on other. On the other hand, k-step FV gives a much better overall correlation between validation and testing errors and a much better solution based on validation is picked in Fig. 3(b) (the small circles).

3.2   Time Series Classification

To evaluate the validation methods on classification tasks we use a classical Japanese Vowels dataset[5]. This benchmark comes as 270 training samples and 370 testing samples, where each sample consists of varying length 12 LPC cepstrum coefficients taken from nine male speakers. The goal of this task is to classify a speaker based on his pronunciation of vowel /ae/. In the training set, there are 30 samples for each user, while in the testing set, this number varies from 29 to 88.

We note that 0 test set misclassifications have been previously achieved with ESNs as reported in [10]. Therefore we shift our efforts to achieve better results on models that reportedly have been performing sub-optimally. It has been reported that models that only store the last state vector $\mathbf{x}(n)$ for every speaker have at best been able to achieve 8 test misclassifications. We replicate the model used in [10], and set up a grid search as described in Sect. 3.1. We run an 18-fold cross-validation, and use a validation length of 15 instances. As the dataset consists of a permutable set of sequences (the order does not matter), we only test the classical SV and the standard k-fold CV. We run the experiments 5 times having different random initializations of the ESN weights. For each of the 5 experiments we run grid search separately and report aggregated results.

We also do individual regularization for each validation split. When doing cross-validation, each split on each regularization degree candidate is evaluated. $\mathbf{W}^{out}$s of each split with their individual best regularization degrees are averaged. We refer to this variation as "IReg Averaged". In another option, the model was retrained on the whole data using the average of the best regularization degrees. We refer to this variation as "IReg Retrained".

The NRMSE errors, misclassifications and their standard deviations are presented in Table 3. We see that all the CV variations outperform all the SV variations. We also see that the individual regularization "IReg" further slightly improves both validation and testing errors, as well as misclassification, which is not surprising. In all validation schemes, except the "CV IReg Averaged", there was at least one model produced that was able to achieve 2 test misclassifications.

---

[5] Publicly available at https://archive.ics.uci.edu/ml/datasets/Japanese+Vowels.

Table 3. Average results on Japanese Wovels task

Method	Final	Validation error	Test error	Misclassifications
SV	As is	0.504 ± 0.017	0.491 ± 0.005	5.0 ± 1.5
	Retrained		0.486 ± 0.003	4.8 ± 1.6
CV	Averaged	0.493 ± 0.004	0.472 ± 0.008	4.2 ± 1.8
	Retrained		0.468 ± 0.006	4.4 ± 2.1
CV IReg	Averaged	0.489 ± 0.004	**0.468** ± 0.009	4.4 ± 1.2
	Retrained		0.470 ± 0.008	**3.8** ± 1.8

4   Discussion

In this contribution we have proposed and motivated different cross-validation schemes for ESNs, have introduced a space- and time-efficient algorithm for doing this, and empirically investigated their effects on several real-world datasets.

The component that dominates the time complexity of the already quite fast ESN training remains constant (does not scale up with k) in our proposed method of doing k-fold cross-validation. The component that does scale linearly with k starts dominating only in some not very common situations, in particular when k is very large. Thus in typical situations k-fold cross-validation of ESNs can be done for virtually the same time complexity as a simple single validation. The time savings are also less evident when the data are short, but then they are also less pertinent. The methods can also have the same space complexity as a simple single validation.

This further sets apart the speed of ESN training from error backpropagation based recurrent neural network training methods, where cross-validation could also be used in principle.

We have demonstrated the proposed validation schemes for classical ESNs, but they apply to other reservoir types as well with the same time and space complexity savings. Namely, the time and space complexity of running the reservoirs is added to the ones of training instead of multiplying them by k in a k-fold/step cross-validation. The time complexity savings on reservoir running could also hold for other readout types.

We have also empirically investigated the benefits of the proposed validation schemes for ESNs on several different datasets. There is no single winner among the validation schemes. The results highly depend on the nature of the data. Overall our experiments show that typically cross-validation predicts testing errors more accurately and produces more robust results. It also can use scarce data more sparingly. AV and FV validation schemes can apparently select for good "forward-predicting" models when the generating process is "evolving" and not quite stationary. SV can also have its benefits, as it can be seen as the last fold of AV. For stationary and ample, well-representative data cross-validation might not be crucial.

How the final trained model is produced from the cross-validation results is also very important.

Acknowledgments. This research was supported by the Research, Development and Innovation Fund of Kaunas University of Technology (grant No. PP-91K/19).

References

1. Jaeger, H.: The "echo state" approach to analysing and training recurrent neural networks. Technical report GMD Report 148, German National Research Center for Information Technology (2001)
2. Jaeger, H., Haas, H.: Harnessing nonlinearity: predicting chaotic systems and saving energy in wireless communication. Science **304**(5667), 78–80 (2004)
3. Jaeger, H.: Echo state network. Scholarpedia **2**(9), 2330 (2007)
4. Lukoševičius, M., Jaeger, H.: Reservoir computing approaches to recurrent neural network training. Comput. Sci. Rev. **3**(3), 127–149 (2009)
5. Stone, M.: Cross-validatory choice and assessment of statistical predictions. J. Roy. Stat. Soc.: Ser. B (Methodol.) **36**(2), 111–133 (1974)
6. Lukoševičius, M.: A practical guide to applying echo state networks. In: Montavon, G., Orr, G.B., Müller, K.-R. (eds.) Neural Networks: Tricks of the Trade. LNCS, vol. 7700, pp. 659–686. Springer, Heidelberg (2012). https://doi.org/10.1007/978-3-642-35289-8_36
7. Yildiz, I.B., Jaeger, H., Kiebel, S.J.: Re-visiting the echo state property. Neural Netw. **35**, 1–9 (2012)
8. Bergmeir, C., Hyndman, R.J., Koo, B.: A note on the validity of cross-validation for evaluating autoregressive time series prediction. Comput. Stat. Data Anal. **120**, 70–83 (2018)
9. Taylor, J.W.: Short-term electricity demand forecasting using double seasonal exponential smoothing. J. Oper. Res. Soc. **54**(8), 799–805 (2003)
10. Jaeger, H., Lukoševičius, M., Popovici, D., Siewert, U.: Optimization and applications of echo state networks with leaky-integrator neurons. Neural Netw. **20**(3), 335–352 (2007)

Workshop: Reservoir Computing - Physical Implementations

Echo State Property of Neuronal Cell Cultures

Tomoyuki Kubota[1]($\boxtimes$), Kohei Nakajima[2], and Hirokazu Takahashi[2]

[1] Department of Advanced Interdisciplinary Studies, Graduate School of Engineering, The University of Tokyo, 4-6-1 Komaba, Meguro-ku, Tokyo 153-8904, Japan
t.kubota@ne.t.u-tokyo.ac.jp
[2] Graduate School of Information Science and Technology, The University of Tokyo, 7-3-1 Hongo, Bunkyo-ku, Tokyo 113-8654, Japan

Abstract. Physical reservoir computing (PRC) utilizes the nonlinear dynamics of physical systems, which is called a reservoir, as a computational resource. The prerequisite for physical dynamics to be a successful reservoir is to have the echo state property (ESP), asymptotic properties of transient trajectory to driving signals, with some memory held in the system. In this study, the prerequisites in dissociate cultures of cortical neuronal cells are estimated. With a state-of-the-art measuring system of high-dense CMOS array, our experiments demonstrated that each neuron exhibited reproducible spike trains in response to identical driving stimulus. Additionally, the memory function was estimated, which found that input information in the dynamics of neuronal activities in the culture up to at least 20 ms was retrieved. These results supported the notion that the cultures had ESP and could thereby serve as PRC.

Keywords: Neuronal cell culture · Physical reservoir computing · Echo state property · Memory capacity

1 Introduction

Neural activity does not have permanently stable or periodic states and shows transient responses, with states moving constantly from one to the next [5,30]. For example, when a locust perceives an odor, neurons in the antenna lobe transit from low to high frequency states. Subsequently, states are maintained for a while, after which they return to the original state. The trajectory of the state transition is reproducible, but it changes depending on the stimulus duration [7].

Such a reproducible response is described by common-signal-induced synchronization or generalized synchronization [18]. When a drive subsystem and a response subsystem are represented by the states x and y, respectively, the response subsystem given the time series input $h(x)$ from the drive subsystem, converges to the state where y is represented by $y = \Phi(x)$, draws a repeatable trajectory for the same time series input [18]. Jaeger called such a trajectory a

© Springer Nature Switzerland AG 2019
I. V. Tetko et al. (Eds.): ICANN 2019, LNCS 11731, pp. 137–148, 2019.
https://doi.org/10.1007/978-3-030-30493-5_13

transient attractor [11] and the property of converging to this trajectory an echo state property (ESP) [13,20]. Reservoir computing (RC) [14,19,31] generates reproducible responses by generalized synchronization between the drive and response subsystem such as a recurrent neural network (RNN) composed of sigmoid or spiking neurons, and emulates desired outputs by linear regression with the reproducible responses. RC can not only be applied in engineering, such as in machine learning, but can also be used to estimate computational capabilities of a dynamical system by analyzing equations executed in the response subsystem, i.e., in the reservoir. For example, by letting a reservoir learn past inputs, we can examine how accurately and how long the inputs would be retained in the reservoir [12].

Furthermore, physical reservoir computing (PRC), which uses a physical system as a reservoir, has been reported in recent years. The framework of RC at least requires generalized synchronization between the drive subsystem and response subsystem to construct a successful reservoir, and to date, various dynamical systems have been utilized as reservoirs, and their computational capabilities have been evaluated by many researchers, as follows: Fernando et al. showed that ripples changing temporally in a bucket have linear separability [8]. Recently, the computational capabilities of delay-based dynamical systems were analyzed by applying time-multiplexing to electronic and laser systems subject to delay feedback and thereby constructing a pseudo network [1,4]. Furthermore, Nakajima et al. applied forces to a device mimicking an octopus arm and investigated its computational capabilities from a dynamical system's point of view [21–26].

As discussed previously, PRC has been applied to various dynamical systems to evaluate their computational capabilities; however, in real neural circuits, synaptic plasticity works according to the long-term input history, and the network structure between neurons is constantly updated and the response of the circuits would alter accordingly [9,16]. Furthermore, computational capabilities of RNNs such as timing capacity, which is the maximum delay after a pulse input that the network produces, have been analyzed [17], whereas PRC with cultures has not been achieved. On the other hand, Dranias et al. evaluated short-term memory combining transient dynamics of cell cultures and support vector machine (SVM), and showed that the cultures discriminated input applied 1 s or more before [6]. However, as SVM was adopted for the readout, responses retrieved from the culture were nonlinearly transformed and one can not distinguish nonlinearity of the culture's responses from of SVM. Therefore, the characteristics of neuronal cultures as reservoirs have not been fully revealed.

In this paper, we constructed a reservoir using a cell culture from rat cortices and stimulated it two times with an identical electrical stimulation trace to investigate whether the reservoir met the prerequisite for RC, i.e., whether it has an ESP. Additionally, we examined their expressive capability of information by measuring memory functions.

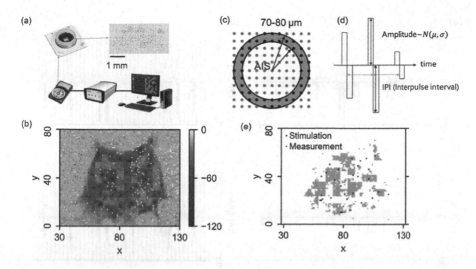

Fig. 1. Composing a reservoir with a rat cortical cell culture. (a) Micro-electrode array (MEA) system. (b) The mean extracellular action potential (EAP) amplitudes of spontaneous activity plotted on the electrode array at 46 days in vitro (DIV). Black points represent electrodes in the vicinity of axon initial segments (AIS). (c) The axons were located by sequentially stimulating electrodes that were included in the area having a radius 70–80 μm from the AIS. The black points are the stimulated electrodes. (d) The waveform of bipolar pulse stimulation. The amplitude follows a normal distribution, and the interpulse interval (IPI) is 10, 20, or 30 ms. (e) The selected electrodes used for measurement (gray) and stimulation (black).

2 Materials and Methods

2.1 Cell Culture

All experiments were approved by the ethical committee of the University of Tokyo and followed "Guiding Principles for the Care and Use of Animals in the Field of Physiological Science" by the Physiological Society of Japan.

Techniques for cell culture have been developed to maintain the culture and conduct experiments for a long time [10,28]. Embryonic rat cortices were dissected from E18 rat and used for cortical cell cultures. The cortices were dissociated in 2 mL of 0.25% trypsin-ethylenediaminetetraacetic acid (Trypsin-EDTA, Life Technologies), from which cells were isolated by trituration, and 38,000 cells were seeded on each MEA (MaxWell Biosystems; Fig. 1(a)). For cell adhesion, 5 mL of 0.05% Polyethyleneimine (PEI; Sigma-Aldrich) and 5 μL of 0.02 mg/mL Laminin (Sigma-Aldrich) were used before plating the cells. After 24 h from plating the cells, the plating media [3] were changed to the growth media [28]. The plating media were composed of Neurobasal 850 μL (Life Technologies), 10% horse serum (HyClone), 0.5 mM GlutaMAX (Life Technologies), and 2% B27 (Life Technologies). The growth media were composed of DMEM 850 μL (Life Technologies), 10% horse serum (HyClone), 0.5 mM GlutaMAX (Life Tech-

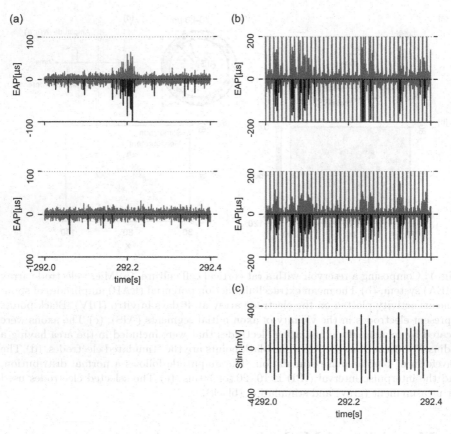

Fig. 2. Detected extracellular action potentials. The position of the measurement electrode (x, y) is (77, 35). In (a) and (b), the gray and black lines represent filtered signals and signals detected as spikes, respectively. (a) Extracellular action potentials of spontaneous activity measured before (upper) and after (lower) the stimulation. (b) Extracellular action potentials evoked by two-trial identical stimulations. Filtered signals saturated when electrical stimulation was conducted. (c) The stimulation signal trace. The mean and standard deviation of stimulation amplitude were 200 mV and 50 mV, respectively. The interpulse interval was 10 µs.

nologies), and 1 mM sodium pyruvate (Life Technologies). All experiments were conducted in an incubator at 37 °C and 5% CO_2. The MEAs were sealed with a lid to prevent water evaporation and invasion of bacteria and fungus.

2.2 Electrode Selection

We calculated the average spike amplitudes during a spontaneous activity at all electrodes, and selected electrodes on which the axon places as stimulation electrodes. The MEA had 26,400 electrodes, which place 17.5 µm apart and are arranged in a 120 × 220 grid. An extracellular action potential (EAP) has a larger negative amplitude in the vicinity of an axon initial segment (AIS) than around other segments [2], and the EAP amplitude of spontaneous activity at each electrode were obtained to search for electrodes near neurons among 26,400 electrodes. We first obtained 20-s voltage traces at 26,400 electrodes, from which negative EAPs x_n were detected. From these x_n, the average of the EAP amplitudes for each electrode were calculated to create a 120×220 map. We smoothed the electrode map using the Gaussian filter, searched local minima on the map, and regarded them as electrodes near AISs. Furthermore, to search for the axon of the AIS, all electrodes in an area with a radius of 70–80 μm from the AIS electrodes were stimulated one at a time. As shown in Fig. 1(c), each electrode was stimulated by 20 bipolar pulses with 50-ms interpulse interval (IPI) and 300-mV amplitude. Spike detection was applied to the voltage traces to obtain band-pass filtered signals. The signal traces for 20 trials were averaged and electrodes whose mean amplitude could be less than −200 μV were selected as the stimulation electrodes.

The AIS electrodes and electrodes with a high firing rate were chosen as the measurement electrodes. The MEA could simultaneously utilize up to 1,024 of 26,400 electrodes. The remaining electrodes, except the stimulation and AIS electrodes, were connected to the electrodes with a higher firing rate. To measure the number of spikes at each electrode, voltage traces in spontaneous activity were measured for 20 s each.

2.3 Spike Detection

The 6th Butterworth bandpass filter and zero phase IIR filter were applied to the voltage traces to extract 300–3000 Hz components. For stimulating electrodes, artifacts caused by stimulation were removed by eliminating traces ±2 ms from the stimulation times. The extracted signals were divided into positive components x_p and negative components x_n, and their standard deviations were calculated as follows [29]:

$$\sigma = \text{median}\left\{ \frac{|x|}{0.6745} \right\}. \tag{1}$$

If the amplitude of extracted signal exceeded 4σ ($x_p > 4\sigma_p, |x_n| > 4\sigma_n$), the value of spike train was set to one; otherwise, it was set to zero. There is a 1–2 ms period called absolute refractory period, in which spikes do not occur even when the neuron is stimulated. As the measurement frequency was 20 kHz, the above spike train was separated by a 1-ms time bin, and if one or more spikes appeared in one bin, the modified spike train was set to one; otherwise it was set to zero.

2.4 Electrical Stimulation

When investigating ESP and memory capacity, we gave bipolar pulse stimuli (Fig. 1(c)) of the amplitude that followed the normal distribution $\mathcal{N}(\mu, \sigma^2)$ with mean μ and standard deviation σ, and of the IPI of $T = 10, 20,$ and 30 ms.

2.5 Spike Metric

The distance between two spike trains were evaluated with Victor's spike metric [32]. Two vectors $a \in \mathbf{N}^{n_a}$ and $b \in \mathbf{N}^{n_b}$ ($n_a \leq n_b$) hold spike times in the first and second trials, respectively, and the total cost to convert a to b is calculated. Although there are many conversion paths, the one with the minimum cost among them is the distance $G(a, b)$. To convert the vector, there are three operations: adding, deleting, and moving spikes, whose costs are one, one, and $q \Delta t$, respectively. Note that Δt is the absolute difference between the times before and after moving the spike, and q is a weight parameter and was set to 10 ms. As the distance $G(a, b)$ depends on n_a, n_b, the normalized distance $d(a, b)$ was obtained as follows:

$$d(a, b) = \frac{G(a, b) - \Delta n}{n_a}. \tag{2}$$

When Victor's metric converts a to b, spikes are added $\Delta n(= n_b - n_a)$ times. Since the spike addition cost is one, there will always be a cost Δn. The remaining n_a spikes are transformed by moving, or adding and deleting, and $d(a, b)$ represent the cost averaged over n_a operations.

2.6 Composing a Reservoir

The interpulse interval was T ($= 10, 20, 30$ ms), whereas the spike detection interval was 1 ms. Since the two intervals must be the same to compose a reservoir, the spike count in a T-width time bin was used as a state of node.

When a dynamical system has computational capabilities, the system converges to the same trajectory without depending on the initial values [13,20]. To wash out the initial transient, the time series data after converging was divided into training data ($i = 1, \cdots, N$) and test data ($i = N + 1, \cdots, 2N$).

Ridge regression was adopted for training readout weights. The readout weights w and the output $\hat{y}$ were calculated by Eqs. (3) and (4), respectively.

$$\hat{w} = \arg\min_{w} \left\{ \frac{1}{N} \sum_{i=1}^{N} (y_i - w \cdot s_i)^2 + \alpha\|w\|_2^2 \right\}, \tag{3}$$

$$\hat{y}_i = \hat{w} \cdot s_i, \tag{4}$$

where y_i was the desired output in the i-th time bin, $\| \cdot \|_2$ representing the Euclidean norm, and α was a hyperparameter that adjusts the degree of normalization. When one uses Ridge regression, the Akaike information criteria AIC is expressed by the following equation:

$$AIC = \frac{1}{N\hat{\sigma}^2} \left(\sum_{i=N+1}^{2N} (\hat{y}_i - y_i)^2 + 2\hat{\sigma}^2 df \right), \tag{5}$$

$$df = \sum_{i=1}^{M+1} \frac{e_i}{e_i + \alpha}, \tag{6}$$

where $\hat{\sigma}^2 = \frac{1}{N-p-1}\sum_{i=N+1}^{2N}(\hat{y}_i|_{\alpha=0} - y_i)^2$, and $e_i(i = 1, 2, \cdots, M+1)$ were the eigenvalues of the matrix $[s_1 \cdots s_N]^T[s_1 \cdots s_N]$. We obtained p pairs of (α, w) during training and p pairs of (α, AIC) during test, with which α was chosen such that the AIC would be minimized.

2.7 Memory Capacity

A measure for short-term memory of a reservoir is memory capacity (MC), which quantifies the past input held in the reservoir by giving the temporally uncorrelated time-series input ν_i to the reservoir and emulating the past inputs. The memory function f_k ($k = 1, 2, \cdots$) represents the accuracy with which ν_{i-k} the input delayed by k time bin are stored in the reservoir and examining f_k provides how accurately and how long the input is held. The memory capacity C is represented by the sum of f_k:

$$f_k = \max_w \frac{\left(\sum_i(\nu_{i-k} - \bar{\nu})(\hat{y}_i - \bar{\hat{y}})\right)^2}{\sum_i(\nu_i - \bar{\nu})^2 \cdot \sum_i(\hat{y}_i - \bar{\hat{y}})^2} \qquad (7)$$

$$C = \sum_k f_k \qquad (8)$$

3 Results and Discussions

3.1 Echo State Property

Spontaneous activities and evoked responses were alternately measured for 300 s per trial. Figure 2(a) and (b) show the filtered signals of spontaneous activity and evoked responses at 292.0–292.4 s, respectively. Figure 2(c) represents the stimulation voltage trace applied to all stimulation electrodes. While the spike frequency in spontaneous activity was low and the spike times were uniformly distributed, the spikes in evoked response densely concentrated around specific times, which corresponded with times when the amplitude was large and indicated that strong stimuli induced spikes at specific times.

Figure 3 shows the difference in the number of spikes, Δn, and the normalized spike train distance, d, with 1-s time bin. Δn converged in approximately 30 s from beginning of stimulation whereas d fluctuated a little before Δn converged; therefore, the spike trains converged in approximately 30 s, which suggests that short-term plasticity occurred in the culture. The spikes in each neuron might converge because of facilitation and depression caused by repeated spikes occurrence.

Furthermore, the spike train distances of spontaneous activities and evoked responses with 300-s time bin are shown in Fig. 4(a) and (b), respectively, where the distances were calculated for each activity and plotted on the electrode map. In addition, the averaged distances for 80 electrodes of all the measurement electrodes are shown in Fig. 4(c). The average distances of spontaneous activity were close to two, which indicated that there were many pairs whose spike time

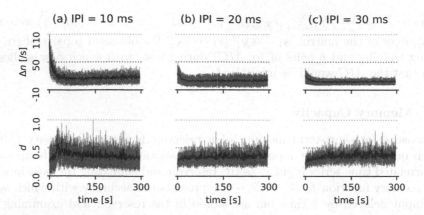

Fig. 3. The average spike train distance changes in time ($\mu = 200\,\mathrm{mV}$). The upper and lower figures show the absolute difference in the number of spikes Δn and the normalized distance d. The figures (a), (b), and (c) show the distances when the inter-pulse interval is 10, 20, and 30 ms, respectively. The black lines and shade are the mean distances and the standard deviation over the measurement electrodes, respectively.

differences were more than 10 ms because the cost to delete and add the spike ($\Delta t > 10\,\mathrm{ms}$) is two. However, the average distances of evoked responses were smaller than those of spontaneous activity, which suggests that the spikes with a time difference of 10 ms or less increased in evoked responses. The distances in evoked responses being smaller than those in spontaneous activity indicated that the cell culture had the ESP.

Additionally, Fig. 4(d) shows the average distances for the mean amplitude μ and IPI of the stimulus. As μ and IPI increased, the average distance decreased. Spike trains were measured in some trials when a bipolar pulse stimulus with 2.4-V amplitude was injected, and the reproducibility between spikes was examined [15], where spikes occurred in 10 ms or more after stimulation and the spike patterns matched better as the spike times were closer to the stimulation. In contrast, when stimulation was repeatedly applied to the cell culture, the spike train distance got smaller as the IPI was larger, which might imply that when multiple stimulations were given, the spike timing was well-matched to a time close to 30 ms from the stimulation.

3.2   Memory Functions

We constructed a reservoir with nodes whose state was the spike number within a time bin T ($= 10$, 20, and 30 ms) and calculated the memory functions. Figure 5 shows the memory functions of the cortical cell culture. ρ_k represents the correlation coefficient of ν_{i-k} with $\hat{y}_i$ that was calculated with $\boldsymbol{w}$ of the memory function f_k. ρ_k (Fig. 5(a) $k = 1, 2$, (b) $k = 1$, and (c) $k = 1$) showed weak positive correlations, which proved that the number of spikes was determined by stimulation up to 20 ms. It has been reported that spikes for 20 ms from the

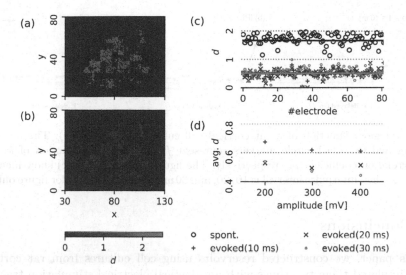

Fig. 4. Echo state property of a rat cortical cell culture. (a) Spike train distances of spontaneous activities on the electrode array. (b) Spike train distances of evoked activities on the electrode array. The interpulse interval (IPI) was 10 ms. (c) Spike train distances of each electrode. 80 of all measurement electrodes are shown here. The colored lines show the mean distances of all electrode distances. In (c) and (d), black circle, blue plus, green cross, and red circle markers stand for spontaneous activity and evoked response with 10, 20, and 30-ms IPI stimulations, respectively. In (b) and (c), mean and standard deviation of stimulation amplitude were 300 mV and 50 mV, respectively. (d) Distances averaged over all measurement electrodes except for stimulation ones. The standard deviation of stimulation amplitude was 50 mV. (Color figure online)

stimulus are caused by the stimulus. These results showed that the number of spikes was determined not only by the stimulus just before but by stimuli up to 20 ms before. A part of stimulus information up to 20 ms before might be retained by excitatory postsynaptic potentials (EPSPs) of pyramidal neurons. If a next EPSP occurs before the previous EPSP returns to the base line, the EPSP is added up. The addition of EPSP is linear when the interval is 30 ms or more, and nonlinear when it is 30 ms or less [27]. As the stimulus is stronger, the stimulated neuron fire more frequently and the postsynaptic neurons retain more nonlinear EPSPs. In other words, the number of spikes of postsynaptic neurons might have contributed to the stimulus strength information held in the reservoir.

However, the stimulation amplitudes older than 20 ms were not retained. Dranias et al. optically stimulated cell cultures where neurons were transfected with ChannelRhodopsin-2, and showed that SVM with time series data of firing rate could distinguish optical stimulations applied 1 s or more before [6]. In our study, since the stimulation interval was 10–30 ms, further studies are needed to examine what occurs in the culture when this interval is longer.

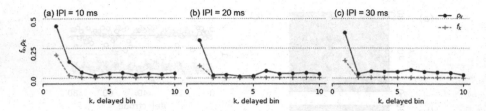

Fig. 5. Memory function of a rat cortical cell culture ($\mu = 400\,\mathrm{mV}$). The red plus markers and the black circle markers represent f_k, the memory function of k, and the correlation coefficient ρ_k, respectively. The figures (a), (b), and (c) show memory functions when interpulse interval is 10, 20, and 30 ms, respectively. (Color figure online)

4   Conclusions

In this paper, we constructed reservoirs using cell cultures from rat cortices and stimulated them two times with an identical electrical stimulation trace to measure the distance between two spike trains. Additionally, we examined their expression of information by measuring memory functions. The results were as follows:

- When the axons of spontaneously active neurons were stimulated, the spike train distances of evoked activity averaged at all electrodes were smaller than the distance of spontaneous activity.
- As the amplitude and IPI of stimulation increased, the spike train distances decreased.
- When the electrical stimulation whose amplitude followed the normal distribution was applied and the reservoir was constructed with the nodes represented by the firing rate, the memory function became relatively high at bins delayed by one and two.

Acknowledgments. This paper is based on results obtained from a project (Exploration of Neuromorphic Dynamics towards Future Symbiotic Society) commissioned by NEDO, KAKENHI grant (17K20090), AMED (JP18dm0307009) and Asahi Glass Foundation. We thank Hitachi UTokyo Laboratory, Hitachi, Ltd. for fruitful discussions. K. N. was supported by JST PRESTO Grant Number JPMJPR15E7, Japan and KAKENHI No. JP18H05472, No. 16KT0019, and No. JP15K16076.

References

1. Appeltant, L., et al.: Information processing using a single dynamical node as complex system. Nat. Commun. **2**, 468 (2011). https://doi.org/10.1038/ncomms1476
2. Bakkum, D.J., et al.: The axon initial segment is the dominant contributor to the neuron's extracellular electrical potential landscape. Adv. Biosyst. **3**(2), 1800308 (2019). https://doi.org/10.1002/adbi.201800308

3. Brewer, G.J., Torricelli, J., Evege, E., Price, P.: Optimized survival of hippocampal neurons in B27-supplemented neurobasal^{TM}, a newserum-free medium combination. J. Neurosci. Res. **35**(5), 567–576 (1993). https://doi.org/10.1002/jnr.490350513
4. Brunner, D., Soriano, M.C., Mirasso, C.R., Fischer, I.: Parallel photonic information processing at gigabyte per second data rates using transient states. Nat. Commun. **4**, 1364 (2013). https://doi.org/10.1038/ncomms2368
5. Buonomano, D.V., Maass, W.: State-dependent computations: spatiotemporal processing in cortical networks. Nat. Rev. Neurosci. **10**(2), 113 (2009). https://doi.org/10.1038/nrn2558
6. Dranias, M.R., Ju, H., Rajaram, E., VanDongen, A.M.: Short-term memory in networks of dissociated cortical neurons. J. Neurosci. **33**(5), 1940–1953 (2013). https://doi.org/10.1523/JNEUROSCI.2718-12.2013
7. Durstewitz, D., Deco, G.: Computational significance of transient dynamics in cortical networks. Eur. J. Neurosci. **27**(1), 217–227 (2008). https://doi.org/10.1111/j.1460-9568.2007.05976.x
8. Fernando, C., Sojakka, S.: Pattern recognition in a bucket. In: Banzhaf, W., Ziegler, J., Christaller, T., Dittrich, P., Kim, J.T. (eds.) ECAL 2003. LNCS, vol. 2801, pp. 588–597. Springer, Heidelberg (2003). https://doi.org/10.1007/978-3-540-39432-7_63
9. Goel, A., Buonomano, D.V.: Timing as an intrinsic property of neural networks: evidence from in vivo and in vitro experiments. Philos. Trans. R. Soc. B: Biol. Sci. **369**(1637), 20120460 (2014). https://doi.org/10.1098/rstb.2012.0460
10. Hales, C.M., Rolston, J.D., Potter, S.M.: How to culture, record and stimulate neuronal networks on micro-electrode arrays (meas). JoVE (J. Vis. Exp.) (39), e2056 (2010). https://doi.org/10.3791/2056
11. Jaeger, H · Identification of behaviors in an agent's phase space. Citeseer (1995)
12. Jaeger, H.: Short term memory in echo state networks, vol. 5. GMD-Forschungszentrum Informationstechnik (2001)
13. Jaeger, H.: Tutorial on training recurrent neural networks, covering BPPT, RTRL, EKF and the "echo state network" approach, vol. 5. GMD-Forschungszentrum Informationstechnik Bonn (2002)
14. Jaeger, H., Haas, H.: Harnessing nonlinearity: predicting chaotic systems and saving energy in wireless communication. Science **304**(5667), 78–80 (2004). https://doi.org/10.1126/science.1091277
15. Jimbo, Y., Kawana, A., Parodi, P., Torre, V.: The dynamics of a neuronal culture of dissociated cortical neurons of neonatal rats. Biol. Cybern. **83**(1), 1–20 (2000). https://doi.org/10.1007/PL00007970
16. Johnson, H.A., Goel, A., Buonomano, D.V.: Neural dynamics of in vitro cortical networks reflects experienced temporal patterns. Nat. Neurosci. **13**(8), 917 (2010). https://doi.org/10.1038/nn.2579
17. Laje, R., Buonomano, D.V.: Robust timing and motor patterns by taming chaos in recurrent neural networks. Nat. Neurosci. **16**(7), 925 (2013). https://doi.org/10.1038/nn.3405
18. Lu, Z., Hunt, B.R., Ott, E.: Attractor reconstruction by machine learning. Chaos: Interdisc. J. Nonlinear Sci. **28**(6), 061104 (2018). https://doi.org/10.1063/1.5039508
19. Maass, W., Natschläger, T., Markram, H.: Real-time computing without stable states: a new framework for neural computation based on perturbations. Neural Comput. **14**(11), 2531–2560 (2002). https://doi.org/10.1162/089976602760407955

20. Manjunath, G., Jaeger, H.: Echo state property linked to an input: exploring a fundamental characteristic of recurrent neural networks. Neural Comput. **25**(3), 671–696 (2013). https://doi.org/10.1162/NECO_a_00411

21. Nakajima, K.: Muscular-hydrostat computers: physical reservoir computing for octopus-inspired soft robots. In: Shigeno, S., Murakami, Y., Nomura, T. (eds.) Brain Evolution by Design. DCA, pp. 403–414. Springer, Tokyo (2017). https://doi.org/10.1007/978-4-431-56469-0_18

22. Nakajima, K., Hauser, H., Kang, R., Guglielmino, E., Caldwell, D.G., Pfeifer, R.: Computing with a muscular-hydrostat system. In: 2013 IEEE International Conference on Robotics and Automation, pp. 1504–1511. IEEE (2013). https://doi.org/10.1109/ICRA.2013.6630770

23. Nakajima, K., Hauser, H., Kang, R., Guglielmino, E., Caldwell, D.G., Pfeifer, R.: A soft body as a reservoir: case studies in a dynamic model of octopus-inspired soft robotic arm. Front. Comput. Neurosci. **7**, 91 (2013). https://doi.org/10.3389/fncom.2013.00091

24. Nakajima, K., Hauser, H., Li, T., Pfeifer, R.: Information processing via physical soft body. Sci. Rep. **5**, 10487 (2015). https://doi.org/10.1038/srep10487

25. Nakajima, K., Hauser, H., Li, T., Pfeifer, R.: Exploiting the dynamics of soft materials for machine learning. Soft Robot. **5**(3), 339–347 (2018). https://doi.org/10.1089/soro.2017.0075

26. Nakajima, K., Li, T., Hauser, H., Pfeifer, R.: Exploiting short-term memory in soft body dynamics as a computational resource. J. R. Soc. Interface **11**(100), 20140437 (2014). https://doi.org/10.1098/rsif.2014.0437

27. Nettleton, J.S., Spain, W.J.: Linear to supralinear summation of AMPA-mediated EPSPs in neocortical pyramidal neurons. J. Neurophysiol. **83**(6), 3310–3322 (2000)

28. Potter, S.M., DeMarse, T.B.: A new approach to neural cell culture for long-term studies. J. Neurosci. Methods **110**(1–2), 17–24 (2001). https://doi.org/10.1016/S0165-0270(01)00412-5

29. Quiroga, R.Q., Nadasdy, Z., Ben-Shaul, Y.: Unsupervised spike detection and sorting with wavelets and superparamagnetic clustering. Neural Comput. **16**(8), 1661–1687 (2004). https://doi.org/10.1162/089976604774201631

30. Rabinovich, M., Huerta, R., Laurent, G.: Transient dynamics for neural processing. Science **321**(5885), 48–50 (2008). https://doi.org/10.1126/science.1155564

31. Verstraeten, D., Schrauwen, B., d'Haene, M., Stroobandt, D.: An experimental unification of reservoir computing methods. Neural Netw. **20**(3), 391–403 (2007). https://doi.org/10.1016/j.neunet.2007.04.003

32. Victor, J.D., Purpura, K.P.: Nature and precision of temporal coding in visual cortex: a metric-space analysis. J. Neurophysiol. **76**(2), 1310–1326 (1996). https://doi.org/10.1152/jn.1996.76.2.1310

Overview on the PHRESCO Project: PHotonic REServoir COmputing

Jean-Pierre Locquet[1,2,3,4,5](✉) [iD] and PHRESCO Partners[1,2,3,4,5]

[1] Katholieke Universiteit Leuven, Leuven, Belgium
jeanpierre.locquet@kuleuven.be
[2] Universiteit Gent, Ghent, Belgium
[3] The Leibniz Institute for Innovative Microelectronics, Frankfurt, Germany
[4] IBM Research GmbH, Rüschlikon, Switzerland
[5] The CentraleSupélec, Gif-sur-Yvette, France

Abstract. PHRESCO is an EU-H2020 funded project that was running for four years and will be ending in September 2019. PHRESCO focused on the development of efficient cognitive computing into a specific silicon-based technology by co-designing a new reservoir computing chip, including innovative electronic and photonic components that will enable major breakthrough in the field. So far, a first-generation reservoir with 18 nodes and integrated readout was designed, fabricated, characterized and a training method has been developed. Additionally, large efforts of the consortium were dedicated to the design of the second-generation chip consisting of larger networks (60 nodes), with an on-chip readout and novel training approaches. This short abstract provides key information on the status of the work achieved and discuss further the potential exploitation routes and the key barriers that still need to be removed to bring the technology to a higher maturity level. A part of the exit strategy of PHRESCO is to identify potential future cooperation with interested stakeholders who are willing to co-develop the PHRESCO technology together with the PHRESCO partners for bringing it to an exploitable or marketable system. This abstract lays down the foundations for potential exploitation activities with interested stakeholders.

Keywords: Reservoir computing · Cognitive computing · Machine learning

1 Introduction

New computing paradigms are required to feed the next revolution in Information and Communication Technology (ICT). Machines that can learn, but also handle vast amount of data, need to be invented. In order to achieve this goal and still reduce the energy footprint of ICT, fundamental hardware innovations must be done. A physical implementation natively supporting new computing methods is required. Most of the time, CMOS is used to emulate e.g. neuronal behavior, but is intrinsically limited in power efficiency and speed. Reservoir computing (RC) is one of the concepts [1] that

---

PHRESCO partners formed by Coordinator of the PHRESCO project.

© Springer Nature Switzerland AG 2019
I. V. Tetko et al. (Eds.): ICANN 2019, LNCS 11731, pp. 149–155, 2019.
https://doi.org/10.1007/978-3-030-30493-5_14

has proven its efficiency to perform tasks where traditional approaches fail. It is also one of the rare concepts of an efficient hardware realization of cognitive computing into a specific, silicon-based technology and of an efficient neural network paradigm suitable for time-series processing. Small RC systems have been demonstrated using optical fibers and bulk components. In 2014, optical RC networks based on integrated photonic circuits were demonstrated by one of the PHRESCO partners [2].

The PHRESCO project aims to bring photonic reservoir computing to the next level of maturity. A new RC chip was designed, including innovative electronic and photonic components that enabled major breakthrough in the field. The consortium of the project is composed by the Katholieke Universiteit Leuven (KUL) in Belgium, who is the coordinator, the Universiteit Gent (UGent) in Belgium, the Leibniz Institute for Innovative Microelectronics (IHP) in Germany, IBM Research GmbH in Switzerland and the CentraleSupélec (CS) in France.

In the following sections we will present key information about the major development achieved during the lifetime of the project, and shed light on the next actions needed to bring the technology to a higher maturity level (e.g. exploitable systems for specific markets). The abstract is organized along the following topics:

1. Motivation for developing the PHRESCO technology from market, business and societal challenges perspectives;
2. An overview on the PHRESCO concept and results achieved;
3. The technical barriers that still need to be removed;
4. Conclusions and outlook for future work and cooperation.

2   Motivation for Developing the PHRESCO Technology

Millions of networked sensors are being embedded in the physical world in devices such as mobile phones, smart meters, automobiles, industrial machines etc., which sense, create and communicate data. Social media sites, smartphones, and other consumer devices have allowed billions of individuals around the world to contribute to the amount of "Big data" available today. According to some estimates [3], the amount of data produced worldwide is doubling every two years; it is expected to increase from 4.4 zettabytes (or 4.4 trillion gigabytes) in 2013 to 44 zettabytes in 2020. Big data analytics have the potential to identify efficiencies that can be made in a wide range of sectors, and to lead to innovative new products and services, competitiveness and economic growth. Moreover, given the intrinsic temporal nature of involved data, for which the RC approach (as a paradigm for modeling RNNs) is chosen and identified as suitable, ultra-efficient photonic implementation could take great advantage thereof.

There are growing demands from many different market sectors to analyze "Big data" and utilize it to generate value (i.e. health care, transport, insurance, marketing, intelligence agencies, modelling of complex systems, etc.). For instance, modern medicine collects huge amounts of information about patients through imaging technology (scans, MRI), genetic analysis (DNA microarrays), and other forms of diagnostic equipment. By applying data analysis or mining to data sets for large numbers of patients, medical researchers are gaining fundamental insights into the genetic and environmental causes of diseases, and creating more effective means of diagnosis. It has been envisioned that the potential annual value to the US health care with the use of "Big Data" is \$300 billion, two-thirds would be in the form of reducing expenditures by about 8%. Another example is the public sector administration in Europe where the use of "Big Data" has a potential annual value of \$250 billion, \$150 billion of which would come from only operational efficiency improvements. In the transport area, navigation companies collect billions of traffic measurement points daily; using this data to reduce congestion could result, by 2020, in worldwide savings of US\$500 billion in time and fuel, and 380 megatons of CO_2 emissions.

The huge potential of Big Data cannot be fully unlocked using traditional computing technologies based on the Von-Neumann architecture. Faster and smarter computing concept will need to be developed to unleash the full potential of Big Data.

Reservoir computing (RC) is one of the concepts that has proven its efficiency to perform tasks where traditional approaches fail. As a simple description for RC, the input data are forced in a so-called reservoir to interact with themselves, and this interaction results in an output pattern that can be more easily classified. Although there is a clear search for **new computing paradigms**, very few examples show a hardware implementation of it. Most of the time, the physical implementation is performed in CMOS to emulate e.g. neuronal behavior, and is intrinsically limited in power efficiency and speed. RC is one of the rare concepts of efficient hardware realizations of cognitive computing, shown to fit natively into a specific technology, namely integrated photonics.

In 2014, optical RC networks based integrated photonic circuits were realized by the Photonics Research Group at Ghent University who is a partner of PHRESCO. In their ground-breaking work the optical reservoir is a passive network of silicon photonic waveguides, interferometers, and splitters. The PHRESCO projects capitalized on this discovery, and aimed at bringing photonic reservoir computing to the next level of maturity. A new RC chip was co-designed, including innovative electronic and photonic components that will enable major breakthrough in the field. The main building blocks of the project consist of: (i) scale up the optical RC systems with the development of two generations of chip consisting of larger networks (18 and 60 nodes), with an on-chip readout and novel training approaches (ii) build an all-optical chip based on the unique electro-optical properties of new materials (iii) implement new learning algorithms to exploit the capabilities of the RC chip. The hardware integration of beyond state-of-the-art components with novel system and algorithm design will pave the way towards a new era of optical, cognitive systems capable of handling huge amount of data at ultra-low power consumption.

3 PHRESCO's Concept and Results Achieved

3.1 Overview on the PHRESCO Concept

In order to develop compact and multifunctional photonic networks integrated on silicon substrates, optically active elements such as non-linear elements, weightining elements, amplifiers and detectors were implemented for the first time into the photonic reservoir computing (PhRC) that was scaled up to reach 60 nodes (4 times larger than previously fabricated). Therefore, an all-optical PhRC was built to boost the speed of the system while strongly reducing their power consumption. The project overview and concept is illustrated in Fig. 1. Moreover, the system layout and learning algorithms such as training methods were developed and investigated.

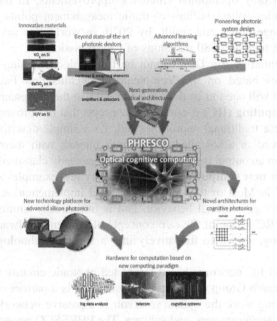

Fig. 1. Overview and concept of the PHRESCO project.

3.2 Overview on the Achieved Results

In the first phase of the PHRESCO project, a first generation reservoir with 18 nodes and integrated readout was designed and fabricated. The chip was designed as a 2×9 photonics swirl reservoir integrated with 3 readout technologies: a silicon readout, a vanadium oxide (VO2) readout and a barium titanate (BaTiO3) readout. The first set of prototype chips were produced and characterised by transmission experiments. The first end-to-end measurements of the reservoir states were attempted. Particle Swarm Optimization training method has been used to simulate the training of a 2×9 reservoir in order to establish a baseline, as well as a first fall-back training method for our prototypes. For a more efficient and fast training, a method based on the inversion of the output nonlinearity of the reservoir has been developed.

At the level of the individual components, the extension of the current function-alities of the new proposed materials was studied. Regarding the VO_2 technology, optical and electro-optical characterization of VO_2-Si photonic structures (1st genera-tion of test chips) was done. However, it was not possible to stabilize a non-volatile switching of the VO_2 layers. On the other hand, the performance of $BaTiO_3$ based programmable weights was studied. Electro-optical characterization of $BaTiO_3$-Si devices confirmed the Pockels effect and the reversible and non-volatile switching behavior of the BaTiO₃ layers. An experimental platform has been designed and developed for characterization of non-linear behavior on $BaTiO_3$ thin films. The test experiments show that polarized light can be coupled and detected before and after the sample respectively.

During the 2nd phase of the PHRESCO project, the main efforts were focused on the co-integration of different components on the photonic chips, i.e. weighting elements, amplifier and nonlinear elements. Regarding the weighting elements, we established a procedure to erase the history of a weight, independent of the previous settings. In the case of VO_2, we have demonstrated a tunable transverse electric pass polarizer based on hybrid VO_2/Si devices showing the potential of VO_2 as an active material for opto-electronic devices, see Fig. 2 [4].

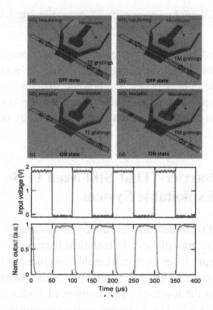

Fig. 2. (Top) concept art of the proposed TE pass polarizer as a function of the input polarization and VO_2 state. (Bottom) electrical driving signal applied to the microheater and photo detected signal at the output of the polarizer for TM polarization.

In parallel, large efforts of the consortium was dedicated to the design of the second-generation chip, see Fig. 3. This part of the work included simulations to optimize the performance of components, architecture layout, best input strategy, best way to readout and combine the different states and power budget studies. The available prototype was upscaled to larger networks (60 nodes), with an on-chip readout and novel training approaches, and the scalability and cascadability of these RC networks were investigated. From the analysis of performance of the reservoir it could be concluded that a chaining architecture can be optimal in order to leverage the performance of electro-optical and fully optical passive photonic reservoir computing systems. The work is currently ongoing and the experimental results will be presented in a separate paper.

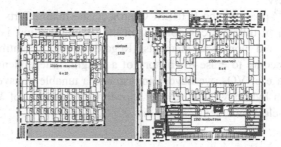

Fig. 3. Low-loss reservoir, 2nd generation mask layout

At the level of the individual components, we greatly progressed in the fabrication of III/V based amplifiers, which showed optical gain at wavelength 1300 nm, and could successfully couple light between Si-photonic structures and III/V devices placed at different photonic layers. Finally, we also examined the non-linearity properties in the BaTiO3 layer by using an experimental setup specifically designed for the study of the photorefractive effect.

4 The Technical Barriers That Still Need to Be Removed to Move to an Exploitable System

Although the PHRESCO project has advanced considerably the state of the art of reservoir computing, there are still several technical barriers that need to be removed in order to bring the concept closer to a marketable product. Below the list of the most important barriers that need to be removed in a future work in collaboration with interested stakeholders in the topic: (1) The number of nodes need to be scaled further up (at least >1000 nodes); (2) The footprint of the demonstrator needs to be reduced with novel designs (currently the reservoirs in PHRESCO are already in the range of 1×1 cm); (3) Better photonic nonlinear elements need to be design (while keeping losses low). Tackling those barriers with stakeholders who are interested to co-develop the PHRESCO technologies further, will be essential to demonstrate the concept on a system level suitable for specific market applications.

5 Conclusions and Outlook for Future Work and Cooperation

To date, PHRESCO have generated a number of innovations that were translated into patents and high impact scientific publications, mainly related to materials, components and the concept/modelling of the demonstrator. Although it's expected that functional demonstrators will be obtained by the end of the project at a small prototype level, obviously there will be still a need to develop those demonstrators further before they can be introduced into the market. This abstract gives an overview on the status of the developed technologies so far, and the barriers that still need to be removed. This abstract lays down the foundations for a potential cooperation with stakeholders who could be interested to co-develop the concept further to a system level targeting specific applications. The market applications that are targeted are cloud-based analytics of big and unstructured data, cognitive signal processing, healthcare and financial sector.

References

1. Verstraeten, D., et al.: An experimental unification of reservoir computing methods. Neural Netw. **20**(3), 391–403 (2007). https://doi.org/10.1016/j.neunet.2007.04.003
2. Vandoorne, K., et al.: Experimental demonstration of reservoir computing on a silicon photonics chip. Nat. Commun. **5**, 3541 (2014). https://doi.org/10.1038/ncomms4541
3. Turner, V.: The digital universe of opportunities (2014). https://www.emc.com/leadership/digital-universe/2014iview/index.htm
4. Sanchez, L., et al.: Experimental demonstration of a tunable transverse electric pass polarizer based on hybrid VO2/silicon technology. Opt. Lett. **43**(15), 3650–3653 (2018). https://doi.org/10.1364/OL.43.003650

Classification of Human Actions in Videos with a Large-Scale Photonic Reservoir Computer

Piotr Antonik[1,2](✉) , Nicolas Marsal[1,2] , Daniel Brunner[3] ,
and Damien Rontani[1,2]

[1] LMOPS EA 4423, CentraleSupélec, 57070 Metz, France
{piotr.antonik,damien.rontani}@centralesupelec.fr
[2] Université de Lorraine, CentraleSupélec, LMOPS, 57000 Metz, France
[3] FEMTO-ST Institute/Optics Department,
CNRS & University Bourgogne Franche-Comté, 25030 Besançon, France

Abstract. The identification of different types of human actions in videos is a major computer vision task, with capital applications in e.g. surveillance, control, and analysis. Deep learning achieved remarkable results, but remains hard to train in practice. Here, we propose a photonic reservoir computer for recognition of video-based human actions. Our experiment comprises off-the-shelf components and implements an easy-to-train neural network, scalable up to 16,384 nodes, and performing with a near state-of-the-art accuracy. Our findings pave the way towards photonic information processing systems for real-time video processing.

Keywords: Photonic reservoir computing ·
Computer vision · Human action classification

1 Introduction

The recognition of human actions has recently become one of the most popular research areas in the field of computer vision [11], driven by the potential applications in various areas such as surveillance, control, and analysis [7]. The complexity of the task stems from the numerous problems, such as background clutter, partial occlusion, different scales, viewpoints, lighting, and appearance [10]. Deep learning has achieved remarkable results in this field [11], but presents several complications, such as the need for (very) large training datasets, the non-trivial tuning of the hyperparameters, and time- and energy-consuming training process, which commonly requires dedicated high-end hardware (GPU).

In this work, we propose a photonic neural network for video processing. Optical computing promise a high level of parallelism in e.g. optical communications. Therefore, neural networks could heavily benefit from parallel signal

P. Antonik—Supported by AFOSR (grants No. FA-9550-15-1-0279 and FA-9550-17-1-0072), Région Grand-Est, and the Volkswagen Foundation via the NeuroQNet.

© Springer Nature Switzerland AG 2019
I. V. Tetko et al. (Eds.): ICANN 2019, LNCS 11731, pp. 156–160, 2019.
https://doi.org/10.1007/978-3-030-30493-5_15

transmission, which is one of the strong suits of photonics. Our optical approach could thus allow one to build high-speed and energy-efficient photonic computing devices.

Our experimental system implements the reservoir computing paradigm – a set of methods for designing and training artificial recurrent neural networks [4,6]. A typical reservoir consists of randomly connected fixed network with random input coupling coefficients. Only the output weights are optimised, which reduces the training process to solving a system of linear equations [5]. The RC algorithm has been successfully applied to channel equalisation, chaotic series forecasting, and phoneme recognition. Its simplicity makes it well suited for electronic, opto-electronic, and all-optical analogue implementations (see [8] for a recent review).

2   Results

The system is tested on the well-known KTH database [9], consisting of video recordings of 25 subjects performing 6 different motions (walking, jogging, running, boxing, hand waving, and hand clapping). At the preprocessing stage, we use the histograms of oriented gradients (HOG) algorithm [2] to preprocess individual video frames and extract relevant spatial and shape information. The photonic reservoir computer classifies the 6 motions given the resulting HOG features, as illustrated in Fig. 1.

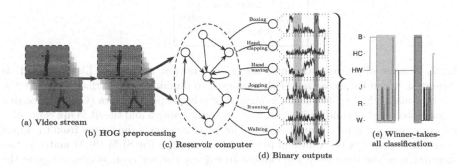

Fig. 1. Scheme of principle of the proposed video processing system. The video input stream (a) undergoes a preprocessing stage (b), where the HOG algorithm is applied to each individual frame. The resulting features are fed into the reservoir computer (c), trained to classify each individual frame. This is achieved by defining 6 binary output nodes (d), one for each action class, that are trained to output 1 for a frame of the corresponding class and 0 for the others. Target outputs are shown in blue. The frame-wise classification (e) is obtained by selecting the node with the maximum output, i.e. the winner-takes-all approach. The final decision for a video sequence is given by the class attributed to the most frames of the sequence. (Color figure online)

The proposed experimental setup, inspired by [1,3] and schematised in Fig. 2, is based on the phase modulation of a spatially extended planar wave by means of a spatial light modulator (SLM). The scheme's notable parallelisation potential allows to implement large neural networks, which is vital for successfully solving challenging tasks in computer vision. The proposed experimental setup can realise a reservoir of 16,384 nodes and could, in principle, be scaled up to as high as 262,144 neurons. The input, the recurrence of the network, and the output layer are implemented digitally to increase the flexibility of the setup. The digital part also dictates the processing speed and allows to classify 2 video frames per second with large reservoirs (16,384 nodes) and up to 7 frames per second with small reservoirs (1,024 nodes).

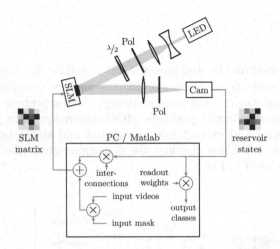

Fig. 2. Scheme of the experimental setup. The output of a green LED (532 nm) is collimated, expanded, polarised (Pol), and used to illuminate the surface of the spatial light modulator (SLM). The latter is imaged by a high-speed camera (Cam.) through a second polariser (Pol) and an imaging lens. Both the camera and the SLM are controlled by a computer, running a Matlab script. The latter generates the inputs from the input videos, and computes the values of pixels to be loaded on the SLM (SLM matrix). The computer uses the data from the camera to extract the reservoir states, compute the outputs and generate the output classes. (Color figure online)

We investigate different network sizes from 1,024 to 16,384 nodes, both numerically and experimentally, and obtain a classification accuracy as high as 92% (see Fig. 3), comparable to the state-of-the-art rates 90.7%–95.6% achieved with far more complex and demanding architectures implemented on noiseless digital processors [11]. This work thus shows that a challenging computer vision task can be efficiently solved with a photonic reservoir computer.

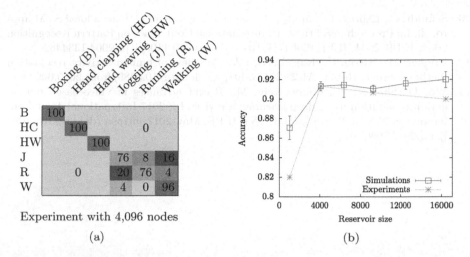

Experiment with 4,096 nodes

(a)   (b)

Fig. 3. (a) Experimental confusion matrix, giving the percentage of actions of class i classified into the class j. Hand gestures (B, HC, HW) are perfectly recognised, while fast spatial motions (J, R, W) are more challenging to differentiate. **(b)** Performance of our photonic reservoir computer on the human action classification task. The error bars on the numerical results show the variability of the accuracy (standard deviation) with 5 different input masks. Experimental variability could not be measured because of the long experimental runtime.

References

1. Bueno, J., et al.: Reinforcement learning in a large-scale photonic recurrent neural network. Optica **5**(6), 756 (2018). https://doi.org/10.1364/optica.5.000756
2. Dalal, N., Triggs, B.: Histograms of oriented gradients for human detection. In: 2005 IEEE Computer Society Conference on Computer Vision and Pattern Recognition (CVPR). IEEE (2005). https://doi.org/10.1109/cvpr.2005.177
3. Hagerstrom, A.M., Murphy, T.E., Roy, R., Hövel, P., Omelchenko, I., Schöll, E.: Experimental observation of chimeras in coupled-map lattices. Nat. Phys. **8**(9), 658–661 (2012). https://doi.org/10.1038/nphys2372
4. Jaeger, H.: Harnessing nonlinearity: predicting chaotic systems and saving energy in wireless communication. Science **304**(5667), 78–80 (2004). https://doi.org/10.1126/science.1091277
5. Lukoševičius, M., Jaeger, H.: Reservoir computing approaches to recurrent neural network training. Comput. Sci. Rev. **3**(3), 127–149 (2009). https://doi.org/10.1016/j.cosrev.2009.03.005
6. Maass, W., Natschläger, T., Markram, H.: Real-time computing without stable states: a new framework for neural computation based on perturbations. Neural Comput. **14**(11), 2531–2560 (2002). https://doi.org/10.1162/089976602760407955
7. Moeslund, T.B., Granum, E.: A survey of computer vision-based human motion capture. Comput. Vis. Image Underst. **81**(3), 231–268 (2001). https://doi.org/10.1006/cviu.2000.0897
8. der Sande, G.V., Brunner, D., Soriano, M.C.: Advances in photonic reservoir computing. Nanophotonics **6**, 3 (2017). https://doi.org/10.1515/nanoph-2016-0132

9. Schuldt, C., Laptev, I., Caputo, B.: Recognizing human actions: a local SVM approach. In: Proceedings of the 17th International Conference on Pattern Recognition 2004, ICPR 2004. IEEE (2004). https://doi.org/10.1109/icpr.2004.1334462

10. Vrigkas, M., Nikou, C., Kakadiaris, I.A.: A review of human activity recognition methods. Front. Robot. AI **2** (2015). https://doi.org/10.3389/frobt.2015.00028

11. Wu, D., Sharma, N., Blumenstein, M.: Recent advances in video-based human action recognition using deep learning: a review. In: 2017 International Joint Conference on Neural Networks (IJCNN). IEEE, May 2017. https://doi.org/10.1109/ijcnn.2017.7966210

A Power-Efficient Architecture
for On-Chip Reservoir Computing

Stijn Sackesyn[1,2]($\boxtimes$) (iD), Chonghuai Ma[1,2] (iD), Andrew Katumba[1,2] (iD),
Joni Dambre[3] (iD), and Peter Bienstman[1,2] (iD)

[1] Photonics Research Group, Department of Information Technology,
Ghent University - imec, Ghent, Belgium
stijn.sackesyn@ugent.be
[2] Center for Nano- and Biophotonics (NB-Photonics),
Ghent University, Ghent, Belgium
[3] IDLab, Department of Information Technology,
Ghent University - imec, Ghent, Belgium

Abstract. Reservoir computing is a neuromorphic computing paradigm
which is well suited for hardware implementations. In this work, an
enhanced reservoir architecture is introduced as to lower the losses
and improve mixing behaviour in silicon photonic reservoir computing
designs.

Keywords: Reservoir computing · Silicon photonics ·
Hardware implementation

1 Introduction

Reservoir computing is a machine learning technique in which a nonlinear
dynamical system is used for computation. It was originally implemented as
an efficient way to train a neural network [2] but it has grown to a commonly
used method for classification and regression tasks. The dynamical system, also
called the reservoir, is held unchanged during the procedure and during training.
Only the weights used to linearly combine the reservoir states are optimized. To
keep the recurrent network unchanged and train only on the level of the reservoir
states makes reservoir computing a computationally cheap method.

2 Silicon Photonic Reservoir Computing

Although reservoir computing was originally invented in computer science as a
software solution to bypass the computational cost of optimizing a neural network, it is perfectly suited to be implemented in various hardware platforms.
These implementations do not suffer from classical digital computer bottlenecks
and are by nature more convenient to operate neuromorphic computing schemes.

© Springer Nature Switzerland AG 2019
I. V. Tetko et al. (Eds.): ICANN 2019, LNCS 11731, pp. 161–164, 2019.
https://doi.org/10.1007/978-3-030-30493-5_16

One such hardware implementation that is especially suited for reservoir computing is silicon photonics. Silicon photonics is a CMOS-compatible platform in which waveguides, splitters and combiners are used to guide light through a silicon chip. It has the advantages of being compact, inexpensive to produce in high volumes and having a mature fabrication process. For reservoir computing, an additional advantage that comes free of costs is that the computation happens in the optical domain, which can improve the reservoir richness as each signal is in essence two-dimensional. Optics also supports much higher bandwidths than electronics and in principle, one can exploit many nonlinear processes in photonics.

3   The Four-Port Architecture

Until now, silicon photonic reservoir computing approaches typically employ the swirl architecture [1,4] as initially defined in [3] and illustrated in Fig. 1a. Even though it indeed introduces the necessary dynamics, there is still room for improvements in terms of losses and mixing. Concerning losses, the swirl architecture fundamentally suffers from modal radiation losses at each 2×1 combiner (for example node 7). These losses are inherent for non-symmetrical reciprocal splitting devices as on average there is a 50% modal mismatch between the two input channels. In terms of mixing, some nodes are partially withdrawn from the dynamics, only consisting of one input and one output (the corner nodes). In general, the most interesting behaviour in swirl reservoirs will be at the inside, by definition of the architecture, while it could be beneficial to also bring the outer layers more into play.

In this paper, we present a new architecture, the four-port architecture, which reduces the problems concerning excessive losses by avoiding 2×1 devices and

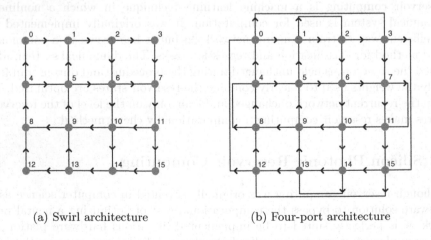

(a) Swirl architecture   (b) Four-port architecture

Fig. 1. Reservoir architectures on a topological level.

employing 2×2 devices exclusively instead, as illustrated in Fig. 1b. It is compared with the swirl architecture in simulations in terms of losses and in terms of power-uniformity.

These simulations show that the four-port architecture indeed suffers less losses and thus has a better energy-efficiency as all input power at a node is redirected to one of the two output channels instead of radiating away. Not only does this avoid undesired losses, but this strategy also contributes to a more uniform power distribution and additional mixing between states which were topologically far apart in the swirl architecture by exploiting extra output ports of the 2×2 devices in an smart way. A better power uniformity does not only increase the richness of the dynamics, it also facilitates measuring the states in an eventual chip as more nodes will be measurable above a certain threshold.

4   Conclusion

The four-port architecture was designed which, in terms of losses and in terms of connectivity within the reservoir, finds itself superior to the swirl architecture.

Measurements on fabricated chips existing of the presented four-port architecture are ongoing and preliminary results will be presented in the scope of this submission. Figure 2 shows a small part of the device under test.

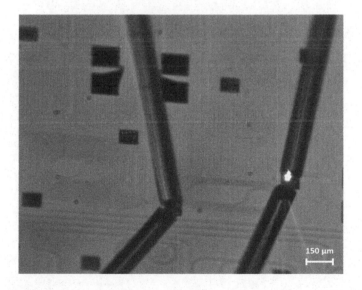

Fig. 2. Photograph of an optical measurement on the silicon chip designed within the scope of PHRESCO. Grating couplers are used to guide light from an optical fiber to the integrated structures on the chip and vice versa.

Acknowledgements. This work was supported in part by the EU project PHRESCO H2020-ICT-2015-688579 and in part by the Research Foundation Flanders (FWO) under Grant 1S32818N.

References

1. Katumba, A., Freiberger, M., Bienstman, P., Dambre, J.: A multiple-input strategy to efficient integrated photonic reservoir computing. Cogn. Comput. **9**(3), 307–314 (2017). https://doi.org/10.1007/s12559-017-9465-5
2. Lukoševičius, M., Jaeger, H.: Reservoir computing approaches to recurrent neural network training. Comput. Sci. Rev. **3**(3), 127–149 (2009). https://doi.org/10.1016/j.cosrev.2009.03.005
3. Vandoorne, K., Dambre, J., Verstraeten, D., Schrauwen, B., Bienstman, P.: Parallel reservoir computing using optical amplifiers. IEEE Trans. Neural Netw. **22**(9), 1469–1481 (2011). https://doi.org/10.1109/TNN.2011.2161771
4. Vandoorne, K., et al.: Experimental demonstration of reservoir computing on a silicon photonics chip. Nat. Commun. **5**, 1–6 (2014). https://doi.org/10.1038/ncomms4541

Time Series Processing with VCSEL-Based Reservoir Computer

Jean Benoit Héroux[1]($\boxtimes$), Naoki Kanazawa[1], and Piotr Antonik[2]

[1] IBM Research—Tokyo, Kawasaki, Kanagawa 212-0032, Japan
heroux@jp.ibm.com
[2] Chaire Photonique, LMOPS EA 4423, CentraleSupélec & Université de Lorraine,
57000 Metz, France

Abstract. Reservoir computing architectures offer important benefits for the implementation of a neural network in a physical medium, as the weighted interconnections between the internal nodes are random and fixed. Experimental results on a time-delay photonic reservoir computer based on directly modulated Vertical Cavity Surface Emitting Lasers and multi-mode fiber couplers are presented. The neuron is made of photodiode, non-linear amplifier and laser chips. The NARMA10 chaotic time-series task is performed with a configuration having 25 virtual nodes operating at 1 GS/s. Experimental and simulated error ranges are in good agreement, which is promising for an expansion to a more elaborate system. The potential of this scheme for the realization of a photonic reservoir cluster device operating at very high speed with low power and a small footprint with a large number of interacting physical and virtual neurons is discussed.

Keywords: Time-delay reservoir · Photonics · Reservoir cluster

1 Introduction

Recurrent neural networks in which part or all of the weights are randomly assigned have attracted much attention recently as a powerful platform for the realization of complex tasks that alleviates the heavy resource requirements for training that are typical of fully optimized deep neural network structures. Reservoir computers are a class of random neural networks in which only the weights between the network and the output layer are trained, via a linear regression. While this scheme, in which an input signal is non-linearly transformed to better recognize hidden features, is increasingly used as an algorithm in high performance computers typically for the analysis or generation of sequential data streams, another active area of research is the realization of reservoir computers via a physical medium that provides the non-linearity and random node interaction. In this last situation, the intent is to build a new kind of computer that

Supported by the New Energy Development Organization (NEDO). P.A. acknowledges financial support by the founders of the Chaire Photonique.

© Springer Nature Switzerland AG 2019
I. V. Tetko et al. (Eds.): ICANN 2019, LNCS 11731, pp. 165–169, 2019.
https://doi.org/10.1007/978-3-030-30493-5_17

does not rely on a traditional von-Neumann architecture and avoids the binary transformation of input information from the physical world for processing. The payoff for the realization of such a solution will be a very power efficient device that can process data streams at very high speed with a small footprint without connection to a server.

While various physical media have been investigated so far including water waves, springs, magnetic materials and biological tissues, photonic reservoir implementations have attracted particular interest in the past ten years or so [1]. Promising attempts have been made to realize an optical device by the manipulation of a light beam traveling in free space or with a signal propagating into single-mode waveguides forming an optical integrated circuit with a physical node array, but challenges related to optical loss and noise control remain limiting factors. A useful scheme that has been proposed that can alleviate these issues is time multiplexing of the signal input [2]: instead of a spatial array of nodes, the time response of a single physical neuron excited by the input signal and a delayed output feedback is acquired to obtain an almost mathematically equivalent system performing complex tasks with virtual, time-varying nodes.

So far, technology mostly inherited from the telecommunications industry including edge-emitting lasers and single mode optical fibers with a length in the hundreds of meter range has been leveraged for the experimental demonstration of time-delay photonic reservoirs. The implementation of a similar scheme in a compact form factor will require further refinement and design choices. A large number of virtual nodes is desirable to perform a task with few errors, which translates however into a long feedback loop for a given signal input rate.

Another area in which important progress has been realized in recent years is related to short range optical data transfer inside high performance computers and data centers, where the requirements are different from those for telecommunication applications [3]: for so-called optical interconnects, the aggregated data rate per unit area (channel density) around a circuit chip as well as the energy required for the transfer of a single bit of information and the fabrication cost per channel are of critical importance. The components of choice for light emission are then directly-modulated Vertical Surface Emitting Lasers (VCSEL), which are low cost, low power and typically made in 12 or 24 channel chip arrays with a 250 μm pitch. For ease of connectivity with low optical loss and high misalignment tolerance, multi-mode waveguides with a relatively large core size typically around 50 μm are used, and the channel distance varies from the centimeter to the hundreds of meter range, i.e., longer than all-electrical connections on a board but shorter than a long-haul telecommunication link.

In this work, we evaluate a prototype time-delay reservoir computer based on multi-mode optical interconnect components and present experimental results for the realization of a time-series benchmark task. The study of new algorithms based on an aggregation of randomly interconnected neural networks is currently a topic of interest, and an optical technology that is a-priori optimized for multi-channel, high data density is a natural match to this research direction.

2   Experimental Details and Results

The optical reservoir computer that we built experimentally is described in more details in Ref. [4] and shown in Fig. 1. For the input, a data stream is programmed into an FPGA and subsequently amplified to drive a VCSEL emitting at an 850 nm wavelength. The emitted light is coupled into a multi-mode graded-index optical fiber to reach a photodiode, and the electrical output is amplified to drive a second identical VCSEL. The assembly formed by the photodiode, amplifier and second VCSEL is the neuron of the system. Its outgoing signal is split into a delayed feedback component that is combined with that of the first VCSEL and an output that is detected by a second photodiode and amplified for readout with a time-sampling oscilloscope.

The formalism describing this time-multiplexed reservoir is well-known, and the node response (corresponding to the optical output power of VCSEL2) is given by

$$x_i(n+1) = \begin{cases} F_{NL}[\alpha x_{i-1}(n) + \beta m_i u(n+1)] & 2 \leq i \leq i_{max} \\ F_{NL}[\alpha x_{N+i-1}(n-1) + \beta m_i u(n+1)] & i = 1 \end{cases} \quad (1)$$

where F_{NL} is the nonlinear response of the neuron, $u(n)$ is the serial signal input, m_i is the scaling step mask randomly chosen in the range $[-1:1]$ and kept fixed for each symbol, and $x_i(n)$ is the physical node state for virtual node i at symbol n. The signal rate determined by the FPGA is 1 GS/s and the feedback loop length is approximately 4 m, which results in $i_{max} = 25$ virtual nodes. The time duration of each mask step m_i is 1 ns. The input and output attenuation factors β and α are determined by the optical loss in the fiber splitter, combiner and couplers as well as the gain in the neuron (Amp2), and are adjusted experimentally to remain close to the edge of the stability.

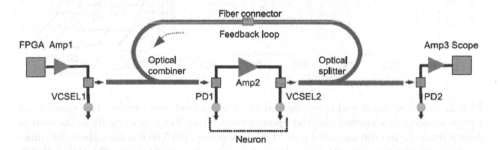

Fig. 1. Experimental setup of the photonic reservoir computer. Optical fibers are shown in blue, and the two coupling arms are linked together by a connector. Forward biases are added to the modulated signals driving VCSEL1 and VCSEL2. A 2 V reverse bias is applied to the photodiode chip PD1. A commercial high speed photoreceiver is used for the combination PD2 and Amp3. (Color figure online)

The task that we present here is the generation of the well-known NARMA10 time-series from a randomly generated data pattern in the interval [0:0.5] with a linear probability distribution. Six series of 800 symbols were programmed and recorded after processing. One of them was used for training, and the other five for test. Figure 2(a) shows a section of the ideal input and experimental output from the reservoir. To each symbol step is applied the random mask modulation for the 25 nodes, and four symbols are shown. The output contains the memory feedback component, so that it does not have the same shape as the input which confirms good operation.

The reservoir was trained offline to reproduce the NARMA10 time series from the experimental data. A matrix form was obtained by averaging the data points of each step, and the performance was evaluated by the usual normalized root mean square error (NRMSE). For comparison purposes, a calculation with a similar input was also performed numerically with a hyperbolic tangent non-linear node, i.e., a simple matrix treatment.

Experimental and simulated error ranges for the five test data sets are shown by the two vertical lines in Fig. 2(b) in red and blue respectively. Results are encouraging considering that there is some overlap between the two lines even though many parameters in the experimental system are not yet optimized. At the moment, the largest sources of noise are the input FPGA device and the oscilloscope, so that performance is not intrinsically limited by the reservoir itself. The two-dimensional graph in the figure shows a numerical matrix calculation to illustrate how results can be improved with more virtual nodes and a larger training set, which will be realized with an improved FPGA system in the future.

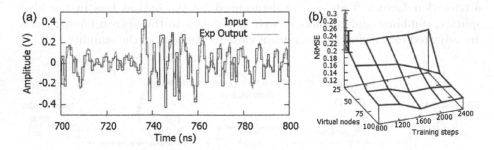

Fig. 2. (a) Ideal input and experimental reservoir output step curves. The signal from the oscilloscope is recorded with data points every 50 ps. The non-linearity in the system comes from the neuron amplifier and VCSEL response. (b) Vertical lines show the range of errors for the five test data sets. Experimental values (red) are [0.264, 0.282, 0.223, 0.314, 0.249] and numerical values (blue) are [0.210, 0.221, 0.245, 0.207, 0.231]. An ideal result gives NRMSE = 0, whereas for a completely uncorrelated output NRMSE = 1. For the surface graph, irregularities are due to the finite amount of random data for the calculation (six data sets for each data point). (Color figure online)

3   Outlook

In conclusion, we present a VCSEL-based photonic reservoir performing a non-linear time-series prediction task with good performance considering the limited number of virtual nodes and input training data. By leveraging a short range and low power optical data transfer technology that is both well-understood and still improving, we expect to build a robust platform for the experimental realization of high speed reservoirs.

The potential of this architecture has two main aspects. First, we would like to eventually use all 12 channels of the VCSEL and photodiode chip arrays by adding bonded wire connections and using multi-core fiber ribbons. With these modifications, we could easily realize a system with 12 physical nodes interconnected with a large number of virtual nodes to obtain a playground for a reservoir cluster. The exact configuration that brings the most benefit will have to be determined first by simulation. Second, the experimental setup could be integrated on a board using polymer waveguides and micro-mirrors to realize a centimeter-scale edge-computing device, as this technology is currently being developed for high performance computers. Applications in which the input is intrinsically optical, such as a distorted signal recovery, would be particularly appealing.

References

1. Tanaka, G., et al.: Recent advances in physical reservoir computing: a review. Neural Netw. **115**, 100–123 (2019)
2. Appeltant, L., et al.: Information processing using a single dynamical node as complex system. Nat. Comm. **2**, 468 (2011)
3. Héroux, J.B., et al.: Energy-efficient 1060-nm optical link operating up to 28 Gb/s. J. Lightwave Technol. **33**, 733–740 (2015)
4. Héroux, J.B., Kanazawa, N., Nakano, D.: Delayed feedback reservoir computing with VCSEL. In: Cheng, L., Leung, A.C.S., Ozawa, S. (eds.) ICONIP 2018. LNCS, vol. 11301, pp. 594–602. Springer, Cham (2018). https://doi.org/10.1007/978-3-030-04167-0_54

Optoelectronic Reservoir Computing Using a Mixed Digital-Analog Hardware Implementation

Miguel C. Soriano$^{(\boxtimes)}$ ⓘ, Pau Massuti-Ballester, Jesús Yelo, and Ingo Fischer ⓘ

Instituto de Física Interdisciplinar y Sistemas Complejos, IFISC (CSIC-UIB),
Campus Universitat Illes Balears, 07122 Palma de Mallorca, Spain
miguel@ifisc.uib-csic.es
http://ifisc.uib-csic.es/

Abstract. Optoelectronic systems have proven to be an attractive platform for the realization of hardware-based Reservoir Computing (RC). These unconventional computers can perform nonlinear prediction of chaotic timeseries and generate arbitrary input/output functions, after training. One of the main advantages of a delay-based optoelectronic reservoir is that it only requires a single nonlinear hardware node and a feedback loop. To implement the RC scheme experimentally, we use photonic and electronic components and an FPGA to drive the system. The resulting optoelectronic RC is a hybrid analog and digital system with great versatility. We show that this set-up can perform a chaotic timeseries prediction task, in which the output is computed online, with low prediction errors. Ultimately, this testbed system will allow to test several strategies to improve the prediction performance, including the addition of multiple delays and the tailoring of the input mask for noise mitigation.

Keywords: Reservoir Computing · Opto-electronic system · Nonlinear time-series prediction

1 Introduction

Reservoir computing (RC) refers to a family of neuro-inspired information processing techniques that combine the advantages of recurrent neural networks with a simple training procedure [1,2]. Photonic hardware implementations of RC, in particular those based on delay-systems, have achieved state-of-the-art performances in many different tasks ranging from speech recognition [3,4] to chaotic time-series prediction [5,6]. The advantage of delay-based RC approaches is that they only require a single nonlinear hardware node [7].

Here, we present a novel hardware implementation of optoelectronic reservoir computing for the online prediction of chaotic time series. This approach is based on two previous experimental implementations. On the one hand, it uses

© Springer Nature Switzerland AG 2019
I. V. Tetko et al. (Eds.): ICANN 2019, LNCS 11731, pp. 170–174, 2019.
https://doi.org/10.1007/978-3-030-30493-5_18

as the dynamical nonlinear node a Mach-Zehnder modulator driven in a non-linear transient regime, similar to the one discussed in [5]. On the other hand, the experimental set-up includes a Field Programmable Gate Array (FPGA) to implement the output layer online, as it has been introduced in [6].

2   Experimental Setup

The classical RC scheme contains three main parts: Input layer, Reservoir and Output layer. Figure 1 illustrates the delay-based RC approach, in which the reservoir is made of a single nonlinear hardware node and multiple virtual nodes. These virtual nodes set the dimension of the reservoir and are defined as equidistant temporal positions (with spacing θ) along the delay loop (of length τ). In turn, the input data is multiplied by a random connectivity mask that maps the input to the different virtual nodes of the reservoir via time-multiplexing. The information within the Reservoir is extracted through a weighted linear combination, with trained weights, constituting the Output layer.

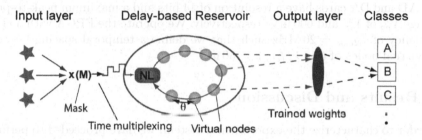

Fig. 1. Schematic representation of delay-based RC. The virtual nodes in the reservoir are defined as equidistant temporal positions along the delay loop [7].

We present the experimental realization in Fig. 2. The optoelectronic system includes a semiconductor laser source, a Mach-Zehnder modulator, a photodetector, filters, amplifiers, an FPGA, analog to digital (AD) and digital to analog (DA) converters. The FPGA is a designed network of logic block circuits to be configured by the user with a hardware description language. In the experimental set-up, the FPGA (ALTERA Cyclone IV) is an essential component that takes care of timing the operations, the input mask multiplication, the delay-feedback loop, and the trained weights multiplication. As a result, this system allows the online computation of the output values.

The Mach-Zehnder (MZ) modulator in Fig. 2 provides a sin^2 nonlinearity, which can be defined as follows:

$$P_{out} = P_{pl}sin^2(2\pi\frac{U}{V_\pi} + \phi_b), \tag{1}$$

where P_{out} is the output power of the MZ, P_{pl} is the power of the pump laser, U is the radio-frequency input of the MZ (input to be processed and sent by the

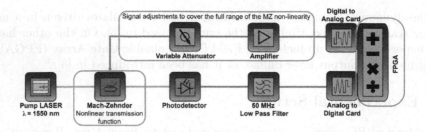

Fig. 2. Block representation of the optoelectronic reservoir computer with mixed digital and analog hardware.

FPGA), $V_\pi = 2.7$ V is the voltage needed to go over one period of the sin^2, and ϕ_b is the operating bias phase of the MZ.

The detection includes a photodetector, filters and the AD card. This detection apparatus defines a band-pass frequency bandwidth with $f_{min} = 100$ KHz and $f_{max} = 20$ MHz, having an optical to electric conversion factor of 1.6 V/mW. The AD and DA cards have a resolution of 14 bits and a maximum peak-to-peak voltage V_{pp} of 1 V and 600 mV, respectively. We operate the FPGA with a clock frequency of $f_{clock} = 20$ MHz such that we define a temporal spacing between the virtual nodes of $\theta = 1/f_{clock} = 50$ ns.

3 Results and Discussion

In order to characterize the experimental system, we have proceeded to perform a chaotic time-series prediction task. For this benchmark task, the input comes from the chaotic dynamics of a Mackey-Glass model (see [8] for more details on the nonlinear time-series prediction task). We have used 4000 samples for this task, divided in 3000 samples for training the system and 1000 samples for the test. The input connectivity mask consists of 6 discrete levels randomly distributed over time and we have considered a reservoir with $N = 500$ virtual nodes. Finally, we have set the delay to $\tau = (N+1)\theta = 25.05$ μs and we operate the Mach-Zehnder with a bias phase of $\phi_b = \pi/4$.

After the output weights have been optimized using a linear regression on the training data, we load these trained weights onto the FPGA. We then run the optoelectronic RC injecting the test sequence and performing the online prediction within the FPGA. We present the experimental results for the one step ahead prediction of the Mackey-Glass chaotic time-series in Fig. 3. We characterize the performance of the optoelectronic RC by computing the normalized mean square error (NMSE). For these operating conditions, we find a NMSE = 0.0032. This low prediction error is comparable to a previous optoelectronic RC implementation (NMSE = 0.0036 [8]) and one order of magnitude lower than other all-optical RC implementations (NMSE = 0.019 [9] and NMSE = 0.042 [10]), validating the current set-up for RC.

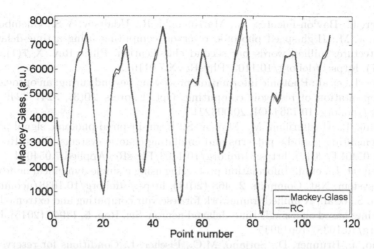

Fig. 3. Prediction of the Mackey-Glass chaotic time series. The original data is depicted in blue and the prediction of the optoelectronic RC is depicted in red. (Color figure online)

In summary, we have implemented an optoelectronic reservoir computer as a hybrid between a digital and an analog part, which achieves a successful prediction of a chaotic time series. The analog part takes care of the nonlinearity, while the digital part takes care of the input pre-processing and the computation of the output. This duality creates a versatile system that allows to test different strategies that have, so far, been unexplored in photonic hardware systems to improve the prediction performance. These strategies include, but are not limited to, the addition of multiple delays within the FPGA and tailoring the input mask. In particular, with multiple delays, we can explore the transition of the current ring topology to more complex network topologies [11]. Finally, this system also has the potential to be operated in a closed loop (connect the output back to the input) autonomous form as it was demonstrated in [6].

References

1. Verstraeten, D., Schrauwen, B., D'Haene, M., Stroobandt, D.: An experimental unification of reservoir computing methods. Neural Netw. **20**(3), 391–403 (2007). https://doi.org/10.1016/j.neunet.2007.04.003
2. Lukoševičius, M., Jaeger, H.: Reservoir computing approaches to recurrent neural network training. Comput. Sci. Rev. **3**(3), 127–149 (2009). https://doi.org/10.1016/j.cosrev.2009.03.005
3. Brunner, D., Soriano, M.C., Mirasso, C.R., Fischer, I.: Parallel photonic information processing at gigabyte per second data rates using transient states. Nat. Commun. **4**, 1364 (2013). https://doi.org/10.1038/ncomms2368

4. Larger, L., Baylón-Fuentes, A., Martinenghi, R., Udaltsov, V.S., Chembo, Y.K., Jacquot, M.: High-speed photonic reservoir computing using a time-delay-based architecture: million words per second classification. Phys. Rev. X **7**(1), 011015 (2017). https://doi.org/10.1103/PhysRevX.7.011015

5. Larger, L., et al.: Photonic information processing beyond turing: an optoelectronic implementation of reservoir computing. Opt. Express **20**(3), 3241–3249 (2012). https://doi.org/10.1364/OE.20.003241

6. Antonik, P., Haelterman, M., Massar, S.: Brain-inspired photonic signal processor for generating periodic patterns and emulating chaotic systems. Phys. Rev. Appl. **7**(5), 054014 (2017). https://doi.org/10.1103/PhysRevApplied.7.054014

7. Appeltant, L., et al.: Information processing using a single dynamical node as complex system. Nat. Commun. **2**, 468 (2011). https://doi.org/10.1038/ncomms1476

8. Ortín, S., et al.: A unified framework for reservoir computing and extreme learning machines based on a single time-delayed neuron. Sci. Rep. **5**, 14945 (2015). https://doi.org/10.1038/srep14945

9. Bueno, J., Brunner, D., Soriano, M.C., Fischer, I.: Conditions for reservoir computing performance using semiconductor lasers with delayed optical feedback. Opt. Express **25**(3), 2401–2412 (2017). https://doi.org/10.1364/OE.25.002401

10. Bueno, J., et al.: Reinforcement learning in a large-scale photonic recurrent neural network. Optica **5**(6), 756–760 (2018). https://doi.org/10.1364/OPTICA.5.000756

11. Hart, J.D., Schmadel, D.C., Murphy, T.E., Roy, R.: Experiments with arbitrary networks in time-multiplexed delay systems. Chaos **27**(12), 121103 (2017). https://doi.org/10.1063/1.5016047

Comparison of Feature Extraction Techniques for Handwritten Digit Recognition with a Photonic Reservoir Computer

Piotr Antonik[1,2], Nicolas Marsal[1,2], Daniel Brunner[3],
and Damien Rontani[1,2(✉)]

[1] LMOPS EA 4423, CentraleSupélec, Université Paris-Saclay, 57070 Metz, France
{piotr.antonik,damien.rontani}@centralesupelec.fr
[2] Université de Lorraine, CentraleSupélec, LMOPS, 57000 Metz, France
[3] FEMTO-ST Institute/Optics Department,
CNRS & University Bourgogne Franche-Comté, 25030 Besançon, France

Abstract. Reservoir computing is a bio-inspired computing paradigm for processing time-dependent signals. Its photonic implementations have received much interest recently, and have been successfully applied to speech recognition and time-series forecasting. However, few works have been devoted to the more challenging computer vision tasks. In this work, we use a large-scale photonic reservoir computer for classification of handwritten digits from the MNIST database. We investigate and compare different feature extraction techniques (such as zoning, Gabor filters, and HOG) and report classification errors of 1% experimentally and 0.8% in numerical simulations.

Keywords: Photonic reservoir computing ·
Handwritten digit classification · Feature extraction · MNIST dataset

1 Introduction

For the past decade, deep learning [10] achieved tremendous success in various computer vision tasks. A deep neural network learns the target task by automatically extracting the most relevant features from the input images or video frames [2]. The key reasons of the deep learning success are the increasing amounts of annotated data available for supervised learning, and the significant advances in computing architectures, especially with the emergence of the general purpose graphical units (GPUs). These major advances, on the other hand, bring out the disadvantages of deep learning approaches, that remain hard to train in practice on regular computers, as they require (very) large datasets and significant computing power (and thus an important energy consumption).

Supported by AFOSR (grants No. FA-9550-15-1-0279 and FA-9550-17-1-0072), Région Grand-Est, and the Volkswagen Foundation via the NeuroQNet.

© Springer Nature Switzerland AG 2019
I. V. Tetko et al. (Eds.): ICANN 2019, LNCS 11731, pp. 175–179, 2019.
https://doi.org/10.1007/978-3-030-30493-5_19

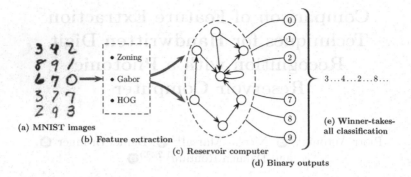

(a) MNIST images

(b) Feature extraction
- Zoning
- Gabor
- HOG

(c) Reservoir computer

(d) Binary outputs

3...4...2...8...

(e) Winner-takes-all classification

Fig. 1. Operating principle of the proposed setup. The images from the MNIST database (a) undergo the feature extraction stage (b), where different algorithms under investigations are applied. The resulting features are fed into the reservoir computer (c) with 10 binary output nodes (d), one for each digit. The nodes are trained to output 1 for the digit associated with the node and 0 for the other digits. The final classification (e) is obtained by selecting the node with the maximum output, i.e. the winner-takes-all approach.

The alternative to deep learning in image classification is the use of much simpler classification systems – e.g. decision trees [12], support vector machines [4], or shallow neural networks – combined with feature extraction techniques, hand-tailored to the target task, as illustrated in Fig. 1. The main difficulty of this approach is to find the most relevant features to feed into the classifier, which is the main focus of the present work.

2   Results

In this work, we investigate and compare three popular feature extraction techniques [1]:

- The zoning technique [8] is a statistical region-based method that aims to get the local characteristics of the image. In practice, pixel densities are computed in smaller zones of the image, with variable zones' dimensions.
- Gabor filters [6] are a powerful tool in texture analysis. They seek for specific frequency content in a fixed direction and a given region of the image.
- The histograms of oriented gradients (HOG) [5] are widely used in computer vision for object detection, based on computing spatial gradients in the image.

The classification stage is based on the reservoir computing paradigm (RC) – a set of methods for designing and training artificial recurrent neural networks [9,11]. A typical reservoir consists of a randomly-connected, fixed network with a trainable output layer. The simplicity of RC makes it well suited for electronic, opto-electronic, and all-optical analogue implementations (see [13] for a recent review).

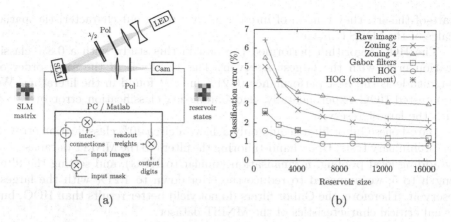

(a)   (b)

Fig. 2. (a) Scheme of the experimental setup. The output of a green LED (532 nm) is collimated, expanded, polarised (Pol), and used to illuminate the surface of the spatial light modulator (SLM). The latter is imaged by a high-speed camera (Cam.) through a second polariser (Pol) and an imaging lens. Both the camera and the SLM are controlled by a computer, that generates the inputs from the MNIST dataset (including the feature extraction stage), processes the camera images and computes the output digits. (b) Performance of the reservoir computer on the MNIST task with different feature extraction techniques. (Color figure online)

In this work, we consider an experimental setup inspired by [3,7] and schematised in Fig. 2a. The scheme, based on the phase modulation of a spatially extended planar wave by means of a spatial light modulator (SLM), allows to implement large reservoir computers (up to 16,384 nodes in this experiment), which is essential for successfully solving image classification tasks. The input, the recurrence of the network, and the output layer are implemented digitally to increase the flexibility of the setup.

Figure 2b presents the numerical (grey) and experimental (red) results. We tested all feature extraction methods with different reservoir sizes from 1,024 up to 16,384 nodes. In all cases, the classification error decreases with the number of neurons. We used the performance on raw images as benchmark, that is, unprocessed pixels were used as inputs to the reservoir computer. Raw images were classified with a 6.4% error with the smallest reservoir (1,024 nodes) and 2.1% with the largest reservoir (16,384 nodes).

The zoning technique is the simplest approach considered here, that consists in averaging pixel values over windows of fixed size. We tested two variants of this method, with windows 2×2 (Zoning 2) and 4×4 (Zoning 4). Zoning 2 technique allows to slightly improve the results. Since its effect can be seen as dimensionality reduction, this result indicates that the default 28×28 MNIST images contain some undesirable information that can be removed to improve performance. Zoning 4 technique, on the other hand, performs worse than the raw images, which indicates that too much information is lost when averaging over 4×4 windows. While those two techniques do not yield error rates comparable to state-of-the-art, they remain of interest as they bring out characteristic spatial scales of the MNIST images.

The HOG algorithm performed the best in this study, with a 0.8% classification error with the largest reservoir. This results is of the same order of magnitude as the best performance (0.21% in [14]) found in the literature. We also tested HOG experimentally, and obtained a classification error of 1.03% with the largest reservoir.

Bare-bones Gabor filters could only achieve a circa 40% classification error in the preliminary tests. Upon hand-tailoring the filters to the MNIST database, i.e. by adding local pooling, normalisation (similar to HOG), and setting the filter length to 5, we managed to reduce the error down to 1.21% with the largest reservoir. Therefore, the Gabor filters do not yield better results than HOG, but reveal critical characteristics of the MNIST dataset.

References

1. Bahi, H.E., Mahani, Z., Zatni, A., Saoud, S.: A robust system for printed and handwritten character recognition of images obtained by camera phone. Technical report (2015). http://www.wseas.org/multimedia/journals/signal/2015/a045714-403.pdf
2. Bengio, Y., Courville, A., Vincent, P.: Representation learning: a review and new perspectives. IEEE Trans. Pattern Anal. Mach. Intell. **35**(8), 1798–1828 (2013). https://doi.org/10.1109/tpami.2013.50
3. Bueno, J., et al.: Reinforcement learning in a large-scale photonic recurrent neural network. Optica **5**(6), 756 (2018). https://doi.org/10.1364/optica.5.000756
4. Cortes, C., Vapnik, V.: Support-vector networks. Mach. Learn. **20**(3), 273–297 (1995). https://doi.org/10.1007/bf00994018
5. Dalal, N., Triggs, B.: Histograms of oriented gradients for human detection. In: 2005 IEEE Computer Society Conference on Computer Vision and Pattern Recognition (CVPR). IEEE (2005). https://doi.org/10.1109/cvpr.2005.177
6. Daugman, J.G.: Two-dimensional spectral analysis of cortical receptive field profiles. Vis. Res. **20**(10), 847–856 (1980). https://doi.org/10.1016/0042-6989(80)90065-6
7. Hagerstrom, A.M., Murphy, T.E., Roy, R., Hövel, P., Omelchenko, I., Schöll, E.: Experimental observation of chimeras in coupled-map lattices. Nat. Phys. **8**(9), 658–661 (2012). https://doi.org/10.1038/nphys2372
8. Hussain, A.B.S., Toussaint, G.T., Donaldson, R.W.: Results obtained using a simple character recognition procedure on Munson's handprinted data. IEEE Trans. Comput. **C–21**(2), 201–205 (1972). https://doi.org/10.1109/tc.1972.5008927
9. Jaeger, H.: Harnessing nonlinearity: predicting chaotic systems and saving energy in wireless communication. Science **304**(5667), 78–80 (2004). https://doi.org/10.1126/science.1091277
10. LeCun, Y., Bengio, Y., Hinton, G.: Deep learning. Nature **521**(7553), 436–444 (2015). https://doi.org/10.1038/nature14539
11. Maass, W., Natschläger, T., Markram, H.: Real-time computing without stable states: a new framework for neural computation based on perturbations. Neural Comput. **14**(11), 2531–2560 (2002). https://doi.org/10.1162/089976602760407955

12. Quinlan, J.R.: Induction of decision trees. Mach. Learn. **1**(1), 81–106 (1986). https://doi.org/10.1007/bf00116251
13. der Sande, G.V., Brunner, D., Soriano, M.C.: Advances in photonic reservoir computing. Nanophotonics **6**(3) (2017). https://doi.org/10.1515/nanoph-2016-0132
14. Wan, L., Zeiler, M., Zhang, S., Cun, Y.L., Fergus, R.: Regularization of neural networks using dropconnect. In: Dasgupta, S., McAllester, D. (eds.) Proceedings of the 30th International Conference on Machine Learning. Proceedings of Machine Learning Research, vol. 28, pp. 1058–1066. PMLR, Atlanta, Georgia, USA, 17–19 June 2013 (2013)

Polarization Dynamics of VCSELs Improves Reservoir Computing Performance

Jeremy Vatin[1,2]([✉]), Damien Rontani[1,2], and Marc Sciamanna[1,2]

[1] Chaire Photonique, CentraleSupélec and Université Paris-Saclay,
2 rue Edouard Belin, 57070 Metz, France
jeremy.vatin@centralesupelec.fr
[2] Université Lorraine and CentraleSupélec, LMOPS EA-4423 Laboratory,
2 rue Edouard Belin, 57070 Metz, France

Abstract. We present a new architecture of time delay-reservoir reservoir computer based on the polarization dynamics of a VCSEL. This architecture achieves state of the art performance. Moreover, implementing polarization rotated feedback instead of a polarization preserving feedback allows further improvement to the computing performance.

This enhancement of the reservoir computing performance is demonstrated on a telecommunication task: nonlinear channel equalization. Numerical simulations show better performance while using the polarization rotated feedback compared to isotropic feedback. This first insight is also proved experimentally and experiment and theory agree not only qualitatively, but also quantitatively.

Keywords: Time-delay reservoir computing · Semiconductor laser · Vertical-cavity surface-emitting laser

Reservoir computing is based on artificial neural network. It consists in, instead of training the links between nodes, only training a readout layer thanks to linear regression. This allows applying this specific technique of machine learning on physical components having physical links between each two nodes that cannot be tuned. However, this approach still has limitation for connecting a larger number of nodes. The time-delay reservoir computing solves this challenge, allowing to use virtual nodes rather than physical ones [1]. It involves one nonlinear physical node, which is submitted to a delay line. Virtual nodes are then spread along the delay-line. The main advantage is that the number of nodes can be easily increased by considering a longer feedback loop.

Several systems based on this approach have already been proposed: optoelectronic [3] or all optical [2,4].

Supported by Ministère de l'Enseignement Supérieur de la Recherche et de l'Innovation; Région Grand-Est; Département Moselle; European Regional Development Fund (ERDF); Metz Métropole; Airbus GDI Simulation; CentraleSupélec; Fondation CentraleSupélec.

© Springer Nature Switzerland AG 2019
I. V. Tetko et al. (Eds.): ICANN 2019, LNCS 11731, pp. 180–183, 2019.
https://doi.org/10.1007/978-3-030-30493-5_20

The reservoir computer we are studying is an all-optical architecture, using a vertical-cavity surface-emitting laser (VCSEL) as a physical node, optically injected and with optical feedback. A scheme of our architecture is proposed Fig. 1.

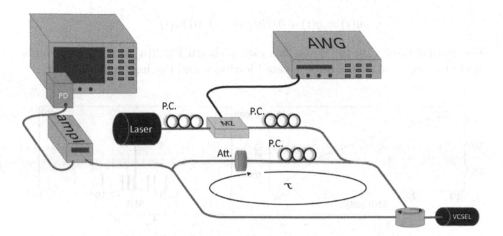

Fig. 1. Schema of the system. AWG: Arbitrary waveform generator, MZ: Mach-Zehnder modulator, P.C.: polarization controller, att: attenuator, ampl: amplifier, PD: photo-diode

The reservoir itself is composed of a VCSEL and a feedback loop, in which the polarization of the emitted light can be tuned thanks to a polarization controller (P.C.). The length τ of the feedback loop is 39 ns, hence a processing rate of 25.6 MHz. 390 virtual nodes are distributed along the delay-line. The input is made optically: The VCSEL is injected with another laser. This other laser is externally modulated with a Mach-Zehnder modulator. The modulation contains the input data to be processed in the reservoir. The output signal is finally sent in an amplifier to reach a power readable by the photodiode, and recorded with the oscilloscope. We use as an output of the reservoir the total optical power, that means the sum of the power of both main and depressed polarization mode. The training and the testing are made with a computer after the record.

This system exploits the specific polarization dynamics of the VCSEL to enhance the computational performance. We proved theoretically that using a polarization rotated feedback allows doubling the memory capacity of the system and therefore improving the performance on computational tasks such as the nonlinear channel equalization [5]. This task consists of reconstructing a signal $u(i)$ distorted in a nonlinear telecommunication channel in order to recover the original signal $d(i)$, which is composed of four different symbols randomly taken in $\{-3, -1, 1, 3\}$. The nonlinear channel under consideration is simulated thanks to the following equations: First the original signal is linearly filtered according to:

$$q(i) = 0.08d(i+2) - 0.12d(i+1) + d(i) + 0.18d(i-1)$$
$$- 0.1d(i-2) + 0.091d(i-3) - 0.05d(i-4) + 0.04d(i-5) \qquad (1)$$
$$+ 0.03d(i-6) + 0.01d(i-7).$$

It is then modified using a nonlinear function:

$$u(i) = q(i) + 0.026q(i)^2 - 0.011q(i)^3. \qquad (2)$$

The symbol error rate (SER) on this task is shown Fig. 2(a), and comparison is made between the polarization rotated feedback and the isotropic feedback.

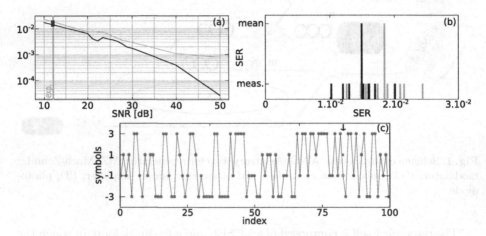

Fig. 2. (a) Performance of the simulated VCSEL-based reservoir depending on the noise in the output layer for isotropic feedback (blue) and rotated feedback (purple), green squares are the expected performance of the experimental reservoir due to the level of noise. (b) Performance of the experimental reservoir: The smaller strips are the values of the SER over different measurement series and the bigger one are the mean SER over those series, for the isotropic feedback (blue) and rotated feedback (purple). (c) Example of experimental signal reconstruction: the crosses are the estimated output and the circles the target values. The arrow points at the only error in this series. (Color figure online)

Figure 2(a) shows that the performance is highly sensitive to the noise in the output layer. This noise is a white gaussian noise that simulates the one added by the amplifier in the output layer. This is the main source of noise in the system. Added to that, it proves that tuning the polarization in the feedback loop significantly improves the performance of the reservoir. This last result is also verified experimentally as shown in Fig. 2(b). The measured SER is consistent with the one predicted theoretically for the same amount of noise. The signal is reconstructed with a remaining SER of 1.5×10^{-2} in the case of rotated feedback as depicted in Fig. 2(c), compared to 2×10^{-2} in the case of isotropic feedback. As shown on Fig. 2(a), the performance can be improved by increasing the SNR. This could be done by replacing the amplifier in the output layer by a less noisy one, or increasing the power of the VCSEL.

In summary, the use of a VCSEL as a physical node adds a new degree of freedom that can be tuned: The polarization of the light. Using a polarization rotated feedback configuration allows improving the performance of our time-delay reservoir as found both in the numerical simulations, and the experiment.

References

1. Appeltant, L., et al.: Information processing using a single dynamical node as complex system. Nat. Commun. **2**, 468 (2011). https://doi.org/10.1038/ncomms1476
2. Duport, F., Schneider, B., Smerieri, A., Haelterman, M., Massar, S.: All-optical reservoir computing. Opt. Express **20**(20), 22783–22795 (2012). https://doi.org/10.1364/OE.20.022783
3. Larger, L., et al.: Photonic information processing beyond turing: an optoelectronic implementation of reservoir computing. Opt. Express **20**(3), 3241–3249 (2012). https://doi.org/10.1364/OE.20.003241
4. Nguimdo, R.M., Verschaffelt, G., Danckaert, J., der Sande, G.V.: Fast photonic information processing using semiconductor lasers with delayed optical feedback: role of phase dynamics. Opt. Express **22**(7), 8672–8686 (2014). https://doi.org/10.1364/OE.22.008672
5. Vatin, J., Rontani, D., Sciamanna, M.: Enhanced performance of a reservoir computer using polarization dynamics in VCSELs. Opt. Lett. **43**(18), 4497–4500 (2018). https://doi.org/10.1364/OL.43.004497

Reservoir-Size Dependent Learning in Analogue Neural Networks

Xavier Porte$^{(\boxtimes)}$ ⓘ, Louis Andreoli, Maxime Jacquot ⓘ, Laurent Larger ⓘ,
and Daniel Brunner ⓘ

FEMTO-ST/Optics Dept., UMR CNRS 6174, Univ. Bourgogne Franche-Comté,
15B avenue des Montboucons, 25030 Besançon Cedex, France
`javier.porte@femto-st.fr`

Abstract. The implementation of artificial neural networks in hardware substrates is a major interdisciplinary enterprise. Well suited candidates for physical implementations must combine nonlinear neurons with dedicated and efficient hardware solutions for both connectivity and training. Reservoir computing addresses the problems related with the network connectivity and training in an elegant and efficient way. However, important questions regarding impact of reservoir size and learning routines on the convergence-speed during learning remain unaddressed. Here, we study in detail the learning process of a recently demonstrated photonic neural network based on a reservoir. We use a greedy algorithm to train our neural network for the task of chaotic signals prediction and analyze the learning-error landscape. Our results unveil fundamental properties of the system's optimization hyperspace. Particularly, we determine the convergence speed of learning as a function of reservoir size and find exceptional, close to linear scaling. This linear dependence, together with our parallel diffractive coupling, represent optimal scaling conditions for our photonic neural network scheme.

Keywords: Neural networks · Reservoir computing · Nonlinear optics

1 Introduction

Nowadays neural networks (NNs) still remain extensively emulated by traditional computers, which posts important challenges in terms of parallelization, energy efficiency and overall computing speed. Ultimately, full hardware integration of NNs, where nonlinear nodes, network connectivity and optimization through learning are implemented via dedicated functionalities of the substrate is desirable. Optics-based solutions like optoelectronic [6,8] or all-photonic [1,4,9] neural networks are of particular interest because they can avoid parallelization bottlenecks.

In the context of hardware implementation, reservoir computing (RC) appeared as a particularly well suited approach to train and operate NNs [5,7].

© Springer Nature Switzerland AG 2019
I. V. Tetko et al. (Eds.): ICANN 2019, LNCS 11731, pp. 184–192, 2019.
https://doi.org/10.1007/978-3-030-30493-5_21

The convenience of RC originates in a training that is restricted to optimization of the readout weights, leaving the input as well as the internal connectivity between neurons unaffected. However, in physically implemented NNs the training itself is conditioned by hardware structure. Here, one of the fundamental questions is how an error landscape is explored by a given learning algorithm when applied to a hardware NN. Therefore, understanding the topology of the cost function and its potential convexity or presence of local minima is of major importance.

We experimentally implement RC in an optoelectronic analogue NN and train it to predict chaotic time series via a greedy algorithm, analogously to [3]. We study in detail the error-landscape, which we also refer as cost-function, differentiating those features caused by topology from those originated by noise. The mapped landscape is rich in features, on average follows an exponential topology and contains numerous local minima with comparable good-performance. We demonstrate that by using our greedy algorithm learning converges systematically. Moreover, we address the particular question of how the NN rate of convergence and the prediction error depend on the network size. This is the first time that the fundamental characteristics of greedy learning are explored in a noisy physically implemented NN.

2   Neural Network Concept and Training

Our optoelectronic NN is composed of up to 961 neurons whose state is encoded in the pixels of a spatial-light modulator (SLM). The neurons are connected among themselves via diffractive optical coupling, which is inherently parallel and scalable [2].

2.1   Experimental Setup

The experimental implementation is schematically illustrated in Fig. 1(a). A laser diode of intensity $|E_i^0|^2$ illuminates the SLM, where the neurons' states x_i are encoded. The SLM is imaged on the camera (CAM) after passing through the polarizing beam splitter (PBS) and twice through the diffractive optical element (DOE). The information detected by the camera is used to drive the SLM, realizing the NN recurrent connectivity. The dynamical evolution of the recurrent NN is given by

$$x_i(n+1) = \alpha|E_i^0|^2 cos^2\left[\beta \cdot \alpha\left|\sum_j^N W_{i,j}^{DOE}E_j(n)\right|^2 + \gamma W_i^{inj}u(n+1) + \theta_i\right], \quad (1)$$

where $E_j(n)$ is the optical electric field for each neuron, β is the feedback gain, γ is the input injection gain, α is an empirical normalization parameter, and θ_i are the phase offsets for each node. After optimization, operational parameters $[\beta, \gamma, \alpha, \theta_i]$ are kept constant. The control PC reads the camera output and sets the new state of the SLM following Eq. 1. Recurrency is stablished by previous neuron state $E_j(n)$, the external information is $u(n+1)$, W^{DOE} is the recurrent neurons' internal coupling and W^{inj} is the information injection matrix with random, independent and uniformly distributed weights between 0 and 1.

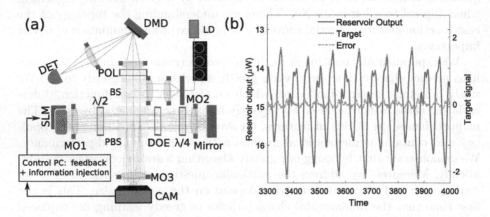

Fig. 1. (a) Schematic of our recurrent neural network. (b) NN performance for the Mackey-Glass chaotic time series prediction task: the target chaotic time series, the reservoir's output and the prediction error are depicted in orange, blue and green, respectively. (Color figure online)

Following the RC principle, the NN training is restricted only to the modification of the readout weights. For that, the neurons of the recurrent NN, i.e. the SLM pixels, are imaged on a digital micro-mirror display (DMD) and focused on the surface of a photodiode. The output of the neural network, as measured by the photodiode is

$$y^{out}(n+1) \propto \left| \sum_i^N W_i^{DMD}(E_i^0 - E_i(n+1)) \right|^2, \tag{2}$$

where $W_{i=1...N}^{DMD}$ are the optically implemented readout weights. The DMD mirrors can be flipped only between two positions, $\pm12°$. Thus, the readout weights are strictly Boolean and physically correspond to the orientation of the mirrors towards or away from the photodiode. By choosing which mirrors are directed towards the detector, we choose the set of active neurons that contribute to computing. Once trained, the DMD is turned into a passive device, operating without bandwidth limitation or energy consumption [3].

2.2   Training with a Greedy Algorithm

Learning in our system optimizes the Boolean readout weights $W_{i=1...N,k}^{DMD}$ during successive learning epochs k such that the output gradually approximates the desired response.

Our greedy learning algorithm explores the error-landscape by favoring the selection of readout weights not tested yet. The vector W_k^{select} is calculated at each iteration, giving a new value $W_k^{select} = rand(N) \cdot W^{bias}$ to each readout weight position. Here, W^{bias} is a vector randomly initialized at the epoch $k = 1$ and the function $rand(N)$ creates N random numbers uniformly distributed between 0 and 1. At every learning epoch, the algorithm chooses a new DMD position l_k as the position of the maximum value of W^{select}, i.e. $l_k = max(W_k^{select})$, and then changes its Boolean readout weight $W_{l_k,k+1}^{DMD} = \neg\, W_{l_k,k}^{DMD}$.

For each epoch k the mean square error (NMSE) ϵ_k is calculated. The error is defined as function of the normalized output $\tilde{y}^{out}(n+1)$ and the target signal $\mathcal{T}$, both normalized by the standard deviation and subtracted their offset:

$$\epsilon_k = \frac{1}{T} \sum_{n=1}^{T} \left(\mathcal{T}(n+1) - \tilde{y}_k^{out}(n+1) \right)^2, \tag{3}$$

where T is the length of the chaotic time series used for training. We train for one-step-ahead prediction, and the target signal $\mathcal{T}(n+1)$ is the injection signal one step ahead $u(n+2)$. The calculation of the reward $r(k)$ is based on the performance evolution in comparison to the previous epoch as

$$r(k) = \begin{cases} 1 & \text{if } \epsilon_k < \epsilon_{k-1} \\ 0 & \text{if } \epsilon_k \geq \epsilon_{k-1}. \end{cases} \tag{4}$$

The modified configuration of the DMD is kept depending on reward $r(k)$:

$$W_{l_k,k}^{DMD} = r(k)W_{l_k,k}^{DMD} + (1 - r(k))W_{l_k,k-1}^{DMD}. \tag{5}$$

If performance has not improved with respect to previous configuration, the DMD mirror is flipped back.

In order to favor the selection of readout weights not yet tested, we implement

$$W^{bias} = \frac{1}{N} + W^{bias}, \qquad W_{l_k}^{bias} = 0, \tag{6}$$

where the values of W^{bias} are increased by $1/N$ at each epoch, setting afterwards the value assigned to the current l_k position to zero. Consequently, the bias for a previously modified weight increases linearly, approaching unity after $k = N$ learning iterations.

Figure 1(b) shows an example of the NN performance after training. The task is prediction of Mackey-Glass chaotic time series. The reservoir's output $y(n)$ is accurately predicting the next step of the chaotic time series $u(n+1)$.

3   Results on Learning and Error Landscape

Two hundred points of the chaotic Mackey-Glass sequence are used as training signal, and the same sequence is repeated for every learning epoch k. At each epoch, the greedy learning maximally modifies the value of a single readout-weight entry, hence varies the system's position within the error-landscape position by distance one. This is the maximal Hamming distance associated to every learning step. An optimization path is a descent trajectory from the readout weight's starting configuration towards a minimum.

The optimized and system parameters are $\beta = 0.8$, $\gamma = 0.25$, $\theta_0 = 0.44\pi$, $\theta_0 + \Delta\theta_0 = 0.95\pi$, $\mu = 0.45$, $N = 961$. The normalization parameter α represents necessary attenuation such the camera is not saturated.

3.1   Learning and Topology

The descending trend of learning for this system was already introduced in [3]. Here we want to explore the systematic characteristics of greedy learning. Since the optimization path in the error-landscape is randomized and subject to experimental noise, we study the statistical variability of learning by measuring twenty different learning curves with identical network parameters. In order to restrict our findings to the properties of our learning-routine, we start all measurements at an identical position in the error-landscape, i.e. with an identical initial $W_{i,1}^{DMD}$.

The results are depicted in Fig. 2(a). Twenty individual learning curves are presented in gray and their average is plotted in red. We can observe that all the curves converge towards a common minimum, after which all curves will slightly increase. The average value of this minimum is $\bar{\epsilon}_{k_{opt}} = (14.2 \pm 1.5) \cdot 10^{-3}$. The green line is the exponential fit to the average of the 20 learning curves. The blue line illustrates the system's testing error, which we determined using a set of 9000 data-points not used during the training sequence. We observe that the testing error $\epsilon_{test} = 15 \cdot 10^{-3}$ matches excellently with the learning error. From this result, we can conclude that no over-fitting is present, which we attribute to the role of noise in our analogue experimental NN.

As shown in more detail in the inset of Fig. 2(a), individual learning curves follow different trajectories, ranging from a rather smooth descent to paths including steep jumps. The large variability among the first learning epoch can be attributed to experimental system's noise because the initial DMD configuration is identical for all curves. However, the local variability during the learning process decreases and can therefore be related to particular topological properties of the various error-landscape explorations rather than to noise.

In order to further study the optimization paths' topology and the global characteristics of descent trajectories, we calculate the gradient of the average descent. At each learning epoch the relative change in local error is $\delta\epsilon/\delta k = \epsilon^{min} - \epsilon_k$, where ϵ^{min} is the previously smallest error achieved by the system. We define $\delta\epsilon^+/\delta k$ ($\delta\epsilon^-/\delta k$) which contains all positive (negative) $\delta\epsilon/\delta k$. At each learning epoch we calculated the average $\delta\epsilon^+/\delta k$ and $\delta\epsilon^-/\delta k$, which correspond to the average positive and negative gradients of the error landscape at each learning epoch k. Data for the average positive (negative) gradients are shown as red (blue) in Fig. 2(b).

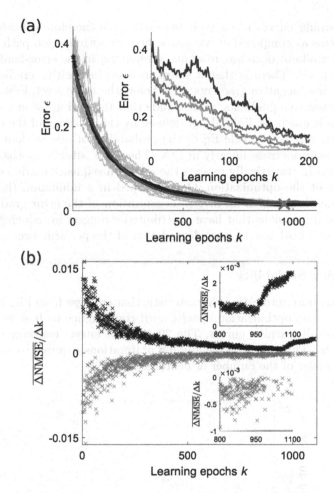

Fig. 2. (a) Learning performance for the Mackey-Glass chaotic time series prediction. Red asterisks are the average of the background 20 gray curves and the green line is its exponential fit. The inset illustrates the strong initial local variability depicting the first 200 epochs for five exemplary learning curves. (b) Average of positive (red) and negative (blue) error gradients of the 20 learning curves displayed in panel (a). Insets: Zooms of the two curves around the epoch $k = 950$. (Color figure online)

Both, red and blue curves in Fig. 2(b) exponentially decrease, which one could expect given that they represent the derivative of the exponentially decaying learning curves. This means that, on average, the error landscape curvature follows a decreasing exponential. Moreover, qualitatively different behaviour can be observed between both gradients when concentrating on the last learning epochs, cf. insets in Fig. 2(b). While the negative gradients follow a noisy convergence towards zero, the positive gradients experience a sudden change of trend after the learning epoch $k \simeq 950$. This iteration corresponds to the average epoch when the learning curves reach their minimum. We therefore consider the NN learning process as composed of two parts: first an optimization path where the error and its gradient decrease, reaching a minimum in the error-landscape and becoming trapped. There is then a sharp increase in positive gradients, while at the same time negative gradients drop below the noise level. From the clear trends of the positive gradients, we conclude that the optimization path around the minimum is not noise limited but defined by the topology of the error landscape. In fact, due to the rule Eq. 6, the probability of one readout weight to be modified again increase linearly in $1/N$. Therefore, after N epochs, the selection of position l_k statistically repeats the selection sequence carried out during the first part of the optimization. Once trapped in a minimum, the sequence of tested dimensions reveal an inverse organization of the error gradient. Consequently, the dimensions that have contributed strongest in reducing the error when optimized, lead now to least degradation of the performance.

3.2 Learning Scalability

We now focus on an interesting characteristic that emerges from Fig. 2(a), where we found that the optimization epochs until convergence to best performance are similar for all learning curves. The 20 different curves converge on average in ~961 epochs. Thus, the average number of iterations required to optimize the NN is of the order of the number of neurons.

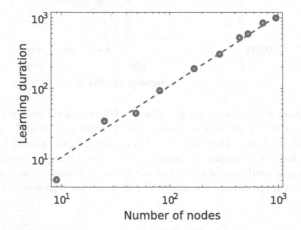

Fig. 3. Scaling of the optimal learning epoch's in function of the number of neurons. Red dashed line shoes the polynomial fit to the data, obtaining a coefficient of 1.08 that indicates close-to-linear scaling. (Color figure online)

We test now if this particular ratio maintains when modifying the size of the NN. Figure 3 shows the results of the optimal learning epoch for different network sizes ranging from 9 to 961 nodes. Impressively, the experimental results (blue circles) have an almost perfect linear distribution over three orders of magnitude. The slope of the linear fit in logarithmic scale is 1.08. Crucially, the prediction performance continuously improves for that larger NNs, hence optimization of all nodes is relevant also for the largest system. The 9 neurons network performs ∼50 worse than the network with 961 neurons.

4 Conclusions

Our work addresses fundamental questions about the size-dependent performance of analogue NNs and about the topology of their learning-error landscape.

We have investigated the features of greedy learning in analogue NNs. We have shown that applying our one-Boolean-step exploration algorithm, learning systematically converges towards similar minima. Nevertheless, different learning curves do not follow the same paths but topologically distinct trajectories. This suggests that all the optimization paths are ultimately trapped in distinct local minima with comparable prediction performance.

We have also experimentally demonstrated that the duration and effectiveness of the training are clearly correlated to the NN size. In particular, the number of epochs required for optimal learning scales almost perfect linearly with the NN size. This is a crucial finding that combines with the inherent parallel nature of diffractive coupling to boost the scalability our photonic NN approach.

Acknowledgements. This work has been supported by the EUR EIPHI program (Contract No. ANR-17-EURE-0002), by the BiPhoProc ANR project (No. ANR-14-OHRI-0002-02), by the Volkswagen Foundation NeuroQNet project and the ENERGETIC project of Bourgogne Franche-Comté. X.P. has received funding from the European Union's Horizon 2020 research and innovation programme under the Marie Sklodowska-Curie grant agreement No. 713694 (MULTIPLY).

References

1. Brunner, D., Soriano, M.C., Mirasso, C.R., Fischer, I.: Parallel photonic information processing at gigabyte per second data rates using transient states. Nat. Commun. **4**, 1364 (2013). https://doi.org/10.1038/ncomms2368
2. Brunner, D., Soriano, M.C., Van der Sande, G. (eds.): Photonic Reservoir Computing: Optical Recurrent Neural Networks. DeGruyter, Berlin (2019)
3. Bueno, J., et al.: Reinforcement learning in a large-scale photonic recurrent neural network. Optica **5**(6), 756 (2018). https://doi.org/10.1364/OPTICA.5.000756. http://arxiv.org/abs/1711.05133. https://www.osapublishing.org/abstract.cfm?URI=optica-5-6-756
4. Duport, F., Schneider, B., Smerieri, A., Haelterman, M., Massar, S.: All-optical reservoir computing. Opt. Express **20**(20), 22783 (2012). https://doi.org/10.1364/oe.20.022783
5. Jaeger, H., Haas, H.: Harnessing nonlinearity: predicting chaotic systems and saving energy in wireless communication. Science **304**(5667), 78–80 (2004). https://doi.org/10.1126/science.1091277. http://www.ncbi.nlm.nih.gov/pubmed/15064413

6. Larger, L., et al.: Photonic information processing beyond turing: an optoelectronic implementation of reservoir computing. Opt. Express **20**(3), 3241–9 (2012). https://doi.org/10.1364/OE.20.003241. http://www.osapublishing.org/viewmedia.cfm?uri=oe-20-3-3241&seq=0&html=true

7. Maass, W., Natschlager, T., Markram, H.: Real-time computing without stable states: a new framework for neural computation based on perturbations. Neural Comput. **14**(11), 2531–2560 (2002). https://doi.org/10.1162/089976602760407955

8. Paquot, Y., et al.: Optoelectronic reservoir computing. Sci. Rep. **2** (2012). https://doi.org/10.1038/srep00287

9. Van Der Sande, G., Brunner, D., Soriano, M.C.: Advances in photonic reservoir computing. Nanophotonics **6**(3), 561–576 (2017). https://doi.org/10.1515/nanoph-2016-0132

Transferring Reservoir Computing: Formulation and Application to Fluid Physics

Masanobu Inubushi(✉) ⓘ and Susumu Goto ⓘ

Graduate School of Engineering Science, Osaka University,
1-3 Machikaneyama, Toyonaka, Osaka 560-8531, Japan
inubushi@me.es.osaka-u.ac.jp

Abstract. We propose a transfer learning for reservoir computing, and verify the effectivity of the proposed methods for the standard inference task of the Lorenz system. Applying the proposed methods to an inference task of fluid physics, we show the inference accuracy is drastically improved compared with the conventional reservoir computing method if available training data size is highly limited.

Keywords: Transfer learning · Reservoir computing · Turbulence

1 Introduction

Reservoir computing is a machine learning method effective for sequential tasks. As with other supervised learning, reservoir computing needs a lot of training data; however, in many practical cases, it is not easy or costly to obtain it. To overcome this problem for the reservoir computing, we here propose to employ an approach by *transfer learning*.

Transfer learning is a machine learning method to tackle "the problem of retaining and applying the knowledge learned in one or more tasks to efficiently develop an effective hypothesis for a new task" [5]. For instance, if we have a well-trained classifier to discern images of apples from those of other fruits, it would be possible to 'reuse' the classifier to discern images of strawberries. The idea of transfer learning is to utilize *similarity* between different tasks, e.g., the pattern similarity between images of apples and strawberries, to obtain the learning model efficiently. See [6] for the detailed definition and applications.

As an example of the unavailability problem of training data, here we consider a common problem underlying machine-learning-based modeling of fluid dynamics. Modeling fluid dynamics, in particular turbulence, is of great importance for a wide variety of engineering applications. Recently, many researchers have shown that the machine learning based modelings of turbulence is effective [7]. However, these modeling needs, as a training data, a long time turbulence data, e.g., velocity fields, with high resolution and high accuracy. Generating

© Springer Nature Switzerland AG 2019
I. V. Tetko et al. (Eds.): ICANN 2019, LNCS 11731, pp. 193–199, 2019.
https://doi.org/10.1007/978-3-030-30493-5_22

such turbulence data is itself not easy and costly, and thus, this is the common problem underlying turbulence modeling based on machine learning.

In this paper, we propose transfer learning of reservoir computing. In particular, we introduce two methods: transfer method (simple reuse method) and transfer learning method. The effectivity of the proposed methods is verified for the standard inference task of the Lorenz chaos. Then, we apply them to an inference problem of fluid physics toward resolving the above-mentioned common problem underlying turbulence modeling based on machine learning.

2   Transfer Learning Formula for Reservoir Computing

We briefly review the echo state network (ESN) model as a conventional reservoir computing introduced by Jaeger [3]. The goal of reservoir computing is to predict/infer a sequence of variable $\{y(t)\}_{t>T}$ from a sequence of data input $\{s(t)\}_{t>T}$, by using a given training data $\mathcal{D} = \{s(t), y(t)\}_{t=1}^{T}$. ESN consists of nodes and recurrent random connections. The state variable x_i of the i-th node in ESN is governed by a signal-driven dynamical system: $x_i(t+1) = \phi\left[\sum_{j=1}^{N} J_{ij} x_j(t) + \sum_{j=1}^{M} v_{ij} s_j(t)\right]$ where N is the number of nodes, J_{ij} and v_{ij} are the random connections, and s_j is the input signal. The prediction/inference by ESN is given by $\hat{y}(t) = \sum_{i=1}^{N} w_i^* x_i(t)$ where the readout weight $\{w_i^*\}_{i=1}^{N}$ is determined by $\partial_{w_j}\mathcal{E}(w) = 0$ $(j = 1, \cdots, N)$. Here, $\mathcal{E}(w) = \langle(y(t) - \hat{y}(t))^2\rangle_T = \langle(y(t) - \sum_{i=1}^{N} w_i x_i(t))^2\rangle_T$ denotes the mean squared error where $\langle a(t)\rangle_T := \frac{1}{T}\sum_{t=1}^{T} a(t)$. The explicit form of the readout weight is $w^* = C^{-1}z$ where $z_j := \langle y(t) x_j(t)\rangle_T$ and $C_{ij} := \langle x_i(t) x_j(t)\rangle_T$.

Let us consider the transfer learning. To this end, we assume that a training data $\mathcal{D}' = \{s'(t), y'(t)\}_{t=1}^{T'}$ is given and it is similar to the data set $\mathcal{D}$ in some sense. Here we propose to 'reuse' the trained readout weight $\{w_i^*\}_{i=1}^{N}$ with a correction weight $\{\delta w_i\}_{i=1}^{N}$ which is determined by the training data $\mathcal{D}' = \{s'(t), y'(t)\}_{t=1}^{T'}$, i.e., $\{w_i^* + \delta w_i\}_{i=1}^{N}$. In the literature of transfer learning [6], the training data $\mathcal{D}$ and $\mathcal{D}'$ are referred to as *source* and *target* domain data, respectively. As with the conventional method, the correction weight is determined by minimizing the mean squared error as follows: $\partial_{\delta w_j}\mathcal{E}(\delta w) = 0$, where

$$\mathcal{E}(\delta w) = \left\langle\left(y'(t) - \sum_{i=1}^{N}(w_i^* + \delta w_i)x_i'(t)\right)^2\right\rangle_{T'} + \lambda\|\delta w\|_2, \tag{1}$$

and x_i' is the state variable of ESN driven by the input signal s' in the target domain. We refer to the parameter λ (≥ 0) similar to l_2 regularization as the *transfer rate*. The explicit form of the correction weight is

$$\delta w^* = [C' + \lambda I]^{-1} z', \tag{2}$$

where $z_j' := \langle y'(t) x_j'(t)\rangle_{T'} - [C'w^*]_j = \langle y'(t) x_j'(t)\rangle_{T'} - [C'C^{-1}z]_j$ and $C_{ij}' := \langle x_i'(t) x_j'(t)\rangle_{T'}$.

Remark 1. If the transfer rate is zero, $\lambda = 0$, the above formula reduces to the conventional reservoir computing method which is just a supervised learning by using the target data $\mathcal{D}'$ only, i.e., transferring no knowledge from source domain. On the other hand, in the limit of large transfer rate, $\lambda \to \infty$, we obtain $\|\delta w^*\|_2 \to 0$, since the variable z' does not depend on the transfer rate λ. Therefore, in this limit, the above formula reduces to a method to simply reuse the weight w^* trained by the source data $\mathcal{D}$ only, i.e., learning no knowledge from the target domain and just transferring the learned method. This method will be referred to as the transfer method. In conclusion, the above formula of the transfer learning constitutes a one-parameter family of learning methods which connects the conventional reservoir computing method ($\lambda = 0$) and the transfer method ($\lambda \to \infty$).

In the following sections, we compare the conventional method ($\lambda = 0$) and the proposed methods: the transfer method ($\lambda \to \infty$) and the transfer learning method ($0 < \lambda < \infty$). The transfer rate λ is determined so that the transfer learning method is the most effective for the small data size.

3   Application to Inference Tasks

We apply the proposed methods to inference tasks. In order to see the effectivity of the proposed methods, we apply them to an inference task of the Lorenz chaos as a test case. Then, we apply them to an inference task of the fluid physics.

For the reservoir computing, the number of nodes is $N = 100$, the elements of the recurrent connection matrix are independently and identically drawn from the Gaussian distribution with mean zero and variance $1/N$, and $\phi[x] = \tanh x$ is used for the activation function (see [2] for the details of ESN).

3.1   Lorenz Chaos

We employ the Lorenz system $dx/dt = \sigma(y - x)$, $dy/dt = x(\rho - z) - y$, $dz/dt = xy - rz$ as a chaotic dynamical system. The task is to infer the variable $z(t)$ from the variable $x(t)$, which is a standard task for reservoir computing [4]. We fix the parameters σ, r to $\sigma = 10, r = 8/3$ and change the parameter ρ. As for the source domain, we use data set $\mathcal{D} = \{x(t), z(t)\}_{t=1}^T$ that is generated by the numerical integration of the Lorenz equation with $\rho = 28$ (so-called the classical parameter values). As for the target domain, we use data set $\mathcal{D}' = \{x(t), z(t)\}_{t=1}^{T'}$ that is generated by the numerical integration of the Lorenz equation with $\rho = 30$.

By using the data set $\mathcal{D}$, we calculate the weight w^* at $\rho = 28$. Then, we use the weight w^* for the inference task at $\rho = 30$ without training (transfer method). Moreover, we calculate the correction weight δw^* with the formula (2) for the data set $\mathcal{D}'$, and use for the inference task at $\rho = 30$ (transfer learning method).

We show the generalization (mean square) error E in Fig. 1(a). The horizontal axis is the length T' of the data set $\mathcal{D}'$. Roughly speaking, 'period' of circling

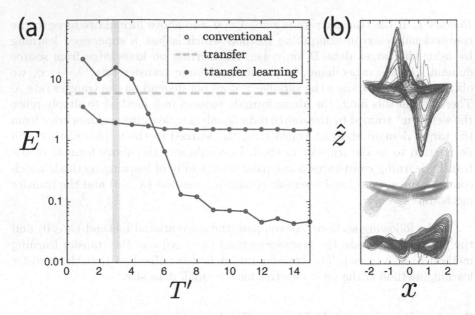

Fig. 1. The inference task of the Lorenz system. (a) Dependence of generalization error on the length of the training data $\mathcal{D}'$. The blue open circles, the yellow broken line, and the red closed circles are the results with the conventional method, transfer method, and transfer learning method ($\lambda = 10^{-2}$), respectively. (b) The results of the inference task in the case of $T' = 3$. The results with the conventional method, transfer method, and transfer learning method are shown in the upper (blue), middle (yellow), and bottom (red) part of the figure, respectively. The gray line depicts the projection of the Lorenz attractor. (Color figure online)

along the one side of the 'wing' of the Lorenz attractor is $O(1)$. The blue open circles are results for the conventional method, i.e., the weight w^* is calculated only with the data set $\mathcal{D}'$. The generalization error decreases for longer T'. The yellow broken line denotes results with the transfer method, which does not use $\mathcal{D}'$ and thus the error does not change. The red closed circles denote results with the transfer learning method with the transfer rate $\lambda = 10^{-2}$. When the available training data size is highly limited (i.e., T' is small), which is the main subject of this paper, the transfer learning is the most effective method for the inference task. In addition, the transfer method outperforms the conventional method in the small T' region.

The results of the inference by using the small data set $T' = 3$ are depicted in Fig. 1(b). The horizontal axis is $x(t)$, which is calculated with the Lorenz equation, and the vertical axis is $\hat{z}(t)$ (shifted appropriately), which is inferred from the variable $x(t)$. The results with the conventional method, transfer method, and transfer learning method are shown in the upper (blue), middle (yellow), and bottom (red) part of the figure, respectively. The gray line depicts the projection of the Lorenz attractor to $x - z$ plane, i.e., the correct data for the inference task.

The conventional method fails to infer the variable z and the resulting 'orbit' is quite different from the Lorenz attractor; however, the proposed methods give the better results, which correspond to the error values shown in Fig. 1(a).

For the inference task of the variable of the Lorenz system, we show that the proposed methods are effective. However, there is no reason for the similarity in the data set $\mathcal{D}$ and $\mathcal{D}'$. If we focus on the similarity in the data set $\mathcal{D}$ and $\mathcal{D}'$, in particular the similarity based on the physical mechanism, the proposed methods become more efficient as we will show in the next section.

3.2 Navier-Stokes Turbulence

We apply the transfer learning formula introduced in Sect. 2 to an inference task for fluid physics, in particular the Navier-Stokes turbulence. The task is to infer the spatial average of the energy dissipation rate $\epsilon(t)$ from a sequence of the average kinetic energy $K(t)$, i.e., the training data is $\{s(t), y(t)\}_{t=1}^{T'} = \{K(t), \epsilon(t)\}_{t=1}^{T'} (= \mathcal{D}')$. Although the energy dissipation rate plays a central role in the statistical theory and modeling of turbulence, the direct measurement of the energy dissipation rate is difficult, while that of the kinetic energy is relatively easy. Therefore, here we consider the inference of the energy dissipation rate for a turbulent flow from the measurement of the kinetic energy for the same flow.

The goal is to infer the energy dissipation rate for a realistic turbulent flow, i.e., turbulence at *high* Reynolds numbers; however, as mentioned in Sect. 1, generating turbulent flow data including $\mathcal{D}'$ at a high Reynolds number is costly. Thus, we assume that (i) the length T' of the data $\mathcal{D}'$ is not sufficient for the training of the reservoir computing. We also assume that (ii) training data $\mathcal{D}$ at a *low* Reynolds number is available and sufficient for the training of the reservoir computing. In the terminology of transfer learning, $\mathcal{D}$ at the low Reynolds number corresponds to the source domain data, while $\mathcal{D}'$ at the high Reynolds number corresponds to the target domain data.

The training data $\mathcal{D}$ and $\mathcal{D}'$ are calculated by the direct numerical simulation of the Navier-Stokes equation with the Fourier spectral method. We use the training data $\mathcal{D}$ at $R_\lambda \simeq 50$ for the low Reynolds number case and the training data $\mathcal{D}'$ at $R_\lambda \simeq 120$ for the high Reynolds number case, where R_λ is the Taylor micro-scale Reynolds number.

Figure 2 shows the results of the transfer learning for the reservoir computing applied to the inference task of the energy dissipation rate. As described in Sect. 2, the readout weight w^* is determined by the training data $\mathcal{D}$ at the low Reynolds number with the sufficiently long period (assumption (ii)). We show the generalization (mean squared) error E in Fig. 2, which quantifies the accuracy of the inference. The horizontal axis is the length of the data set T' normalized by a characteristic time scale (eddy turnover time τ). The blue open circles represent the generalization error with the conventional reservoir computer trained by the data set $\mathcal{D}'$ only, and thus transfer learning is not employed (i.e., $\lambda = 0$). For longer available training data, the inference is more accurate; however, in the case that only limited training data is available, e.g., $T'/\tau = 200$, the inference accuracy is drastically degraded.

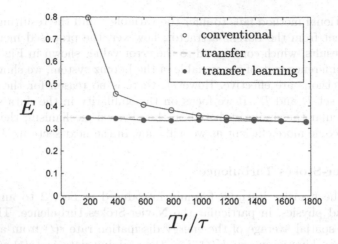

Fig. 2. The inference task of the energy dissipation rate of the Navier-Stokes turbulence. Dependence of generalization error on the length of the trining data $\mathcal{D}'$ normalized by a characteristic time scale of turbulence (specifically, eddy turnover time τ). The blue open circles and red closed circles are the results with the conventional and transfer learning method ($\lambda = 10^{-3}$), respectively. The yellow broken line denotes the results with the transfer method. (Color figure online)

On the other hand, the red closed circles represent the generalization error with the transferred reservoir computer by using the formula (2) with $\lambda = 10^{-3}$. Even if the training data is not available sufficiently (the assumption (i)), the inference accuracy is not degraded compared with the conventional method. The yellow broken line shows the generalization error with the trained weight w^* at the low Reynolds number (the transfer method). The transfer method is effective for this inference task. This is explained by the similarity between the data $\mathcal{D}$ and $\mathcal{D}'$ because of the dynamics of the *energy cascade* in turbulence [1].

4   Conclusion

We introduce the transfer learning for reservoir computing and derive the formula. The effectivity of the proposed methods is verified for the standard inference task of the Lorenz system. Furthermore, we apply the proposed method to the fluid physics, in particular the Navier-Stokes turbulence, which is important in a wide variety of physics and engineering applications. The inference accuracy of the energy dissipation rate is drastically improved compared with the conventional reservoir computing method if available training data size is highly limited. Behind the strong similarity in the data $\mathcal{D}$ and $\mathcal{D}'$ of the turbulence, there exists the energy cascade mechanism. We emphasize that, by focusing such a physical mechanism, the transfer method can be effective.

The reservoir computing is the low-computational-cost method of machine learning, and the transfer learning is the machine learning method working well with the small amount of data. These two methods are both simple, easy and inexpensive machine learning, and therefore, the concepts of reservoir computing and transfer learning are well suited to each other. In fact, our results suggests the reservoir computing combined with the transfer learning offers an effective machine learning method. For a future study, transfer learning from numerical simulation data (source domain) to real experiment data (target domain) may be interesting.

References

1. Inubushi, M., Goto, S.: Inference of the energy dissipation rate of turbulence by machine learning (in preparation)
2. Inubushi, M., Yoshimura, K.: Reservoir computing beyond memory-nonlinearity trade-off. Sci. Rep. **7**(1), 10199 (2017)
3. Jaeger, H.: The "echo state" approach to analysing and training recurrent neural networks-with an erratum note. Bonn Ger.: Ger. Natl. Res. Cent. Inf. Technol. GMD Tech. Rep. **148**(34), 13 (2001)
4. Lu, Z., Pathak, J., Hunt, B., Girvan, M., Brockett, R., Ott, E.: Reservoir observers: model-free inference of unmeasured variables in chaotic systems. Chaos: Interdisc. J. Nonlinear Sci. **27**(4), 041102 (2017)
5. Silver, D., et al.: Best of NIPS 2005: highlights on the 'inductive transfer: 10 years later' workshop (2006)
6. Weiss, K., Khoshgoftaar, T.M., Wang, D.: A survey of transfer learning. J. Big data **3**(1), 9 (2016)
7. Wu, J.L., Xiao, H., Paterson, E.: Physics-informed machine learning approach for augmenting turbulence models: a comprehensive framework. Phys. Rev. Fluids **3**(7), 074602 (2018)

Special Session: Artificial Intelligence in Medicine

Special Session: Artificial Intelligence in Medicine

In conventional settings, diagnosis of diseases and further medical and surgical treatment choices are mostly subjective, based on limited assessment of patient phenotypes. In the era of petabytes of clinical data and detailed patient information, from genomics to histological or volumetric images, gathered over entire lifespans, AI and big data analysis are instrumental in providing decision support and to understanding patient treatments.

In this session, we discussed state-of-the-art AI methods to analyze multimodal patient data, understand their responses to medication and treatments, and to predict the risk of diseases prior to onset. Our focus was on topics related to methods for clinical diagnosis and outcome prediction, identification of increased-risk patient categories, ethics and security of healthcare data used by AI, and methods for computational pathology.

From the large number of submitted papers, and together with our dedicated reviewers, we selected the four most interesting contributions for oral presentation. Our speakers reported advances in diagnosing breast cancer, clinical text mining, and the use of EEG for predicting certain disorders. In addition to the accepted papers, Dr. Shadi Albarqouni, senior research scientist at the Technische Universität München, was invited to present his work on deep learning for medical imaging.

Organization

Chairs

Carsten Marr   Helmholtz Zentrum München (GmbH), Germany
Narges Ahmidi   Helmholtz Zentrum München (GmbH), Germany
                          Johns Hopkins University, USA
Tingying Peng   Helmholtz Zentrum München (GmbH), Germany

Program Committee and the Reviewers

Stefan Bartzsch   Helmholtz Zentrum München (GmbH), Germany
Felix Bork   Technische Universität München, Germany
Mustafa Büyüközkan   Helmholtz Zentrum München (GmbH), Germany
Nikolaos Chlis   Helmholtz Zentrum München (GmbH), Germany
Dominik Jüstel   Helmholtz Zentrum München (GmbH), Germany
Gabriel Maicas   University of Adelaide, Australia
Michael Menden   Helmholtz Zentrum München (GmbH), Germany
Bjoern Menze   Technische Universität München, Germany
Molly O'brian   Johns Hopkins University, USA
Ario Sadafi   Helmholtz Zentrum München (GmbH)
                          and Technische Universität München, Germany
Benjamin Schubert   Helmholtz Zentrum München (GmbH), Germany

Towards Deep Federated Learning in Healthcare
(Abstract of Talk of the Invited Speaker)

Shadi Albarqouni

Technische Universität München, Germany

Deep Learning (DL) has emerged as a leading technology in computer science for accomplishing many challenging tasks. This technology shows an outstanding performance in a broad range of computer vision and medical applications. However, this success comes at the cost of collecting and processing a massive amount of data, which are in healthcare often inaccessible due to privacy issues.

Federated Learning is a new technology that allows training DL models without sharing the data. Using Federated Learning, DL models at local hospitals share only the trained parameters with a centralized DL model, which is, in return, responsible for updating the local DL models as well. Yet, a couple of well-known challenges in the medical imaging community, e.g., heterogeneity, domain shift, scarify of labeled data and handling multi-modal data, might hinder the utilization of Federated Learning.

In this talk, a couple of proposed methods, to tackle the challenges above, will be presented paving the way to researchers to integrate such methods into the privacy-preserved federated learning.

Investigation of EEG-Based Graph-Theoretic Analysis for Automatic Diagnosis of Alcohol Use Disorder

Wajid Mumtaz[1,2,3](✉) [iD], Lukáš Vařeka[1] [iD], and Roman Mouček[1] [iD]

[1] NTIS - New Technologies for the Information Society,
Faculty of Applied Sciences, University of West Bohemia,
Univerzitní 8, 306 14 Pilsen, Czech Republic
wajidmumtaz@gmail.com, {lvareka,moucek}@kiv.zcu.cz
[2] Center for Intelligent Signal and Imaging Research (CISIR),
Department of Electrical and Electronic Engineering,
Universiti Teknologi PETRONAS, 32610 Seri Iskandar, Perak, Malaysia
[3] Department of Electrical Engineering,
School of Electrical Engineering and Computer Science,
National University of Sciences and Technology, H-12, Islamabad, Pakistan

Abstract. Abnormal functional connectivity (FC) has been commonly observed during alcohol use disorder (AUD). In this work, FC analysis has been performed by incorporating EEG-based graph-theoretic analysis and a machine learning (ML) framework. Brain FC was quantified with synchronization likelihood (SL). Undirected graphs for each channel pair were constructed involving the SL measures. Furthermore, the graph-based features such as minimum spanning tree, distances between nodes, and maximum flow between the graph nodes were computed, termed as *EEG data matrix*. The matrix was used as input data to the ML framework to classify the study participants. The ML framework was validated with data acquired from 30 AUD patients and an age-matched group of 30 healthy controls. In this study, the classifiers such as SVM (accuracy = 98.7%), Naïve Bayes (accuracy = 88.6%), and logistic regression (accuracy = 89%) have shown promising discrimination results. The method was compared with two existing methods that also involve resting-state EEG data. The first method reported a classification accuracy of 91.7% while utilizing the time-based features such as Approximate Entropy (ApEn), Largest Lyapunov Exponent (LLE), Sample Entropy (SampEn), and four other Higher Order Spectra (HOS) features [1]. The second method reported 95.8% accuracy involving wavelet-based signal energy [2]. Since the study has utilized a small sample size, the generalization could not be possible. The FC-based graph-theoretic analysis in combination with ML methods could be used as an endophenotype for screening AUD patients.

Keywords: Machine learning for biomedical systems ·
Support vector machine · Time series analysis ·
EEG-based graph-theoretic analysis · EEG-based synchronization likelihood

© Springer Nature Switzerland AG 2019
I. V. Tetko et al. (Eds.): ICANN 2019, LNCS 11731, pp. 205–218, 2019.
https://doi.org/10.1007/978-3-030-30493-5_23

1 Introduction

Alcohol Use Disorder (AUD) is a chronic, compulsive, and relapsing mental condition. AUD patients may lose their control over alcohol intake [3]. Hence, severe alcohol consumption could become a social issue [4]. Moreover, the toxic effects of alcohol may damage other body parts and could become dangerous to health [5]. Unfortunately, severe alcohol consumption affects people from all ages including both the school-age students and older people. According to the World Health Organization (WHO), AUD effects include behavioral, cognitive, psychosocial and social aspects [3].

Clinical practitioners screen AUD patients involving a clinical questionnaire such as AUDIT (Alcohol Use Disorder Identification Test) [6]. However, the questionnaire is subjective that may involve either subjective responses from the patients or subjective assessments by the practitioners. Therefore, the development of an objective method that could improve the screening of AUD patients would be beneficial. In this context, electroencephalogram (EEG)-based objective methods can help by classifying an under-observation subject as either suffering from AUD condition or not. The EEG modality is useful in many clinical applications such as diagnosis of epilepsy [7, 8], autism spectrum disorder [9] and quantification of sleep stages [10].

Resting-state EEG-based brain functional connectivity (FC) analysis provides the means to study brain functional states. For example, the synchronization likelihood (SL), an FC measure, could implicate discriminant features for different brain disorders such as depression [11–14] and Alzheimer's disease [15]. In addition, EEG-based graph-theoretic analysis provides the means to study brain structural information. In addition, it has been revealed that a normal brain exhibits a small-world architecture [16]. Any abnormality in the brain structure could implicate a neurological disorder or vice versa. A variety of neurological and psychiatric diseases alters connectivity within brain networks [17]. The graph theory has been utilized to delineate functional and structural network alterations in neurologic disorders such as Alzheimer's disease [18–20], and psychiatric disorders such as schizophrenia [21, 22]. EEG has a high temporal resolution suitable for assessing brain connectivity and network measures. Hence, the EEG-based graph-theoretic analysis involving brain FC could provide a mechanism to study the relationship between the structure and function of the brain.

This study aims to develop a ML-based framework for automatic screening of AUD patients involving graph-theoretic analyses. Therefore, in this paper, the EEG-based graph-theoretic analysis involving brain functional connectivity (FC) to screen AUD patients is presented. To the best of our knowledge, no such study has investigated the EEG-based network theoretic analysis for screening AUD patients. The study has performed quantification of FC while computing SL values between different sensor pairs at sparse locations [23].

The rest of the paper is organized as follows: Sect. 2 details on the methodology including the information of study participants, experiment design and data acquisition, noise reduction, the theoretical background of SL, feature selection and classification methods, and validation. Section 3 elaborates on the study's results. Finally, Sect. 4 provides the discussion and conclusion, respectively.

2 Methodology

2.1 Study Participants and Experiment Design

In this research, the study participants (volunteers) were recruited from the clinic Bingkor, Sabah, east Malaysia after ethics approval from the ethics committee of University Malaya, Malaysia. The sample size i.e., 30 participants in each group was computed according to the sample size calculation procedure [39, 40]. Equation 1 shows the mathematical notation for the sample size formula:

$$n = \frac{P(1-P)\left(Z_{1-\frac{\alpha}{2}}\right)^2}{e^2}$$

$$n = \frac{0.87(1-0.87)(1.645)^2}{(0.1)^2} \cong 30$$

(1)

where P = 87%, α = 0.05, e = 10%. The 'P' value corresponds to the possible classification accuracy value. The 'e' corresponds to the maximum allowable error.

The experimental procedure was briefed to the study participants; they signed informed consent. The first group of study participants included thirty AUD patients (mean 55.4 ± 12.87 years) and passed the diagnostic criteria defined by the Alcohol Use Disorders Identification Test (AUDIT) [24]. For example, the participants with the AUDIT score greater than seven were categorized as AUD patients [25]. The second study group involved thirty age-matched healthy controls (mean 42.67 ± 15.90 years) who were assessed for any neurological disorder and were found healthy.

Figure 1 shows the 10–20 electrode placements [26] in a 19 sensors EEG cap with linked-ear (LE) as a reference. However, in this study, the EEG data were re-referenced according to infinity reference (IR) based on the reference electrode standardization technique (REST) [27]. The use of REST reference has been recommended by many

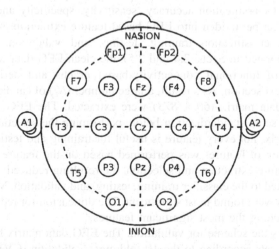

Fig. 1. EEG sensors located according to the 10/20 system

EEG studies such as [34]. Moreover, the EEG data were recorded with the amplifier configurations with two types of filters: (1) 0.5 Hz to 70 Hz, and (2) a 50 Hz notch filter. The sampling rate was set to 256 samples per second.

The recording of experimental data involved both the qualitative information (AUDIT questionnaires) and the quantitative data (EEG data). In this study, the AUDIT scores were recorded from the patients and considered as gold standard during EEG analysis. On the other hand, the resting-state EEG data involved 5 min of eyes closed (EC) and 5 min of eyes open (EO) sessions.

2.2   EEG Noise Reduction

EEG data are often confounded with electro-ocular (EOG) potentials, eye blinks, eye movements; electromyographic (EMG) potentials, muscular activities, and heartbeats. EOG and EMG artefacts could be characterized by high amplitude and high frequency bursts, respectively. In this study, the multiple source eye correction (MSEC) method [28] was employed to correct the artefacts. A detailed discussion of the method is out-of-scope here. However, a brief elaboration is provided here.

The technique was implemented in the standard brain electric source analysis (BESA) software [29]. The method is semi-automatic and involved the construction of artifact topographies based on the artifacts marked, manually. For this purpose, at-least one minutes of EEG data should be screened for artifacts. The artifacts founded during this one-minute screening of EEG data were averaged and implicated into a template artifact and formed the artifact topography. This artifact topography and a suitable head-model were used to reconstruct the EEG data while successfully removing the artifacts.

2.3   EEG Analysis Involving the ML Scheme

In this study, the EEG analysis involved an application of a machine learning (ML) framework that provides measures to quantify the performance of the proposed features in term of classification accuracy, sensitivity, specificity and the f-measure. The ML scheme can be divided into EEG-based feature extraction, selection of most relevant features, classification training, testing, and validation. During feature extraction (as elaborated in Sects. 2.4 and 2.5), the clean EEG data were subjected to the computation of functional connectivity-based graphs and derived features as explained in the next section. As a result, a large number (N_c) of candidate features that formed an EEG data matrix (60×873) were extracted. The EEG data matrix was subjected to the z-score standardization before performing the selection of features. In the EEG data matrix, not every feature is useful for training and testing the classifier. Hence, the selection of features was performed based on the feature ranking method [30, 31]. After feature extraction and selection, the resulting reduced dimensional data matrix was subjected to the classifier training, testing, and validation. More specifically, the top 15 features were found most significant. The dimension for reduced data matrix was 60×15 including the most significant features.

Figure 2 shows the scheme for validation. The EEG data matrix has been divided into test and train sets according to the 10-fold cross-validation (CV).

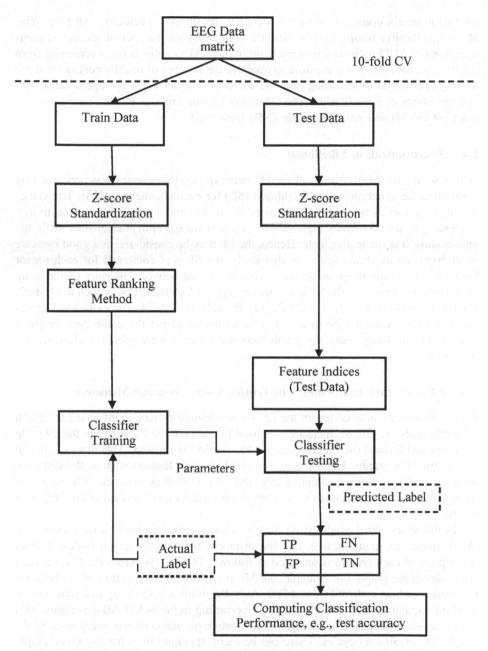

Fig. 2. The proposed ML framework for validation

In this manuscript, the classification performance metrics were classification accuracy, sensitivity, specificity, and f-measures. In this study, the average classification accuracies were reported. Moreover, these metrics were determined based on the confusion matrix computation of the accuracy, sensitivity, specificity, and f-measure. More specifically, *sensitivity* is a measure to quantify the number of successful identifications of AUD patients who were diagnosed as AUD patients (after screening from the doctor). *Specificity* is a measure to quantify the number of healthy controls that are healthy controls (after screening from the doctor). *Accuracy* is the average of sensitivity and specificity of classification. The classifiers logistic regression (LR), support vector machine (SVM) and naïve Bayesian (NB) were used.

2.4   Synchronization Likelihood

In this study, the quantification of the FC between two times series was performed by computing the synchronization likelihood (SL) for each channel pair [23]. The values for SL may vary between 0 (minimum connectivity) and 1 (maximum connectivity). For example, the SL shows high values such as 1 during epileptic seizures while the values show 0 at the resting state. Hence, the SL may be considered as a good measure to discriminate the brain states. In this study, the SL was computed for each sensor location. The graph-theoretic features were computed between all pairs of the scalp electrodes. For example, the SL at O1 was computed for channel pairs such as O1-Fp2; O1-F4; O1-F8; O1-T4; O1-T6; O1-P4; O1-P8; O1-O2; O1-C4. The extracted features were arranged column-wise in a matrix to further construct the undirected weighted graphs. In this study, only the graph-theoretic features were used for classification purposes.

2.5   EEG Feature Extraction: The Graph-Based Theoretic Measures

Figure 3 provides an overview of the EEG-based feature extraction involving the graph theoretic analysis. The SL measure was used to compute the FC between the 19 scalp locations and formed the synchronization likelihood (SL) matrix. This matrix is useful to construct the graphs. Furthermore, the graph-derived features such as the distances between nodes, minimum spanning tree and maximum flow between different nodes were computed. These features were arranged in a matrix form, termed as the *EEG data matrix*.

In this study, three graph-based measures were computed such as maximum flow (MF), minimum spanning tree, and the distances between the graph nodes. A brief description of each feature is provided as follows. The MF quantifies the link between the nodes of the graph. For example, the MF is equal to zero if no flow exists between two nodes such as node at Fp1 and Fp2. Also, the minimum spanning tree provides the minimum spanning tree T of the graph G. According to the MATLAB description [35] "For connected graphs, a spanning tree is a subgraph that connects every node in the graph but contains no cycles. There can be many spanning trees for any given graph. By assigning a weight to each edge, the different spanning trees are assigned a number for the total weight of their edges. The minimum spanning tree is then the spanning tree whose edges have the least total weight. For graphs with equal edge weights, all spanning trees are minimum spanning trees, since traversing n nodes requires n-1 edges." Moreover, the distances between the graph nodes quantifies the shortest path distances between node pairs in a graph.

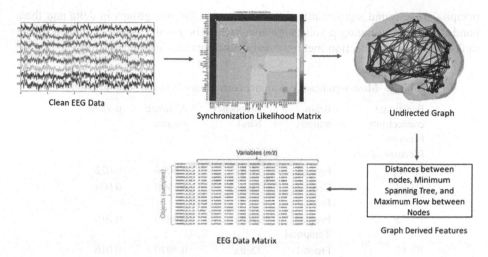

Fig. 3. Graph derived feature extraction involving the EEG data

2.6 EEG Data Matrix and z-score Standardization

The z-score standardization was performed to standardize the values in the EEG data matrix. Equation 2 shows the mathematical formula for the z-score standardization:

$$z = \frac{x - \mu}{\sigma} \tag{2}$$

where the mean (μ) and standard deviation (σ) of each feature vector (x) were used to compute z, i.e., the standard value for each feature.

2.7 Rank-Based Feature Selection Method

In this study, the feature selection involved a rank-based feature selection method based on the receiver-operating curve (ROC) [30, 31]. The method computed the area under the curve (AUC) for each feature termed as z-value. A high z-value (equal or near 0.5) corresponded to the ability of a feature to discriminate within classes. The z-value represents the relevance of feature with the class label. For example, a feature with a z-value that is 0 means it is not a good feature as for as classification is concerned. On the contrary, a feature with a z-value that is 0.5 implicates good classification ability, accordingly.

3 Results

3.1 Significant EEG Features

The most significant FC pairs were identified by the feature selection method. Table 1 shows the most significant FC pairs. The brain regions such as frontal, temporal, and occipital were found significantly different between the two groups in delta and theta bands. The corresponding p-values are also listed. The p-values were computed by the rank-based feature selection method cited in the previous section [36–38].

Table 1. Most significant functional connectivity between scalp locations

Functional connectivity between Electrode pairs	Brain regions	Frequency band	Absolute z-value	p-value
Fp1-Fp2	Frontal	Delta	0.5000	0.022
C3-F4	Central-Frontal	Delta	0.4990	0.008
F7-T3	Frontal-Temporal	Delta	0.4980	0.010
F3-T7	Frontal-Temporal	Delta	0.4970	0.016
F7-F8	Frontal-Frontal	Theta	0.4960	0.022
T4-T5	Temporal	Theta	0.4949	0.022
F8-T3	Frontal-Temporal	Theta	0.4939	0.008
F3-T4	Frontal-Temporal	Delta	0.4929	0.010
F4-T7	Frontal-Temporal	Theta	0.4919	0.002
F8-Fp1	Frontal	Delta	0.4909	0.045
T3-F4	Temporal-Frontal	Theta	0.4899	0.0021
C4-T3	Central-Temporal	Delta	0.4879	0.016
F8-T3	Frontal-Temporal	Theta	0.4869	0.022
T4-T3	Temporal	Delta	0.4859	0.013
P3-Fp1	Parietal-Frontal	Theta	0.4848	0.022

Table 2 shows the LR classification for individual features and their integration. The feature *minimum spanning tree* provides results such as (*accuracy = 88.7%, sensitivity = 83.66%, specificity = 93.6%, and f-measure = 87%*). The feature *maximum flow between nodes* provides the second highest accuracy. Finally, the integration of these features provides results such as the *accuracy = 89%, sensitivity = 85.3%, specificity = 91.5%, and f-measure = 87%*.

Table 2. LR classification while discriminating AUD patients and healthy controls

EEG features	Accuracy	Sensitivity	Specificity	F-measure
Minimum Span Tree (MST)	88.7%	83.6%	93.6%	87%
Distances (D) between nodes	74.85%	80.3%	68.6%	74%
Maximum flow (MF) between nodes	76.1%	80.3%	79.0%	75%
Integration of MST, D, and MF	89%	85.3%	91.5%	87%

Table 3 shows the NB classification for individual features and their integration. The feature *minimum spanning tree* provides results such as (*accuracy = 87.6%, sensitivity = 94%, specificity = 81.9%, and f-measure = 0.89*). The feature *maximum flow between nodes* provides the second highest accuracy. Finally, the integration of these features provides results such as the *accuracy = 88.6%, sensitivity = 96%, specificity = 81.5%, and f-measure = 87%*.

Table 3. NB classification while discriminating AUD patients and healthy controls

EEG features	Accuracy	Sensitivity	Specificity	F-measure
Minimum Span Tree (MST)	87.6%	94%	81.9%	89%
Distances (D) between nodes	83%	80.6%	85.6%	82%
Maximum flow (MF) between nodes	64.8%	54%	74.8%	57%
Integration of MST, D, and MF	88.6%	96%	81.4%	87%

Table 4 shows the SVM classification for individual features and their integration. The feature *minimum spanning tree* provides results such as (*accuracy = 90%, sensitivity = 91.9%, specificity = 87%, and f-measure = 89%*). The feature *maximum flow between nodes* provides the second highest accuracy. Finally, the integration of these features provides results such as the *accuracy = 91.7%, sensitivity = 95.3%, specificity = 89.4%, and f-measure = 92%*.

Table 4. SVM classification while discriminating AUD patients and healthy controls

EEG features	Accuracy	Sensitivity	Specificity	F-measure
Minimum Span Tree (MST)	90%	91.9%	87.0%	89%
Distances (D) between nodes	84.3%	78.6%	88.8%	84%
Maximum flow (MF) between nodes	76.5%	73.8%	81.3%	59%
Integration of MST, D, and MF	98.7%	95.3%	98.9%	98%

This is a proof of concept study and classification was employed to determine the significance of the features. It was hypothesized that the connectivity matrix and the graph-based features could better represent the brain functional and structural information, respectively. For example, the minimum spanning tree was computed based on the connectivity matrix, i.e., the SL matrix (the matrix provides functional connectivity between different EEG channels). In this study, among other features, the MST was selected as a feature in order to investigate its usefulness during classification and the results have proved this hypothesis.

The EEG features employed in this study provide two types of real-world perspectives. Firstly, this study focused on investigating the discriminability of the features rather than studying their associated neurobiology with the AUD. In this perspective, the study resulted into high classification results that signified the robustness of the features. A variety of neurological and psychiatric diseases alters connectivity within brain networks [17]. Hence, EEG features derived from connectivity matrix (i.e., SL) such as MST, MF and D were suitable to study the functional and structural network alterations. Secondly, the results signify the importance of the spatial locations of the AUD and healthy brains as mentioned in Table 1. The frontal and temporal areas were identified as the most significant brain regions. This finding is in accordance with the studies reporting the affected brain areas during AUD [43].

4 Discussion

This manuscript aims to develop an ML-based framework involving graph-based theoretic measures. The method serves as an objective method for screening AUD patients from healthy controls, termed as a decision support system [32]. In this manuscript, the primary finding is that the EEG-based graph-based theoretic analysis and the subsequently derived features have shown to be promising in differentiating AUD patients and healthy controls. Hence, they can be termed as EEG-based biomarkers for screening AUD patients [33]. Comparing with our previous study [41], that focuses on depressed patients and investigates different EEG reference choices, the present study employed AUD patients aiming at investigating the effectiveness of graph-theoretic analysis. The second study [42] focuses on FC measures only; however, in the present study, the FC matrix was further analyzed, and the graph-theoretic measures were computed. Hence, the present study provides enhanced analysis than the previous two studies and adds new results to the knowledge-base.

In this study, we hypothesized that the EEG-based graph-theoretic measures could provide a mechanism to study the relationship between the structure and functional connectivity of the brain. (The SL quantify the brain functional connectivity and the minimum spanning tree quantify the brain structural information). Hence, both the brain functional and structural information is quantified by the MST features and this idea was endorsed by the classification because the resulting classifiers were implicated into higher discrimination efficiencies. In our investigation, we are comparing these measures between the AUD and control groups.

Moreover, the EEG-based graph-based theoretic analysis provides integration of brain structural and functional information. The graph provides details about the brain structure and the SL measure provides functional connectivity. The conventional screening methods typically employ the use of questionnaire-based or symptom-based methods for the assessment of AUD. This approach inherently based on subjective analysis could be confounded due to human errors. On the contrary, the employment of

EEG-based automatic method could improve the screening by providing additional objective evidence on the diagnosis.

In this paper, we have extended our previous work to a next level where we shall employ the graph theoretic approach on EEG data implicated into graph-based features [42]. Previously, the EEG feature extraction involved the computation of synchronization likelihood features. The maximum classification accuracy was reported, i.e., 98%. In addition, the method has been compared with two more methods from the literature that also involve resting-state EEG data. The comparison was performed by implementing the methods while utilizing the common dataset acquired in this study. The first method reported a classification accuracy of 91.7% while utilizing the time-based features such as Approximate Entropy (ApEn), Largest Lyapunov Exponent (LLE), Sample Entropy (SampEn), and four other Higher- Order Spectra (HOS) features [1]. In addition, the second method reported 95.8% accuracy involving wavelet-based signal energy [2]. The features will further differentiate the AUD patients and healthy controls involving the classification methods.

In comparison with the previous studies, higher accuracy is achieved. This could be due to the integration of both structural and functional information. As the classification results have been presented, there might be a concern that the results could be due to classification overfitting. To avoid this concern, we have adopted the following precautions.

- The EEG noise can confound the result; therefore, a proper check on the clean EEG data is necessary. In this study, the data histograms are used to check the presence of any kind of outliers. It is found that the data are appropriate for classification purposes.
- The gender distribution could harm the classification results due to gender bias. Therefore, equal sample sizes in both groups of gender are selected.
- In this study, the incorporation of the classifier with three different structures has proved the validity of our data, too.
- The 10-fold cross validation is suitable to make sure that the over-fitting can be avoided.

The study is confounded with a few limitations. First, the effect of medication cannot be ruled out completely. To reduce this factor, the AUD patients are asked to be medication free at-least two weeks before the first EEG recording. Second, lifestyle factors including the appetite, sleep patterns, cannot be controlled since the patients are out-patients. Finally, despite the promising classification findings, the small sample size poses a constraint that the results should not be generalized to a wider population. Therefore, the validation of the proposed method on larger population sizes is suggested.

5 Conclusion

In this study, the brain functional connectivity-based graph-theoretic analysis has been proposed for objective assessment and screening of AUD patients. More specifically, the integration of both structural and functional information has proved feasible to discriminate the AUD patients from healthy controls based on resting-state EEG data. The automatic discrimination provides a clinical tool to objectively screen the AUD patients among of population of potential patients. However, the study has been confounded with small sample sizes. This further warrants a caution during generalization of these results and further replication of into a larger population of the study sample.

Acknowledgment. This publication was supported by the project LO1506 of the Czech Ministry of Education, Youth and Sports under the program NPU I. The research work was supported by the HiCoE grant for CISIR, Ministry of Education, Malaysia.

References

1. Acharya, U.R., et al.: Automated diagnosis of normal and alcoholic EEG signals. Int. J. Neural Syst. **22**(03), 1250011 (2012). https://doi.org/10.1142/S0129065712500116
2. Faust, O., Yu, W., Kadri, N.A.: Computer-based identification of normal and alcoholic EEG signals using wavelet packets and energy measures. J. Mech. Med. Biol. **13**(03), 1350033 (2013). https://doi.org/10.1142/S0219519413500334
3. American Psychiatric Association: Diagnostic and Statistical Manual of Mental Disorders (DSM-5®). American Psychiatric Publishing, Washington, D.C. (2013)
4. Kendler, K.S., et al.: Divorce and the onset of alcohol use disorder: a Swedish population-based longitudinal cohort and co-relative study. Am. J. Psychiatry **174**(5), 451–458 (2017). https://doi.org/10.1176/appi.ajp.2016.16050589
5. Juhás, M., et al.: Deep grey matter iron accumulation in alcohol use disorder. NeuroImage **148**, 115–122 (2017). https://doi.org/10.1016/j.neuroimage.2017.01.007
6. Saunders, J.B., et al.: Development of the alcohol use disorders identification test (AUDIT): WHO collaborative project on early detection of persons with harmful alcohol consumption-II. Addiction **88**(6), 791–804 (1993). https://doi.org/10.1111/j.1360-0443.1993.tb02093.x
7. Acharya, U.R., et al.: Deep convolutional neural network for the automated detection and diagnosis of seizure using EEG signals. Comput. Biol. Med. (2017). https://doi.org/10.1016/j.compbiomed.2017.09.017
8. Saini, J., Dutta, M.: An extensive review on development of EEG-based computer-aided diagnosis systems for epilepsy detection. Netw.: Comput. Neural Syst. 1–27 (2017). https://doi.org/10.1080/0954898x.2017.1325527
9. Srengers, J., et al.: O132 an EEG-based decision-support system for diagnosis and prognosis of autism spectrum disorder. Clin. Neurophysiol. **128**(9), e221 (2017). https://doi.org/10.1016/j.clinph.2017.07.143
10. Memar, P., Faradji, F.: A novel multi-class EEG-based sleep stage classification system. IEEE Trans. Neural Syst. Rehabil. Eng. **26**(1), 84–95 (2018). https://doi.org/10.1109/TNSRE.2017.2776149
11. Anand, A., et al.: Activity and connectivity of brain mood regulating circuit in depression: a functional magnetic resonance study. Biol. Psychiatry **57**(10), 1079–1088 (2005). https://doi.org/10.1016/j.biopsych.2005.02.021
12. Bae, J.N., et al.: Dorsolateral prefrontal cortex and anterior cingulate cortex white matter alterations in late-life depression. Biol. Psychiatry **60**(12), 1356–1363 (2006). https://doi.org/10.1016/j.biopsych.2006.03.052

13. Fingelkurts, A.A., et al.: Impaired functional connectivity at EEG alpha and theta frequency bands in major depression. Hum. Brain Mapp. **28**(3), 247–261 (2007). https://doi.org/10.1002/hbm.20275

14. Hamilton, J.P., Gotlib, I.H.: Neural substrates of increased memory sensitivity for negative stimuli in major depression. Biol. Psychiatry **63**(12), 1155–1162 (2008). https://doi.org/10.1016/j.biopsych.2007.12.015

15. Stam, C., et al.: Disturbed fluctuations of resting state EEG synchronization in Alzheimer's disease. Clin. Neurophysiol. **116**(3), 708–715 (2005). https://doi.org/10.1016/j.clinph.2004.09.022

16. Smit, D.J., et al.: Heritability of "small-world" networks in the brain: a graph theoretical analysis of resting-state EEG functional connectivity. Hum. Brain Mapp. **29**(12), 1368–1378 (2008). https://doi.org/10.1002/hbm.20468

17. Catani, M., Ffytche, D.H.: The rises and falls of disconnection syndromes. Brain **128**(10), 2224–2239 (2005). https://doi.org/10.1093/brain/awh622

18. He, Y., Chen, Z., Evans, A.: Structural insights into aberrant topological patterns of large-scale cortical networks in Alzheimer's disease. J. Neurosci. **28**(18), 4756–4766 (2008). https://doi.org/10.1523/JNEUROSCI.0141-08.2008

19. Stam, C.J., et al.: Small-world networks and functional connectivity in Alzheimer's disease. Cereb. Cortex **17**(1), 92–99 (2006). https://doi.org/10.1093/cercor/bhj127

20. Stam, C., et al.: Graph theoretical analysis of magnetoencephalographic functional connectivity in Alzheimer's disease. Brain **132**(1), 213–224 (2008). https://doi.org/10.1093/brain/awn262

21. Alexander-Bloch, A., et al.: The discovery of population differences in network community structure: new methods and applications to brain functional networks in schizophrenia. Neuroimage **59**(4), 3889–3900 (2012). https://doi.org/10.1016/j.neuroimage.2011.11.035

22. Fornito, A., et al.: Schizophrenia, neuroimaging and connectomics. Neuroimage **62**(4), 2296–2314 (2012). https://doi.org/10.1016/j.neuroimage.2011.12.090

23. Stam, C., Van Dijk, B.: Synchronization likelihood: an unbiased measure of generalized synchronization in multivariate data sets. Physica D: Nonlinear Phenom. **163**(3), 236–251 (2002). https://doi.org/10.1016/S0167-2789(01)00386-4

24. Babor, T.F., et al.: AUDIT: the alcohol use disorders identification test: guidelines for use in primary health care. World Health Organization (2001)

25. Bush, K., et al.: The AUDIT alcohol consumption questions (AUDIT-C): an effective brief screening test for problem drinking. Arch. Intern. Med. **158**(16), 1789–1795 (1998). https://doi.org/10.1001/archinte.158.16.1789

26. Jasper, H.H.: The ten twenty electrode system of the international federation. Electroencephalogr. Clin. Neurophysiol. **10**, 371–375 (1958)

27. Qin, Y., Xu, P., Yao, D.: A comparative study of different references for EEG default mode network: the use of the infinity reference. Clin. Neurophysiol. **121**(12), 1981–1991 (2010). https://doi.org/10.1016/j.clinph.2010.03.056

28. Berg, P., Scherg, M.: A multiple source approach to the correction of eye artifacts. Electroencephalogr. Clin. Neurophysiol. **90**(3), 229–241 (1994). https://doi.org/10.1016/0013-4694(94)90094-9

29. Hoechstetter, K., Berg, P., Scherg, M.: BESA research tutorial 4: distributed source imaging. BESA Research Tutorial, pp. 1–29 (2010)

30. Liu, H., Motoda, H.: Computational Methods of Feature Selection. CRC Press, Boca Raton (2007)

31. Mamitsuka, H.: Selecting features in microarray classification using ROC curves. Pattern Recogn. **39**(12), 2393–2404 (2006). https://doi.org/10.1016/j.patcog.2006.07.010

32. Sadegh-Zadeh, K.: Clinical decision support systems. In: Sadegh-Zadeh, K. (ed.) Handbook of Analytic Philosophy of Medicine. PM, vol. 119, pp. 705–722. Springer, Dordrecht (2015). https://doi.org/10.1007/978-94-017-9579-1_20

33. Mumtaz, W., et al.: Review on EEG and ERP predictive biomarkers for major depressive disorder. Biomed. Signal Process. Control 22, 85–98 (2015). https://doi.org/10.1016/j.bspc.2015.07.003

34. Chella, F., Pizzella, V., Zappasodi, F., Marzetti, L.: Impact of the reference choice on scalp EEG connectivity estimation. J. Neural Eng. 13(3), 036016 (2016). https://doi.org/10.1088/1741-2560/13/3/036016

35. Elowitz, M.B., Leibler, S.: A synthetic oscillatory network of transcriptional regulators. Nature 403(6767), 335 (2000). https://doi.org/10.1038/35002125

36. Theodoridis, S., Koutroumbas, K.: Pattern Recognition, pp. 341–342. Academic Press, Cambridge (1999)

37. Liu, H., Motoda, H.: Feature Selection for Knowledge Discovery and Data Mining. Kluwer Academic Publishers, Norwell (1998)

38. Ross, D.T., et al.: Systematic variation in gene expression patterns in human cancer cell lines. Nat. Genet. 24(3), 227–235 (2000). https://doi.org/10.1038/73432

39. Fosgate, G.T.: Practical sample size calculations for surveillance and diagnostic investigations. J. Vet. Diagn. Invest. 21(1), 3–14 (2009). https://doi.org/10.1177/104063870902100102

40. Zhou, X.-H., McClish, D.K., Obuchowski, N.A.: Statistical Methods in Diagnostic Medicine, vol. 569. Wiley, Hoboken (2009)

41. Mumtaz, W., Malik, A.S.: A comparative study of different EEG reference choices for diagnosing unipolar depression. Brain Topogr. 31(1), 875–885 (2018). https://doi.org/10.1007/s10548-018-0651-x

42. Mumtaz, W., Kamel, N., Ali, S.S.A., Malik, A.S.: An EEG-based functional connectivity measure for automatic detection of alcohol use disorder. Artif. Intell. Med. 84(1), 79–89 (2018). https://doi.org/10.1016/j.artmed.2017.11.002

43. Mumtaz, W., Vuong, P.L., Xia, L., Malik, A.S., Rashid, R.B.A.: Automatic diagnosis of alcohol use disorder using EEG features. Knowl.-Based Syst. 105, 48–59 (2016). https://doi.org/10.1016/j.knosys.2016.04.026

EchoQuan-Net: Direct Quantification of Echo Sequence for Left Ventricle Multidimensional Indices via Global-Local Learning, Geometric Adjustment and Multi-target Relation Learning

Rongjun Ge[1,2,4,7], Guanyu Yang[1,2,4], Chenchu Xu[7], Jiulou Zhang[1,2,4],
Yang Chen[1,2,3,4(✉)], Limin Luo[1,2,4], Cheng Feng[5], Heye Zhang[6], and Shuo Li[7]

[1] Laboratory of Image Science and Technology,
School of Computer Science and Engineering, Southeast University, Nanjing, China
{gerj16,chenyang.list}@seu.edu.cn
[2] Key Laboratory of Computer Network and Information Integration,
Southeast University, Ministry of Education, Nanjing, China
[3] School of Cyber Science and Engineering, Southeast University, Nanjing, China
[4] Centre de Recherche en Information Biomedicale Sino-Francais, Rennes, France
[5] Department of Ultrasound, The 3rd People's Hospital of Shenzhen,
Shenzhen, China
[6] School of Biomedical Engineering, Sun Yat-Sen University, Guangzhou, China
[7] Department of Medical Imaging, Western University, London, Canada

Abstract. Accurately quantifying multidimensional indices of the left ventricle (LV) in 2D echocardiography (echo) is clinically significant for cardiac disease diagnosis. However, the challenges of the frequently missing information, the high geometric variability and the uncertain multidimensional indices relation hinder its automated analysis development. Here, we propose an EchoQuan-Net to directly quantify LV in echo sequence from the multidimension, covering length and width for 1D, area for 2D, and volume for 3D. The net consist of three components: (1) Global-Local Learning to capture contextual information in the cardiac cycle for each frame, with global information from whole sequence and local information from the individual frame; (2) Geometric Adjustment to promote a canonical region of interest for LV, with translation, rotation and scale invariant; (3) Multi-target Relation learning to promote joint quantification for LV multidimensional indices, with sparse latent regression. The experiments reveal that EchoQuan-Net gains high accuracy, with mean accuracy error of 3.14 mm, 3.10 mm, 276 mm^2 and 13.5 ml for length, width, area and volume. The results show great potential of our method in clinical cardiac function assessment.

Keywords: Echo · Direct quantification ·
Global and local information · Geometric invariant · Multi-target

© Springer Nature Switzerland AG 2019
I. V. Tetko et al. (Eds.): ICANN 2019, LNCS 11731, pp. 219–230, 2019.
https://doi.org/10.1007/978-3-030-30493-5_24

1   Introduction

Accurately quantifying multidimensional indices of the left ventricle in 2D echocardiography (echo) sequence is of great clinical significance for diagnosis of cardiac disease [1–3]. (1) The 2D echo of medical ultrasound imaging is the essential modality in the assessment of various cardiac conditions, risk stratification, and prognostication [2], because of its non-invasiveness, low-cost, portability and accessibility that promote a real-time visual record of cardiac structures and activity [4]. (2) The multidimensional indices, including length and width of 1D, area of 2D, volume of 3D, together enable a comprehensive evaluation for LV size and systolic function [1,5]. Length is distance from apex to middle mitral valve plane for major axis, width is perpendicular to the major axis, at one-third of the major axis from the mitral valve plane for minor axis, area is measured from blood pool for region, and volume is the stereoscopic size. (3) Therefore, the achievement of automatic direct quantification is urgent for fast and accurate LV assessment [6]. Clinically, LV indices quantification is a time-consuming, subjective and non repeatable laborious work. It requires experts to manually zoom into the echo to catch LV, then further detect the biological sites (e.g., apex, plane of mitral valve, etc.), determine LV cavity, and compute modified Simpson's method, then finally measure them (Fig. 1(a1)–(a3)). Moreover, the segmentation-based method though provides limited quantification (only area), it is still an open problem for precise segmentation in 2D echo due to low signal-to-noise ratio, edge dropout, shadows, indirect relation between pixel intensity and the physical property of the tissue, and anisotropy of ultrasonic imaging [7], and needs extra user interaction [6], which may prevent it from efficient clinical application.

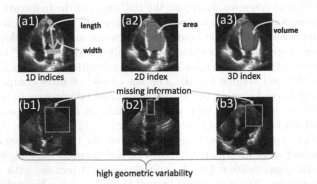

Fig. 1. (1) Multidimensional indices: length and width of 1D (a1), area of 2D (a2), and volume of 3D (a3). (2) Frequently missing information in box (b1-3) due to ultrasound imaging modality, and high geometric variability (b1-3) caused by subjective imaging acquisition.

Direct quantification of LV without segmentation [6,8–10] has been extensively studied in recent decades thanks to its efficiency and effectiveness, but it

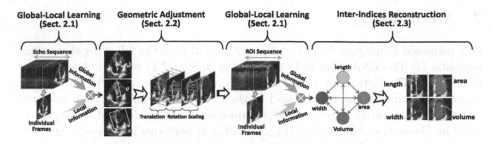

Fig. 2. The EchoQuan-Net directly achieves multidimensional quantification for LV in echo sequence, with composed of three effective components: (1) Global-Local Learning for contextual information in the cardiac cycle, (2) Geometric Adjustment for translation, rotation and scale invariant, (3) Multi-target relation learning for joint quantification of LV multidimensional indices.

mainly focuses on indices in single short-axis view cardiac magnetic resonance, and is still a big ambition to be repeated in 2D echo for multidimensional quantification due to the challenges: (1) the frequently missing information (box in Fig. 1(b1-3)) in individual echo frames, such as edge dropouts, etc., caused by the drawback of the ultrasound imaging modality; (2) the high geometric variability (Fig. 1(b1-3)) in position, orientation and scale of LV, led by the subjective imaging acquisition [4]; (3) the uncertain relation inter multidimensional indices, impeding the joint learning for multi-targets.

Here, an EchoQuan-Net is proposed to directly quantify LV in echo sequence from the multidimension, covering length and width for 1D, area for 2D, and volume for 3D, as shown in Fig. 2. The net is composed of three components: (1) Global-Local Learning for contextual information in the cardiac cycle, (2) Geometric Adjustment for translation, rotation and scale invariant, and (3) Multi-target Relation Learning for joint quantification of multidimensional indices. In the EchoQuan-Net, the **Global-Local Learning** combines sequence global information and individual frame local information, so that each frame is embedded with inter-frame relation, gets robust feature representation with the context of spatia-temporal information across the cardiac cycle, and meanwhile fixes missing information. The **Geometric Adjustment**, following the Global-Local Learning on echo image sequence, predicts the geometric transformation parameters of LV center, orientation and scale, then makes translation, rotation and scaling on various subjectively acquired echo for canonical LV ROI, where LV is centered, major axis is perpendicular to horizon and LV region is uniform. The **Multi-target Relation Learning**, following the Global-Local Learning on LV ROI sequence, builds a latent space for common representation of multi-target, then utilize a sparse matrix to mode intrinsic physiological interrelation, so that all multidimensional indices are regularized with each other for a joint quantification.

2   Methodology

Our proposed EchoQuan-Net (Fig. 2) is integrated with the following three components: (1) The Global-Local Learning (Fig. 3, Sect. 2.1) makes hybrid utilization of 3D convolution on global cardiac cycle sequence, and 2D convolution on each local individual frames, and further fuses the global and local information, to get contextual information for frames in echo sequence and LV sequence. (2) The Geometric Adjustment (Fig. 4, Sect. 2.2) performs a successive geometric transformation of translation, rotation and scaling on whole image with its predicted transformation parameters, to provide canonical ROI. (3) The Multi-target Relation Learning (Fig. 5, Sect. 2.3) builds a latent space for common representation of multi-target, and uses sparse matrix for capturing interrelation with $l_{2,1}$-norm based sparse learning, to promote a joint quantification.

2.1   Global-Local Learning for Spatio-Temporal Contextual Information

The proposed Global-Local Learning (Figs. 2 and 3) captures spatio-temporal contextual information across the cardiac cycle, so that promotes accurate spatial parameters and indices quantification for each frame by inter-frame relation.

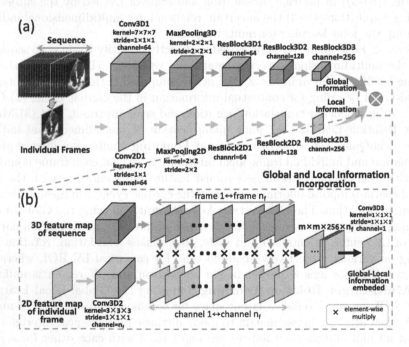

Fig. 3. Global-Local Learning captures spatio-temporal contextual information across the cardiac cycle, for each frame.

The Global-Local Learning deploys a set of 3D convolution process with the whole cardiac sequence as input to get global information of the whole cycle. Meanwhile, it performs the successive 2D convolution processes on the individual frame to get local information of each frame in the sequence. And the 3D/2D ResBlocks used after maxpooling 3D/2D are developed from ResNet to promote faster training [11].

Then, to fuse the global and local information, the Global-Local Learning firstly utilizes a 3D convolution to extend individual frames feature maps of $m \times m \times 256$ to $m \times m \times 256 \times n_f$. Secondly, the extended feature map of individual frames is arranged along channels: $1 \leftrightarrow n_f$, to align with the feature map of sequence arranged along frames: $1 \leftrightarrow n_f$, as the corresponding inter-action coefficient between frames. So that each local individual frame captures its contextual information with all frames in the global sequence. Finally, a 2D convolution of $1 \times 1 \times 1$ integrates the extended channels: $1 \leftrightarrow n_f$, to conclude and embed the spatio-temporal contextual information for each frame.

2.2   Geometric Adjustment for Translation, Rotation and Scale Invariant

The proposed Geometric Adjustment (Figs. 2 and 4) enables canonical ROI for LV in echo sequence, to achieve translation, rotation and scale invariant for the quantification. It goes through two stages of transformation parameters prediction and geometric transformation.

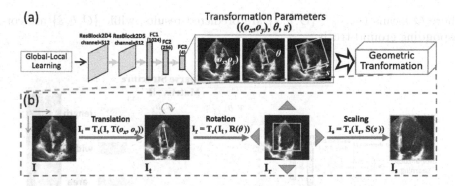

Fig. 4. The Geometric Adjustment make transformation parameters prediction (a) and geometric transformation (b) to achieve the canonical ROI of LV, which is equipped with translation, rotation and scale invariant for quantification, meanwhile other chambers interference reduction.

To make accurate prediction of LV center (o_x, o_y), orientation θ and scale s in each view, the Geometric Adjustment (Fig. 4(a)) utilizes successive convolutional ResBlocks to extract visual discriminative feature after getting global-local information of cardiac activity. Then, it abstracts and regresses transformation parameters by fully-connected (FC) layers with task-specific activation function in the last layer: $tanh(\cdot)$ for center, $\pi * tanh(\cdot)$ for orientation, and $0.5 * (1 + tanh(\cdot))$ for scale. The task-specific activation function effectively avoids and fixes obviously ridiculous transformation in early stage for explicitly learning and speeding up training.

As shown in Fig. 4(b), to transform the echo images into the canonical ROI, the Geometric Adjustment deploys a set of affine map to translate, rotate and scale, with the affine matrixes constructed from the predicted transformation parameters: $T(o_x, o_y) = \begin{bmatrix} 1 & 0 & o_x \\ 0 & 1 & o_y \\ 0 & 0 & 1 \end{bmatrix}$ as translation matrix, $R(\theta) = \begin{bmatrix} cos(\theta) & -sin(\theta) & 0 \\ sin(\theta) & cos(\theta) & 0 \\ 0 & 0 & 1 \end{bmatrix}$ as rotation matrix, and $S(s) = \begin{bmatrix} s & 0 & 0 \\ 0 & s & 0 \\ 0 & 0 & 1 \end{bmatrix}$ as scaling matrix, together with the bilinear interpolation to calculate pixel value. This process is inspired by [12,13] and [14] of improving performance for classification and segmentation.

Consequently, the echo images are converted into the canonical ROI as I_s in Fig. 4(b), which is equipped with translation, rotation and scale invariant for quantification, meanwhile other chambers interference reduction. The cost function for the Geometric Adjustment is formulated as:

$$\mathcal{L}_{Geo} = \underbrace{\left\| O - \hat{O} \right\|_2}_{Translation} + \underbrace{\left\| (\theta - \hat{\theta} + \pi) mod\, 2\pi - \pi \right\|_1}_{Rotation} + \underbrace{\left\| s - \hat{s} \right\|_1}_{Scaling} \qquad (1)$$

where O means (o_x, o_y), $\{O, \theta, s\}$ are predicted results, while $\{\hat{O}, \hat{\theta}, \hat{s}\}$ are corresponding ground truth.

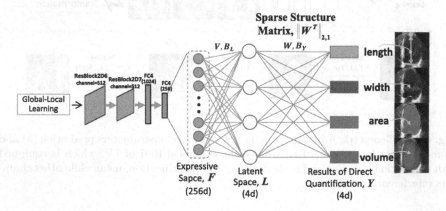

Fig. 5. The Multi-target Relation Learning effectively captures intrinsic physiological interrelation among multidimensional indices of LV for joint quantification.

2.3 Multi-target Relation Learning for Joint Quantification of Multidimensional Indices

The proposed Multi-target Relation Learning (Figs. 2 and 5) effectively extracts intrinsic physiological interrelation among multidimensional indices of LV to directly enable joint quantification, via a latent space model and a $l_{2,1}$-norm based sparse structure matrix.

The latent variables $L \in \mathbb{R}^4$ in the latent space are got from the expressive space F via a linear representer theorem [15] extracts higher level concepts to build a common representation for multiple regression outputs, formulated as $L = VF + B_L$. The expressive space F is obtained through Global-Local learning, visual structure feature capture and abstract, to disentangle complex nonlinear relationship between image appearance and the multidimensional indices.

The sparse structure matrix is performed on the latent variables to construct quantification results $Y \in \mathbb{R}^4$, with regularization of $l_{2,1}$-norm [16] to explicitly model multi-target correlation within multidimensional indices, formulated as $Y = WL + B_Y$, where $\left\| W^T \right\|_{2,1}$ needs be minimized. It promote common information cross related indices with the l_2 norm, and enable sparse selection of the most relevant information with the l_1 norm for each index.

Thus, the relation among the multidimensional indices is explicitly learned and maintained. The optimization object for the Multi-target Relation Learning is to minimize the indices loss together with the regularization term, as:

$$\mathcal{L}_{Ind} = \left\| Y - \hat{Y} \right\|_1 + \left\| W^T \right\|_{2,1} \tag{2}$$

3 Experiments

3.1 Data Set and Details

Data Set: A data set of 1000 2D echos of 50 patients from 2 hospitals is used to evaluate the performance of EchoQuan-Net. (1) Each patient provides 20 frames in a cardiac cycle. All images are resized to 256×256 to be fed into the net. (2) All ground truth are annotated by two cardiac radiologists with double-check.

Data Augmentation: To improve the generalization, the data set is augmented to 4000 images with three strategies: (1) random rotation $-15° \leftrightarrow 15°$), (2) random zooming ($0.9 \leftrightarrow 1.1$ times), and (3) random rotation + zooming.

Implementation Details: The network is optimized with the Adam (initial learning rate $= 10^{-4}$). 10-fold cross-validation is used for performance assessment. The performance of the network is measured by calculating the mean absolute error (MAE), as:

$$MAE = \frac{1}{50 \times 20} \sum_{sub=1}^{50} \sum_{f=1}^{20} \left| \hat{y}_{sub,f}^t - y_{sub,f}^t \right| \tag{3}$$

where $t \in \{length, width, area, volume\}$ which means indices type, or $t \in \{o_x, o_y, \theta, s\}$ which means geometric transformation parameters. $1 \leq sub \leq 50$ and $1 \leq f \leq 20$, in which sub represents subject, f means frames.

Table 1. The proposed method achieves the best performance compared to the other configurations, gaining the lowest MAE for each index.

	Length (mm)	Width (mm)	Area (mm^2)	Volume (ml)
No Global-Local Learning	3.65 ± 3.27	3.46 ± 3.19	343 ± 318	17.0 ± 14.9
No Geometric Adjustment	3.98 ± 3.50	3.89 ± 3.43	374 ± 334	22.7 ± 20.2
No M-tR Learning*	3.74 ± 3.45	3.52 ± 3.28	336 ± 306	18.2 ± 17.1
Proposed method	$\mathbf{3.14 \pm 2.69}$	$\mathbf{3.10 \pm 2.76}$	$\mathbf{276 \pm 245}$	$\mathbf{13.5 \pm 11.6}$

*M-tR Learning: Multi-target Relation Learning

3.2 Results and Analysis

Our EchoQuan-Net (the 4th row in Table 1) achieves excellent quantification performance with MAE as low as 3.14 mm for length and 3.10 mm for width of 1D, 276 mm^2 for area of 2D, and 13.5 ml for volume of 3D. And the acquired quantified results on an example subject is given in Fig. 6 to illustrate that EchoQuan-Net effectively, multidimensionally and continually quantifies the cardiac changes across the cardiac cycle.

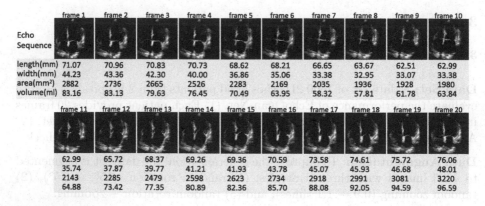

Fig. 6. The proposed EchoQuan-Net is able to effectively, multidimensionally and continually quantifies the cardiac changes across the cardiac cycle.

Advantages of Global-Local Learning. The comparison of the 1st and 4th rows in Table 1 illustrates that performance is improved by 13.97%, 10.40%, 19.53, and 20.59% for length, width, area and volume, with Global-Local Learning, compared to the situation of being degenerated to the normal local learning with only 2D ResBlocks. The Global-Local Learning has superiority to learn the spatio-temporal contextual information in cardiac activity for each frame, so that inter-frame relation is embedded to fix frequently missing information in individual frame with other frames. It also promotes a more accurate transformation parameters prediction (Table 2) with tiny MAE of (3.43 mm, 3.48 mm) for translation, 7.14° for rotation and 0.0463 for scaling, thanks to capturing the continuous motion of LV with contextual information across the sequence.

Table 2. Global-Local Learning promotes a better prediction on geometric transformation parameters.

	o_x (mm)	o_y (mm)	θ (°)	s
No Global-Local Learning	3.95 ± 3.62	4.02 ± 3.69	11.47 ± 8.636	0.0574 ± 0.0452
Proposed method	$\mathbf{3.43 \pm 2.92}$	$\mathbf{3.48 \pm 3.01}$	$\mathbf{7.14 \pm 4.28}$	$\mathbf{0.0463 \pm 0.0323}$

Advantages of Geometric Adjustment. The comparison of the 2nd and 4th rows in Table 1 proves that Geometric Adjustment improves the performance by 21.11%, 20.31%, 26.20%, and 40.53% for length, width, area and volume, compared to the situation of no Geometric Adjustment that makes LV quantification from the whole echo image of high geometric variability. The Geometric Adjustment transforms the echo image into the canonical ROI of LV as Fig. 7, so that guarantees the translation, rotation and scale invariant for the generalization, as well as reduces the inference from the other chambers for coarse-to-fine.

echo image   canonical ROI   echo image   canonical ROI   echo image   canonical ROI   echo image   canonical ROI

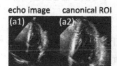 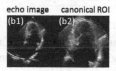 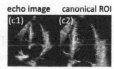 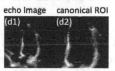

Fig. 7. Geometric Adjustment enables the canonical ROI for LV, guaranteeing the translation, rotation and scale invariant.

Advantages of Multi-target Relation Learning. The comparison of the 3rd and 4th rows in Table 1 shows the performance improvement of 16.04%, 11.93%, 17.86, and 25.82% for length, width, area and volume, by introducing Multi-target Relation Learning, compared to the situation of no Multi-target Relation Learning with fully-connected layer regression instead. Besides, Fig. 8 further shows that Multi-target Relation Learning promotes a better convergence for quantification. All of these effectiveness is beneficial from that the embedded relation mines the intrinsic physiological constraint among the indices for efficient learning of the multidimensional quantification set.

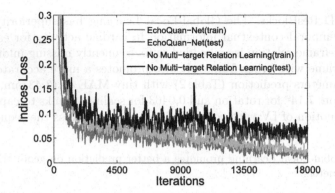

Fig. 8. The Multi-target Relation Learning effectively captures intrinsic physiological interrelation among multidimensional indices of LV for better convergence of joint quantification.

Table 3. The proposed method achieves best quantification performance compared to the existing methods, with the lowest MAE for each index.

	Length (mm)	Width (mm)	Area (mm²)	Volume (ml)
Multi-features+RF	3.86 ± 3.48	3.23 ± 2.91	323 ± 266	18.4 ± 15.7
SDL+AKRF	3.73 ± 3.05	3.21 ± 2.82	280 ± 236	18.9 ± 15.2
MCDBN+RF	3.93 ± 3.38	3.18 ± 3.00	312 ± 255	17.6 ± 14.9
Indices-Net	3.29 ± 2.42	4.27 ± 3.37	354 ± 338	16.1 ± 14.4
U-Net	N/A	N/A	387 ± 296	N/A
Proposed method	$\mathbf{3.14 \pm 2.69}$	$\mathbf{3.10 \pm 2.76}$	$\mathbf{276 \pm 245}$	$\mathbf{13.5 \pm 11.6}$

Comparison with the Existing Method. As shown in Table 3, our proposed method outperforms all the competitors with performance averagely improved by 14.78%, 9.34%, 15.65% and 23.656% on length, width, area and volume. Furthermore, (1) during the comparison with the existing direct quantification methods of Multi-features+RF [8], SDL+AKRF [9], MCDBN+RF [10] and Indices-Net [6], our framework is fed with the echo image covering all four chambers including LV, left atrium, right ventricle and right atrium, while the competitors need manually cropped ROI that mainly covers LV. For echo image, this is not an easy work, unlike cardiac magnetic resonance which has relatively fixed intersection point [6,8,10]. (2) Towards a full-automatic clinical tool, our framework comprehensively outputs all four indices directly, while the segmentation method U-Net [17] can only provide area without user interaction. Moreover, several post processing, such as maximum connected region extraction, is also deployed for U-Net to better collect area.

4 Conclusion

In this paper, we propose an EchoQuan-Net to achieve direct quantification of Echo sequence for LV multidimensional indices. It works through three effective components to handle the grave challenges in the echo LV direct quantification. (1) The Global-Local Learning combines sequence global information and individual frame local information to capture spatio-temporal contextual information in cardiac activity for each frame; (2) The Geometric Adjustment transform the subjectively acquire echo image into a canonical ROI of LV for translation, rotation and scale invariant; (3) The Multi-target Relation Learning captures intrinsic physiological interrelation among multidimensional indices of LV to promote a cooperatively and comprehensively joint quantification. Consequently, it gains promising results on all multidimensional indices, revealing its great potential as a clinical tool for cardiac function assessment. Meanwhile, our method also has great possibilities to be generally developed for handling the other multi-target learning issues in the sequence data, since it is a unified network that comprehensively considers the spatio-temporal analysis, the geometric variability and the inter-targets relation.

Acknowledgments. This work was supported by the Postgraduate Research & Practice Innovation Program of Jiangsu Province (No. KYCX17_0104); the China Scholarship Council (No. 201706090248); the National Natural Science Foundation (No. 61871117, 61828101); the States Key Project of Research and Development Plan (No. 2017YFA0104302, 2017YFC0109202 and 2017YFC0107900); and the Science and Technology Program of Guangdong (No. 2018B030333001).

References

1. Schiller, N.B., Shah, P.M., Crawford, M., DeMaria, A., Devereux, R., Feigenbaum, H., et al.: Recommendations for quantitation of the left ventricle by two-dimensional echocardiography. J. Am. Soc. Echocardiogr. **2**(5), 358–367 (1989). https://doi.org/10.1016/S0894-7317(89)80014-8
2. Kjaergaard, J., Petersen, C.L., Kjaer, A., Schaadt, B.K., Oh, J.K., Hassager, C.: Evaluation of right ventricular volume and function by 2D and 3D echocardiography compared to MRI. Eur. J. Echocardiogr. **7**(6), 430–438 (2006). https://doi.org/10.1016/j.euje.2005.10.009
3. Margossian, R., et al.: The reproducibility and absolute values of echocardiographic measurements of left ventricular size and function in children are algorithm dependent. J. Am. Soc. Echocardiogr. **28**(5), 549–558 (2015). https://doi.org/10.1016/j.echo.2015.01.014
4. Abdi, A.H., Luong, C., Tsang, T., Allan, G., Nouranian, S., Jue, J., et al.: Automatic quality assessment of echocardiograms using convolutional neural networks: feasibility on the apical four-chamber view. IEEE Trans. Med. Imaging **36**(6), 1221–1230 (2017). https://doi.org/10.1109/TMI.2017.2690836
5. Pascual, M., et al.: Effects of isolated obesity on systolic and diastolic left ventricular function. Heart **89**(10), 1152–1156 (2003). https://doi.org/10.1136/heart.89.10.1152

6. Xue, W., Islam, A., Bhaduri, M., Li, S.: Direct multitype cardiac indices estimation via joint representation and regression learning. IEEE Trans. Med. Imaging **36**(10), 2057–2067 (2017). https://doi.org/10.1109/TMI.2017.2709251

7. Carneiro, G., Nascimento, J.C., Freitas, A.: The segmentation of the left ventricle of the heart from ultrasound data using deep learning architectures and derivative-based search methods. IEEE Trans. Image Process. **21**(3), 968–982 (2012). https://doi.org/10.1109/TIP.2011.2169273

8. Zhen, X., Wang, Z., Islam, A., Bhaduri, M., Chan, I., Li, S.: Direct estimation of cardiac bi-ventricular volumes with regression forests. In: Golland, P., Hata, N., Barillot, C., Hornegger, J., Howe, R. (eds.) MICCAI 2014. LNCS, vol. 8674, pp. 586–593. Springer, Cham (2014). https://doi.org/10.1007/978-3-319-10470-6_73

9. Zhen, X., Islam, A., Bhaduri, M., Chan, I., Li, S.: Direct and simultaneous four-chamber volume estimation by multi-output regression. In: Navab, N., Hornegger, J., Wells, W.M., Frangi, A.F. (eds.) MICCAI 2015. LNCS, vol. 9349, pp. 669–676. Springer, Cham (2015). https://doi.org/10.1007/978-3-319-24553-9_82

10. Zhen, X., Wang, Z., Islam, A., Bhaduri, M., Chan, I., Li, S.: Multi-scale deep networks and regression forests for direct bi-ventricular volume estimation. Med. Image Anal. **30**, 120–129 (2016). https://doi.org/10.1016/j.media.2015.07.003

11. He, K., Zhang, X., Ren, S., Sun, J.: Deep residual learning for image recognition. In: Proceedings of the IEEE Conference on Computer Vision and Pattern Recognition, pp. 770–778 (2016). https://doi.org/10.1109/CVPR.2016.90

12. Jaderberg, M., Simonyan, K., Zisserman, A.: Spatial transformer networks. In: Advances in Neural Information Processing Systems, pp. 2017–2025 (2015). arXiv:1506.02025

13. Dai, J., He, K., Sun, J.: Instance-aware semantic segmentation via multi-task network cascades. In: Proceedings of the IEEE Conference on Computer Vision and Pattern Recognition, pp. 3150–3158 (2016). https://doi.org/10.1109/CVPR.2016.343

14. Vigneault, D.M., Xie, W., Ho, C.Y., Bluemke, D.A., Noble, J.A.: Ω-net (omega-net): fully automatic, multi-view cardiac MR detection, orientation, and segmentation with deep neural networks. Med. Image Anal. **48**, 95–106 (2018). https://doi.org/10.1016/j.media.2018.05.008

15. Kimeldorf, G.S., Wahba, G.: A correspondence between Bayesian estimation on stochastic processes and smoothing by splines. Ann. Math. Stat. **41**(2), 495–502 (1970). https://doi.org/10.1214/aoms/1177697089

16. Ding, C., Zhou, D., He, X., Zha, H.: R 1-PCA: rotational invariant L_1-norm principal component analysis for robust subspace factorization. In: Proceedings of the 23rd International Conference on Machine Learning, pp. 281–288. ACM (2006). https://doi.org/10.1145/1143844.1143880

17. Ronneberger, O., Fischer, P., Brox, T.: U-Net: convolutional networks for biomedical image segmentation. In: Navab, N., Hornegger, J., Wells, W.M., Frangi, A.F. (eds.) MICCAI 2015. LNCS, vol. 9351, pp. 234–241. Springer, Cham (2015). https://doi.org/10.1007/978-3-319-24574-4_28

An Attention-Based ID-CNNs-CRF
Model for Named Entity Recognition
on Clinical Electronic Medical Records

Ming Gao(iD), Qifeng Xiao(iD), Shaochun Wu$^{(\boxtimes)}$(iD), and Kun Deng(iD)

Department of Intelligent Information Processing, Shanghai University,
Shanghai 200444, China
{qywtgm950120,hurricane,scwu,dengkun}@shu.edu.cn

Abstract. Named Entity Recognition (NER) on Clinical Electronic
Medical Records (CEMR) is a fundamental step in extracting disease
knowledge by identifying specific entity terms such as diseases, symptoms, etc. However, the state-of-the-art NER methods based on Long
Short-Term Memory (LSTM) fail to fully exploit GPU parallelism under
the massive medical records. Although a novel NER method based on
Iterated Dilated CNNs (ID-CNNs) can accelerate network computing, it
tends to ignore the word-order feature and semantic information of the
current word. In order to enhance the performance of ID-CNNs-based
models on NER tasks, an attention-based ID-CNNs-CRF model which
combines word-order feature and local context is proposed. Firstly, Position Embedding is utilized to fuse word-order information. Secondly, ID-
CNNs architecture is used to rapidly extract global semantic information.
Simultaneously, the attention mechanism is employed to pay attention
to the local context. Finally, we apply the CRF to obtain the optimal tag
sequence. Experiments conducted on two CEMR datasets show that our
model outperforms traditional ones. The F1-scores of 94.55% and 91.17%
are obtained respectively on these two datasets and both are better than
LSTMs-based models.

Keywords: Clinical electronic records · Named entity recognition ·
Convolutional Neural Network

1 Introduction

With the rapid development of the medical industry, data mining for CEMR
plays an important role in precision medicine. The state-of-the-art NER methods use Long Short-Term Memory (LSTM) to extract features and then employ
Conditional Random Field (CRF) to obtain the optimal tag sequence [1–4]. However, the temporal structure of LSTMs-based models is usually computationally
expensive and inefficient, especially when faced with massive medical records.
Recent works [5,6] attempt to apply Convolutional Neural Networks (CNN)
to NER. Nevertheless, the performance of CNNs-based models is poorer than

© Springer Nature Switzerland AG 2019
I. V. Tetko et al. (Eds.): ICANN 2019, LNCS 11731, pp. 231–242, 2019.
https://doi.org/10.1007/978-3-030-30493-5_25

LSTMs-based models for its neglect of global semantic information. Recently, the Iterated Dilated CNNs (ID-CNNs) [7] is proposed to efficiently aggregate broad context. But this model ignores the significance of word-order feature and local context in the text.

To address these issues, we proposed an attention-based ID-CNNs-CRF model. We first introduce Position Embedding to capture word-order information. Then, the attention mechanism is applied to the ID-CNNs-CRF model because of its good performance in NLP tasks, which enables the enhancement of the influence of critical words. Finally, we apply the CRF to obtain the optimal tag sequence. Experimental results on two CEMR datasets demonstrate that the proposed model achieves better prediction performance with a high F1-score than those of baseline methods.

The remainder of the paper is structured as follows: Sect. 2 introduces the related work. Section 3 describes the details of the proposed method. Section 4 demonstrates the proposed methodology with a series of experiments. Finally, the conclusions of this study are given in Sect. 5.

2　Related Work

Named entity recognition (NER) of CEMR designed to identify critical entities of interest is a basic step in medical information extraction. Traditionally, many simple but straightforward methods, such as rule-based and heuristic-search-based methods, have been utilized to identify critical medical entities [8]. These methods tend to achieve low recall value due to the inability to cover all medical entities. With the expansion of medical data, these time-consuming and laborious original methods seem to be clumsy.

NER methods based on statistical machine learning includes Support Vector Machine (SVM) [9], Hidden Markov Model (HMM) [10,11], Maximum Entropy Hidden Markov Model (MEHMM) [12] and Conditional Random Field (CRF) [13]. GuoDong et al. [11] presented an HMM NER system to deal with the special phenomena in the biomedical domain. Kaewphan et al. [14] utilized a CRF-based machine learning system named Nersuite to achieve an F-score of 88.46% on Gellus corpus. The CRF methods have decent performance but rely heavily on the selection of features.

In recent years, deep learning methods have been developed for NER. LSTM architectures that are capable of learning the long-term dependencies have been put forward [15]. Rondeau et al. [3] utilized NeuroCRF to obtain an F1-score of 89.28% on the WikiNER dataset. However, they are inefficient because of their sequential processing on sentences. To alleviate this problem, Emma et al. [5] applied ID-CNNs architecture to speed up the processing of the network. However, these models tend to ignore the word-order feature and local context compared to LSTM-based models.

In this paper, in order to promote the performance of the ID-CNNs-CRF model, Position Embedding is utilized to introduce word-order information and attention mechanism is applied to focus on those critical words by assigning different weights.

3   Method

In this section, we construct a variety of features as model input and then describe the proposed attention-based ID-CNNs-CRF model in detail.

3.1   The Construction of Features

In order to get richer semantic features, we construct a module for extracting Word representation. Word representation is a concatenation of Word Embedding, Char Embedding, POS Embedding, and Position Embedding. The extraction module of word representation is shown in Fig. 1.

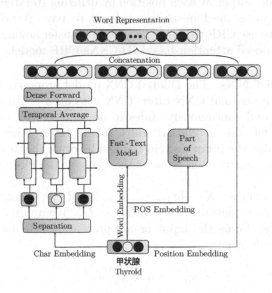

Fig. 1. The extraction module of the word representation. The one on the far left is char embedding, the middle one is word embedding, and the two on the right are part of speech embedding and position embedding.

Word Embedding. We choose Fast-Text [16] instead of Word2vec [17] to characterize words because it is not only fast in the massive text but also relieves the problem of out-of-vocabulary. In order to obtain high-quality word characterization, a large amount of medical record data is crawled from the Internet [18] as a training corpus of the word embedding.

Char Embedding. To obtain a high-quality word representation and avoid word segmentation errors, we take the character of the word as a sequence, and then get the character level representation through the bidirectional recurrent neural network.

POS Embedding. The performance of NLP tasks relies heavily on word representation, especially linguistic features. We utilize Jieba [19] (a Chinese word segmentation tool) to segment each sentence and get part of speech (such as nouns, verbs, etc).

Position Embedding. To obtain the position information, we number each position (each position corresponds to a number) and then introduce a certain position embedding for each word by random embedding.

3.2 Attention-Based ID-CNNs-CRF Model

Firstly, we obtain word representation according to the description in Sect. 3.1 and then obtain the output at each position by utilizing Iterated Dilated CNNs. Secondly, the attention mechanism is employed to pay attention to the local context. Finally, we use CRF to learn the rules of transfer among labels. Figure 2 illustrates the proposed attention-based ID-CNNs-CRF model.

Iterated Dilated CNNs. The Dilated CNN model proposed by Yu et al. [20] is different from the normal CNN filter. CNN acts on a continuous position of the input matrix and continuously slides to do convolution and pooling. The Dilated CNN model adds a dilation width to the filter. When it acts on the input matrix, it skips the input data in the middle of all the dilation width, and the size of the filter matrix itself remains the same, so the filter achieves a wider input.

For the input sequence $X = \{x_1, x_2, ..., x_t, ..., x_n\}$, we denote the jth dilated convolutional layer of dilation with width δ as D_δ^j. We apply a dilation-1 convolution D_1^0 to transforms the input as a representation i_t and take it as the begining of the stack layers.

$$c_t^{(0)} = i_t = D_1^{(0)} x_t \tag{1}$$

next, the stacked layers are represented as follows:

$$c_t^{(j)} = r(D_{2^{L_c-1}}^{(j-1)} c_t^{(j-1)}) \tag{2}$$

where $c_t^{(j)}$ denotes the output of the jth dilated convolutional layer of dilation, L_c denotes the layers of dilated convolutions and r denotes the activation function of *Relu* activation function.

Each block has three layers of expansion convolution (excluding input), so the output of each block is $c_t^{(3)}$. Figure 3 illustrates the architecture of a block. Our model stacks four blocks of the same architectures and each of that is a three-layer Dilated convolution layer with the dilation width of $[1, 1, 2]$. We define these three layers of expansion convolution as a block B.

$$b_t^{(1)} = B(x_t) \tag{3}$$

$$b_t^{(k)} = B(b_t^{(k-1)}) \tag{4}$$

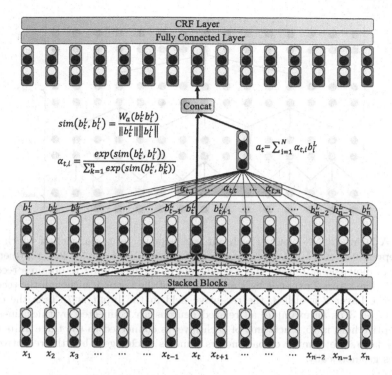

Fig. 2. Attention-based ID-CNN-CRF architecture. We stack 4 Dilated CNN blocks, each as shown in Fig. 3. To simplify the drawing, we use *Stacked Blocks* to represent all the layers in the middle. For the output b_t^L at position t (the orange unit with four circles), we calculate its similarity with all the output units and take them as weights $\{\alpha_{t,1}, ..., \alpha_{t,i}, ..., \alpha_{t,n}\}$. The output is multiplied by the corresponding weight and then summed as the attention vector a_t. Concatenate a_t with b_t^L and add a full connection layer to map it into the category space. Finally, the optimal tag sequence is obtained by CRF. (Color figure online)

Attention Mechanism. Our model can rapidly aggregate broad context by using Iterated Dilated CNNs. However, the extracted broad context usually ignores the importance of the current word for the current tag. To alleviate this problem, we apply the attention mechanism to our model. Attention mechanism has recently become popular in image processing and natural language processing. Luo et al. [21] utilized an attention mechanism to enforce tagging consistency across multiple instances of the same token in a document. Bharadwaj et al. [22] introduced the attention mechanism to enhance their model performances. However, their attention mechanism focuses on which encoded elements contribute to the generation of the current unit. Different from them, We apply the attention mechanism to focus on the related tokens in the sequence.

In the attention layer, we calculate the attention weight by Eq. 5 between the current token and all tokens in the sequence as the projection matrix and

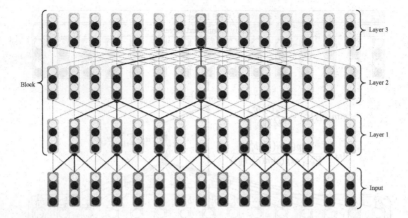

Fig. 3. A Dilated CNN block. The Input is a sequence of texts, each of which is the word representation of Fig. 1 (15 words are entered). The circles just represent the concept of dimensions. Layer 1 has a coefficient of expansion of 1 to obtain a receptive field of 3. The Layer 2 expands by 1 on the basis of the Layer 1, although the receptive field in the Layer 2 is still 3, the receptive field in the Input is 7. Take the orange unit of Layer 3 as an example, all related neurons are connected by thick lines, and the receptive field is 15. Each unit of each layer is connected to the three units of the previous layer, the number of parameters remains unchanged, but the receptive field increases geometrically.

normalize it with the softmax function.

$$\alpha_{t,i} = \frac{exp(sim(b_t^L, b_i^L))}{\sum_{k=1}^{n} exp(sim(b_t^L, b_k^L))} \tag{5}$$

where b_t^L is the final output position t in Eq. 4, sim denotes the similarity between two vectors. The similarity between two vectors is measured by cosine similarity according to Eq. 6.

$$sim(b_t^L, b_i^L) = \frac{W_a(b_t^L b_i^L)}{\|b_t^L\| \, \|b_i^L\|} \tag{6}$$

where W_a is a weight matrix which is learned in the training process. We take the weight vector computation of the unit at position b_t^L as an example to explain the detailed computation process. According to Eq. 7, we can achieve the coefficient of each output in the attention layer, then we calculate the output a_t under this attention coefficient.

$$a_t = \sum_{i=1}^{N} \alpha_{t,i} b_i^L \tag{7}$$

then, the output of the current position and the output of the attention layer are concatenated as the output of this module.

$$o_t = W_o[a_t : b_t^L] \tag{8}$$

where W_o is a weight matrix that maps the output to the category space. Finally, like the general NER method, the CRF layer is added for the final sequence labeling.

Linear Chain CRF. The Linear Chain CRF combines the advantages of the maximum entropy model and the hidden Markov model. It considers the transfer relationships among the labels, enabling the acquisition of the optimal label sequence in the sequence labeling.

We consider E to be the matrix of scores output by our model. The ith column is the vector o_i obtained by the Eq. 8. The element E_{i,y_i} of the matrix is the score of the tag y_i of the ith token in the sentence. We introduce a tagging transition matrix T. The element T_{y_{i-1},y_i} of the matrix is the score of transition from tag y_{i-i} to y_i. This transition matrix will be trained as the parameter of our network.

Therefore, for a sentence $S = \{w_1, w_2, w_3, ..., w_n\}$, the score of its prediction sequence $y = \{y_1, y_2, y_3, ..., y_n\}$ can be expressed as Eq. 9:

$$s(S, y) = \sum_{i=1}^{n} E_{i,y_i} + \sum_{i=1}^{n-1} T_{y_{i-1},y_i} \tag{9}$$

under the condition of the given sentence S, the probability of sequence label y is as follows:

$$p(y|S) = \frac{e^{s(S,y)}}{\sum_{\tilde{y} \in Y_X} e^{s(S,\tilde{y})}} \tag{10}$$

during the training process, the likelihood function of the marker sequence is:

$$log(p(y|S)) = s(S, y) - log(\sum_{\tilde{y} \in Y_X} e^{s(S,\tilde{y})}) \tag{11}$$

where Y_X represents all possible sets of markers, and a valid output sequence can be obtained by the likelihood function. The set of sequences with the highest overall probability is output by Eq. 12 when predicting the optimal label:

$$y^* = \arg\max_{\tilde{y} \in Y_X} s(X, \tilde{y}) \tag{12}$$

The loss function is minimized by backpropagation during training and the Viterbi algorithm is applied to find the tag sequence with maximum probability during testing.

4 Experiments and Analysis

4.1 Datasets

Both CCKS2017 [23] and CCKS2018 [24] (China Conference on Knowledge Graph and Semantic Computing) have published NER tasks for recognizing

Table 1. Statistics on different categories of entities in the data set.

CCKS2017				CCKS2018			
Entity	Count	Train	Test	Entity	Count	Train	Test
Symptom	7831	6846	1345	Symptom	3055	2199	856
Check	9546	7887	1659	Description	2066	1529	537
Treatment	1048	853	195	Operation	1116	924	192
Disease	722	515	207	Drug	1005	884	121
Body	10719	8942	1777	Body	7838	6448	1390
Total	29866	24683	5183	Total	15080	11984	3096

entities (such as Body, Symptom, etc.) on CEMR. Table 1 lists the statistics for different categories of entities in the datasets. CCKS2017 Task 2 (Electronic Medical Record Named Entity Recognition) provides a total of 800 real electronic medical record data to identify five types of entities. Each clinical electronic case is divided into four domains including general items, Medical history, diagnosis, and discharge. The CCKS2018 CNER Task 1 provides 600 annotated corpus as training data set to identify five types of entities.

To effectively train and evaluate the proposed model, we split the dataset into the training set and test set by a ratio of 4:1, In addition, we collected nearly 13,496 medical records from the Internet [18] as training corpora for word vectors.

4.2   Parameter Setting

The parameters of our model are listed in Table 2. We obtain word representation with word embedding dim 256, char embedding dim 64, POS embedding dim 64 and position embedding dim 128. Filter width, num filter, dilation and block number of Iterated Dilated CNNs module are set to 3, 256, [1, 1, 2] and 4 respectively. Other parameters can also be seen in Table 2.

4.3   Comparison with Other Methods

To fully validate the effectiveness of our model, we compare our attention-based ID-CNNs-CRF model to previous state-of-the-art methods on the two datasets described in Sect. 4.1. As shown in Table 3, our attention-based ID-CNNs-CRF outperforms the prior methods for most metrics. Compared with the ID-CNNs-CRF, our method obtains better performance (improvements of 5.95%, 7.48% and 7.08% in Precision, Recall, and F1-score, respectively). This demonstrates that the position embedding and attention mechanism have a huge performance improvement for the ID-CNNs model since it can fuse word-order information and pay attention to the local context. In addition, our model outperforms the Bi-LSTM-CRF model, showing that our attention-based ID-CNNs-CRF is also an effective token encoder for structured inference.

Table 2. Parameter setting for modules

Module	Parameter name	Value
Word representation	Word embedding dim	256
	Char embedding dim	64
	POS embedding dim	64
	Position embedding dim	128
Iterated Dilated CNNs	Filter width	3
	Num filter	256
	Dilation	[1, 1, 2]
	Block number	4
Other	Learning rate	1e−3
	Dropout	0.5
	Gradient clipping	5
	Batch size	64
	Epoch	40

Table 3. Results of different NER models on CCKS2017 and CCKS2018 data sets.

Model	CCKS2017			CCKS2018		
	Prec.(%)	Rec.(%)	F(%)	Prec.(%)	Rec.(%)	F(%)
CRF	89.18	81.60	85.11	92.67	72.10	77.58
LSTM-CRF	87.73	87.00	87.24	82.61	81.70	82.08
Bi-LSTM-CRF	94.73	93.29	93.97	90.43	90.49	90.44
ID-CNNs-CRF	88.20	87.15	87.47	81.27	81.42	81.34
Attention-based ID-CNNs-CRF (Ours)	94.15	94.63	94.55	91.11	91.25	91.17

Our method is not only a better token encoder than the Bi-LSTM-CRF but it is also faster. Table 4 lists test time on the test set, compared to the Bi-LSTM-CRF. Our test set contains 512 pieces of data, and each model is tested 5 times for average. The average test time for our model and Bi-LSTM-CRF is 12.81 and 15.62, respectively. The model we proposed is 22% faster than the Bi-LSTM-CRF. Compared with the ID-CNNs-CRF, the speed of our model is slightly worse, but our model outperforms it by a wide margin.

Table 4. Comparison of test time.

Model (512 test data)	Time(s)	Speed
Bi-LSTM-CRF	15.62	1.0×
ID-CNNs-CRF	11.96	1.31×
Attention-ID-CNNs-CRF	12.81	1.22×

4.4 Comparison of Entity Category

To comprehensively compare the recognition capabilities in different entity categories, we enumerate and compare the F1-score of the three main models (Bi-LSTM-CRF, ID-CNN-CRF without position embedding and attention-based ID-CNNs-CRF model we proposed) in five categories. Detailed experimental results are shown in Table 5.

Table 5. Comparison of entity category on CCKS2017 and CCKS2018 data sets.

CCKS2017						
Model	Body	Check	Disease	Signs	Treatment	Avg
Bi-LSTM-CRF	94.81	96.40	87.79	95.82	88.10	93.97
ID-CNNs-CRF	90.89	88.28	79.67	91.04	81.27	87.47
Attention-ID-CNNs-CRF	95.38	97.79	86.55	96.91	87.64	94.55
CCKS2018						
Model	Body	Symptom	Operation	Drug	Description	Avg
Bi-LSTM-CRF	92.59	93.12	87.43	82.86	86.33	90.44
ID-CNNs-CRF	84.37	85.81	76.59	80.42	84.61	81.34
Attention-ID-CNNs-CRF	95.18	94.47	85.86	80.06	91.99	91.17

As shown in Table 5, the average F1-score of our model is improved by 0.58% (0.73% in CCKS2018) compared to the Bi-LSTM-CRF while maintaining 1.22× (Table 4) faster test time speed. The performance of the ID-CNNs-CRF is worse than the other two models in each category. For most categories, the model we proposed is significantly better than the Bi-LSTM-CRF. For example, it has a slightly high F1-score (2.59% on Body and 5.66% on Description in CCKS2018) than the Bi-LSTM-CRF model. But on a few categories, the Bi-LSTM-CRF has a higher F1-score (1.24% on Disease and 0.46% on Treatment in CCKS2017) than our model. We conjecture that it might due to the imbalanced data distribution (the number of different entities varies greatly) because the entity number of these categories are all small (it can be seen in Sect. 4.1).

5 Conclusion

In this paper, we propose an attention-based ID-CNNs-CRF model for NER on CEMR. Firstly, word representation combined with position embedding is used for the input of our model to capture the word-order information. Secondly, we stack four Dilated CNN blocks to obtain broad semantic information to make up for the shortage of CNN-based model in the language field. Then, the attention mechanism is applied to pay more attention to the characteristics of the current words and increase the performance of the model. Finally, the CRF is utilized to obtain the optimal tag sequence. The experiments on two CEMR datasets demonstrate that the attention-based ID-CNNs-CRF is superior to state-of-the-art methods with faster test time. There is no significant improvement in our model in the number of entities with fewer samples. Therefore, our future work is to study how to improve the recognition of these entities with fewer samples.

References

1. Lample, G., Ballesteros, M., Subramanian, S., Kawakami, K., Dyer, C.: Neural architectures for named entity recognition. arXiv preprint arXiv:1603.01360 (2016)
2. Ma, X., Hovy, E.: End-to-end sequence labeling via bi-directional lstm-cnns-crf. arXiv preprint arXiv:1603.01354 (2016). https://doi.org/10.18653/v1/p16-1101
3. Rondeau, M.-A., Su, Y.: LSTM-based NeuroCRFs for named entity recognition. In: INTERSPEECH, pp. 665–669 (2016). https://doi.org/10.21437/interspeech.2016-288
4. Rei, M., Crichton, G.K., Pyysalo, S.: Attending to characters in neural sequence labeling models. arXiv preprint arXiv:1611.04361 (2016)
5. Collobert, R., Weston, J., Bottou, L., Karlen, M., Kavukcuoglu, K., Kuksa, P.: Natural language processing (almost) from scratch. J. Mach. Learn. Res. **12**(Aug), 2493–2537 (2011). https://doi.org/10.1016/j.chemolab.2011.03.009
6. Wang, C., Wei, C., Bo, X.: Named entity recognition with gated convolutional neural networks (2017). https://doi.org/10.1007/978-3-319-69005-6_10
7. Strubell, E., Verga, P., Belanger, D., McCallum, A.: Fast and accurate sequence labeling with iterated dilated convolutions. arXiv preprint arXiv:1702.02098 138 (2017)
8. Hirschman, L., Morgan, A.A., Yeh, A.S.: Rutabaga by any other name: extracting biological names. J. Biomed. Inform. **35**(4), 247–259 (2002). https://doi.org/10.1016/s1532-0464(03)00014-5
9. Han, X., Ruonan, R.: The method of medical named entity recognition based on semantic model and improved SVM-KNN algorithm. In: 2011 Seventh International Conference on Semantics, Knowledge and Grids, pp. 21–27. IEEE (2011). https://doi.org/10.1109/skg.2011.24
10. Collier, N., Nobata, C., Tsujii, J.-I.: Extracting the names of genes and gene products with a hidden Markov model. In: Proceedings of the 18th Conference on Computational linguistics-Volume 1, pp. 201–207. Association for Computational Linguistics (2000). https://doi.org/10.3115/990820.990850
11. GuoDong, Z., Jian, S.: Exploring deep knowledge resources in biomedical name recognition. In: JNLPBA Workshop, pp. 96–99. Association for Computational Linguistics (2004). https://doi.org/10.3115/1567594.1567616
12. Chieu, H.L., Ng, H.T.: Named entity recognition with a maximum entropy approach. In: Proceedings of the Seventh Conference on Natural Language Learning at HLT-NAACL 2003-Volume 4, pp. 160–163. Association for Computational Linguistics (2003). https://doi.org/10.3115/1119176.1119199
13. Leaman, R., Islamaj Doğan, R., Lu, Z.: DNorm: disease name normalization with pairwise learning to rank. Bioinformatics **29**(22), 2909–2917 (2013). https://doi.org/10.1093/bioinformatics/btt474
14. Kaewphan, S., Van Landeghem, S., Ohta, T., Van de Peer, Y., Ginter, F., Pyysalo, S.: Cell line name recognition in support of the identification of synthetic lethality in cancer from text. Bioinformatics **32**(2), 276–282 (2015). https://doi.org/10.1093/bioinformatics/btv570

15. Zhu, Q., Li, X., Conesa, A., Pereira, C.: GRAM-CNN: a deep learning approach with local context for named entity recognition in biomedical text. Bioinformatics **34**(9), 1547–1554 (2017). https://doi.org/10.1093/bioinformatics/btx815

16. Joulin, A., Grave, E., Bojanowski, P., Mikolov, T.: Bag of tricks for efficient text classification. arXiv preprint arXiv:1607.01759 (2016). https://doi.org/10.18653/v1/e17-2068

17. Le, Q.V., Mikolov, T.: Distributed representations of sentences and documents. arXiv preprint arXiv:1405.4053 (2014)

18. http://case.medlive.cn/all/case-case/index.html?ver=branch

19. https://github.com/fxsjy/jieba

20. Yu, F., Koltun, V.: Multi-scale context aggregation by dilated convolutions. arXiv preprint arXiv:1511.07122 (2015)

21. Luo, L., et al.: An attention-based BiLSTM-CRF approach to document-level chemical named entity recognition. Bioinformatics **34**(8), 1381–1388 (2017). https://doi.org/10.1093/bioinformatics/btx761

22. Bharadwaj, A., Mortensen, D., Dyer, C., Carbonell, J.: Phonologically aware neural model for named entity recognition in low resource transfer settings. In: Proceedings of the 2016 Conference on Empirical Methods in Natural Language Processing, pp. 1462–1472 (2016). https://doi.org/10.18653/v1/d16-1153

23. Li, J., Zhou, M., Qi, G., Lao, N., Ruan, T., Du, J. (eds.): CCKS 2017. CCIS, vol. 784. Springer, Singapore (2017). https://doi.org/10.1007/978-981-10-7359-5

24. Zhao, J., Harmelen, F., Tang, J., Han, X., Wang, Q., Li, X. (eds.): CCKS 2018. CCIS, vol. 957. Springer, Singapore (2019). https://doi.org/10.1007/978-981-13-3146-6

Deep Text Prior: Weakly Supervised Learning for Assertion Classification

Vadim Liventsev[1,2](✉) , Irina Fedulova[1] , and Dmitry Dylov[2]

[1] Philips Innovation Labs RUS, Bolshoy Boulevard 42, bld. 1, Moscow 121205, Russia
{Vadim.Liventsev,Irina.Fedulova}@philips.com
[2] Skolkovo Institure of Science and Technology,
Bolshoy Boulevard 30, bld. 1, Moscow 121205, Russia
{Vadim.Liventsev,D.Dylov}@skoltech.ru

Abstract. The success of neural networks is typically attributed to their ability to closely mimic relationships between features and labels observed in the training dataset. This, however, is only part of the answer: in addition to being fit to data, neural networks have been shown to be useful priors on the conditional distribution of labels given features and can be used as such even in the absence of trustworthy training labels. This feature of neural networks can be harnessed to train high quality models on low quality training data in tasks for which large high-quality ground truth datasets don't exist. One of these problems is assertion classification in biomedical texts: discriminating between positive, negative and speculative statements about certain pathologies a patient may have. We present an assertion classification methodology based on recurrent neural networks, attention mechanism and two flavours of transfer learning (language modelling and heuristic annotation) that achieves state of the art results on MIMIC-CXR radiology reports.

Keywords: Assertion classification · Natural language processing ·
Biomedical texts · Deep learning · Transfer learning ·
Weakly supervised learning

1 Introduction

Consider the following excerpt from a radiology report:

INDICATION: Evaluate for *pneumonia*

What does this sentence say about the health of this patient?

A This patient has *pneumonia*
B This patient does not have *pneumonia*
C Nothing

© Springer Nature Switzerland AG 2019
I. V. Tetko et al. (Eds.): ICANN 2019, LNCS 11731, pp. 243–257, 2019.
https://doi.org/10.1007/978-3-030-30493-5_26

What about the following sentences?

IMPRESSION: No evidence of acute *pneumonia.*

FINDINGS: Developing *pneumonia* should be excluded by repeat image.

IMPRESSION: Effusions represent area of atelectasis, although *pneumonia* could also have this appearance.

This task[1] is known as *assertion classification* [1]. Given a sentence that mentions a certain condition, the classifier is tasked to discriminate between 3 types of assertions:

1. *positive* - the author states that the patient *has* the condition
2. *negative* - the author asserts that the patient does *not* have the condition
3. *speculative*[2] - the author mentions the condition, but does not assert anything as to whether the patient has it

This problem is an important bottleneck preventing efficient information extraction from clinical texts. Whenever one needs to turn a textual description of a patient's condition into a structured table summarizing the information known about said patient's health, the following pipeline is typically applied [2]:

1. Find all medical conditions mentioned in the report in any way.
2. Treat the report as a sequence of *assertions* where an *assertion* is defined as a sentence and a *concept of interest marker* that highlights the medical condition that will potentially (if the assertion about it is not speculative) end up in the summary. The examples above use *italics* to represent the *concept of interest marker.*
3. Classify assertions.
4. Produce two tables: conditions this patient is known to have and conditions this patient is known to not have.

Steps 2 and 4 are trivial and step 1 can be achieved by checking every word in a sentence against a comprehensive dictionary of medical terms such as UMLS [3], a method implemented in Apache Ctakes [4], QuickUMLS [5] and MetaMap [6].

Step 3, however, is an open problem [7] and an increasingly important one due to the advent of deep learning in medical image classification [8] and segmentation [9]. Most methods [10] leverage large datasets of medical images enriched with annotations to train deep convolutional neural networks that predict the annotation based on the image. The annotations take form of either a segmentation map for every image indicating which pixels belong to an organ or a pathology or simply one class per image, i.e. *healthy, pneumothorax* or *atelectasis.* Due to the nature of record-keeping in hospitals, these annotations often have to be

---

[1] It's C, B, C and probably C.
[2] Uzuner et. al. [1] refer to this type as *alter-assertion.*

derived from freetext reports written by medical professionals which requires a lot of manual labour or an efficient methodology for automated assertion classification. In particular, this is the case for MIMIC-CXR [11] and CheXPert [12]: newly available datasets of chest X-rays that are the largest to date and have the potential to enable a new wave of research in medical image analysis, such as DualNet [13].

For the assertion classification problem itself, training data is even scarcer. The only dataset of sentences annotated with *concept of interest markers* and *assertion classes* available to the scientific community, *2010 I2B2 Challenge* [7], contains 394 annotated reports and 877 unannotated reports for training. MIMIC-CXR, on the other hand, contains 206574 unannotated reports. In this paper, we propose a method for semi-supervised learning that can take advantage of large corpora, but only requires annotations for a small subset of them.

2   Related Work

2.1   Assertion Classification

Rule-Based Approaches. Most available solutions are (rather unsurprisingly, given the data deficit) rule-based and focus on detecting negation, but not speculation. NegEx [14] uses hand-crafted templates, *negation cues*, to detect situations where the *concept of interest* is negated. DEEPEN [15], Negtool [16] and NegBio [17], as well as similar systems for Russian [18] and Dutch [19] do the same, but employ dependency parsing with LinkParser [20], SPECIALIST [21] or Stanford Parser [22] and apply negation cues to the parse tree. See "Negation's Not Solved" [23] for a comprehensive comparison of rule-based negation detectors (Fig. 1).

```
{} <{dependency:/nmod:of|nmod:for/} ({lemma:/evidence/} >{dependency:/neg/} {})
```

Fig. 1. A sample negation cue from NegBio. Detects the phrase "No evidence of/for X"

Uzuner et al. [1] as well as Apache Ctakes [4] extend this approach to detecting speculative assertions. Others [24,25] formulate the task as detecting *negation and speculation* scopes: substring of the assertion invalidated by a negation or speculation cue.

The advantage of rule-based approaches is that they don't require training labels and, at least in theory, are immune to dataset bias [26]. However, NegBio [17] was designed with MIMIC-CXR in mind and our experiments show that it underperforms on the *i2b2* dataset and [12] noticed that it underperforms on CheXPert.

Statistical Approaches. Statistical approaches to assertion classification were largely enabled by *2010 I2B2 Challenge* [7]. The leaders of the challenge [27–29] represented assertions as vectors of complicated hand-crafted features like *lexical categories* of words in the sentence and even *"cue scope*: a feature that indicated the number of negation and speculation cue scopes that enclosed the concept in question." [28]. These features were used to train support vector machines [30]. So one can view this SVM as the final step in the pipeline which is mostly rule-based.

Unlike rule-based systems, statistical ones are prone to dataset bias: they are not guaranteed to work on a different dataset without retraining. In this case though even retraining these models on a different dataset can be problematic, since they use features hand-crafted for the *I2B2 Challenge*. Besides, as *I2B2 Challenge* didn't involve a lot of unannotated texts, the systems that were developed for it don't focus on harnessing unlabelled data in any way.

2.2   Weakly Supervised Learning

More generally, one can view assertion classification as the task of weakly supervised learning [31]. One can manually annotate some (but not all) assertions from the dataset and employ *learning with incomplete supervision*. Alternatively, one can use one of the rule-based classifiers, annotate the dataset with them and train the system on those annotations - *learning with inexact supervision*.

There is a family of methods collectively referred to as *semi-supervised learning* [32]: the core idea is to build a model of the distribution of the large unlabelled dataset X, either explicitly modelling it as a probability distribution (Bayesian methods) or implicitly via clustering or representation learning [33] and then use this model when training the model for $y|x$ where $x \in X$, $y \in Y$ and Y is the set of labels.

Active learning [34] reduces the necessary label subset by carefully selecting which data points from X to include in it. Annotation is done in epochs, after every epoch a model for $y|x$ is trained and the data points that the model finds hardest to classify (low confidence) are annotated in the next epoch. This is often done implicitly: several different models are trained [35] and their disagreement with each other is considered a sign of low confidence, but approaches that use a model with explicit confidence levels exist as well [36].

MentorNet [37] applies curriculum learning [38,39] to learning with inexact supervision. In curriculum learning, the loss function being minimized contains an extra term that indicates whether the model is supposed to pay attention to a certain part of the dataset at a certain stage of optimization. This curriculum term can be represented as a neural network as well and learned jointly with the main model. In our problem setting, where only a fraction of training labels is trustworthy, the networks can be jointly trained on this subset, then the curriculum network (MentorNet) is fixed and used to train the main model on inexact labels.

3 Methodology

Let A be a corpus of assertions. Every assertion is a 3-tuple $(s, i, c) \in A$ where

- s is a sentence
- $i = (\textbf{start}, \textbf{end})$ is a concept of interest marker indicating which part of the represents the object of the assertion. Assertion classification can be viewed as asking a question "Does this sentence affirm the concept of interest, negate it or neither?".
- c is some metadata about the context of the sentence. A human reader faced with a sentence always takes into account cues that they found before (or even after) the sentence - the context. We found it important for neural networks to take it into account as well.

There is an unknown ground truth function $t(s, i, c)$ that maps an assertion to its correct class c: *affirmative, negative* or *speculative* Two functions are known:

1. $t_a(s, i, c)$ is equal to $t(s, i, c)$ for all (s, i, c) in its domain A_t, but its domain is much smaller. A_t represent the assertions that have been manually annotated and can be exploited for machine learning. A_t is a small subset of A, for most assertions in A $(s, i, c) \in A$ it is undefined.
2. $h(s, i, c)$ is a mediocre approximation of $t(s, i, c)$, but it is defined for all (s, i, c) for which t is defined. It represents a rule-based algorithm for assertion classification.

There is also a practically unlimited corpus of unrelated texts T. Be it a literary corpus or Wikipedia, we can use unrelated texts to learn general patterns of English language.

Our goal is to use A, t_a and h to come up with $f(s, i, c)$ that approximates $t(s, i, c)$ as closely as possible. Or, in mathematical terms

$$\min_f L(f, t) \tag{1}$$

where L is some loss function that measures dissimilarity between functions f and t.

The naive solution would be to pick a parametric model $m_w(s, i, c)$ and optimize it's parameters w to mimic $t_a(s, i, c)$. However, since the domain of $t_a(s, i, c)$ is a small subset of A this approach skips a lot of valuable information that could be learned from A and T. To mitigate this problem we employ 2 flavours of transfer learning:

1. *Word vectors.* T can be used to derive compact representations of words and sentences and then use them as input to f. In particular, one can train a neural language model on T to predict a word based on its context and use intermediate representations of words produced by this neural network. We use FastText [40] word vectors trained on Wikipedia and ELMo [41] contextualized word representations trained on One Billion Words Benchmark [42].

2. *Heuristic annotation.* We can train neural network $m_w(s, i, c)$ in 2 steps: first optimize m_w to mimic $h(s, i, c)$ and use the obtained weights w as initialization for the second step: learning to mimic $t_a(s, i, c)$. Because $h(s, i, c)$ is defined for all $(s, i, c) \in A$, this way we utilize the entire dataset of assertions to find a good initialization that would help gradient descent to avoid local minima in the second step. Our results (see Sect. 5) also let us speculate that some neural networks $m_w(s, i, c)$ inherently represent a subspace of the space of possible functions $f(s, i, c)$ that contains functions similar to $t(s, i, c)$ and just training m_w on h is enough for it to outperform h.

3.1   Representations

Up to this point, we've been deliberately vague about which structures actually represent s, i and c to emphasize generality of the approach.

In our implementation, sentence s and concept of interest marker i are jointly represented as a sequence of word vectors v_i where $v_{1..n}$ is a FastText or ELMo representation of the word and v_{n+1} is 0 or 1 indicating whether the concept of interest includes this word.

Context c is represented as a one-hot encoded section label vector. Clinical texts (see Fig. 3 for an example) typically follow a specific structure with clearly delimited sections. Moreover, the sections are mostly the same in all the reports (*medical history, findings, impression*). The section where the assertion occurred can be very important for its interpretation so we extract them from the reports and represent as a one-hot encoded vector.

Assertion class $t(s, i, c)$ is represented as 2 numbers from $[0; 1]$ interval:

1. *Reality confidence.* Is the doctor asserting anything as to whether this patient has this pathology or not?
2. *Reality.* Is the doctor asserting the presence or the absence of this pathology in the patient?

Thus an *affirmative* assertion is $(1, 1)$, a *negative* one $(1, 0)$ and a *speculative* one is $(1, x)$ where x could be any number. We introduce a loss function that accounts for the fact that when reality confidence is low, predicted reality doesn't matter -*confidence adjusted mean squared error*:

$$\text{CMSE}(\text{TC}, \text{PC}, \text{TR}, \text{PR}) = (\text{TC} - \text{PC})^2 + \text{TC} * (\text{TR} - \text{PR})^2 \tag{2}$$

where TC stands for *true confidence*, PC - for *predicted confidence*, TR for *true reality* and PR for *predicted reality*.

This approach was chosen over one-hot encoding assertion classes (representing them with 3 numbers: *affirmative, negative* and *speculative*) because, intuitively, the task of assertion classification should break down into negation detection and speculation detection. This approach also lets us specify *confidence levels*: for some examples in A, $t_a(s, i, t)$ can be, sat, $(0.8, 1)$ if the author of the assertion expresses uncertainty.

3.2 Models

We use several models $m_w(s, i, c)$ that fall into 2 classes.

One is attention-based neural network inspired by [43] with 1, 3 or 8 (n) attention heads. Three 3×3 successive convolutional layers calculate n attention masks, every attention mask is a vector of the same length as the sentence. Every word vector is multiplied by its corresponding number between 0 and 1 in the attention mask and the results are summed up. Then n sums (for each attention mask) and the metadata vector are concatenated and fed into a feed-forward neural network.

The other model is based on LSTM [44] - a recurrent neural network with a hidden and cell state. It consists of 3 neural networks:

1. LSTM cell with hidden size of 8, 128, 512 or 1024
2. A feed-forward network with 2 hidden layers that maps the context vector c to the initial hidden and cell states of LSTM
3. A feed-forward network with input size 256 and hidden sizes 128, 32 and 8 that maps the final cell and hidden states to the desired output

3.3 Training Schedule

We compare 3 training schedules (Fig. 2):

1. *Classic.* Initialize w randomly, then minimize $L(m_w(s, i, c), t_a(s, i, c))$. This schedule is fast, but misses out on most data in A
2. *Deep Prior.* Initialize w randomly, minimize $L(m_w(s, i, c), h(s, i, c))$.
3. *Transfer.* Use the weights obtained with *Deep Prior* as initialization. Minimize $L(m_w(s, i, c), t_a(s, i, c))$.

Both neural models are trained using Adam optimizer [45] with CMSE as the loss function. We initialize models weights w randomly with Xavier initialization [46] and shuffle training batches every epoch. As a result, model training is not deterministic. To account for that and obtain robust results, we use *nested cross-validation* [47]. In *Classic* and *Transfer* schedules, the ground truth dataset is split into 3 folds then one by one each fold is used as the *testing* dataset while the rest is used as the *training-validation dataset*. The same procedure is then applied to the *training-validation dataset*: it is split into 3 folds and 3 models are trained: each model uses 1 fold as the *validation dataset* and the rest as the *training dataset* A_t. The purpose of the validation dataset is twofold[3]:

1. We use early stopping to abort training whenever the exponential moving average of accuracy on the validation dataset starts decreasing.
2. After 3 instances of a model (w_1, w_2 and w_3) are trained on a train-validation dataset, the best one (according to accuracy on the validation set) is picked as the finalist model. The results of 3 finalists are averaged to represent the quality of $m_w(s, i, c)$ generally

---

[3] pun intended.

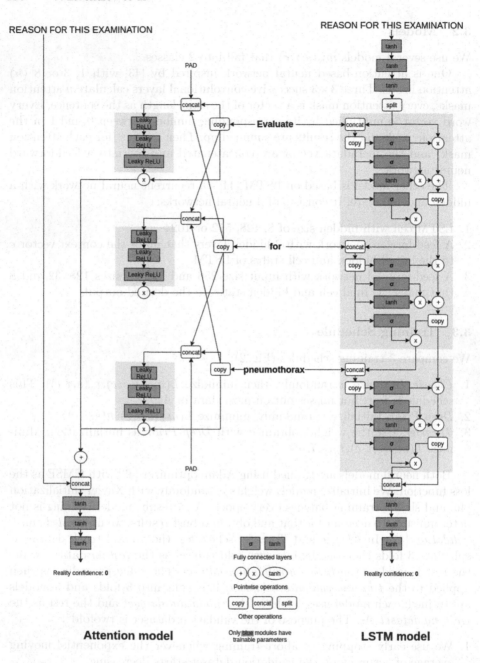

Fig. 2. Neural networks used

In *Deep Prior* schedule, A_t is not used for training, so it's entirety is used as a *testing-validation dataset*. Given that *Deep Prior* schedule takes the longest time (it goes over all assertions in A), we decided against cross-validating it.

4 Datasets

4.1 I2B2 Dataset

I2B2 Dataset was published as part of 2010 I2B2 Challenge [7] and includes 394 reports with assertions annotated as either *present, absent, possible, conditional, hypothetical* or *referring to another patient* as well as 877 unannotated reports for training and a holdout testing set of 256 reports.

In terms of our methodology, *conditional, hypothetical, potential* and *referring to another patient* assertions are *speculative*, *present* means *affirmative* and *absent* means *negative*.

Since this dataset doesn't contain a lot of unannotated reports, it isn't a good fit for Deep Text Prior. However, since this is the golden standard in assertion classification and in order to check for dataset bias [26], we did test our methodology on it.

Apache Ctakes [4] was used as $h(s, i, c)$ and the provided annotations were used as $t_a(s, i, c)$.

4.2 MIMIC-CXR Dataset

The main dataset this work is validated on is MIMIC-CXR [11]: the largest chest X-ray dataset available to date. In addition to X-ray images, MIMIC-CXR contains text descriptions in the form of radiology reports by medical professionals. It contains almost 70000 high quality texts (no misspellings and artifacts like OCR errors sometimes found in text corpora) of similar structure: sections like "Indication", "Findings" and "Impression". Not every section is present in every report, but there is a limited superset that all sections fall into. Moreover, sections are very clearly delimited: see Fig. 3.

```
[**Hospital 9**] MEDICAL CONDITION:
  64 year old immunocompromised women with persistent cough/SOB and fluid
  overload
REASON FOR THIS EXAMINATION:
  ?pna, pleural effusions
```

```
                              FINAL REPORT
CHEST RADIOGRAPH

INDICATION:  Immunocompromised woman, shortness of breath.

COMPARISON:  [**2192-12-8**].

FINDINGS:  As compared to the previous radiograph, there is no relevant
change.  The lung volumes have increased.  The monitoring and support devices
are all unchanged.  Unchanged scarring at the left and right lung bases but no
newly appeared parenchymal opacity.  Unchanged size of the cardiac silhouette.
```

Fig. 3. A sample radiology report from MIMIC-CXR

We obtain an assertion dataset A by

- Splitting the report into sections and extracting section titles c with several convoluted regular expressions.
- Splitting each section into sentences s
- Finding all pathologies in said sentence with QuickUMLS [5] and defining a respective concept of interest marker i.

We obtained 2 ground truth assertion datasets A_{t1} and A_{t2} by manually annotating assertions from A making heavy use of confidence levels (i.e reality confidence that isn't 0 or 1). We noticed that a lot of assertions in MIMIC-CXR are identical (sentence "*No pneumothorax*" occurs over 1000 times). Hence as A_{t1} (MIMIC-CXR-FREQ) we annotated 300 of the most common sentences in A to capture the distribution of sentences in the dataset as closely as possible. However, sentence length has been shown ot inversely correlate with sentence frequency [48] and MIMIC-CXR is no exception: the most frequent sentences in the dataset are the shortest ones. This means that A_{t1} contains the simplest assertions to classify and is a bad benchmark for a classifier since the most challenging samples are missing. To mitigate this, we also annotated 100 of the longest sentences in A (A_{t2}, MIMIC-CXR-LONG).

The state of the art system currently used to classify assertions in MIMIC-CXR is *NegBio* [17]. Unfortunately, NegBio only detects negation, but not speculation, so we introduced an expanded version: *NegBio+* that, in addition to NegBio's built-in rules, applied one more heuristic: if the sentence occurs in "history", "impression" or "medical condition" section, the assertion is considered to be real, i.e. the author is actually describing the state of the patient. If the section title is "indication" or "reason for this examination", it is considered speculative as these sections typically list what pathologies the patient has to be tested for, not the ones she has. This is not strictly true: there are, for instance, affirmative assertions in "Indication", but it's a good first-order approximation. *NegBio+* was used as $h(s, i, c)$ to train the models.

5   Results

See Tables 1, 2 and 3 for model accuracy. See appendix for f1 scores.

Based on these results we argue that the *attention model with 8 attention heads* over ELMo word vectors pretrained on heuristic annotations and fine-tuned on ground truth is currently the best tool for weakly supervised assertion classification. It outperforms everything else for the most challenging sentences in the dataset (*MIMIC-CXR-LONG*) and is one of the best models for the simple sentences (*MIMIC-CXR-FREQ*). It lags behind on the *I2B2 challenge*, but that is to be expected since this model requires a lot of pretraining data.

We demonstrate that pretraining models on heuristic annotations before they are trained on ground truth can help avoid local minima and greatly improve model accuracy (see Table 4). This is especially true for smaller models (with a low number of trainable parameters) like `lstm8` and `attention2`.

Table 1. Accuracy on I2B2 Challenge

Vectors	Model	ctakes	0.796	0.796	0.796
			Classic	Deep Prior	Transfer
elmo	attention1		0.625	0.704	0.756
	attention8		0.621	0.706	0.738
	lstm1024		0.798	0.755	0.866
	lstm128		0.846	0.747	0.868
	lstm512		**0.858**	0.733	**0.858**
	lstm8		0.621	0.624	0.673

Table 2. Accuracy on MIMIC-CXR-FREQ

Vectors	Model	NegBio+	0.834	0.834	0.834
			Classic	Deep Prior	Transfer
elmo	attention1		0.615	0.803	0.932
	attention3		0.863	0.225	0.950
	attention8		0.919	0.576	0.944
	lstm8		0.639	0.878	0.898
	lstm128		0.927	0.873	0.967
	lstm512		0.912	0.873	**0.975**
	lstm1024		0.785	0.876	0.939
fasttext	attention1		0.610	0.429	0.870
	attention3		0.944	0.325	0.773
	attention8		0.778	0.441	0.838
	lstm8		0.276	0.388	0.058
	lstm128		0.705	0.914	0.929
	lstm512		0.748	0.381	0.841
	lstm1024		0.578	0.309	0.796

Table 3. Accuracy on MIMIC-CXR-LONG

Vectors	Model	NegBio+	0.71	0.71	0.71
			Classic	Deep Prior	Transfer
elmo	attention3		0.800	0.770	0.683
	attention8		0.833	0.710	**0.843**
	lstm128		0.716	0.690	0.750
	lstm512		0.763	0.710	0.691
	lstm8		0.721	0.700	0.821
fasttext	attention3		0.722	0.680	0.821
	attention8		0.810	0.750	0.739
	lstm128		0.686	0.690	0.722
	lstm512		0.862	0.700	0.788
	lstm8		0.830	0.690	0.694

Table 4. Heuristic pretraining effect on MIMIC-CXR-FREQ

Vectors	Model	Classic	Transfer
elmo	attention1	0.615	+0.317
	attention3	0.863	+0.097
	attention8	0.919	+0.025
	lstm8	0.639	+0.249
	lstm128	0.927	+0.040
	lstm512	0.912	+0.063
	lstm1024	0.785	+0.154
fasttext	attention1	0.610	+0.260
	attention8	0.778	+0.050
	lstm512	0.748	+0.097
	lstm1024	0.578	+0.218

Another tool responsible for the demonstrated results is word vectors that include contextual information. Table 5 demonstrates the advantage of ELMo over Fasttext approach that represents words with the same vectors irregardless of the context.

Finally, we can see (Table 6) that LSTM over ELMo models demonstrate an interesting effect: when they are optimized to mimic the heuristic engine $h(s, i, c)$, they end up better at classifying ground truth assertions than the engine itself. This suggests that they are effective regularizers: they inherently represent a subspace of the solution space that's more likely to be contain good solutions. This is consistent with Deep Image Prior [49], a similar result in Computer Vision.

This also suggests a generalization of the method beyond assertion classification in medical texts: any problem that is usually solved via rule-based systems due to lack of training labels can be tackled by

Table 5. Contextualized embeddings effect on MIMIC-CXR-FREQ

Model	Schedule	fasttext	elmo
lstm8	Classic	0.276	+0.363
	Deep Prior	0.388	+0.490
	Transfer	0.058	+0.848
lstm512	Classic	0.748	+0.164
	Deep Prior	0.381	+0.492
	Transfer	0.841	+0.134
lstm1024	Classic	0.578	+0.207
	Deep Prior	0.309	+0.567
	Transfer	0.796	+0.143
attention1	Classic	0.610	+0.164
	Deep Prior	0.429	+0.374
	Transfer	0.870	+0.062

Table 6. Deep prior effect on MIMIC-CXR-FREQ

Vectors	Model	Advantage over NegBio+
elmo	lstm8	+0.044
	lstm128	+0.039
	lstm512	+0.039
	lstm1024	+0.042

1. Generating labels with the best system currently available
2. Pre-training a parametrized model on those labels
3. Fine-tuning it on a small dataset of true labels

Acknowledgments. The authors would like to acknowledge Artem Shelmanov and Ilya Sochenkov for sharing their expertise in natural language processing, mentorship and support.

Appendix: F1 Scores

Throughout the paper, we use accuracy as the metric for our results. For completeness sake, micro-averaged f1 scores are attached here (Tables 7, 8 and 9).

Table 7. F1 scores on MIMIC-CXR-FREQ

	NegBio+	0.808	0.808	0.808
Vectors	Model	Classic	Deep Prior	Transfer
elmo	attention1	0.623	0.788	0.905
	attention3	0.864	0.474	0.927
	attention8	0.870	0.690	0.903
	lstm1024	0.651	0.816	0.939
	lstm128	0.918	0.853	0.903
	lstm512	0.877	0.853	0.951
	lstm8	0.465	0.857	0.790
fasttext	attention1	0.590	0.573	0.762
	attention3	0.493	0.519	0.682
	attention8	0.783	0.591	0.643
	lstm1024	0.611	0.483	0.765
	lstm128	0.609	0.860	0.885
	lstm512	0.747	0.530	0.804
	lstm8	0.302	0.546	0.480

Table 8. F1 scores on MIMIC-CXR-LONG

	NegBio+	0.612	0.612	0.612
Vectors	Model	Classic	Deep Prior	Transfer
elmo	attention3	0.471	0.410	0.448
	attention8	0.625	0.519	0.628
	lstm128	0.414	0.454	0.436
	lstm512	0.517	0.437	0.576
	lstm8	0.591	0.445	0.655
fasttext	attention3	0.456	0.448	0.621
	attention8	0.618	0.414	0.459
	lstm128	0.418	0.440	0.420
	lstm512	0.562	0.445	0.469
	lstm8	0.624	0.440	0.384

Table 9. F1 scores on I2B2 Challenge

Vectors	Model	Classic	Deep Prior	Transfer
	ctakes	0.678	0.678	0.678
elmo	attention3	0.471	0.410	0.448
	attention8	0.625	0.519	0.628
	lstm128	0.414	0.454	0.436
	lstm512	0.517	0.437	0.576
	lstm8	0.591	0.445	0.655
fasttext	attention3	0.456	0.448	0.621
	attention8	0.618	0.414	0.459
	lstm128	0.418	0.440	0.420
	lstm512	0.562	0.445	0.469
	lstm8	0.624	0.440	0.384

References

1. Uzuner, Ö., Zhang, X., Sibanda, T.: Machine learning and rule-based approaches to assertion classification. J. Am. Med. Inform. Assoc. **16**(1), 109–115 (2009)
2. Goff, D.J., Loehfelm, T.W.: Automated radiology report summarization using an open-source natural language processing pipeline. J. Digit. Imaging **31**(2), 185–192 (2018)
3. Bodenreider, O.: The unified medical language system (UMLS): integrating biomedical terminology. Nucleic Acids Res. **32**(suppl-1), D267–D270 (2004)
4. Chute, C.G., et al.: Mayo clinical text analysis and knowledge extraction system (cTAKES): architecture, component evaluation and applications. J. Am. Med. Inform. Assoc. **17**(5), 507–513 (2010). https://doi.org/10.1136/jamia.2009.001560
5. Soldaini, L., Goharian, N.: Quickumls: a fast, unsupervised approach for medical concept extraction. In: MedIR Workshop, sigir (2016)
6. Aronson, A.R.: Effective mapping of biomedical text to the UMLS Metathesaurus: the MetaMap program. In: Proceedings of the AMIA Symposium, p. 17. American Medical Informatics Association (2001)
7. Uzuner, Ö., South, B.R., Shen, S., DuVall, S.L.: 2010 i2b2/va challenge on concepts, assertions, and relations in clinical text. J. Am. Med. Inform. Assoc. **18**(5), 552–556 (2011)
8. Miranda, E., Aryuni, M., Irwansyah, E.: A survey of medical image classification techniques. In: 2016 International Conference on Information Management and Technology (ICIMTech), pp. 56–61, November 2016.https://doi.org/10.1109/ICIMTech.2016.7930302
9. Lai, M.: Deep learning for medical image segmentation. arXiv preprint arXiv:1505.02000 (2015)
10. Litjens, G., et al.: A survey on deep learning in medical image analysis. Med. Image Anal. **42**, 60–88 (2017)
11. Johnson, A.E., et al.: MIMIC-CXR: a large publicly available database of labeled chest radiographs. arXiv preprint arXiv:1901.07042 (2019)

12. Irvin, J., et al.: CheXpert: a large chest radiograph dataset with uncertainty labels and expert comparison. arXiv preprint arXiv:1901.07031 (2019)
13. Rubin, J., Sanghavi, D., Zhao, C., Lee, K., Qadir, A., Xu-Wilson, M.: Large scale automated reading of frontal and lateral chest x-rays using dual convolutional neural networks. arXiv preprint arXiv:1804.07839 (2018)
14. Chapman, W.W., Bridewell, W., Hanbury, P., Cooper, G.F., Buchanan, B.G.: A simple algorithm for identifying negated findings and diseases in discharge summaries. J. Biomed. Inform. **34**(5), 301–310 (2001)
15. Mehrabi, S., et al.: DEEPEN: a negation detection system for clinical text incorporating dependency relation into NegEx. J. Biomed. Inform. **54**, 213–219 (2015)
16. Enger, M., Velldal, E., Øvrelid, L.: An open-source tool for negation detection: a maximum-margin approach. In: Proceedings of the Workshop Computational Semantics Beyond Events and Roles, pp. 64–69 (2017)
17. Peng, Y., Wang, X., Lu, L., Bagheri, M., Summers, R.M., Lu, Z.: NegBio: a high-performance tool for negation and uncertainty detection in radiology reports. CoRR abs/1712.05898 (2017). http://arxiv.org/abs/1712.05898
18. Shelmanov, A., Smirnov, I., Vishneva, E.: Information extraction from clinical texts in Russian. In: Computational Linguistics and Intellectual Technologies: Papers from the Annual International Conference "Dialogue", vol. 14, pp. 537–549 (2015)
19. Afzal, Z., Pons, E., Kang, N., Sturkenboom, M.C., Schuemie, M.J., Kors, J.A.: ContextD: an algorithm to identify contextual properties of medical terms in a Dutch clinical corpus. BMC Bioinform. **15**(1), 373 (2014)
20. Sleator, D.D., Temperley, D.: Parsing English with a link grammar. arXiv preprint cmp-lg/9508004 (1995)
21. McCray, A.T., Srinivasan, S., Browne, A.C.: Lexical methods for managing variation in biomedical terminologies. In: Proceedings of the Annual Symposium on Computer Application in Medical Care, p. 235. American Medical Informatics Association (1994)
22. Chen, D., Manning, C.: A fast and accurate dependency parser using neural networks. In: Proceedings of the 2014 Conference on Empirical Methods in Natural Language Processing (EMNLP), pp. 740–750 (2014)
23. Wu, S., et al.: Negation's not solved: generalizability versus optimizability in clinical natural language processing. PLoS One **9**(11), e112774 (2014)
24. Apostolova, E., Tomuro, N., Demner-Fushman, D.: Automatic extraction of lexico-syntactic patterns for detection of negation and speculation scopes. In: Proceedings of the 49th Annual Meeting of the Association for Computational Linguistics: Human Language Technologies: Short Papers-Volume 2, pp. 283–287. Association for Computational Linguistics (2011)
25. Zou, B., Zhou, G., Zhu, Q.: Tree kernel-based negation and speculation scope detection with structured syntactic parse features. In: Proceedings of the 2013 Conference on Empirical Methods in Natural Language Processing, pp. 968–976 (2013)
26. Torralba, A., Efros, A.A.: Unbiased look at dataset bias (2011)
27. de Bruijn, B., Cherry, C., Kiritchenko, S., Martin, J., Zhu, X.: NRC at i2b2: one challenge, three practical tasks, nine statistical systems, hundreds of clinical records, millions of useful features
28. Clark, C., et al.: Determining assertion status for medical problems in clinical records

29. Demner-Fushman, D., Apostolova, E., Islamaj Dogan, R., et al.: NLM's system description for the fourth i2b2/va challenge. In: Proceedings of the 2010 i2b2/VA Workshop on Challenges in Natural Language Processing for Clinical Data, Boston, MA, USA: i2b2 (2010)
30. Cortes, C., Vapnik, V.: Support-vector networks. Mach. Learn. **20**(3), 273–297 (1995)
31. Zhou, Z.H.: A brief introduction to weakly supervised learning. Natl. Sci. Rev. **5**(1), 44–53 (2017)
32. Olivier Chapelle, B.S., Zien, A.: Semi-Supervised Learning. Adaptive Computation and Machine Learning Series. MIT Press, Cambridge (2010)
33. Bengio, Y., Courville, A., Vincent, P.: Representation learning: a review and new perspectives. IEEE Trans. Pattern Anal. Mach. Intell. **35**(8), 1798–1828 (2013)
34. Settles, B.: Active learning. Synth. Lect. Artif. Intell. Mach. Learn. **6**(1), 1–114 (2012)
35. Hanneke, S., et al.: Theory of disagreement-based active learning. Found. Trends® Mach. Learn. **7**(2–3), 131–309 (2014)
36. Zhang, C., Chaudhuri, K.: Beyond disagreement-based agnostic active learning. In: Advances in Neural Information Processing Systems, pp. 442–450 (2014)
37. Jiang, L., Zhou, Z., Leung, T., Li, L.J., Fei-Fei, L.: MentorNet: learning data-driven curriculum for very deep neural networks on corrupted labels. arXiv preprint arXiv:1712.05055 (2017)
38. Kumar, M.P., Packer, B., Koller, D.: Self-paced learning for latent variable models. In: Advances in Neural Information Processing Systems, pp. 1189–1197 (2010)
39. Jiang, L., Meng, D., Zhao, Q., Shan, S., Hauptmann, A.G.: Self-paced curriculum learning. In: Twenty-Ninth AAAI Conference on Artificial Intelligence (2015)
40. Bojanowski, P., Grave, E., Joulin, A., Mikolov, T.: Enriching word vectors with subword information. CoRR abs/1607.04606 (2016). http://arxiv.org/abs/1607.04606
41. Peters, M.E., et al.: Deep contextualized word representations. arXiv preprint arXiv:1802.05365 (2018)
42. Chelba, C., et al.: One billion word benchmark for measuring progress in statistical language modeling. arXiv preprint arXiv:1312.3005 (2013)
43. Vaswani, A., et al.: Attention is all you need. CoRR abs/1706.03762 (2017). http://arxiv.org/abs/1706.03762
44. Hochreiter, S., Schmidhuber, J.: Long short-term memory. Neural Comput. **9**(8), 1735–1780 (1997)
45. Kingma, D.P., Ba, J.: Adam: a method for stochastic optimization. CoRR abs/1412.6980 (2014). http://arxiv.org/abs/1412.6980
46. Glorot, X., Bengio, Y.: Understanding the difficulty of training deep feedforward neural networks. In: Proceedings of the Thirteenth International Conference on Artificial Intelligence and Statistics, pp. 249–256 (2010)
47. Varma, S., Simon, R.: Bias in error estimation when using cross-validation for model selection. BMC Bioinform. **7**(1), 91 (2006)
48. Sigurd, B., Eeg-Olofsson, M., Van De Weijer, J.: Word length, sentence length and frequency - Zipf revisited. Studia Linguistica **58**(1), 37–52 (2004). https://doi.org/10.1111/j.0039-3193.2004.00109.x
49. Ulyanov, D., Vedaldi, A., Lempitsky, V.: Deep image prior. In: Proceedings of the IEEE International Conference on Computer Vision and Pattern Recognition (2018)

Inter-region Synchronization Analysis Based on Heterogeneous Matrix Similarity Measurement

Hengjin Ke[1] , Dan Chen[1(✉)] , Lei Zhang[1], XinHua Zhang[2], Xianzeng Liu[2], and Xiaoli Li[3]

[1] School of Computer Science, Wuhan University, Wuhan 430072, China
`dan.chen@whu.edu.cn`
[2] Neurology, Peking University People's Hospital, Beijing 100044, China
[3] State Key Laboratory of Cognitive Neuroscience and Learning and IDG/McGovern Institute for Brain Research, Beijing Normal University, Beijing 100875, China

Abstract. EEG synchronization is an essential tool to understand mechanisms of communication between brain regions. Despite numerous successes along this direction, grand challenges still remain: (1) to establish the relation between treatment outcome and the synchronization patterns amongst brain regions and (2) to correctly quantify the synchronization amongst brain regions with different electrode placement topologies. As for this problem, we propose an approach to inter-region synchronization analysis based on Heterogeneous Matrix Similarity Measurement to avoid information loss of the results on similarity. It is measured by deriving the bridge matrix that quantifies the distance from the source matrix to the target one both of arbitrary dimensions. The similarity measures are then used to examine the relation between brain region interactions (directional) and treatment outcome. Experiments have been performed on EEG collected from patients received anti-depression treatment with both non-effective and effective outcomes. Experimental results indicate that for the non-effective group, 97% (95%) *similarity* values from **right temporal** to central (from **left temporal** to central) are greater than the average *similarity* value of all patients; The new correlation measures can quantify the directional synchronization between brain regions and form an indicator for anti-depression treatment outcome with a high-level confidence. The approach also holds potentials in sophisticated applications involving clustering and dimensionality reduction.

Keywords: Synchronization pattern analysis · Brain region · Matrix Similarity Measurement · EEG · Anti-depression

1 Introduction

Neural synchronization dynamics embedded in multivariate EEG is explicitly related to brain disorders and the corresponding treatment outcomes, such as

© Springer Nature Switzerland AG 2019
I. V. Tetko et al. (Eds.): ICANN 2019, LNCS 11731, pp. 258–272, 2019.
https://doi.org/10.1007/978-3-030-30493-5_27

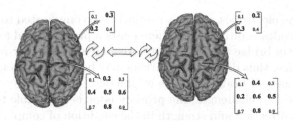

Fig. 1. Inter-region correlation analysis of heterogeneous matrices (regions). Most existing approaches give the same value when measuring the regions with **different distributions** (dotted in bold style).

major depressive disorder (MMD) and anti-depression treatment [1]. These can be characterized by particular synchronization patterns amongst functional brain regions [13], i.e., how intensively these are coupled and which (driver) drives the other(s) (response). Numerous methods have emerged for the synchronization measurement of bivariate EEG typically including spectrum-based coherence and mutual information (MI) [6]. Coherence is linear and not capable enough of characterizing the non-linear dynamics of EEG. MI is particularly salient for discrimination and robustness to noises [4]. Maximal Information Coefficient (MIC) has developed as the best bivariate synchronization measurement for analyses [20] in terms of non-linearity and robustness to noise (Fig. 1).

Synchronization manifests a key feature for establishing communication between different regions of the brain at various scales. Numerous attempts have been performed to quantify the relation between treatment outcomes and the synchronization patterns including time-frequency analysis [9], information theory [18], and the dominant complex network methods [3]. The latter methods exploit the complex network theory to treat EEG channels and the paired connections amongst them as functional connectivity matrices [21,24]. It can successfully discriminate the ongoing EEG of healthy individuals and those of subjects with brain disorders [15].

The above theories and methods largely ignore the topology of the connectivity matrices, i.e., the distribution of correlation pattern in the virtual space. This loss of structural information of the inter-region correlation may lead to severe distortion of the analysis. For example, data elements in two matrices can have the same values but different distributions (see the below), the two matrices denoting totally different dynamics will be treated as the same by existing methods:

Connectivity matrices corresponding to brain regions are naturally heterogeneous to each other as the number of electrodes placed within different brain regions is indefinite for commonly used EEG systems, such as the international 10–20 system and Geodesic Sensor Net. Existing matrix-based synchronization measurement approaches only apply to **isomorphic matrices**, i.e., they can be flattened as vectors of the same size. Furthermore, the common objective of synchronization measurement is to construct a network amongst all regions corresponding to the electrodes, which is routinely too complicated to explain as the number of electrodes is often large. There exists a pressing need for an approach that can explicitly bridge the treatment outcomes and the inter-region synchronization dynamics, thus distinct biomarkers may be available to understand the treatment outcomes of brain disorders in practice.

To tackle the two challenges, the primary issue is to be able to measure the synchronization direction and strength in the condition of complicated topology. The measurement should emphasize asymmetry, and the permutation conditional mutual information is such an example [18]. This study develops a new "inter-region" synchronization analysis approach based on **similarity measurement** (see Sect. 3).

Experiments have been performed to validate the effectiveness of Heterogeneous Matrix Similarity Measurement (HMSM) (Sect. 4.1) by measuring the similarity of images with various scales and layouts. After that, a case study has been performed using the similarity measures to examine the relation between brain region interactions and treatment outcome (Sect. 4). The case study examined an EEG dataset of 17 patients with depression (authorization attained from the Peking University People's Hospital). The measures manifest an appropriate biomarker for its statistically significant capability to distinct the effective group from the non-effective group.

The main contributions of this study include:

1. A topology-sensitive method has been developed to quantify the correlations of different regions of a complex system. The method gains this capability by measuring the similarity between matrices of arbitrary dimensions. The method significantly outperforms the deep learning counterparts.
2. The solution proved useful to reveal the relation between the treatment outcomes and brain region dynamics. It exhibits a biomarker for the excellent performance in treatment outcome evaluation for anti-depression treatment.

2   Related Work

This study focuses on the similar measurement of matrices denoting inter-region correlations. Existing methods are mainly constructed based on **information entropy** and **graph theory**. To further understand functional connectivity in the brain, Li et al. presents a methodology based on permutation analysis and conditional mutual information to identify the coupling direction between neuronal signals recorded from different brain areas [18]. To examine the neurophysiologic connectivity measures, Leuchter et al. proposed a weighted network analysis to examine resting state functional connectivity and achieved 81% accuracy on average [17]. Based on the resulted correlation measures, patients were clustered with the development of diagnostic classifiers (biomarkers) of high (82–93%) sensitivity and specificity for depression subtypes [8]. To enable accurate and early diagnosis of depression, Mumtaz proposed a machine learning framework using synchronization likelihood (SL) features as input. It can discriminate MDD patients and healthy controls with high accuracy, sensitivity,

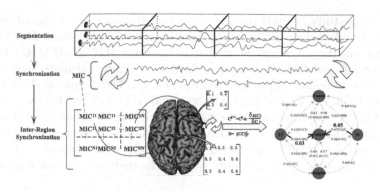

Fig. 2. Overview of the proposed approach.

specificity, and f-measure [19]. Recently, deep learning methods have boomed. Dosovitskiy [7] proposed a deep perceptual similarity metrics to generate high-resolution images from compressed abstract representations. Instead of computing distances in the image space, the features of the image were first extracted by deep autoencoder and then Euclidean distance was computed. This metric reflects the perceptual similarity of images much better on two examples. The graph theory approach calculates the similarity in terms of node degrees, clustering coefficients, and distances. However, it relies on linear mathematical operations such as sums, which neglect the influence of spatial information. In particular, it obtained the same value when measuring the regions with **different distributions** in the case of a fixed partition, such as in this paper. The same situation for information entropy because it also relies on probability calculation (counting).

Most existing approaches to similarity measurement could effectively measure the similarity between heterogeneous matrices with different values. However, they all lack the capability of topology measures. In contrast to the existing work, we aim to find a solution with the capability of quantifying the similarity of square matrices with arbitrary dimensions with topological features highlighted.

3   Inter-region Synchronization Pattern Analysis Based on Heterogeneous Matrix Similarity Measurement

The proposed approach operates in three phases (see Fig. 2): (1) Calculation of the Correlation Matrix based on Maximal Information Coefficient ($CMMIC$), (2) **Similarity measurement** of heterogeneous synchronization matrices. After that, the inter-region synchronization measures can be obtained, which forms the basis of treatment outcome evaluation.

3.1   Maximal Information Coefficient and Extension

MIC is the best bivariate synchronization measurement for analyses in terms of nonlinearity and robustness to noises [20]. In view of this desirable feature of MIC, this study extends the MIC measure to quantify the global synchronization of multivariate EEG per brain region, which combines MIC with a correlation matrix, i.e., Correlation matrix based on MIC ($CMMIC$). Given a 10–20 international EEG system, five $CMMIC$s will be calculated corresponding to five brain regions (Frontal: Fp1, Fp2, F3, F4; Left temporal (LT): F7, T3, T5; Central: C3, C4, Fz, Cz, Pz; Right temporal (RT): F8, T4, T6; Occipital: P3, P4, O1, O2). A $CMMIC$ can be formalized as the matrix where each element in the matrix $MIC_{ij}(i, j = 1, ..., n)$ denotes the synchronization strength between channel i and j. CMMIC can express the MIC value of all channels in the whole brain, and can also express the MIC value of channels in a certain region.

3.2 Theory and Design

This subsection first presents the theory of Heterogeneous Matrix Similarity Measurement (HMSM) (Sect. 3.2) followed by the convergence proof (Sect. 3.2). The theory of HMSM is proposed (Sect. 3.2).

Theories of Heterogeneous Matrix Similarity Measurement. In this study, the HMSM forms the theoretic basis of analysis of inter-region synchronization pattern. In particular, the inter-region synchronization can be quantified by similarity measure between two $CMMIC$ matrices (X and Y) corresponding to a pair of brain regions. These matrices are routinely heterogeneous with different dimensions. To the best of our knowledge, existing methods for similarity measurement of matrices will not apply by their nature. One may tentatively reshape or tailor the matrices to have the same dimensions first, but this will inevitably lead to information loss and the reliability of the results cannot be guaranteed. To address this problem, the HMSM aims to measure similarity between two heterogeneous matrices directly.

The formal definitions for similarity measurement calculation of two heterogeneous matrices are as follows:

Definition 1 *Heterogeneous Matrix Similarity Measurement (HMSM):*
Two heterogeneous square matrices X and Y, to find a similarity coefficient matrix C such that:

$$Y \approx CXC^T, s.t. \, X, Y \neq 0 \tag{1}$$

where X (source) and Y (target) are both non-zero square matrices with arbitrary dimensions $s(x)$ and $s(y)$ respectively; The coefficient matrix C has $s(y)$ rows and $s(x)$ columns; HMSM denotes the similarity of the heterogeneous matrices directly.

Definition 2 *Direction: The direction $X \rightarrow Y$ is defined as from X to Y if they satisfy Definition 1.*

Next, the coefficient matrix C is derived via the HMSM algorithm iteratively. C is updated at each iteration based on the latest C and the inputs X and Y in accordance with Eq. 1. The cost function should be defined first to quantify the quality of the approximation of C. It can be constructed considering the similarity between the residual error of Y and CXC^T:

$$E = |Y - CXC^T|^2 \qquad (2)$$

Then the optimization problem becomes:

- **Optimization Problem.** Minimize $|Y - CXC^T|^2$ with respect to C.

The above equations (e.g., Eq. 1) explicitly involve non-linear relations amongst indefinite matrices, which determine that C cannot be directly solved through linear matrix operations, i.e., there is no analytical solution to C but only approximation solution. This study adopts Gradient Descent (GD) for solving C as it (1) guarantees convergence [2] and (2) needs no analytical solution demanded by counterparts such as the Least Square methods. The coefficient matrix C can then be derived as follows:

First, normal distribution applies to the residual error E to keep generality, and the maximum likelihood function is defined as:

$$L(C) = \prod_{i,j} \frac{1}{\sqrt{2\pi}\sigma_{ij}} exp\left(-\frac{E_{ij}^2}{2\sigma_{ij}}\right) \qquad (3)$$

After Logarithm of both sides ($\propto$ denotes removing constant term):

$$lnL(C) \propto \sum_{ij} ln\frac{1}{\sqrt{2\pi}\sigma_{ij}} - \sum_{ij} \frac{1}{2\sigma_{ij}}[Y - CXC^T]^2 \qquad (4)$$

To maximize the Logarithm function in Eq. 4, the objective function should be minimized with assuming that the variance of noise at each data point is the same:

$$J(C) = \frac{1}{2}tr[Y - CXC^T][Y - CXC^T]^T \qquad (5)$$

where $tr(.)$ denotes the trace of a matrix.

Solve the optimization problem in GD approach:

$$\frac{\partial J(C)}{\partial C} = Y^T CX^T + YCX + CXC^T[XC^T + X^T C^T]^T \qquad (6)$$

Then the coefficient matrix C is updated with learning rate lr:

$$C^{(g+1)} = C^{(g)} - lr\frac{\partial J(C)}{\partial C} \qquad (7)$$

where g is the index of the iteration; C is said solved once the threshold of residual error being reached.

Finally, the distance between the source and the target matrices ($CMMICs$ in the context of this study) is calculated as:

$$D = \|CC^T\|_{\mathcal{F}}^2 = \frac{1}{s(CC^T)}\sqrt{tr(CC^T * CC^T)}$$
$$= \frac{1}{s(y)}\sqrt{tr(CC^T * CC^T)} \tag{8}$$

where $tr(.)$ is the trace of a matrix; Coefficient $\frac{1}{s(y)}$ is calculated to eliminate the influence incurred by the different numbers of elements of the source and the target matrices; The normalization (denoted as $\mathcal{F} - Norm$) is designed slightly different from the conventional Frobenius norm; Although the Heterogeneous Matrix Similarity Measure (D) is different from the conventional similarity between isomorphic matrices as it is not symmetrical, it manifests a new type of similarity because of the property of directional inverse. Given that C is the coefficient matrix of $X \to Y$, $C^\dagger$ is the coefficient matrix of $Y \to X$, and vice versa; where $C^\dagger$ is the Moore–Penrose inverse of C:

$$Y = CXC^T \Leftrightarrow C^\dagger Y (C^T)^\dagger = C^\dagger CXC^T (C^T)^\dagger = X. \tag{9}$$

The synchronization strength is defined as follows, which is negatively correlated with the distance between the source and target matrices (D):

$$sync(D) = \frac{2e^{-D}}{1 + e^{-D}} = \frac{2}{1 + e^D} \ (s.t \ D \geq 0) \tag{10}$$

Obviously, Eq. 10 defines a monotone decreasing function and the value range of $sync(D)$ is $(0, 1]$, which can be seen as the synchronization measure between two matrices.

Design of the Inter-region Synchronization Analysis. Brain can be partitioned into five regions by the 10–20 international system. In this study, the HMSM applies in inter-region synchronization analysis. Given two heterogeneous $CMMICs$ corresponding to two different brain regions (X for the source and Y for the target), the synchronization measure from the source to the target is then derived by calculating the coefficient matrix C (Eq. 8). The Algorithm 1 is designed to implement the HMSM theory.

Initialization is extremely important to the performance and correctness (trapped in local minimum or not) in multi-way analyses approaches [23], and the derivation of the coefficient matrix C is exactly the case. In this study, the calculation of C intensively relies on the two heterogeneous matrices (X and Y), which makes this task even more challenging. An initialization method is designed for this purpose, namely Random Mean Deviation Initialization ($RMDI$), which takes the two matrices into account at the same time.

Suppose $\bar{x}$ and $\bar{y}$ are the mean values of the source(target) matrix $X(Y)$, the initial coefficient matrix (ICM) applies the norm distribution with the mean value $\bar{y} - \bar{x}$ and variance σ^2 ($\sigma = 1$ for the boundness of MIC):

$$ICM \sim N(\bar{y} - \bar{x}, \sigma^2) \tag{11}$$

Algorithm 1. The algorithm for Calculating the coefficient matrix

---

Input:
   Region Matrix X; Region Matrix Y;Maximum Iterations mi;Learning Rate lr;
Output:
   Coefficient Matrix C;
1: $error=0;mi=1e+6;lr = 10^{-7};C = $ initialization$(X, Y);epoch = 5;$
2: **for** $i = 1; i < mi; i++$ **do**
3:    newlr = guessLearningRate(error) according to the inverse function based on residuals;
4:    $[error, C^{(g+1)}] = getTrainingError(C^g, X, Y, newlr);$
      $C^{(g+1)} = C^g - newlr * [Y^T C X^T + Y C X + C X C^T [X C^T + X^T C^T]^T]$ error $=$
      $|Y - C^{g+1} X [C^{g+1}]^T|^2$
5:    apply momentum on newlr and $C^{(g+1)}$.
6:    StopCriteria(error, epoch);
7: **end for**
8: **return** C;

The inverse function based on residuals in this study is defined as below with initial learning rate(ILR) as $1e - 7$:

$$lr = \begin{cases} ILR & error \le 0.05 \text{ ,} \\ 10 * ILR & 0.05 < error \le 0.1, \\ 50 * ILR & 0.1 < error \le 0.5, \\ 100 * ILR & error > 0.5. \end{cases}$$

Convergence. In this section, the theoretical proof of convergence for GD algorithm is addressed.

Theorem 1. *The quadratic matrix function g: $R^{m \times n} \to S^n$ defined by*

$$g(X) = XAX^T + A, \tag{12}$$

where $A \in S^m, X \in S^n$, is convex when $A \succeq 0$.

Proof. The function $h : S^n \to R$ defined by $h(Y) = -log \, det(-Y)$ is convex and increasing on $dom(h) = -S^n$. By the composition theorem [5], it can be concluded that

$$f(X) = -log \, det(-(X^T AX + C)) \tag{13}$$

is convex on

$$domf = \{X \in R^{m \times n} | X^T AX + C \prec \mathbf{0}\} \tag{14}$$

Based on the theory of trigonometric inequality for norm [5], the objective function 2 is convex.

With GD algorithm, the local optimal solution of convex optimization is the global optimal solution [5].

Table 1. Comparing with deep learning method on topological measure. The distance and its corresponding similarity are denoted as "[distance (similarity)]."

Approach	A → B	A → C	A → D
autoencoder-dense	0.0718(1.0)	88.4259(0.007)	89.0186(0.0)
autoencoder-cnn	0.0175(1.0)	15.7862(0.022)	16.1353(0.0)
autoencoder-resnet	8.0E−4(1.0)	0.0275(0.016)	0.0279(0.0)
our	16.9116(1.0)	21.7188(0.722)	34.1752(0.0)

4   Experiments and Results

Experiments had first been performed in order to demonstrate the performance of HMSM on similarity measurement in terms of sensitivity to topology on a public image dataset. Secondly, a case study had been made to examine the effectiveness of the proposed inter-region analysis approach on a real EEG dataset via evaluation of the anti-depression treatment outcome.

4.1   Experiments on Topology-Sensitive Similarity Measurement

This experiments aimed to evaluate the capability of measuring topology-sensitive information of the proposed HMSM approach on a public image database[1]. Each image (The entire image will be used as the source and target matrix in this experiment, and it will not be divided into regions) was first converted to a $2D$ grey image with size (286×286) by rgb2gray (Matlab function) and referred to as the original image in the rest of this section (Sub-plot (A) of Fig. 3). Its three variants were created as follows: Variant B: It was produced by adding Gaussian noise on the original image, and the residual error was 14.3 (Sub-plot (B) of Fig. 3); Variant C: It was produced by shrinking the original image to 272×272 (Sub-plot (C) of Fig. 3); Variant D: It was produced by shuffling some portions of Variant 2 in the month area (Sub-plot (D) of Fig. 3).

The similarities of the original image to the three variants were calculated (Eqs. 8 and 10). The results were presented in Fig. 3 as the values on the arrows. For this task, the similarity of $A \to B$ is the highest (1.0) followed by that of $A \to C$ (0.584), and that of $A \to D$ is the least (0). Obviously, the **similarities** (as defined in Eq. 10) of the above images to the original one are descending.

The dataset contained 1521 images. The corresponding $HMSMs$ for the images were calculated. The similarity of $A \to B$ is maximal (1), and the similarity of $A \to D$ is minimal (0). The average similarity of $A \to C$ is 0.722 with variance of 0.038. This was not a surprise as the noise processing with smaller residual error still retained most information of the original image in variant B; In contrast, variant C "compressed" 10% information of the original one; These was a change in topology from variant C to variant D, and this enlarged the dissimilarity compared with A only with a few exceptions.

---

[1] https://www.bioid.com/facedb/.

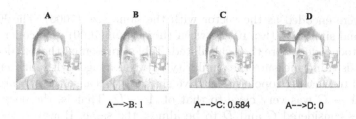

A—>B: 1   A-->C: 0.584   A-->D: 0

Fig. 3. Similarity between the image and its variants. The similarity (S) from the source image (A) to target image (B) denotes as "A → B : S".

The experimental results in this example indicated that the proposed HMSM-based approach could properly measure the similarity between heterogeneous matrices and was sensitive to the difference in topology.

The experiments were designed to compare HMSM with deep learning methods (CNN [14] or ResNet [11]) in topology-sensitive similarity measurements and the results were presented in Table 1. The above images (A, B, C, D) were first compressed to the same size (1000) by four corresponding autoencoders respectively. An autoencoder learns to compress data from the input layer into a shortcode (encoder network), and then uncompress that code into something that closely matches the original data (decoder network). We only describe the processing of the original image due to layout restrictions, all the variants shared the same network architecture except the input layer and output layer. The input of "autoencoder-dense" are the flattened vectors of the original image (A) and its variates (B, C, D), while the input of "autoencoder-CNN" and "autoencoder-resnet" are [4, (143, 143)] of A, [4, (143, 143)] of B, [4, (136, 136)] of C and [4, (136, 136)]of D respectively, where [4, (143, 143)] etc, presents the [filter, (input_shape)]. The architecture of three networks was described as follows (except input and output layer): "autoencoder-dense": The encoder network consists of two dense layers (5000 and 1000). The decoder network consists of two dense layers (1000 and 5000). "autoencoder-CNN": The encoder network consists of three same convolutional layers (receptive field: 3 × 3, filter:4) and two dense layers (5000 and 1000). The decoder network consists of two dense layers (1000 and 5000) and three deconvolutional layers with a receptive field (3 × 3) and filter 4. "autoencoder-resnet": The encoder layer followed the configuration of Resnet-16 [11], except that we only use one dense layer (1000). In the decoder network, In symmetrical position, we use deconvolutional layer, upsampling layer instead of convolutional layer and pooling layer (maxpooling, avgpooling), respectively, while other layers remain the same. The above three models were trained using SGD and applied a very small weight decay to keep the model's training error low [16]. The activation ReLU was used in all parameter layers except "autoencoder-resnet". Weight initialization was performed conforming to that proposed in [10] and batch normalization applied to the network [12]. The objective was to minimize the mean squared error in the model which was trained in terms of different variates (A variate corresponded to a model) with 40 epochs. For each image (e.g. 286 × 286), the corresponding variates were encoded as the vector with the same size (1000). The Euclidean distance and similarity (Eq. 10) between the output (1000) of encoder in terms of four autoencoders were then calculated. Table 1 presented that all deep learning methods could measure the similarity between images with different pixels, but **failed** to measure topology-sensitive similarity because the synchronization value of $A \rightarrow C$ was very close to that of $A \rightarrow D$. That is, the deep learning approaches considered C and D to be almost the same. However, the HMSM can effectively measure similarity on both tasks because there is an appropriate synchronization value from source image to its variants.

4.2   Inter-region Synchronization Analysis for Treatment Outcome Evaluation

Experiments had also been performed to examine the effectiveness of the proposed inter-region analysis approach on a real EEG dataset via evaluation of the anti-depression treatment outcome.

Data Description. The EEG dataset was constructed from 17 patients with major depressive disorder (MDD) in Peking University People's Hospital. EEG were recorded simultaneously through twenty different channels (Fp1, Fp2, F3, F4, C3, C4, P3, P4, O1, O2, F7, F8, T3, T4, T5, T6, Fz, Cz, Pz, ECG) in 256 Hz with 19 electrodes and one Electrocardiograph. All MMD subjects received antidepressant treatment (escitalopram oxalate tablets 10 mg) and were selected by experts in the hospital. Among those, seven subjects were diagnosed non-effective and the rest effective, and those were separated into the non-effective and the effective groups respectively.

Treatment Outcome Evaluation. For this round of experiments, the time window was first set as 4096 (16 s), the whole sample space is split into 5210 segments (effective: 3120, non-effective: 2090). All synchronization strengths ($sync(D)$ in Eq. 10) belonging to effective and non-effective groups were calculated with each denoting the $sync(D)$ (uni-directional) from a brain region to another (**directed edge**) in a particular time window, i.e., 3120 $sync(D)$s of the effective group and 2090 $sync(D)$s of the non-effective group for a given edge. For each group per directed edge (grouped edge), a K-means (cluster center: 2) algorithm was applied to separate the group edge. The average synchronization strength and its standard deviation (in bracket) per directed edge were calculated (Eq. 10) and were shown as the italic number associated with each edge in Fig. 4. With running the T-test between two groups, a small p-value (<0.05) in Fig. 4 illustrated that the synchronization values between respondents and non-respondents are statistically different.

In this study, the synchronization measures of the effective group were higher than those of the non-effective group. The two most significant cases were right temporal (RT) to central (non-effective 0.66 vs. effective 0.49) and left to central (non-effective 0.67 vs. effective 0.53).

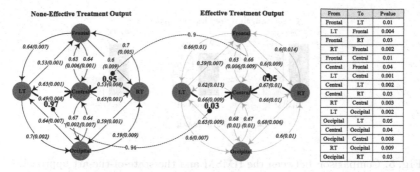

From	To	Pvalue
Frontal	LT	0.01
LT	Frontal	0.004
Frontal	RT	0.03
RT	Frontal	0.002
Frontal	Central	0.01
Central	Frontal	0.04
LT	Central	0.001
Central	LT	0.002
Central	RT	0.03
RT	Central	0.003
LT	Occipital	0.002
Occipital	LT	0.05
Central	Occipital	0.04
Occipital	Central	0.008
RT	Occipital	0.009
Occipital	RT	0.03

Fig. 4. The directionality index (direction and corresponding strength) on brain region. The nodes are the brain region and the edges are the directionality index. The table illustrated the P-value after running T-test from one brain region to another between two groups (respondents and non-respondents).

This set of experiments was conducted by comparing our HMSM approach against the deep learning approaches such as CapsulesNet [22] and ResNet-16 [11] upon the same dataset. We followed the parameter configuration of CapsulesNet [22] and ResNet-16 [11] respectively, except that the input was changed to the size of our synchronization matrix (20×20), and the output dense layer was changed to 2 with activation softmax since it was a binary classification. The above two models were trained using SGD and applied a very small weight decay to keep the model's training error low [16]. Weight initialization was performed conforming to that proposed in [10] and batch normalization applied to the network [12]. The objective was to minimize the mean squared error in the model with epochs 40. The two models processed the whole brain CMMICs, which differed from the previous clustering of HMSM in that its CMMIC was the correlation matrix of the whole brain (See Sect. 3.1) rather than the division matrix of regions, as the initial inputs in model training. After shuffling the whole sample space, CMMICs were divided into training sets, validation sets and test sets. A 5-fold cross validation algorithm was employed to evaluate the training performance of the classifier with training and validation sets. The performance of the classification was reported with the test sets.

As shown in Fig. 5, HMSM and Resnet-16 achieved excellent performance. However, HMSM and CapsuleNet were robust across different training set because the performance had not declined significantly, even if the percentage of test set increased to 50%.

In summary, these results indicate that (1) on average, the synchronization strength of non-effective group is lower than effective group in most cases, and (2) Synchronization direction of inter-region plays an important role in evaluation, and (3) characterization of direction strength can provide useful information to differentiate the effective anti-depression treatment outcome from non-effective group, which can serve as a biomarker (LT/RT to central).

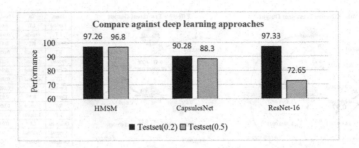

Fig. 5. Comparison between the HMSM and the state-of-the-art approaches.

5   Conclusions and Future Work

This study proposed an HMSM theory for distance calculation between heterogeneous matrices and developed a new inter-region approach to quantifying the synchronization amongst brain regions. First, experiments were performed on one public image dataset to measure the similarity between the images and their variants in different dimensions and layouts. The preliminary results indicated that the proposed approach could properly quantify the similarity of heterogeneous matrices and reflect the difference incurred by variations in topology. Second, the similarity measures were used to examine the relation between brain region interactions (directional) and treatment outcome. Experimental results indicate that: for the non-effective group, 97% (95%) *similarity* values from **right temporal** to central (from **left temporal** to central) are greater than the average *similarity* value of all patients; the measures can form a biomarker for anti-depression treatment outcome. Furthermore, the HMSM theory and algorithm hold potentials in clustering, dimensionality reduction, and sophisticated applications involving quantification of relation between images.

The approach currently only concerns pairwise brain regions and does not directly apply to multiple regions. Future work will consider incorporating the correlation matrix theory to extend the HMSM to directly quantifying multiple brain regions.

Acknowledgement. This work was supported in part by the National Natural Science Foundation of China (No. 61772380), Foundation for Innovative Research Groups of Hubei Province (No.2017CFA007), and Major Project for Technological Innovation of Hubei Province (No. 2019AAA044).

References

1. Abdallah, C.G., et al.: Ketamine treatment and global brain connectivity in major depression. Neuropsychopharmacology **42**, 1210–1219 (2017). https://doi.org/10.1038/npp.2016.186

2. Baird, L., Moore, A.: Gradient descent for general reinforcement learning. In: Proceedings of the 1998 Conference on Advances in Neural Information Processing Systems II, pp. 968–974. MIT Press, Cambridge (1999). https://doi.org/10.1145/1514274.1514279
3. Bassett, D.S., Sporns, O.: Network neuroscience. Nat. Neurosci. **20**, 353–364 (2017). https://doi.org/10.1038/nn.4502
4. Bonita, J.D., et al.: Time domain measures of inter-channel EEG correlations: a comparison of linear, nonparametric and nonlinear measures. Cogn. Neurodyn. **8**, 1–15 (2014). https://doi.org/10.1007/s11571-013-9267-8
5. Boyd, S., Vandenberghe, L.: Convex Optimization. Cambridge University Press, London (2004)
6. Chen, D., Li, X., Cui, D., Wang, L., Lu, D.: Global synchronization measurement of multivariate neural signals with massively parallel nonlinear interdependence analysis. IEEE Trans. Neural Syst. Rehabil. Eng. **22**(1), 33–43 (2014). https://doi.org/10.1109/TNSRE.2013.2258939
7. Dosovitskiy, A., Brox, T.: Generating images with perceptual similarity metrics based on deep networks. In: Advances in Neural Information Processing Systems 29 (NIPS), pp. 658–666. Curran Associates Inc., Barcelona (2016). http://papers.nips.cc/paper/6158-generating-images-with-perceptual-similarity-metrics-based-on-deep-networks.pdf
8. Drysdale, A.T., et al.: Resting-state connectivity biomarkers define neurophysiological subtypes of depression. Nat. Med. **23**, 28–38 (2017). https://doi.org/10.1038/nm.4246
9. Ghuman, A.S., McDaniel, J.R., Martin, A.: A wavelet-based method for measuring the oscillatory dynamics of resting-state functional connectivity in MEG. NeuroImage **56**(1), 69–77 (2011). https://doi.org/10.1016/j.neuroimage.2011.01.046
10. He, K., Zhang, X., Ren, S., Sun, J.: Delving deep into rectifiers: surpassing human-level performance on ImageNet classification. In: IEEE International Conference on Computer Vision (ICCV 2015), vol. 1502, pp. 1026–1034, February 2015. https://doi.org/10.1109/ICCV.2015.123
11. He, K., Zhang, X., Ren, S., Sun, J.: Deep residual learning for image recognition, vol. abs/1512.03385, Las Vegas, NV, USA, June 2016. https://doi.org/10.1109/CVPR.2016.90
12. Ioffe, S., Szegedy, C.: Batch normalization: accelerating deep network training by reducing internal covariate shift. In: Bach, F., Blei, D. (eds.) Proceedings of the 32nd International Conference on Machine Learning. Proceedings of Machine Learning Research, PMLR, Lille, France, 07–09 July 2015, vol. 37, pp. 448–456 (2015). http://proceedings.mlr.press/v37/ioffe15.html
13. Kandel, E.R., Markram, H., Matthews, P.M., Yuste, R., Koch, C.: Neuroscience thinks big (and collaboratively). Nat. Rev. Neurosci. **14**, 659–664 (2013). https://doi.org/10.1038/nrn3578
14. Ke, H., Chen, D., Li, X., Tang, Y., Shah, T., Ranjan, R.: Towards brain big data classification: Epileptic EEG identification with a lightweight VGGNet on global MIC. IEEE Access **6**, 14722–14733 (2018). https://doi.org/10.1109/ACCESS.2018.2810882
15. Kim, D.J., et al.: Disturbed resting state EEG synchronization in bipolar disorder: a graph-theoretic analysis. NeuroImage: Clin. **2**(Suppl. C), 414–423 (2013). https://doi.org/10.1016/j.nicl.2013.03.007
16. Krizhevsky, A., Sutskever, I., Hinton, G.E.: ImageNet classification with deep convolutional neural networks. Commun. ACM **60**(2), 2012 (2012). https://doi.org/10.1145/3065386

17. Leuchter, A.F., Cook, I.A., Hunter, A.M., Cai, C., Horvath, S.: Resting-state quantitative electroencephalography reveals increased neurophysiologic connectivity in depression. PLOS One **7**(2), 1–13 (2012). https://doi.org/10.1371/journal.pone.0032508
18. Li, X., Ouyang, G.: Estimating coupling direction between neuronal populations with permutation conditional mutual information. NeuroImage **52**(2), 497–507 (2010). https://doi.org/10.1016/j.neuroimage.2010.05.003
19. Mumtaz, W., Ali, S.S.A., Yasin, M.A.M., Malik, A.S.: A machine learning framework involving EEG-based functional connectivity to diagnose major depressive disorder (MDD). Med. Biol. Eng. Comput. **56**(2), 233–246 (2018). https://doi.org/10.1007/s11517-017-1685-z
20. Reshef, D.N., et al.: Detecting novel associations in large datasets. Science **334**(6062), 1518–1524 (2011). https://doi.org/10.1126/science.1205438
21. Rubinov, M., Sporns, O.: Complex network measures of brain connectivity: uses and interpretations. NeuroImage **52**(3), 1059–1069 (2010). https://doi.org/10.1016/j.neuroimage.2009.10.003
22. Sabour, S., Frosst, N., Hinton, G.E.: Dynamic routing between capsules. In: Advances in Neural Information Processing Systems 30 (NIPS), pp. 3859–3869 (2017). http://papers.nips.cc/paper/6975-dynamic-routing-between-capsules.pdf
23. Smilde, A., Bro, R., Geladi, P.: Multi-way Analysis. Wiley, West Sussex (2004)
24. Stam, C.: Functional connectivity patterns of human magnetoencephalographic recordings: a 'small-world' network? Neurosci. Lett. **355**(1), 25–28 (2004). https://doi.org/10.1016/j.neulet.2003.10.063

Bi-ResNet: Fully Automated Classification of Unregistered Contralateral Mammograms

Runze Wang, Yanan Guo, Wendao Wang, and Yide Ma[✉]

Lanzhou University, Lanzhou 730000, Gansu, China
{wangrz17,guoyn16,wangwd17,ydma}@lzu.edu.cn

Abstract. Motivated by the fact that the contralateral mammograms can provide the symmetrical difference of the left and right breasts to assist identify the breast cancer, we propose a bilateral residual neural network (Bi-ResNet) that can automatically classify the normality/abnormality based on the unregistered contralateral whole mammograms. Specifically, the parallel ResNet network is designed to simultaneously process a group of contralateral mammograms and respectively capture the discriminative representations from the left and right mammograms, and the concatenation strategy in the final is used to integrate the differentiated features for the abnormal classification task. The proposed Bi-ResNet can achieve reproducible and similar results based on different backbones and is superior to traditional contralateral analysis methods in both automation and performance. Finally, our proposed Bi-ResNet is greatly demonstrated on the publicly available DDSM dataset, a total of 10480 images, yielding the highest AUC of 0.908 on the abnormal classification task. Through the massive experiments, we deem our model is stable and robust, and has the potential to be recommended to clinical application in the future.

Keywords: Contralateral whole mammograms ·
Residual neural network · Classification

1 Introduction

Breast cancer, as the most frequently diagnosed cancer and the leading cause of cancer death in women, is responsible for 25% of all cancer cases and 15% of cancer-related deaths amongst females worldwide [20]. Widely believed the early detection and diagnosis can effectively decrease the mortality rate associated with breast cancer, greatly improving the prospects of survival [4,22]. Currently, mammography works as one of the most effective and reliable diagnostic methods in the early stage of breast cancers [15] and is becoming more and more popular in the clinical application.

In traditional clinical interpretation, mammogram diagnosis is usually performed by highly professional trained radiologists. However, manual analysis

© Springer Nature Switzerland AG 2019
I. V. Tetko et al. (Eds.): ICANN 2019, LNCS 11731, pp. 273–283, 2019.
https://doi.org/10.1007/978-3-030-30493-5_28

of mammograms is not only time-consuming and laborious but also subject to inter-observer variability [26]. Therefore, Mammogram-motivated computer-aided diagnosis (CAD) system was recommended as "the second opinion" to release radiologists from the heavy workload and improve diagnostic accuracy [9], which has received extensive attention in the past decades. Almost all of those clinical CAD systems can be roughly divided into single-view and multi-view based methods. Recently, research on breast cancer CAD systems based on single-view is moving in a more favorable direction [1], due to the advances in machine learning, especially deep learning. As an essential part of the CAD system, the single-view classification algorithm of mammograms based on deep learning has been widely studied [6,8,14,18,25,28,30]. However, all these methods ignore the strong association between multiple mammographic views, at the same time, the normality of a single mammogram does not guarantee the health of the subject. In this context, multi-view based analysis methods emerged as the times require. The mainstream of the multi-view CAD classification models is based on the craniocaudal (CC) and mediolateral oblique (MLO) views acquired from the same breast. Recently, Carneiro et al. [2] and Dhungel et al. [7] have taken the ipsilateral CC and MLO views and the corresponding masks of masses and microcalcifications into account, improving the classification ability to some extent. But the normality in unilateral CC and MLO mammograms still cannot provide a confirmed diagnosis (normal/abnormal) for the subject, similar to the single-view based methods. Therefore, other considerations should be introduced.

In clinical practice, radiologists typically detect inconspicuous lesions by global-to-local symmetry difference analysis by comparing the mammograms of left and right breasts. Motivated by this, some researches started to exploit the CAD model based on contralateral mammograms (contralateral mammograms refer to images acquired from both breasts and that each of the contralateral images can include multiple-views) instead of the single or unilateral ones. In 2010, Wang et al. [27] utilized feature-based contralateral asymmetry to classify low/high risk mammograms; In 2012, Zheng et al. [29] used the contralateral density asymmetry to classify high risk patients; In 2013, Rodriguez-Rojas et al. [23] developed a feature-based genetic algorithm to classify low/high risk mammograms using contralateral asymmetry and Tan et al. [24] exploited feature based contralateral asymmetry SVM to performed near-term breast cancer prediction; In 2014, Martí et al. [19] proposed temporal registration based random forest-LOOCV to classify the MLO-view contralateral mammograms as normal/abnormal; recently, Celaya-Padilla et al. [3] proposed an automated registered contralateral subtraction to perform normal/abnormal classification, etc. Indeed, these methods effectively improve the recognition performance of breast abnormalities through a variety of contralateral analysis. But it is should be noted that methods above heavily rely on the hand-crafted features that are always domain-specific, for different databases, the performance varies greatly. In addition, the separate design of features and classifiers is often non-automatic or semi-automatic due to the need for complex multi-step operations, and it is prone to sub-optimal results [7]. Currently, it sounds that the end-to-end network based on deep learning can resist the interference introduced by human factors.

In this paper, an extended version of the deep residual network (ResNet)[10], called Bilateral ResNet (Bi-ResNet), is proposed, and it is able to accept unregistered contralateral mammograms as input and unify the learning of contralateral features and classifier in an end-to-end supervised training fashion for normal/abnormal classification. The proposed Bi-ResNet achieved a fully automated classification AUC of 0.908 on the most commonly used DDSM dataset [12], including 5240 pairs of contralateral mammograms, without bells and whistles. This competitive result shows that our proposed Bi-ResNet has the great potential to be used in breast screening programs.

This paper is organized as follows. Section 2 introduces the dataset and proposed method, Sect. 3 describes the experimental results and discussion; conclusions are presented in Sect. 4.

2   Materials and Methods

2.1   Dataset

The DDSM database includes 2620 mammography screening cases, of which 695 are normal cases, and abnormal cases include benign without callback, benign and malignant, a total of 1925 cases. Every case contains CC and MLO views of both breasts, for a total of 10480 mammograms. We randomly selected 80% normal cases and abnormal cases, respectively, to form a training set, and the remaining cases form a test set. Let $D = \{(X^{(i)}, y^{(i)})_j\}_{j=1}^{|D|}$ represent the dataset, where $i \in \{CC, MLO\}$ indexes the two views (CC and MLO), $X = \{X_{Left}, X_{Right}\}$ represent the contralateral mammograms (see in Fig. 1)

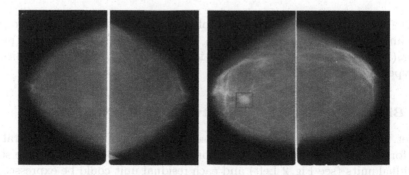

Fig. 1. A set of normal (Left) and abnormal (Right) contralateral mammograms. Obviously, there is no obvious discrepancy in the normal contralateral mammograms; on the contrary, significant visual difference occur in the abnormal contralateral mammograms, where the suspicious lesion marked by red box in the left breast is not visible in the right breast. (Color figure online)

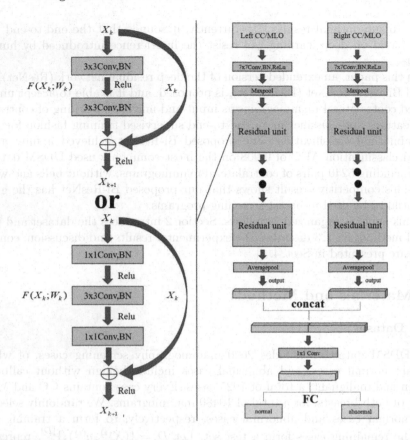

Fig. 2. The proposed Bi-ResNet for fully automated classification of unregistered contralateral mammograms. Upper left: a residual unit for ResNet-18/34. Lower left: a residual unit for ResNet-50/101/152.

and $y \in \{0, 1\}$ denotes the class label of a pair of contralateral mammograms that can be either normal or abnormal. The latter refers to those with proven cancer (i.e. masses, calcification, distortions, etc.) and benign abnormalities of any type.

2.2   Bilateral Residual Neural Network (Bi-ResNet)

ResNet, proposed by He et al. [10] to ease the training of the deeper neural network for a stronger feature representation capability, consists of multiple stacks of residual units (see Fig. 2 Left) and each residual unit could be expressed by:

$$X_{k+1} = H(X_k) + F(X_k; W_k) \tag{1}$$

where $k \in \{1, \cdots, K\}$ refers to the kth residual unit, X_k and W_k separately represent the input feature maps and the set of weights for the kth residual unit, $F(.)$ is a function represented by several combinations of a convolutional layer [16,17], a batch normalization (BN)[13] and a rectilinear unit (ReLU)[21], and $H(X_k) = X_k$ is an identity mapping [11]. Recursive calculation by (1), the output at the location K within the ResNet can be represented as:

$$X_K = X_k + \sum_{k=1}^{K-1} F(X_k; W_k) \tag{2}$$

Our proposed Bi-ResNet (see Fig. 2 Right) could be considered as the integration of two identical ResNet for processing contralateral mammograms, where we concatenate the output from the last average pool of each ResNet, followed by a 1×1 convolutional layer and a fully connected layer containing two nodes, one denoting normal and the other abnormal, which is expressed as the function $G(.)$ in the following formula:

$$\hat{y} = G(X_{Left,L}, X_{Right,L}; W_{Bi}) \tag{3}$$

where the $W_{Bi} = [W_{Left,l}, W_{Right,l}, W_{conv\&fc}]$ indicate the weights of Bi-ResNet, with the $W_{conv\&fc}$ denotes the weights of the final 1×1convolutional layer and fully connected layer, $W_{Left,l}$ the weights of the left mammogram (CC/MLO) and $W_{Right,l}$ the weights of the right mammogram (CC/MLO, the same as the left). The $X_{Left,L}$ and $X_{Right,L}$ respectively represent the outputs from the last average pool of each ResNet and are denoted as follows:

$$X_{Left,L} = X_{Left,l} + \sum_{l=1}^{L-1} F(X_{Left,l}; W_{Left,l}) \tag{4}$$

$$X_{Right,L} = X_{Right,l} + \sum_{l=1}^{L-1} F(X_{Right,l}; W_{Right,l}) \tag{5}$$

For training the Bi-ResNet, we calculate the cross entropy loss function for the predicted probability and ground truth, and adapt the stochastic gradient descent (SGD) optimizer to minimize the loss in an end-to-end training manner. The loss function could be expressed as follows:

$$\ell(W_{Bi}) = -\sum_{j=1}^{|D|} \sum_{i \in \{CC,MLO\}} y(i,j) log\, \hat{y}(i,j) \tag{6}$$

Under the constraint of the above loss function, and driven by a large amount of contralateral training data with its corresponding tags, our proposed Bi-ResNet can learn contralateral data representations with multiple levels of abstraction to achieve accurate classification. Intuitively, when the input contralateral mammograms is normal, the contralateral differences of features extracted by the Bi-ResNet are relatively small. While when the input belongs to the abnormal case, the contralateral features presents a competitive relationship, thereby highlighting the abnormality, which is similar to the radiologist's contralateral image diagnosis process. Then we fuse the bilateral features through a simple 1×1 convolutional layer and perform normal/abnormal classification based on feature differences through a fully connected layer.

3 Experiments

3.1 Experimental Configurations

Our proposed Bi-ResNet are implemented based on the deep learning framework PyTorch, and the hardware platform is i5 Intel CPU with 8 GB RAM and GTX1080 GPU with 8 GB on-chip memory. The parameters of SGD are configured as follows: the initial learning rate is set to 0.001 and decay 0.1 for every 5 epoch, the momentum is set to 0.9, and the weight decay is set to 0.0001. The batch size is set to 8 according to the GPU memory. Both branches of the proposed Bi-ResNet are initialized with the ResNet's pre-training parameters on ImageNet [5] database and then all of the Bi-ResNet is fine-tuned on the DDSM database. It is worth mentioning that the ImageNet images have three channels (RGB), while mammograms are grayscale. Therefore, we replicated every single-channel mammogram image three times and stacked them into a three-channel image. To avoid overfitting, we apply a series of data augmentation methods including flipping horizontally, flipping vertically, random rotation to generate more training samples.

3.2 Experiment Results and Discussion

Comparison with Standard ResNet Series Network. To evaluate the effectiveness of our proposed Bi-ResNet, we use the standard ResNet as a comparative experiment and keep the training and test sets consistent with Bi-ResNet. It is worth mentioning that the standard ResNet only accepts single input image, so for normal cases, we randomly select one of the pair of contralateral mammograms as input, while for the abnormal cases, we select the ipsilateral mammogram as input. Moreover, in order to verify the robustness of Bi-ResNet, we conduct a set of comparative experiments on each network of the ResNet series. The results are shown in Fig. 3.

Obviously, the Bi-ResNet-101 has achieved the highest classification performance on the DDSM dataset with an AUC of 0.908. At the same time, the classification AUC of Bi-ResNet-18/34/50/101/152 is 0.032, 0.024, 0.03, 0.045 and 0.032 higher than the standard ResNet-18/34/50/101/152, respectively, with an average increment of 0.033, which greatly proves the validity and universality of our approach.

Comparison with Precious Similar Methods. Table 1 presents several similar approaches that incorporate a various type of contralateral analysis, for objectively evaluating the performance of the proposed Bi-ResNet. In addition

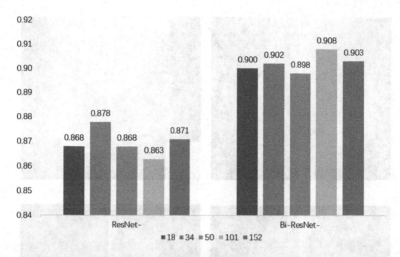

Fig. 3. The normal/abnormal classification AUC of ResNet-18/34/50/101/152 and proposed Bi-ResNet-18/34/50/101/152.

Table 1. Performance comparison between several similar contralateral analysis approaches.

Methodologies	Database	AUC
Rodriguez-Rojas et al. [23]	Private database (200 cases)	0.880
Martí et al. [19]	Private database (264 images)	0.760
Zheng et al. [29]	Private database (451 cases)	0.761
Wang et al. [27]	Private database (200 cases)	0.750
Tan et al. [24]	Private database (994 cases)	0.720
Celaya-Padilla et al. [3]	DDSM-Lumisys (88 cases)/BCDR (64 cases)	0.738/0.767
Bi-ResNet-101	Entire DDSM(2620 cases)	0.908

to our method, the other methods listed in the Table 1 almost require manual intervention and multi-step operations, not an end-to-end fully automated solution. Moreover, most of the other methods use a private database with a small amount of data, however, our method is verified in the most commonly used public database DDSM, and our data volume is much higher than others. It should be specially stated that the recently published [3] also exploit the contralateral mammograms to perform normal/abnormal (referred to as healthy/unhealthy in [3]) classification. However, their approach requires a series of complex steps such as image preprocessing, contralateral image registration, asymmetric feature extraction, and feature selection, etc. Nevertheless, our proposed Bi-ResNet only needs to provide the input contralateral mammograms and corresponding label, without any intermediate steps, and the network automatically captures the required contralateral asymmetry or difference information to complete the normal/abnormal classification with an AUC of 0.908. Obviously, our proposed

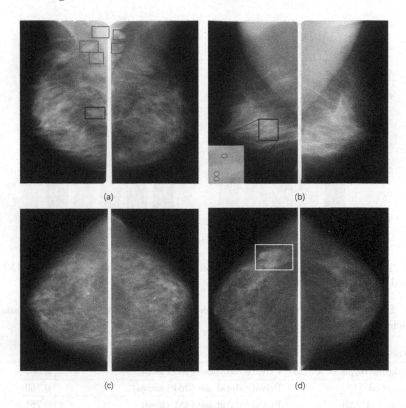

Fig. 4. Some misclassification examples from the test set. (a) and (b) are abnormal examples that were misclassified as normal, and (c) and (d) are normal examples that were misclassified as abnormal. (Color figure online)

Bi-ResNet obtains state-of-the-art results both in terms of automation and classification accuracy.

Misclassified Cases Analysis. Figure 4 shows several misclassification examples, Where (a) and (b) represent several false negative examples, and (c) and (d) represent some false positive examples. The red and green boxes in (a) mark the mass and the lymph nodes, respectively, and it is not difficult to find that the shape of the lymph nodes is similar to the mass. It should be specially stated that although we did not show it in Fig. 4, we can still find the existence of lymph nodes in many false positive examples, so we think that the lymph nodes in the pectoral muscle may cause some adverse interference to our model. The region marked by the red box in (b) is the microcalcifications of clustered distribution. First, the microcalcifications account for an extremely small proportion of the entire breast region and even experienced radiologists may overlook these microcalcifications. To make matters worse, the multi-stage pooling process of convolutional neural networks undoubtedly lose some information, which further causes our model lacking sensitivity to microcalcifications. Example (c) represents a considerable part of false positive examples, where their commonality is the BI-RADS category of breast density is usually C (heterogeneously dense) or D (extremely dense). High-density breast tissue has high-luminance properties similar to masses, leading to possible misclassification. Although the expert comment in case (d) is normal, it is undeniable that the area marked by the yellow box is indeed invisible in the contralateral mammogram and its morphology is also suspected of lesions, which may be the reason why our network misclassified the example.

4   Conclusion

In this paper, a novel end-to-end Bi-ResNet is proposed for the fully automated normal/abnormal classification of unregistered contralateral whole mammograms. Comparing with single-view classification based on ResNet series networks, the Bi-ResNet analysis strategy has achieved significant improvement, with a highest and average AUC increment of 0.045 and 0.033, respectively. This result shows that the strategy can achieve reproducible and similar results even with the different backbone networks and proves the effectiveness and robustness of the strategy. Moreover, experiments prove that our proposed Bi-ResNet surpasses traditional contralateral mammograms analysis approaches in terms of the automation degree and classification performance. Finally, the Bi-ResNet-101 is validated in the most commonly used pubic mammographic database DDSM with the highest AUC of 0.908, which indicate that the model could classify normal contralateral mammograms from any other type of cancer or benign abnormality and has the potential to be used in the breast screening programs. In addition, we believe that the proposed Bi-ResNet strategy may inspire other CAD tasks where symmetrical differences within or between organs are sought, such as lung, brain, and prostate images, etc.

In the future work, we will consider developing a pectoral muscle segmentation network based on deep learning semantic segmentation network to exclude the pectoral muscle region so as to eliminate the interference of lymph nodes on the model. In addition, a breast density classification network may be introduced to alleviate a large number of false positive problems caused by high breast density images.

Acknowledgements. This work is jointly supported by the National Natural Science Foundation of China (Nos. 61175012 and 61201421) and Natural Science Foundation of Gansu Province (No. 18JR3RA288).

References

1. Burt, J.R., et al.: Deep learning beyond cats and dogs: recent advances in diagnosing breast cancer with deep neural networks. Br. J. Radiol. **91**(1089), 20170545 (2018)
2. Carneiro, G., Nascimento, J., Bradley, A.P.: Automated analysis of unregistered multi-view mammograms with deep learning. IEEE Trans. Med. Imaging **36**(11), 2355–2365 (2017)
3. Celaya-Padilla, J.M., et al.: Contralateral asymmetry for breast cancer detection: a CADx approach. Biocybern. Biomed. Eng. **38**(1), 115–125 (2018)
4. Communities, C.O., Communities, S.O.O.: Health statistics : atlas on mortality in the European Union. Office for Official Publications of the European Communities (2010)
5. Deng, J., Dong, W., Socher, R., Li, L.J., Li, K., Fei-Fei, L.: ImageNet: a large-scale hierarchical image database (2009)
6. Dhungel, N., Carneiro, G., Bradley, A.P.: A deep learning approach for the analysis of masses in mammograms with minimal user intervention. Med. Image Anal. **37**, 114–128 (2017)
7. Dhungel, N., Carneiro, G., Bradley, A.P.: Fully automated classification of mammograms using deep residual neural networks. In: 2017 IEEE 14th International Symposium on Biomedical Imaging (ISBI 2017), pp. 310–314. IEEE (2017)
8. Gallego-Posado, J., Montoya-Zapata, D., Quintero-Montoya, Q.: Detection and diagnosis of breast tumors using deep convolutional neural networks. In: Research Group on Mathematical Modeling School of Mathematical Sciences Universidad EAFIT Medellín (2016)
9. Giger, M.L., Karssemeijer, N., Schnabel, J.A.: Breast image analysis for risk assessment, detection, diagnosis, and treatment of cancer. Annu. Rev. Biomed. Eng. **15**, 327–357 (2013)
10. He, K., Zhang, X., Ren, S., Sun, J.: Deep residual learning for image recognition. In: Proceedings of the IEEE Conference on Computer Vision and Pattern Recognition, pp. 770–778 (2016)
11. He, K., Zhang, X., Ren, S., Sun, J.: Identity mappings in deep residual networks. In: Leibe, B., Matas, J., Sebe, N., Welling, M. (eds.) ECCV 2016. LNCS, vol. 9908, pp. 630–645. Springer, Cham (2016). https://doi.org/10.1007/978-3-319-46493-0_38
12. Heath, M., Bowyer, K., Kopans, D., Moore, R., Kegelmeyer, W.P.: The digital database for screening mammography. In: Proceedings of the 5th International Workshop on Digital Mammography, pp. 212–218. Medical Physics Publishing (2000)
13. Ioffe, S., Szegedy, C.: Batch normalization: accelerating deep network training by reducing internal covariate shift. arXiv preprint arXiv:1502.03167 (2015)
14. Jadoon, M.M., Zhang, Q., Haq, I.U., Butt, S., Jadoon, A.: Three-class mammogram classification based on descriptive CNN features. BioMed Res. Int. **2017**, 11 (2017)
15. Kooi, T., et al.: Large scale deep learning for computer aided detection of mammographic lesions. Med. Image Anal. **35**, 303–312 (2017)
16. Krizhevsky, A., Sutskever, I., Hinton, G.E.: ImageNet classification with deep convolutional neural networks. In: Advances in Neural Information Processing Systems, pp. 1097–1105 (2012)
17. LeCun, Y., Bengio, Y., et al.: Convolutional networks for images, speech, and time series. Handb. Brain Theory Neural Netw. **3361**(10), 1995 (1995)

18. Lévy, D., Jain, A.: Breast mass classification from mammograms using deep convolutional neural networks. arXiv preprint arXiv:1612.00542 (2016)

19. Martí, R., Díez, Y., Oliver, A., Tortajada, M., Zwiggelaar, R., Lladó, X.: Detecting abnormal mammographic cases in temporal studies using image registration features. In: Fujita, H., Hara, T., Muramatsu, C. (eds.) IWDM 2014. LNCS, vol. 8539, pp. 612–619. Springer, Cham (2014). https://doi.org/10.1007/978-3-319-07887-8_85

20. Torre, L.A., Bray, F., Siegel, R.L., Ferlay, J., Lortet-Tieulent, J., Jemal, A.: Global cancer statistics, 2012. CA Cancer J. Clin. **65**, 69–90 (2015)

21. Nair, V., Hinton, G.E.: Rectified linear units improve restricted Boltzmann machines. In: Proceedings of the 27th International Conference on Machine Learning (ICML-10), pp. 807–814 (2010)

22. Fact Sheet No. 297: Cancer. World Health Organization, France (2009)

23. Rodriguez-Rojas, J., Garza-Montemayor, M., Trevino-Alvarado, V., Tamez-Pena, J.G.: Predictive features of breast cancer on Mexican screening mammography patients. In: Medical Imaging 2013: Computer-Aided Diagnosis, vol. 8670, p. 867023. International Society for Optics and Photonics (2013)

24. Tan, M., Zheng, B., Ramalingam, P., Gur, D.: Prediction of near-term breast cancer risk based on bilateral mammographic feature asymmetry. Acad. Radiol. **20**(12), 1542–1550 (2013)

25. Tsochatzidis, L., Costaridou, L., Pratikakis, I.: Deep learning for breast cancer diagnosis from mammograms—a comparative study. J. Imaging **5**(3), 37 (2019)

26. Wang, H., et al.: Breast mass classification via deeply integrating the contextual information from multi-view data. Pattern Recogn. **80**, 42–52 (2018)

27. Wang, X., Lederman, D., Tan, J., Wang, X.H., Zheng, B.: Computerized detection of breast tissue asymmetry depicted on bilateral mammograms: a preliminary study of breast risk stratification. Acad. Radiol. **17**(10), 1234–1241 (2010)

28. Wu, E., Wu, K., Cox, D., Lotter, W.: Conditional infilling GANs for data augmentation in mammogram classification. In: Stoyanov, D., et al. (eds.) RAMBO/BIA/TIA -2018. LNCS, vol. 11040, pp. 98–106. Springer, Cham (2018). https://doi.org/10.1007/978-3-030-00946-5_11

29. Zheng, B., Sumkin, J.H., Zuley, M.L., Wang, X., Klym, A.H., Gur, D.: Bilateral mammographic density asymmetry and breast cancer risk: a preliminary assessment. Eur. J. Radiol. **81**(11), 3222–3228 (2012)

30. Zhu, W., Lou, Q., Vang, Y.S., Xie, X.: Deep multi-instance networks with sparse label assignment for whole mammogram classification. In: Descoteaux, M., Maier-Hein, L., Franz, A., Jannin, P., Collins, D.L., Duchesne, S. (eds.) MICCAI 2017. LNCS, vol. 10435, pp. 603–611. Springer, Cham (2017). https://doi.org/10.1007/978-3-319-66179-7_69

Pattern Recognition for COPD Diagnostics Using an Artificial Neural Network and Its Potential Integration on Hardware-Based Neuromorphic Platforms

Pouya Soltani Zarrin[1(✉)] [iD] and Christian Wenger[1,2] [iD]

[1] IHP–Leibniz-Institut fuer innovative Mikroelektronik,
15236 Frankfurt (Oder), Germany
soltani@ihp-microelectronics.com
[2] Brandenburg Medical School, 16816 Neuruppin, Germany

Abstract. This paper presents the development of an Artificial Neural Network (ANN) model for pattern recognition for Chronic Obstructive Pulmonary Disease (COPD) diagnosis. Recent advancements in healthcare devices and the availability of numerous medical data have facilitated the management of chronic diseases. In addition, machine learning tools have made the management and diagnostic of a chronic illness more efficient by converting collected medical data from biosensors into meaningful clinical information. However, securing sensitive medical data, collected from patients, is still a challenging task. Hardware-based neural networks address this data safety concern by on-chip processing of acquired data, without cloud communications. Therefore, the presented ANN model was designed to comply with the intrinsic structure of neuromorphic platforms for future integrations.

Keywords: Pattern recognition · Neuromorphic learning ·
COPD diagnostics

1 Introduction

With the expansion of Internet-of-Things (IoT) and wearable technologies in healthcare, constant and remote monitoring of patients has become a reality. Recent advances in point-of-care medical devices have facilitated the early detection, prevention, and treatment of various diseases [1]. In addition, the availability of numerous clinical data, thanks to medical IoT, has paved the way towards the better management of chronic disease [2]. For example, various technologies have widely been used for monitoring blood glucose levels in diabetic patients, heart activity tracking in elderly, and gait pattern observations of Parkinson's diseased patients [1]. However, without analytical insight, the collected data from medical sensors are merely raw data with low clinical value. For instance, in our previous work, a portable biosensor for the management of COPD in home-care environments was presented [3]. The developed biosensor was capable to

© Springer Nature Switzerland AG 2019
I. V. Tetko et al. (Eds.): ICANN 2019, LNCS 11731, pp. 284–288, 2019.
https://doi.org/10.1007/978-3-030-30493-5_29

characterize the viscosity of saliva samples for diagnostic purposes. However, viscosity properties of saliva samples is one parameter out of various parameters which are required for COPD detection. In addition, the ambient conditions such as temperature considerably affect the viscosity of samples, causing diagnostic complexities. In other words, two identical samples collected from a single patient could be diagnosed as diseased and healthy in low and high temperatures, respectively. This issue is not a measurement error (to be addressed using a temperature compensation sensor), but an environmental parameter affecting the samples. Therefore, various parameters including ambient conditions, patients medical background, smoking history, gender, and age are required for the disease diagnosis. As a result, upon viscosity measurements by the developed biosensor, a sophisticated diagnostic algorithm, by concurrent consideration of all essential parameters, is required for the detection of COPD. Therefore, machine learning tools, or more specifically pattern recognition methods, could make the diagnostic procedure more efficient by converting the collected data from medical sensors into meaningful clinical information. Moreover, machine learning can be used for identifying diagnostic links between symptoms and diseases, which have been previously unknown, and providing treatment plans and recommendations to the healthcare specialists. Therefore, the objective of this work was to develop an ANN for pattern recognitions for COPD diagnosis. Nevertheless, the end goal of this project is to transfer the modeled ANN onto our previously developed neuromorphic platform [4]. As reported in [4], a new learning algorithm for neuromorphic systems was implemented based on the inherent stochasticity of CMOS integrated HfO2-based Resistive Random Access Memory (RRAM) devices. For the implementation of the learning algorithm, the RRAM array was integrated into a mixed-signal neuromorphic circuit with software-based neurons [4]. The hardware-based approach provides many advantages over the ANN such as low energy consumption. In addition, the main advantage of the hardware-based platforms, which is the main incentive behind the current work, is the possibility of securing sensitive medical data by processing them on-chip, without requiring any cloud communication or backend post-processing. Although, the advantages of neuromorphic platforms for medical applications are briefly discussed in the last section of the paper, the focus of this extended abstract is to present the hardware-compatible simulation model using an ANN.

2   Methods

Prior to modeling the ANN, required clinical parameters for diagnosing COPD and their respected ranges for diseased and healthy subjects were identified. Apart from the sensor measurements on the viscosity of saliva samples, the ambient temperature, patients smoking background, cytokine level, pathogen load, mucin combinations, gender, and age are among the eight fundamental parameters required for the diagnosis of COPD [5]. Since, at this stage of the project, actual medical records for training the neural network were not available, synthetic data were generated for thousand subjects. For clinical relevancy, synthetic

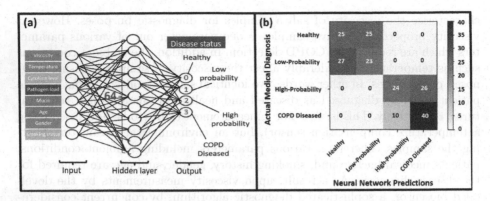

Fig. 1. (a) The structure of the ANN developed for pattern recognition for COPD diagnostics; (b) ANN predictions for the test subjects compared to their actual status.

data for various parameters were created with respect to the reported ranges in previous studies [5–8]. For example, patients with severe COPD have shown a lower cytokine level of TNFα compared with controls (COPD: 0–20 pg/ml and controls: 0–40 pg/ml) [6]. Additionally, viscosity of saliva was investigated to be greater in COPD samples (1.6–2 mPa s) compared to healthy controls (1–1.5 mPa s) [3,7,8]. Further information regarding clinical parameters and their ranges for different stages of COPD is available in the literature [5–8]. By taking into account the literature information for synthesizing data, first 300 subjects were considered as COPD diseased patients and were labeled as 3; whereas, the last 300 were chosen to be healthy subjects with the label 0. Labels 2 and 1 indicated patients (200 subjects for each label) with high and low probabilities of containing COPD, respectively. In a similar manner, 200 additional data for test subjects were created to evaluate the ANN's performance after training it. Generated data for train and test samples were normalized separately, to the range of 0–1, to improve network's performance. As shown in Fig. 1(a), a dense ANN with a single hidden layer was modeled on the JupyterLab environment using Python and Keras. The input layer of the network consisted of 8 neurons, considering the abovementioned clinical parameters. Considering the intrinsic structure and characteristics of the neuromorphic platform (64 × 64 grid structure of the memristor), a hidden layer with 64 neurons and a sigmoid activation function was modeled. The output layer, with a sigmoid activation function, consisted of 4 neurons for the four potential outcomes. Adam optimization algorithm, with a 0.0001 learning rate, and the cross entropy error function were used for training and optimizing the network parameters. A dropout, with 20% probability, was implemented for overfitting prevention. After modeling the ANN, the generated data were fed into the network for 150 training epochs with a batch size of 10. During training the network, 10% of the data were split for validation purposes. The trained network was used to predict the labels of the generated test data.

3   Results and Discussions

The developed ANN provided accuracies of 65% and 72% for the network and validation performances, respectively. The synthetic nature of the generated data, used for training the network, could possibility be the main reason behind ANN's limited accuracy. Using actual medical records, therefore, will potentially improve the accuracy in the future. After the training procedure, generated test data were fed into the network for performance evaluations. The ANN was able to precisely predict the status of 112 samples (out of 200), as presented in Fig. 1(b). Moreover, no diseased patient was diagnosed as healthy or vice versa, representing ANN's reliability for medical applications.

Machine learning provides promising results for diagnosing various diseases and identifying trends in medical data. It fills the gap between patients, in remote locations, and the medical staff by providing accurate and real-time predictions on their health status or disease progress. Furthermore, it assists the health-care professionals by recommending treatment plans and medical prescriptions. Machine learning will play a pivotal role in the future for assisting patients with degenerative conditions, management of chronic conditions, monitoring patients rehabilitation progress, and predicting critical and emergency health conditions. Although machine learning is of a great importance for healthcare, various drawbacks and challenges of this approach have limited its real-world applications. The main drawback of this method is the security risk related to handling sensitive medical data which concerns patients' privacy. Precise security regulations need to be considered for designing short-range (device to smartphone via blue-tooth) and long-range (smartphone to backend using internet) communications for transferring data from medical devices to the backend. In addition, storage of the collected medical data in a single database increases the potential risk for malicious attacks [1]. Furthermore, acquired data require pre-processing and data curation–compression to comply with the broadband communications specifications, prior to their extraction from the device towards the backend. Moreover, enormous computational power with significant energy consumption is required for post-processing and analyzing the collected data at the backend to provide meaningful and real-time information to users and the medical staff.

The abovementioned limitations of machine learning in healthcare can possibly be addressed using neuromorphic platforms. A neuromorphic unit offers a hardware-based imitation of neural networks by using actual electrical components as neurons and synapses. By bringing the data post-processing from the backend onto the chip, real-time analysis of data in a less time consuming manner with a smaller time delay is feasible [9]. Furthermore, securing sensitive medical data on a single chip, without external communications, is far more practical. In addition, the energy-efficient neuromorphic platforms are relatively better immune against false operations and, therefore, offer a large fault tolerance for medical applications [10]. Therefore, the integration of the presented ANN model, for diagnosing COPD, on our previously developed hardware-based neuromorphic platform is the next goal of this work.

References

1. Baker, S.B., Xiang, W., Atkinson, I.: Internet of things for smart healthcare: technologies, challenges, and opportunities. IEEE Access **5**, 26521–26544 (2017). https://doi.org/10.1109/ACCESS.2017.2775180
2. Fogel, A.L., Kvedar, J.C.: Artificial intelligence powers digital medicine. NPJ Digit. Med. **1**, 5 (2018). https://doi.org/10.1038/s41746-017-0012-2
3. Zarrin, P.S., Jamal, F.I., Roeckendorf, N., Wenger, C.: Development of a portable dielectric biosensor for rapid detection of viscosity variations and its in vitro evaluations using saliva samples of COPD patients and healthy control. Healthcare **7**, 11 (2019). https://doi.org/10.3390/healthcare7010011
4. Wenger, C., et al.: Inherent stochastic learning in CMOS integrated HfO2 arrays for neuromorphic computing. IEEE Electron Device Lett. **40**, 639–642 (2019). https://doi.org/10.1109/LED.2019.2900867
5. Csikesz, N.G., Gartman, E.J.: New developments in the assessment of COPD: early diagnosis is key. Int. J. Chronic Obstructive Pulm. Dis. **9**, 277 (2014). https://doi.org/10.2147/COPD.S46198
6. Barreiro, E., et al.: Cytokine profile in quadriceps muscles of patients with severe COPD. Thorax **63**, 2 (2008). https://doi.org/10.1136/thx.2007.088575
7. Soltani Zarrin, P., Jamal, F.I., Guha, S., Wessel, J., Kissinger, D., Wenger, C.: Design and fabrication of a BiCMOS dielectric sensor for viscosity measurements: a possible solution for early detection of COPD. Biosensors **8**, 78 (2018). https://doi.org/10.3390/bios8030078
8. Verdugo, P.: Supramolecular dynamics of mucus. Cold Spring Harb. Perspect. Med. **2**, 11 (2012). https://doi.org/10.1101/cshperspect.a009597
9. Heinis, T., Schmuker, M.: Neuromorphic hardware as database co-processors: potential and limitations. In: EDBT 2019, Lisbon, pp. 694–697 (2019). https://doi.org/10.5441/002/edbt.2019.89
10. Jeong, D.S., Kim, K.M., Kim, S., Choi, B.J., Hwang, C.S.: Memristors for energy-efficient new computing paradigms. Adv. Electron. Mater. **2**, 9 (2016). https://doi.org/10.1002/aelm.201600090

Quantifying Structural Heterogeneity of Healthy and Cancerous Mitochondria Using a Combined Segmentation and Classification USK-Net

Manish Mishra[1,3], Sabine Schmitt[2], Hans Zischka[2], Michael Strasser[3], Nassir Navab[1,4], Carsten Marr[3], and Tingying Peng[1,3(✉)]

[1] Computer Aided Medical Procedures (CAMP), Technical University of Munich, Munich, Germany
tingying.peng@tum.de
[2] Institute for Molecular Toxicology and Pharmacology, Helmholtz zentrum muenchen, Oberschleißheim, Germany
[3] Institute of Computational Biology (ICB), Helmholtz zentrum muenchen, Oberschleißheim, Germany
[4] Computer Aided Medical Procedures (CAMP), Johns Hopkins University, Baltimore, USA

Abstract. Mitochondria are the main source of cellular energy and thus essential for cell survival. Pathological conditions like cancer, can cause functional alterations and lead to mitochondrial dysfunction. Indeed, electron micrographs of mitochondria that are isolated from cancer cells show a different morphology as compared to mitochondria from healthy cells. However, the description of mitochondrial morphology and the classification of the respective samples are so far qualitative. Furthermore, large intra-class variability and impurities such as mitochondrial fragments and other organelles in the micrographs make a clear separation between healthy and cancerous samples challenging. In this study, we propose a deep-learning based model to quantitatively assess the status of each intact mitochondrion with a continuous score, which measures its closeness to the healthy/tumor classes based on its morphology. This allows us to describe the structural transition from healthy to cancerous mitochondria. Methodologically, we train two USK networks, one to segment individual mitochondria from an electron micrograph, and the other to softly classify each image pixel as belonging to (i) healthy mitochondrial, (ii) cancerous mitochondrial and (iii) non-mitochondrial (image background & impurities) tissue. Our combined model outperforms each network alone in both pixel classification and object segmentation. Moreover, our model can quantitatively assess the mitochondrial heterogeneity within and between healthy samples and different tumor types, hence providing insightful information of mitochondrial alterations in cancer development.

© Springer Nature Switzerland AG 2019
I. V. Tetko et al. (Eds.): ICANN 2019, LNCS 11731, pp. 289–298, 2019.
https://doi.org/10.1007/978-3-030-30493-5_30

Keywords: Segmentation · Classification ·
Convolutional neural network · Deep learning · Cancer ·
Mitochondria · USK-net

1   Introduction

Mitochondria are essential for multiple critical functions in healthy cells [1]. They generate up to 90% of the cellular ATP, are essential for several anabolic and catabolic processes, serve as a calcium reservoir and are involved in the regulation of apoptosis. In cancerous cells, these functions are either impaired or altered to meet the specific demands of cellular maintenance and reproduction of cancer. Often, mitochondrial oxidative phosphorylation is affected and ATP production is reduced to 50% (known as Warburg hypothesis [2]), mitochondrial regulation of cell death breaks up, and cancerous mitochondria are involved in reprogramming adjacent stromal cells to optimise the cancer cell environment [3]. Furthermore, different cancer cell types undergo different bioenergetic alterations, some to more glycolytic and others to more oxidative states, leading to a highly complex pattern of mitochondrial alteration in cancer [3].

Mitochondrial dysfunctions in cancer is mirrored by a pronounced change in the morphology of these organelles. Distinct morphological features can be found in mitochondria that are isolated from liver cancer cell lines as compared to healthy liver tissue (as shown in Fig. 1). Moreover, considerable structural variability is found between different cell lines, which might be related to specific mitochondrial functional differences [4]. So far, descriptions of mitochondrial structure peculiarities remain largely qualitative and subject to individual expert interpretation, which limits a systematic examination of potential structural-functional correlation. A quantitative and objective structure-based assessment of mitochondria is thus required for a better understanding of mitochondrial dysfunctions in different cancerous environments and for the development of mitochondria targeting therapies.

Deep learning approaches are known to be versatile and adaptive to new environments and excel on a couple of recent biomedical challenges, like the classification of skin cancer [5] or the prediction of mutations from histopathological slides [6]. A first deep learning application in the mitochondrial classification was proposed [7], which trained a convolutional neural network (CNN) to classify each image patch of the micrographs into healthy or tumour classes. Though it achieves relatively high classification accuracy, this patch-based approach cannot filter out the confounding effects of non-mitochondrial structures in the images and its performance drops dramatically when applied to images with large proportions of impurities. Since high quality images with minimal amount of impurities are experimentally extremely challenging, and for some cell types nearly impossible, the application of [7] method is limited. Therefore, in the present study, we develop a joint segmentation-classification model to derive a continuous score for each intact tumour mitochondrion. With this mitochondrion-specific score, we can quantify mitochondrial heterogeneity within and between healthy

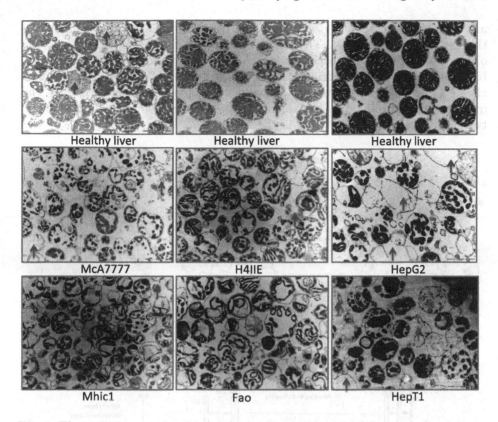

Fig. 1. Electron micrographs of isolated mitochondria from healthy mouse liver tissue (green boxes) and different liver tumor cell lines (red boxes, bars equal 1 /μ m). Intraclass variation is found for both healthy and tumor samples: While some mitochondria from healthy liver cells resemble cancerous structures (indicated by the red arrow) there are also mitochondria in cancer samples that resemble the healthy morphology (indicated by the green arrow in H4IIE). Furthermore, the images are corrupted by impurities and fragments from mitochondria and other organelles (indicated by blue arrows). (Color figure online)

tissue and different tumour subtypes, and describe the structural transition from healthy to tumorous tissue.

2   Method

We employ a joint segmentation-classification framework to derive structural score for each mitochondrion (Fig. 2). For each electron micrograph (Fig. 2a), a pixel classification module generates three probability maps for the healthy mitochondrial, cancerous mitochondrial, and non-mitochondrial classes (Fig. 2b). In parallel, an object segmentation module identifies each mitochondrion as a region of interest (Fig. 2c). Both networks use a U-shape sparse kernel structure, and are termed ClassUSK and SegUSK, respectively, and differ in output layer and the loss function (Fig. 2e). The outputs of both networks are overlaid on the original image: the transparent green, red and blue represents pixel-level probability of healthy mitochondrial, cancerous mitochondrial and non-mitochondrial classes, whilst the white outlines are segmented object contours (Fig. 2d). The quantification of individual mitochondrial structure combines the results of the two networks, with details given in the Subsect. 2.3.

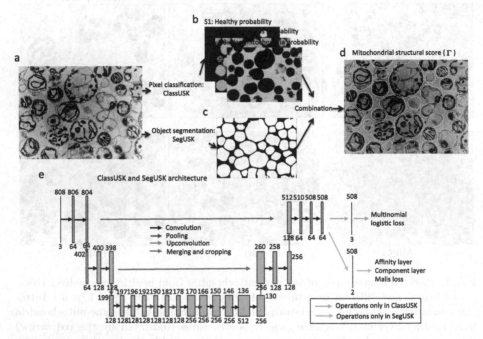

Fig. 2. Schematic plot of our joint segmentation-classification framework to derive structural score for each intact mitochondrion. For each electron micrograph (a), a pixel classification module, ClassUSK, generates three probability maps to quantify the probability of each pixel that belong to the healthy mitochondrial, cancerous mitochondrial, and non-mitochondrial classes (b). An parallel object segmentation module, SegUSK, identifies each mitochondrion as a region of interest (c). Both networks share the same network architecture but differ in output layer and the loss function (e). The outputs of the two networks are overlaid on the original image: the transparent green, red and blue represents pixel-level probability of the three classes respectively, whilst the white outlines are segmented object contours (d). (Color figure online)

The USK-net proposed by [8] is a combination of the U-shape subnetwork (U-net) [9,10] and sparse kernel subnetwork (SK-net) [11]. Like U-net, the USK-net has a contraction path which use convolution and pooling layers with down-sampling for efficient feature learning, and an expansion path which use upsampling and deconvolution to locate the learned feature back to the original image space. The shortcut between the contraction path and expansion path preserves edge information and hence to contribute for an accurate segmentation on the object boundaries. The bottom of USK-net is composed of a SK-net, which uses sparse kernel technique for an efficient feature learning with location information. Compare to U-net and SK-net architectures of similar input and output size, USK-net has only 25% of U-net and SK-net weights and hence is much easier to be trained, particularly with limited amount of data [8].

2.1 Pixel Classification

We classify each pixel into three classes using ClassUSK: (i) healthy mitochondrial, (ii) cancerous mitochondrial and (iii) non-mitochondrial structures. We use a soft-max layer as our output layer so that we can obtain a continuous score that measures the probability of one pixel belonging to each class, mis-classified rather than a single label. One key factor of a successful classification is that the context size considered for around each pixel in its classification should be large enough to cover an entire mitochondrion, as it is difficult to differentiate a fraction of the inner structure of a intact mitochondrion from impurities that include a broken mitochondrial fragment (as shown in Fig. 1). In the original USK-net structure proposed in [8], the context size is 180×180 pixels, which is not sufficient for our application, as some mitochondria in specific cancer cell lines (e.g. Fao) have a very large size. Hence, we use two downsampling and upsampling layers to increase the context size of our USK-net to 300×300 pixels (Fig. 2e).

2.2 Object Segmentation

Since our aim is to derive one score per mitochondrion, we have to segment individual mitochondria as objects of interest. A simple way to obtain segmentation is to invert the classification result of the non-mitochondrial structures (class iii). This approach however is error-prone in practice, since e.g. as small amount of mis-classified pixels between two closely neighboring mitochondria could merge them as a single mitochondrion, and mis-classified pixels inside one mitochondrion would break into two or more individual mitochondria (an evaluation is provided in the Results section). Since this kind of mis-segmentation is usually difficult to be restored with post-processing techniques (such as morphological filtering), we train a separate network to achieve an end-to-end segmentation.

The architecture of our SegUSK is very similar to ClassUSK, but with an affinity layer and a component layer after soft-max layer (Fig. 2e). The affinity layer explores neighboring pixels and compute connectivity among pixels. The component layer is based on the OpenCV flood-filling algorithm and outputs separated connected components from a foreground-background labeled ground truth. The Malis (maximum affinity learning of image segmentation) loss function takes the output of the affinity and component layers and attributes the error map between the prediction and the ground truth for backpropagation (refer to [8] and [12] for details of affinity layer, component layer and Malis loss calculation). Malis loss has recently been used for neuron segmentation on electron microscopy images and improves upon other state-of-the-art segmentation techniques [13]. Here we only use SegUSK to segment mitochondria from background, regardless of its tissue type.

2.3 Mitochondrion-Specific Structural Score

With ClassUSK, we can calculate a normalised per-pixel score that measures the structural transition from healthy to cancerous class by:

$$\Gamma_{\text{pixel}} = \frac{e^{s_2 - s_1}}{e^{s_1 - s_2} + e^{s_2 - s_1}} \tag{1}$$

where s_1 and s_2 are the probabilities belonging to healthy class and cancerous class, respectively. A weighted average of the per-pixel scores within the segmented region of a single mitochondrion serves as a measure of its closeness to the two classes, yielding:

$$\Gamma_{\text{mitochondrion}} = \frac{\sum_{\text{pixel} \in \text{mitochondrion}} w_{\text{pixel}} * \Gamma_{\text{pixel}}}{\sum_{\text{pixel} \in \text{mitochondrion}} w_{\text{pixel}}}, \quad w_{\text{pixel}} = 1 - s_3 \tag{2}$$

$$\text{only if} \quad \frac{1}{n_{\text{pixel}}} \sum_{\text{pixel} \in \text{mitochondrion}} s_3 > 0.5 \tag{3}$$

where the individual mitochondria segmentation is provided by the prediction of SegUSK and s_3 is the probability belonging to non-mitochondrial class. We add the condition Eq. (3) here to filter out the mistakes made by SegUSK in segmenting impurities as mitochondria (Fig. 2d).

2.4 Acquisition of Mitochondrial Images and Network Training

Mitochondria from rat liver were isolated according to Schulz et al. [14] and from cultured cells (McA 7777, MH1C1, H4IIE, Fao, HepG2 and HepT1) according to Schmitt et al. [15]. Electron microscopy (EM) of the isolated mitochondria was done as described previously [16]. Briefly, samples were fixed in 2.5% glutaraldehyde, postfixation and prestaining was done with osmium tetroxide. After dehydration with ethanol and propylene oxide, samples were embedded in Epon. Ultrathin sections were stained with uranylacetate and lead citrate and examined with an EM 10 CR transmission electron microscope (Zeiss, Germany).

Our dataset comprises 163 images from of healthy samples mitochondria and 240 images from of cancer samples mitochondria (100 McA7777, 66 H4IIE, 56 MH1C1, 9 Fao, 8 HepG2 and 11 HepT1). Each image has a size of 1032×1376 pixel ($20000\times$ magnification). We annotate 60 healthy images and 60 tumour images (randomly sampled from different tumour cell lines) by manually outlining the contours of each mitochondrion. All mitochondria from healthy images are considered as healthy while all mitochondria from tumour images be cancerous, regardless of their morphology. Among these annotated images, we use 45 healthy and 45 tumour images to train our networks and reserve the remaining 15 healthy and 15 tumour images to test networks. In the preprocessing step we augment and normalize our data using caffe neural tool [8] in an on-the-fly fashion. The entire training process takes around 6 h for each network using three GPUs and the testing time is 0.02 s per image.

3   Results and Discussion

3.1   Evaluation of ClassUSK, SegUSK and CombUSK

We evaluate ClassUSK, SegUSK and our combination framework (termed CombUSK here) on our reserved test dataset (30 images, 15 healthy tissue, 15 tumour). We first consider the classification accuracy on pixel level. As shown in Fig. 3a, both ClassUSK and CombUSK gain a high pixel classification accuracy of $87.8 \pm 4.2\%$ (mean $\pm$ s.d.) and $88.1 \pm 4.2\%$ respectively. By contrast, the performance of SegUSK (mitochondria vs non-mitochondria) is poorer, $80.3 \pm 5.7\%$, as it tends to predict impurities in the background into the foreground, i.e. mitochondrial class. The evaluation criteria for object segmentation is based on counting the number of correctly segmented mitochondria (Dice coefficient $>80\%$), divided by the average of total number of ground-truth mitochondria and that of automatically segmented mitochondria (similar to the definition of the Dice coefficient, but number based rather than pixel based). For ClassUSK, mitochondrial segmentation is generated through labeling the connected components of the inverted label map of the predicted non-mitochondrial class. This approach performs poorly when mitochondria are nearby and touching, since ClassUSK tends to merge them into a single mitochondrion (shown in the exemplary segmentation of Fig. 3b, red arrows). In comparison, although SegUSK performs better in separating nearby mitochondria and generating more clear boundaries, it cannot differentiate impurities from real mitochondria, probably as both have similar shape features. The best performance an astonishing accuracy of $96.4 \pm 3.3\%$ (less than one mitochondrion error per image), is achieved with CombUSK, as it effectively overcomes the disadvantages of both networks.

3.2   Heterogeneity Quantification of Mitochondrial Structures

Our structural score, $\Gamma_{mitochondria}$, does not only allow us to assign single mitochondria to healthy and cancerous states, but also to states in between the two extremes. In Fig. 4, we show exemplary images of mitochondria for different ranges of our score. The structural differences suggest that the transition from healthy to cancerous mitochondria is not an instant switching, but rather a continuous process. This is quantified by our structural score: the intensely stained mitochondrial matrix tends to connect the inner-membrane on both sides in mitochondria with a low score. These connections begin to break up when the score increases and the matrix shrinks into isolated dots in mitochondria with a high score. Distributions of scores for different samples exhibit the clear differences between healthy and cancerous cells: the majority of healthy mitochondria have a score under 0.2 (Fig. 4), while most of cancerous mitochondria have a score above 0.8. An exception is the HepT1 tumour cell line, which shows an increased heterogeneity in structural scores as compared to other tumour cell lines. This is an inspring finding as HepT1 is known to be resistant to chemotherapy. The varying mitochondrial structural patterns identified by our structural score might reflect the resistance against single chemotherapeutics.

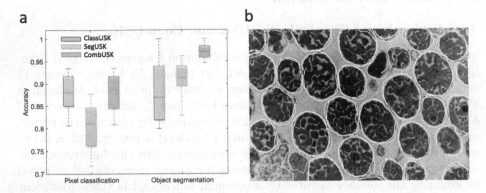

Fig. 3. CombUSK gives rise to highly accurate mitochondria segmentation. (a) Evaluation of ClassUSK, SegUSK and CombUSK on 30 test images. (b) Exemplary mitochondrial segmentation (ground-truth: white outlines, red: ClassUSK, green: SegUSK, blue: CombUSK, note that Cyan outlines are due to an overlay effect of SegUSK and CombUSK). Red arrow indicates a merged segmentation of three closely-attached mitochondria by ClassUSK. Green arrow indicates an mis-segmentation of impurities as mitochondria by SegUSK. (Color figure online)

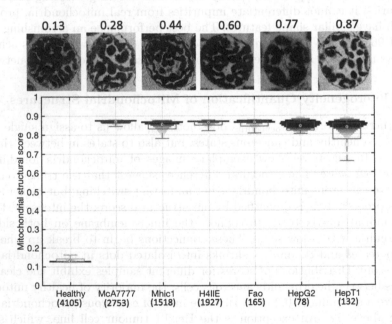

Fig. 4. Quantifying structural heterogeneity of healthy and cancerous mitochondria. A score from 0 to 1 represents the structural transition from healthy to cancerous mitochondria, with six exemplary mitochondria shown on the top.

4   Conclusion

In this study, we develop a combined segmentation-classification approach to automatically segment mitochondria in electron microscopy images and to quantify the structural transition from healthy to tumor samples using a continuous score from 0 to 1. We demonstrate that the combination of the two parallel networks, ClassUSK and SegUSK, achieves superior results than using each network alone in both pixel classification and object segmentation, hence leading to the generation of a more reliable mitochondrial structural score. Using this mitochondrion-specific score, we can quantify mitochondrial structural heterogeneity of healthy tissue and different tumour subtypes. The presented study demonstrates that deep learning is not only a passive learning tool that simply repeats human annotations, but may have the potential to derive objective quantification beyond human observations and to provide new insights to biologists in cancer research.

References

1. Alirol, E., Martinou, J.C.: Mitochondria and cancer: is there a morphological connection? Oncogene **25**(34), 4706–4716 (2006)
2. Warburg, O.: On the origin of cancer cells. Science **123**(3191), 309–314 (1956)
3. Wallace, D.C.: Mitochondria and cancer. Nat. Rev. Cancer **12**(10), 685–698 (2012)
4. Smith, R.A.J., Hartley, R.C., Cochemé, H.M., Murphy, M.P.: Mitochondrial pharmacology. Trends Pharmacol. Sci. **33**(6), 341–352 (2012)
5. Esteva, A., et al.: Dermatologist-level classification of skin cancer with deep neural networks. Nature **542**(7639), 115–118 (2017)
6. Coudray, N., et al.: Classification and mutation prediction from non-small cell lung cancer histopathology images using deep learning. Nat. Med. **24**(10), 1559–1567 (2018)
7. Mishra, M., et al.: Structure-based assessment of cancerous mitochondria using deep networks. In: ISBI, pp. 545–548. IEEE (2016)
8. Tschopp, F.: Efficient convolutional neural networks for pixelwise classification on heterogeneous hardware systems. CoRR abs/1509.03371 (2015)
9. Ronneberger, O., Fischer, P., Brox, T.: U-Net: convolutional networks for biomedical image segmentation. In: Navab, N., Hornegger, J., Wells, W.M., Frangi, A.F. (eds.) MICCAI 2015. LNCS, vol. 9351, pp. 234–241. Springer, Cham (2015). https://doi.org/10.1007/978-3-319-24574-4_28
10. Falk, T., et al.: U-Net: deep learning for cell counting, detection, and morphometry. Nat. Methods **16**(1), 67–70 (2019)
11. Li, H., Zhao, R., Wang, X.: Highly efficient forward and backward propagation of convolutional neural networks for pixelwise classification, December 2014
12. Turaga, S.C., Briggman, K.L., Helmstaedter, M., Denk, W., Seung, H.S.: Maximin affinity learning of image segmentation. CoRR abs/0911.5372 (2009)
13. Funke, J., et al.: Large scale image segmentation with structured loss based deep learning for connectome reconstruction. IEEE Trans. Pattern Anal. Mach. Intell. **24**, 1669–1680 (2018)

14. Schulz, S., et al.: A protocol for the parallel isolation of intact mitochondria from rat liver, kidney, heart, and brain. In: Posch, A. (ed.) Proteomic Profiling. MMB, vol. 1295, pp. 75–86. Springer, New York (2015). https://doi.org/10.1007/978-1-4939-2550-6_7

15. Schmitt, S., Eberhagen, C., Weber, S., Aichler, M., Zischka, H.: Isolation of mitochondria from cultured cells and liver tissue biopsies for molecular and biochemical analyses. In: Posch, A. (ed.) Proteomic Profiling. MMB, vol. 1295, pp. 87–97. Springer, New York (2015). https://doi.org/10.1007/978-1-4939-2550-6_8

16. Zischka, H., et al.: Electrophoretic analysis of the mitochondrial outer membrane rupture induced by permeability transition. Anal. Chem. 80(13), 5051–5058 (2008)

Breast Cancer Classification on Histopathological Images Affected by Data Imbalance Using Active Learning and Deep Convolutional Neural Network

Bogdan Kwolek[1,2]([✉]), Michał Koziarski[1,2], Andrzej Bukała[2],
Zbigniew Antosz[2], Bogusław Olborski[2], Paweł Wąsowicz[2], Jakub Swadźba[2,3],
and Bogusław Cyganek[1,2]

[1] Faculty of Computer Science, Electronics and Telecommunications,
AGH University of Science and Technology, 30 Mickiewicza, 30-059 Kraków, Poland
{bkw,cyganek}@agh.edu.pl
[2] Diagnostyka Consilio, 93-357 Łódź, Poland
sekretariat.consilio@diag.pl
[3] Department of Laboratory Medicine, Faculty of Medicine,
Andrzej Frycz Modrzewski Kraków University, 30-705 Kraków, Poland

Abstract. In this work, we propose an algorithm for training deep neural networks for classification of breast cancer in histopathological images affected by data unbalance with support of active learning. The output of the neural network on unlabeled samples is used to calculate weighted information entropy. It is utilized as uncertainty score for automatic selecting both samples with high and low confidence. A number of low confidence samples that are selected in each iteration is manually labeled by pathologist. A threshold that decays over iteration number is used to decide which high confidence samples should be concatenated with manually labeled samples and then used in fine-tuning of convolutional neural network. The neural network can optionally be trained using weighted cross-entropy loss to better cope with bias towards the majority class.

1 Introduction

In the last decade, a significant effort has been put forth for breast cancer recognition from histological slides. Histological slides allow the pathologist to distinguish between the normal tissue, non-malignant (benign) tissue, and malignant lesions. Currently, substantial efforts are devoted to recognize the two fundamental types of breast cancer with Computer Aided Diagnosis (CAD) [1]. Various computer-based approaches for analysis of histological images have been proposed to support pathologists in quantifying morphological features [2], detecting malignant lesions [3], and predicting prognosis for breast cancer [1].

Prior to a visual analysis by pathologist the tissue samples are collected during biopsy and then stained with Hematoxylin and Eosin (H&E). Afterwards,

© Springer Nature Switzerland AG 2019
I. V. Tetko et al. (Eds.): ICANN 2019, LNCS 11731, pp. 299–312, 2019.
https://doi.org/10.1007/978-3-030-30493-5_31

in traditional approach the pathologist examines microscopic images of the tissue samples from the biopsy with different magnification factors. To make the correct diagnosis, the doctor investigates various image features including patterns, textures, and different morphological properties [4]. Different magnification factors are inherent in analysis of histological images, and require panning, zooming, focusing, and the whole diagnosis process is very time consuming and tiresome. The diagnosis results are influenced by many subjective factors. As a consequence, such manual process sometimes leads to erroneous or insufficient diagnosis for breast cancer identification. Manual classification of histological images is laborious for pathologists, prone to inconsistencies, expensive and time-consuming. In some cases, detailed analysis of a single case could require several slides with multiple stainings. Moreover, pathologists undergo pressure to handle large volumes of cases while providing a larger amount of information in the pathology reports. In order to minimize risks associated with improper diagnosis of cancer, as well as to provide a support for pathologists in preparing reports, various image processing and recognition techniques have been elaborated for analyzing pathological images at microscopic resolution [5]. Unfortunately, traditional computer signal processing and computer vision techniques are not able to meet the requirements and to fulfill the expectations of clinicians.

Owing to advancement of digital imaging techniques, a remarkable progress in histological image processing and recognition has been made [1]. Modern whole slide image (WSI) scanners can process entire tissue slices and deliver high-resolution images. WSIs are very large in size and contain huge information. Such images are characterized by small inter-class variance and large intra-class variance. Moreover, features extracted from similar histological images with unlike magnification are usually very different. Thus, automatic classification of breast cancer pathological images is a challenging task.

The approaches to automatic classification of breast caner pathological images can be divided into two groups: methods based on feature engineering and classical machine learning, and methods based on feature/deep learning. In contrast to traditional approaches, which rely on hand-crafted features, recent algorithms learn useful features directly from the training image patches by the optimization of the loss function. The history of extracting handcrafted features for breast cancer recognition image classification is a long one [2]. In [6], a multiple magnifications-based framework for breast cancer on histopathological image classification has been proposed. The authors utilized various joint color-texture features and classifiers, and demonstrated that suitable feature-classifier combinations can largely outperform relevant methods. Before 2017, the system proposed in [6] outperformed in terms of recognition accuracy all machine learning based approaches. A review of current deep learning based approaches to histopathological image analysis can be found in [7]. Several methods were proposed for classification of histopathological images, and they mainly differ in architectures of convolutional neural networks, data augmentation, etc.

As noted in a recent survey [7], the problem of insufficient labeled images is very important in the area of histopathological image analysis. Most of the approaches to cope with this issue fall into one of the following categories: (1) increasing the number of examples with annotations, (2) utilization of models/parameters for/from other tasks, or (3) exploitation of weak label or unlabeled data. Some research has been done on methods belonging to first two categories. In particular, several data augmentation and transfer learning based method were elaborated to improve learning of neural networks for histological image classification. However, little work has been done on exploitation of weak label or unlabeled data. One of the main reasons is that for histopathological image classification this is very hard and challenging task. One of the reasons that significant effort has only been devoted to classification of histopathological images is that almost all research group are working on freely available datasets, like BreaKHis [8], and thus they underestimate the costs and difficulties associated with data annotation. As previously mentioned, WSIs are very large in size, contain huge information and manual labeling of such images, which requires a highly qualified pathologists, is very time consuming. In [9] an interactive machine-learning system for digital pathology has been proposed. The proposed framework utilizes active learning to direct user feedback, making classifier training efficient and scalable in datasets containing huge amount of histologic objects.

In this work, we propose an algorithm for training deep neural networks for classification of breast cancer in histopathological images with support of active learning (AL). Instead of random selection, AL methods typically actively select samples with lowest confidence as the most valuable samples to add them to the query and finally train the model incrementally [10]. Randomly selecting samples instead of actively choosing samples establishes a lower bound. In the proposed method, both samples with high and low confidence are included in the query. We utilize information entropy as uncertainty score for automatic selection of both samples with high and low confidence. A threshold that decays over iteration number is used to decide which high confidence samples should be concatenated with manually labeled samples and then used in fine-tuning of a convolutional neural network. A *weighted* entropy [11] is calculated on the basis of prediction of CNN, which is tuned in every iteration in such a way. The pool of labeled samples is updated with newly labeled samples by the pathologist. Such high confidence samples for the labeling are selected automatically by the algorithm on the basis of weighted entropy after selecting the uncertain samples.

The contribution of this work is as follows: first, we propose an improved deep convolutional neural network model to achieve accurate and precise classification or grading of breast cancer pathological images. Meanwhile, online data augmentation, transfer learning and fine-tuning strategies are employed to avoid model overfitting effectively. Second, we propose an active learning scheme for fine-tuning a deep residual convolutional neural network on unbalanced data. Finally, experimental results based on freely available pathological image dataset show that the performance of our method is better or at least comparable with recent state-of-the-art methods for breast cancer classification on histopathological images, with good robustness and generalization ability.

2   Breast Cancer Classification on Histopathological Images

During analysis of the stained tissue, pathologists examines overall architecture of tissue, along with nuclei layout, density and variability. The diagnosis process using H&E stained biopsies is not trivial, and the average diagnostic compatibility between pathologists is about 75% [12]. Most of the WSI scanners that are presently in the use carry out slide scanning at ×20 or ×40 magnification with spatial resolution in order of $0.5\,\mu$/pixel and $0.25\,\mu$/pixel, respectively. One of the major difficulties in breast cancer histopathology image analysis is variability of appearance, which is mostly the result of variations in the conditions of the tissue preparation and staining processes. The color appearance can significantly vary due to differences in fixation and in staining processes. A typical histopathology slide comprises a tissue area of about $15 \times 15\,mm$. Considering the resolutions on which the slides are scanned, the scanned images have size of up to several gigapixels. Taking into consideration that classification of larger images requires far larger number of parameters, as well as that typical WSI image can consists of as many as tens of billions of pixels, the WSI scans are divided into patches of size a few hundred pixels times a few hundred pixels, see also Fig. 1, which are then analyzed independently. The mentioned figure depicts sample histopathological images [8] with two fundamental types of breast cancer: benign and malignant.

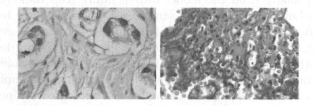

Fig. 1. Sample histopathological images with two fundamental types of breast cancer: benign (left) and malignant (right).

In the area of digital pathology, there are available some freely available datasets that contain hand-annotated histopathological images and corresponding labels. In [13], a performance comparison between four machine learning algorithms, including decision tree (DT), Naïve Bayes (NB), k-nearest neighbors (KNN) and support vector machine (SVM) on the Wisconsin Breast Cancer dataset [14], which consists of 699 instances (458 benign and 241 malignant cases). Experimental results demonstrated that the SVM classifier achieves the highest accuracy of 97.13% with 10-fold cross-validation. BreaKHis dataset [8] contains 7 909 histopathological images of breast cancer from 82 patients. The authors employed six different feature descriptors and four different classical machine learning methods, including 1-NN (1 Nearest Neighbor), QDA

(Quadratic Discriminant Analysis), RF (Random Forest), and SVM with the Gaussian kernel function to carry out classification of benign and malignant tumors. The classification accuracy is between 80 and 85% in 5-fold cross-validation. Although classical machine learning methods demonstrated great usefulness in digital pathology, present deep learning-based methods outperform traditional ones. In [15] a modified AlexNet [16] convolutional neural network improved classification accuracy by 4–6% on BreaKHis dataset. A CNN-based method proposed in [17] can classify breast cancer histopathological images independently of their magnifications. Two different architectures were studied: a single task CNN to predict malignancy, and a multi-task CNN to predict both malignancy and image magnification level simultaneously. Evaluations were carried out on the BreaKHis dataset, and the experimental results were competitive with state-of-the-art results achieved by classical machine learning methods. In [18] a pre-trained ResNet_V1_152 [19] has been applied to carry out diagnosis of benign and malignant tumors on BreaKHis as well as multi-class classification of various subtypes of histopathological images of breast cancer. This deep CNN achieved an accuracy of 98.7% and 96.4% for binary classification and multi-class classification, respectively. It is worth noting that although there are 7 909 histopathological images in the BreaKHis dataset, the number of images is far from enough for effectively using current deep learning techniques.

Another difficulty in breast cancer classification on histopathological images is class imbalance [20]. It can be observed in several histopathological benchmark datasets, including BreaKHis. As underlined in [20], it is largely unclear to what extent the data imbalance affects the performance of deep learning algorithms for histopathological image analysis, and what techniques [21] should be applied to learn from unbalanced data. Overall, class unbalance is very important problem, which is, however, frequently neglected in many evaluations on BreaKHis dataset. Moreover, several studies reports only the classification accuracy, which is sometimes badly chosen as the only metric to judge the classification performance on the imbalanced data.

Significant effort has been devoted to classification of histopathological images. Several research group performed evaluations on freely available datasets with histopathological images, and thus they were not involved in manual data annotation, which is very costly and time consuming task. Active learning is a machine learning technique, which is generally effective when the acquisition cost of label data is substantial [7]. It can be used to support supervised learning for automatic choosing the most valuable unlabeled sample(s) (i.e. the one(s) that could led to improved classification performance when labeled properly and included in training data) and display it for manual labeling by pathologists. Most active learning focuses on selecting examples from a so-called area of uncertainty, i.e. space that is nearest to the model's decision boundary, which for a binary classification problem can be expressed as: $x' = \arg\max_x \min_y P(y|x), y \in \{0,1\}$. However, simple selecting samples from an unlabeled pool with considerable data imbalance may pose some practical difficulties. The greater fraction of examples in the majority class may result in a model preferring one class over another. If labels of samples selected by an active learning algorithm are considered as random variables, the class imbalance would result in preference for majority examples in the training dataset subset, i.e. over-representation. Unless properly treated such an over-representation may result in a model with predictive preference for the majority class when labeling. This important issue motivated us to explore active learning on unbalanced data for cancer classification on histopathological images by the use of convolutional neural networks.

3   The Algorithm

The aim of active learning algorithms is to attain best possible performance of the learned model with as few labeled samples as possible [10]. Standard AL algorithms run in an iterative manner and in each learning step usually select only a few of the most informative samples, i.e. samples that have quite low prediction confidence, and frequently engage the user to label the recommended data. The algorithm stops when a predefined stopping criteria is met. In case of unbalanced data, conventional classification algorithms are often biased towards the majority class because their loss functions attempt to optimize quantities, which do not take into account the data distribution. In the worst case, particularly when the dataset is severely unbalanced, minority examples can be treated as outliers of the majority class and ignored. The learning algorithm simply produces a trivial classifier with tendency to classify every example as the majority class. An approach proposed in [22] performs active learning using both majority & clearly classified samples and minority & most informative samples. However, it is unable to cope with unbalanced data due to reasons mentioned above.

We use weighted information entropy [11] as uncertainty score. Weighted entropy is a measure of information supplied by a probabilistic experiment whose elementary events are characterized both by their objective probabilities and by some qualitative (objective or subjective) weights associated with the events. We characterize each event x_i by $\{p_i, w_i\}$, $i = 1, \ldots, n$, $\sum_{i=1}^{n} p_i = 1$, $w_i \leq 0$, where p_i is the probability of the event x_i and weight w_i quantifies qualitative aspect of x_i. The weighted entropy of $\{p_i, w_i\}$, $i = 1, \ldots, n$ is defined as:

$$H_w(w_1, w_2, \ldots, w_n; p_1, p_2, \ldots, p_n) = -\sum_{k=1}^{n} w_k p_k \log p_k \qquad (1)$$

In our approach the weights w_i are determined on the basis of class probabilities. We give higher weight to minority class and lower weight to majority class.

The pseudo-code of the algorithm is listed below. The input arguments are as follows: max_it – maximum number of iterations, unc_samp_size – number of the most uncertain samples that are selected in each iteration for manual labeling by the pathologist, delta – initial threshold value that is used in selection of the most confident samples, delta_decay – smaller than zero factor to decrease the value of delta if the classifier performs better, x_init, y_init –

Algorithm 1. Active learning on unbalanced data

Input: `max_it`, `unc_samp_size`, `delta`, `delta_decay`, `x_init`, `y_init`, `x_pool`, `-`, `W`

```
 1: PL = x_init, y_init
 2: uratio = num(y_init==0) / num(y_init==1)
 3: PU = x_pool, -

 4: train(PL, W, uratio)
 5: W = load_weights

 6: for i in range(max_it):
 7:     y_pred_prob = predict(PU, W)
 8:     un_idx = get_uncertain_samples( y_pred_prob, unc_samp_size, uratio)
 9:     y_pool[un_idx]           # perform manual labeling of PU[un_idx]
10:     PL = append(PL, PU[un_idx])
11:     hc_idx = get_high_confid_samples(y_pred_prob, delta, uratio)
12:     hc = hc_idx - un_idx # remove samples also selected as uncertain
13:     PH = PU[hc]
14:     ptrain = concatenate(PL, PH)
15:     train(ptrain, W)          # optionally: train(ptrain, W, uratio)
16:     PU = delete(PU[un_idx])
17:     W = load_weights
18:     delta = delta * delta_decay
19:     uratio = num(y_pred_prob==0) / num(y_pred_prob==1)
20:     acc = evaluate(x_test, y_test, W)
21: return W
```

initial training pool consisting of samples and corresponding labels, `x_pool` – unlabeled pool of data samples, `W` – weights of the pre-trained neural network. The factor `uratio`, see line #2 in Algorithm 1, is used to express unbalance in dataset, and it can be determined as ratio of number of samples in each class. The output of the predictor, see line #7 in Algorithm 1, is used to calculate the weighted entropy, which in turn is used to determine the most uncertain samples, see line #8, as well as high confidence samples, see line #11. The number of uncertain samples selected in each iteration from the unlabeled pool `PU` depends on value of the predefined constant `unc_samp_size`. After selecting such samples, see index `un_idx`, a pathologist manually assigns labels to the recommended samples. The labeled pool `PL` is progressively updated in each iteration by `PU[un_idx]`, see line #10. The discussed pool `PL` is concatenated with high confidence pool `PH` and then used to fine-tune the neural network, i.e. to update weights `W`. This means that the samples from the unlabeled pool `PU` are progressively fed into the convolutional neural network. Depending on the option, it can be trained using commonly used categorical cross-entropy or weighted categorical cross-entropy in order to better cope with unbalanced data. The indexes of samples from `PU` to be included in `PH` are calculated on the basis of the method `get_high_confid_samples`, which selects samples whose weighted entropy is smaller than `delta`, see line #11. The `delta` variable is updated in every iteration, see line #18. In this way, more and more confident samples are selected in subsequent iterations. The `uratio` factor is calculated on the basis of the predictions `y_pred_prob`. Since the `get_high_confid_samples` method can select samples that were previously selected by `get_uncertain_samples` method, the algorithm excludes samples with indexes `un_idx`, which were labeled by the pathologist. During training of the neural network in a predefined number of epochs, the best weights are stored, and then loaded before evaluation of the accuracy, as well as calculating the prediction in the next iteration. The pool of unlabeled samples `PU` is updated at the end of each iteration. To overcome the influence from the imbalanced histopathological images in subclasses, the minority class can be additionally balanced by turning images up and down, right and left, and rotating them counterclockwise by 90 and 180°.

4   Experimental Results and Discussion

We investigated the performance of various convolutional neural networks in breast cancer classification on histopathological images. It is well known that learning hyperparameters have a great influence on the performance of the trained CNN model, particularly the learning rate. Thus, in addition to investigations on various models of neural networks for active learning on unbalanced data, we devoted a considerable attention to selecting the learning hyperparameters for training. Due to limited amount of training images with breast cancer and model over-fitting risks we investigated techniques for reducing the number of CNN parameters as well as various data augmentation methods. The evaluations were realized on BreaKHis dataset consisting of 2480 images in benign class and 5429 images in malignant class. We randomly selected 6418 images for training subset, 802 images for validation and 689 images for test subset.

4.1   Breast Cancer Classification on Histopathological Images Using Deep Convolutional Neural Networks

At the beginning we investigated transfer learning of pre-trained VGG16 convolutional neural network for breast histopathology image classification. We utilized VGG16 with weights learned on imageNet dataset without the top layer. We extracted the features delivered by the VGG without the top layer and stored them for the future use. Next, on features with shape $(7, 7, 512)$ we trained a convolutional neural network consisting of 256 filters in the first layer and relu activation, dropout layer and output layer with sigmoid activation. The network has been trained using binary cross-entropy loss and RMSprop algorithm with learning rate set to 2e-5. The batch size has been set to 32 and training was in 200 epochs. The classification accuracy on test data was equal to 87.16%. Afterwards, we investigated fine-tuning of VGG16 with on-line data augmentation. The weights of VGG16 network without the top layer were frozen. After flattening the last layer of such a base network, we added a dense layer consisting of 256 neurons with relu activation, dropout set to 0.5 and an output neuron with activation set to sigmoid. The network has been trained using binary cross-entropy loss and RMSprop algorithm with learning rate set to 2e-5. During training we executed online data augmentation (rotation, horizontal/vertical shift, image flip). The batch size has been set to 32 and training was in 30 epochs. The classification accuracy on the test data was equal to 85.7%. Finally, the block5_conv1 layer has been set as trainable and complementary training of the network with such a trainable layer has been done in 100 epochs. The classification accuracy on the test dataset improved to 96.5%.

In the next stage we investigated resNet neural networks. Similarly to experiments with the VGG network, the fully connected layer of the pre-trained network has been replaced with a new fully connected layer. The network was pre-trained using the same parameters on the same training data as the VGG16 neural network. Next, the res5a_branch2a layer has been set as trainable and complementary training of the network with such a trainable layer has been done in 60 epochs. During training the network achieved 99.0% accuracy on the validation data. On test data the accuracy was equal to 97.8%.

In the last stage of this part of experiments we investigated pre-trained resNet18 neural network with weights learned on the imageNet dataset. The fully connected layer of the network has been replaced with a new fully connected layer with 256 neurons and the network has been pre-trained on the same training data using identical parameters and online data augmentation. The best classification accuracy that we obtained was about 75%. Finally, the zero_padding2d_18 layer has been set as trainable and complementary training of the network with such a trainable layer has been done in 60 epochs. On the test data the classification accuracy improved to 91.5%.

The above experiments demonstrated that by setting in the base neural network the last layer as trainable and then extending such a base network about a dense layer, pre-trained in advance and fine-tuning the neural network in several epochs can lead to substantial improvement in classification performance. Bearing in mind that resNet18 neural network has far smaller number of training parameters in comparison to VGG16 and resNet50 neural networks, and thus the fine-tuning can be done in far shorter time, and particularly that our focus was on developing best strategies for active learning for breast cancer classification on histopathological images affected by data unbalance and not just experimenting with neural network architectures to obtain improvements in classification accuracy, the resNet18 neural network has been selected for further investigations on active learning algorithms.

4.2 Breast Cancer Classification on Histopathological Images Using Active Learning and Deep Convolutional Neural Network

In the next stage of the experiments we investigated active learning algorithms for breast cancer classification on histopathological images affected by data unbalance. The active learning algorithm was based in resNet18 neural network with weights learned on the imageNet dataset. The weights of the network without the top layer were frozen. The fully connected layer of the network has been replaced with a new fully connected layer consisting of 256 neurons with

relu activation and subsequent dropout set to 0.5. The output of the network
was softmax layer. This means that the zero_padding2d_18 layer that in the
previous experiment was set as trainable, and thanks to which the classification
accuracy considerably improved, in the discussed evaluations has been frozen.
From the training data, twenty percent of training samples were selected for
initial training of the neural network. The neural network has been initially
trained in five epochs, and size of batch with data shuffling equal to 32. To better cope with imbalanced data, it has been trained using *weighted* categorical
cross-entropy and RMSprop algorithm with learning rate set to 2e-5. The class
weights were determined on the basis of data labels in the initial data. The best
weights obtained in the training were stored and then utilized to initialize the
network in the subsequent iterations, see line #4 − 5 in Algorithm 1. The active
learning has been performed in nine iterations. In each iteration a complementary training of the network in five epochs has been carried out. The number of
uncertain samples, i.e. samples labeled in each iteration by pathologist has been
set to 300.

Table 1 presents results that were achieved in experimental evaluations. The
presented results are averages of scores from ten independent runs of each considered algorithm with unlike weights initializations. First row contains results
that were achieved by a baseline active learning algorithm with samples selected
randomly for annotation by a pathologist. The discussed algorithm does not
use the high-confident samples as pseudo-annotated data. Second row contains
scores that were obtained by active learning algorithm using samples that were
selected on the basis of weighted categorical cross-entropy. In a similar way to
the previous algorithm, it does not use the high-confident samples as pseudo-
annotated data. As we can observe, this algorithm achieves better classification
performance. Third row contains scores that were obtained by algorithm using
samples that were selected on the basis of categorical cross-entropy, and in which
high-confident samples were utilized as pseudo-annotated data. The accuracy,
recall and F1-scores are better in comparison to scores achieved by previously
discussed algorithms. The last row contains results achieved by the proposed
algorithm, which employs samples that are selected on the basis of weighted
categorical cross-entropy, and in which high-confident samples are utilized as
pseudo-annotated data. The accuracy, recall and F1-scores are superior to scores
achieved by algorithms discussed previously. Particularly, owing to using the

Table 1. Classification performance on BreaKHis dataset using active learning: rs -
random sampling, hc - high confidence samples, wce - weighted cross-entropy.

Algorithm	Accuracy	Precision	Recall	F1-score
rs, hc-no	0.9258	0.9416	0.9517	0.9467
wce, hc-no	0.9428	0.9652	0.9517	0.9583
wce-no, hc	0.9467	0.9644	0.9584	0.9613
wce, hc	0.9507	0.9612	0.9678	0.9644

weighted cross-entropy and high-confident samples as pseudo-annotated data our algorithm achieves the smallest number of false negatives, i.e. it achieves the highest recall. This is highly desirable property because practically it is very dangerous and costly to miss an image with malignant while it is positive.

Figure 2 illustrates sample results that were obtained in one of the mentioned above experiments, in which we evaluated the classification performance of active learning using entropy (c.f. green curves) and weighted entropy (c.f. blue curves). Left plot on Fig. 2 depicts the evolution of size of PL pool over iteration number as well as evolution of size of PH pool (consisting of high confidence samples) vs. iteration number. As we explained in Sect. 3, the concatenated PL and PH samples were used to train the CNN. As previously mentioned, the number of PL samples increases about 300 in each iteration, whereas the number of PH samples depends on their uncertainty score referred to the **delta** parameter, and therefore can be different in each iteration, see also green and blue curves on the left plot on Fig. 2. Right plot on Fig. 2 presents classification accuracies vs. iteration number. As we can observe, on the initial training pool consisting of only labeled data, the algorithm based on entropy achieved 86% accuracy, whereas algorithm based on weighted entropy achieved 83.9% accuracy. In seventh iteration the entropy-based algorithm learned on 3035 samples from PL pool and 3146 training samples from PH pool, and achieved on such training data 93.3% classification accuracy, whereas weighted entropy-based algorithm learned on 3035 samples from PL pool and 3335 samples from PH pool and achieved 94.62% classification accuracy. In all remaining experiments, the classification accuracies achieved by weighted entropy-based active learning in iterations #3 − 7 were higher in comparison to accuracies achieved by entropy-based active learning. As we can observe on Fig. 2, in next iterations the increase of the classification accuracy was not so high despite larger number of the PL samples and smaller proportion of PH data in total training data fed to the CNN. In tenth iteration the classification accuracy on the test data was equal to 95.06% for entropy-based algorithm and 95.21% for weighted-entropy based algorithm. The classification accuracies are far larger than 91.5% accuracy achieved by the resNet18 neural network with **zero_padding2d_18** set as trainable, c.f. results in Subsect. 4.2. Comparing results achieved by resNet18 neural network with the fully connected layer replaced by new fully connected layer and fine-tuned as in most relevant work, c.f. Subsect. 4.2, and results obtained with active learning, we can observe considerable improvement of classification accuracy. The increase of the classification accuracy from about 75% to 95% has been achieved owing to use our techniques for training neural networks on histopathological images affected by unbalanced data.

Since calculations of precision and recall do not make use of the true negatives, precision-recall analysis is useful in cases where there is an imbalance in the samples between the two classes. Figure 3 illustrates the precision-recall that has been obtained on the basis of results produced by weighted entropy based-algorithm. The average precision score is equal to 0.99.

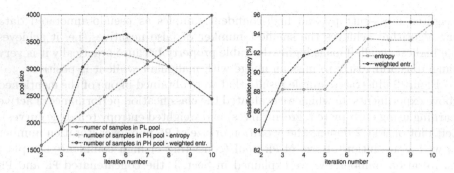

Fig. 2. Number of samples in PL and PH pool vs. iteration number (left). Classification accuracy vs. iteration number (right). (Color figure online)

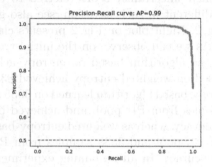

Fig. 3. Precision-recall curve.

The algorithm has been implemented in python language using Keras/ TensorFlow frameworks. The training has been realized on TitanX GPU.

5   Conclusions

The proposed AL-based algorithm for breast cancel classification on unbalanced histopathological datasets considerably reduces the label effort from pathologists, without significantly sacrificing the accuracy. Thanks to the use of weighted cross-entropy in the loss function during training the CNN, as well as weighted entropy both for selecting uncertain samples and determining high confidence samples the proposed algorithms achieves high classification accuracy, high average precision score as well as precision-recall tradeoff, and it is less biased towards the majority class.

Acknowledgments. This research was co-financed by the European Regional Development Fund in the Intelligent Development 2014-2020 Programme, within the grant "The system of automatic analysis and recognition of histopathological images" supported by the National Center for Research and Development: grant no. POIR.01.01.01-00-0861/16-00, and Diagnostyka Consilio.

References

1. Chen, J.M., et al.: Computer-aided prognosis on breast cancer with hematoxylin and eosin histopathology images: a review. Tumor Biol. **39**(3) (2017)
2. Veta, M., Pluim, J., van Diest, P., Viergever, M.: Breast cancer histopathology image analysis: a review. IEEE Trans. Biomed. Eng. **61**(5), 1400–1411 (2014)
3. Filipczuk, P., Fevens, T., Krzyzak, A., Monczak, R.: Computer-aided breast cancer diagnosis based on the analysis of cytological images of fine needle biopsies. IEEE Trans. Med. Imaging **32**(12), 2169–2178 (2013)
4. Aswathy, M., Jagannath, M.: Detection of breast cancer on digital histopathology images: present status and future possibilities. Inform. Med. Unlocked **8**, 74–79 (2017)
5. McCann, M.T., Ozolek, J.A., Castro, C.A., Parvin, B., Kovacevic, J.: Automated histology analysis: opportunities for signal processing. IEEE Signal Process. Mag. **32**(1), 78–87 (2015)
6. Gupta, V., Bhavsar, A.: Breast cancer histopathological image classification: is magnification important? July 2017
7. Komura, D., Ishikawa, S.: Machine learning methods for histopathological image analysis. Comput. Struct. Biotechnol. J. **16**, 34–42 (2018)
8. Spanhol, F., Oliveira, L., Petitjean, C., Heutte, L.: A dataset for breast cancer histopathological image classification. IEEE Trans. Biomed. Eng. **63**(7), 1455–1462 (2016)
9. Nalisnik, M., et al.: Interactive phenotyping of large-scale histology imaging data with HistomicsML. Sci. Rep. **7** (2017). Article no. 14588
10. Huang, S., Jin, R., Zhou, Z.: Active learning by querying informative and representative examples. IEEE Trans. PAMI **36**(10), 1936–1949 (2014)
11. Guiasu, S.: Weighted entropy. Rep. Math. Phys. **2**(3), 165–179 (1971)
12. Elmore, J., et al.: Diagnostic concordance among pathologists interpreting breast biopsy specimens. JAMA **313**(11), 1122–1132 (2015)
13. Asri, H., Mousannif, H., Moatassime, H.A., Noel, T.: Using machine learning algorithms for breast cancer risk prediction and diagnosis. Procedia Comput. Sci. **83**, 1064–1069 (2016)
14. Wolberg, W., Street, W., Mangasarian, O.: Breast Cancer Wisconsin (Diagnostic) Data Set (1993). https://archive.ics.uci.edu/ml/datasets/Breast+Cancer+Wisconsin+(Diagnostic). Accessed 31 Mar 2019
15. Spanhol, F.A., Oliveira, L.S., Petitjean, C., Heutte, L.: Breast cancer histopathological image classification using convolutional neural networks, July 2016
16. Krizhevsky, A., Sutskever, I., Hinton, G.E.: ImageNet classification with deep convolutional neural networks (2012)
17. Bayramoglu, N., Kannala, J., Heikkilä, J.: Deep learning for magnification independent breast cancer histopathology image classification, December 2016
18. Motlagh, M.H., et al.: Breast cancer histopathological image classification: a deep learning approach. bioRxiv (2018)
19. He, K., Zhang, X., Ren, S., Sun, J.: Deep residual learning for image recognition, June 2016
20. Koziarski, M., Kwolek, B., Cyganek, B.: Convolutional neural network-based classification of histopathological images affected by data imbalance. In: Bai, X., et al. (eds.) FFER/DLPR-2018. LNCS, vol. 11264, pp. 1–11. Springer, Cham (2019). https://doi.org/10.1007/978-3-030-12177-8_1

21. He, H., Garcia, E.A.: Learning from imbalanced data. IEEE Trans. Knowl. Data Eng. **21**(9), 1263–1284 (2009)
22. Wang, K., Zhang, D., Li, Y., Zhang, R., Lin, L.: Cost-effective active learning for deep image classification. IEEE TCSVT **27**(12), 2591–2600 (2017)

Measuring the Angle of Hallux Valgus Using Segmentation of Bones on X-Ray Images

Konrad Kwolek[2]($\boxtimes$), Henryk Liszka[2,4], Bogdan Kwolek[1], and Artur Gądek[2,3]

[1] AGH University of Science and Technology, 30 Mickiewicza, 30-059 Krakow, Poland
[2] Department of Orthopedics and Rehabilitation, University Hospital, Krakow, Poland
kwolekonrad@gmail.com, ortopediasu@gmail.com
[3] Department of Orthopedics and Physiotherapy, Jagiellonian University Collegium Medicum, Krakow, Poland
[4] Department of Anatomy, Jagiellonian University Collegium Medicum, Krakow, Poland

Abstract. Hallux valgus is a common feet problem. A hallux valgus deformity is when there is medial deviation of the first metatarsal and lateral deviation of the great toe. In this work, we introduce an algorithm for automatic recognition of hallux valgus on X-ray images with feet. The bones are segmented on the basis of U-Net convolutional neural network. The neural network has been trained on thirty manually segmented images by an orthopedist. We present both qualitative and quantitative segmentation results on ten test images. We present algorithms for great toe extraction and hallux valgus angle (HVA) estimation. The HVA is estimated as the angle between two lines fitted to big toe skeleton. We compare results that were obtained manually, by computer-assisted programs that are used by radiologists, and by the proposed algorithm.

1 Introduction

The term "Deep Learning" (DL) was introduced to Machine Learning (ML) in 1986 [1] and latterly utilized for Neural Networks (ANN) in 2000 [2]. Recent progress in Artificial Intelligence (AI) aiming at the imitation of human cognition by a computer is driven by advances in ML, in which algorithms learn from data without human intervention. Architectures and ANNs falling into DL category are composed of multiple layers to discover most discriminative features in data with multiple levels of abstraction [3]. The advancements in artificial intelligence and deep-learning have been enabled by the availability and proper use of labeled big data, along with markedly increased computing power and developments in learning algorithms. Recent progress in artificial intelligence and deep learning [4] led to remarkable development in many areas. Artificial intelligence has recently surpassed human performance in several domains [5]. Some jobs and routine works might soon be automated, and there is great hope for that in healthcare.

© Springer Nature Switzerland AG 2019
I. V. Tetko et al. (Eds.): ICANN 2019, LNCS 11731, pp. 313–325, 2019.
https://doi.org/10.1007/978-3-030-30493-5_32

AI will allow for much more personalized medicine [6] and bring revolution in the use of large medical datasets [7].

Computers are revolutionizing the field of diagnostic imaging [8]. It appears that in near future the computers will supersede the radiologists that perform routine analyses, and in so doing considerably reduce human errors [9] or at least create a new generation of systems with radiologists-in-the-loop [10]. In [11] a neural network architecture that learns highly relevant disease features for knee osteoarthritis diagnosis from plain radiographs has been proposed. In another work [12], a DenseNet with 121 layers has been fine-tuned for the classification of miscellaneous chest pathologies on the basis of X-ray images, achieving radiologist-level classification performance in identifying pneumonia.

In roadmap for implementation of AI in radiology [10], automated image segmentation, lesion detection, measurement, labeling and comparison with historical images were identified as key areas. Deep learning has recently been applied to lung segmentation in chest radiographs [13]. The segmentation model is based on the U-Net architecture [14]. Recently, in [15] an integrated computer-aided diagnosis system for digital X-ray mammograms via detection and segmentation has been proposed.

Delineation of bone contours from X-ray radiographs can be very useful for radiologists and orthopedists for preoperative planning [16]. However, little work has been done in this domain until now. Automatic delineation of bone contours from X-ray radiographs is a very challenging task due to complex structure of the bones in X-ray images [16]. Images may be contaminated by noise, sampling artifacts, spatial aliasing, have insufficient contrast or resolution, or intensities depending on the X-ray device such that the boundaries of the regions of interest become indistinguishable or disconnected. In [17], a mean-shift segmentation is executed for initial segmentation, followed by an adaptive region merging process relying on the maximal similarity between regions. Active shape model (ASM) is frequently used algorithm in bone segmentation. In [16], ASM has been utilized to extract the distal femur and proximal tibia in knee radiographs of varying image quality. A spectral clustering method based on the eigensolution of an affinity matrix has been applied for denoising X-ray images. Finally, ASM-based segmentation has been employed on denoised X-ray images. In [18] an algorithm for bone segmentation on the basis of prior shape and a straightened boundary image (SBI) has been proposed. To the best of our knowledge, convolutional neural networks and deep learning were not used for automatic bone delineation or segmentation on X-ray images.

A hallux valgus deformity is when there is medial deviation of the first metatarsal and lateral deviation of the great toe. Hallux valgus is a common feet problem [19]. In [20] the results achieved by the Hallux Angles App and the iPinPoint App with goniometer measurements were compared. The angular measurements were performed with the smartphone parallel to the X-ray film. Afterwards, a target line has been lined up along the diaphysis of 1 of the bones (metatarsal or phalange) utilized as a reference. Next, the device has been rotated, in the same plane parallel to the X-ray, to align the target line to the second reference bone segment, and the angle formed between them has been determined. To the best of our knowledge, no work devoted to automatic hallux valgus recognition on X-ray images has been done until now.

Numerous studies have reported the reliability of different radiographic measurements evaluating a hallux valgus deformity [21,22]. These methods intend to make measurements cheaper, faster, or more straight-forward. However, manual or computer-assisted measurements methods to estimate angles for accessing the severity of hallux valgus deformity have been shown to be prone to errors and to be time consuming.

In this work we introduce an algorithm for automatic recognition of hallux valgus on X-ray images with feet. We propose an algorithm for foot bone segmentation on X-ray images. The bones are segmented on the basis of U-Net convolutional neural network. The neural network has been trained on thirty manually segmented images by an orthopedist. We present both qualitative and quantitative segmentation results on ten test images. In addition to the separated bones, the test dataset also contains values of HVA angles, which were determined by both the radiologist and the orthopedist. We present algorithms for great toe extraction and hallux valgus angle (HVA) estimation. The HVA has been estimated on the basis of lines fitted to skeleton representing the segmented big toe. We compare the results that were obtained manually, by computer-assisted programs that are used by radiologists, and by the proposed algorithm. The most similar work is [23], in which a deep learning method for bone suppression in a single chest radiograph has been proposed. The contribution of this work is an algorithm for automatic recognition of hallux valgus on X-ray images. To best of our knowledge, this is seminal paper on application of such techniques in orthopedic. In [24] fine-tined deep networks were applied to classify X-ray images. The networks achieved at least 90% accuracy in identifying laterality, body part, and exam view, whereas accuracy for fractures was about 83% for the best performing neural network. Our work significantly goes beyond the standard classification of X-ray images and introduces deep-learning based X-ray image analysis for orthopedics.

2   Radiographic Measurements of Hallux Angles

Hallux valgus is a common feet deformity that reveals itself by lateral deviation of the first toe and progressive subluxation of the first metatarsophalangeal joint. It affects approximately 23% to 35% of adults [25] and is associated with foot pain, impaired gait, and increased fall risk of elderly. As digital radiography is replacing standard radiography, computers are becoming a more common way to view and measure the HV angle. Another alternative is to use visual comparison with a categorical grading scale, e.g. the Manchester scale [26], which relies on visual comparison on the basis of four standardized photographs with increasing HV severity. Although such scales acknowledged their usefulness for classifying the severity of foot deformity, they have rather quite limited clinical applicabilities to estimate the HV progression, i.e. when incremental measurements of the progression of HV deformity are needed. Naturally, one can measure the HV angles on the basis of foot photographs taken by a smartphone. However, in such an approach, each one must take into account the measuring errors resulting from different thickness of soft tissue (thickening of the epidermis), and in particular corns of the epidermis (*clavus*). A precise and valid measurement of HV angle requires information about bones.

In clinical practice, HVA and IMA angles are used to select surgical treatments, to evaluate the results of surgical treatment, etc. Hallux valgus angle has the best reliability [27]. Several methods for measuring the HVA and IMA were proposed in the literature [21]. However, the most reliable and valid method has not been selected until now. According to Schneider et al. [28] a method in which first metatarsal axis is a line drawn from the center of the head of the first metatarsal through the center of the base of the first metatarsal provides the best preoperative and postoperative measurement accuracy and can be recommended for clinical use. Recently, a new method based on the most medial prominent points has been proposed [22]. Figure 1 illustrates determining the HVA angle. As described in [21], first metatarsal axis line is drawn from the center of the head of the first metatarsal head through the center of the base of the first metatarsal.

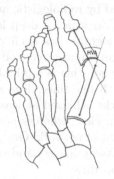

Fig. 1. X-ray image featuring the hallux valgus angle (HVA) measurement.

3   Framework for Bone Segmentation on X-Ray Images

At the beginning we discuss segmentation in biomedical image analysis. Then we present bone segmentation on X-ray images. Afterwards we present our dataset with bone annotations for training and evaluating deep models and algorithms for hallux valgus recognition. Finally, we present the neural network for bone segmentation on X-ray radiographs.

3.1   Image Segmentation

Segmentation is an active research topic in the field of biomedical image analysis as it is a common preprocessing and evaluation tool for a number of medical applications. Traditional methods are based on active shape models, clustering, and interactive methods. Recent methods are based on convolutional neural networks. Neural networks for image segmentation are different from networks for classification because they should predict a class for each pixel of the input image instead of class for the whole input image. Thus, such networks need to output not only what is in the input, but also where. The most basic segmentation architectures can be based on CNN models like AlexNet or VGG, adapted to perform dense prediction. The adjustment of such CNN can be done by substituting fully connected layers in the CNN by convolutional ones and adding a spatial loss. In [29], a skip-architecture that combines high-level representation from deep decoding layers with the appearance representation from shallow encoding layers to deliver a detailed segmentation has been introduced. The introduced Fully Convolutional Network (FCN) owes its name to its architecture, which is built only on layers, such as convolution, pooling and learnable upsampling. An upsampling path following the downsampling path allows obtaining detailed segmentation. The upsampling path is used to provide precise localization (where), whereas the downsampling one is utilized to extract and interpret the context (what). In order to recover the fine-grained spatial information lost in pooling (downsampling) layers, skip connections were used. Ronneberger et al. [14] employed such skip connections in the proposed U-Net architecture. The U-Net network consists of a contracting path to capture context and a symmetric expanding path that allows obtaining precision in the location. Through concatenating the output from encoder layers with their upsampled counterparts in the decoder, this network is capable of passing the spatial information that is lost in the encoding back to the decoder. Additionally, owing to executing convolutions on such concatenated layers, the decoder can learn far more spatially precise output and then propagate it to layers of higher resolution. Since its introduction three years ago, the U-Net architecture has been used to create deep learning models for segmenting nerves in ultrasound images, lungs in CT scans and X-ray images. In the subsequent subsections we preset the proposed framework for bone segmentation on X-ray images.

3.2 Bone Segmentation on X-Ray Images

Automatic delineation of bones from X-ray radiographs is a challenging task due to complex structure of the bones in X-ray images [16]. Images may be contaminated by noise, sampling artifacts, spatial aliasing, have insufficient contrast or resolution, or intensities. Usually, images have different sizes and differ considerably by intensity ranges, see Fig. 2. Frequently, feet occupy different parts of the images. The images might contain some text, e.g. labels or remarks. Moreover, bones can contain surgical nails, rods, pins that are routinely used in internal fixation of fractures, see also 2nd image on Fig. 2. Delineation of bone contours on X-ray radiographs is associated with distinguishing soft tissue from bones. To evaluate HVA, hallux sesamoids exclusion is necessary. However, the boundaries of the regions of interest are not easy to distinguish on X-ray radiographs.

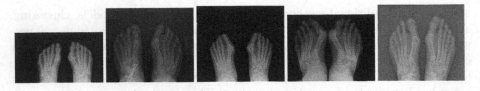

Fig. 2. Example samples of foot images on x-ray radiographs.

3.3   Dataset

The dataset contains forty annotated X-ray images. Thirty images are in training part of the dataset and ten images are in test part. They were split randomly into train and test subsets. On each X-ray image the bones were manually delineated by an orthopedist. Figure 3 depicts sample examples of the annotated images, see also corresponding images on Fig. 2. For training and evaluation of neural networks for bone segmentation both X-ray and annotated images were rescaled to size 256×256 pixels. In order to make possible determining the HVA angles and to compare results, the dataset contains also X-ray images of width equal to 256 pixels and the image height resulting from the height/width ratio for the original X-ray image. For images in the test subset the HVA angles were determined both by the radiologist and orthopedist. The angles were determined on images with the original resolution, i.e. on images not downscaled before the measurement.

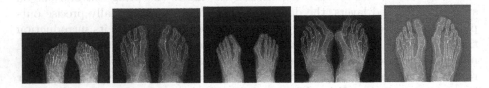

Fig. 3. Segmented images from Hallux dataset, see corresponding images on Fig. 2. The illustrated annotations are stored in vectorized graphics format and a better viewing can be obtained by zooming this figure.

3.4   Neural Network for Bone Segmentation on X-Ray Radiographs

The network architecture has been based on the U-Net comprising a down-sampling (encoding) path and an up-sampling (decoding) path as shown in Fig. 4. The down-sampling path is built on five convolutional blocks. Every block has two convolutional layers with filters of size of 3×3, stride of 1 in both directions. For the down-sampling, max pooling with stride 2×2 is applied on the end of every blocks except the last block. In the up-sampling path, every block begins with a deconvolutional layer with filter size of 3×3 and stride of 2×2, which doubles the size of feature maps in both directions but decreases the number

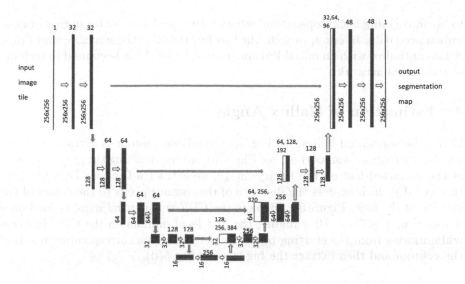

Fig. 4. Architecture of U-Net used for bone segmentation.

of feature maps by two. In every up-sampling block, two convolutional layers reduce the number of feature maps arising by concatenation of deconvolutional feature maps and the feature maps from corresponding block in the encoding path. Unlike from the original U-Net architecture [14], we utilize zero padding to keep the output dimension for all the convolutional layers of both down-sampling and up-sampling path. Finally, a 1×1 convolutional layer is employed to reduce the number of feature maps to two in order to reflect the bone and background areas.

The pixel-wise cross-entropy has been used as the loss function for bone structure segmentation. The binary cross-entropy can be expressed as:

$$\mathcal{L}_{CE} = -\frac{1}{N} \sum_{i=1}^{N} [y_i \log(\hat{y}_i) + (1 - y_i) \log(1 - \hat{y}_i)] \tag{1}$$

where N stands for the number of training samples, y is true value and $\hat{y}$ denotes predicted value.

The Sørensen–Dice (Dice) coefficient is a statistic utilized for comparing the similarity of two samples. It equals twice the number of elements common to both data sets divided by the sum of the number of elements in each set. The Dice similarity is the same as F1-score. In our approach we employed a Dice loss that can be expressed as follows:

$$\mathcal{L}_{Dice} = -\frac{\sum_{i=1}^{N} p_i g_i}{\sum_{i=1}^{N} p_i^2 + \sum_{i=1}^{N} g_i^2} \tag{2}$$

where the sums run over the N pixels of the predicted segmentation image $p_i \in P$ and ground truth one $g_i \in G$. The advantage of cross-entropy is that it is easier to optimize using backpropagation, whereas Dice performs better in case of class imbalanced data. In our approach, the loss function $\mathcal{L}$ is the sum $\mathcal{L}_{CE}$ and $\mathcal{L}_{Dice}$. Adam optimizer with an initial learning rate of 3×10^{-4} has been used in training of the neural network.

4   Estimation of Hallux Angle

Given the segmented bones, see Fig. 5(a), the bone contour is extracted. After- wards, the contour is smoothed, see Fig. 5(b). After extracting the gravity centers of the extracted feet on the binary image, we extract a Contour Distance Fea- ture (CDF) which express the distance of the contour pixels to the center of the gravity of the foot. Figure 5(c) depicts the CDF versus pixel number for bones depicted on Fig. 5(a). After finding the first local minima on the CDF function while moving from the starting point, we determine the corresponding pixels of the contour and then extract the big toe, see Fig. 5(d).

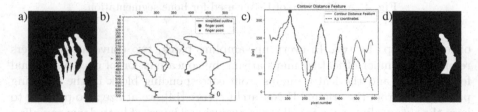

Fig. 5. Big toe extraction. Segmented bones (a), smoothed outline (b), Contour Dis- tance Feature (c), extracted big toe (d).

Given the extracted big toe we performed the image skeletonization using morphological thinning, i.e. shrinking the image until the area of interest is 1 pixel wide. The process of skeleton construction, which is also known as medial axis or symmetric axis extraction, is pretty sensitive to noise on the shape boundary. Small perturbations of the shape boundary lead to redundant skele- ton branches that may seriously change the topology of the skeleton graph. Such unnecessary branches are eliminated in an operation called skeleton pruning. In our algorithm we employed the geodesic distance transform [30] to determine the longest continuous path in the thinned image. In this way we robustly eliminated spurs and artifacts from the skeleton. Next, inflection point on curve extracted in such a way has been determined to split the skeleton into two parts corre- sponding to main segments of the big toe. After skipping a few begin as well the end points we fit a line by linear regression (by minimizing the sum of squared errors between the best fit line and data points belonging to the toe segment). The line parameters were determined on the basis of the following equations:

$$m = \frac{\sum_{i=1}^{n} x_i \sum_{i=1}^{n} y_i - n \sum_{i=1}^{n} x_i y_i}{(\sum_{i=1}^{n} x_i)^2 - n \sum_{i=1}^{n} x_i{}^2} \tag{3}$$

$$b = \frac{\sum_{i=1}^{n} x_i y_i \sum_{i=1}^{n} x_i - n \sum_{i=1}^{n} y_i \sum_{i=1}^{n} x_i{}^2}{(\sum_{i=1}^{n} x_i)^2 - n \sum_{i=1}^{n} x_i{}^2} \tag{4}$$

where n stands for number of pixels on skeleton segment. The angle between two fitted lines to skeleton parts estimates the HVA value.

5   Experimental Results

At the beginning we implemented several U-Net architectures and evaluated them in terms of mean IoU (Intersection over Union) and mean Dice metrics. Various loss functions, among others weighted cross-entropy, Dice Loss, sum of the cross-entropy and the Dice-loss, Generalised Dice Loss, pixel-wise Wasserstein distance were investigated at this stage of experiments. Variety of optimizers and learning parameters were investigated as well. This way we selected the architecture, c.f. Subsect. 3.4, which achieved the best results. Figure 6 depicts segmentation results for images selected from the test subset of the Hallux dataset. On all images from test part of Hallux dataset the mean IoU was equal to 0.935 and mean Dice was equal to 0.966. As we can observe, promising results were achieved using the proposed framework for bone segmentation on X-ray images. It is worth nothing that hallux sesamoids, which are quite often not easy to distinguish from the bone on radiographic images have been distinguished fairly well, i.e. not segmented as the bone.

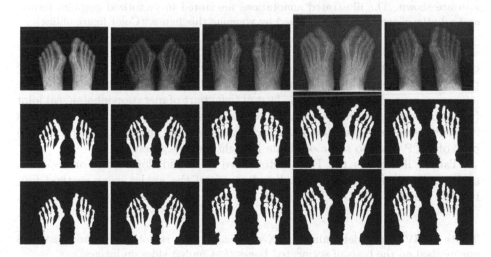

Fig. 6. X-ray images (1st row), manual bone segmentation by orthopedist (2nd row), bone segmentation by our network (3rd row). From left to right the images #6, 7, 9, 10 and 12 are shown.

Figure 7 shows lines that were fitted to skeletons of the segmented big toes by the U-Net. The lines were extracted on the basis of Eqs. (3) and (4). As we

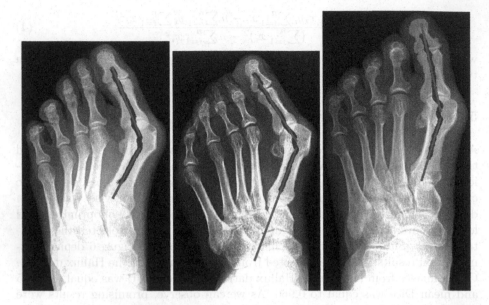

Fig. 7. Fitting lines to the segmented big toe by the U-Net. For better visualization, the image intensity values of depicted X-ray images were adjusted. The segmentation masks generated by the U-Net were overlaid transparently on the enhanced X-ray images (yellowish color). The skeleton of the segmented big toe mask was marked with blue, whereas the fitted lines were marked with red. From left to right the images #6, 7, 9 are shown. The illustrated annotations are stored in vectorized graphics format and a better viewing can be obtained by zooming this figure. (Color figure online)

can observe (on the zoomed figure), the bones are segmented by the U-Net quite precisely. The sesamoids bones are extracted quite properly by the U-Net. On the left image we can observe that the lateral (left) part of metatarsophalangeal joint (MTP) is somewhat oversegmented, whereas the medial (right) part of MTP is insensibly undersegmented. On the right image we can observe that lateral (left) part of MTP close to sesamoid bone is slightly oversegmented.

Table 1 illustrates the HVA determined by a radiologist, the orthopedist, and our algorithm. The orthopedist determined the angles using method four in [28]. RadiAnt DICOM Viewer [31] has been used to determine HVA values.

Table 1. HVA [deg] determined on X-ray images by a radiologist, orthopedist, and our method on the basis of segmented bones (feet on left sides on images).

	im_6	im_7	im_9	im_10	im_12	im_15	im_16	im_20	im_21	im_27
Radiologist	45	49	20	39	35	38	52	38	23	15
Orthopedist	42	45	21	41	36	37	50	37	25	16
Our method	50.6	53.4	29	47	43.7	39.3	55.5	38.8	28	21.9

A radiologist and the orthopedist estimated the angles on images of original resolution, whereas our software determined the HVA angles on images with width equal to 256 pixels and the image height resulting from the height/width ratio for the original X-ray image. The orthopedist determined also HVA angles on segmented images by our U-Net network. The following HVA values [deg] were obtained: 43, 49, 21, 41, 35, 39, 52, 36, 24.5 and 16 for images with indexes listed in Table 1. As we can observe, the difference between results obtained by the radiologist and our algorithm in not greater than nine degrees. The images with indexes mentioned above belong to test part of Hallux dataset and were not used to train the neural network.

Future work will include estimation of HVA angles on images with 512 columns. To achieve this a considerable work is needed to collect proper number of X-ray images for training a neural network, manually segment such images and to design a network architecture for segmentation of bones on images with higher resolution. Various strategies and algorithms for estimation of HVA angles on X-ray images will be investigated as well.

The network was trained in 150 epochs using 16 images per batch, the Adam solver and the proposed loss function. Momentum was set to 0.9, while a learning rate of 1×10^{-4} was used and a weight decay of 5×10^{-4}. While learning of the neural network the data were augmented using image rotation ($\pm 10°$), horizontal and vertical shift (shift range set to 0.1), zoom (range set to 0.2), rescale and horizontal flip. The training of neural networks has been performed on TitanX GPU. The neural network has been implemented in Python using Keras framework with TensorFlow backend. The algorithms for big toe extraction and HVA estimation have been implemented in Matlab and Python.

6   Conclusions

We introduced algorithm for recognition and measurement of hallux valgus on X-ray images. We prepared a dataset with manually extracted bones for training neural networks and evaluating the accuracy of HVA angle estimation. We presented HVA values that were determined by radiologist, orthopedist and our algorithm. The experimental results are promising in terms of quality of segmentation and HVA accuracy. The novelty includes a new approach to measure the HVA, modified U-Net for bone segmentation, big toe extraction and HVA estimation.

References

1. Dechter, R.: Learning while searching in constraint-satisfaction-problems. In: Proceedings of the Fifth AAAI National Conference on Artificial Intelligence, pp. 178–183. AAAI Press (1986)
2. Aizenberg, I., Aizenberg, N., Vandewalle, J.: Multi-Valued and Universal Binary Neurons: Theory, Learning and Applications. Kluwer Academic Publishers, Norwell (2000)

3. LeCun, Y., Bengio, Y., Hinton, G.: Deep learning. Nature **521**(7553), 436–444 (2015)
4. Liu, W., Wang, Z., Liu, X., Zeng, N., Liu, Y., Alsaadi, F.E.: A survey of deep neural network architectures and their applications. Neurocomputing **234**, 11–26 (2017)
5. Schroff, F., Kalenichenko, D., Philbin, J.: FaceNet: a unified embedding for face recognition and clustering. In: IEEE Conference on Computer Vision and Pattern Recognition (CVPR), pp. 815–823 (2015)
6. Fogel, A., Kvedar, J.: Artificial intelligence powers digital medicine. NPJ Digit. Med. **1**(1), 5 (2018)
7. Topol, E.J.: High-performance medicine: the convergence of human and artificial intelligence. Nat. Med. **25**(1), 44–56 (2019)
8. Mayo, R.C., Leung, J.: Artificial intelligence and deep learning - radiology's Next frontier? Clin. Imaging **49**, 87–88 (2018)
9. Fazal, M.I., Patel, M.E., Tye, J., Gupta, Y.: The past, present and future role of artificial intelligence in imaging. Eur. J. Radiol. **105**, 246–250 (2018)
10. Liew, C.: The future of radiology augmented with artificial intelligence: a strategy for success. Eur. J. Radiol. **102**, 152–156 (2018)
11. Tiulpin, A., Thevenot, J., Rahtu, E., Lehenkari, P., Saarakkala, S.: Automatic knee osteoarthritis diagnosis from plain radiographs: a deep learning-based approach. Sci. Rep. **8**(1) (2018)
12. Rajpurkar, P., et al.: CheXNet: radiologist-level pneumonia detection on chest X-Rays with deep learning. CoRR abs/1711.05225 (2017)
13. Islam, J., Zhang, Y.: Towards robust lung segmentation in chest radiographs with deep learning. CoRR abs/1811.12638 (2018)
14. Ronneberger, O., Fischer, P., Brox, T.: U-Net: convolutional networks for biomedical image segmentation. In: Navab, N., Hornegger, J., Wells, W.M., Frangi, A.F. (eds.) MICCAI 2015. LNCS, vol. 9351, pp. 234–241. Springer, Cham (2015). https://doi.org/10.1007/978-3-319-24574-4_28
15. Ribli, D., Horváth, A., Unger, Z., Pollner, P., Csabai, I.: Detecting and classifying lesions in mammograms with deep learning. Sci. Rep. **8**, 1 (2018)
16. Wu, J., Mahfouz, M.R.: Robust X-ray image segmentation by spectral clustering and active shape model. J. Med. Imaging **3** (2016)
17. Stolojescu-Crisan, C., Stefan, H.: An interactive X-ray image segmentation technique for bone extraction. In: International Work-Conference on Bioinformatics and Biomedical Engineering, pp. 1164–1171 (2014)
18. Mohammadi, H.M., de Guise, J.A.: Enhanced X-ray image segmentation method using prior shape. IET Comput. Vision **11**(2), 145–152 (2017)
19. Liszka, H., Gądek, A.: Results of scarf osteotomy without implant fixation in the treatment of hallux valgus. Foot Ankle Int. **39**(11), 1320–1327 (2018)
20. Dinato, M., de Faria Freitas, M., Milano, C., Valloto, E., Ninomiya, A.F., Pagnano, R.G.: Reliability of two smartphone applications for radiographic measurements of hallux valgus angles. J. Foot Ankle Surg. **56**(2), 230–233 (2017)
21. Srivastava, S., Chockalingam, N., Fakhri, T.E.: Radiographic measurements of hallux angles: a review of current techniques. Foot **20**(1), 27–31 (2010)
22. Heineman, N., Chhabra, A., Zhang, L., Dessouky, R., Wukich, D.: Point vs. traditional method evaluation of hallux valgus: interreader reliability and intermethod performance using X-ray and MRI. Skeletal Radiol. **48**(2), 251–257 (2019)
23. Yang, W., et al.: Cascade of multi-scale convolutional neural networks for bone suppression of chest radiographs in gradient domain. Med. Image Anal. **35**, 421–433 (2017)

24. Olczak, J., et al.: Artificial intelligence for analyzing orthopedic trauma radiographs. Acta Orthop. **88**(6), 581–586 (2017)
25. Wülker, N., Mittag, F.: The treatment of hallux valgus. Deutsches Ärzteblatt Int. **109**(49), 857–868 (2012)
26. Garrow, A.P., Papageorgiou, A., Silman, A.J., Thomas, E., Jayson, M.I.V., Macfarlane, G.J.: The grading of hallux valgus. The Manchester scale. J. Am. Podiatr. Med. Assoc. **91**(2), 74–78 (2001)
27. Lee, K.M., Ahn, S., Chung, C.Y., Sung, K., Park, M.: Reliability and relationship of radiographic measurements in hallux valgus. Clin. Orthop. Relat. Res. **470**(9), 2613–2621 (2012)
28. Schneider, W., Csepan, R., Knahr, K.: Reproducibility of the radiographic metatarsophalangeal angle in hallux surgery. J. Bone Joint Surg. Am. **85–A**, 494–499 (2003)
29. Long, J., Shelhamer, E., Darrell, T.: Fully convolutional networks for semantic segmentation. In: IEEE Conference on Computer Vision and Pattern Recognition (CVPR), pp. 3431–3440 (2015)
30. Soille, P.: Morphological Image Analysis: Principles and Applications, 2nd edn. Springer, Heidelberg (2003). https://doi.org/10.1007/978-3-662-05088-0
31. RadiAnt: Radiant DICOM Viewer

Human Body Posture Recognition
Using Wearable Devices

Junxiu Liu[1], Mingxing Li[1], Yuling Luo[1(✉)], Su Yang[2], and Senhui Qiu[1]

[1] School of Electronic Engineering, Guangxi Normal University, Guilin 541004, China
yuling0616@mailbox.gxnu.edu.cn
[2] School of Computing, Engineering and Intelligent Systems, Ulster University,
Derry BT48 7JL, Northern Ireland, UK

Abstract. Recently, the activities of elder people are monitored to support them live independently and safely, where the embedded hardware systems such as wearable devices are widely used. It is a research challenge to deploy deep learning algorithms on embedded devices to recognize the human activities, with the hardware constraints of limited computing resources and low power consumption. In this paper, human body posture recognition methods are proposed for the wearable embedded systems, where back propagation neural network (BPNN) and binary neural network (BNN) are employed to classify the human body postures. The BNN quantizes the synaptic weights and activation values to +1 or −1 based on the BPNN, and is able to achieve a good trade-off between the performance and cost for the embedded systems. In the experiments, the proposed methods are deployed on embedded device of Raspberry Pi 3 for real application of body postures recognition. Results show that compared with BPNN, the BNN can achieve a better trade-off between classification accuracy and cost including required computing resource, power consumption and processing time, e.g. it uses 85.29% less memory, 8.86% less power consumption, and has 5.19% faster classification speed. Therefore, the BNN is more suitable for deployment to resource constrained embedded hardware devices, which is of great significance for the application of human body posture recognition using wearable devices.

Keywords: Human body posture recognition ·
Back propagation neural network · Binary neural network ·
Wearable devices

1 Introduction

With the increasing of life expectancy and a growing rate of older people, the ageing of the population has become a global challenge [1]. In order to safeguard the older people to live independently and safely, in particular for those with chronic diseases, developing new technologies for living assistance has become a potential solution [2]. Human activity recognition is a well-recognised method to

© Springer Nature Switzerland AG 2019
I. V. Tetko et al. (Eds.): ICANN 2019, LNCS 11731, pp. 326–337, 2019.
https://doi.org/10.1007/978-3-030-30493-5_33

support the care of the elderly [3]. The long-term monitoring of human activities (especially for people with chronic diseases and special needs) using wearable devices is one of the most promising solution [4]. The decision tree, a machine learning method, is used to classify the human activity in the approach of [5] where a self-collected database is used for testing. In the meantime, neural networks are becoming increasingly popular to recognise the human body postures [6]. For example, the neural networks are used for classification in a body postures recognition dataset [7]. Large neural network is computing-intensive which often requires parallel GPUs for calculations. This consumes a lot of computing power, especially while running on the embedded platforms, which limits its potential applications. Some algorithms such as matrix decomposition [8], weight quantization [9], network pruning [10] and binary networks [11] are proposed to address this problem. They are based on two solutions: one is the intuitive solution, i.e. increase the computing capacity and the memory for the embedded computing devices; and the other solution is to compress the neural network which can fit in to the embedded devices. This work aims to minimise the power consumption and required computing resource of neural network classifier which runs on a wearable device. The application of human body posture recognition is used as test-bench and the neural network is optimised for implementation on the embedded hardware system. The contributions in this work are in two-fold: (1). The body postures recognition dataset (BPRD) is preprocessed, and then used for deep neural network training. The traditional back propagation neural network (BPNN) and a binary neural network (BNN) are used to classify the human body postures in the BPRD, where the weights and activation values are binarized for the BNN; (2). The neural network models are deployed and evaluated on the embedded devices. Experimental results show that the BNN has relatively simple architecture and can yield good performance with much less energy consumption compared with BPNN.

The rest of the paper is organised as follows: related works are introduced in Sect. 2. Section 3 presents the research methods and Sect. 4 provides the experimental results and analysis. The conclusion and future work are given in Sect. 5.

2   Related Works

The wearable devices have been studied and used for decades. For example, sensors are added to wearable computers that enable them to adjust the functionality to suit the human activities and conditions [12]. Some new sensors are designed for body wear [13], e.g. a novel, non-polluting integrated sensor worn on the wrist. These sensors provide information of human activities. Recently, wearable sensors are used to collect large amounts of data, where machine learning algorithms are used to the data analytics [14]. However most of the algorithms are not deployed to embedded hardware systems such as wearable devices, which is a challenge and needs to be further explored [15]. Deep learning algorithm is an advanced machine learning algorithm. If it is deployed to embedded devices, the resulting huge memory consumption and computational burden become an issue [16], which prevents the scalability of neural network in the embedded devices. Increasing the memory of hardware devices leads to extra cost, therefore compressing the neural network model becomes an appealing solution. For example, vector quantization methods are used to compress deep convolutional neural networks to reduce the size of model [17]. After that, the Binary-Connect was proposed [18] to quantize the weights to only +1 or −1. The BNN [11] is also a promising algorithm which quantifies the weights and activation values of the traditional neural network model into +1 and −1. Moreover, multiply and addition operations for the fixed point numbers can be simplified using bit operations, such as shifting and XOR operations, which can reduce the power consumption of hardware deep neural networks. Based on the quantization weights, redundant operations are reduced for the neural network in the approach of [19]. For the BNN, to improve the accuracy is one challenge. The training strategy, regularization and the angle of activation approximation are used in the approach of [20] to improve the accuracy of BNN. There are two network models which have different numbers of binarized hidden layers, and they are evaluated on the MNIST and CIFAR10 datasets [20]. The AlexNet is binarized in the approach [21], which achieves a good trade-off between compressed network and network recognition accuracy, but the algorithm is not deployed on wearable devices.

In the past few years, the research of human body postures recognition has shown a good overall recognition performance. Machine learning [5] and deep learning [22] have been applied to recognise human body postures. Therefore, human body postures recognition is considered as a potential technology of electronic health monitoring and alert system. Developing independent and cost-efficient computing applications, which can support home care and enable doctors, families and patients to collaborate together, has become a challenge [2]. The research of human body posture recognition may be conducive to the development of assisting technology to support the care of the elderly, chronic patients and people with special needs. In this work, traditional and binarized neural networks are employed to recognise the human body postures. Various parameters of these network models are evaluated and analysed. Experimental results show that the BNN can achieve a trade-off between the performance and cost, and has the potential to be used for the data analytics in embedded hardware systems.

3 Methodology

This section presents the used methods including the BPNN and BNN, and the work procedures and related algorithms for training neural networks are presented.

The BPNN is a multilayer feed forward network with an error back propagation algorithm [23]. It uses the cost function and gradient descent method to obtain the synaptic weights. Functions such as measuring the smallest sum of squared errors and cross entropy are used for the training. The BNN is a simplified version of BPNN, where the weights and activations of are quantified to +1 or −1. There are two ways to quantify them including (a). Deterministic binarization. A binary operation converts real weight values to two possible values. A simple binary operation is based on the sign function, e.g.

$$X_b = \begin{cases} +1, & if \ X \geq 0 \\ -1, & if \ X < 0, \end{cases} \tag{1}$$

where X_b is a binary value (synaptic weight or activation value), X is the real value; and (b). Stochastic binarization. The binarization operation averages the discretization of the input synaptic weight of the hidden layer. Stochastic binarization is an accurate quantization process, which is given by

$$X_b = \begin{cases} +1, & p = \sigma(x) \\ -1, & p = 1 - p, \end{cases} \tag{2}$$

where p is probability value, $\sigma(x)$ is a probabilistic function which is given by

$$\sigma(x) = clip(\frac{1+x}{2}, 0, 1) = max(0, min(0, \frac{1+x}{2})), \tag{3}$$

where $clip()$ is a cut function, which limits the input value ranging from 0 to 1. The stochastic binarization is an alternative method, which requires to generate random values and is difficult to achieve in hardware systems. Therefore, this study mainly investigates the deterministic binarization.

The BNN reduces the computational cost and speeds up the reasoning process. Its training uses binary weights and activation values to calculate the parameter gradient, but the updated parameters are floating-point weights. Floating-point weights are one of the necessary conditions for stochastic gradient descent. The BNN propagates gradient through the discretization. The derivative of deterministic binary symbolic function is 0 everywhere, which obviously can't satisfy the condition of back propagation. When the gradient is propagated, this study adopts the quantized deterministic binary sign function so that the gradient is fixed to 1, and the back propagation of the binarized neural network can be satisfied. In this process, for $y = Sign(x)$, its relaxed post-function is $g_x = g_y 1_{(|x| \leq 1)}$. When x belongs to the range of $(-1, 1)$, the gradient is fixed at 1. It can be described as a piecewise activation function $Htanh(x) = Clip(x, -1, 1) = max(-1, min(1, x))$.

In this study, BNN is also used in conjunction with the stochastic gradient descent algorithm. The data pre-transmission weights and activation values are binarized, but floating-point parameters are used when updating weights. This work also employed batch normalization, which not only can accelerate the network training, but also reduce the overall impact of the weight scale.

Batch normalization integrating scattered points is a method to optimize neural networks. Because the output values of each layer have different mean and variance, the distribution of the output data are different. To obtain an effective training, the data are batch normalized at each layer. It can reduce the disappearance of the gradient and accelerate the convergence rate. This work uses dropout, $L1$ and $L2$ regularization [24], which deal with network over-fitting. The activation values of each layer are normalized to the region where the mean value is 0 and the variance is 1. In this way, most of the data are transformed from the region with gradient 0 to the region with large gradient in the middle of the activation function, thus avoiding the problem of gradient vanishing. As the gradient gets bigger, the network is updated faster.

For batch normalization of the data, the average value of the input data is calculated by

$$M = \frac{1}{n} \sum_{i=1}^{n} Z_i, \tag{4}$$

where $Z = \{Z_1, Z_2...Z_n\}$ is sum of input weights for each layer of network and M is the average of input weights. Then the variance of the input weights is calculated by

$$V = \frac{1}{n} \sum_{i=1}^{n} (Z_i - M)^2, \tag{5}$$

where V is the variance of the input weights. Data Z_i to be normalized is calculated by

$$\overline{Z}_i = \frac{Zi - M}{\sqrt{V + \varepsilon}}, \tag{6}$$

where ε is a constant number and has a stabilizing effect. $P = \{P_1, P_2...P_n\}$ is the value after batch normalization, which is calculated by

$$P_i = \gamma \overline{Z}_i + \beta, \tag{7}$$

where γ is scale and β is shift.

Figure 1 is an example which shows the data distribution after batch normalization of each layer. The normalized data is distributed to a reasonable range, avoiding the disappearance of gradients and effectively accelerating the convergence of neural networks.

4   Experimental Results

In this section, firstly, the dataset used in this approach is introduced. Then the performances of BPNN and the BNN are evaluated base on the metrics of memory, power consumption, computing time etc.

4.1   Dataset

A public dataset of BPRD [5] is used to test the neural networks in this study. There are 5 different body movement measurements (classes), including standing up, standing, sitting down, sitting and walking. The dataset are collected from 4

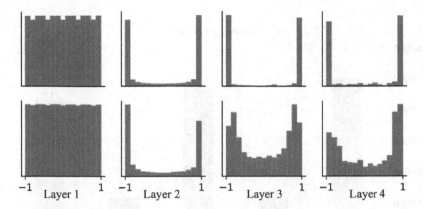

Fig. 1. Batch normalization. Top and bottom are the results without and with using the batch normalization, respectively.

subjects with different age range and gender, and each subject was tested for 2 hours. The dataset contains 165,633 samples in total where each sample has 12 features. Although the number of subjects is relatively small, the observations (i.e. the subjects) are diverse enough. The tri-axial ADXL335 accelerometers are attached to the waist, left thigh, right arm and right ankle for each subject. Each sensor collected three acceleration values in x, y, z directions, which are used as the input features for the classifier. According to analysis in the approach of [25], the data collected by the tri-axial ADXL335 accelerometers is pre-processed and grouped in this work. In order to remove redundant information, the Mark Hall's algorithm is used as the feature selection algorithm. It is configured to adopt the "Best First" approach, which uses a greedy strategy based on backtracking. Figure 2 shows the sample size for each class of the dataset used in this work. The 75- and 31-year-old subjects are male, and the 46- and 28-year-old subjects are female. The number indicates the sample size which shows the diversity of the dataset.

The numbers of samples for postures of sitting down and standing up are much smaller than other 3 postures. As the unbalanced data can affect the classification results, a data balancing technique is used in this work [26], i.e. 1.5 times oversampling is performed on the postures of sitting down and standing up, and 0.5 times undersampling is performed on other postures. The data processing is used to delete the redundant information, correct the errors and provide the consistency of the data. To eliminate the dimensionality influence between characteristics, data standardisation is used. After this, each characteristic is in the same order of magnitude and the speed to find the optimal solution is accelerated. In this study, a data standardization method is used, which is given by $X' = (X - X_{min})/(X_{max} - X_{min})$, where X' is standardised data, and X is the original data. X_{max} and X_{min} are largest, lowest values, respectively. The result shows that the recognition accuracy of the final recognition is improved by 2%.

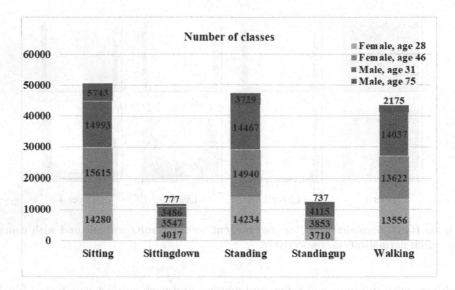

Fig. 2. Number of samples in each action in the BPRD dataset.

In this work, the dataset is randomly divided into 5 equal parts, and then the 5-fold cross-validation test mode is used. When training a neural network, the number of training iterations is 5,000.

4.2   Performance Analysis

In this approach, the initial weights are random numbers between −1 and 1, with variance of 1 and mean of 0. The cross-entropy function is used as the cost function, which is given by

$$C = -\frac{1}{n} \sum_x \sum_j [y_j ln a_j^L + (1 - y_j) ln(1 - a_j^L)], \tag{8}$$

where n is the total number of training data and the sum is the corresponding target output on all training inputs x. y is the target value on the output neuron, a is the actual output value, L is the total number of layers of the network and $\sum_j$ is the sum of all the output neurons.

In this work, the optimizer uses a stochastic gradient descent algorithm. During the training stage, the BPNN model is first trained and the network structure is established. As shown in Table 1, various training experiments have been carried out in this work and the classification effect of the final 4-layer network structure is best. Then the number of neurons in each hidden layer is tuned, where the best performance is obtained when the neurons in the hidden layer are 24 and 36 respectively. Other parameters also need to be tuned in this work, including learning rate and small batch data. In this work the learning rate is changed during the training process, which is firstly set to 0.01, then is reduced

Table 1. Accuracy and training time by using different network structures.

Over fitting algorithms	Structures	Acc. (%)	Time (min)
L2	$12 + 24 + 36 + 5$	99.00	38.75
	$12 + 36 + 24 + 5$	98.50	38.41
	$12 + 24 + 36 + 24 + 5$	98.65	39.37
Dropout	$12 + 24 + 36 + 5$	99.10	20.19
	$12 + 36 + 24 + 5$	98.95	20.66
	$12 + 24 + 36 + 24 + 5$	98.95	25.74
N/A	$12 + 24 + 36 + 5$	99.30	37.23
	$12 + 36 + 24 + 5$	99.00	36.29
	$12 + 24 + 36 + 24 + 5$	98.85	37.51

by 10% for each training iteration of 100 times. This allows a quick converge in the beginning and to achieve an optimal result at a later stage. Moreover, the number of training iterations is 5,000 and the number of small batch data is set to 1,000. The $L1$, $L2$ normalization and dropout algorithms are used to deal with the overfitting. Table 1 shows the comparison results using the $L2$ normalization and dropout algorithms. By using the $L2$ normalization algorithm, the recognition accuracy of the network is reduced, and the training time is almost unchanged. The recognition accuracy by using the dropout algorithm is almost unchanged, but some training time can be reduced and the retention rate of dropout is generally set to 0.9. To reduce the training time of the large number of samples, 10% of all samples are randomly selected to train the network structure in this work. The highest recognition accuracy obtained with the 10% training network of all samples is 99.30%, retaining the structure of the neural network at this time. This neural network structure is then used to train all samples with a final recognition accuracy of 98.00%.

Before the network is binarized, the optimal result of the traditional BPNN is obtained first. The BNN is further built upon the optimized BPNN, where the binarization of the activation value can be viewed as an additional layer. Especially, in this study the floating-point parameters are used to update the weights. The classification accuracy of the proposed method using BNN is 81.00%. Although the classification accuracy decreases, the memory required by the BNN is greatly reduced. Once the models of BPNN and BNN are trained, they are deployed into embedded hardware devices. Raspberry Pi 3 (Model B) is used, which has a 1.2 GHz 64-bit quad-core Broadcom BCM2837 ARMv8 CPU and is very suitable for wearable devices due to the compact size. By observing the size of the network, the BNN is a more compact network compare with BPNN, which is an advantage for the applications in embedded devices with limited resources. The BPNN and BNN models are evaluated based on the hardware systems with 20,000 samples, where the processing times and power consumptions are recorded. To measure power consumption, the UT658B USB

Table 2. Power consumptions of the BPNN and BNN (Raspberry Pi).

State	BPNN	BNN
Size of models	1.70 MB	0.25 MB
Bare metal	5.27 V, 0.25 A, 1.3175 W	
With network connection, no application	5.27 V, 0.29 A, 1.5283 W	
With network connection and fan, no application	5.27 V, 0.38 A, 2.0026 W	
With network connection and fan, application is running	5.27 V, 0.64 A 3.3728 W 135 s, 184.977 J	5.27 V, 0.63 A 3.3201 W 128 s, 168.640 J

power monitor is used to measure the voltage and current [27,28]. In this work, the operating frequency of Raspberry Pi is 600 MHz. Table 2 shows that the model size and the power consumption results of the BPNN and BNN. The size of the BPNN and the BNN model is 1.70 MB and 0.25 MB, where the latter is 85.29% smaller than the former. The bare metal state means that only the Raspberry Pi CPU is running without external interfaces and devices. Also, the power consumption with network connection and fan (heat dissipation) are also provided. For the BPRD dataset, the BPNN needs 135 s to complete 20,000 predictions, and BNN only needs 128 s which has 5.19% improvement than the former. For the power consumption, the BPNN consumes 184.977 J and BNN consumes 168.640 J. The BNN saves about 8.86% energy consumption compared with the BPNN when performing the same task.

To further investigate the trade-off between accuracy and power consumption of the BNN, various approaches are included in Table 3 for illustration. The neural networks are optimized for the datasets including MNIST, CIFAR10, ImageNet and BPRD. Two BNN models with one, two hidden layers (MLP-1, MLP-2) are evaluated on the MNIST and CIFAR10 datasets [21]. The BNN is also evaluated on the ImageNet dataset [20]. Compared to MLP-1, the MLP-2 model has one more hidden layer. Its recognition accuracy is reduced on the MNIST or CIFAR10 datasets, but the size of the network is compressed and power consumption is reduced. For the ImageNet dataset, the AlexNet model is binarized to obtain the BNN model, and the recognition accuracy is also reduced from 80.2% to 50.4%, but the neural model is compressed. It can be seen that when the neural network is binarized, the recognition accuracy is reduced [20,21] and this work is same, where a trade-off between performance and cost should be achieved. In this work, although the recognition accuracy of BNN is lower than BPNN, other performances (processing time, memory and power consumptions) are actually better. To do classification for one sample in the dataset, BNN takes 6.40 ms, which saves 0.35 ms comparing to BPNN (a reduction of 5.19%), and BNN consumes 8.43 mWs, which is 0.82 mWs less than BPNN (a reduction of 8.86%). Further, the size of the BNN is equivalent to 14.71% of BPNN.

Table 3. The processing time, memory requirement and power consumption of different approaches. The processing time and power consumption were calculated based on one sample in this work as same as in the approach of [21].

Approach	Mode l	Acc. (%)	Time (ms)	Mem. (KB)	Enrg. (mWs)
MNIST [21]	MLP-1	91.54	17.35	14.73	5.37
	MLP-2	84.65	9.17	13.533	5.37
CIFAR10 [21]	MLP-1	52.30	21.29	13.84	4.37
	MLP-2	41.80	19.65	14.00	2.31
ImageNet [20]	AlexNet	80.20	N/A	$232 \times 1{,}024$	N/A
	BNN	50.40	N/A	$22.6 \times 1{,}024$	N/A
This work	BPNN	98.00	6.75	$1.70 \times 1{,}024$	9.25
	BNN	81.00	6.40	$0.25 \times 1{,}024$	8.43

The advantage of the BNN is less processing time, memory requirement and power consumption which are beneficial for the deployments of embedded hardware systems with limited memory resources such as the wearable devices.

5 Conclusions

In this work, human body posture recognition methods are proposed which are based on the embedded wearable devices. The BPNN and BNN models are employed to recognize human postures which achieve the classification accuracies of 98.00% and 81.00%, respectively. Results show that although the BNN has a lower recognition accuracy than the BPNN, the BNN has a compact network model and can save 8.86% power consumption, 5.19% processing time and 85.29% memory requirement. It has demonstrated a better trade-off between the recognition accuracy and power consumption than the BPNN, and it is more suitable for the neural network running on the embedded hardware systems with limited memory resources such as wearable devices. Future work will investigate the methods to improve the accuracy of the BNN and further reduce the memory requirement.

Acknowledgments. This research was partially supported by the National Natural Science Foundation of China under Grant 61603104, the Guangxi Natural Science Foundation under Grants 2017GXNSFAA198180 and 2016GXNSFCA380017, the funding of Overseas 100 Talents Program of Guangxi Higher Education under Grant F-KA16035, and 2018 Guangxi One Thousand Young and Middle-Aged College and University Backbone Teachers Cultivation Program.

References

1. Hubbard, R., et al.: The ageing of the population:implications for multidisciplinary care in hospital. Age Ageing **33**(5), 479–482 (2004). https://doi.org/10.1093/ageing/afh164
2. Hong, Y.J., Kim, I.J., Ahn, S.C., Kim, H.G.: Activity recognition using wearable sensors for elder care. In: International Conference on Future Generation Communication and Networking, pp. 302–305 (2008). https://doi.org/10.1109/FGCN.2008.165
3. Lara, O.D., Labrador, M.A.: A survey on human activity recognition using wearable sensors. IEEE Commun. Surv. Tutor. **15**(3), 1192–1209 (2013). https://doi.org/10.1109/SURV.2012.110112.00192
4. Naddeo, S., Verde, L., Forastiere, M., De Pietro, G., Sannino, G.: A real-time m-health monitoring system: an integrated solution combining the use of several wearable sensors and mobile devices. In: International Joint Conference on Biomedical Engineering Systems and Technologies, pp. 545–552 (2017). https://doi.org/10.5220/0006296105450552
5. Ugulino, W., Cardador, D., Vega, K., Velloso, E., Milidiú, R., Fuks, H.: Wearable computing: accelerometers' data classification of body postures and movements. In: Advances in Artificial Intelligence, pp. 52–61 (2012). https://doi.org/10.1007/978-3-642-34459-6_6
6. Kumari, P., Mathew, L., Syal, P.: Increasing trend of wearables and multimodal interface for human activity monitoring: a review. Biosens. Bioelectron. **90**(1), 298–307 (2017). https://doi.org/10.1016/j.bios.2016.12.001
7. Hu, F., Wang, L., Wang, S., Liu, X., He, G.: A human body posture recognition algorithm based on BP neural network for wireless body area networks. China Commun. **13**(8), 198–208 (2016). https://doi.org/10.1109/CC.2016.7563723
8. Denil, M., Shakibi, B., Dinh, L., Ranzato, M., de Freitas, N.: Predicting parameters in deep learning. In: International Conference on Neural Information Processing Systems, pp. 2148–2156 (2013). https://doi.org/10.14288/1.0165555
9. Ieee, S.M., Ieee, F., Sze, V., Chen, Y.H., Yang, T.J., Emer, J.S.: Efficient processing of deep neural networks: a tutorial and survey. Proc. IEEE **105**(12), 2295–2329 (2017). https://doi.org/10.1109/JPROC.2017.2761740
10. Han, S., et al.: EIE: efficient inference engine on compressed deep neural network. In: International Symposium on Computer Architecture ISCA, pp. 243–254 (2016). https://doi.org/10.1109/ISCA.2016.30
11. Liang, S., Yin, S., Liu, L., Luk, W., Wei, S.: FP-BNN: binarized neural network on FPGA. Neurocomputing **275**(1), 1072–1086 (2018). https://doi.org/10.1016/j.neucom.2017.09.046
12. Farringdon, J., Moore, A., Tilbury, N., Church, J., Biemond, P.: Wearable sensor badge and sensor jacket for context awareness. In: International Symposium on Wearable Computers(ISWC), pp. 107–113 (1999). https://doi.org/10.1109/ISWC.1999.806681
13. Poh, M.Z., Swenson, N.C., Picard, R.W.: A wearable sensor for unobtrusive, long-term assesment of electrodermal activity. IEEE Trans. Biomed. Eng. **57**(5), 1243–1252 (2010). https://doi.org/10.1109/TBME.2009.2038487
14. Janidarmian, M., Fekr, A.R., Radecka, K., Zilic, Z.: A comprehensive analysis on wearable acceleration sensors in human activity recognition. Sensors **17**(3), 529–555 (2017). https://doi.org/10.3390/s17030529

15. Barkallah, E., Freulard, J., Otis, M.J., Ngomo, S., Ayena, J.C., Desrosiers, C.: Wearable devices for classification of inadequate posture at work using neural networks. Sensors 17(9), 1–24 (2017). https://doi.org/10.3390/s17092003
16. Wong, A., Shafiee, M.J., Li, F., Chwyl, B.: Tiny SSD: a tiny single-shot detection deep convolutional neural network for real-time embedded object detection. In: Canadian Conference on Computer and Robot Vision, pp. 95–101 (2018). https://doi.org/10.1109/crv.2018.00023
17. Lei, W., Chen, H., Wu, Y.: Compressing deep convolutional networks using k-means based on weights distribution. In: International Conference on Intelligent Information Processing, pp. 1–6 (2017). https://doi.org/10.1145/3144789.3144803
18. Rastegari, M., Ordonez, V., Redmon, J., Farhadi, A.: XNOR-Net: ImageNet classification using binary convolutional neural networks. In: Leibe, B., Matas, J., Sebe, N., Welling, M. (eds.) ECCV 2016. LNCS, vol. 9908, pp. 525–542. Springer, Cham (2016). https://doi.org/10.1007/978-3-319-46493-0_32
19. Ardakani, A., Condo, C., Ahmadi, M., Gross, W.J.: An architecture to accelerate convolution in deep neural networks. IEEE Trans. Circuits Syst. I: Regular Pap. 65(4), 1349–1362 (2018). https://doi.org/10.1109/TCSI.2017.2757036
20. Wei, T., Gang, H., Liang, W.: How to train a compact binary neural network with high accuracy? In: AAAI Conference on Artificial Intelligence (AAAI), pp. 2625–2631 (2017)
21. McDanel, B., Teerapittayanon, S., Kung, H.T.: Embedded binarized neural networks. In: International Conference on Embedded Wireless Systems and Networks (EWSN), pp. 168–173 (2017)
22. Münzner, S., Schmidt, P., Reiss, A., Hanselmann, M., Stiefelhagen, R., Dürichen, R.: CNN-based sensor fusion techniques for multimodal human activity recognition. In: International Symposium on Wearable Computers (ISWC), pp. 158–165 (2017). https://doi.org/10.1145/3123021.3123046
23. Li, J., Cheng, J.H., Shi, J.Y., Huang, F.: Brief introduction of back propagation (BP) neural description of BP algorithm in mathematics. Adv. Comput. Sci. Inf. Eng. 169(1), 553–558 (2012). https://doi.org/10.4314/wsa.v31i2.5199
24. Ng, A.: Feature selection, L1 vs. L2 regularization, and rotational invariance. In: International Conference on Machine Learning (ICML), pp. 1–8 (2004). https://doi.org/10.1145/1015330.1015435
25. Kwolek, B., Kepski, M.: Improving fall detection by the use of depth sensor and accelerometer. Neurocomputing 168(1), 637–645 (2015). https://doi.org/10.1016/j.neucom.2015.05.061
26. Mollineda, R.A.: Surrounding neighborhood-based SMOTE for learning from imbalanced data sets. Prog. Artif. Intell. 1(4), 347–362 (2012). https://doi.org/10.1007/s13748-012-0027-5
27. Kaup, F., Gottschling, P., Hausheer, D.: PowerPi: measuring and modeling the power consumption of the raspberry Pi. In: Local Computer Networks (LCN), pp. 236–243 (2014). https://doi.org/10.1109/LCN.2014.6925777
28. Astudillo-Salinas, F., Barrera-Salamea, D., Vazquez-Rodas, A., Solano-Quinde, L.: Minimizing the power consumption in raspberry Pi to use as a remote WSN gateway. In: Latin-American Conference on Communications (LATINCOM), pp. 1–5 (2016). https://doi.org/10.1109/LATINCOM.2016.7811590

Collaborative Denoising Autoencoder for High Glycated Haemoglobin Prediction

Zakhriya Alhassan[1,2](✉), David Budgen[1], Ali Alessa[3], Riyad Alshammari[4],
Tahani Daghstani[4], and Noura Al Moubayed[1]

[1] Department of Computer Science, Durham University, Durham, UK
`zakhriya.n.alhassan@durham.ac.uk`
[2] Computing and Information Technology, University of Jeddah,
Jeddah, Kingdom of Saudi Arabia
[3] Computer Science and Engineering, University of Bridgeport, Bridgeport, USA
[4] CPHHI, King Saud bin Abdulaziz University for Health Sciences; KAIMRC,
Ministry of the National Guard Health Affairs, Riyadh, Kingdom of Saudi Arabia

Abstract. A pioneering study is presented demonstrating that the presence of high glycated haemoglobin (HbA1c) levels in a patient's blood can be reliably predicted from routinely collected clinical data. This paves the way for performing early detection of Type-2 Diabetes Mellitus (T2DM). This will save healthcare providers a major cost associated with the administration and assessment of clinical tests for HbA1c. A novel collaborative denoising autoencoder framework is used to address this challenge. The framework builds an independent denoising autoencoder model for the high and low HbA1c level, which extracts feature representations in the latent space. A baseline model using just three features: patient age together with triglycerides and glucose level achieves 76% F1-score with an SVM classifier. The collaborative denoising autoencoder uses 78 features and can predict HbA1c level with 81% F1-score.

Keywords: Healthcare · Machine learning · Deep learning ·
Collaborative denoising autoencoder · Clinical data · Diabetes ·
Type-2 Diabetes Mellitus · Glycated haemoglobin · HbA1c ·
KAIMRC dataset

1 Introduction

The haemoglobin when joined with the glucose within the blood, it forms glycated haemoglobin, referred to as HbA1c [24,30]. Glycated haemoglobin (HbA1c) is the basis for one of the important blood tests used to indicate the status of glucose levels in the blood [24,30].

The level of HbA1c is strongly related to the average glucose concentration in the blood and the life span of the red blood cells. The normal red blood cells of the human body can last for two to three months before being reproduced.

© Springer Nature Switzerland AG 2019
I. V. Tetko et al. (Eds.): ICANN 2019, LNCS 11731, pp. 338–350, 2019.
https://doi.org/10.1007/978-3-030-30493-5_34

Hence the level of HbA1c can indicate the average level of blood glucose over the whole period of the life span of the red blood cells [25,31]. This can provide physicians with an important long-term measure for blood glucose levels [1].

The traditional method of measuring the blood glucose levels is by using the fasting plasma glucose (FPG) test. The FPG test can provide a measure of short-term blood glucose level but requires the patients to undertake an overnight fasting prior to the test. However, the HbA1c blood test can provide an overall average of the blood glucose level over the preceding two to three months (long-term) and can be taken without the patient having to fast the night before the test. Thus, HbA1c test is an attractive option for both patients and practitioners for measuring glucose levels in the blood [1]. Recent studies have showed that HbA1c test can be used as indicator for diagnosing diabetes (T2DM) especially when combined with vital signs such as Body Mass Index (BMI) [15,21].

The International Expert Committee (IEC), with members from the American Diabetes Association (ADA), the European Association for the Study of Diabetes, and the International Diabetes Federation [6,11], recommends HbA1c test to evaluate the adults with a high risk of diabetes [1]. HbA1c levels are also related to chronic complications [10]. The IEC recommends effective clinical intervention for those patients with a HbA1c level of 6.5% or more.

Non-diabetic people with an elevated HbA1c level are also at an increased risk of cardiovascular disease [1,23]. The HbA1c test helps with predicting which patients are likely to develop T2DM in the future [12]. It also helps physicians decide how frequently patients need to undertake clinical screening for T2DM [12]. The earlier the diagnosis of T2DM, the better chance there is of delaying (and possibly preventing) long-term complications [13].

Research has shown that reducing the HbA1c level can significantly reduce the possibility of developing serious complications for diabetic patients. Lowering the HbA1c level by 1% for diabetic patients can help reduce the risk of developing heart failure by 16%, cataracts by 19%, retinopathy and kidney disease by 25% and death caused by vascular diseases by 43% [18,34]. Hence, a close monitoring of HbA1c levels is recommended for diabetic patients and also for those with potential for developing diabetes [23].

In this paper, we introduce the first study that employs machine learning models to predict the level of HbA1c using patient's data. The study uses routinely collected data from hospital patients to predict the level of HbA1c. We introduce a novel deep learning based framework to predict the level of HbA1c using a significantly large and unique clinical dataset (KAIMRC dataset). According to the IEC classification of the test level, we formulate the risk of the HbA1c test classification problem as a binary classification problem (patients with less than a 6.5% HbA1c level being coded as low HbA1c, and for 6.5% or more being coded as high HbA1c).

We suggest that the outcomes of this work should encourage more clinical investigations of how far other lab tests may be predictable from the HbA1c level. By predicting high and low levels of HbA1c, we also expect that this work will help reduce the need for HbA1c clinical assessments, resulting in reducing the cost and time needed to perform the test. Finally, and most importantly, it will help identify patients who are at risk of developing diabetes by using their hospital history visit records, and hence planning for preventive interventions.

To the best of our knowledge, there are no studies that have investigated the prediction of HbA1c levels using any form of machine learning by making use of patient visit data. The main contributions of this paper are: (I) introduces a novel collaborative autoencoder framework; (II) investigates the use of the routinely collected clinical patient data as predictors for HbA1c levels; (III) presents a unique and large dataset that contains the HbA1c level for 14,609 different patient visits.

2   Related Work

With the help of Electronic Health Records (EHR), clinical data has developed into an interesting frontier for machine learning research. In recent years, machine learning models have shown powerful capabilities of analysing and understanding complex clinical data in a variety of medical applications. Autoencoders had several successes in diverse areas of applications, and especially recently with the development of deep variations [2,14]. In the medical field, autoencoders were mainly used to analyse medical imaging data including: removing the noise [16], data analytics [33], and outlier detection [3,7].

From a clinical perspective, there have been several studies investigating the trend of HbA1c levels. McCarter et al. [28] studied the association of the HbA1c levels for T2DM patients over clinical variables. Conversely, Nathan et al. [29] were able to demonstrate promising results for calculating clinical variables, specifically, the average glucose level, from the HbA1c levels. Kazemi et al. [22] added complications, such as retinopathy, to the clinical variables to analyse the trend of the HbA1c levels. The above studies used the statistical linear model to achieve this task and the data was mostly collected from diabetic patients. Other work by Rose et al. [32] showed a correlation between the mean blood glucose level and the HbA1c level. The correlation coefficients were found to be between 0.71 to 0.86. However, the result can be significantly affected by the time of day (before or after meals) at which the blood glucose level was measured. A very recent study in 2018 by Wells et al. [36] discussed the use of mathematical equations for HbA1c prediction, obtaining an accuracy of 77% using only non-diabetic patients records.

The above studies used statistical and mathematical approaches to investigate the correlation between HbA1c levels and clinical variables. However, they did not explore the prediction power of the HbA1c levels using machine learning techniques. Our approach investigates the use of a novel deep learning technique to predict the HbA1c level for diabetic and non-diabetic patients using only routinely collected clinical data.

3 Dataset

KAIMRC is one of the leading institutions in health research in the Middle East. The KAIMRC[1] dataset was collected by Ministry of National Guard Health Affairs (NGHA) from three main national guard hospitals in Saudi Arabia[2].

Table 1. Statistics of HbA1c KAIMRC dataset

Characteristic	Overall
Number of patient visits	13,317
Number of features	78
Number of different health conditions	99
Number of patient visit types	4
Number of discharge types	8

The KAIMRC dataset contains a full history of patients for the period between 2010 and 2015. In addition, it contains 41 million time-stamped lab test readings, such as Blood Urea Nitrogin (BUN), cholesterol (Chol) and Mean Corpuscular Hemoglobin (MCH). It also holds time-stamped data on vital signs, such as BMI and Hypertension. Other complementary features were also collected during each visit, such as visit type (in-patient, or emergency), gender, patient age, service type (such as Cardiology, Neurology or Endocrinology), length of stay (LOS) and discharge type [4].

Predicting the HbA1c level using only general clinical data is very challenging. There are many factors that affect HbA1c level and stability such as improved diet and physical exercises. Changes in patient lifestyle is known to have a significant effect on the level of HbA1c. Furthermore, the significant amount of missing data in most of the medical datasets forms another major challenge. Figure 1 demonstrates the challenge of separating the data between the two classes: high and low level HbA1c, by visualising a two dimensional projection of the data using t-SNE [27].

---

[1] Access to the KAIMRC dataset can be obtained upon an official request to KAIMRC.
[2] Western, Central and Eastern regions of Saudi Arabia.

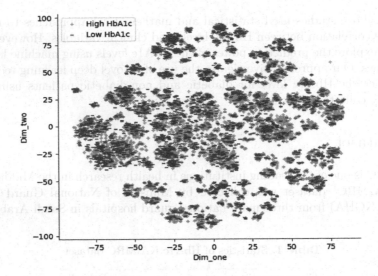

Fig. 1. Projection of the row data onto two dimensional space using t-SNE

3.1 Data Pre-processing

Each patient visit is described by a set of measures. These measures are represented as episodes. Episodes contain the data of irregularly collected vital signs and lab readings. In addition to this, the patient details and visit details (e.g., gender, age, visit type and service provided) are integrated into the episodes. For a patient with an in-patients visit type, only the data for the first day was considered. Cases with values of less than 0.1 of HbA1c are considered to be erroneous readings and have been excluded. This resulted in reducing the dataset size from 14,609 down to 13,317 cases (Table 1).

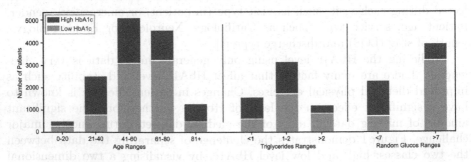

Fig. 2. Classes distribution over patients age, random glucose and triglycerides.

We use 78 features for our analysis: gender, age, service, specialty, visit type and 73 vital signs and lab results. Some features are collected frequently on an hourly basis, such as vital signs. In these cases, the average value for the readings on that day is used instead. 58% of the KAIMRC dataset is labelled as high level: patients with a 6.5 HbA1c level or more, while the remaining 42% are labelled as low level (less than a 6.5). An integer encoding method was used to encode the values of categorical features such as age and gender. Data standardization is used to change the distribution of the features' values so that they are centered on 0 and a standard deviation of 1.

We measured the correlation between the HbA1c level and the 78 features using Pearson Correlation Coefficient (PCC) [8]. The result shows positive linear correlation for 44 features and zero or negative correlation for the remaining 34 features. There are three features, age, Triglycerides (Trig) and Random Glucose (Glur), with correlation between 0.2 and 0.3. Figure 2 shows the class distribution with regards to these three features.

In general, mining clinical data is made difficult by several problems such as missing data, variety of lengths, and irregularity. For instance, the percentage of missing values for Triglycerides and Random Glucose is 54% and 51% respectively. As clinical data is sensitive, we avoided using techniques to interpolate this problem.

4   Methods

To increase the separability between the classes, high vs low HbA1c levels, the framework generates new features from each class separately by modelling directly the data that belongs to a given class. This is motivated by the success of pre-training in deep learning models [17, 35], however we use a separate model per class to reduce the within-class noise and increase between-class separability. These two models "collaborate" by combining their outputs together to form the input to a third classification model.

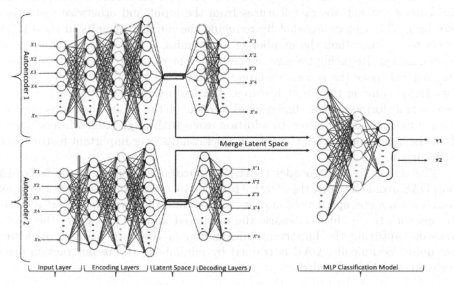

Fig. 3. Proposed collaborative-denoising autoencoders framework

Figure 3 demonstrates the collaborative denoising autoencoders (Col-DAE) framework piloted here. Autoencoder1 models the low HbA1c level class, while Autoencoder2 models the high HbA1c level class. The features of latent space of both models are then merged and fed into the MLP classification model. The MLP model is trained to predict the level of HbA1c. To classify a sample, we feed it to both pre-trained autoencoder models with their outputs merged and fed to the MLP model for prediction.

4.1 Denoising Autoencoders

Autoencoders fall under the umbrella of representative learning with deep neural networks. The goal of an Autoencoder (AE) is to learn an abstract representation of the data presented at its input. The input can be reconstructed from that representation. Hence the desired output of the AE is the input itself [9,35], making it a completely unsupervised deep learning method.

Given an input x the AE model tries to reconstruct it as r. An AE network consists of an encoder $h = f(x)$ network that transforms x from the input space to the latent space, and a decoder network which does the opposite transformation to reconstruct the original input using the decoder function $r = g(h)$ [5,26]. The encoder and decoder are implemented as deep neural networks, that are optimised simultaneously using standard stochastic gradient descent algorithms.

The AE's main task is to minimise the reconstruction error between the input x and the reconstructed variable r using loss functions:

$$L(x, g(f(x))) \tag{1}$$

One of the main challenges with training an AE is over-fitting to the training data, also known as the identity function problem [19]. This occurs when the AE cannot extract abstract features from the input and otherwise memorises the data. This can be avoided by reducing the number of units in the hidden layers to be fewer than the number of input units, applying dropouts, or using regularisation. Reducing the size of the hidden layers maps the input to a lower dimensional space (known as the latent space). This forces the network to learn correlations among the input features. The features at the latent space represent a transformed lower dimensional version of the input that conserves its most important information. In addition, even with large hidden layers (larger than the input dimension), the autoencoder can discover important features and structures in the data.

The Denoising Autoencoder (DAE) is an extended version of the basic AE [35]. DAE avoids learning the identity function by trying to reconstruct the input data x from a corrupted version of it, $\tilde{x}$. Decoding the corrupted input requires the network to be able to denoise the corrupted input. This pushes the model towards capturing the important information in the data without modelling the noise. Technically, DAE is trained by minimising the reconstruction error between x and the decoded

$$\tilde{x} : L(x, g(f(\tilde{x}))) \tag{2}$$

Here deep denoising autoencoders are used to model a sequences of patient observations as input $x = (x_1, x_2, ...x_n)$. Mean Squared Error (MSE) is used to calculate the reconstruction error:

$$MSE = \frac{1}{n} \sum_{i=1}^{n} (y - r)^2 \qquad (3)$$

4.2   Models and Experimental Setup

Our framework (Col-DAE) consists of two denoising autoencoders and one s Multi-Layer Perceptron (MLP) model. The first DAE (Autoencoder1) models the high HbA1c level while the second DAE (Autoencoder2) models the low HbA1c level. Each DAE model has three hidden layers for the encoder, as shown in Fig. 3. Prior to the encoder, an isotropic Gaussian distributed noise is added to the input layer [35]. The number of neurons for the layers in the encoder are 90, 120 and 130 respectively. Each DAE model consists of two hidden layers for the decoder. The first decoding layer (90 neurons) takes the embeddings in the latent space as an input while the output layer has the same size as the input. Tanh activation function is used in all encoder and decoder layers.

An MLP model is used as the classification model. It consists of three dense layers. The first merges the latent spaces from the two DAEs with 260 neurons (130 from each DAE). The second and third layers of the MLP classification model has 70 and 32 neurons respectifly. Relu activation function is used in both layers and Sigmoid for the output layer used for the two classes prediction.

The DAEs (Autoencoder 1 and Autoencoder 2 in Fig. 3) are pre-trained and validated to reconstruct the input using 80% and 10% of the data respectively. The remaining 10% of the data is kept as the test set. The MLP classification model is trained and validated using the training and validation sets along with the associated HbA1c labelled levels. Each test sample is fed to both DAEs and the embedding outputs from both DAEs are merged to form the input vector of the MLP classification model. These tasks (pre-train the DAEs, train the MLP and testing the Col-DAE model) are repeated 10 times using a 10-folds cross-validation approach to ensure that all data points are included in testing the model. The DAEs of the Col-DAE model are trained for 100 epochs using an Adam optimiser with Mean Squared Error as the loss function. The MLP of the Col-DAE model uses same optimiser, loss function and number of epochs for training.

4.3   Comparative Models

We have compared our results against popular base-line models: Support Vector Machines (SVM) and Logistic Regression (LR). We also compare our results to deep learning approaches such as MLP and Autoencoders [20, 26]. We explored all models with different feature sizes (using the top three correlated features, the 44 features with positive correlation and the 78 originally collected). Because deep learning approaches work with high dimensional data, the MLP, DAE and

Col-DAE models were not experimented using three features. For the purpose of fair comparison, these models were employed using the same data pre-processing, training and testing techniques used for the Col-DAE model.

In addition to the accuracy, we report F1, F1 Weighted, Recall and Precision measures to evaluate the performance of the proposed models. The Col-DAE model was experimented on using different combination of regularisers, activation functions, dropout rates, learning rates, and optimisers. We only report the results with the best performance as per the reported measures.

5   Results

The performance metrics for predicting the level of HbA1c, obtained using the compared models: SVM, LR, MLP, DAE and Col-DAE with different feature sizes, are presented in Table 2. All models show better performance when using the 78 features despite the negative linear correlation for 34 features except for DAE models. We report an F1-score of 73.34% and 65.63% F1-Weighted measures for DAE. However, the DAE models show clear signs of over-fitting. The large difference between the two measures is explained by the 79.08% recall and 68.48% precision. This is being ascribed to biased classification behaviour. The DAE model performs the lowest among all models.

The rest of the models (SVM, LR, MLP and Col-DAE) do not show any bias towards any of the classes. The SVM and MLP models with 78 features achieved competitive performance with an F1-score of 78.84% and 79.72%. However, the SVM model achieved promising performance, using three features only, with 76% F1-score.

Table 2. Performance of classifiers for HbA1c risk prediction

Model	Features size	F1-score	Accuracy	F1-weighted	Recall	Precision
SVM	3	0.7609	0.7172	0.7162	0.7724	0.7499
	44	0.7817	0.7292	0.7239	0.8324	0.7370
	78	0.7884	0.7398	0.7356	0.8322	0.7492
LR	3	0.7168	0.6526	0.6476	0.7543	0.6830
	44	0.7394	0.6859	0.6832	0.7648	0.7159
	78	0.7574	0.7113	0.7098	0.7733	0.7424
MLP	44	0.7857	0.7465	0.7456	0.7973	0.7752
	78	0.7972	0.7588	0.7573	0.8146	0.7827
DAE	44	0.7334	0.6650	0.6563	0.7908	0.6848
	78	0.7301	0.6563	0.6445	0.7985	0.6737
Col-DAE	44	0.7974	0.7626	0.7620	0.8032	0.7929
	78	**0.8109**	**0.7760**	**0.7751**	0.8240	0.7987

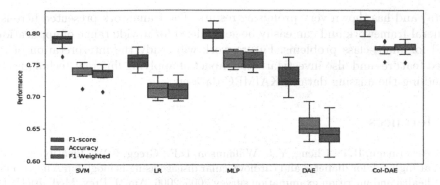

Fig. 4. Box plot of the detailed performance for the proposed models.

Table 2 shows that the Col-DAE model using 78 features achieved better results than the compared base-line models, with 81.09% F1-score and 77.60% of accuracy. Figure 4 summarises the 10-folds results of the reported measures achieved by the proposed models. Figure 4 shows small variation between the folds and especially in F1-score which demonstrates the consistency of the Col-DAE's performance. The differences between the obtained F1-scores for Col-DAE with 78 features and the comparative models are statistically significant (p-value is 0.006 with SVM, 0.005 with LR and DAE and 0.028 with MLP).

6 Discussion and Conclusion

Our framework is trained using patient clinical data from patients visiting the hospitals for a variety of health conditions. Despite the large number of missing values, the SVM achieved 76% F1-score using three features. The Col-DAE outperformed the base-line classifiers and achieved 81% for F1-score using 78 features. Due to the lack of similar studies and related work using machine learning, the accuracy achieved in this paper could not be compared to any previous work. Hence, this forms the baseline accuracy for studies aiming at predicting of HbA1c level in the future.

The outcome of this work is significant for enabling physicians to make preventative intervention decisions in order to successfully manage the risk of high level HbA1c. The replication of this work using other hospital clinical datasets can ultimately help provide improved healthcare services to patients and reduce the cost and time needed to assess the HbA1c test. This can help identifying patients who are at a high risk of developing T2DM, and has the potential to be used as an early warning indicator for developing serious health complications.

We introduced here the collaborative denoising autoencoders (Col-DAE). The framework uses denoising autoencoders to model separately the high and low level HbA1c data. The latent spaces of both models are then merged and passed to a MLP model for decision making. This framework was utilised for a complex classification challenge (HbA1c level prediction from routinely collected clinical data) and has shown very promising results. The framework presented here is a general framework and can easily be generalised for a wide range of applications and for multi-class problems. Future work will study the interpretation of the used features and also investigate the impact of applying different techniques for handling the missing data in KAIMRC dataset.

References

1. Ackermann, R.T., Cheng, Y.J., Williamson, D.F., Gregg, E.W.: Identifying adults at high risk for diabetes and cardiovascular disease using hemoglobin A1c: national health and nutrition examination survey 2005–2006. Am. J. Prev. Med. **40**(1), 11–17 (2011)
2. Al Moubayed, N., Breckon, T., Matthews, P., McGough, A.S.: SMS spam filtering using probabilistic topic modelling and stacked denoising autoencoder. In: Villa, A.E.P., Masulli, P., Pons Rivero, A.J. (eds.) ICANN 2016. LNCS, vol. 9887, pp. 423–430. Springer, Cham (2016). https://doi.org/10.1007/978-3-319-44781-0_50
3. Alhassan, Z., Budgen, D., Alshammari, R., Daghstani, T., McGough, A.S., Al Moubayed, N.: Stacked denoising autoencoders for mortality risk prediction using imbalanced clinical data. In: International Conference on Machine Learning and Applications. IEEE (2018)
4. Alhassan, Z., McGough, A.S., Alshammari, R., Daghstani, T., Budgen, D., Al Moubayed, N.: Type-2 diabetes mellitus diagnosis from time series clinical data using deep learning models. In: Kůrková, V., Manolopoulos, Y., Hammer, B., Iliadis, L., Maglogiannis, I. (eds.) ICANN 2018. LNCS, vol. 11141, pp. 468–478. Springer, Cham (2018). https://doi.org/10.1007/978-3-030-01424-7_46
5. An, J., Cho, S.: Variational autoencoder based anomaly detection using reconstruction probability. Spec. Lect. IE **2**, 1–18 (2015)
6. Association, A.D., et al.: Diagnosis and classification of diabetes mellitus. Diabetes Care **37**(Supplement 1), S81–S90 (2014)
7. Baur, C., Wiestler, B., Albarqouni, S., Navab, N.: Deep autoencoding models for unsupervised anomaly segmentation in brain MR images. arXiv preprint arXiv:1804.04488 (2018)
8. Benesty, J., Chen, J., Huang, Y., Cohen, I.: Pearson correlation coefficient. In: Cohen, I., Huang, Y., Chen, J., Benesty, J. (eds.) Noise Reduction in Speech Processing, pp. 1–4. Springer, Heidelberg (2009). https://doi.org/10.1007/978-3-642-00296-0
9. Bengio, Y., Simard, P., Frasconi, P.: Learning long-term dependencies with gradient descent is difficult. IEEE Trans. Neural Netw. **5**(2), 157–166 (1994)
10. Bonora, E., Tuomilehto, J.: The pros and cons of diagnosing diabetes with A1c. Diabetes Care **34**(Supplement 2), S184–S190 (2011)
11. Committee, I.E., et al.: International expert committee report on the role of the A1c assay in the diagnosis of diabetes. Diabetes Care **32**(7), 1327–1334 (2009)
12. Edelman, D., Olsen, M.K., Dudley, T.K., Harris, A.C., Oddone, E.Z.: Utility of hemoglobin A1c in predicting diabetes risk. J. Gen. Intern. Med. **19**(12), 1175–1180 (2004)
13. Federation ID: IDF diabetes atlas (2017). http://www.diabetesatlas.org
14. Gao, S., Zhang, Y., Jia, K., Lu, J., Zhang, Y.: Single sample face recognition via learning deep supervised autoencoders. IEEE Trans. Inf. Forensics Secur. **10**(10), 2108–2118 (2015)

15. Gerstein, H.C., et al.: Annual incidence and relative risk of diabetes in people with various categories of dysglycemia: a systematic overview and meta-analysis of prospective studies. Diabetes Res. Clin. Pract. **78**(3), 305–312 (2007)
16. Gondara, L.: Medical image denoising using convolutional denoising autoencoders. In: 2016 IEEE 16th International Conference on Data Mining Workshops (ICDMW), pp. 241–246, December 2016
17. Goodfellow, I., Bengio, Y., Courville, A.: Deep Learning. MIT Press, Cambridge (2016)
18. UK Prospective Diabetes Study Group, et al.: Tight blood pressure control and risk of macrovascular and microvascular complications in type 2 diabetes: UKPDS 38. BMJ: Br. Med. J. **317**(7160), 703 (1998)
19. Hinton, G.E., Zemel, R.S.: Autoencoders, minimum description length, and Helmholtz free energy. In: Advances in Neural Information Processing Systems, p. 3 (1994)
20. Hochreiter, S., Schmidhuber, J.: Long short-term memory. Neural Comput. **9**(8), 1735–1780 (1997)
21. Kälsch, J., et al.: Normal liver enzymes are correlated with severity of metabolic syndrome in a large population based cohort. Sci. Rep. **5**, 13058 (2015)
22. Kazemi, E., Hosseini, S.M., Bahrampour, A., Faghihimani, E., Amini, M.: Predicting of trend of hemoglobin A1c in type 2 diabetes: a longitudinal linear mixed model. Int. J. Prev. Med. **5**(10), 1274 (2014)
23. Khaw, K.T., Wareham, N., Bingham, S., Luben, R., Welch, A., Day, N.: Association of hemoglobin A1c with cardiovascular disease and mortality in adults: the European prospective investigation into cancer in Norfolk. Ann. Intern. Med. **141**(6), 413–420 (2004)
24. Koenig, R.J., Peterson, C.M., Jones, R.L., Saudek, C., Lehrman, M., Cerami, A.: Correlation of glucose regulation and hemoglobin A1c in diabetes mellitus. N. Engl. J. Med. **295**(8), 417–420 (1976)
25. Larsen, M.L., Hørder, M., Mogensen, E.F.: Effect of long-term monitoring of glycosylated hemoglobin levels in insulin-dependent diabetes mellitus. N. Engl. J. Med. **323**(15), 1021–1025 (1990)
26. LeCun, Y., Bengio, Y., Hinton, G.: Deep learning. Nature **521**(7553), 436–444 (2015)
27. Maaten, L.V.D., Hinton, G.: Visualizing data using t-SNE. J. Mach. Learn. Res. **9**(Nov), 2579–2605 (2008)
28. McCarter, R.J., Hempe, J.M., Chalew, S.A.: Mean blood glucose and biological variation have greater influence on HbA1c levels than glucose instability: an analysis of data from the diabetes control and complications trial. Diabetes Care **29**(2), 352–355 (2006)
29. Nathan, D.M., et al.: Translating the A1c assay into estimated average glucose values. Diabetes Care **31**, 1473–1478 (2008)
30. Peterson, K.P., Pavlovich, J.G., Goldstein, D., Little, R., England, J., Peterson, C.M.: What is hemoglobin A1c? An analysis of glycated hemoglobins by electrospray ionization mass spectrometry. Clin. Chem. **44**(9), 1951–1958 (1998)
31. Pradhan, A.D., Rifai, N., Buring, J.E., Ridker, P.M.: Hemoglobin A1c predicts diabetes but not cardiovascular disease in nondiabetic women. Am. J. Med. **120**(8), 720–727 (2007)
32. Rose, E., Ketchell, D.S.: Does daily monitoring of blood glucose predict hemoglobin A1c levels? Clinical Inquiries, 2003 (MU) (2003)

33. Shin, H., Orton, M., Collins, D.J., Doran, S., Leach, M.O.: Autoencoder in time-series analysis for unsupervised tissues characterisation in a large unlabelled medical image dataset. In: 2011 10th International Conference on Machine Learning and Applications and Workshops, vol. 1, pp. 259–264, December 2011
34. Stratton, I.M., et al.: Association of glycaemia with macrovascular and microvascular complications of type 2 diabetes (UKPDS 35): prospective observational study. BMJ **321**(7258), 405–412 (2000)
35. Vincent, P., Larochelle, H., Lajoie, I., Bengio, Y., Manzagol, P.A.: Stacked denoising autoencoders: learning useful representations in a deep network with a local denoising criterion. J. Mach. Learn. Res. **11**(Dec), 3371–3408 (2010)
36. Wells, B.J., et al.: Predicting current glycated hemoglobin values in adults: development of an algorithm from the electronic health record. JMIR Med. Inform. **6**(4), e10780 (2018)

Special Session: Informed and Explainable Methods for Machine Learning

Special Session: Informed and Explainable Methods for Machine Learning

Although the latest advancements in deep learning have pushed the boundaries of artificial intelligence and already resulted in successful solutions for very challenging problems, two major practical aspects, which are of utmost importance in a variety of industry applications, remain mostly uncovered. The first aspect is related to lack of training data (i.e. having thin data scenarios), which due to the Vapnik-Chervonenkis theory is a crucial issue to train well-generalizing deep models with millions of adjustable parameters. Whereas, the second aspect is related to the lack of traceability as the typical connectionist models can be seen as black boxes since their inner computations become difficult-to-interpret with the increased complexity. This special session aimed at bringing together lead researchers from industry and research to concentrate on methods to incorporate knowledge into the state-of-the-art machine learning models to reduce the need for massive training datasets and investigate a variety of methods to obtain interpretable solutions.

Organization

Christian Bauckhage   Fraunhofer Center for Machine Learning,
                             University of Bonn, Germany
Rafet Sifa   Fraunhofer IAIS, Germany
Anna Ladi   Fraunhofer IAIS, Germany
Fabian Hadiji   Goedle.io, Germany

On Chow-Liu Forest Based
Regularization of Deep Belief Networks

Alex Sarishvili[1,3]([✉]) [iD], Andreas Wirsen[1,3] [iD], and Mats Jirstrand[2,3] [iD]

[1] Fraunhofer ITWM, Kaiserslautern, Germany
{alex.sarishvili,andeas.wirsen}@itwm.fraunhofer.de
[2] Fraunhofer-Chalmers Centre, Göteborg, Sweden
mats.jirstrand@fcc.chalmers.se
[3] Fraunhofer Center for Machine Learning, Munich, Germany

Abstract. In this paper we introduce a methodology for the simple integration of almost-independence information on the visible (input) variables of restricted Boltzmann machines (RBM) into the weight decay regularization of the contrastive divergence and stochastic gradient descent algorithm. After identifying almost independent clusters of the input coordinates by Chow-Liu tree and forest estimation, the RBM regularization strategy is constructed. We show an example of a sparse two hidden layer Deep Belief Net (DBN) applied on the MNIST data classification problem. The performance is quantified by estimating misclassification rate and measure of manifold disentanglement. Approach is benchmarked to the full model.

Keywords: Probabilistic graphs · Restricted Boltzman machines · Chow-Liu trees

1 Introduction

One way of modeling complex high dimensional probability distributions is to use a large number of relatively simple probabilistic models and to combine them, e.g. Gaussian mixtures. If sufficiently many models are included in the mixture, it is possible to approximate complicated smooth distributions with arbitrarily accuracy [7]. In high dimensional spaces the building of mixtures is very inefficient and therefore it is not recommended to follow this approach. Instead one should apply the multiplication of simple probability distributions and renormalization. Of course the product of Gaussians can not approximate arbitrarily well the smooth multivariate distributions like the mixtures of Gaussians, but if the individual distributions are slightly complicated and if latent (hidden) variables are introduced, the product of distributions can do this job

This work was developed in Fraunhofer Cluster of Excellence "Cognitive Internet Technologies".

© Springer Nature Switzerland AG 2019
I. V. Tetko et al. (Eds.): ICANN 2019, LNCS 11731, pp. 353–364, 2019.
https://doi.org/10.1007/978-3-030-30493-5_35

very well. Individual models of this product approximations (factorizations) are called "experts".

Unsupervised neural networks assume unlabelled data to be generated from a neural network structure. They have been applied extensively to pattern recognition and representation learning. The most basic one is the restricted Boltzmann machine (RBM) [14]. A RBM is one way to approximate the mentioned multivariate probability distribution functions (pdf).

On the other hand side, it is known that these directed graphs can represent immoral dependence structures properly. These very important structures have the property that, the parents of a node are not connected to each other. However, a good graphical model for a pdf of visible units, that does not include latent variables and is a directed acyclic graph, would contain very large number of nodes in a Bayesian network. A large number of parents mirrors into the exponentially growing number of parameters in a Bayesian network. Therefore, this kind of modeling is computationally and statistically not feasible [10].

In our approach we approximate the pdf by a tree (Chow-Liu tree) and integrate the gained information, regarding the interdependencies between the data into the regularization term of a RBM. We will demonstrate on the MNIST data set that integrating a-priori dependency information into the learning procedure of the RBM's in the first layer of a DBN (Deep Belief Net) improves the classification performance of the approach.

In the first step, we factorize the input vector using Chow-Liu trees, prune the estimated tree by removing insignificant edges and finally incorporate the obtained information about almost independent groups of visible coordinates into the regularization term of the RBM learning algorithm. Regularization made by weight decay is a main instrument to avoid overfitting in machine learning approaches. However, especially in RBMs weight decay has further benefits, in particular the receptive fields of the hidden units become more smooth and more interpretable by shrinking useless weights [8].

In this paper we construct a two layer DBN which is a composition of two RBM's. In section two we describe the graphical models. Particularly we begin with the RBMs and then continue with the Chow-Liu graphs and the Chow-Liu Forests. In the third section we show the natural way of integrating the independence information given by Chow-Liu forest into the learning algorithm of the RBM. In section four we demonstrate the benefits of the approach on the MNIST benchmark. We will end our paper with a summary.

2   Graphical Models

In this section we describe the graphical models required within our approach. Particularly we begin with RBM's, continue with the Chow-Liu graphs and the Chow-Liu Forests.

RBM is a bipartite undirected graph, which consists of d visible units $V = (V_1, ..., V_d)$ and m hidden units $H = (H_1, ..., H_m)$. It is known that in a RBM the hidden units are capturing dependencies between visible units. the use of these hidden (latent) variables allows the description of complex distributions over the visible variables by means of simple conditional distributions. That means that the hidden units are conditionally independent of one another given the visible nodes and vice versa. This can easily be seen by u-separation and makes inference especially easy.

In a binary RBM the random visible v and hidden h variables are taking values in $(v, h) \in \{0, 1\}^{d+m}$. The joint probability distribution $p(v, h)$ (the Gibb's distribution), marginal and conditional pdf's of visible and hidden units are given by the following formulas [6]):

$$p(v, h) = \frac{1}{Z} e^{-E(v,h)}, \quad p(v) = \frac{1}{Z} \sum_h p(v, h),$$

$$p(h|v) = \frac{e^{-E(v,h)}}{\sum_h p(v, h)}, \tag{1}$$

where the energy function E is linear in parameters i.e.

$$E(v, h) = -\sum_{i=1}^{m} \sum_{j=1}^{d} w_{ij} h_i v_j - \sum_{j=1}^{d} b_j v_j - \sum_{i=1}^{m} c_i h_i. \tag{2}$$

and the normalization term:

$$Z = \sum_{v,h} e^{-E(v,h)}. \tag{3}$$

For all $i \in \{1, ..., m\}$ and $j \in \{1, ..., d\}$ the weight w_{ij} is real valued and associated with the edge between units v_j and v_i. b_j and c_i are bias parameters. In terms of probability the special structure of the RBM graph, where there are no connections between visible units and no connections between different hidden units. That means that the hidden variables are independent given the state of the visible variables and vice versa. In the rest of this paper we will concentrate on this property of the RBM: The visible (input) variables are independent conditionally given the state of the hidden variables. Before we do that approximate the factorization of the pdf of the visible variables independently from the hidden variables using Chow Liu forests (a tree shaped Bayesian network) [15].

Formally a Bayesian network is a pair $B = (\mathcal{G}, \Theta)$, where $\mathcal{G} = (v, E)$ is a directed acyclic graph with vertices $v = (v_1, ..., v_d)$ and edges E. A Bayesian network defines unique probability distribution function over v:

$$P(v) = \prod_{i=1}^{d} P(v_i | Pa_i(\mathcal{G})) \tag{4}$$

The conditional probability for each $i = 1, ..., d$, $P(v_i | Pa_i(\mathcal{G}))$ represents the probability of v_i given its parents Pa_i in $\mathcal{G}$. Learning the structure of such a network is super-exponential in the number of parameters and the task is NP-hard [5,12]. However, in certain cases a Bayesian network which is a directed acyclic graph can be written as a Gibb's distribution. The procedure is called moralization of a Bayesian network and is described in more details in [11]. However, the moralization procedure is much more simple for trees. We can see that the moral graph of a tree-structured Bayesian network is obtained by removing the arrows from the edges of the graph.

In contrary to the classical Bayesian network-based factorization of the probability distribution functions, Chow and Liu proposed a method constructing the distribution for which the corresponding Bayesian networks are tree structured. Chow and Liu have shown that in order to minimize the KL-divergence between constructed and true discrete probability distribution, it is enough to maximize the total mutual information of the edges of the tree [3]. The mutual information is defined in the following formula:

$$I(V_i, V_j) = \sum_{V_i} \sum_{V_j} P(V_i, V_j) \log \frac{P(V_i, V_j)}{P(V_i)P(V_j)}$$

The result of Chow and Liu can be accomplished by calculating mutual information values between node pairs and solving the minimum spanning tree problem e.g. by Kruskal's algorithm. The whole procedure has complexity $\mathcal{O}\left(d^2 B^2 N\right)$, where B is the maximum number of values that one coordinate of the discrete random variable V can take. N is the number of data available for the analysis and d is the dimensionality of the input vector. In contrast to that the exhaustive search for optimal tree is ruled out, because the huge number of possible trees which is d^{d-2} (see [16]). The approximation of Chow-Liu is therefore restricted to a product of second order marginal distributions.

However, the Maximum Likelihood (ML) favors richer models more and the outputs are trees anyway. Therefore, the Chow-Liu tree estimation algorithm (like Kruskal's algorithm) does not produce a consistent estimate. In this paper we used the CLThresh algorithm which has a thresholding mechanism to prune weak edges from the Chow-Liu tree. In [15] it has been shown that the CLThresh algorithm is consistent if the true distribution is forest-structured.

It has to be mentioned that there are other algorithms which are generating tree-structured distributions as well, and are able to reflect more complex dependencies (e.g., thin junction trees, [1]). However, these algorithms have higher time complexity than the Chow-Liu algorithm and they do not guarantee optimality within the model class for the structure that is learned. Furthermore, Chow-Liu Trees have good properties of probabilistic inference as the tree is equal to its junction tree.

3   Restricted Sparce Boltzmann Machines ($RSBM$)

In this section we show a natural way of integrating the independence information given by Chow-Liu forest into the learning algorithm of the RBM.

The marginal probability distribution function for the visible and hidden variables of a RBM are:

$$p(h|v) = \prod_{i=1}^{m} p(h_i|v), \quad p(v|h) = \prod_{j=1}^{d} p(v_j|h) \tag{5}$$

Considering Eqs. (1), (2) and (5) we obtain for binary visible and hidden units well known log-likelihood. We call the parameters by $\theta = W, c, b$.

$$\log(p(v|\theta)) = \log \sum_{h} e^{-E(v,h)} - \log \sum_{h,v} e^{-E(v,h)} \tag{6}$$

and after simple computations the gradient is:

$$\frac{\partial \log(p(v|\theta))}{\partial \theta} = \underbrace{\sum_{h,v} p(h,v) \frac{\partial E(v,h)}{\partial \theta}}_{:=s} - \sum_{h} p(h|v) \frac{\partial E(v,h)}{\partial \theta} \tag{7}$$

(7) shows the difference between the expectations of the energy gradient under the model distribution and the conditional distribution of the hidden variables given training sample v. The second term can be calculated efficiently since it factorizes nicely. The first term approximation is much more harder and can also be expressed by:

$$s = \sum_{v} p(v) \sum_{h} p(h|v) \frac{\partial E(v,h)}{\partial \theta} \tag{8}$$

The second term in (8) equals to the second term in (7) and can be calculated similarly. The first term is the marginal of the visible units on which we want to focus.

Now we assume that the Chow-Liu tree algorithm based factorization of the probability distribution function of the input (visible) variables v and the following thresholding algorithm, which generates a forest, have been accomplished. As a result we get a forest $F_{CL} = (v, E_{CL})$. Here E_{CL} is the set of edges of the graph. The result of this two algorithms is a partition of v in clusters of coordinates. We denote the set of these clusters with C_v.

$$C_v \equiv \{\{(A_1), ..., (A_g)\} | A_i \subset V, A_i \cap A_j = \emptyset, \forall i, j \in \{1, ..., g\}\}. \tag{9}$$

Now we can modify the (4) by the following. Let $k_{j(1)}, ..., k_{j(n_j)}$ be the indexes of the visible units in a cluster $j \in \{1, ..., g\}$ then:

$$P(v) = \prod_{j=1}^{g} \prod_{i=k_{j(1)}}^{k_{j(n_j)}} P(v_i|Pa_i(F_{CL})) \tag{10}$$

The conditional probability for each v_i, $P(v_i|Pa_i(F_{CL}))$ represents the probability of v_i given its parents Pa_i in F_{CL}.

Considering that the graph of an RBM has only connections between the layers of visible and hidden variables and not between the variables in the same layer, the visible variables are independent to each other given the hidden variables and vice versa. On the other hand side according to the CLThresh factorization of the $p(v)$ by (10) where we get separate product terms for each cluster, the variables between the clusters from $\mathcal{C}_v$ are independent to each other without conditioning on the hidden variables. We now integrate this finding into the learning algorithm of the RBM. Especially, we insert it into the first term of the likelihood gradient in (8).

We know that (see for details [6]) the marginal of visible random variables can be computed easily because of node restrictions between visible and hidden units, for a binary RBM:

$$p(v) = p(A_1, A_2, ..., A_g) = \frac{1}{Z} \sum_{i=1}^{m} p(v, h_i)$$

$$= \frac{1}{Z} \underbrace{\prod_{j=1}^{d} e^{b_j v_j}}_{J1} \underbrace{\prod_{i=1}^{m} \left(1 + e^{c_i + \sum_{j=1}^{d} w_{ij} v_j}\right)}_{J2} \tag{11}$$

(Z is given in (3)) That means that the marginalized RBM is a product of expert models in which a number of experts for individual components are combined multiplicatively. In the first term of the product ($J1$) in (11) only experts for individual visible variables are combined. However, in the second product term ($J2$) we have m experts which are acting on all visible variables simultaneously. Now we concentrate on ($J2$) in (11). For one arbitrary expert $k \in \{1, ..., m\}$ its contribution to ($J2$) from (11) ex_k is:

$$ex_k = 1 + e^{c_k + \sum_{j=1}^{d} W_{kj} v_j} = 1 + e^{c_k} e^{w_{k1} v_1} e^{w_{k2} v_2} ... e^{w_{kd} v_d} \tag{12}$$

Now assume we have a Chow-Liu forest with two trees with equal number of nodes, $A_1 = \{v_1, ..., v_{d/2}\}$ and $A_2 = \{v_{d/2+1}, ..., v_d\}$ i.e. $A_1 \perp\!\!\!\perp A_2$. Therefore the Term ex_k and consequently ($J2$) in (11) according to the (10) can be approximated by two product terms with variables form each cluster. e.g. setting weights $w_{k1}, ..., w_{kd/2}$ to zero we obtain for the k-th expert:

$$ex_k = 1 + e^{c_k} e^{w_{kd/2+1} v_{d/2+1}} ... e^{w_{kd} v_d} \tag{13}$$

That means the expert k is acting only on the variables from set A_2. If we divide the m experts in two groups in this manner, where the first group is acting on A_1 and the second group on A_2, we obtain the desired factorization:

$$p(v) = p(A_1, A_2) = \frac{1}{Z} \sum_{i=1}^{m} p(v, h_i) =$$

$$\approx \frac{1}{Z} \prod_{j=1}^{d} e^{b_j v_j} \prod_{i=1}^{m/2} \left(1 + e^{c_i + \sum_{j=1}^{d/2} w_{ij} v_j}\right) \prod_{i=m/2+1}^{m} \left(1 + e^{c_i + \sum_{j=d/2+1}^{d} w_{ij} v_j}\right) \tag{14}$$

Inserting the modified marginal of the visible random variables from the (14) into (8) and the result into (7) we obtain the modified gradient of the likelihood.

To penalize the corresponding weights and consequently to obtain this modified gradient of the likelihood we modified the weight decay regularization technique (which is already implemented in any RBM learning algorithm). Weight decay regularization limits the capacity of the learning machine and consequently reduces overfitting. L^r regularization adds the term $\Lambda \circ ||W||_r^r$ to the negative log-likelihood term (see Eq. 6) of an RBM. The most popular choices of r are $r = 2$ ridge regression, or $r = 1$ Lasso (which we used).

In the next we want to give some ideas regarding the control of the weight decay parameter matrix Λ by the CLThresh hyperparameter β. For that reason we define the ϵ-independence:

Definition 1. *Let $F_{CL} = (v, E_{CL})$ be a CL-forest. Two sets of random variables $A_i, A_j \in v$ are ϵ-independent $A_i \overset{\epsilon}{\underset{\sim}{\perp\!\!\!\perp}} A_j$, if for a small $0 < \epsilon \ll 1$ and any $v_{k(i)} \in A_i$, $v_{k(j)} \in A_j$*

$$\max_{k(i),k(j)} I(v_{k(i)}.v_{k(j)}) \leq \epsilon \tag{15}$$

I is the mutual information.

The following Lemma 1 shows that any two clusters of random variables which have been generated by the CLThresh algorithm are automatically ϵ-independent to each other.

Lemma 1. *Any pair of the clusters of variables $A_i, A_j \subset C_v, \forall i, j \in \{1, ..., g\}$ are ϵ-independent to each other with $\epsilon = N^{-\beta}$. Here N is the number of training samples and the β is the CLThresh hyperparameter.*

Proof. The proof of the Lemma 1 follows from the way the forest are generated out of the CL-tree in the CLThresh algorithm (see [15]) and the data processing inequality (DPI) [4]. Assume the connection between CL-tree nodes $v_i \in A_i$ and $v_j \in A_j$ have been removed after CLThresh. Then for any $v_k \in A_j$, $I(v_j, v_k) \leq I(v_i, v_j) \leq \epsilon$ by DPI. $\qquad\square$

To connect ϵ-independence with the classical one, we formulate the following lemma:

Lemma 2. *If $A_i, A_j \subset C_v \overset{Lemma 1}{\Longrightarrow} A_i \overset{\epsilon}{\underset{\sim}{\perp\!\!\!\perp}} A_j$. Then for any realization of $v_i \in A_i$ and $v_j \in A_j$*

$$|P(v_i = r_i, v_j = r_j) - P(v_i = r_i)P(v_j = r_j)| \leq \epsilon$$

Proof. Because of the Lemma 1 we know that $A_i \overset{\epsilon}{\underset{\sim}{\perp\!\!\!\perp}} A_j$. Let $I(r_i, r_j) = \epsilon$ knowing that $I(r_1, r_2) = D_{KL}(P(r_1, r_2)||P(r_1)p(r_2))$. Here D_{KL} is the Kullback-Leibler Divergence (KLD). The total variation distance (TVD) between the probabilities is defined as $\Delta(P_1, P_2) = 1/2 \sum_x (P_1(x) - P_2(x))$. Pinskers inequality [13] relates the KLD and the TVD: $\Delta(P_1, P_2) \leq \sqrt{\ln(2)/2D_{KL}(P_1, P_2)}$.

For any realization $v_i = r_i$ and $v_j = r_j$ we obtain from the Eq. 15:

$$\delta = |P(v_i = r_i, v_j = r_j) - P(v_i = r_i)P(v_j = r_j)|$$
$$\leq 2\Delta(P(v_i.v_j), P(v_i)P(v_j))$$
$$\leq \sqrt{\frac{ln(2)}{2}D_{KL}(P(v_i.v_j), P(v_i)P(v_j))} = \sqrt{\frac{ln(2)}{2}I(v_i.v_j)}$$

From Lemma 1 follows:

$$I(v_i, v_j) \leq \epsilon = N^{-\beta}, \forall v_i \in A_i, v_j \in A_j$$

That means:

$$\delta \leq \sqrt{\frac{ln(2)}{2}I(v_i, v_j)} \leq I(v_i, v_j) \leq \epsilon.$$

□

Lemma 2 formalizes the intuition that the CLThresh parameter β controls the value of the regularization matrix Λ at corresponding places. High value of β implies that CLThresh algorithm prunes fewer connections in the CL-Tree. That means the ϵ value is smaller which implies that the ϵ-independence is stronger, therefore the regularization parameter matrix Λ can be higher for weights under consideration (i.e. in (16) all $w.$ weights)

The generalization of the method to the factorizations with more than two tree containing forests, with different number of visible variables is straightforward. We only have to determine the number of experts for each tree of the forest. The simplest way is to say that the number of experts of a particular tree should be proportional to the number of variables in this tree. E.g. for a two tree forest with 2 variables in the first, 4 variables in second and the total number of hidden variables equal to 9 we obtain the following weight matrix:

$$W = \begin{bmatrix} W_{11} & W_{12} & w_{13} & w_{14} & w_{15} & w_{16} \\ W_{21} & W_{22} & w_{23} & w_{24} & w_{25} & w_{26} \\ W_{31} & W_{32} & w_{33} & w_{34} & w_{35} & w_{36} \\ w_{41} & w_{42} & W_{43} & W_{44} & W_{45} & W_{46} \\ w_{51} & w_{52} & W_{53} & W_{54} & W_{55} & W_{56} \\ w_{61} & w_{62} & W_{63} & W_{64} & W_{65} & W_{66} \\ w_{71} & w_{72} & W_{73} & W_{74} & W_{75} & W_{76} \\ w_{81} & w_{82} & W_{83} & W_{84} & W_{85} & W_{86} \\ w_{91} & w_{92} & W_{93} & W_{94} & W_{95} & W_{96} \end{bmatrix} \tag{16}$$

Here the capital $W.$ weights are regularized by standard λ setting (in our case 0.0001, see also Table 1 and the small $w.$ are regularised by a $\lambda = 0.01$). On Fig. 1 the corresponding sparce RBM is illustrated. In general the distribution of experts to the identified groups can be described by the following equation.

$$|H_i| = \left\lfloor m \times |A_i| \times \frac{1}{d} \right\rfloor \tag{17}$$

where m is the number of hidden units, d is the number of visible variables, $|A_i|, i = 1, ..., g$ is the number of visible variables in the corresponding group and $|H_i|$ is the number of hidden variables for the group $i \in \{1, ..., g\}$

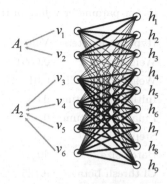

Fig. 1. Sparce RBM. The thin connections indicate the insignificant dependencies identified by an CL-Tree

4   Example

Within this section we demonstrate the benefits of the approach described before using the MNIST data benchmark. To see the effect of the sparcification of the first layer RBM weight matrix on the size of the used training data set we applied the following steps. In the first step we defined 1000 image increments beginning with the 30000 images initially included into the training data sample. In the second step we balanced each training data set in order to have the same number of digit representatives. We decided to take only a 2 layer DBN with 500 and 100 units in each layer. The hyper parameters of the training are set to the default values given in the paper of [8]. Furthermore, we applied the standard preprocessing: we removed pixels with no variation, and pixels with variation in only 0.01% of the images. After all, we get 482 out of 784 dimensional input of pixel binary values. The hyper parameter settings are summarized in Table 1. The hyper parameter "weight decay CL" is the developed Chow-Liu Forest based weight decay parameter. The hyper-parameter CLthresh Beta is used in the CLthresh algorithm (see for details [15] Formula (10) on page 1625). Beside the standard on failure rate based classification performance estimation we investigated the ability of the approach to disentangle the manifolds in the data set. In [2] the manifold disentanglement perspective of the deep learning is presented. Manifold perspective means that the deeper layers can extract the underlying factors of variations that define the structure of the data geometrically. After the learning process the propagation of signals through the layers of the machine manifolds of different classes are unfolded so that at the last layer a linear discriminant function is enough to classify the digits in the MNIST data set. In the following we will show the manifold disentanglement performance of sparce RBM in the first layer of the DBN. We estimated the Manifold disentanglement by the a(=0.01)-trimmed mean (see [2]):

Table 1. Hyper-parameter values of the DBN

Hyperparameters	Value
Weight decay	0.0001
Weight decay CL	0.01
Initial momentum	0.05
Final momentum	0.95
Step ratio	0.5
Batch size	40
CLthresh beta	0.125

$$G(i,j) = \frac{|G_M(i,j) - G_E(i,j)|}{|G_E(i,j)|}, \quad \forall i,j \in \{1,...,N\}$$

$$Dis := \frac{1}{N(1-2a)} \sum_{i=aN+1}^{N-aN} \tilde{G}(i,j)$$

(18)

Here, $G_M(i,j)$ is the sum of all edge lengths on the shortest path traveled on the manifold surface between i and j (Geodesic Distance), $G_E(i,j)$ is the Euclidean distance between two points i and j, $\tilde{G}(i,j)$ is in increasing order sorted $G(i,j)$. Small value of Dis indicates the more disentangled manifolds in the data space. The disentanglement have been calculated based on the best model (we made 10 runs) for each data set. The following Figs. 2 and 3 are showing the results.

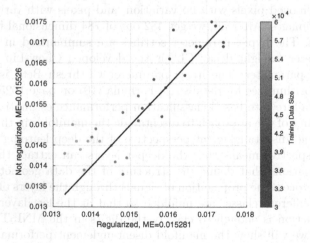

Fig. 2. Mean misclassification rate (over 10 independent runs) of the DBN with and without CL-regularization on different size training data sets. Mean (ME) over all data siezes for regularized approach was 1,52% and for not regularized approach 1,55%.

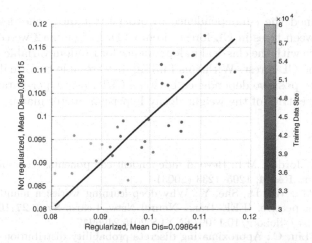

Fig. 3. Manifold disentanglement performance of the RBM in first DBN layer calculated with CL-regularization and without for different training data sample sizes by the formula (18).

We can see on the Fig. 3 that the disentanglement performance (see (18)) of the regularizesd DBN is slightly better than the not-regularized DBN by maintaining the mean classification performance which is shown on the Fig. 2. Especially for small and large training data sets the regularization showed the better disentanglement and classification performances (See the upper right and lover left figure segments on the Figs. 3 and 2). For all training set sizes we were able to regularize close to 50% of all weights, e.g. for full data model we regularized 117637 out of 242000 weights (500 hidden units times 484 visible variables (pixels)) in the first RBM. The regularization power was 100 times higher then for normally regularized weights see also the Table 1.

5   Summary

We demonstrated that the Chow-Liu forest based preprocessing and integrating the gained information into the learning process of the first RBM is a way to reduce the DBN complexity without loosing the classification performance and the disentanglement property of the machine. The overall performance of the approach was around 1,3% errors on the official MNIST test data set. If we compare the result with the result in the paper [9] of 1,25% errors where a DBN with 3 layers with 500,500,2000 units in each layer has been chosen. We see that the proposed much more simpler model, is able to came close to this performance by using a DBN with only 2 RBM layers with 500 and 100 hidden units plus a linear map and extremely regularized over 45% of weights in the first RBM.

Furthermore we don't used any hyperparameter search algorithms and further ideas to improve the results like e.g. local elastic deformations. Therefore the approach can be improved in several ways. In the theoretical part of the paper we defined the ϵ-independence in order to quantify the degree of independence between trees in CL forest (Lemma 1). In Lemma 2 we connected the ϵ-independence with the classical independence for random variable clusters constructed by the CL-forest. With this finding we were able to formalize the intuition about the positive dependence between CLThresh hyper-parameter β and corresponding entries of the weight decay hyper-parameter matrix Λ.

References

1. Bach, F.R., Jordan, M.I.: Beyond independent components: trees and clusters. J. Mach. Learn. Res. **4**, 1205–1233 (2003)
2. Brahma, P.P., Wu, D., She, Y.: Why deep learning works: a manifold disentanglement perspective. IEEE Trans. Neural Netw. Learn. Syst. **27**(10), 1997–2007 (2016). https://doi.org/10.1109/TNNLS.2015.2496947
3. Chow, C., Liu, C.: Approximating discrete probability distributions with dependence trees. IEEE Trans. Inf. Theory **14**, 462–467 (1968)
4. Cover, T.M., Thomas, J.A.: Elements of Information Theory. Wiley Interscience (2006). https://doi.org/10.1002/047174882X
5. Dasgupta, S.: The Sample Complexity of Learning Fixed-Structure Bayesian Networks. Kluwer Academic Publishers, Boston (1999). https://doi.org/10.1023/A:1007417612269
6. Fischer, A., Igel, C.: An introduction to restricted Boltzmann machines. In: Alvarez, L., Mejail, M., Gomez, L., Jacobo, J. (eds.) CIARP 2012. LNCS, vol. 7441, pp. 14–36. Springer, Heidelberg (2012). https://doi.org/10.1007/978-3-642-33275-3_2
7. Hinton, G.E.: Training products of experts by minimizing contrastive divergence. Neural Comput. **14**(8) (2002). https://doi.org/10.1162/089976602760128018
8. Hinton, G.E.: A practical guide to training restricted Boltzmann machines. In: Montavon, G., Orr, G.B., Müller, K.-R. (eds.) Neural Networks: Tricks of the Trade. LNCS, vol. 7700, pp. 599–619. Springer, Heidelberg (2012). https://doi.org/10.1007/978-3-642-35289-8_32
9. Hinton, G.E., Osindero, S., Teh, Y.: A fast learning algorithm for deep belief nets. Neural Comput. **18**(7), 27–54 (2006). https://doi.org/10.1162/neco.2006.18.7.1527
10. Koller, D., Friedman, N.: Probabilistic Graphical Models, Principles and Techniques. MIT Press, Cambridge (2009). https://doi.org/10.1017/S0269888910000275
11. Koller, D., Friedman, N., Getoor, L., Taskar, B.: Graphical Models in a Nutshell. MIT Press, Cambridge (2007). https://doi.org/10.1.1.146.2935
12. Meek, C.: Finding a path is harder than finding a tree. J. Artif. Intell. Res. **15** (2001). https://doi.org/10.1613/jair.914
13. Pinsker, M.S.: On estimation of information via variation. Probl. Inf. Transm. **41**(2), 71–75 (2005). https://doi.org/10.1007/s11122-005-0012-8
14. Salakhutdinov, R., Mnih, A., Hinton, G.E.: Restricted Boltzmann machines for collaborative filtering. In: International Conference on Machine Learning 24 (2007). https://doi.org/10.1145/1273496.1273596
15. Tan, V.Y.F., Anandkumar, A., Willsky, A.: Learning high-dimensional markov forest distributions: analysis of error rates. J. Mach. Learn. Res. **12**, 1617–1653 (2011)
16. West, D.B.: Introduction to Graph Theory, 2nd edn. Prentice Hall, Upper Saddle River (2001)

Prototypes Within Minimum Enclosing Balls

Christian Bauckhage[1,2,3](✉) (iD), Rafet Sifa[1,2], and Tiansi Dong[3]

[1] Fraunhofer Center for Machine Learning, Sankt Augustin, Germany
[2] Fraunhofer IAIS, Sankt Augustin, Germany
{christian.bauckhage,rafet.sifa,tiansi.dong}@iais.fraunhofer.de
[3] B-IT, University of Bonn, Bonn, Germany

Abstract. We revisit the kernel minimum enclosing ball problem and show that it can be solved using simple recurrent neural networks. Once solved, the interior of a ball can be characterized in terms of a function of a set of support vectors and local minima of this function can be thought of as prototypes of the data at hand. For Gaussian kernels, these minima can be naturally found via a mean shift procedure and thus via another recurrent neurocomputing process. Practical results demonstrate that prototypes found this way are descriptive, meaningful, and interpretable.

1 Introduction

The problem of characterizing data in terms of prototypes commonly arises in contexts such as clustering, latent component analysis, manifold identification, or classifier training [15,16,19,20,23]. Ideally, prototype identification should be computationally efficient and yield representative and interpretable results either for meaningful downstream processing or for assisting analysts in their decision making. In this paper, we discuss a two-stage approach based on kernel minimum enclosing balls and their characteristic functions that meets all these requirements.

Minimum enclosing balls (MEBs) are central to venerable techniques such as support vector clustering [3], or support vector data description [17]. More recently, balls have shown remarkable success in structuring deep representation learning [6,7,14]. Here, we revisit the kernel MEB problem and discuss how to solve it using simple recurrent neural networks. Resorting to recent work [1] which showed that recurrent neural networks can accomplish Frank-Wolfe optimization [9], we show how the Frank-Wolfe algorithm allows for finding MEBs and how this approach can be interpreted in terms of reservoir computing.

The solution to the kernel MEB problem consists in a set of support vectors that define the surface of a ball in a high dimensional feature space. Its interior, too, can be characterized in terms of a function of its support vectors. Local minima of this function coincide with representative and easily interpretable prototypes of the given data and we show that, for balls computed using Gaussian kernels, these minima are naturally found via generalized mean shifts [4,10].

© Springer Nature Switzerland AG 2019
I. V. Tetko et al. (Eds.): ICANN 2019, LNCS 11731, pp. 365–376, 2019.
https://doi.org/10.1007/978-3-030-30493-5_36

Since the mean shift procedure, too, can be interpreted in terms of reservoir computing, the approach we present in this paper constitutes an entirely neuro-computing based method for prototype extraction.

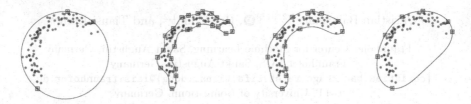

(a) Euclidean MEB   (b) Gaussian kernel MEBs for growing scale parameters

Fig. 1. A 2D data set, its Euclidean MEB, and several Gaussian kernel MEBs. Squares indicate which data points support the surface of the corresponding ball

2   Minimum Enclosing Balls in Data- and Feature Space

In order for our presentation to be self-contained, we begin with a brief review of the minimum enclosing ball (MEB) problem.

Given a data matrix $X = [x_1, \ldots, x_n] \in \mathbb{R}^{m \times n}$, the minimum enclosing ball problem asks for the smallest Euclidean m-ball $\mathcal{B}(c, r)$ with center $c \in \mathbb{R}^m$ and radius $r \in \mathbb{R}$ that contains each of the given data points x_i.

Understood as an inequality constrained convex minimization problem, the *primal MEB problem* is to solve

$$c_*, r_* = \underset{c, r}{\mathrm{argmin}} \quad r^2$$
$$\text{s.t.} \quad \left\| x_i - c \right\|^2 - r^2 \leq 0 \qquad i = 1, \ldots, n. \tag{1}$$

Evaluating the Lagrangian and Karush-Kuhn-Tucker conditions for (1) yields the corresponding *dual MEB problem*

$$\mu_* = \underset{\mu}{\mathrm{argmax}} \quad \mu^\mathsf{T} z - \mu^\mathsf{T} X^\mathsf{T} X \mu$$
$$\text{s.t.} \quad \begin{aligned} \mu^\mathsf{T} 1 &= 1 \\ \mu &\succeq 0 \end{aligned} \tag{2}$$

where $\mu \in \mathbb{R}^n$ is a vector of Lagrange multipliers, $0, 1 \in \mathbb{R}^n$ denote vectors of all zeros and ones, and the entries of $z \in \mathbb{R}^n$ are given by $z_i = x_i^\mathsf{T} x_i$.

The Karush-Kuhn-Tucker conditions further reveal that, once (2) has been solved, center and radius of the sought after ball amount to

$$c_* = X \mu_* \tag{3}$$

$$r_* = \sqrt{\mu_*^\mathsf{T} z - \mu_*^\mathsf{T} X^\mathsf{T} X \mu_*}. \tag{4}$$

Note that the given data points enter the problem in (2) only in form of inner products with other data points because $\boldsymbol{X}^\mathsf{T}\boldsymbol{X}$ is an $n \times n$ Gramian with entries $(\boldsymbol{X}^\mathsf{T}\boldsymbol{X})_{ij} = \boldsymbol{x}_i^\mathsf{T}\boldsymbol{x}_j$ and $\boldsymbol{z} = \text{diag}[\boldsymbol{X}^\mathsf{T}\boldsymbol{X}]$. The dual thus allows for invoking the kernel trick where inner products are replaced by non-linear kernel functions so as to implicitly solve the problem in a high dimensional feature space.

Hence, letting $K : \mathbb{R}^m \times \mathbb{R}^m \to \mathbb{R}$ be a Mercer kernel, we introduce $\boldsymbol{K} \in \mathbb{R}^{n \times n}$ where $K_{ij} = K(\boldsymbol{x}_i, \boldsymbol{x}_j)$ and $\boldsymbol{k} \in \mathbb{R}^n$ such that $\boldsymbol{k} = \text{diag}[\boldsymbol{K}]$ and obtain the *kernel MEB problem*

$$\boldsymbol{\mu}_* = \underset{\boldsymbol{\mu}}{\text{argmax}} \ \boldsymbol{\mu}^\mathsf{T}\boldsymbol{k} - \boldsymbol{\mu}^\mathsf{T}\boldsymbol{K}\boldsymbol{\mu}$$

$$\text{s.t.} \quad \begin{aligned} \boldsymbol{\mu}^\mathsf{T}\boldsymbol{1} &= 1 \\ \boldsymbol{\mu} &\succeq \boldsymbol{0}. \end{aligned} \tag{5}$$

Once (5) has been solved, the radius of the minimum enclosing ball in feature space can be computed analogously to (4), namely

$$r_* = \sqrt{\boldsymbol{\mu}_*^\mathsf{T}\boldsymbol{k} - \boldsymbol{\mu}_*^\mathsf{T}\boldsymbol{K}\boldsymbol{\mu}_*}. \tag{6}$$

However, the center of the feature space ball cannot be computed similarly since (3) does not lend itself to the kernel trick. Nevertheless, computing

$$\boldsymbol{c}_*^\mathsf{T}\boldsymbol{c}_* = \boldsymbol{\mu}_*^\mathsf{T}\boldsymbol{K}\boldsymbol{\mu}_*. \tag{7}$$

still allows for checking whether or not an arbitrary $\boldsymbol{x} \in \mathbb{R}^m$ resides within the kernel MEB of the given data. This is because the inequality $\|\boldsymbol{x} - \boldsymbol{c}_*\|^2 \le r_*^2$ can be rewritten as

$$K(\boldsymbol{x}, \boldsymbol{x}) - 2\,\boldsymbol{\kappa}^\mathsf{T}\boldsymbol{\mu}_* + \boldsymbol{\mu}_*^\mathsf{T}\boldsymbol{K}\boldsymbol{\mu}_* \le \boldsymbol{\mu}_*^\mathsf{T}\boldsymbol{k} - \boldsymbol{\mu}_*^\mathsf{T}\boldsymbol{K}\boldsymbol{\mu}_* \tag{8}$$

where $\boldsymbol{\kappa} \in \mathbb{R}^n$ in the second term on the left has entries $\kappa_i = K(\boldsymbol{x}, \boldsymbol{x}_i)$.

Figure 1 compares the Euclidean minimum enclosing ball of a set of 2D data to kernel minimum enclosing balls computed using Gaussian kernels

$$K(\boldsymbol{x}_i, \boldsymbol{x}_j) = \exp\left(-\frac{\|\boldsymbol{x}_i - \boldsymbol{x}_j\|^2}{2\lambda^2}\right) \tag{9}$$

with different scale parameters λ. In order to visualize the surfaces of the feature space balls in the original data space, we considered the function

$$f(\boldsymbol{x}) = \sqrt{K(\boldsymbol{x}, \boldsymbol{x}) - 2\,\boldsymbol{\kappa}^\mathsf{T}\boldsymbol{\mu}_* + \boldsymbol{\mu}_*^\mathsf{T}\boldsymbol{K}\boldsymbol{\mu}_*} - \sqrt{\boldsymbol{\mu}_*^\mathsf{T}\boldsymbol{k} - \boldsymbol{\mu}_*^\mathsf{T}\boldsymbol{K}\boldsymbol{\mu}_*} \tag{10}$$

and highlighted the contour for which $f(\boldsymbol{x}) = 0$. Note that $f(\boldsymbol{x})$ can be seen as a characteristic function of the corresponding MEB $\mathcal{B}$, because $f(\boldsymbol{x}) \le 0 \Leftrightarrow \boldsymbol{x} \in \mathcal{B}$ and $f(\boldsymbol{x}) > 0 \Leftrightarrow \boldsymbol{x} \notin \mathcal{B}$.

Finally, we note that those data points $\boldsymbol{x}_i$ which support the surface of an MEB $\mathcal{B}$ in data- or in feature space are easily identified. This is because only if $\boldsymbol{x}_i$ resides on the surface of the ball will its Lagrange multiplier μ_{i*} exceed zero; for points inside the ball the inequality constraints in (2) or (5) are inactive and their multipliers vanish. Below, we will refer to points whose multipliers exceed zero as the *support vectors* $\boldsymbol{s}_j$ of $\mathcal{B}$.

Algorithm 1. Frank-Wolfe algorithm for (11)

---

initialize a feasible point in Δ^{n-1}, for instance

$$\mu_0 = \tfrac{1}{n}\mathbf{1}$$

for $t = 0, \ldots, t_{\max}$ **do**
    determine the step direction

$$\nu_t = \underset{\nu \in \Delta^{n-1}}{\operatorname{argmin}} \ -\nu^\mathsf{T} \nabla \mathcal{D}(\mu_t)$$

update the current estimate

$$\mu_{t+1} = \mu_t + \tfrac{2}{t+2}\left[\nu_t - \mu_t\right]$$

---

3   Neural Computation of Minimum Enclosing Balls

Next, we discuss how the Frank-Wolfe algorithm [9] solves the kernel MEB problem and how this approach can be interpreted in terms of reservoir computing.

Observe that the kernelized dual Lagrangian $\mathcal{D}(\mu) = \mu^\mathsf{T} k - \mu^\mathsf{T} K \mu$ in (5) is concave so that $-\mathcal{D}(\mu) = \mu^\mathsf{T} K \mu - \mu^\mathsf{T} k$ is convex. We may therefore rewrite the maximization problem in (5) in terms of a minimization problem

$$\mu_* = \underset{\mu \in \Delta^{n-1}}{\operatorname{argmin}} \ \mu^\mathsf{T} K \mu - \mu^\mathsf{T} k \tag{11}$$

where we also exploited that the non-negativity and sum-to-one constraints in (5) require any feasible solution to reside in the standard simplex $\Delta^{n-1} \subset \mathbb{R}^n$.

Written as in (11), our problem is clearly recognizable as an instance of a convex minimization problem over a compact convex set and we note that the Frank-Wolfe algorithm provides a simple iterative solver for this setting.

Algorithm 1 shows how it specializes to our context: Given an initial guess μ_0 for the solution, each iteration of the algorithm determines which $\nu_t \in \Delta^{n-1}$ minimizes the inner product $\nu^\mathsf{T} \nabla \mathcal{D}(\mu_t)$ and applies a conditional gradient update $\mu_{t+1} = \mu_t + \eta_t (\nu_t - \mu_t)$ where the step size $\eta_t = \tfrac{2}{t+2} \in [0, 1]$ decreases over time. This way, updates will never leave the feasible set and the efficiency of the algorithm stems from the fact that it turns a quadratic problem into a series of simple linear problems.

Next, we build on recent work [1] and show how Frank-Wolfe optimization for the kernel MEB problem can be implemented by means of rather simple recurrent neural networks.

For the gradient of the negated dual Lagrangian $-\mathcal{D}(\mu)$, we simply have $-\nabla \mathcal{D}(\mu) = 2 K \mu - k$ so that each iteration of the Frank-Wolfe algorithm has to compute

$$\nu_t = \underset{\nu \in \Delta^{n-1}}{\operatorname{argmin}} \ \nu^\mathsf{T} \left[2 K \mu_t - k\right]. \tag{12}$$

The objective function in (12) is linear in ν and needs to be minimized over a compact convex set. Since minima of a linear functions over compact convex sets are necessarily attained at a vertex, the solution of (12) must coincide with a vertex of Δ^{n-1}. Since the vertices of the standard simplex in $\mathbb{R}^n$ correspond to the standard basis vectors $e_j \in \mathbb{R}^n$, we can cast (12) as

$$\nu_t = \operatorname*{argmin}_{e_j \in \mathbb{R}^n} e_j^\mathsf{T} \left[2 \, K \mu_t - k \right] \approx g_\beta (2 \, K \mu_t - k). \tag{13}$$

where $g_\beta(x)$ introduced in the approximation on the right of (13) represents the vector-valued softmin operator. Its i-th component is given by

$$\left(g_\beta(x) \right)_i = \frac{e^{-\beta x_i}}{\sum_j e^{-\beta x_j}} \tag{14}$$

and we note that

$$\lim_{\beta \to \infty} g_\beta(x) = \operatorname*{argmin}_{e_j \in \mathbb{R}^n} e_j^\mathsf{T} x = e_i. \tag{15}$$

Based on the relaxed optimization step in (13), we can therefore rewrite the Frank-Wolfe updates for our problem as

$$\mu_{t+1} = \mu_t + \eta_t \left[\nu_t - \mu_t \right] \tag{16}$$
$$= (1 - \eta_t) \, \mu_t + \eta_t \, \nu_t \tag{17}$$
$$\approx (1 - \eta_t) \, \mu_t + \eta_t \, g_\beta (2 \, K \mu_t - k). \tag{18}$$

Choosing an appropriate parameter β for the softmin function, the non-linear dynamical system in (18) mimics the Frank-Wolfe algorithm up to arbitrary precision and can therefore solve the kernel MEB problem.

From the point of view of neurocomputing this is of interest because, the system in (18) is algebraically equivalent to the equations that govern the internal dynamics of the simple recurrent architectures known as echo state networks [11]. In other words, we can think of this system in terms of a reservoir of n neurons whose synaptic connections are encoded in the matrix $2 \, K$. The system evolves with fixed inputs inputs k and its non-linear readout happens according to (6) and (7). The step size η_t assumes the role of the leaking rate of the reservoir. Since η_t decays towards zero, neural activities will stabilize and the system is guaranteed to approach a fixed point $\mu_* = \lim_{t \to \infty} \mu_t$.

What is further worth noting about the reservoir governed by (18) is that its synaptic connections and constant input are determined by the training data for the problem under consideration. Understanding the MEB problem as a learning task, both could be seen as a form of short term memory. At the beginning of a learning episode, data is loaded into this memory and used to determine support vectors. At the end of a learning episode, only those data points and activities required for decision making, i.e. those x_i and μ_i for which $\mu_i > 0$, need to be persisted in a long term memory to be able to compute the characteristic function in (10).

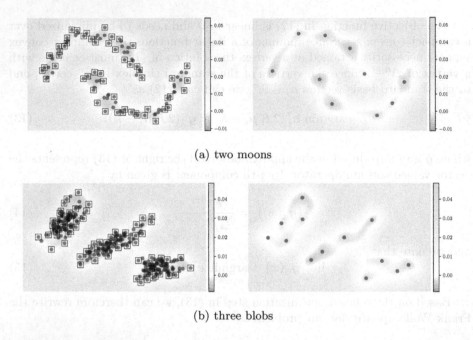

(a) two moons

(b) three blobs

Fig. 2. Two additional 2D data sets. Squares highlight support vectors s_j of Gaussian kernel MEBs; the color coding indicates the characteristic function $f(x)$ in (10). The panels on the right show local minima of $f(x)$. Points that minimze the function $f(x)$ can be understood as prototypes for the given data

4   Neural Reduction of Support Vectors to Prototypes

Figure 2 show two more 2D data sets for which we computed Gaussian kernel MEBs. Squares highlight support vectors, the coloring indicates values of the characteristic function $f(x)$ in (10), and blue dots represent its local minima. Both examples illustrate that (i) the number of support vectors of a kernel MEB is typically smaller than the number of data points the support vectors are computed from, (ii) the number of local minima of the characteristic function is typically smaller than the number of support vectors, and (iii) points where the characteristic function achieves a minimum constitute characteristic prototypes for the given data. Curiously, however, we are not aware of any prior work where minimizers of $f(x)$ have considered as prototypes before. Next, we therefore discuss a simple recurrent procedure for how to compute them.

Solving the kernel MEB problem yields a vector of Lagrange multipliers whose non-zero entries indicate support vectors of $\mathcal{B}$. As the multipliers of all other data points equal zero, the characteristic function in (10) can be evaluated using only the support vectors and their multipliers.

Hence, letting $l \leq n$ denote the number of support vectors of $\mathcal{B}$, we next collect all of the support vectors of $\mathcal{B}$ in a matrix $S = [s_1, \ldots, s_l] \in \mathbb{R}^{m \times l}$ and consider a vector $\sigma \in \mathbb{R}^l$ of their multiplies. Furthermore, introducing a reduced kernel matrix $Q \in \mathbb{R}^{l \times l}$ where $Q_{ij} = K(s_i, s_j)$ and kernel vector $q \in \mathbb{R}^l$ such that $q = \mathrm{diag}[Q]$, allows us to rewrite the characteristic function in (10) as

$$f(x) = \sqrt{K(x, x) - 2\kappa^\mathsf{T}\sigma + \sigma^\mathsf{T}Q\sigma} - \sqrt{\sigma^\mathsf{T}q - \sigma^\mathsf{T}Q\sigma} = \sqrt{d(x)} - r_* \quad (19)$$

where the entries of $\kappa \in \mathbb{R}^l$ now amount to $\kappa_j = K(x, s_j)$ and where function $d : \mathbb{R}^m \to \mathbb{R}$ computes the squared feature space distance between x and the center of $\mathcal{B}$.

Writing the characteristic function like this and observing that, on the outside of $\mathcal{B}$, the distance function $d(x)$ will grow beyond all bounds, it is clear that the problem of estimating local minimizers of $f(x)$ is equivalent to the problem of estimating those $x \in \mathcal{B}$ for which the gradient of $d(x)$ vanishes.

Assuming that $K(\cdot, \cdot)$ is a Gaussian kernel such as in (9), we have $K(x, x) = 1$ so that the gradient of $d(x)$ becomes

$$\nabla d(x) = -\frac{2}{\lambda^2} \sum_{j=1}^{k} \sigma_j K(x, s_j) [x - s_j]. \quad (20)$$

Equating the right hand side to 0 provides

$$x = \frac{\sum_j \sigma_j K(x, s_j)s_j}{\sum_j \sigma_j K(x, s_j)} = \frac{\sum_j \sigma_j \kappa_j s_j}{\sum_j \sigma_j \kappa_j} = \frac{S\Sigma\kappa}{\kappa^\mathsf{T}\sigma} = S\Sigma D^{-1}\kappa \quad (21)$$

where we introduced two diagonal matrices $\Sigma = \mathrm{diag}[\sigma]$ and $D = \mathrm{diag}[(\sigma\kappa^\mathsf{T})1]$, respectively. But this result is to say that local minima of $f(x)$ correspond to weighted means or convex combinations of the support vectors of $\mathcal{B}$.

We also recognize (21) as an extension of classical mean shift updates [4,10] (where there are no scaling parameters σ_j). Hence, when started with $x_0 \in \mathbb{R}^m$, the following process with step size $\gamma_t \in [0, 1]$ will find the nearest minimizer of the characteristic function

$$\kappa_t = \mathrm{vec}\big[K(x_t, s_j)_j\big] \quad (22)$$

$$D_t = \mathrm{diag}\big[(\sigma\kappa_t^\mathsf{T})1\big] \quad (23)$$

$$x_{t+1} = (1 - \gamma_t)x_t + \gamma_t S\Sigma D_t^{-1}\kappa_t. \quad (24)$$

Looking at (24), we recognize these dynamics as yet another variant of the internal dynamics of a reservoir of neurons and note that, for $\gamma_t = 1$, the updates in (24) become the mean shift updates in (21). In other words, mode seeking via mean shifts can be seen as yet another form of neurocomputing.

Letting $x_0 \leftarrow s_j$ be a copy of one of the support vectors in S and starting mode seeking at this point will identify the minimizer closest to this support vector. Repeating this process for all the support vectors of $\mathcal{B}$ will thus collapse them into another, usually smaller, set of points that can be understood as prototypes of the given data. Collecting these in a matrix $P \in \mathbb{R}^{m \times p}$ where $p \leq l \leq n$ therefore provides a reduced representation for a variety of downstream processing steps.

Table 1. Sample mean and prototypes extracted from the CBCL face data

$\bar{x}$	k-means prototypes	MEB prototypes
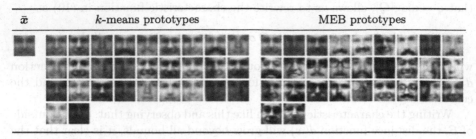		

5   Practical Examples

In order to provide illustrative examples for the performance of our approach, we next present results obtained in experiments with three standard benchmark data sets: The MIT CBCL face database[1] contains intensity images of different faces recorded under various illumination conditions, the well known MNIST database [12] consist of intensity images of ten classes of handwritten digits, and the recently introduced MNIST-Fashion data [22] contains intensity images of fashion items again sampled from ten classes.

For each experimental setting, we vectorized the designated training samples which left us with a data matrix $X \in \mathbb{R}^{361 \times 2429}$ for the CBCL data and matrices $X \in \mathbb{R}^{361 \times 6000}$ for each class in two MNIST data sets.

In each experiment, we computed the sample mean $\bar{x} = \frac{1}{n} X \mathbf{1}$ as a reference prototype and normalized the data in X to zero mean and unit variance before running our procedure. Scale parameters λ for the Gaussian MEB kernels were determined using the method in [8] and reused during mean shift computation; the activation function for neural MEB computation was set to g_∞.

An favorable property of MEB-based prototype identification is that it does not need manual specification of the number p of prototypes. Minimum enclosing ball computation and mean shift on the resulting support vectors automatically identify appropriate numbers l and p of support vectors and prototypes. Hence, after having obtained p MEB-based prototypes for each data matrix, we also ran k-means clustering for $k = p$ in order to provide an intuitive baseline comparison.

The rightmost column of Table 1 shows the $p = 29$ MEB-based prototypes we found for the CBCL face data; the center column of the table shows cluster prototypes resulting from k-means clustering for $k = 29$ and the single image in the leftmost column depicts the overall sample mean for comparison.

The cluster means in the center column represent average faces which are smoothed to an extent that makes it difficult to discern characteristic features. The MEB prototypes, on the other hand, show distinguishable and therefore

---

Table 2. Sample mean and prototypes extracted from MNIST digit data

interpretable visual characteristics. In other words, these prototypes reveal that the CBCL data contains pictures of faces of people of pale or dark complexion, of people wearing glasses, sporting mustaches, or having been photographed under varying illumination conditions.

What is further worth noting is that several of the MEB-based prototypes coincide with given images or, put differently, with actual data points. This phenomenon is known from latent factor models such as archetypal analysis [2,5,19] or CUR decompositions [13,18,21] and usually considered beneficial for interpretability [16]. The fact that we observe it here suggests that, for real world data, some of the support vectors of a kernel minimum enclosing ball themselves constitute minima of the corresponding characteristic function so that the above mean shift procedure will not reduce them any further. Since support vectors reside on the boundary of a given data set, this also explains the apparent variety among the MEB-based prototypes. While this also holds for prototypes extracted via archetypal analysis or CUR decompositions, the prototypes resulting from our approach do not exclusively coincide with extremal data points. In fact, some of them resemble the overall sample mean or the local means found via k-means. In contrast to archetypal analysis, CUR decompositions, or k-means clustering, we therefore observe that our MEB-based approach produces extremal *and* central prototypes simultaneously.

Tables 2 and 3 show examples of results obtained from the MNIST data sets. These are apparently analogous to the results we just discussed and therefore corroborate that our approach identifies prototypes that cover a wide variety of aspects of a data set.

Table 3. Sample mean and prototypes extracted from MNIST fashion data

$\bar{x}$	k-means prototypes	MEB prototypes

6   Conclusion

The problem of extracting representative prototypes from a given set of data frequently arises in data clustering, latent component analysis, manifold identification, or classifier training. Methods for this purpose should scale well and yield meaningful and interpretable results so as to assist downstream processing or decision making by analysts. In this paper, we proposed a two-stage approach based on kernel minimum enclosing balls and their characteristic functions. Our approach can be efficiently computed and empirical results suggest that it yields notably distinct prototypes that are therefore interpretable. Contrary to established techniques for clustering or factor analysis, our method yields central and extremal prototypes alike.

From the point of view of neurocomputing, our approach in interesting in that it can be computed using simple recurrent neural networks. Building on recent work in [1], we showed that kernel minimum enclosing balls can be computed using architectures akin to those found in reservoir computing. We also showed that, if kernel minimum enclosing balls are determined w.r.t. Gaussian kernels, the problem of further reducing the support vectors of a ball naturally leads to a variant of the mean shift procedure which can be understood as a form of recurrent neural computation, too.

References

1. Bauckhage, C.: A neural network implementation of Frank-Wolfe optimization. In: Lintas, A., Rovetta, S., Verschure, P.F.M.J., Villa, A.E.P. (eds.) ICANN 2017. LNCS, vol. 10613, pp. 219–226. Springer, Cham (2017). https://doi.org/10.1007/978-3-319-68600-4_26
2. Bauckhage, C., Thurau, C.: Making archetypal analysis practical. In: Denzler, J., Notni, G., Süße, H. (eds.) DAGM 2009. LNCS, vol. 5748, pp. 272–281. Springer, Heidelberg (2009). https://doi.org/10.1007/978-3-642-03798-6_28
3. Ben-Hur, A., Horn, D., Siegelmann, H., Vapnik, V.: Support vector clustering. J. Mach. Learn. Res. **2**(Dec), 125–137 (2001)
4. Cheng, Y.: Mean shift, mode seeking, and clustering. IEEE Trans. Pattern Anal. Mach. Intell. **17**(8), 767–776 (1995). https://doi.org/10.1109/34.400568
5. Cutler, A., Breiman, L.: Archetypal analysis. Technometrics **36**(4), 338–347 (1994). https://doi.org/10.1080/00401706.1994.10485840
6. Dong, T., et al.: Imposing category trees onto word-embeddings using a geometric construction. In: Proceedings ICLR (2019)
7. Dong, T., Wang, Z., Li, J., Bauckhage, C., Cremers, A.: Triple classification using regions and fine-grained entity typing. In: Proceedings AAAI (2019)
8. Evangelista, P.F., Embrechts, M.J., Szymanski, B.K.: Some properties of the Gaussian kernel for one class learning. In: de Sá, J.M., Alexandre, L.A., Duch, W., Mandic, D. (eds.) ICANN 2007. LNCS, vol. 4668, pp. 269–278. Springer, Heidelberg (2007). https://doi.org/10.1007/978-3-540-74690-4_28
9. Frank, M., Wolfe, P.: An algorithm for quadratic programming. Naval Res. Logist. **3**(1–2), 95–110 (1956)
10. Fukunaga, K., Hostetler, L.: The estimation of the gradient of a density function with applications in pattern recognition. IEEE Trans. Inf. Theory **21**(1), 32–40 (1975). https://doi.org/10.1109/TIT.1975.1055330
11. Jäger, H., Haas, H.: Harnessing nonlinearity: predicting chaotic systems and saving energy in wireless communication. Science **304**(5667), 78–80 (2004). https://doi.org/10.1126/science.1091277
12. LeCun, Y., Boottou, L., Bengio, Y., Haffner, P.: Gradient-based learning applied to document recognition. Proc. IEEE **86**(11), 2278–2324 (1998). https://doi.org/10.1109/5.726791
13. Mahoney, M., Drineas, P.: CUR matrix decompositions for improved data analysis. PNAS **106**(3), 697–702 (2009). https://doi.org/10.1073/pnas.0803205106
14. Ruff, L., et al.: Deep one-class classification. In: Proceedings ICML (2018)
15. Schleif, F.M., Gisbrecht, A., Tino, P.: Supervised low rank indefinite kernel approximation using minimum enclosing balls. Neurocomputing **318**(Nov), 213–226 (2018). https://doi.org/10.1016/j.neucom.2018.08.057
16. Sifa, R.: An overview of Frank-Wolfe optimization for stochasticity constrained interpretable matrix and tensor factorization. In: Kůrková, V., Manolopoulos, Y., Hammer, B., Iliadis, L., Maglogiannis, I. (eds.) ICANN 2018. LNCS, vol. 11140, pp. 369–379. Springer, Cham (2018). https://doi.org/10.1007/978-3-030-01421-6_36

17. Tax, D., Duin, R.: Support vector data description. Mach. Learn. **54**(1), 45–46 (2004). https://doi.org/10.1023/B:MACH.0000008084.60811.49
18. Thurau, C., Kersting, K., Bauckhage, C.: Deterministic CUR for improved large-scale data analysis: an empirical study. In: Proceedings SDM. SIAM (2012)
19. Thurau, C., Kersting, K., Wahabzada, M., Bauckhage, C.: Descriptive matrix factorization for sustainability: adopting the principle of opposites. Data Min. Knowl. Discov. **24**(2), 325–354 (2012). https://doi.org/10.1007/s10618-011-0216-z
20. Tsang, I., Kwok, J., Cheung, P.M.: Core vector machines: fast SVM training on very large data sets. J. Mach. Learn. Res. **6**(Apr), 363–392 (2010)
21. Wang, S., Zhang, Z.: Improving CUR matrix decompositions and the Nyström approximation via adaptive sampling. J. Mach. Learn. Res. **14**(1), 2729–2769 (2010)
22. Xiao, H., Rasul, K., Vollgraf, R.: Fashion-MNIST: a novel image dataset for benshmarking machine learning algorithms. arXiv:1708.07747 [cs.LG] (2017)
23. Zhang, K., Kwok, J.: Clustered Nyström method for large scale manifold learning and dimension reduction. IEEE Trans. Neural Netw. **21**(10), 1576–1587 (2010). https://doi.org/10.1109/TNN.2010.2064786

Exploring Local Transformation Shared Weights in Convolutional Neural Networks

Rohan Ghosh$^{(\boxtimes)}$ and Anupam K. Gupta

National University of Singapore, Singapore, Singapore
rghosh92@gmail.com, anupamkg@vt.edu

Abstract. We explore the methods which add domain-specific transformation knowledge to the weights of a Convolutional Neural Network (CNN), through transformation-based weight-sharing. Often, such *network augmentations* have demonstrated superior parameter-efficiency and better generalization abilities. We focus on two conjugate methods in literature which increase the degree of weight-sharing within CNNs, showcasing opposing philosophies. The first, transformation shared max-pooling (TSMP), explicitly injects local visual transformation invariance into network representations. The second, transformation shared kernel extension (TSKE), is simply TSMP without max-out, and therefore avoids explicit injection of local invariances. Elaborate empirical validation of these methods is presented on variations of MNIST, which evaluate various aspects of these methods. Mainly, we find that for small test-data distortions (rotation, scale and elastic), TSKE outperforms TSMP, whereas for large test-data distortions, TSMP showcases superior performance.

Keywords: Convolutional neural network ·
Invariance to transformations ·
Visual transformation based weight sharing

1 Introduction

Convolutional neural networks have revolutionized the approach to large scale machine learning in computer vision [1]. The convolutional structure is very potent at capturing useful features at various scales, keeping the representation translation equivariant, while slowly injecting translational invariance across layers through pooling operations [2]. Spatial pooling (or selective sampling [3]) throughout the layers adds much needed translation and partial distortion invariance to the internal representations. Furthermore, small convolutional kernel weights ensure less free-parameters for the non-convex optimization, reducing chances of overfitting. Stacking multiple layers enhances richness of representation, capturing high-level and low-level concepts alike [4]. As such, in addition

---

Both authors contributed equally to this work.

© Springer Nature Switzerland AG 2019
I. V. Tetko et al. (Eds.): ICANN 2019, LNCS 11731, pp. 377–390, 2019.
https://doi.org/10.1007/978-3-030-30493-5_37

to CNNs, many other competition winning architectures leverages the convolutional + max-pooling structure [5]. Therefore CNNs represent a significant improvement to deep, fully connected neural networks [6,7], both in design and performance. Although DNNs may exhibit a larger parametric space and higher capacity (complexity) of information representation, previous literature [8] has highlighted its ineffiency of converting this added capacity to achieving a better fit on the training data itself. In contrast, the CNN architecture scales excellently to very large training datasets, and is thus more efficient than a DNN in using its available capacity.

An important concept accompanying the convolution operation native to a 2D CNN, is its weight sharing property across all spatial locations (spatiotemporal sharing in 3D CNNs [9]). This ensures that convolutional weight features acquired from a certain location in a certain image found in the training data, is subsequently searched across all spatial locations. The spatial location of a local feature (e.g. a corner) is subject to change with different visual presentations of an object, including even presentation of different objects. Therefore the convolution operation is quite desirable for learning robust representations.

In recent years, works have surfaced which add to the weight sharing methodology present in a CNN, by accomodating for changes in feature appearance due to visual transformations other than translation [10–13]. Particularly in [11] and [12], the CNN weights are automatically rotated or spatially scaled by certain amounts, to add invariance to spatial scaling or rotations in the data. Such extended weight sharing abilities naturally add to the robustness of the network, as layers demonstrate heightened invariance to the desired visual transformations (see results in [11]).

Such *knowledge transfer* through visual transformations on weights (translation, rotation or spatial scaling) benefits a learner by reducing the number of examples required to learn appropriate weights, and the number of trainable model parameters as well. For instance, in [14], when translational knowledge transfer (convolution) was not applied, the optimization process resulted in weights which are translated versions of each other. With convolution, the returned weights no longer were related through translational shifts, and a richer and sharper class of weights resulted. Translational repetition (i.e. convolution) of the kernels therefore naturally avoids kernel pairs which are related to each other through translation. Similarly, one should expect the same for other transformations, such as spatial scaling and rotation.

2   Motivation

In that direction, we analyse and compare two different extensions to a conventional CNN architecture, both of which incorporate transformation-aware knowledge transfer on the network weights. In the first method, local kernels are either spatially scaled or rotated, thereafter their responses max-pooled. We call this method kernel Transformation Shared Max-Pooling (TSMP). In the second method, instead of max-pooling the responses of rotated/scaled kernels at each location, we simply append those responses, and effectively increase the number of weights in each layer. We denote this method as the transformation shared kernel extension (TSKE, used in [11,12]). The ideological difference between TSMP and TSKE lies in how each considers transformation invariance. TSMP focuses on invariance to transformations, as the max-pooling operation channels the best response across local changes of scale/rotation. TSKE on the other hand does not discard kernel responses from transformed weights, instead adding to the outputs at each layer, and therefore does not explicitly incorporate invariance.

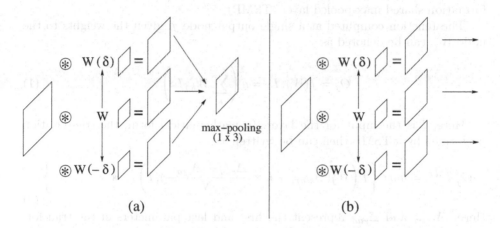

(a)   (b)

Fig. 1. Schematic of architectural variations on CNN. (a) shows the Transformation-Shared Max-Pooling (TSME) architectural variation of CNN. In this architecture, inavariance to scale/rotation is incorporated by max-pooling of responses across the learned and transformed kernals (by scaling or rotating), (b) shows the Transformation-Shared Kernal Extension (TSKE) architectural variation of CNN. In this architecture, instead of max-pooling across transformed kernels, they are apended with learned kernels thus increasing the number of convolutional kernels in each layer.

3   Methods

3.1   Method A: Kernel Transformation Shared Max-Pooling

As we previously note, this method is simply a generalization of the weight extended,locally scale-invariant architecture proposed in [11]. The transformations analyzed in this work are limited to spatial scaling and rotation. Here we formally describe the computations within the architecture, with the appropriate parameter denominations, which will serve us well in explaining the experimental procedures described later.

Figure 1(a) demonstrates the layer topology corresponding to the same. In addition to reproducing weights across various translations (convolution) required for inducing translational invariance, the weights are reproduced across another transformation (scaling or rotation) to make the computations invariant to that particular local transformation. This is done by transforming the weights themselves through that particular transformation, by treating them as images. For spatial scaling, the weights are increased or decreased in kernel size after either upsampling or downsampling them respectively. For rotations, the weights are simply rotated clockwise or anticlockwise to produce rotated copies. Subsequently, to ensure a higher degree of invariance to the transformation, a max-pooling layer is introduced which only channels the maximum response at each pixel location from the responses obtained across scale or rotation changes. For sake of simplicity for reference purposes, we denote such a layer as the transformation shared max-pooled layer (TSMP).

The function computed at a single output node j, given the weights to the node W_j, can be denoted as

$$O_j = f(W_j, I) = g\left(\sum_k W_{kj} I_k\right). \tag{1}$$

Here, I is the input for the layer. In mathematical form, the computation performed by a TSMP then can be written as

$$O_j^{TSMP} = max\left\{ f\left(W_j(\Delta_{start} + k \times \frac{\Delta_{end} - \Delta_{start}}{N}), I\right) \; : \; k = 0, ..., N\right\}. \tag{2}$$

Here, Δ_{start} and Δ_{end} represent the first and last parameters of the transformations applied on the weights W_j. Note that for all experiments reported in this work, we only control Δ_{start} and Δ_{end}. N is chosen each time such that the step-size of the transformation parameters is fixed (0.1 for scaling, 15° for rotation). Thus, we have $(\Delta_{end} - \Delta_{start})/N = 0.1$, for 2D scaling transformations, and $(\Delta_{end} - \Delta_{start})/N = 15$ for rotational transformations. $W_j(\delta)$ are the generated weights after the particular transformation has been applied on kernels in W_j. Each TSMP output node records the maximum response of the convolutional kernel across its transformed variations (scaling or rotation in this work), at each spatial location. Notice that setting $\Delta_{start} = \Delta_{end} = 0$, would essentially turn this into a normal CNN output node.[1]

3.2   Method B: Transformation Shared Kernel Extension

In this method, we wish to avoid explicitly encoding invariance to transformations. Thus, we introduce a slight tweak to that architecture, where now, instead of max-pooling the responses from transformed kernels, we simply append those responses (see Fig. 1(b)). In mathematical notation, we can express this addition to the weight space as follows. Given the current set of weights W, at a certain layer in a CNN, we generate a modified, larger set weights

---

[1] Therefore, Δ_{start} and Δ_{end} allows one to control the extent of invariance in each subsequent layer, to the transformation intended.

$$W_{TSKE} = \left\{ W(\Delta_{start} + k \times \frac{\Delta_{end} - \Delta_{start}}{N}) : k = 0, ..., N \right\}. \qquad (3)$$

In the above example, the weights are extended by either rotation or spatial scaling, but in practice we can also use both set of transformations to increase the number of kernels.

4 Training and Testing Protocols

In the analyses that follow, we use variations of MNIST, which we create by adding rotational (MNIST-rot), spatial scaling (MNIST-Scale) and elastic distortions (MNIST-elastic). There are publicly available variations of MNIST which contain scale and rotation transformations on the data, but they encompass a fixed range of transformation parameters. In this work, we customise the extent of the applicable transformations, to cater to our intended analyses, and thus generate these variations by ourselves. In what follows, we use various proportions of the MNIST training data (with or without variations) for training our networks, but in each case, we use the full extent of the MNIST testing data samples (with or without variations). Furthermore, all the CNNs used below contain the exact same number of weight parameters, with a fixed kernel size of 3×3, although the number of actually *trainable* parameters may vary. This common architecture contains two convolutional layers followed by a fully connected dense layer.

Note that all subsequent estimates of accuracy and invariance reported are the average values across six random folds of training data applicable to each case.

5 Results

5.1 Results: Kernel Transformation Shared Max-Pooling (TSMP)

5.1.1 Effect of TSMP Layer on Test Accuracy

In this section, we discuss results of layer-specific testing for kernel tranformations for both scale-shared an rotation-shared architectures. We only add the TSMP convolution to one of the layers of our 2-layer CNN. Note that all accuracy measures and invariance scores are the average values reported after six folds of CNN training. The method used for computing invariance scores is the same as [2].

1. **For best performance, layer with kernel transformations were different for scale-sharing and rotation-sharing architectures.** For TSMP-Scale, kernel transformations in layer 2 (L2) resulted in best performance with test accuracy for L2 exceeding by 1.57% & 5.06% over L1 for MNIST-orig and -scale test datasets. For rotation-shared architecture, higher performance was obtained with kernel transformations in L1 instead. This resulted in increase in performance by 0.55% & 1.23% for MNIST-orig and -rot test datasets.

2. **High transformation-specific invariance scores are not necessarily an indicator of high performance.** Even though the transformation (scale & rotation) specific invariance scores for L1 and L2 are same, gap in test performance on MNIST-orig and transformed test datasets (MNIST-scale & MNIST-rot) is considerable.
3. **We find that general elastic deformations based invariance scores are a better predictor of performance than transformation specific invariance scores.** As an example, in case of TSMP-Rotation models, rotation specific invariance scores for L2 are identical irrespective of kernel transformations done in L1 or L2. However, the corresponding elastic distortion L2 invariance scores show a reduction when TSMP-Rot is used in layer 2.

On the evidence of this analysis, TSMP-Scale is added to the second layer of the CNN, and TSMP-Rot at the first layer, for all subsequent analyses.

Table 1. Performance comparison between CNN and its TSMP variant for different TSMP layers. All the networks were trained on 2% of MNIST-orig train data. The networks were tested on original and transformed variants of MNIST-orig test data as well as local elastic distortions. Along with test accuracy, transformation specific invariance scores for both layers of the network are also shown. For TSMP-Scale, $\Delta_{start} = 0.8$ and $\Delta_{end} = 1.3$, and for TSMP-Rot, $\Delta_{start} = -15°$ and $\Delta_{end} = 15°$.

Classifier	TSMP layer	Test-data	Accuracy	Invariance measure (scaling)		Invariance measure (elastic distortions)	
				L1	L2	L1	L2
TSMP-Scale	1	MNIST-orig	93.97	23.33	20.00	37.03	33.33
TSMP-Scale	1	MNIST-scale	64.71	23.33	20.00	37.03	33.33
TSMP-Scale	2	MNIST-orig	95.54	21.11	26.66	37.96	44.44
TSMP-Scale	2	MNIST-scale	69.77	21.11	26.66	37.96	44.44
				Invariance measure (rotation)		Invariance measure (elastic distortions)	
TSMP-Rot	1	MNIST-orig	94.92	25.92	18.51	38.88	33.33
TSMP-Rot	1	MNIST-rot	84.23	25.92	18.51	38.88	33.33
TSMP-Rot	2	MNIST-orig	94.37	23.61	18.51	37.03	31.48
TSMP-Rot	2	MNIST-rot	83	23.61	18.51	37.03	31.48

5.1.2 Effect of Datasize and Invariance Variation When Trained on MNIST-orig

Here, we discuss the results of data-size and invariance level variation (Δ in Eq. 2) on the test accuracy, for baseline CNN (normal CNN), scale-shared TSMP, and rotation-shared TSMP, when trained on MNIST-orig dataset and tested on MNIST-orig, MNIST-scale and MNIST-rot datasets. The results are shown in Table 2, which includes reduction in error values (in percentage), corresponding to the best performing classifier for each data-size.

1. **TSMP-Scale performs better than normal CNN on MNIST-orig and MNIST-scale. The percentage reduction in error rate is similar across variable training data sizes.** The best accuracy achieved by scale-sharing architectures (40% MNIST-orig train data) demonstrate an error improvement over normal CNN by 15% and 22% on MNIST-orig and -scale test datasets respectively. This shows that, in this case, the benefits incorporating such local-transformation invariance do not decrease with larger training datasize.

2. **Rotation-sharing architectures, however, lagged behind normal CNN for both original and rotation transformed test datasets across variable datasizes.** The gap in test error between Normal CNN and rotation-shared architecture was significant ($\approx$15%) for small training datasize (2%). This shows that rotational sharing (with 3×3 kernel size, can be detrimental to performance).

3. **With larger training data size, incorporating more invariance (higher Δ) led to better performance for TSMP-Scale on MNIST-orig-test.** As an example, TSMP-Scale (0.3:3.6) architecture exceeds TSMP-Scale (0.8:1.3) architecture in test accuracy by 0.1%, which is significant because the accuracy is already high in that case.

Table 2. Performance comparison between CNN and its TSMP variant on original and transformed variants of test MNIST across varying datasizes. In addition to comparing the effect of training datasize on test accuracy, the effect of increase in transformation-specific (scale or rotation) invariance is also compared. The extent of invariance (Δ_{start}:Δ_{end}) is shown in the first column. All the networks are trained on MNIST-orig train data.

Classifier	Test-data	2% training	5% training	10% training	40% training
Normal CNN	orig	0.9508	0.9706	0.9804	0.9889
Normal CNN	scale	0.6632	0.7135	0.7544	0.8024
Normal CNN	rot	0.8502	0.8883	0.913	0.935
TSMP-Scale (0.8:1.3)	orig	0.9538	0.9727	0.9809	0.9896
TSMP-Scale (0.5:2.1)	orig	0.9447	0.9698	0.9807	0.9902
TSMP-Scale (0.3:3.6)	orig	0.944	0.9663	0.9818	0.9906
Error improvement %	orig	(+6.09%)	(+7.14%)	(+7.14%)	(+15.31%)
TSMP-Scale (0.8:1.3)	scale	0.6773	0.7359	0.7644	0.8171
TSMP-Scale (0.5:2.1)	scale	0.7059	0.7458	0.7859	0.8455
TSMP-Scale (0.3:3.6)	scale	0.7014	0.7415	0.7786	0.8255
Error improvement %	scale	(+12.67%)	(+11.27%)	(+12.82%)	(+21.81%)
TSMP-Rot (−15:15)	orig	0.9442	0.9675	0.9789	0.9885
TSMP-Rot (−30:30)	orig	0.9081	0.9626	0.9767	0.9875
TSMP-Rot (−45:45)	orig	0.7351	0.9577	0.9764	0.9876
Error improvement %	orig	(−13.41%)	(−10.55%)	(−7.65%)	(−3.60%)
TSMP-Rot (−15:15)	rot	0.8267	0.8693	0.9047	0.9293
TSMP-Rot (−30:30)	rot	0.7588	0.8655	0.9006	0.9211
TSMP-Rot (−45:45)	rot	0.5779	0.8418	0.8908	0.9225
Error improvement %	rot	(−15.68%)	(−17.00%)	(−9.54%)	(−8.76%)

5.1.3   Effect of Datasize and Invariance Variation When Trained on Transformed MNIST Variants

Unlike the previous section, where all the architectures were trained on different proportions of MNIST-orig, here the architectures are trained on different proportion of transformation-inclusive MNIST-scale and MNIST-rot train datasets. The resuls are shown in Table 3.

1. **TSMP-Scale showed higher performance in comparison to normal CNN when trained and tested on MNIST-scale across all variable datasizes.** The best performance was exhibited by scale-shared (0.3:3.6) architecture that achieved test accuracy of 98.08%, achieving an improvement of 11.9% in error percentage over normal CNN.
2. **TSMP-Rotation again performed worse than normal CNN, when trained and tested on rotation transformed data.** The percentage difference in test error between normal CNN and best performing TSMP-Rotation architectures ranged from 35% (small training datasize) to 6% (large training datasize).
3. **Increase in extent of kernel transformations boosts performance for scale but curtails performance for rotation. However, with increase in training datasize the performance gap shows a clear reduction.** For scale, increase in $\Delta_{end} - \Delta_{start}$, resulted in performance improvement over normal CNN ranging from 1.9% (2% train data) to 0.12% (40% train data) when trained and tested of MNIST-scale. However, in case of rotation, increase in the transformation range resulted in performance drop ranging

Table 3. Performance comparison between CNN and its TSMP variant on original and transformed variants of train and test MNIST across varying datasizes. In addition to comparing the effect of training datasize on test accuracy, the effect of increase in transformation-specific (scale or rotation) invariance is also compared. The extent of invariance ($\Delta_{start}:\Delta_{end}$) is shown in the first column.

Classifier	Test data	Training data	2% training	5% training	10% training	40% training
Normal CNN	scale	scale	87.80	93.99	95.98	97.82
Normal CNN	rot	rot	91.91	95.17	96.40	97.83
TSMP-Scale (0.8:1.3)	scale	scale	89.26	94.49	96.20	97.96
TSMP-Scale (0.5:2.1)	scale	scale	90.53	94.64	96.56	98.00
TSMP-Scale (0.3:3.6)	scale	scale	91.16	95.36	96.71	98.08
Error improvement	scale	scale	(+27.54%)	(+22.79%)	(+18.15%)	(+11.92%)
TSMP-Rot (−15:15)	rot	rot	89.06	94.17	95.97	97.70
TSMP-Rot (−30:30)	rot	rot	86.25	94.47	95.73	97.55
TSMP-Rot (−45:45)	rot	rot	70.20	93.94	95.66	97.56
Error improvement	rot	rot	(−35.29%)	(−14.49%)	(−11.94%)	(−5.99%)

from 18.86% (2% train data) to 0.14% (40% train data) from MNIST-rotation train and test data.[2]

5.2 Performance Comparison Between Normal CNN and Variants (TSMP & TSKE)

This section compares performance of TSMP, TSKE and normal CNN architectures when trained and tested on MNIST-orig, -scale and -rot datasets. The training datasize is fixed at 2% of training data. The kernel extensions (TSKE) are done in first layer. The parameters $(\Delta_{start}, \Delta_{end})$ are $(0.8, 1.3)$ for TSMP-Scale and $(-15, 15)$ for TSMP-Rot. For fair comparison we use the same range of transformation in TSKE-Scale and TSKE-Rot. However, for TSKE-Scale, we only extend the kernels for $s = 0.8$ and $s = 1.3$, not the scale factors in between. As such, TSKE-Scale enjoys one-third the number of trainable parameters in the first layer, as two-thirds are simply the transformation extended kernels (in either direction). We ensure that the weight configurations are the same for all networks, in all layers. The results are summarized in Table 4

1. **We find TSMP and TSKE peformances to be comparable to each other, except in the case of MNIST-Rot testing, where TSKE-Rot shows significant improvement on TSMP-Rot.** For orig/rot training/testing, TSKE shows an accuracy improvement of 1.5% over TSMP.
2. **TSMP & TSKE architectures showed best performance except in the cases when MNIST-rot was the training dataset. In those cases, normal CNNs exhibit better performance.** For MNIST-rot training and MNIST-orig and MNIST-rot testing, TSKE-Rotation (best transformation augmented architecture) lags behind the normal CNN in test accuracy by 0.41% and 0.28% respectively.
3. **Scale augmented architectures (TSMP-Scale and TSKE-Scale) exhibited better performance than normal CNN when tested on transformed datasets without any transformation-specific training. Thus, exhibiting transformation independent distortion invariance.** For MNIST-orig training and MNIST-rot and -scale testing, TSKE-Scale and TSMP-Scale fared better than normal CNN by 1.38% and 1.36% respectively.
4. **When trained and tested on scale-transformed MNIST, TSKE-Scale comes out at the top. This shows that explicitly building transformation-specific invariance (TSMP-Scale) may not result in best performance.** TSKE-Scale achieves higher than normal CNN and TSMP-Scale by 2.16% and 0.19% respectively.

---

[2] We find that with TSMP-Rot, as partially seen in Table 1, higher range of transformation invariance improves rotation invariance measures, but drastically affects general distortion invariance (elastic). We hypothesize that this may be the primary reason behind higher rotation negatively affecting test accuracy.

Table 4. Performance comparison between CNN and variants (TSMP & TSKE) on original and transformed variants of MNIST. Only 2% of original or transformed data is used for training. For TSKE, kernal extensions are done in layer 1.

Training	Testing	TSKE-Rot	TSKE-Scale	Normal CNN	TSMP-Scale	TSMP-Rot
orig	orig	94.97	95.34	94.8	95.38	94.42
orig	rot	84.05	NA	83.57	NA	82.67
orig	scale	NA	66.71	64.37	67.73	NA
scale	scale	NA	89.45	87.29	89.26	NA
rot	rot	91.41	NA	91.69	NA	89.06

5.3   Performance Comparison Between CNN. TSMP & TSKE, with Varying Test Data Distortion Types

Here, we compare between the two architectures described in this paper namely TSMP and TSKE for spatial scale (Table 5(a)), rotation (Table 5(b)) and local elastic deformations (Table 5(c)). Note that the training data used here is 2% of the original MNIST-train, and is the same for all classifiers in each validation split (total of 6). All distortions are added to the entire original MNIST-test data. For adding elastic deformations, we refer the reader to [15], where the

Table 5. Performance comparison between CNN and variants (TSMP & TSKE) for different distortion types. (a) shows the test accuracy achieved by CNN and TSMP & TSKE variants for scale distortions for different level of scale changes, denoted by s, (b) shows the test accuracy achieved by CNN and TSMP & TSKE variants for rotational distortions. θ is the rotational distortion angle, and (c) shows the test accuracy achieved by CNN and TSMP & TSKE variants for elastic distortions. α is the elastic distortion level. For all tables, left to right represents increasing distortion magnitude on the MNIST-test data. All the networks were trained on 2% of MNIST-orig data and tested on MNIST-transformed test data.

Classifier	$s = 0.8$	$s = 0.6$	s=0.4
Normal CNN	0.7134	0.2102	0.1281
TSMP-Scale	0.7003	0.2735	0.1454
TSKE-Scale	0.7323	0.2008	0.1357

(a)

Classifier	$\theta = 15°$	$\theta = 45°$	$\theta = 75°$
Normal CNN	0.8129	0.2828	0.1624
TSMP-Rot	0.7299	0.2082	0.1229
TSKE-Rot	0.8279	0.2981	0.157

(b)

Classifier	$\alpha = 15$	$\alpha = 45$	$\alpha = 75$
Normal CNN	0.8826	0.6077	0.3403
TSMP-Scale	0.88	0.6345	0.3706
TSMP-Rot	0.8535	0.6085	0.3551
TSKE-Scale	0.8866	0.6097	0.337
TSKE-Rot	0.8912	0.6168	0.3396

(c)

deformation method used two parameters $\sigma = 4$ and $\alpha = 34$. The parameter α controls the magnitude of the elastic deformation (in terms of displacement). We keep σ constant at 4, and instead vary α. For rotational distortion, we rotate the images by angles 15°, 45° and 75°, and report the classification accuracies for each case. Similarly, for spatial scale distortion, we scale the images by $s = 0.8$, $s = 0.6$ and $s = 0.4$, to show the effect of increasing scale differences between testing and training on performance.

1. **At low distortional levels, kernel extensions (TSKE) showed better promise than kernel shared (TSMP) and normal CNN architectures.** For elastic distortion ($alpha = 15$) TSKE-Rot was the best performing architecture and achieved test accuracies of 1.12% and 0.86% in excess of TSMP-Scale and normal CNN respectively. Simiarly for spatial scale ($s = 0.8$) and rotation ($\theta = 15°$), TSKE architecture showed best performance that exceeded TSMP and normal CNN architectures by 3.2% & 1.89% for spatial scale and 9.8% & 1.5% for rotation.

2. **For greater distortion levels, however, kernel shared architectures (TSMP) showed best performance except for rotational distortions.** For elastic distortion and spatial scale, TSMP-scale showed best performance that exceeded TSKE and normal CNN architectures by 3.1% & 3.03% at $alpha = 75$ and 0.97% & 1.73% at $s = 0.4$ respectively. In case of rotation, however, normal CNN showed best performance that exceeded TSMP and TSKE architectures by 3.95% & 0.54% at $\theta = 75°$. This shows that at high distortion levels, explicitly building invariance promotes better performance, except for rotation.

6   Discussions and Future Directions

In this section, we analyze this study from the perspective of recent developments in the field towards transformation augmentations in convolutional neural networks. The underlying idea behind the TSMP framework used in our experiments is invariance to transformations [11,12], whereas the TSKE framework is roughly aligned with the philosophy of equivariance to transformations [16,17]. Note that an equivariant representation is identified by the exact predictability of the feature output, in response to transformations in the input. Although TSKE enjoys this property in the first layer, the equivariance principle is not exploited beyond the first layer.

Recent work [16,18] has explored equivariant weight sharing for rotational transformations, but using group convolutions [16], which is different from TSKE. Another strategy of achieving equivariant representations (for scale) can be observed in [10], where multiple CNNs for differently scaled versions of the input are simultaneously trained, and the final feature vectors across all scales are concatenated. The TSKE module discussed in this work (with scale and rotational transformations) can be additionally incorporated into such networks for better performance and more efficient learning of kernels within a CNN.

One of the consistent findings in this work is the inability of the rotation-shared architectures to improve upon the conventional CNN itself. Although a successful rotation invariant TSMP-like architecture was proposed in [12], the filter sizes used in that work were much larger (35×35) than in this work (3×3). Our findings indicate that using the same rotation shared property with filters of small size does not show improvements in performance, rather negatively affecting accuracy. One of the main reasons for these observations is that rotation of small sized images is very susceptible to interpolation errors [18]. For such small filters, exact rotation to any angle other than (0, 90, 180, 270) degrees will introduce artifacts, which can degrade the quality of the subsequently max-pooled feature maps. A workaround to the problem of exact rotation and scale in small filters was proposed in [18] and recently in [19], which use steerable filters from the family of circular harmonics (rotation) and log-radial harmonics (scale) respectively.

The results in this paper demonstrate that data augmentation remains essential for ensuring good generalization performance of CNNs. Currently, the state-of-the-art in many vision tasks is held by Auto-Augment [20], which learns an optimal data augmentation policy on each dataset. Therefore, augmentation remains a crucial part of achieving good performance on most datasets. The methods reported in this paper, along with previous discussed works in literature, represent a step towards avoiding data augmentation, by replacing them with network augmentations. Importantly, including in this study, we observe that network augmentations have the potential to improve the generalization ability of the architectures, even in the data augmentation regime [11, 16–18].

7   Conclusions

This paper presents an elaborate analysis of a particular class of improved CNN architectures. These architectures employ transformation knowledge based kernel sharing to incorporate transformation-specific invariance. To that end, we explored two specific variants, namely, Transformation-Shared Max Pooling (TSMP) and Transformation-Shared Kernel Extension (TSKE). These methods only differ in the way transformation-specific information is incorporated in the architecture. While TSMP explictly injects local invariances, TSKE instead preserves responses from the transformed weights. We carried detailed analysis on the performance of these competing architectures in contrast to the baseline CNN model. The analysis reported included: (1) effect of the TSMP layer, (2) effect of training data size and invariance level in the presence and absence of training data augmentation and (3) effect of increasing local and global distortions of the test data. We summarize the major findings here: (1) Higher invariance in TSMP-Scale architectures usually led to better performance, but the trend was reversed for TSMP-Rot, (2) Performance improvements from TSMP-Scale (on normal CNN) were consistent for varying training data size, but only when original MNIST-Train was used for training, (3) TSKE outperformed TSMP when distortion levels (rotation/scale/elastic) in the test data is on the lower side, even with slightly fewer trainable weights, but (4) For larger distortions, however, TSMP architectures fared better.

References

1. Krizhevsky, A.: ImageNet classification with deep convolutional neural networks alex. J. Geotech. Geoenvironmental Eng. **12**, 04015009 (2015)
2. Goodfellow, I., Lee, H., Le, Q.V., Saxe, A., Ng, A.Y.: Measuring invariances in deep networks. Adv. Neural Inf. Process. Syst. **22**, 646–654 (2009)
3. Long, J., Shelhamer, E., Darrell, T.: Fully convolutional networks for semantic segmentation. In: 2015 IEEE Conference on Computer Vision and Pattern Recognition (CVPR), pp. 3431–3440, June 2015
4. Zeiler, M.D., Fergus, R.: Visualizing and understanding convolutional networks. In: Fleet, D., Pajdla, T., Schiele, B., Tuytelaars, T. (eds.) ECCV 2014. LNCS, vol. 8689, pp. 818–833. Springer, Cham (2014). https://doi.org/10.1007/978-3-319-10590-1_53
5. Bengio, Y.: Deep learning of representations: looking forward. In: Dediu, A.-H., Martín-Vide, C., Mitkov, R., Truthe, B. (eds.) SLSP 2013. LNCS (LNAI), vol. 7978, pp. 1–37. Springer, Heidelberg (2013). https://doi.org/10.1007/978-3-642-39593-2_1
6. Hornik, K., Stinchcombe, M., White, H.: Multilayer feedforward networks are universal approximators. Neural Netw. **2**, 359–366 (1989)
7. Hinton, G.E., Osindero, S., Teh, Y.W.: A fast learning algorithm for deep belief nets. Neural Comput. **18**, 1527–1554 (2006)
8. Dauphin, Y. N., Bengio, Y.: Big neural networks waste capacity, arXiv e-prints, arXiv:1301.3583, January 2013
9. Ji, S., Xu, W., Yang, M., Yu, K.: 3D convolutional neural networks for human action recognition. IEEE Trans. Pattern Anal. Mach. Intell. **35**, 221–231 (2013)
10. Xu, Y., Xiao, T., Zhang, J., Yang, K., Zhang, Z.: Scale-invariant convolutional neural networks, arXiv e-prints, arXiv:1411.6369, November 2014
11. Kanazawa, A., Sharma, A., Jacobs, D.W.: Locally scale-invariant convolutional neural networks. NIPS (2014)
12. Marcos, D., Volpi, M., Tuia, D.: Learning rotation invariant convolutional filters for texture classification. In: Proceedings - International Conference on Pattern Recognition (2017)
13. Marcos, D., Kellenberger, B., Lobry, S., Tuia, D.: Scale equivariance in CNNs with vector fields. ICML (2018)
14. Kavukcuoglu, K., et al.: Learning convolutional feature hierarchies for visual recognition. In: Advances in Neural Information Processing Systems 23 (2010)
15. Simard, P.Y., Steinkraus, D., Platt, J.C.: Best practices for convolutional neural networks applied to visual document analysis. In: Proceedings of the International Conference on Document Analysis and Recognition, ICDAR (2003)
16. Cohen, T., Welling, M.: Group equivariant convolutional networks. In: Balcan, M.F., Weinberger, K.Q. (eds.) Proceedings of the 33rd International Conference on Machine Learning Research, PMLR, New York, USA, 20–22 June 2016, vol. 48, pp. 2990–2999, June 2016
17. Worrall, D.E., Garbin, S.J., Turmukhambetov, D., Brostow, G.J.: Harmonic networks: deep translation and rotation equivariance. In: The IEEE Conference on Computer Vision and Pattern Recognition (CVPR), July 2017

18. Weiler, M., Hamprecht, F.A., Storath, M.: Learning steerable filters for rotation equivariant CNNs. In: 2018 IEEE/CVF Conference on Computer Vision and Pattern Recognition, pp. 849–858 (2017)
19. Ghosh, R., Gupta, A.K.: Scale steerable filters for locally scale-invariant convolutional neural networks. ICML (2019)
20. Cubuk, E.D., Zoph, B., Mane, D., Vasudevan, V., Le, Q.V.: AutoAugment: learning augmentation policies from data, arXiv e-prints, arXiv:1805.09501, May 2018

The Good, the Bad and the Ugly: Augmenting a Black-Box Model with Expert Knowledge

Raoul Heese[1,2]([✉]), Michał Walczak[1,2], Lukas Morand[3], Dirk Helm[3], and Michael Bortz[1,2]

[1] Fraunhofer Center for Machine Learning, Sankt Augustin, Germany
[2] Fraunhofer ITWM, Fraunhofer-Platz 1, 67663 Kaiserslautern, Germany
raoul.heese@itwm.fraunhofer.de
[3] Fraunhofer Institute for Mechanics of Materials IWM,
Wöhlerstraße 11, 79108 Freiburg, Germany

Abstract. We address a non-unique parameter fitting problem in the context of material science. In particular, we propose to resolve ambiguities in parameter space by augmenting a black-box artificial neural network (ANN) model with two different levels of expert knowledge and benchmark them against a pure black-box model.

Keywords: Expert knowledge · Mixture of experts · Custom loss

1 Introduction

A central aspect of the physical description of elastoplastic solids are stress-strain relationships. They describe the deformation behavior of a material as a reaction to an external load. In this manuscript, we consider the exponential hardening model [1]

$$R(\varepsilon, \mathbf{p}) = \frac{\gamma_1}{\beta_1}(1 - e^{-\beta_1 \varepsilon}) + \frac{\gamma_2}{\beta_2}(1 - e^{-\beta_2 \varepsilon}). \tag{1}$$

The hardening stress R depends on the accumulated plastic strain ε and the material model parameters $\mathbf{p} \equiv (\gamma_1, \gamma_2, \beta_1, \beta_2)$.

The so-called *parameter identification problem* [4] entails finding material parameters $\mathbf{p}$ for a given stress-strain curve $\mathbf{C} = \{R_1(\varepsilon_1), \ldots, R_S(\varepsilon_S)\}$ evaluated at a discretized strain interval $\varepsilon_1, \ldots, \varepsilon_S$. Thus, to solve this problem, we search for a suitable inverse mapping

$$E : \mathcal{C} \longmapsto \mathcal{P} \quad \text{with} \quad E(\mathbf{C}) = \mathbf{p} \tag{2}$$

from the domain of stress-strain curves $\mathcal{C}$, to the domain of material parameters $\mathcal{P}$. A summary of typical solution strategies can be found in Ref. [4] and references therein. However, all of these approaches only consider injective backward

© Springer Nature Switzerland AG 2019
I. V. Tetko et al. (Eds.): ICANN 2019, LNCS 11731, pp. 391–395, 2019.
https://doi.org/10.1007/978-3-030-30493-5_38

mappings. In contrast, Eq. (1) is symmetric with respect to the permutation $(\gamma_1 \mapsto \gamma_2, \gamma_2 \mapsto \gamma_1, \beta_1 \mapsto \beta_2, \beta_2 \mapsto \beta_1)$ and therefore Eq. (2) is clearly non-injective.

In the following, we will present three different models for Eq. (2) based on ANNs and compare their performance. In particular, we will demonstrate that incorporating expert knowledge about the forward mapping into the model can improve the results.

2   Data

The test and training data consists of stress-strain curves $\mathbf{C} \in \mathcal{C}$ generated from the forward mapping $R(\varepsilon, \mathbf{p})$ with the same equally-spaced discretization for $S = 20$. We use $\mathbf{C_p}$ to denote the stress-strain curve for the parameters $\mathbf{p} \in \mathcal{P}$. Since a uniform sampling in $\mathcal{P}$ will lead to a very unbalanced distribution of curves in $\mathcal{C}$, we choose our data in such a way that all curves of the sample have an approximately even distance to their nearest neighbor in $\mathcal{C}$. For this purpose, we define the distance between two curves $\mathbf{C_p}$ and $\mathbf{C_{p'}}$ as

$$d(\mathbf{C_p}, \mathbf{C_{p'}}) \equiv \frac{1}{\varepsilon_S - \varepsilon_1} \int_{\varepsilon_1}^{\varepsilon_S} |R(\varepsilon, \mathbf{p}) - R(\varepsilon, \mathbf{p'})| \, d\varepsilon. \tag{3}$$

Summarized, our test and training data contains curves sampled for parameters in the sets $\mathcal{P}_{\text{test}} \subset \mathcal{P}$ and $\mathcal{P}_{\text{train}} \subset \mathcal{P}$, respectively.

3   Models

We consider three models with different network architectures, each of them based on a different perspective on the problem. The input of all models is the twenty-dimensional vector of hardening stresses $(R_1, \ldots, R_{20})$ associated with a stress-strain curve $\mathbf{C}$. Their output is a four-dimensional parameter vector $\mathbf{p}$.

(1) The bad: The first model, $\hat{E}_{\text{bad}}$, is a fully-connected MLP as shown in Fig. 1a with a conventional least squares loss function. Such kind of models have already been applied successfully to parameter identification problems with injective backward mappings [4]. However, we expect them to fail for our non-injective problem since the model will in fact be trained to predict the mean of the ambiguities in the data as proven in Ref. [3]. We therefore regard this model as a naive black-box approach [2] to the problem, which can be used as a worst-case limit for the other models. Thus this model serves as the "bad" candidate among our competitors.

(2) The good: The second model, $\hat{E}_{\text{good}}$, is a fully-connected MLP of the same architecture as $\hat{E}_{\text{bad}}$, but with a custom loss function

$$L_{\text{good}} = \sum_{\mathbf{p} \in \mathcal{P}_{\text{batch}}} \sum_{i=1}^{S} \left[R(\varepsilon_i, \mathbf{p}) - R(\varepsilon_i, |\hat{E}_{\text{good}}(\mathbf{C_p})|) \right]^2 \tag{4}$$

depending on the current training batch $\mathcal{P}_{\text{batch}} \subset \mathcal{P}_{\text{train}}$. Thus, our expert knowledge about the forward mapping is directly incorporated into the loss function and allows us to circumvent parameter ambiguities. This model can consequently be seen as a grey-box approach [2] to the problem. Since MLPs are comparably easy to train and the choice of loss function allows us to exploit expert knowledge efficiently, we consider this model as the "good" candidate.

(3) The ugly: The third model, $\hat{E}_{\text{ugly}}$, is a mixture of experts (MOE) network [5] as shown in Fig. 1a with two experts and one gate. For this model we use expert knowledge about the symmetry of the forward mapping to set the number of experts, but make no further use of the specific form of Eq. (1). We can therefore consider this model as an enhanced black-box approach. On the one hand, MOE networks can handle complicated data ambiguities and are even able to resolve them in a useful manner. On the other hand, they are much more difficult to train than regular MLP networks due to higher complexity. Therefore, this model represents the "ugly" candidate in our benchmark.

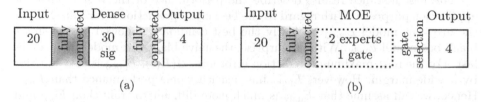

(a)   (b)

Fig. 1. ANN architectures. (a) For the naive black-box model $\hat{E}_{\text{bad}}$ and the grey-box model $\hat{E}_{\text{good}}$ consisting of three fully-connected layers. In total, there are 754 trainable model parameters for each of the two models. (b) For the enhanced black-box model $\hat{E}_{\text{ugly}}$ with 1,988 trainable model parameters. Each expert consists of one dense layer with 30 neurons and a tanh activation function, the gate has a single layer with 10 neurons and a tanh activation function.

4   Results

To test the model performance, we train each of the three models $N = 100$ times with different random seeds for the network initialization. For each training instance i, we calculate the mean prediction error

$$\delta(E, i) \equiv \frac{1}{|\mathcal{P}_{\text{test}}|} \sum_{\mathbf{p} \in \mathcal{P}_{\text{test}}} d(\mathbf{C_p}, \mathbf{C}_{E(\mathbf{C_p})}) \tag{5}$$

and the maximum prediction error

$$\Delta(E, i) \equiv \max_{\mathbf{p} \in \mathcal{P}_{\text{test}}} d(\mathbf{C_p}, \mathbf{C}_{E(\mathbf{C_p})}) \tag{6}$$

for the model E with respect to the test data $\mathcal{P}_{\text{test}}$. We summarize the benchmark results in Table 1, where we show the averages $\langle \delta(E, i) \rangle$ and $\langle \Delta(E, i) \rangle$ and the standard deviations $\sigma[\delta(E, i)]$ and $\sigma[\Delta(E, i)]$, respectively, over all N training instances for all three models. We also list the best instance results $\min_i \delta(E, i)$ and $\min_i \Delta(E, i)$.

Table 1. Benchmark results: performance of the three models in solving the parameter identification problem. Smaller values are better, the best results are highlighted in bold. The grey-box approach $\hat{E}_{\text{good}}$ is clearly superior

Model	Metric					
	$\langle \delta(E, i) \rangle$	$\langle \Delta(E, i) \rangle$	$\sigma[\delta(E, i)]$	$\sigma[\Delta(E, i)]$	$\min_i \delta(E, i)$	$\min_i \Delta(E, i)$
$\hat{E}_{\text{bad}}$	20.27	320.77	2.24	97.38	15.57	155.17
$\hat{E}_{\text{good}}$	**0.88**	**9.48**	**0.11**	**17.52**	**0.73**	**6.21**
$\hat{E}_{\text{ugly}}$	15.75	2,397.69	3.75	1,589.80	9.49	663.74

The best instance results describe the performance of the best models for application purposes. With regard to the best mean prediction error $\min_i \delta(E, i)$, the grey-box model $\hat{E}_{\text{good}}$ is clearly the best candidate, followed by the smart black-box model $\hat{E}_{\text{ugly}}$. Without surprise, the naive black-box model $\hat{E}_{\text{bad}}$ comes last. For the best maximum prediction error $\min_i \Delta(E, i)$, $\hat{E}_{\text{good}}$ is also superior by a wide margin. However, $\hat{E}_{\text{ugly}}$ has a much worse performance than $\hat{E}_{\text{bad}}$. Hence, we can assume that $\hat{E}_{\text{ugly}}$ is much more difficult to train than $\hat{E}_{\text{bad}}$ and particularly prone to overfitting.

The other statistics also allow to quantify the difficulty of the training process. Although the best expected mean prediction error $\langle \delta(E, i) \rangle$ of $\hat{E}_{\text{ugly}}$ is better than for $\hat{E}_{\text{bad}}$, its best maximum prediction error $\langle \Delta(E, i) \rangle$ is much worse, what again indicates overfitting. Additionally, the corresponding variances are much higher for $\hat{E}_{\text{ugly}}$ in comparison with the other candidates, which underpins our assumption of a much more difficult training process.

Summarized, we find that it can be a very useful strategy to directly incorporate the expert knowledge about the forward mapping of a parameter identification problem in the loss function of the model. This approach simplifies the training process in comparison with a MOE model and furthermore leads to a superior overall prediction performance.

The conceptional idea presented here can also be used as an origin for further studies. One could examine in how far the choice of the training and test data sets influences the results. A further point to consider is the robustness to noise in the data (which occurs for experimentally measured stress-strain curves). Finally, the method can also be easily adapted to parameter identification problems in other fields of application.

References

1. Helm, D.: Stress computation in finite thermoviscoplasticity. Int. J. Plast **22**(9), 1699–1727 (2006)
2. Kroll, A.: Grey-Box Models: Concepts and Application, pp. 42–51. IOS Press, Amsterdam (2000)
3. Bishop, C.M.: Mixture density networks. Technical report, Neural Computing Research Group (1994)
4. Mahnken, R.: Identification of Material Parameters for Constitutive Equations, chap. 19, pp. 637–355. American Cancer Society (2004)
5. Yuksel, S.E., Wilson, J.N., Gader, P.D.: Twenty years of mixture of experts. IEEE Trans. Neural Netw. Learn. Syst. **23**, 1177–1193 (2012)

Hierarchical Attentional Hybrid Neural Networks for Document Classification

Jader Abreu(iD), Luis Fred(iD), David Macêdo$^{(\boxtimes)}$(iD), and Cleber Zanchettin$^{(\boxtimes)}$(iD)

Centro de Informática, Universidade Federal de Pernambuco,
50.740-560, Recife, PE, Brazil
{jaoa,lfgs,dlm,cz}@cin.ufpe.br

Abstract. Document classification is a challenging task with important applications. The deep learning approaches to the problem have gained much attention recently. Despite the progress, the proposed models do not incorporate the knowledge of the document structure in the architecture efficiently and not take into account the contexting importance of words and sentences. In this paper, we propose a new approach based on a combination of convolutional neural networks, gated recurrent units, and attention mechanisms for document classification tasks. We use of convolution layers varying window sizes to extract more meaningful, generalizable and abstract features by the hierarchical representation. The proposed method in improves the results of the current attention-based approaches for document classification.

Keywords: Text classification · Attention mechanisms · Document classification · Convolutional neural networks

1 Introduction

Text classification is one of the most classical and important tasks in the machine learning field. The document classification, which is essential to organize documents for retrieval, analysis, and curation, is traditionally performed by classifiers such as Support Vector Machines or Random Forests. As in different areas, the deep learning methods are presenting a performance quite superior to traditional approaches in this field [5]. Deep learning is also playing a central role in Natural Language Processing (NLP) through learned word vector representations. It aims to represent words in terms of fixed-length, continuous and dense feature vectors, capturing semantic word relations: similar words are close to each other in the vector space.

In most NLP tasks for document classification, the proposed models do not incorporate the knowledge of the document structure in the architecture efficiently and not take into account the contexting importance of words and sentences. Much of these approaches do not select qualitative or informative words

---

J. Abreu and L. Fred—Contributed equally and are both first authors.

© Springer Nature Switzerland AG 2019
I. V. Tetko et al. (Eds.): ICANN 2019, LNCS 11731, pp. 396–402, 2019.
https://doi.org/10.1007/978-3-030-30493-5_39

and sentences since some words are more informative than others in a document. Moreover, these models are frequently based on recurrent neural networks only [6]. Since CNN has leveraged strong performance on deep learning models by extracting more abundant features and reducing the number of parameters, we guess it not only improves computational performance but also yields better generalization on neural models for document classification.

A recent trend in NLP is to use attentional mechanisms to modeling information dependencies without regard to their distance between words in the input sequences. In [6] is proposed a hierarchical neural architecture for document classification, which employs attentional mechanisms, trying to mirror the hierarchical structure of the document. The intuition underlying the model is that not all parts of a text are equally relevant to represent it. Further, determining the relevant sections involves modeling the interactions and importance among the words and not just their presence in the text.

In this paper, we propose a new approach for document classification based on CNN, GRU [4] hidden units and attentional mechanisms to improve the model performance by selectively focusing the network on essential parts of the text sentences during the model training. Inspired by [6], we have used the hierarchical concept to better representation of document structure. We call our model as Hierarchical Attentional Hybrid Neural Networks (HAHNN). We also used temporal convolutions [2], which give us more flexible receptive field sizes. We evaluate the proposed approach comparing its results with state-of-the-art models and the model shows an improved accuracy.

2   Hierarchical Attentional Hybrid Neural Networks

The HAHNN model combines convolutional layers, Gated Recurrent Units, and attention mechanisms. Figure 1 shows the proposed architecture. The first layer of HAHNN is a pre-processed word embedding layer (black circles in the Fig. 1). The second layer contains a stack of CNN layers that consist of convolutional layers with multiple filters (varying window sizes) and feature maps. We also have performed some trials with temporal convolutional layers with dilated convolutions and gotten promising results. Besides, we used Dropout for regularization. In the next layers, we use a word encoder applying the attention mechanism on word level context vector. In sequence, a sentence encoder applying the attention on sentence-level context vector. The last layer uses a Softmax function to generate the output probability distribution over the classes.

We use CNN to extract more meaningful, generalizable and abstract features by the hierarchical representation. Combining convolutional layers in different filter sizes with both word and sentence encoder in a hierarchical architecture let our model extract more rich features and improves generalization performance in document classification. To obtain representations of more rare words, by taking into account subwords information, we used FastText [3] in the word embedding initialization.

We investigate two variants of the proposed architecture. There is a basic version, as described in Fig. 1, and there is another which implements a TCN

[2] layer. The goal is to simulate RNNs with very long memory size by adopting a combination of dilated and regular convolutions with residual connections. Dilated convolutions are considered beneficial in longer sequences as they enable an exponentially larger receptive field in convolutional layers. More formally, for a 1-D sequence input $x \in \mathbb{R}^n$ and a filter $f : \{0, ..., k-1\} \to \mathbb{R}$, the dilated convolution operation F on element s of the sequence is defined as

$$F(s) = (x *_d f)(s) = \sum_{i=o}^{k-1} f(i) \cdot \mathbf{x}_{s-d \cdot i} \qquad (1)$$

where d is the dilatation factor, k is the filter size, and $s - d \cdot i$ accounts for the past information direction. Dilation is thus equivalent to introducing a fixed step between every two adjacent filter maps. When $d = 1$, a dilated convolution reduces to a regular convolution. The use of larger dilation enables an output at the top level to represent a wider range of inputs, expanding the receptive field.

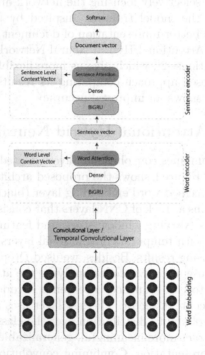

Fig. 1. Our HAHNN architecture include an CNN layer after the embedding layer. In addition, we have created a variant which includes a temporal convolutional layer [2] after the embedding layer.

The proposed model takes into account that the different parts of a document have no similar relevant information. Moreover, determining the relevant sections involves modeling the interactions among the words, not just their isolated presence in the text. Therefore, to consider this aspect, the model includes two levels of attention mechanisms [1]. One structure at the word level and other at the sentence level, which let the model pay more or less attention to individual words and sentences when constructing the document representation.

The strategy consists of different parts: (1) A word sequence encoder and a word-level attention layer; and (2) A sentence encoder and a sentence-level attention layer. In the word encoder, the model uses bidirectional GRU to produce annotations of words by summarizing contextual information from both directions. The attention levels let the model pay more or less attention to individual words and sentences when constructing the representation of the document [6].

Given a sentence with words $w_{it}, t \in [0, T]$ and an embedding matrix W_e, a bidirectional GRU contains the forward $\overrightarrow{GRU\,f}$ which reads the sentence s_i from w_{i1} to w_{iT} and a backward $\overleftarrow{GRU\,f}$ which reads from w_{iT} to w_{i1}:

$$x_{it} = W_e w_{it}, t \in [1, T], \tag{2}$$

$$\overrightarrow{h_{it}} = \overrightarrow{GRU}(x_{it}), t \in [1, T], \tag{3}$$

$$\overleftarrow{h_{it}} = \overleftarrow{GRU}(x_{it}), t \in [T, 1]. \tag{4}$$

An annotation for a given word w_{it} is obtained by concatenating the forward hidden state and backward hidden state, i.e., $h_{it} = [\overrightarrow{h_{it}}, \overleftarrow{h_{it}}]$, which summarizes the information of the whole sentence. We use the attention mechanism to evaluates words that are important to the meaning of the sentence and to aggregate the representation of those informative words into a sentence vector. Specifically,

$$u_{it} = \tanh(W_w h_{it} + b_w) \tag{5}$$

$$\alpha_{it} = \frac{\exp(u_{it}^{\top} u_w)}{\sum_t \exp(u_{it}^{\top} u_w)} \tag{6}$$

$$s_i = \sum \alpha_{it} h_{it} \tag{7}$$

The model measures the importance of a word as the similarity of u_{it} with a word level context vector u_w and learns a normalized importance weight α_{it} through a softmax function. After that, the architecture computes the sentence vector s_i as a weighted sum of the word annotations based on the weights. The word context vector u_w is randomly initialized and jointly learned during the training process.

Given the sentence vectors s_i, and the document vector, the sentence attention is obtained as:

$$\overrightarrow{h_{it}} = \overrightarrow{GRU}(s_i), i \in [1, L], \tag{8}$$

$$\overleftarrow{h_{it}} = \overleftarrow{GRU}(s_i), i \in [L, 1]. \tag{9}$$

The proposed solution concatenates $h_i = [\overrightarrow{h_i}, \overleftarrow{h_i}]$ h_i which summarizes the neighbor sentences around sentence i but still focus on sentence i. To reward sentences that are relevant to correctly classify a document, the solution again use attention mechanism and introduce a sentence level context vector u_s using it to measure the importance of the sentences:

$$u_{it} = \tanh(W_s h_i + b_s) \tag{10}$$

$$\alpha_{it} = \frac{\exp(u_i^\top u_s)}{\sum_i \exp(u_i^\top u_s)} \tag{11}$$

$$v = \sum \alpha_i h_i \tag{12}$$

In the above equation, v is the document vector that summarizes all the information of sentences in a document. Similarly, the sentence level context vector u_s can be randomly initialized and jointly learned during the training process. A fully connected softmax layer gives us a probability distribution over the classes. The proposed method is openly available in the github repository[1].

3   Experiments and Results

We evaluate the proposed model on two document classification datasets using 90% of the data for training and the remaining 10% for tests. The word embeddings have dimension 200 and we use Adam optimizer with a learning rate of 0.001. The datasets used are the IMDb Movie Reviews[2] and Yelp 2018[3]. The former contains a set of 25k highly polar movie reviews for training and 25k for testing, whereas the classification involves detecting positive/negative reviews. The latter include users ratings and write reviews about stores and services on Yelp, being a dataset for multiclass classification (ratings from 0–5 stars). Yelp 2018 contains around 5M full review text data, but we fix in 500k the number of used samples for computational purposes.

Table 1. Results in classification accuracies.

Method	Accuracy on test set	
	Yelp 2018 (five classes)	IMDb (two classes)
VDNN [7]	62.14	79.47
HN-ATT [6]	72.73	89.02
CNN [5]	71.81	91.34
Our model with CNN	**73.28**	92.26
Our model with TCN	72.63	**95.17**

---

[1] https://github.com/luisfredgs/cnn-hierarchical-network-for-document-classification.

[2] http://ai.stanford.edu/~amaas/data/sentiment/.

[3] https://www.yelp.com/dataset/challenge.

Table 1 shows the experiment results comparing our results with related works. Note that HN-ATT [6] obtained an accuracy of 72,73% in the Yelp test set, whereas the proposed model obtained an accuracy of 73,28%. Our results also outperformed CNN [6] and VDNN [7]. We can see an improvement of the results in Yelp with our approach using CNN and varying window sizes in filters. The model also performs better in the results with IMDb using both CNN and TCN.

3.1 Attention Weights Visualizations

To validate the model performance in select informative words and sentences, we present the visualizations of attention weights in Fig. 2. There is an example of the attention visualizations for a positive and negative class in test reviews. Every line is a sentence. Blue color denotes the sentence weight, and red denotes the word weight in determining the sentence meaning. There is a greater focus on more important features despite some exceptions. For example, the word "loving" and "amazed" in Fig. 2(a) and "disappointment" in Fig. 2(b).

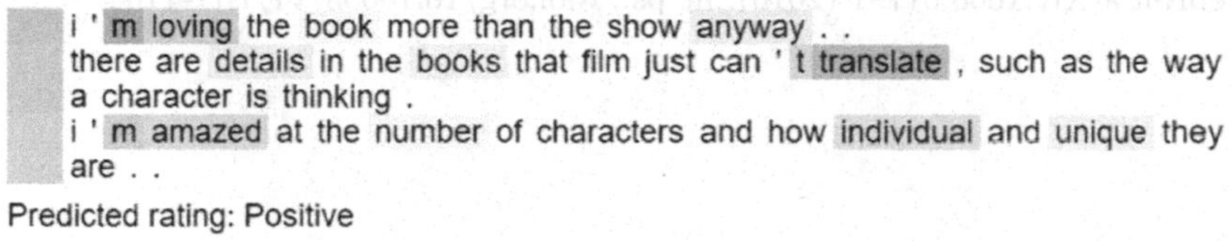

i ' m loving the book more than the show anyway . .
there are details in the books that film just can ' t translate , such as the way
a character is thinking .
i ' m amazed at the number of characters and how individual and unique they
are . .

Predicted rating: Positive

(a) A positive example of visualization of a strong word in the sentence.

the movie was a big disappointment to me . the directing was so bad .
and i think the director thinks the audience are so stupid or retarded to believe
those cheap story lines or cheap jokes or even cheap plot . .

Predicted rating: Negative

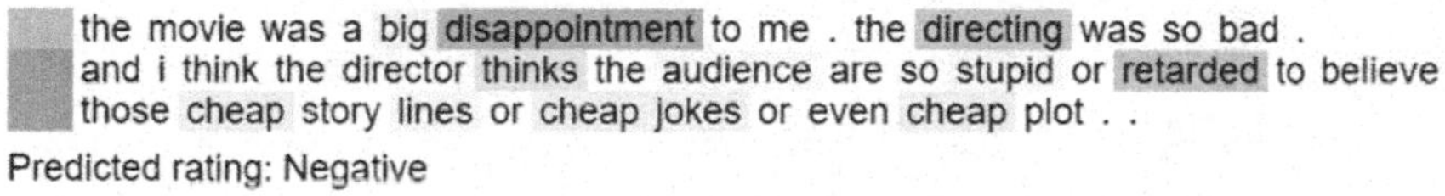

(b) A negative example of visualization of a strong word in the sentence.

Fig. 2. Visualization of attention weights computed by the proposed model (Color figure online)

Occasionally, we have found issues in some sentences, where fewer important words are getting higher importance. For example, in Fig. 2(a) notes that the word "translate" has received high importance even though it represents a neutral word. These drawbacks will be taken into account in future works.

4 Final Remarks

In this paper, we have presented the HAHNN architecture for document classification. The method combines CNN with attention mechanisms in both word and sentence level. HAHNN improves accuracy in document classification by incorporate the document structure in the model and employing CNN's for the extraction of more abundant features.

References

1. Bahdanau, D., et al.: Neural machine translation by jointly learning to align and translate. arXiv preprint arXiv:1409.0473 (2014)
2. Bai, S., Kolter, J.Z., Koltun, V.: An empirical evaluation of generic convolutional and recurrent networks for sequence modeling. arXiv preprint arXiv:1803.01271 (2018)
3. Bojanowski, P., et al.: Enriching word vectors with subword information. arXiv preprint arXiv:1607.04606 (2016). https://doi.org/10.1162/tacl_a_00051
4. Cho, K., et al.: Learning phrase representations using RNN encoder-decoder for statistical machine translation. arXiv preprint arXiv:1406.1078 (2014). https://doi.org/10.3115/v1/D14-1179
5. Kim, Y.: Convolutional neural networks for sentence classification. arXiv preprint arXiv:1408.5882 (2014). https://doi.org/10.3115/v1/D14-1181
6. Yang, Z., et al.: Hierarchical attention networks for document classification. In: Conference of the North American Chapter of the Association For Computational Linguistics: Human Language Technologies, San Diego, CA, USA, pp. 1480–1489 (2016). https://doi.org/10.18653/v1/N16-1174
7. Conneau, A., et al.: Very deep convolutional networks for text classification. arXiv preprint arXiv:1606.01781 (2016). https://doi.org/10.18653/v1/E17-1104

Reinforcement Learning Informed
by Optimal Control

Magnus Önnheim[1,2]([✉]), Pontus Andersson[1,2], Emil Gustavsson[1,2],
and Mats Jirstrand[1,2]

[1] Fraunhofer-Chalmers Centre, Chalmers Science Park, 412 88 Gothenburg, Sweden
magnus.onnheim@fcc.chalmers.se
[2] Fraunhofer Center for Machine Learning, Gothenburg, Sweden

Abstract. Model-free reinforcement learning has seen tremendous
advances in the last few years, however practical applications of pure
reinforcement learning are still limited by sample inefficiency and the
difficulty of giving robustness and stability guarantees of the proposed
agents. Given access to an expert policy, one can increase sample efficiency by in addition to learning from data, and also learn from the
experts actions for safer learning.

In this paper we pose the question whether expert learning can be
accelerated and stabilized if given access to a family of experts which are
designed according to optimal control principles, and more specifically,
linear quadratic regulators. In particular we consider the nominal model
of a system as part of the action space of a reinforcement learning agent.
Further, using the nominal controller, we design customized reward functions for training a reinforcement learning agent, and perform ablation
studies on a set of simple benchmark problems.

Keywords: Reinforcement learning · Expert learning ·
Linear quadratic control · Optimal control · Adaptive control ·
Online learning

Consider a standard problem in optimal control where one wants to find a
sequence of control signals u_t such that the following optimization is solved.

$$\min_u \mathbb{E}\left[\sum_{t=0}^{T} \ell(y_t, u_t, t)\right] \tag{1a}$$

$$x_{t+1} = f(x_t, u_t) + v_t, \quad t = 0, \ldots, T-1, \tag{1b}$$

$$y_t = g(x_t, u_t) + w_t, \quad t = 0, \ldots, T, \tag{1c}$$

This work was developed in Fraunhofer Cluster of Excellence Cognitive Internet Technologies. It has also partially been funded by the Swedish Foundation for Strategic
Research.

© Springer Nature Switzerland AG 2019
I. V. Tetko et al. (Eds.): ICANN 2019, LNCS 11731, pp. 403–407, 2019.
https://doi.org/10.1007/978-3-030-30493-5_40

where ℓ denotes a loss function, f and g denote the system and observation dynamics, and where v_t and w_t denote system and observation noise, respectively. Further, we assume that we are presented with a *nominal* version of (1), where ℓ is a quadratic form, and $f(x, u) = Ax + Bu$, $g(x, u) = Cx$, for some matrices A, B, C, and where v_t, w_t are i.i.d. samples of zero-mean Gaussian distributions with covariance matrices V and W, respectively. In the sequel we will refer to the A, B, C-matrices of (1) whenever we are talking about a nominal model.

Given the above nominal model, it is well-known from control theory that we can design an optimal nominal controller as a linear quadratic regulator (LQR), consisting of a Kalman estimator, with Kalman gains K_t, and linear feedbacks L_t (see e.g. [1]). From the optimal LQR we have a feedback law that explicitly gives the control signal through

$$\hat{x}_{t+1} = A\hat{x}_t + Bu_t + K_t \left[y_t - C \left(A\hat{x}_t + Bu_t \right) \right] \tag{2a}$$

$$u_t = L_t \hat{x}_t. \tag{2b}$$

One can alternatively consider a model-free reinforcement learning approach to solving the problem (1). Given the recent highly impressive successes of model-free reinforcement learning to highly complex domains (e.g. AlphaZero), it is perhaps surprising that such an approach can fail to perform on even simple problems [6], in particularly with regards to sample efficiency and robustness. In the authors' view, this failure is in large part due to an inherent disadvantage of model-free approaches as compared to model-based approaches in the case where good models are available.

Here we consider an indirectly model based approach to solving the problem (1). Given a fixed nominal model, we ask whether it is possible to modify the operation of the nominal controller using a reinforcement learning agent. That is, instead of using a reinforcement learning agent for directly providing actual control signals u_t as actions, we investigate various ways of letting the reinforcement learning agent's actions affect the control law in (2). This requires some care when defining the action space of the agent, and also opens up for designing various reward functions guided by the fixed nominal model, and we perform ablation studies over these design choices. We note the previous similar work done in [3,5], however, to the authors' knowledge, direct manipulation of nominal models seems to be unexplored in the literature.

1   Actions

There are many ways of modifying the operations of the nominal controller, but for brevity we here only discuss what we consider to be illustrative subsets of the full action space, left undefined here. This subset consists of

(a) Perturbations δA_t of the nominal A-matrix.
(b) Perturbations δu_t of the nominal control signal u_t.
(c) Hidden (explained later) perturbations δu_t^h of the nominal control signal u_t.

For completeness, the control law (2) using the possible actions (a)–(c) is

$$\hat{x}_{t+1} = (A + \delta A_t)\hat{x}_t + B(u_t - \delta u_t^h) + L_t \left[y_t - C \left((A + \delta A_t)\hat{x}_t + B(u_t - \delta u_t^h) \right) \right], \quad (3a)$$

$$u_t = K_t \hat{x}_t + \delta u_t + \delta u_t^h, \quad (3b)$$

where the Kalman filter K_t and feedback L_t are adjusted according to the perturbations in the A-matrix. Note the difference that δu_t^h does not affect the state estimation Eq. (3a), whereas δu_t does.

2 Environment

For the observation space we will, again for brevity, only use a rolling window of measurements, that is, the observation o_t at time t that the agent receives is $[y_t, y_{t-1}, \ldots, y_{t-m}]^T$ for a window length m. To facilitate online learning, we will introduce normal shocks to the benchmark problems, simulating control towards a varying reference signal. We thus also extend the size and timing of the normal shocks to the observations. We point out however that the observation space can be extended in many different ways, e.g., by including the nominally estimated states, the nominal value function etc. to the observation.

As rewards we use the following signals:

System loss: $R_t = -\ell(y_t, u_t, t)$,
Innovation: $R_t = -\|y_t - C(A\hat{x}_t + Bu_t)\|^2$, and
Nominalized: $R_t = -\ell(y_t, u_t, t) - \delta R_t^{\text{nom}}$,

as well as a weighted aggregation of the above. *System loss* represents the naïve reward derived from (1), *Innovation* represents modifying the nominal model such that the system estimations becomes correct, *Nominalized* reward represents a reward shaping [4], intended to reduce the variance of stochastic policy gradient estimates as in Generalized Advantage Estimation [7], by factoring out a part of the raw system reward that can be considered as being the responsibility of nominal controller. That is, we may take $\delta R_t^{\text{nom}}(x_t, u_t, x_{t+1}) = \gamma V^{\text{nom}}(x_{t+1}) - V^{\text{nom}}(x_t)$, where $V^{\text{nom}}(x_t)$ denotes the (known) value function of the nominal control policy *assuming the nominal model to be exactly correct*. Concretely we implement an approximation of this by letting

$$\delta R_t^{\text{nom}} = -\ell(\hat{x}_{t+1|t,u_t}) \approx \mathbb{E}_{\pi^{\text{nom}}} \left[-\ell(x_{t+1}, u_t)|x_0, \ldots, x_t, u_0, \ldots, u_{t-1} \right]. \quad (4)$$

3 Experimental results

In view of [6], and the therein demonstrated failure of model-free reinforcement learning approaches to optimal control for even simple problems, we take as benchmark problems perturbations of a discrete-in-time frictionless unit mass double integrator system. The nominal model is thus

$$f^{nom}(x, y) = \begin{bmatrix} 1 & dt \\ 0 & 1 \end{bmatrix} x + \begin{bmatrix} dt^2/2 \\ dt \end{bmatrix} u, \qquad g^{nom}(x, u) = \begin{bmatrix} 1 & 0 \end{bmatrix} u. \quad (5a)$$

We train all agents with a PPO2 algorithm [8], as implemented in [2], with an increased learning rate, and use neural networks to approximate both the value functions and the policy. We train in an online fashion, i.e., we learn from a single trajectory of the system. Further, we induce large random shocks to the system at regular intervals, and all agents are trained using 10000 samples.

Misidentified linear system. The $(2,2)$-component of the A-matrix is replaced by $1 - \mu \in (0,1]$, representing friction.

Piecewise linear system. $f(x,u) = f^{nom}(x,u) + \mathbb{I}_{\|x\|>1}\begin{bmatrix} 0 \\ -\text{sgn}(x)\sin\theta \end{bmatrix}$ corresponding to a mass on plane that at unit distance away from the origin slopes downward at an angle θ.

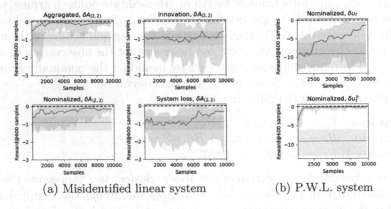

(a) Misidentified linear system   (b) P.W.L. system

Fig. 1. Median reward of 12 agents compared to an optimal controller, evaluated after every 256 samples during training on a set of fixed episodes. Trained agent is in blue, nominal controller is in orange, and shaded regions indicate the 10–90th percentiles. (a) Varying reward signals. (b) Varying action spaces. (Color figure online)

Main results are presented in Fig. 1. Figure 1a shows a clear improvement in sample efficiency using reward nominalization, compared to both raw system loss and innovation rewards. A weighted aggregation appears to show an additional increase in robustness, indicated by relatively narrower error bars. Figure 1b illustrates the importance of choosing the correct action, in the top row the agents' actions enters the feedback loop of the nominal controller, and the action of the agent causes severe problems for the nominal state estimator. On the other hand, when acting invisibly, the agent successfully learns to compensate for the unmodelled nonlinearities using only roughly 1000 samples.

References

1. Glad, T., Ljung, L.: Control Theory. Taylor & Francis, London (2000)
2. Hill, A., et al.: Stable baselines (2019). https://github.com/hill-a/stable-baselines
3. Koryakovskiy, I., Kudruss, M., Vallery, H., Babuška, R., Caarls, W.: Model-plant mismatch compensation using reinforcement learning. IEEE Robot. Autom. Lett. **3**(3), 2471–2477 (2018)
4. Ng, A.Y., Harada, D., Russell, S.: Policy invariance under reward transformations. ICML **99**, 278–287 (1999)
5. Rastogi, D., Koryakovskiy, I., Kober, J.: Sample-efficient reinforcement learning via difference models. In: Machine Learning in Planning and Control of Robot Motion Workshop at ICRA (2018)
6. Recht, B.: A tour of reinforcement learning. AAnnu. Rev. Control. Robot. Auton. Syst. (2018)
7. Schulman, J., Moritz, P., Levine, S., Jordan, M., Abbeel, P.: High-dimensional continuous control using generalized advantage estimation. arXiv preprint arXiv:1506.02438 (2015)
8. Schulman, J., Wolski, F., Dhariwal, P., Radford, A., Klimov, O.: Proximal policy optimization algorithms. arXiv preprint arXiv:1707.06347 (2017)

Explainable Anomaly Detection
via Feature-Based Localization

Shogo Kitamura[✉] 🆔 and Yuichi Nonaka

Research & Development Group, Hitachi, Ltd., Kokubunji-shi, Tokyo, Japan
{shogo.kitamura.hb,yuichi.nonaka.zy}@hitachi.com

Abstract. Deep learning has made remarkable progress in the field of image anomaly detection. An explanation of the validity of the detection is required for practical application. Conventional methods for localizing abnormal regions are mainly based on image reconstruction errors. However, they cannot directly extract specific features from abnormal regions, which limits localization performance. To address this issue, we developed a method for explainable anomaly detection in an unsupervised manner. We trained a feature extractor to extract features that had both the compactness of the normal state and the descriptiveness of the abnormal state for the input images and their reconstructed images. For explainability, our method localized and visualized abnormal regions by accumulating intermediate layers, which led to a significant difference in features extracted from the input image and the reconstructed image. The quantitative results of the defect segmentation and the qualitative results of anomaly localization from experiments on two datasets showed that our method outperformed conventional methods when it came to localizing abnormal regions.

Keywords: Anomaly detection · Deep learning · Explainable · Localization

1 Introduction

To automate the inspection of social infrastructures, industrial products, and so on, the need for vision-based inspection has been increasing. In recent years, image anomaly detection using deep learning has made remarkable progress. Practically, it is difficult to guarantee the validity of the detection if the anomaly detection process uses the black box approach. The ability to explain why an anomaly has been detected is needed.

Many methods have been proposed for localizing abnormal regions in pixelwise based on the reconstruction error of the image [2, 4, 15, 19]. These methods only use normal images because it is difficult to collect abnormal images. A generator that has only been trained to reconstruct normal images cannot correctly reconstruct an image if the test image is abnormal. Using this property, these methods can be used to localize abnormal regions based on the difference

© Springer Nature Switzerland AG 2019
I. V. Tetko et al. (Eds.): ICANN 2019, LNCS 11731, pp. 408–419, 2019.
https://doi.org/10.1007/978-3-030-30493-5_41

between the test image and the reconstructed image. However, methods based on the image reconstruction error do not capture any specific features in abnormal regions, which limits localization performance.

To address this problem, we developed a method for producing explainable anomaly detection in an unsupervised manner. We set up an encoder-decoder based generator which was trained with normal images only. In the inference phase, if an abnormal image was fed into the generator, there was a non-trivial difference between the input and reconstructed images. Our method detected the difference on the feature-level rather than on the pixel-level. To detect differences in features, which are not found in normal images, we trained an extractor by feeding it normal images and an external multiclass dataset. The normal images included the input images and the reconstructed images, and we trained the extractor to have both the compactness for the normal state and the descriptiveness for the abnormal state for the input images and the reconstructed images. We used the extractor to extract features from both the test image and its reconstructed image. For explainability, our method localized and visualized abnormal regions by accumulating intermediate layers, which led to a significant difference in features. We calculated the contribution to the differences based on both the difference between the features and the gradient of this difference. The quantitative results of the defect segmentation and the qualitative results of anomaly localization from experiments on two datasets showed that our method outperformed conventional methods at localizing abnormal regions.

In summary, this paper makes the following three contributions.

- We developed a training framework in which a feature extractor was trained to have both compactness for the normal state and descriptiveness for the abnormal state for the input images and the reconstructed images. To do this, we used a one class dataset created by mixing normal images and their reconstructed images, and an external multiclass dataset for obtaining external knowledge.
- We developed a localization method using intermediate layers that led to a significant difference in the feature vectors extracted from the test image and the reconstructed image.
- We conducted experiments on two datasets that demonstrated the performance of our method for localizing abnormal regions was better than that of conventional methods.

2   Related Work

There are various methods for localizing and visualizing unseen anomalies in images. One of the most common methods is to localize abnormal regions based on image reconstruction errors. Convolutional autoencoder (CAE) [11], variational autoencoder (VAE) [9] and generative adversarial network (GAN) [1,8,14] are often used to reconstruct images. Schlegl et al. [15] used AnoGAN, which optimized the latent variable in GAN for minimizing the difference between the test image and the reconstructed image, for determining whether the image was normal or abnormal by evaluating the residual difference. The GAN was trained with normal images. To reduce the computational cost of the optimal latent variable determination in AnoGAN, Zenati et al. [19] developed a method (EGBAD) using the GAN model [6,7], which estimated latent variables without optimization for each test image. Bergmann et al. [2] developed a method (SSIM-AE) that incorporated structural similarity (SSIM) [18] into the loss function of CAE and the difference metric between the test image and the reconstructed image to improve the accuracy of the image reconstruction and anomaly detection performance. This method, which was based on image reconstruction errors, did not capture any specific features in abnormal regions. This limited the method's localization capability.

When humans look at unknown anomalies, it is possible for them to recognize specific features of the images and accurately identify abnormal regions by using previously obtained knowledge. Other methods use external datasets in combination with images for inspection. Perera et al. [13] used a feature extractor called DOC that had been trained to have both compactness in normal images for inspection and descriptiveness in other multiclass datasets. In the test phase, DOC determined that a feature specific image to normal images was abnormal. Despite this, DOC was not considered for the localization of abnormal regions.

For tasks such as image classification based on convolutional neural networks, Selvaraju et al. [16] proposed a method (Grad-CAM) for localizing regions important for classification using intermediate layers, which gave a high contribution to the classification results. Grad-CAM was not considered for detecting anomalies.

3   Methodology

3.1   Overview

We developed a method for localizing abnormal regions by capturing specific features without using abnormal images in the training process. The overview of our method is shown in Fig. 1.

Training Method: First, normal images were inspected and the images that were reconstructed from these were mixed to create a target dataset of one class. To obtain external knowledge, multiclass datasets such as ImageNet [5] were used as reference datasets. Training was performed ensure both the compactness of the target dataset and the descriptiveness of the reference dataset.

Inference: The extractor, which had a multiclass classification ability, extracted specific features from the test images that were not in the normal images. Then, the abnormal regions were localized based on the extracted features. The trained extractor extracted feature vectors from both the test images and the reconstructed images. Then, the abnormal regions were localized by using the intermediate layers, which led to a significant difference in the feature vectors.

In the following sections, we describe how to set up the generator, how to train the extractor, and how to localize the abnormal regions.

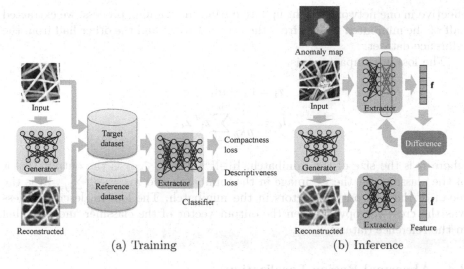

(a) Training   (b) Inference

Fig. 1. Our method

3.2 Generator Setup

We mixed the normal images for inspection and the reconstructed images to create the target dataset. We trained the generator to reconstruct images by feeding it the normal images in advance. Then, we input the normal images into the trained generator and generated the reconstructed images. We could select any model, including CAE [2,11], VAE [9], and GAN [1,6–8,14] as the generator.

3.3 Training the Extractor

We used a model that had been obtained by removing the final layer (the classifier) from an image classification model (such as Alexnet [10] or VGG16 [17]) as the extractor. In the training phase, we connected the classifier to the extractor and trained the extractor in the same way as the DOC had been trained [13]. The loss function is

$$loss = l_D(r) + \lambda l_C(t) \tag{1}$$

where t is the target dataset, r is the reference dataset, $l_C(t)$ is the loss in compactness, $l_D(r)$ is the loss in descriptiveness, and λ is a positive constant for balancing two losses. Both $l_C(t)$ and $l_D(r)$ are calculated from the classifier output.

The loss in compactness was calculated using the target dataset (t). To recognize images in the target dataset as the same class, it positioned the feature vectors extracted from the images close to each other. The loss in descriptiveness was calculated using only the reference dataset (r), which provided the ability to classify multiclass images in the reference dataset. To make both $l_C(t)$ and $l_D(r)$

effective in one network weight update during the training process, we extracted half of the minibatch images from the target dataset and the other half from the reference dataset.

The loss in compactness is

$$\mathbf{z_i} = \mathbf{h_i} - \mathbf{m_i} \tag{2}$$

$$l_C = \frac{1}{nk_h}\sum_{i=1}^{n}\mathbf{z_i}^T\mathbf{z_i} \tag{3}$$

where n is the size of the minibatch, $\mathbf{h_i} \in \mathbb{R}^{k_h} (1 \leq i \leq n)$ is the output vector of the classifier for the ith image in the minibatch and $\mathbf{m_i} = \frac{1}{n-1}\sum_{j \neq i}\mathbf{h_j}$ is the mean of the rest output vectors in the minibatch. The loss in descriptiveness was the cross-entropy between the output vector of the classifier and the label in the reference dataset.

3.4   Abnormal Region Localization

We made improvements to the Grad-CAM [16] to enable localization of the abnormal regions. Grad-CAM is a method of localizing regions that are important for classifying images into different classes. A map $L_{map}^c \in \mathbb{R}^{u \times v}$ for localizing the regions important for image classification is formulated

$$\alpha_t^c = \frac{1}{uv}\sum_i\sum_j\frac{\partial y^c}{\partial A_{ij}^t} \tag{4}$$

$$L_{map}^c = ReLU\left(\sum_t\alpha_t^c A^t\right) \tag{5}$$

where y^c is the output score for the class c of interest, $A^t \in \mathbb{R}^{u \times v}$ is the tth intermediate layer, and α_t^c is the weight for A^t. As described in Eq. (4), α_t^c was calculated using global average pooling on the gradient of y^c with respect to A_{ij}^t. As described in Eq. (5), the localization map L_{map}^c was obtained by taking positive values of the sum of intermediate layers weighted by α_t^c.

Based on the feature vector of the test image, we localized the abnormal regions in the same way. We denoted the feature vector extracted by the extractor from the test image as $\mathbf{f} \in \mathbb{R}^{k_f}$, and we denoted the feature vector extracted from the reconstructed image as $\hat{\mathbf{f}} \in \mathbb{R}^{k_f}$. Unlike y^c in the image classification task, it was not possible to determine an element of interest in $\mathbf{f}$ during the localization of unknown anomalies. To make the elements with large differences between $\mathbf{f}$ and $\hat{\mathbf{f}}$ have a greater effect on the weighting of the intermediate layers, we used global average pooling on the sum of the products in all elements for the difference and the gradient. As described in Eq. (6), the above equation was reformulated as global average pooling was performed on the gradient of the sum of squares in all elements of differences between $\mathbf{f}$ and $\hat{\mathbf{f}}$. By replacing Eqs. (4)

with (6) and calculating Eq. (5), we then obtained an anomaly map that localized the anomaly regions.

$$\alpha_t^c = \frac{1}{uv}\sum_i\sum_j\sum_d (f_d(A) - \hat{f}_d)\frac{\partial}{\partial A_{ij}^t}(f_d(A) - \hat{f}_d)$$

$$= \frac{1}{2uv}\sum_i\sum_j\frac{\partial}{\partial A_{ij}^t}\sum_d (f_d(A) - \hat{f}_d)^2 \tag{6}$$

4   Experiments

4.1   Datasets

We conducted experiments on two different datasets as the target datasets for inspection.

NanoTWICE: The first dataset was NanoTWICE [3], which contained images of nanofibrous materials. The NanoTWICE dataset contained both normal images without defects and abnormal images with defects. Together with the abnormal images, pixel-level ground truth images for the defect regions were prepared. We conducted defect segmentation on NanoTWICE and quantitatively evaluated the anomaly localization performance. Table 1(a) shows the details of the NanoTWICE dataset used in the experiment, and Fig. 2(a) shows an example of a normal image from NanoTWICE used for training. We collected training patches for the extractor by cropping five normal images. We cropped to a randomly determined size in the range of 128×128–320×320 and then resized them to 128×128. We collected patches while moving a 128×128 window with a 30-pixel stride to the test image, created anomaly maps for all patches, and integrated them by averaging the overlapping areas.

SVHN: The second dataset we used was SVHN [12], a real-world dataset of digit images. In the experiment on SVHN, we considered *3* as normal and *8* as abnormal, and we evaluated the performance qualitatively to localize the abnormal regions. Table 1(b) shows details of the SVHN dataset used in the experiment, and Fig. 2(b) shows examples of normal images for training. Since SVHN has a small image size (unlike NanoTWICE), cropping was not conducted, and the entire image was used as the extractor's input.

We used ImageNet [5], a large-scale image dataset, as a reference for obtaining external knowledge. In the experiment on SVHN, we also created another reference dataset using digits other than *3* and *8*, and we compared the localization performance using ImageNet as the reference dataset. Table 2 details the reference datasets used in the experiment.

Table 1. Test datasets

(a) NanoTWICE

(b) SVHN

Item	Value
Training images	5
Training patches	10,000
Patch size	128×128
Test images	40
Patch stride for test	30

Item	Value
Training images	4,200
Image size	32×32
Normal images for test	1,000
Abnormal images for test	200

Table 2. Reference datasets

	ImageNet	SVHN (other digits)
Classes	1,000	8
Images	50,000	33,600

4.2 Experimental Setup

In the experiments on both NanoTWICE and SVHN, we compared EGBAD [19] and SSIM-AE [2]. For EGBAD, we created anomaly maps based on the absolute difference of pixel values between the test image and the reconstructed image. In the case of SSIM-AE, we created anomaly maps based on SSIM values between the test image and the reconstructed image. To evaluate the progress of localizing the abnormal regions from the above methods, we prepared two models with different generators, one with BiGAN [6] used in EGBAD for image reconstruction and one with SSIM-AE. We denoted $Proposed_{GAN}$ and $Proposed_{AE}$ for each model.

In the experiment on NanoTWICE, in addition to the quantitative comparison of the localization performances, we evaluated the relationship between the depth of the intermediate layer used to create the anomaly map and the accuracy of the anomaly map. We used the area under the ROC curve (AUC) for pixel-level ground truth as the evaluation metrics.

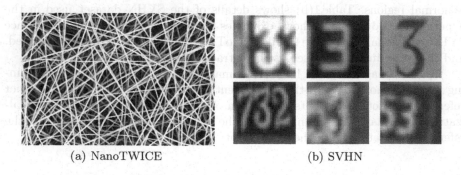

(a) NanoTWICE   (b) SVHN

Fig. 2. Examples of training images

Table 3. Details of extractor training on each target dataset

	NanoTWICE	SVHN
Generator	BiGAN/SSIM-AE	
Reference dataset	ImageNet	ImageNet/other digits
Base model	VGG16	
iterations	20,000	100,000
Optimizer	Momentum SGD $(lr = 5 \times 10^{-5}, momentum = 0.9)$	
λ	0.1	
Trainable layers	Last four layers	
Batch size	32	
CPU	Core i7-8700 3.20GHz	
GPU	GeForce GTX 1080 Ti	
Framework	TensorFlow	

Table 4. Results on NanoTWICE dataset

(a) Comparison to
other methods

(b) Comparison among intermediate
layers used for localization

Method	AUC
EGBAD [19]	0.773
SSIM-AE [2]	0.966
Proposed$_{GAN}$	0.957
Proposed$_{AE}$	**0.974**

Layer	AUC	
	Proposed$_{GAN}$	Proposed$_{AE}$
3rd conv.	0.867	0.909
4th conv.	**0.957**	**0.974**
5th conv.	0.913	0.953

In the experiment on SVHN, we evaluated the relationship between the reference dataset used and the localization performance in addition to the qualitative comparison of localization performance.

Table 3 details the extractor training. We used VGG16 pre-trained with ImageNet as a base model for the extractor, and we trained the extractor using the target dataset and the reference dataset for the last four layers (convolution: one layer, full-connect: three layers).

4.3 Results

Table 4 shows the results of the quantitative evaluation on NanoTWICE. Our method, when incorporating SSIM-AE as the generator, outperformed all other methods, achieving an AUC of 0.974 at pixel-level. We also found that the fourth convolutional layers had the highest localization performance among the intermediate layers. Figure 3 shows two results obtained from localizing the abnormal regions in NanoTWICE. The anomaly maps shown in Fig. 3(b) indicate that the degree of anomaly gets higher as the color gets closer to red. We confirmed that our method correctly localized the abnormal regions that are not contained in

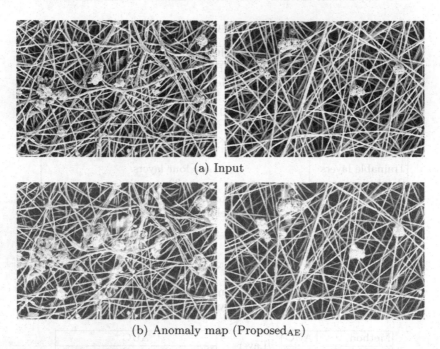

(a) Input

(b) Anomaly map (Proposed$_{AE}$)

Fig. 3. Visualized results of localized abnormal regions on NanoTWICE dataset (Color figure online)

the normal images shown in Fig. 2(a) but are contained in the abnormal images shown in Fig. 3(a).

Figure 4 shows six examples of results obtained from localizing the abnormal regions in SVHN. As with Fig. 3(b), the anomaly maps indicate that the degree of anomaly became higher as the color became closer to red. We confirmed that our method localized the abnormal regions more accurately than other methods. In particular, the difference between *3* and *8* was correctly localized in the results using the reference dataset consisting of other digits, compared with the results of using ImageNet as the reference dataset.

4.4   Discussion

We discuss the results of our quantitative evaluation on NanoTWICE. Table 4(a) shows that in both BiGAN and SSIM-AE, the accuracy of localization is higher when we incorporate them into the proposed method as generators rather than using them alone. Our method, which took external knowledge from an external dataset and extracted specific features from abnormal regions, was more accurate than the methods based only on image reconstruction errors. Table 4(b) shows that the localization accuracy varied depending on the depth of the intermediate layer used to create the anomaly map. We believe this is because the quality of features such as textures, which were easily extracted, differed depending on the depth of the intermediate layer. As the quality of the anomalies depends on

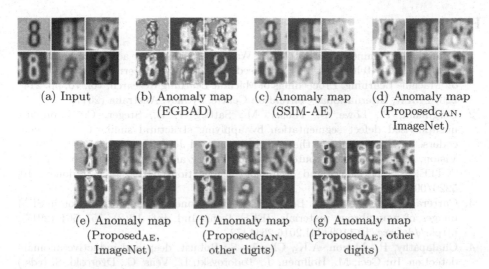

(a) Input   (b) Anomaly map   (c) Anomaly map   (d) Anomaly map
                       (EGBAD)   (SSIM-AE)   (Proposed$_{GAN}$,
                                                                          ImageNet)

(e) Anomaly map   (f) Anomaly map   (g) Anomaly map
  (Proposed$_{AE}$,   (Proposed$_{GAN}$,   (Proposed$_{AE}$, other
   ImageNet)   other digits)   digits)

Fig. 4. Visualized results of localized abnormal regions with SVHN dataset (Color figure online)

the dataset, we need to investigate the relationship between the depth of the intermediate layer and the localization accuracy of the abnormal regions using further datasets.

We discuss the results of our qualitative evaluation on SVHN. As shown in Fig. 4, when using a reference dataset consisting of other digits, the proposed method localizes the abnormal regions more accurately. We believe that the closer the domain of the target dataset and the reference dataset, the more accurate the localization of abnormal regions. To achieve robust anomaly localization for various datasets, we need to explore how to select a reference dataset close to the target dataset.

5   Summary

We developed a method for explainable anomaly detection that localized and visualized abnormal regions. An extractor, which was incorporated in our method, directly extracted specific features from abnormal regions. We used an extractor to extract features from both the test image and its reconstructed image. Our method then localized the abnormal regions by accumulating intermediate layers in the extractor, which led to a significant difference in features. Experimental results on two datasets show that our method outperformed conventional methods at localizing abnormal regions.

References

1. Arjovsky, M., Chintala, S., Bottou, L.: Wasserstein generative adversarial networks. In: Precup, D., Teh, Y.W. (eds.) Proceedings of the 34th International Conference on Machine Learning. Proceedings of Machine Learning Research, vol. 70, pp. 214–223. PMLR, International Convention Centre, Sydney, Australia (2017)
2. Bergmann, P., Löwe, S., Fauser, M., Sattlegger, D., Steger, C.: Improving unsupervised defect segmentation by applying structural similarity to autoencoders. In: Proceedings of the 14th International Joint Conference on Computer Vision, Imaging and Computer Graphics Theory and Applications. pp. 372–380. SCITEPRESS - Science and Technology Publications (2019). https://doi.org/10.5220/0007364503720380
3. Carrera, D., Manganini, F., Boracchi, G., Lanzarone, E.: Defect detection in SEM images of nanofibrous materials. IEEE Trans. Ind. Inf. **13**(2), 551–561 (2017). https://doi.org/10.1109/TII.2016.2641472
4. Chalapathy, R., Menon, A.K., Chawla, S.: Robust, deep and inductive anomaly detection. In: Ceci, M., Hollmén, J., Todorovski, L., Vens, C., Džeroski, S. (eds.) ECML PKDD 2017. LNCS (LNAI), vol. 10534, pp. 36–51. Springer, Cham (2017). https://doi.org/10.1007/978-3-319-71249-9_3
5. Deng, J., Dong, W., Socher, R., Li, L.J., Li, K., Fei-Fei, L.: ImageNet: a large-scale hierarchical image database. In: 2009 IEEE Conference on Computer Vision and Pattern Recognition. pp. 248–255. IEEE, June 2009. https://doi.org/10.1109/CVPR.2009.5206848
6. Donahue, J., Krähenbühl, P., Darrell, T.: Adversarial feature learning. In: International Conference on Learning Representations (ICLR) (2017)
7. Dumoulin, V., et al.: Adversarially Learned Inference. In: International Conference on Learning Representations (ICLR) (2017)
8. Goodfellow, I., et al.: Generative adversarial nets. In: Ghahramani, Z., Welling, M., Cortes, C., Lawrence, N.D., Weinberger, K.Q. (eds.) Advances in Neural Information Processing Systems 27, pp. 2672–2680. Curran Associates, Inc. (2014)
9. Kingma, D.P., Welling, M.: Auto-encoding variational bayes. In: International Conference on Learning Representations (ICLR) (2014)
10. Krizhevsky, A., Sutskever, I., Hinton, G.E.: ImageNet classification with deep convolutional neural networks. In: Pereira, F., Burges, C.J.C., Bottou, L., Weinberger, K.Q. (eds.) Communications of the ACM, vol. 60, pp. 84–90. Curran Associates, Inc., May 2017. https://doi.org/10.1145/3065386
11. Masci, J., Meier, U., Cireşan, D., Schmidhuber, J.: Stacked convolutional auto-encoders for hierarchical feature extraction. In: Honkela, T., Duch, W., Girolami, M., Kaski, S. (eds.) ICANN 2011. LNCS, vol. 6791, pp. 52–59. Springer, Heidelberg (2011). https://doi.org/10.1007/978-3-642-21735-7_7
12. Netzer, Y., Wang, T., Coates, A., Bissacco, A., Wu, B., Ng, A.Y.: Reading digits in natural images with unsupervised feature learning. In: NIPS Workshop on Deep Learning and Unsupervised Feature Learning 2011 (2011)
13. Perera, P., Patel, V.M.: Learning deep features for one-class classification. IEEE Trans. Image Process. **35**, 1 (2019). https://doi.org/10.1109/TIP.2019.2917862
14. Radford, A., Metz, L., Chintala, S.: Unsupervised representation learning with deep convolutional generative adversarial networks. In: International Conference on Learning Representations (ICLR) (2016)

15. Schlegl, T., Seeböck, P., Waldstein, S.M., Schmidt-Erfurth, U., Langs, G.: Unsupervised anomaly detection with generative adversarial networks to guide marker discovery. In: Niethammer, M., et al. (eds.) IPMI 2017. LNCS, vol. 10265, pp. 146–157. Springer, Cham (2017). https://doi.org/10.1007/978-3-319-59050-9_12
16. Selvaraju, R.R., Cogswell, M., Das, A., Vedantam, R., Parikh, D., Batra, D.: Grad-CAM: visual explanations from deep networks via gradient-based localization. In: 2017 IEEE International Conference on Computer Vision (ICCV), pp. 618–626. IEEE, October 2017. https://doi.org/10.1109/ICCV.2017.74
17. Simonyan, K., Zisserman, A.: very deep convolutional networks for large-scale image recognition. In: International Conference on Learning Representations (ICLR) (2015)
18. Wang, Z., Bovik, A.C., Sheikh, H.R., Simoncelli, E.P.: Image quality assessment: from error visibility to structural similarity. IEEE Trans. Image Process. **13**(4), 600–612 (2004). https://doi.org/10.1109/TIP.2003.819861
19. Zenati, H., Foo, C.S., Lecouat, B., Manek, G., Chandrasekhar, V.R.: Efficient GAN-Based Anomaly Detection. CoRR arXiv:1802.06222 (2018)

Bayesian Automatic Relevance Determination for Feature Selection in Credit Default Modelling

Rendani Mbuvha[✉], Illyes Boulkaibet, and Tshilidzi Marwala

School of Electrical and Electronic Engineering, University of Johannesburg,
Johannesburg, South Africa
rendani.mbuvha@wits.ac.za, {ilyesb,tmarwala}@uj.ac.za

Abstract. This work develops a neural network based global model interpretation mechanism - the Bayesian Neural Network with Automatic Relevance Determination (BNN-ARD) for feature selection in credit default modelling. We compare the resulting selected important features to those obtained from the Random Forest (RF) and Gradient Tree Boosting (GTB). We show by re-training the models on the identified important features that the predictive quality of the features obtained from the BNN-ARD is similar to that of the GTB and outperforms those of RF in terms of the predictive performance of the retrained models.

Keywords: Bayesian · Neural networks · Hybrid Monte Carlo · Credit default modelling · Automatic Relevance Determination

1 Introduction

Accurate estimation of an individual's credit risk is of benefit both to the lending institution and the borrower in any credit agreement [3]. The lending institution benefits from increased profits or reduced loss while the borrower benefits through only being involved in transactions which are within their ability of fulfilment.

Literature in machine learning approaches to credit risk modelling is dominated by tree based models and artificial neural networks (ANNs) [1,13].

Xia et al. [12] proposed an Xtreme Gradient boosting (XGboost) tree model for credit scorecard creation. Their results show that XGboost after Bayesian parameter tuning outperforms random forests (RF) and support vector machines (SVMs) based on accuracy and the Area Under the Curve (AUC) measures on five benchmark credit datasets. Sun and Vasarhelyi [11] compare deep neural networks (DNNs) in predicting credit default on a Brazilian Banking Dataset. Their results show that DNNs outperform logistic regression and tree based methods on the AUC performance measure.

© Springer Nature Switzerland AG 2019
I. V. Tetko et al. (Eds.): ICANN 2019, LNCS 11731, pp. 420–425, 2019.
https://doi.org/10.1007/978-3-030-30493-5_42

There has been no attempt to address some of the shortcomings of ANNs when it comes to their applications to credit risk modelling. These shortcomings include the fact that traditional ANNs do not give an indication of which attributes are relevant for the prediction of credit risk and therefore does not aid the transparency of the credit granting process.

Bayesian Neural Networks (BNNs) have numerous advantages over traditional ANNs that are useful in addressing the shortcomings above. These include the ability to produce a predictive distribution of the probability to default. The ability to perform Automatic Relevance Determination (ARD) in BNNs also allows decision makers to infer the relative influences of different attributes on the network outputs.

The main contribution of this work is to provide a systematic empirical comparison of the feature selection performance of BNNs with ARD (BNN-ARD) with that of ensemble tree based methods in the RF and Gradient Tree Boosting (GTB).

2   Bayesian Neural Networks

ANNs can be formulated within the Bayesian Probabilistic framework such that the error function of the network can be viewed as the kernel of the posterior distribution of network parameters [7]. In the Bayesian sense the posterior distribution of the network weight parameters given the observed data and prior distribution of weights can be derived using Bayes formula as [5]:

$$P(\mathbf{w}|\alpha, \beta, \mathcal{H}) = \frac{1}{Z(\alpha, \beta)} \exp\left(-\alpha\left(E_W + \beta E_D\right)\right) \tag{1}$$

$$= \frac{1}{Z_M(\alpha, \beta)} \exp\left(-M(\mathbf{w})\right) \tag{2}$$

Where αE_W is the kernel of the prior distribution on the weights and βE_D is the kernel of the data likelihood.

The distribution functions above are not computationally tractable in closed form and thus we rely on approximate inference techniques such as Hybrid Monte Carlo [10].

3   Hybrid Monte Carlo

HMC is a Markov Chain Monte Carlo technique that generates samples from the exact posterior distribution by simulating a Markov Chain that converges to the true target posterior distribution of the BNN.

HMC uses the gradient information of the neural network to create a vector field around the current state giving it a trajectory towards a high probability next state. The dynamical system formed by the model parameters w and the auxiliary momentum variables p is represented by the Hamiltonian $H(w, q)$ written as follows [8]:

$$H(w, q) = M(w) + K(p) \tag{3}$$

Where $M(w)$ is the negative log likelihood of the posterior distribution in Eq. 2, also referred to as the potential energy. $K(p)$ is the kinetic energy defined by the kernel of a Gaussian [9]. The trajectory vector field is defined by considering the parameter space as a physical system that follows Hamiltonian Dynamics. The dynamical equations governing the trajectory of the chain are then defined by the Hamilton's equations at a fictitious time t as follows [8]:

$$\frac{\partial w_i}{\partial t} = \frac{\partial H}{\partial p_i} \tag{4}$$

$$\frac{\partial p_i}{\partial t} = -\frac{\partial H}{\partial w_i} \tag{5}$$

Due to the discretising errors that arise from when simulating the dynamics above a Metropolis acceptance step is performed in order to accept or reject the new sample proposed by the trajectory [10].

4   Automatic Relevance Determination (ARD)

An Automatic Relevance Determination (ARD) BNN is one where weights associated with each network input belong to a distinct class. The loss function in ARD is as follows [6]:

$$P(w|D, \alpha, \beta, H) = \frac{1}{Z(\alpha, \beta)} \exp\left(\beta E_D + \sum_c \alpha_c E_{W_C}\right) \tag{6}$$

The regularisation hyperparameters for each class of weights can be estimated online during the inference stage by alternating between HMC samples of weights and re-sampling the regularisation parameters. Irrelevant inputs will have high regularisation parameter values meaning weights will be forced to decay to values close to zero.

5   Feature Importance in Tree Based Ensembles

The relative importance of a feature in a tree ensemble is calculated as the average improvement over all internal nodes of the splitting criterion (information gain) as a result of the particular feature being chosen as a vertex [2].

Given a model (RF or GTB) with M trees with each tree labelled T_m, the aggregated feature importance measure for feature l becomes [4]:

$$I_l = \frac{1}{M} \sum_{m=1}^{M} I_l(T_m) \tag{7}$$

Where $I_l(T_m)$ improvement in the splitting criterion in the mth tree when feature l is used as a vertex.

6   Experiment and Data

We use the Taiwan Credit Card Dataset of Yeh and Lien [13] to perform an empirical comparative study on the various feature selection approaches. We train the BNN-ARD, a RF and a GTB to evaluate the quality of the features they select.

7   Results

Figure 1 shows the ARD output of BNN-ARD. Since the BNN-ARD model gives samples of the posterior variances we use the mean posterior variances to infer relative importance. The BNN-ARD model identifies the payment status variables, PAY0, PAY6 and PAY2 as the most important features.

Figure 2 shows the feature importances from the RF and GTB models. Both models agree with BNN-ARD that the payment status in the previous month (PAY0) is the most important feature. The RF model also suggests that AGE, bill amounts (BILL AMT1 and BILL AMT2) and the credit limit (LIMIT BAL) are important features, the GTB suggests that credit limit (LIMIT BAL) and the most recent payment amount (PAY AMT1) have a higher importance than suggested by the other models.

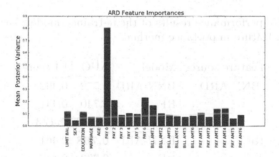

Fig. 1. Mean posterior variances from the BNN-ARD model which indicate the relevance of each attribute.

There is significant disagreement between the models as can be seen from the feature importance plots. In order to further investigate which model yields the most accurate feature ranking we re-train each of the models on a subset of the top 10 most important features as suggested by the respective plots.

Table 1 shows the results of re-training models based on the subset of important features suggested by the respective models. As can be seen in Table 1 there has been marginal deterioration in predictive performance across all models, indicating that majority of the features where noisy and thus did not provide any additional information in the prediction of defaults. While the best performance in terms of AUC is obtained when using BNN-ARD features, the relatively low

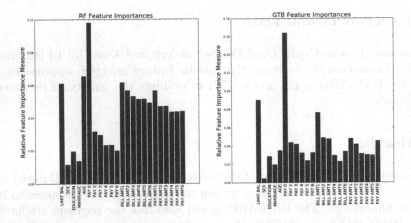

Fig. 2. Feature importance measures from the RF and GTB models

marginal difference in AUC obtained results in the conclusion that the difference in quality of BNN-ARD and GTB features is not statistically significant. The lowest performance across all models is recorded when using RF features. Overall the results in Table 1 show that features obtained from the BNN-ARD are of similar predictive quality to those obtained from the GTB.

Table 1. Predictive performance results of the re-trained models based on the top 10 features from each feature importance method

Feature source	Model	AUC	F1 score
BNN-ARD	BNN-ARD	0.772	0.462
	RF	0.740	0.441
	GTB	**0.778**	**0.473**
RF	BNN-ARD	**0.768**	0.461
	RF	0.733	0.447
	GTB	0.765	**0.464**
GTB	BNN-ARD	0.774	0.455
	RF	0.728	0.435
	GTB	**0.775**	**0.465**

8   Conclusion

We compare the resulting features with those obtained from the popular RF and GTB models. We also show by reducing the feature space based on obtained feature importances that the features obtained from the BNN-ARD result in better predictive performance when compared to those obtained from the RF. The resulting predictive performance of BNN-ARD features is similar to that of features obtained from GTB.

References

1. Baesens, B., Van Gestel, T., Viaene, S., Stepanova, M., Suykens, J., Vanthienen, J.: Benchmarking state-of-the-art classification algorithms for credit scoring. J. Oper. Res. Soc. **54**, 06 (2003)
2. Breiman, L., Ihaka, R.: Nonlinear discriminant analysis via scaling and ACE. University of California, Department of Statistics (1984)
3. Hand, D.J., Henley, W.E.: Statistical classification methods in consumer credit scoring: a review. J. R. Stat. Soc.: Ser. (Stat. Soc.) **160**(3), 523–541 (1997)
4. Hastie, T., Tibshirani, R., Friedman, J.: The Elements of Statistical Learning. Springer Series in Statistics. Springer, New York Inc., Heidelberg (2001)
5. MacKay, D.J.C.: A practical Bayesian framework for backpropagation networks. Neural Comput. **4**(3), 448–472 (1992). ISSN 0899–7667
6. MacKay, D.J.C.: Probable networks and plausible predictions – a review of practical Bayesian methods for supervised neural networks (1995)
7. Mbuvha, R., Jonsson, M., Ehn, N., Herman, P.: Bayesian neural networks for one-hour ahead wind power forecasting. In 2017 IEEE 6th International Conference on Renewable Energy Research and Applications (ICRERA), pp. 591–596, November 2017
8. Neal, R.M.: Bayesian learning via stochastic dynamics. In: Advances in Neural Information Processing Systems, pp. 475–482 (1993)
9. Neal, R.M.: Bayesian Learning for Neural Networks, vol. 118. Springer, Heidelberg (2012). https://doi.org/10.1007/978-1-4612-0745-0
10. Neal, R.M., et al.: MCMC using Hamiltonian dynamics. Handb. Markov Chain. Monte Carlo **2**(11), 2 (2011)
11. Sun, T., Vasarhelyi, M.A.: Predicting credit card delinquencies: an application of deep neural networks. Intell. Syst. Account. Financ. Manag. **25**(4), 174–189 (2018)
12. Xia, Y., Liu, C., Li, Y.Y., Liu, N.: A boosted decision tree approach using Bayesian hyper-parameter optimization for credit scoring. Expert Syst. Appl. **78**, 225–241 (2017). ISSN 0957–4174
13. Yeh, I.-C., Lien, C.: The comparisons of data mining techniques for the predictive accuracy of probability of default of credit card clients. Expert Syst. Appl. **36**(2), 2473–2480 (2009)

TSXplain: Demystification of DNN Decisions for Time-Series Using Natural Language and Statistical Features

Mohsin Munir[1,2]([⊠])(iD), Shoaib Ahmed Siddiqui[1,2](iD), Ferdinand Küsters[3],
Dominique Mercier[1,2], Andreas Dengel[1,2], and Sheraz Ahmed[2]([⊠])(iD)

[1] Technische Universität Kaiserslautern, 67663 Kaiserslautern, Germany
[2] German Research Center for Artificial Intelligence (DFKI) GmbH,
67663 Kaiserslautern, Germany
{mohsin.munir,sheraz.ahmed}@dfki.de
[3] Ingenieurgesellschaft Auto und Verkehr (IAV), 67663 Kaiserslautern, Germany
ferdinand.kuesters@iav.de

Abstract. Neural networks (NN) are considered as black boxes due to the lack of explainability and transparency of their decisions. This significantly hampers their deployment in environments where explainability is essential along with the accuracy of the system. Recently, significant efforts have been made for the interpretability of these deep networks with the aim to open up the black box. However, most of these approaches are specifically developed for visual modalities. In addition, the interpretations provided by these systems require expert knowledge and understanding for intelligibility. This indicates a vital gap between the explainability provided by the systems and the novice user. To bridge this gap, we present a novel framework i.e. Time-Series eXplanation (TSXplain) system which produces a natural language based explanation of the decision taken by a NN. It uses the extracted statistical features to describe the decision of a NN, merging the deep learning world with that of statistics. The two-level explanation provides ample description of the decision made by the network to aid an expert as well as a novice user alike. Our survey and reliability assessment test confirm that the generated explanations are meaningful and correct. We believe that generating natural language based descriptions of the network's decisions is a big step towards opening up the black box.

Keywords: Demystification · Textual explanation ·
Statistical feature extraction · Deep learning · Time-series analysis ·
Anomaly detection

1 Introduction

Deep neural networks (DNN) have become ubiquitous these days. They have been successfully applied in a wide range of sectors including automotive [12],

© Springer Nature Switzerland AG 2019
I. V. Tetko et al. (Eds.): ICANN 2019, LNCS 11731, pp. 426–439, 2019.
https://doi.org/10.1007/978-3-030-30493-5_43

government [27], wearable [21], dairy [16], home appliances [25], security and surveillance [15], health [6], and many more, mainly for regression, classification, and anomaly detection problems [5,7,19,20,32]. The neural network's capability of automatically discovering features to solve any task at hand makes them particularly easy to adapt to new problems and scenarios. However, this capability to automatically extract features comes at the cost of a lack of transparency/intelligibility of their decisions. Since there is no clear indication regarding why the network reached a particular prediction, these deep models are generally referred to as a black box [23].

It has been well established in the prior literature that an explanation of the decision made by a DNN is essential to fully exploit the potential of these networks [1,22]. With the rise in demand for these deep models, there is an increasing need to have the ability to explain their decisions. For instance, big industrial machines cannot be powered down just because a DNN predicted a high anomaly score. It is important to understand the reason for reaching a particular decision, i.e. why the DNN computed such an anomaly score. Significant efforts have been made in the past in order to better understand these deep models [2,23,24,30]. Since most of the prior literature is validated specifically for visual modalities, therefore, a very common strategy is to visualize the learned features [14,24,30,31]. Visual representation of filters and learned features, which might provide important hints to the expert, is already a good step in this direction, but might not be intelligible to the end user. Adequate reasoning of the decision taken increases the user's confidence in the system.

The challenges pertaining to interpretability and visualization of deep models developed for time-series analysis are even more profound. Directly applying the interpretation techniques developed specifically for visual modalities are mostly uninterpretable in time-series data, requiring human expertise to extract useful information out of these visualizations [23]. Therefore, we aim to bridge this gap by generating explanations based on the concrete knowledge extracted out of the statistical features for the decision made by the network, assisting both novice and expert users alike. Statistical features are considered to be transparent which ameliorates the process of gaining user trust and confidence.

This paper presents a novel approach for generating natural language explanations of the classification decisions made by a DNN (system pipeline is shown in Fig. 1). We take a pre-trained DNN and find the most salient regions of the input that were mainly leveraged for a particular prediction. Important data points that contributed towards the final network decision are tracked using the influence computation framework. These points are then combined with different statistical features, which are further used to generate natural language explanations. We specify confidence to the generated explanation by accessing its reliability through sanity check. Our experiments on the classification datasets and a survey show that the generated natural language explanations help in better understanding the classification decision. To the best of the authors' knowledge, it is the very first attempt to generate natural language explanations for classification task on time-series data.

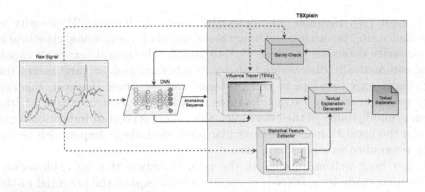

Fig. 1. TSXplain system diagram. Based on a time-series classifier, the anomalous time-series are passed on to the *influence tracer* module which highlights the most influential data points. Furthermore, these influential data points are used along with the point-wise and sequence-wise statistical features to generate textual explanations of the anomalous time-series.

2   Related Work

Over the past few years, there have been numerous advancements in the field of network interpretation and visualizations. For understanding a DNN on an image classification task, Zeiler and Fergus [31] proposed a technique for the visualization of deep convolutional neural networks. Their visualization reveals features in a fully trained network. Highlighting the input stimuli which excites individual feature maps at different layers, helps in understanding what the network has learned. Yosinski et al. [30] introduced DeepVizToolbox which helped in understanding how neural network models work and which computations they perform at the intermediate layers of the network at two different levels. This toolbox visualizes the top images for each unit, forward activation values, deconvolutional highlighting, and preferred stimuli. Simonyan et al. [24] also presented an approach to visualize the convolutional neural networks developed for image classification. Their visualization provides image-specific class saliency maps for the top predicted class.

Bau et al. [4] introduced a framework to interpret the deep visual representations and quantify the interpretability of the learned CNN model. First, they gather a broad set of human-labeled visual concepts and then gather the response of hidden variables to known concepts. To understand a neural network, Mahendran and Vedaldi [17] highlight the encoding learned by the network through inversion of the image representations. They also study the locality of the information stored in the representations. Melis et al. [18] designed self-explaining models where the explanations are intrinsic to the model for the robust interpretability of a network. Bach et al. [2] introduced an approach to achieve pixel-level decomposition of an image classification decision. They generate a heatmap for a well-classified image that segments the pixels belonging to the predicted class.

Current DNN visualizations and interpretations only help an expert to understand and improve the overall process. However, they still lack the actual reasoning why a particular decision has been taken by the learned model. There has been some work in the domain of image captioning where visual attributes are leveraged to support the DNN decision. Guo et al. [8] proposed a textual summarization technique of image classification models. They train a model with the image attributes which are used to support the classification decision. In the same domain, Hendricks et al. [10] proposed a model that predicts a class label and explains the reason for the classification based on the discriminating properties of the visual objects. Kim et al. [13] introduced a textual explanation system for self-driving vehicles. They generate introspective explanations to represent the causal relationships between the system's input and its behavior which is also the target of our study. In most of the aforementioned techniques, a neural network black box is further used to generate descriptions and explanations of another neural network. Despite being a promising direction, this introduces another level of opaqueness into the system.

In the domain of relation extraction, Hancock et al. [9] proposed a supervised rule-based method to train classifiers with natural language explanations. In their framework, an annotator provides a natural language explanation for each labeling decision. Similar work has been presented by Srivastava et al. [26], but they jointly train a task-specific semantic parser and classifier instead of a rule-based parser. These systems, however, rely on a labeled set of training examples that are not available in most of the real-world applications.

3   Methodology

Different visualization and interpretation techniques developed specifically to understand deep models aid an expert in understanding the learning and decision-making process of the network. However, the provided interpretation/visualization cannot be readily understood by a novice user. It is up to the user to draw conclusions about the network's decision with the help of the available information. Many of the existing techniques use a separate deep network that is trained for the generation of explanations using the primary model [8,10,13]. These explanations still suffer from a lack of transparency, as they are also generated by a deep model. Therefore, we approach the problem of generating explanations in a way that significantly improves the intelligibility of the overall process. We leverage the statistical time-series features to provide a concrete natural language based explanation of an anomalous sequence. These features also help in gaining a user's trust because of their lucid nature. The proposed system is composed of different modules as highlighted in Fig. 1. The raw input is first passed on to DNN for classification. If the sequence is classified an anomalous, the whole TSXplain system is activated which is composed of four modules, namely *influence tracer (TSViz)* [23], *statistical feature extractor, sanity check,* and *textual explanation generator*. The *influence tracer* is employed to discover the most salient regions of the input (Sect. 3.1). The

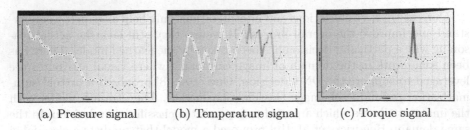

(a) Pressure signal   (b) Temperature signal   (c) Torque signal

Fig. 2. Visualization of the saliency information along with the raw signal values. Influential data points are highlighted in the shades of red, dark being more influential. (Color figure online)

statistical feature extractor module extracts different statistical features from the sequence (Sect. 3.2). The results from previous two modules are passed onto the *textual explanation generator* module in order to come up with a natural language description of the encountered anomaly (Sect. 3.3). Furthermore, we introduce a *sanity check* module to get a coarse estimate of the system's confidence regarding the generated explanation (Sect. 3.4).

3.1   Influence Tracer (TSViz)

The influence tracer module is based on the TSViz framework [23]. The proposed influence tracing algorithm can be used to trace the influence at several different levels. However, we only consider the main influence for our method i.e. the influence of the input on the output. These influences can be effectively computed as the gradients of the output y w.r.t. the input components x_j:

$$I_{input}^j = \left| \frac{\partial y}{\partial x_j} \right| \tag{1}$$

The absolute value of the gradient is used here, as only the magnitude of the influences is of relevance to us. Once the influences have been computed, we use the max-min scaling to scale the values for visualization and further processing as specified by:

$$I_{input,scaled}^j = \frac{I_{input}^j - \min\limits_{j} I_{input}^j}{\max\limits_{j} I_{input}^j - \min\limits_{j} I_{input}^j} \tag{2}$$

The computed influence values are visualized on top of the original signals to provide a hint regarding the encapsulated information. This visualization is presented in Fig. 2, where color-coded data points in each input channel are shown with respect to their influence on the final decision of the DNN. The dark shade represents the high influence of a data point. It can be observed in Fig. 2 that the network decision is mostly based on the small and big spikes in the observed time-series.

Siddiqui et al. [23] discussed the problem of extremely confident predictions for the influence tracing algorithm and suggested a remedy to overcome this issue by imposing regularization on top of the activations itself when training the network. The new objective can be written as:

$$\mathcal{W}^* = \arg\min_{\mathcal{W}} \frac{1}{|\mathcal{X}|} \sum_{(\mathbf{x},y)\in\mathcal{X}\times\mathcal{Y}} \mathcal{L}(\Phi(\mathbf{x};\mathcal{W}),y) + \lambda\|\mathcal{W}\|_2^2 + \beta\|z^L\|_2^2 \qquad (3)$$

where Φ defines the mapping from the input to the output space, $\mathcal{W}$ encapsulates all the parameters of the network, z^L denotes the activation values of the last layer before application of the sigmoid layer, and λ and β denotes hyperparameters controlling the contribution of the regularization terms and the empirical risk. We use the same modified objective to train our network in order to avoid extremely confident predictions.

The *influence tracer* module consumes both the DNN model as well as the raw input. Then, it performs the backward pass through the network from output to input in order to obtain these influence values. The output from this module is consumed by the *textual explanation generator* module (Sect. 3.3).

3.2   Statistical Feature Extractor

This module extracts different statistical features from the input sequence. Since we are dealing with sequential data comprising of time-series, we calculate different point-wise as well as sequence-wise features. These features include, but not limited to, lumpiness, level shift, KL score, number of peaks, and ratio beyond r-sigma (explained below). These features have been previously used by Bandara et al. [3] and have been proposed in [11,29].

1. *Lumpiness:* Initially, daily seasonality from the sequence is removed by dividing it into blocks of n observations. Variance of the variances across all blocks is computed which represent the lumpiness of the sequence.
2. *Level shift:* The sequence is divided into n observations and the maximum difference in mean between consecutive blocks is considered as the level shift. It highlights the block which is different from the rest of the sequence.
3. *KL score:* To calculate this score, the sequence is divided into consecutive blocks of n observations. This score represents the maximum difference in Kullback-Leibler divergence among consecutive blocks. A high score represents high divergence.
4. *Number of peaks:* This feature identifies the number of peaks in a sequence. The sequence is smoothed by a Ricker wavelet for widths ranging from 1 to n. It detect peaks with sufficiently high signal-to-noise ratio.
5. *Ratio beyond r-sigma:* It gives the ratio of data points which are r standard deviations away from the mean of a sequence.
6. *Standard deviation:* This feature represents the standard deviation of a sequence.

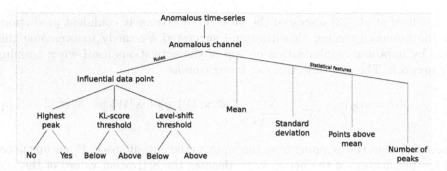

Fig. 3. An example of rules and statistical features used to generate textual explanations for a given anomalous time-series. (Color figure online)

In addition to the above mentioned sequence-wise features, we also use point-wise features including peak, valley, maximum point, minimum point, highest spike, and lowest valley, etc. The fusion of sequence-wise and point-wise features provides vivid characteristics of the highly influential data points.

3.3   Textual Explanation Generator

The influential points determined by the *influence tracer* module along with the features computed by the *statistical feature extractor* module are passed onto this module for the generation of the textual explanations of a given anomalous sequence. This module also receives input from the *sanity check* module (explained in Sect. 3.4) which allows the system to specify its confidence over the generated explanation.

We designed a set of rules to incorporate a range of features. These rules are defined based on the statistical time-series features in a way that explains different characteristics of a given sequence. Based on the classification decision given by the network along with the time-series features, this module provides explanations of the data points which influence the network decision. The explanations are generated channel-wise so that the anomalous data points in each channel can be highlighted. An example of defining rules and an overall hierarchy of this process is shown in Fig. 3. The influential data points are provided by the *influence tracer* module and the rules defined under the 'Influential data point' parent node in Fig. 3 are applied on those data points. The statistical features (mean, standard deviation, points above mean, and number of peaks calculated on whole sequence mentioned on right branch of 'Anomalous channel' parent node in Fig. 3) provide supporting arguments to the defined rules. The values of child nodes (highest peak, KL-score, level-shift, mean, standard deviation, points above mean, and number of peaks) in green are incorporated in the textual explanations. For multi-channel input, the salient data points from an individual channel are passed onto this module.

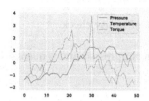
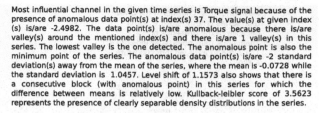

Most influential channel in the given time series is Torque signal because of the presence of anomalous data point(s) at index(s) 37. The value(s) at given index (s) is/are -2.4982. The data point(s) is/are anomalous because there is/are valley(s) around the mentioned index(s) and there is/are 1 valley(s) in this series. The lowest valley is the one detected. The anomalous point is also the minimum point of the series. The anomalous data point(s) is/are -2 standard deviation(s) away from the mean of the series, where the mean is -0.0728 while the standard deviation is 1.0457. Level shift of 1.1573 also shows that there is a consecutive block (with anomalous point) in this series for which the difference between means is relatively low. Kullback-leibler score of 3.5623 represents the presence of clearly separable density distributions in the series.

Sanity Check: The system is fairly confident regarding the provided explanation.

Most influential channel in the given time series is Temperature signal because of the presence of anomalous data point(s) at index(s) 21. The value(s) at given index(s) is/are 2.6410. The data point(s) is/are anomalous because there is/are peak(s) around the mentioned index(s) and there is/are 7 peak(s) in this series.

Sanity Check: The system is not confident regarding the provided explanation.

Fig. 4. Channel-wise natural language explanations generated for the expert users. The second time-series in this figure represents a failure-case where the explanations are not accurate, which is also detected in the sanity check.

We have defined two levels of abstraction for the textual explanations. The first level of explanation is defined for the users who are not interested in detailed explanations or don't have enough knowledge of time-series data (novice user) but would like to get information regarding the most salient regions and channels of the input. The second level of explanation, on the other hand, is defined for the expert users. Such users are generally interested in knowing the details such as, why a network took a particular decision? Sample explanations for the expert users are shown in Fig. 4. The explanations shown in this figure clearly point out the anomalous channel in a given anomalous time-series sequence. Moreover, characteristics of anomalous data points and anomalous channel are also explained in the form of feature values. Figure 5a shows an example of a simple explanation which is generated for the novice users.

3.4   Sanity Check

In order to assess the reliability of a given explanation, a simple sanity check is performed. The output of the *textual explanation generator* module along with the original input is passed onto this module. The data point corresponding to the explanation is suppressed and this masked sequence is fed again to the network for inference. The masking is performed by linear interpolation between the last and the first retained points.

If the removed data points were indeed causal for the network prediction, we expect the prediction to flip. We use this sanity check to compute confidence over the provided explanation. If we observe a flip in the prediction, we assign high confidence to the provided explanation. On the other hand, if the prediction is retained, we assign low confidence to the provided explanation. This check confirms that the generated explanations are referring to the data points which are

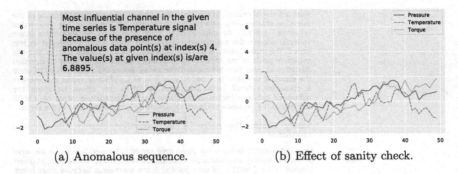

(a) Anomalous sequence.   (b) Effect of sanity check.

Fig. 5. As a result of the sanity check, the anomalous peak is suppressed and sequence in (a) is classified as normal in (b).

actually contributing towards the classification decision taken by the network. Finally, the sanity check output is passed back to the *textual explanation generator* module with the confidence information which is mentioned in Fig. 4 after the generated explanations. In Fig. 4, the second example didn't observe any flip in the prediction after the suppression of the deemed causal point (a failure case), resulting in low system confidence for the provided explanation. Figure 5 visualizes the process of sanity check on an anomalous sequence. In Fig. 5a, the textual explanation highlights an anomalous data point in the temperature signal. The network prediction is flipped after suppressing the mentioned anomalous data point in the temperature signal as shown in Fig. 5b.

4   Experimental Setup and Dataset

To classify a time-series as normal or anomalous, we trained a CNN model with three convolutional layers comprising of 16, 32 and 64 filters respectively, with Leaky ReLU as the activation function, followed by a single dense layer. Since the focus of this paper is on a generation of textual explanations, we selected the hyperparameters (e.g. number of layers, number of filters) based on our experience and did not invest any significant effort into hyperparameter optimization or model selection. It is important to mention that we have chosen a convolutional neural network as our DNN, because CNNs are generally easier to optimize, achieved state-of-the-art results in anomaly detection [20,32], and the base module of TSXplain (*influence tracer* module) is currently based on CNNs. To generate natural language explanations on time-series data, we used the datasets mentioned in this section.

4.1   Machine Anomaly Detection [23]

It is a synthetic time-series classification dataset curated by Siddiqui et al. (2018) [23]. This dataset comprises of 60000 time-series with 50 time-stamps each. Each sequence consists of three channels which represent values from pressure, torque, and temperature sensors. The dataset contains point anomalies in the torque and temperature signals, while the pressure signal is kept intact. A sequence is labeled as anomalous if it contains a point anomaly. The dataset is split into 45000 training sequences with 7505 anomalous sequences, 5000 validation sequences with 853 anomalous sequences, and 10000 test sequences with 1696 anomalous sequences.

4.2   Mammography [28]

This breast cancer screening dataset contains 11183 time-series and it is commonly used for classification purposes. Anomalies at certain points/features make the whole time-series anomalous. The dataset is split into 8000 training time-series with 186 anomalous time-series, 1000 validation time-series with 25 anomalous time-series, and 2183 test time-series with 49 anomalous time series.

4.3   NASA Shuttle [28]

This dataset describes radiator positions in a NASA shuttle with 9 attributes. There are 46464 time-series in this dataset. We split the dataset into 25000 training time-series with 483 anomalous time-series, 5000 validation time-series with 102 anomalous time-series, and 16464 test time-series with 293 anomalous time series.

5   Evaluation and Discussion

In order to completely assess the relevance and correctness of a given explanation, we conducted a survey in which novice and expert users were asked to evaluate the generated explanations. We provided 20 time-series from the *Machine Anomaly Detection* dataset along with the generated explanations to the participants and asked questions related to whether the generated explanation was (i) relevant, (ii) sufficient, (iii) meaningful, (iv) correct, and (v) satisfactory from

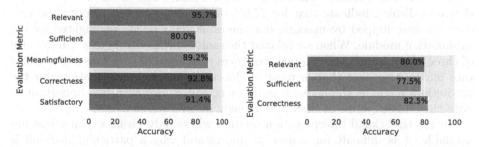

Fig. 6. Summary of evaluation done by expert (left) and novice (right) users on Machine Anomaly Detection dataset [23] explanations.

Table 1. Effect of masking the data points in *Machine Anomaly Detection* dataset which are relevant for the explanation

Window size	Anomalous sequence	Flipped prediction after masking	Percentage flipped
1	1511	1104	73.0%
3	1511	1319	87.3%

the experts. Whereas, the novice users were only questioned about (i), (ii), and (iv). There were 7 expert and 6 novice participants, who provided their binary (agree/disagree) feedback to the aforementioned question category. By analyzing the accumulated feedback of the participants shown in Fig. 6, it is clear that most of the participants considered the provided explanations relevant, meaningful, and correct. The majority of the experts were satisfied with the reasoning of the NN decision provided in the explanation. Although, 20% experts and 22.5% novice participants thought that the provided explanation is not sufficient.

In this study, we are trying to infer the causality through the provided explanations, so it is also important to assess the reliability of the generated explanations. Therefore, we introduced the sanity check module in the system pipeline to obtain a measure of confidence over the provided explanations (as explained in Sect. 3.4). We also computed this confidence estimate over the entire test set of *Machine Anomaly Detection* dataset in order to get an impression regarding the overall reliability of the generated explanations. The cumulative results are presented in Table 1. Since it is important to compute these statistics only over examples where the classification from the network was correct, we were left with 1511 out of 1696 total anomalous sequences in the test set. In the first setup, a masked sequence is generated by suppressing the exact data points for which the explanations have been generated by the *textual explanation generator* module. Since we are suppressing the exact point, we represent this setup with a window size of one. In the second setup, a sequence is masked with a window size of three covering one preceding and one following value in order to cover up any minor misalignment of the most salient region highlighted by the *textual explanation generator* module. In this case, a total of three data points were suppressed. We represent this setup with a window size of three. The results shown in Table 1 indicate that for 73.0% of the anomalous sequences, the predictions were flipped by masking out the exact data points highlighted by the explanation module. When we relaxed the sanity check criteria to a window size of three, the percentage of flipped sequences rose up to 87.3%. This high success rate makes it evident that in most of the cases, plausible explanations for the predictions made by the network could be provided. However, it is important to note that this experiment does not strongly imply causality.

In the traditional interpretation settings where only visual explanations are available, it is difficult for a user to understand why a particular decision is taken by the network just by looking at the plots. However, it is relatively easy to understand the classification reason by reading the explanations provided by

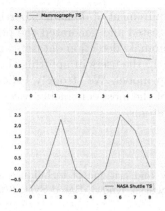

The anomalous data point(s) is/are present at index(s) 3 with value(s) of 2.5460. The data point(s) is/are anomalous because there is/are peak(s) around the mentioned index(s). There is/are 1 peak(s) in the observed time-series and the highest peak is the one detected. The anomalous data point(s) is/are 2.40 standard deviation(s) away from the mean of the series, where the mean is 0.9330 while the standard deviation is 1.0602. There is/are 4 such point(s) which crossed the mean of the series.

The anomalous data point(s) is/are present at index(s) 2, 5, 6 with value(s) of 2.29, -0.01, 2.53. The data point(s) is/are anomalous because there is/are peak(s) around the mentioned index(s). There is/are 2 peak(s) in the observed time-series and the peak at index 6 is highest. The anomalous data point(s) is/are 1.89, -0.01, 2.09 standard deviation(s) away from the mean of the series, where the mean is 0.5718 while the standard deviation is 1.2121. There is/are 4 such point(s) which crossed the mean of the series.

Fig. 7. Sample explanations generated for mammography (top) and NASA shuttle (bottom) time-series.

our system for the corresponding plots in Figs. 4 and 7. We specifically opted for statistical features due to their strong theoretical foundations and transparency. The point-wise features of an anomalous data point help a user in understanding how that data point is different from the rest of the sequence. Whereas sequence-wise features help in highlighting the overall behavior of an anomalous sequence. An end-user can confirm part of the explanation by looking at the plot, which elevates his trust in the system.

One of the major limitations of the TSXplain is its specificity for the task at hand. The computed features and the rule-base are not easily transferable between different tasks. Another limitation is regarding the availability of a suitable set of statistical features for a particular task. We leave these open questions as future work. We would also like to test our system on more complex datasets along with the employment of more sophisticated influence tracing techniques [2] to further enhance the utility of the system.

6   Conclusion

This paper proposes a novel demystification framework that generates natural language based descriptions of the decision made by the deep learning models developed for time-series analysis. The *influence tracer* module identifies the most salient regions of the input. Statistical features from the sequence are simultaneously extracted from the *statistical feature extractor* module. These two results are passed onto the *textual explanation generator* for the generation of the final explanation by the system. We used statistical features over other alternates due to their strong theoretical foundations and transparency which significantly improved the intelligibility and reliability of the provided explanations. A confidence estimate is also provided by the system using the sanity check module which is based on assessment whether the estimated influential point is indeed causal for the prediction. The generated explanations are directly intelligible for both expert and novice users alike evading the requirement of domain expertise to understand the encapsulated information. We tested and generated the proposed framework on different synthetic and real anomaly detection datasets and demonstrated our results. The explanations can help the users in uplifting their confidence regarding the performance of the deep model. We strongly believe that natural language based descriptions are one of the most convincing ways in the long-term for the realization of explainable systems.

References

1. Andrews, R., Diederich, J., Tickle, A.B.: Survey and critique of techniques for extracting rules from trained artificial neural networks. Knowl.-Based Syst. **8**(6), 373–389 (1995)
2. Bach, S., Binder, A., Montavon, G., Klauschen, F., Müller, K.R., Samek, W.: On pixel-wise explanations for non-linear classifier decisions by layer-wise relevance propagation. PLoS ONE **10**(7), e0130140 (2015)
3. Bandara, K., Bergmeir, C., Smyl, S.: Forecasting across time series databases using recurrent neural networks on groups of similar series: A clustering approach. arXiv preprint (2017)
4. Bau, D., Zhou, B., Khosla, A., Oliva, A., Torralba, A.: Network dissection: Quantifying interpretability of deep visual representations. In: Proceedings of the IEEE Conference on Computer Vision and Pattern Recognition, pp. 6541–6549. IEEE (2017)
5. Cai, Y., et al.: A high-performance and in-season classification system of field-level crop types using time-series landsat data and a machine learning approach. Remote Sens. Environ. **210**, 35–47 (2018)
6. Crabtree, B.F., Ray, S.C., Schmidt, P.M., O'Connor, P.T., Schmidt, D.D.: The individual over time: Time series applications in health care research. J. Clin. Epidemiol. **43**(3), 241–260 (1990)
7. Goldstein, M.: Anomaly detection in large datasets. Ph.D. thesis, University of Kaiserslautern, München, Germany, February 2014. http://www.goldiges.de/phd
8. Guo, P., Anderson, C., Pearson, K., Farrell, R.: Neural network interpretation via fine grained textual summarization. arXiv preprint arXiv:1805.08969 (2018)
9. Hancock, B., Bringmann, M., Varma, P., Liang, P., Wang, S., Ré, C.: Training classifiers with natural language explanations. In: Proceedings of the Conference on Association for Computational Linguistics Meeting, vol. 2018, p. 1884. NIH Public Access (2018)
10. Hendricks, L.A., Akata, Z., Rohrbach, M., Donahue, J., Schiele, B., Darrell, T.: Generating visual explanations. In: Leibe, B., Matas, J., Sebe, N., Welling, M. (eds.) ECCV 2016. LNCS, vol. 9908, pp. 3–19. Springer, Cham (2016). https://doi.org/10.1007/978-3-319-46493-0_1
11. Hyndman, R.J., Wang, E., Laptev, N.: Large-scale unusual time series detection. In: 2015 IEEE International Conference on Data Mining Workshop (ICDMW), pp. 1616–1619. IEEE (2015)
12. Kang, M.J., Kang, J.W.: Intrusion detection system using deep neural network for in-vehicle network security. PLoS ONE **11**(6), e0155781 (2016)
13. Kim, J., Rohrbach, A., Darrell, T., Canny, J., Akata, Z.: Textual explanations for self-driving vehicles. In: Ferrari, V., Hebert, M., Sminchisescu, C., Weiss, Y. (eds.) ECCV 2018. LNCS, vol. 11206, pp. 577–593. Springer, Cham (2018). https://doi.org/10.1007/978-3-030-01216-8_35

14. Kuo, C.C.J., Zhang, M., Li, S., Duan, J., Chen, Y.: Interpretable convolutional neural networks via feedforward design. J. Vis. Commun. Image Represent. **60**, 346–359 (2019)
15. Kushwaha, A.K., Dhillon, J.K.: Deep learning trends for video based activity recognition: A survey. Int. J. Sens. Wirel. Commun. Control. **8**(3), 165–171 (2018)
16. Lark, R.M., Nielsen, B.L., Mottram, T.T.: A time series model of daily milk yields and its possible use for detection of a disease (ketosis). Anim. Sci. **69**(3), 573–582 (1999)
17. Mahendran, A., Vedaldi, A.: Understanding deep image representations by inverting them. In: CVPR, pp. 5188–5196 (2015)
18. Melis, D.A., Jaakkola, T.: Towards robust interpretability with self-explaining neural networks. In: Advances in Neural Information Processing Systems, pp. 7775–7784 (2018)
19. Munir, M., Siddiqui, S.A., Chattha, M.A., Dengel, A., Ahmed, S.: FuseAD: Unsupervised anomaly detection in streaming sensors data by fusing statistical and deep learning models. Sensors **19**(11) (2019). https://doi.org/10.3390/s19112451
20. Munir, M., Siddiqui, S.A., Dengel, A., Ahmed, S.: DeepAnT: A deep learning approach for unsupervised anomaly detection in time series. IEEE Access **7**, 1991–2005 (2019)
21. Ordóñez, F., Roggen, D.: Deep convolutional and LSTM recurrent neural networks for multimodal wearable activity recognition. Sensors **16**(1), 115 (2016)
22. Saad, E.W., Wunsch II, D.C.: Neural network explanation using inversion. Neural Netw. **20**(1), 78–93 (2007)
23. Siddiqui, S.A., Mercier, D., Munir, M., Dengel, A., Ahmed, S.: TSViz: Demystification of deep learning models for time-series analysis. IEEE Access **7**, 67027–67040 (2019)
24. Simonyan, K., Vedaldi, A., Zisserman, A.: Deep inside convolutional networks: visualising image classification models and saliency maps. arXiv preprint (2013)
25. Singh, S., Majumdar, A.: Deep sparse coding for non-intrusive load monitoring. IEEE Trans. Smart Grid **9**(5), 4669–4678 (2018)
26. Srivastava, S., Labutov, I., Mitchell, T.: Joint concept learning and semantic parsing from natural language explanations. In: Proceedings of the 2017 Conference on Empirical Methods in Natural Language Processing, pp. 1527–1536 (2017)
27. Trippi, R.R., Turban, E.: Neural Networks in Finance and Investing: Using Artificial Intelligence to Improve Real World Performance. McGraw-Hill, Inc., New York City (1992)
28. Vanschoren, J., Van Rijn, J.N., Bischl, B., Torgo, L.: OpenML: networked science in machine learning. ACM SIGKDD Explor. Newsl. **15**(2), 49–60 (2014)
29. Wang, X., Smith, K., Hyndman, R.: Characteristic-based clustering for time series data. Data Min. Knowl. Discov. **13**(3), 335–364 (2006)
30. Yosinski, J., Clune, J., Fuchs, T., Lipson, H.: Understanding neural networks through deep visualization. In: International Conference on Machine Learning Workshop on Deep Learning (2015)
31. Zeiler, M.D., Fergus, R.: Visualizing and understanding convolutional networks. In: Fleet, D., Pajdla, T., Schiele, B., Tuytelaars, T. (eds.) ECCV 2014. LNCS, vol. 8689, pp. 818–833. Springer, Cham (2014). https://doi.org/10.1007/978-3-319-10590-1_53
32. Zheng, Y., Liu, Q., Chen, E., Ge, Y., Zhao, J.L.: Time series classification using multi-channels deep convolutional neural networks. In: Li, F., Li, G., Hwang, S., Yao, B., Zhang, Z. (eds.) WAIM 2014. LNCS, vol. 8485, pp. 298–310. Springer, Cham (2014). https://doi.org/10.1007/978-3-319-08010-9_33

DeepMimic: Mentor-Student Unlabeled Data Based Training

Itay Mosafi(✉) , Eli (Omid) David , and Nathan S. Netanyahu

Department of Computer Science, Bar-Ilan University, 5290002 Ramat-Gan, Israel
itay.mosafi@gmail.com, mail@elidavid.com, nathan@cs.biu.ac.il

Abstract. In this paper, we present a deep neural network (DNN) training approach called the "DeepMimic" training method. Enormous amounts of data are available nowadays for training usage. Yet, only a tiny portion of these data is manually labeled, whereas almost all of the data are unlabeled. The training approach presented utilizes, in a most simplified manner, the unlabeled data to the fullest, in order to achieve remarkable (classification) results. Our DeepMimic method uses a small portion of labeled data and a large amount of unlabeled data for the training process, as expected in a real-world scenario. It consists of a mentor model and a student model. Employing a mentor model trained on a small portion of the labeled data and then feeding it only with unlabeled data, we show how to obtain a (simplified) student model that reaches the same accuracy and loss as the mentor model, on the same test set, without using any of the original data labels in the training of the student model. Our experiments demonstrate that even on challenging classification tasks the student network architecture can be simplified significantly with a minor influence on the performance, i.e., we need not even know the original network architecture of the mentor. In addition, the time required for training the student model to reach the mentor's performance level is shorter, as a result of a simplified architecture and more available data. The proposed method highlights the disadvantages of regular supervised training and demonstrates the benefits of a less traditional training approach.

1 Introduction

Deep neural networks (DNNs) have been used lately very effectively in many applications, e.g., object detection (as described by the ImageNet challenge [9]), with state-of-the-art performance [16] exceeding human-level capabilities, natural language processing, where text translation using DNNs with attention mechanism [24] has achieved remarkable results, playing highly-complex games (such as chess [8] and Go [31]) at a grandmaster level, generation of realistic-looking images [10], etc.

The recent impressive advancement of deep learning (DL) can be attributed to a number of factors, including: (1) Enhancement of computational capabilities (e.g., using strong graphical processing units (GPUs)), (2) improvement of network architectures, and (3) acquisition of vast amounts of training data. With the

© Springer Nature Switzerland AG 2019
I. V. Tetko et al. (Eds.): ICANN 2019, LNCS 11731, pp. 440–455, 2019.
https://doi.org/10.1007/978-3-030-30493-5_44

growing availability of powerful computational capabilities, much of the research has focused on innovative network architectures for the pursuit of state-of-the-art performance in various problem domains. Some examples include: *transferable architectures* [40], which suggest a method of learning the model architectures directly on the dataset of interest, *fractional max-pooling* [11], which offers a modification to the standard max-pooling in convolutional neural networks (CNNs) [20], and *exponential linear units* (ELUs) [6], which provide a new activation function for improving learning characteristics.

In this paper, we focus mainly on the usage of large amounts of available unlabeled data to form a training method that utilizes these available resources. Specifically, we focus here on a new DeepMimic training methodology, demonstrating its effectiveness with respect to object classification based on the use of CNNs.

Occupied mainly by the performance of DNNs in numerous applications, researchers may tend to overlook various aspects of the learning process, e.g., the specific manner in which supervised learning (i.e., the training of a network using labeled data) is performed. In the case of multi-label classification, each data item is associated with a class label, and is represented by a "one-hot" encoding vector. (The dimension of a one-hot encoding vector is the number of possible classes in the dataset, such that, the correct class index contains '1' and all the other indexes contain '0'.) It is reasonable to assume that a label distribution that is different from the one-hot vector representation might gain extra insight or knowledge about the model, thereby changing significantly the training process.

To explore this idea, we need a meaningful label distribution, which we gain by using the proposed DeepMimic paradigm. In our method, we use a relatively small subset of our data to perform supervised training on our mentor model, while treating the rest of the dataset as unlabeled data, i.e., ignoring the labels completely. Once the mentor is trained, we use it to create a label distribution by outputting the softmax components for each data item in the unlabeled dataset. During the data splitting process, the one-hot labels are used merely to ensure a balanced dataset split. We later show that this might not be actually required, based on our empirical results for the unbalanced dataset, which yield the same accuracy gained for the balanced dataset.

Using the unlabeled data and the label distribution produced by the mentor model we train a student model. We are able to achieve comparable performance to the mentor's, with a student model that is simpler, shallower, and substantially faster. In other words, our method can extract a model's knowledge and successfully transfer it to another model using essentially no labeled data. These remarkable results suggest that the method presented can be used in many applications. One can take advantage of large amounts of unlabeled data and mimic a black-box trained model, without even knowing its architecture or the labeled data used for its training. For example, an individual can purchase a neural network-based product and create a copy of it, which will match the original product's performance with no access to the data used to train the product.

Finally, the student model could result in a substantially simpler architecture. Therefore, we can achieve a much faster inference time, which is very important in production for various real life systems and services.

2   Background

Many real-life problems have led to interesting, innovative DL techniques, such as those pertaining to the mentor-student learning process [37]. These methods suggest a less strict training of the mentor with the overall gain of lowering the risk of overfitting by the student. In [17] a class-distance loss is presented that assists mentor networks in forming densely-clustered vector spaces to make it easy for a student network to learn from. In [12] the authors focus on enhancing the robustness of the student network without sacrificing the performance. Model compression, originally researched in [4], presents a way of compressing the function learned by a complex model into a much smaller, faster model.

The problems addressed in this paper are considered nowadays rather simple, and thus the method should be reestablished on more challenging problems. Furthermore, years ago, when the Internet was much less developed and considerably smaller amounts of data were available, the focus was directed at the ability to generate synthetic data for training and development purposes. With tens of zettabytes ($1000^7 = 10^{21}$ bytes) of data available online, acquiring unlabeled data is no longer an issue. Currently, the main interest is to develop ways of exploiting these data efficiently.

The ability to distill the knowledge of a neural network to create a simpler and more suitable production network [2,7,14,25] is extremely valuable. During this knowledge transfer, the method of training on soft labels, i.e., using a vector of classes (whose probabilities sum up to 1) as labels, seems to provide much more information for the training process compared to the training with one-hot vectors only. This supports the notion that training based on one-hot labels may not be ideal.

Another interesting aspect of soft-label training is its use of regularization [1]. Regularization techniques for preventing overfitting and achieving better generalization consist mainly of *dropout* [15,33], i.e., randomly "shutting down" some of the neurons, *DropConnect* [35], for random cancellation of synapses between neurons in a very similar way to dropout, random noise addition [22], and weight decay [19]. These techniques are also referred to as L_1 and L_2 regularization [26]. Another work is the mixup paper [39], which shows that averaging the training examples and their labels, e.g., creating a new image and its label as a weighted average of the original two images and two one-hot vectors used as labels, to improve the regularization. It is also possible to transfer knowledge from different types of networks, e.g., a recurrent neural network (RNN) to a DNN, as shown in [5].

Mimicking a model's predictions, in order to obtain knowledge has been researched in various aspects. In [27] it is used to transfer knowledge from one domain to another, in order to generalize it and teach a reinforcement learning agent how to behave in multiple tasks simultaneously. In our case, we mimic a mentor model and try to acquire its knowledge as well; yet, we always remain in the same domain and try to maximize the student's performance there. In [30] the authors show that their method can extract the policy of a reinforcement learning agent, and train a new network, which is dramatically smaller and more efficient, while performing at a comparable level of the agent's. Thinner and deeper student models are presented in [29]; the method discussed allows using not only the outputs but the intermediate representations learned by the mentor as hints to improve the training process and the final performance of the student. In [34] it is argued that even though a student model does not have to be as deep as its mentor, it requires the same number of convolutional layers in order to learn functions of comparable accuracy. According to their results, the large gap between CNNs and fully-connected DNNs cannot be significantly reduced, as long as the student model does not consist of multiple convolutional layers.

The difference from our work is that both mentor and student models are trained over the entire dataset, i.e., there are no unique data seen only by the student, as in our case. For now, the state-of-the-art results on any visual tasks are achieved by CNNs, as the classical fully-connected DNNs simply cannot compete with it. Even though the DNN limits can be pushed further [23], they are no match for the CNN architecture which relies on local correlations in a given image. Our method may enable DNNs to overcome this boundary, since the regular training procedures which failed to do so are not used by our method. Note that we can alter the mentor model as we deem fit, and rely on the soft labels it predicts, in order to train a student model, regardless of its architecture.

3   DeepMimic Training

3.1   Data Split

When it comes to available data, our goal is to simulate real-life scenarios. In such a case, we would usually have huge amounts of unlabeled data; these data are considered useless, most of the time, unless used for training autoencoders [3], for example.

In order to simulate such a scenario, we choose a ratio between the mentor's training data and that of the student's, such that there is a sufficient amount of unlabeled data to train the student and a sufficient amount of training data for the mentor model to reach good performance on the test set. We performed this experiment on the following datasets: MNIST [21], CIFAR-10 [18], and Tiny ImageNet [36]. All the training data chosen for the student model are treated as unlabeled data, i.e., we ignore the labels as described in the next section.

The ratio chosen for the data split is 1:4, which produced the best results after testing various split ratios, and considering the need for sufficient training data for the student model (see Fig. 1). All the images are randomly assigned to create balanced datasets in most experiments. In other words, by splitting the data randomly, we ensure that for each image of a certain label in the mentor

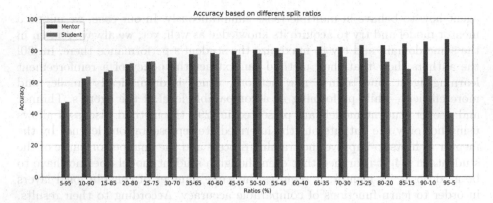

Fig. 1. Accuracy of trained mentor and student on CIFAR-10 test set as a function of dataset split ratio. Plot depicts the influence of different split ratios on the performance of mentor and student. To allocate sufficient training data to the student and still enable the mentor to reach high accuracy, we choose a 20–80 split (used in the experiments reported for the three different datasets). Presumably, similar ratios to 20–80 with smaller student datasets could work as well, but for more complicated problems, a larger dataset available for the student would probably be required; this insight could serve as a rule of thumb for desired ratios between labeled and unlabeled data.

dataset, there are four images of the same label in the student training set. This way both the mentor and the student datasets contain an equal number of images of each label, i.e., the image amount per each class is balanced. In order to simulate scenarios where the available data distribution is unknown and unbalanced, we modified in some of the experiments the student dataset by forcing, e.g., a different number of samples in each class, as described in Sect. 4.3. We did that by either removing a random number of samples from each class in the student dataset or adding a random amount of out-of-domain images to the student training set. Regardless, it seems the student is only bound to its mentor accuracy rate, i.e., even if there were huge amounts of data for the student, it could not be significantly better than its mentor. As for testing, we used the original test set of each dataset, respectively, to test both models. Since the datasets are fairly limited in size, we decided not to split the data to training, validation and testing; instead, we use all of the available data for training and testing.

3.2   Training Method

In the training process, we first start by training the mentor model using its assigned dataset. Regularization methods, such as dropout, were vastly used in order to reach high accuracy on the test set. Considering mainly classification problems, the last layer of each model is a softmax layer, which normalizes the output and provides a distribution for each possible class (with all distributions summing up to 1). The training uses a stochastic gradient descent (SGD) algorithm and the cross-entropy loss. Once the mentor is well trained, we can predict a soft label for each image in the student dataset. By doing that, we generate an estimate for each image while still ignoring all the real labels. We now train the student model, using its assigned data and the soft labels generated by the mentor model. For the student training, we also use SGD and cross entropy loss. In the student model training, regularization is less needed since training on the soft labels creates a very strong generalization in the training process [1]. The student reaches the mentor's accuracy on the test set, in all experiments. Based of the performances of shallow students on test sets, it is clear that the student architecture does not have to be similar to the mentor's, while the performance remains almost identical on the test set. In all the classification tasks we worked on, the reduced student network consistently maintained the mentor's performance.

4 Experiments

4.1 MNIST

MNIST is a relatively simple dataset containing handwritten digit images; it is ideal to perform a "sanity check" on the method. It contains 70,000 (28×28) grayscale images, 60,000 of which for training the model and the remaining 10,000 for testing it.

As mentioned in the previous section, we use 20% of the training set for the mentor training; after it is trained, we use the remaining 80% and the trained mentor model to create the soft label distributions. In the experiments reported below, we tested a student model identical to the mentor model, as well as shallower and more simplified student models. The mentor's accuracy is relative to the amount of data used for training; it is not expected to reach state-of-the-art results with only one fifth of the original training data. This is true for all models trained on a small subset of the standard dataset. As can be seen from Table 1 and Fig. 2, the Mentor and Student-A (i.e., the model with the identical architecture) reach almost identical results (i.e., identical loss and accuracy) on the test set, while all the unlabeled data used for training Student-A are never used to train the Mentor. Student-B reaches very close results, as well, i.e., it is possible to create a rather simplified student model to mimic successfully a mentor without knowing its architecture.

Table 1. Model architectures, test accuracy, and relative accuracy between Students and Mentor for MNIST dataset. Symbols: c-convolutional layer, mp-max pooling layer, fc-fully connected layer, s-softmax layer. θ^n means n consecutive layers of type θ.

Model	Architecture	Accuracy	Relative accuracy
Mentor	$c - mp - c - mp - fc^2 - s$	97.46%	–
Student-A	$c - mp - c - mp - fc^2 - s$	97.38%	99.91%
Student-B	$c - mp - fc^2 - s$	97.17%	99.70%

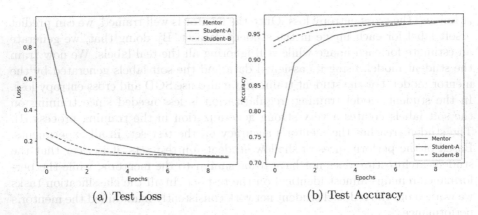

(a) Test Loss   (b) Test Accuracy

Fig. 2. Models' test loss and accuracy for MNIST dataset; even shallower Student-B model reaches almost identical results to the Mentor's.

4.2 CIFAR-10

CIFAR-10 is an established dataset used for object recognition. It consists of 60,000 (32×32) RGB images in 10 classes, with 6,000 images per class. There are 50,000 training images and 10,000 test images in the official data. We used deeper networks for this task; as before, the student networks manage to achieve very good results compared to the mentor's, using various network architectures.

Table 2. Model architectures, test accuracy, and relative accuracy between Students and Mentor for CIFAR-10 dataset. Symbols: c-convolutional layer, mp-max pooling layer, fc-fully connected layer, s-softmax layer. θ^n means n consecutive layers of type θ.

Model	Architecture	Accuracy	Relative accuracy
Mentor	$c^2 - mp - c^2 - mp - c^2 - mp - fc^2 - s$	73.14%	–
Student-A	$c^2 - mp - c^2 - mp - c^2 - mp - fc^2 - s$	73.58%	100.6%
Student-B	$c^2 - mp - c - fc^2 - s$	72.38%	98.96%
Student-C	$c^2 - mp - fc^2 - s$	69.63%	95.2%

As can be seen from Table 2 and Figs. 3 and 4, Student-A matches the Mentor's performance, and Student-B reaches a very high accuracy compared to that of the Mentor (only 0.76% lower), which serves as its only training source. Finally, Student-C still reaches good results (only 3.51% lower than the Mentor's accuracy), despite its substantially shallower architecture.

It might be of interest to observe also the training of a student model on 80% of the data using one-hot labels instead of the mentor's predictions as a simpler mentor model training. There is a limit, of course, to simplifying the model and still obtain better accuracy than the original mentor, while training merely on

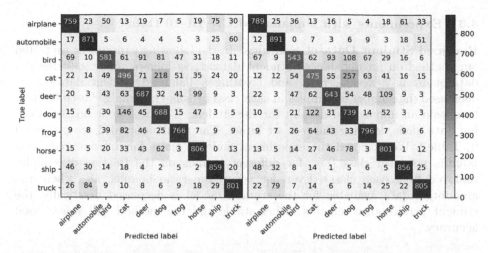

Fig. 3. Confusion matrices of Mentor (LHS) and Student-A (RHS) for CIFAR-10 dataset. Note model's frequency of true and false class predictions for each and every class in test set. This helps understand the degree of confusion in the model, with respect to certain classes, e.g., the model sometimes mistakes a Dog for a Cat and vice versa. A confusion matrix is much more informative than an accuracy measurement. Note that where Mentor tends to make mistakes, so does Student, i.e., they are very similar in all aspects. This best illustrates the successful knowledge transfer from Mentor to Student.

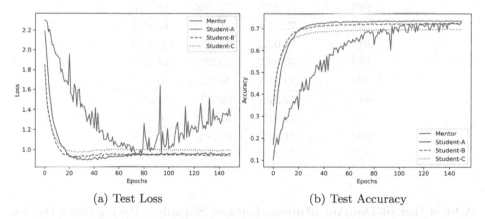

(a) Test Loss   (b) Test Accuracy

Fig. 4. Models' test loss and accuracy over 150 epochs for CIFAR-10 dataset; Students are averaged over multiple runs to show consistent results. In contrast to Mentor's spiky and increasing loss function, Student models remain steady and consistent, owing to the very strong regularization of soft label training.

20% of the data. In our case, Student-B and Student-C reach accuracy rates of 77.22% and 72.64%, respectively, while the original Mentor reaches an accuracy rate of 73.14%. Note that the models described have four times more data to train on with simpler architectures.

4.3   Experiments with Unbalanced CIFAR-10 Data

Reduced Student Dataset Samples. In the following experiment we tested our method on an unbalanced student dataset, as follows. After splitting the dataset by a 20%–80% ratio and creating a balanced student dataset, we decreased the number of samples in each class by some randomly chosen fraction to obtain an unbalanced dataset for the student. This was done on the training data alone, keeping the test set intact. The results obtained are presented in Table 3. We executed the experiment multiple times for different reduction bounds per each class (i.e., for different bounds on the fraction of samples removed from each class). Even for very large ratio bounds, i.e., where the amount of data available for the student model is decreased drastically, the student performance remains rather stable and the method still shows good accuracy.

Table 3. Student accuracy using DeepMimic with unbalanced CIFAR-10 dataset (due to removal of data samples). Each entry is an average over multiple runs. Training is based on Mentor reaching 72.92% accuracy on test set. All models were trained over 150 epochs.

Ratio bound	Student-A	Student-B	Student-C
5%	73.01%	71.74%	69.28%
10%	73.29%	71.38%	68.79%
20%	72.45%	71.34%	68.98%
30%	73.02%	71.32%	68.58%
40%	72.92%	70.73%	68.36%
50%	73.02%	70.72%	68.15%
60%	72.32%	69.77%	67.16%
70%	72.21%	69.35%	66.62%
80%	71.86%	69.36%	66.62%
90%	72.33%	69.64%	66.94%

Added Out-of-Domain Student Dataset Samples. Having shown that an unbalanced dataset for the student model (generated by removing at random large amounts of samples from the balanced dataset) has little effect on the performance, we now demonstrate the effect of adding "out-of-domain" random data to the student dataset, by testing our models on this newly created dataset. Specifically, the student dataset is modified by adding samples whose labels are very different from the categories contained in the CIFAR-10 dataset, so as to ensure non-related data to the student dataset. The labels of the added samples are, for example, Flowers, Food Containers, Fruits and Vegetables, Household Electrical Devices and Furniture, Trees, Insects, and others, taken from the CIFAR-100 dataset. As before, we use for each experiment a specified fraction limit per each class on the number of samples added at random from the other categories. The results are presented in Table 4; as can be seen, the models perform very well, reaching good accuracy with no disruption caused by the addition of out-of-domain data.

Table 4. Student accuracy using DeepMimic with unbalanced CIFAR-10 dataset (due to added data samples). Each entry is an average of multiples runs. Training is based on Mentor reaching 72.92% accuracy on test set. All models were trained over 150 epochs.

Ratio	Student-A	Student-B	Student-C
5%	73.36%	71.2%	68.96%
10%	73.42%	71.45%	69.04%
20%	73.36%	71.68%	69.01%
30%	73.17%	71.49%	69.35%
40%	73.30%	71.37%	69.05%
50%	73.20%	71.53%	68.96%
60%	73.16%	71.58%	69.04%

4.4 Tiny ImageNet

The Tiny ImageNet dataset is the most challenging dataset we have applied our method on. The training data consists of 100,000 (64 × 64) RGB images in 200 classes, with 500 images per class. There are 10,000 images in the validation set and in the test set. As can be seen from Table 5, the architecture used for the networks is much deeper. This makes it possible to demonstrate the effect of removing a substantial amount of layers without having almost a negative impact on the model's performance. Note that Student-B and Student-C have much simpler architectures, yet, their obtained results are very close to the Mentor's.

As can be seen in [38], obtaining over 55% accuracy on the test set is an impressive result; in contrast, a random guess yields only 0.5% accuracy. Therefore, and considering that only a fifth of the original training data is used for training, obtaining over 20% accuracy on the test set for the Mentor is satisfactory, as well. The result demonstrates our method's effectiveness for this dataset. Table 5 and Fig. 5 show that both Student-A and Student-B definitely match the Mentor's performance. Student-C is the shallower model we use. Still, it achieves only 0.85% less accuracy than the Mentor's, attesting to the method's effectiveness and impressive results, even when applied to highly-complex and involved datasets. Figures 6 and 7 contain images classified correctly by both the Mentor and Student-A and images classified differently by the Mentor and Student-A, respectively.

Table 5. Model architectures, test accuracy, and relative accuracy between Students and Mentor for Tiny ImageNet dataset. Symbols: c-convolutional layer, bn-batch norm layer, d-dropout layer, fc-fully connected layer, s-softmax layer. θ^n means n consecutive layers of type θ.

Model	Architecture	Accuracy	Relative accuracy
Mentor	$(c - bn - d)^9 - fc - bn - d - s$	20.45%	–
Student-A	$(c - bn - d)^9 - fc - bn - d - s$	20.47%	100.09%
Student-B	$(c - bn - d)^6 - fc - bn - d - s$	20.51%	100.29%
Student-C	$(c - bn - d)^3 - fc - bn - d - s$	19.60%	95.84%

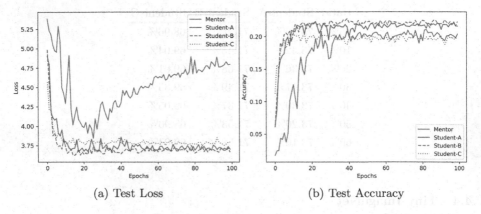

(a) Test Loss   (b) Test Accuracy

Fig. 5. Models' test loss and accuracy over 100 epochs for Tiny ImageNet dataset; Students and Mentor are averaged over multiple runs.

(a) Plate   (b) Ladybug   (c) Candle

(d) Snorkel   (e) Viaduct   (f) Espresso

Fig. 6. Images successfully classified by both Mentor and Student-A.

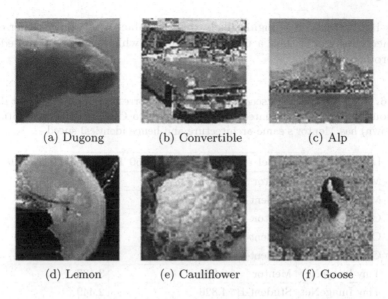

(a) Dugong   (b) Convertible   (c) Alp

(d) Lemon   (e) Cauliflower   (f) Goose

Fig. 7. Images classified successfully by Mentor, but incorrectly by Student-A as: (a) Sea Cucumber, (b) Sports Car, (c) Seacoast, (d) Banana, (e) Brain Coral, and (f) Albatross. Although Mentor and Student are very similar in knowledge, they are not identical.

4.5   Inference Time Measurements

We have tested also comparative inference times (in seconds) for each student model versus its associated mentor, running on the test sets that correspond to the datasets experimented with (see Table 6). Each model was tested on two different GPU architectures, with the results averaged over 100 executions. When using a more complex and deeper network, which is usually the case in real-life scenarios, the time reduction is more significant, and may allow for much faster data processing. Sometimes the student seems to slightly surpasses the mentor; this behavior was observed mostly for student models which are replicas of the mentor, or a student with relatively little reduction in architecture. Determining whether a smaller, albeit less accurate model, should be used versus a larger, more accurate model, is an interesting question. For DNN-based cloud services, the answer would probably be never, as such services usually rely on very strong and expensive hardware, so we would not be limited by any restrictions and just use the most accurate model. However, embedded devices which usually do not rely on strong hardware or stable internet connection, e.g., a cell phone or an IOT (Internet of things) device, are mostly more limited as far as size, memory, and power. The manufacturers would usually develop an extremely small and less powerful hardware, in order to keep the product small, elegant, and rather inexpensive. The mentioned limitations are quite problematic when one is interested in deploying a massive model on a product. In such scenarios, creating a significantly smaller and faster model would enable to deploy it on a smaller hardware, so it is highly likely that manufacturers would rather employ a less accurate model than a more accurate one which cannot be embedded in their products.

Table 6. Inference times (in seconds) on test sets corresponding to different datasets for various models and associated mentors, using two GPU architectures. Student-A (not shown) has Mentor's same architecture and hence identical speed.

Dataset	Model	GeForce Gtx 1050 Ti	GeForce Gtx 1070
MNIST	Mentor	0.551	0.652
MNIST	Student-B	0.399	0.609
CIFAR-10	Mentor	1.637	1.097
CIFAR-10	Student-B	1.275	0.922
CIFAR-10	Student-C	1.129	0.859
Tiny ImageNet	Mentor	6.328	3.137
Tiny ImageNet	Student-B	4.826	2.449
Tiny ImageNet	Student-C	4.089	2.194

5   Conclusion

In this paper, we have presented a novel approach for training deep neural networks. Our DeepMimic method relies on utilizing two models, which are not necessarily identical. We have shown that reducing the student model's complexity has a minor effect on its success rate compared to the mentor's. According to this empirical evidence, it is possible to mimic a black-box mentor model with an unknown architecture and reach the same accuracy. In a series of experiments, we have shown that for both balanced and unbalanced training data available for the student, the method manages to mimic the mentor model successfully. One only needs to exploit large amounts of unlabeled data, which is the expected scenario in real-life situations. Our method raises serious security implications, as one can "duplicate" a proprietary neural network, by creating a copy of it without having access to the original training data. The method presented yields impressive results and exploits large amounts of unlabeled data for training, without having to manually tag them. We have worked solely on CNNs for both the mentor and student models. Our method can be further extended and used to explore the relations between different types of networks, e.g., a fully-connected network and a CNN.

This could prove as a key factor to obtain, extract, and transfer knowledge between different types of networks, thereby pushing further the performance level.

6   Future Work

As can be seen from Table 5, the larger the network, the easier it is to reduce its size more significantly with low reduction in accuracy. In such cases, the effect on the inference time is more noticeable and such compressed networks have an advantage, as shown in Table 6. Therefore, we would prefer to test our method on deeper networks such as VGG [32] and ResNet [13], expecting to create models with even more improved inference times. So far we have experimented mainly with CNNs for classification problems, but it is of interest to explore the effect of DeepMimic in other problem domains, e.g., networks designed for detection and segmentation. Such networks usually perform feature extraction on the input and rely on massive architectures to do so, we expect our method to be very beneficial in these domains.

An additional idea that might lead to a much smaller, yet a more accurate student, is to distill multiple mentor models into a single student model. By doing so, the student training data can be increased by using multiple mentors to generate the data or we could average different mentor predictions to make the student hopefully more accurate.

An interesting work regarding CNN classifiers using low-shot learning is given in [28]. The idea is to enable a model to successfully classify a newly seen category after being presented with merely few training examples. This notion resembles the way human vision works using imprinted weights. The authors use a CNN as an embedding extractor, and after a classifier is trained, the embedding vectors of new low-shot examples are used to imprint weights for new classes in the extended classifier. As a result, the new model is able to classify well examples belonging to a novel category after seeing only a few examples. Combining this work and DeepMimic might be very interesting, in the following sense. While using a mentor model trained on specific categories, upon the arrival of a novel category it might be easier to implant the new category in a student model combining the two processes described in DeepMimic and [28]. It is possible that a student model would adjust more naturally to new categories during the training process itself rather than an already trained model.

References

1. Aghajanyan, A.: SoftTarget regularization: an effective technique to reduce over-fitting in neural networks. In: Proceedings of the IEEE International Conference on Cybernetics, Exeter, UK, pp. 1–5 (2017). https://doi.org/10.1109/CYBConf.2017.7985811
2. Ba, J., Caruana, R.: Do deep nets really need to be deep? In: Advances in Neural Information Processing Systems, Montreal, Quebec, Canada, vol. 27, pp. 2654–2662 (2014)
3. Baldi, P.: Autoencoders, unsupervised learning, and deep architectures. In: Proceedings of the ICML Workshop on Unsupervised and Transfer Learning, Edinburgh, Scotland, vol. 27, pp. 37–50 (2012)

4. Buciluă, C., Caruana, R., Niculescu-Mizil, A.: Model compression. In: Proceedings of the 12th ACM SIGKDD International Conference on Knowledge Discovery and Data Mining (KDD), pp. 535–541, Philadelphia, Pennsylvania. ACM (2006). https://doi.org/10.1145/1150402.1150464
5. Chan, W., Ke, N.R., Lane, I.: Transferring knowledge from a RNN to a DNN. arXiv preprint arXiv:1504.01483 (2015)
6. Clevert, D.A., Unterthiner, T., Hochreiter, S.: Fast and accurate deep network learning by exponential linear units (ELUs). In: International Conference on Learning Representations, San Juan, Puerto Rico (2016)
7. Correia-Silva, J.R., Berriel, R.F., Badue, C., de Souza, A.F., Oliveira-Santos, T.: Copycat CNN: stealing knowledge by persuading confession with random non-labeled data. In: International Joint Conference on Neural Networks, Rio de Janeiro, Brazil, pp. 1–8 (2018). https://doi.org/10.1109/IJCNN.2018.8489592
8. David, O.E., Netanyahu, N.S., Wolf, L.: DeepChess: end-to-end deep neural network for automatic learning in chess. In: Villa, A.E.P., Masulli, P., Pons Rivero, A.J. (eds.) ICANN 2016. LNCS, vol. 9887, pp. 88–96. Springer, Cham (2016). https://doi.org/10.1007/978-3-319-44781-0_11
9. Deng, J., Dong, W., Socher, R., Li, L.J., Li, K., Fei-Fei, L.: ImageNet: A large-scale hierarchical image database. In: Proceedings of the IEEE International Conference on Computer Vision and Pattern Recognition, Miami Beach, Florida, pp. 248–255 (2009). https://doi.org/10.1109/CVPR.2009.5206848
10. Goodfellow, I., et al.: Generative adversarial nets. In: Advances in Neural Information Processing Systems, Montreal, Quebec, Canada, pp. 2672–2680 (2014)
11. Graham, B.: Fractional max-pooling. arXiv preprint arXiv:1412.6071 (2014)
12. Guo, T., Xu, C., He, S., Shi, B., Xu, C., Tao, D.: Robust student network learning. arXiv preprint arXiv:1807.11158 (2018)
13. He, K., Zhang, X., Ren, S., Sun, J.: Deep residual learning for image recognition. In: Proceedings of the Conference on Computer Vision and Pattern Recognition, Las Vegas, Nevada, pp. 770–778 (2016). https://doi.org/10.1109/CVPR.2016.90
14. Hinton, G., Vinyals, O., Dean, J.: Distilling the knowledge in a neural network. arXiv preprint arXiv:1503.02531 (2015)
15. Hinton, G.E., Srivastava, N., Krizhevsky, A., Sutskever, I., Salakhutdinov, R.R.: Improving neural networks by preventing co-adaptation of feature detectors. arXiv preprint arXiv:1207.0580 (2012)
16. Hu, J., Shen, L., Sun, G.: Squeeze-and-excitation networks. arXiv preprint arXiv:1709.01507 (2017). https://doi.org/10.1109/CVPR.2018.00745
17. Kim, S.W., Kim, H.E.: Transferring knowledge to smaller network with class-distance loss. In: International Conference on Learning Representations Workshop (2017)
18. Krizhevsky, A.: Learning multiple layers of features from tiny images. Technical report, University of Toronto (2009)
19. Krogh, A., Hertz, J.A.: A simple weight decay can improve generalization. In: Advances in Neural Information Processing Systems, pp. 950–957 (1992)
20. LeCun, Y., Bottou, L., Bengio, Y., Haffner, P.: Gradient-based learning applied to document recognition. Proc. IEEE 86(11), 2278–2324 (1998). https://doi.org/10.1109/5.726791
21. LeCun, Y., Cortes, C., Burges, C.J.: MNIST handwritten digit database (2010). http://yann.lecun.com/exdb/mnist/
22. Li, Y., Liu, F.: Whiteout: Gaussian adaptive noise regularization in feedforward neural networks. arXiv preprint arXiv:1612.01490 (2016)

23. Lin, Z., Memisevic, R., Konda, K.: How far can we go without convolution: improving fully-connected networks. arXiv preprint arXiv:1511.02580 (2015)
24. Luong, M., Pham, H., Manning, C.D.: Effective approaches to attention-based neural machine translation. In: Empirical Methods in Natural Language Processing, Lisbon, Portugal, pp. 1412–1421 (2015). https://doi.org/10.18653/v1/D15-1166
25. Mosafi, I., David, O.E., Netanyahu, N.S.: Stealing knowledge from protected deep neural networks using composite unlabeled data. In: Proceedings of the International Joint Conference on Neural Networks, Budapest, Hungary (2019)
26. Ng, A.Y.: Feature selection, L_1 vs. L_2 regularization, and rotational invariance. In: Proceedings of the International Conference on Machine Learning, Banff, Alberta, Canada, p. 78 (2004). https://doi.org/10.1145/1015330.1015435
27. Parisotto, E., Ba, J.L., Salakhutdinov, R.: Actor-Mimic: Deep multitask and transfer reinforcement learning. In: International Conference on Learning Representations, San Juan, Puerto Rico (2016)
28. Qi, H., Brown, M., Lowe, D.G.: Low-shot learning with imprinted weights. In: Proceedings of the Conference on Computer Vision and Pattern Recognition, Salt Lake City, Utah, pp. 5822–5830 (2018). https://doi.org/10.1109/CVPR.2018.00610
29. Romero, A., Ballas, N., Kahou, S.E., Chassang, A., Gatta, C., Bengio, Y.: FitNets: hints for thin deep nets. In: International Conference on Learning Representations, San Diego, California (2015)
30. Rusu, A.A., et al.: Policy distillation. In: International Conference on Learning Representations, San Juan, Puerto Rico (2016)
31. Silver, D., et al.: Mastering the game of Go without human knowledge. Nature, 354–359 (2017). https://doi.org/10.1038/nature24270
32. Simonyan, K., Zisserman, A.: Very deep convolutional networks for large-scale image recognition. arXiv preprint arXiv:1409.1556 (2014)
33. Srivastava, N., Hinton, G., Krizhevsky, A., Sutskever, I., Salakhutdinov, R.: Dropout: a simple way to prevent neural networks from overfitting. J. Mach. Learn. Res. **15**, 1929–1958 (2014)
34. Urban, G., et al.: Do deep convolutional nets really need to be deep and convolutional? In: International Conference on Learning Representations, Toulon, France (2017)
35. Wan, L., Zeiler, M., Zhang, S., Le Cun, Y., Fergus, R.: Regularization of neural networks using DropConnect. In: Proceedings of the International Conference on Machine Learning, Atlanta, Georgia, pp. 1058–1066 (2013)
36. Wu, J., Zhang, Q., Xu, G.: Tiny ImageNet Challenge (2017). http://cs231n.stanford.edu
37. Yang, C., Xie, L., Qiao, S., Yuille, A.L.: Knowledge distillation in generations: more tolerant teachers educate better students. arXiv preprint arXiv:1805.05551 (2018)
38. Yao, L., Miller, J.: Tiny ImageNet classification with convolutional neural networks (2015). http://cs231n.stanford.edu
39. Zhang, H., Cisse, M., Dauphin, Y.N., Lopez-Paz, D.: mixup: beyond empirical risk minimization. In: Proceedings of the International Conference on Learning Representations, Vancouver, British Columbia, Canada (2018)
40. Zoph, B., Vasudevan, V., Shlens, J., Le, Q.V.: Learning transferable architectures for scalable image recognition. arXiv preprint arXiv:1707.07012 (2017). https://doi.org/10.1109/CVPR.2018.00907

Evaluation of Tag Clusterings for User Profiling in Movie Recommendation

Guglielmo Faggioli[1]([⊠]), Mirko Polato[1][iD], Ivano Lauriola[1,2][iD], and Fabio Aiolli[1][iD]

[1] Department of Mathematics, University of Padova,
Via Trieste, 63, 35121 Padova, Italy
{gfaggiol,mpolato,aiolli}@math.unipd.it, ivano.lauriola@phd.unipd.it
[2] Bruno Kessler Foundation, Via Sommarive, 18, 38123 Trento, Italy

Abstract. In the web 2.0 era, tags provide an effective mechanism to rapidly annotate and categorize items. However, tags suffer from many problems typically linked to language, like synonymy, polysemy, and ambiguity in general. To overcome this limitation, tag clustering can be used to group tags that represent similar concepts. One of the domains where tag clustering has shown to be particularly useful is the *movie recommendation*, where tags are used to represent users' preferences and affinities. In this context it is not yet available a golden standard that can prove the quality of a clustering technique, especially considering that the final aim is the users' satisfaction rather than an accuracy-like score. To this end, we propose an evaluation criterion for the quality of the resulting clusters based on human judgments.

Keywords: Movie recommendation · Tag clustering · Human judgment · Word embeddings

1 Introduction

Tagging usually refers to the association of a relevant keyword or phrase with an entity. In the Web 2.0 era, tags are widely used to annotate resources, such as, pictures, movies, tweet and so on. Nowadays, collaborative (a.k.a. social) tagging is almost a must-have feature for content sharing web applications enabling users to contribute to the community by adding new free-text keywords (i.e., tags). Tags are useful for resource discovery because they provide meaningful descriptions of resources and help users in finding what they really want. Tags are also very important to profile users. For example, a Recommender System (RS) can take advantage of the available tags to describe the user through the most popular tags that characterize the resources the user liked/consumed in the past. Because of the collaborative nature of social tagging applications, they are dynamic and able to grasp changes in the vocabulary as well as absorbing trends.

© Springer Nature Switzerland AG 2019
I. V. Tetko et al. (Eds.): ICANN 2019, LNCS 11731, pp. 456–468, 2019.
https://doi.org/10.1007/978-3-030-30493-5_45

Although tags bring many benefits, there are several obstacles that may hinder their application in web services. The main issue regards the free-form nature of tagging and the lack of explicit semantics which gives rise to problems, such as, syntactic variations, typographical errors, polysemy, synonym, and hypernym/hyponym. Moreover, different users may use different words to tag the same resource, mainly because users focus their attention on different aspects. For example, in a tagging system for movies, a user can annotate a movie using keywords related to the general plot, while other users can use keywords related to specific scenes. In order to alleviate some of the aforementioned issues, tag clustering has lately attracted the research community interest. Tag clustering can be defined as the process of grouping tags in such a way that members of the same tag cluster are perceived by users as related to each other. Clearly, the underlying meaning of this *relation* highly depends on the context at hand.

The literature offers different approaches for dealing with tag clustering. For example, graph-based approaches [3,4,14] have been hugely applied to tag clustering in the past. While, the two most successful standard clustering methods in this context have been K-Means [12,19,21], and hierarchical agglomerative clustering [8,11]. Especially, the latter is particularly suited for this task because of the intrinsic hierarchical nature of tags. [7] offers a comprehensive quantitative comparison of many tag clustering algorithms. However, besides the merits of each specific algorithm, a very important aspect that must be taken into account is the way tags are represented in the first place. Different representations can lead to highly different clusters even if the underlying algorithm is fixed. Moreover, there is not a Gold Standard to evaluate tag clustering, and different domain applications can require different evaluation procedures.

For these reasons, in this work we compare different tag representation approaches applied in the context of a movie recommender system. Specifically, the task at hand consists in developing a user profile that can be explored by the user. A user profile is expressed in terms of movie-related tags (e.g., popular tags for the movies already seen by the user) where tag clustering is used for grouping semantic-related tags. As mentioned previously, there not exists a recognized best practice to objectively judge the quality of a clustering method. Therefore, since the system is user-oriented, we have investigated with real users how they perceive the quality of different tag clusters, and whether there exist tag representations that lead to better clustering from a human perspective. In the study we fixed as clustering algorithm the hierarchical agglomerative clustering [7,8,11], and we compared two families of representations: (*i*) graph-based (domain-aware) representations, and (*ii*) word-embeddings (domain agnostic), namely, word-2-vec [13] and GloVe [15]. Differently from [7] which are based on objective measures, we compare the representations directly with real users. The study has been carried out by means of a user-study in which a user, given a specific tag, had to select among two clusters (generated with two different tag representations) the one that, in her opinion, best suits the given tag.

In short, results of our analysis show that general-purpose word embeddings pre-trained on large corpora better meet the user preferences.

2   Use Case: A Movie Recommender Systems

A specific use case where the tag clustering algorithm and underlying representations are extremely relevant is the recommender system domain. In Fig. 1 it is possible to observe a GUI (Graphical User Interface) of a system developed to aid the user and explain her the system's knowledge about herself. The idea behind this system is to describe users considering their affection toward a cluster of tags, identified as a latent "topics". The interface shows a polar chart that describes how the system assumes the user to be keen on a specific topic (top left of the figure). The user can observe which tags are part of the clustering that forms the selected topic, by exploring the (top) right-hand side of the interface. Additionally, the user can learn which movies have generated such belief about the topic, by observing movies in the lower part of the GUI. Finally, if the user desires to further explore a specific topic, she can receive recommendations on movies that can be labelled with said topic [6]. It is clear that, for similar systems, being able to correctly cluster tags, means being capable of grasping specific intrinsic aspects of the environment.

3   Representations and Embeddings

To define possible tag representations, we introduce the following notation. We define a tag t any human readable keyword or phrase that can be associated with a resource r. $\mathcal{T}$ represents the set of all possible tags, while $\mathcal{R}$ describes the set of possible resources. Note that, the relation between tags and resources is clearly *many-to-many*. A single resource can be described by a number of different tags, while a tag can be associated to many resources. We define $\mathcal{R}_t$ the set of resources associated to tag t. Several representations have been used to evaluate the tag clustering approach.

These representations are the tags co-occurrences in the resource space (*co-R*), tags co-occurrences in the tag space (*co-T*), word-2-vec (*w2v*) and GloVe.

3.1   Domain Aware Representations

As observed in [3], a typical approach to represent resources and tags' associations, consists in using bipartite graphs. Let $\mathcal{E} = \{(i,j) | (i \in \mathcal{R} \wedge j \in \mathcal{T}) \vee (i \in \mathcal{T} \wedge j \in \mathcal{R})\}$ be the set of edges that links a tag with a resource or viceversa. We define $G(\mathcal{T}, \mathcal{R}, \mathcal{E})$ the bipartite graph between tags and resources (see Fig. 2). Note that, in the specific setting of resource tagging, the bipartite graph is usually undirected, because the connection between tags and resources is typically bidirectional.

Without loss of generality, by using bipartite graphs, we can represent a tag $t \in \mathcal{T}$ in the space of resources as the probability of a random walk with an odd number of steps ending in a specific resource. Formally, the representation $\boldsymbol{\tau} \in \mathcal{S}^{|\mathcal{R}|}$ for tag t, where $\mathcal{S} \equiv \{\mathbf{v} \in \mathbb{R}^{|\mathcal{R}|} \mid \|\mathbf{v}\|_1 = 1\}$, has a component τ_i for each resource r_i, where τ_i is the probability of a Markov chain starting from the

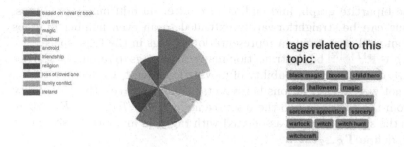

Movies you have seen related to this topic:

About this topic you might also like:

Fig. 1. Screenshot of an application that relies on tag clustering to describe users.

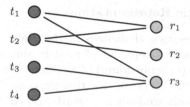

Fig. 2. Example of the tags-resources bipartite graph.

node t of the bipartite graph, and ending in r_i after an odd number of steps. This reasoning can be straightforwardly extended to an even number of steps for the random walk, producing a representation of tags in the tags' space.

Let $\mathbf{T}_{\mathcal{T} \to \mathcal{R}} \in \mathbb{R}^{|\mathcal{T}| \times |\mathcal{R}|}$ be the transition matrix from tags to resources. Entries (i, j) of $\mathbf{T}_{\mathcal{T} \to \mathcal{R}}$ describe the probability of moving form tag t_i to the resource r_j. Since we do not weight associations between tags and resources, the probability of moving from the tag node t to the resource node r is $1/|\mathcal{R}_t|$, $\forall r \in \mathcal{R}_t$, where $\mathcal{R}_t$ describes the set of resources associated with tag t. In an analogously way it is possible to define $\mathbf{T}_{\mathcal{R} \to \mathcal{T}}$.

One-Step Markov Chain: co-R. Consider the case when a single step on the bipartite graph is taken from a tag t to a resource r. In this case, we represent t in the resource space and the representation τ_t for tag t corresponds to the t-th row of $T_{\mathcal{T} \to \mathcal{R}}$. In this kind of representation, we observe how likely is it to observe a resource associated with the tag we want to represent. Computing the similarity with this kind of representation measures how likely is to observe two tags together. A particular aspect that makes this representation vulnerable is the bias toward popular tags: tags that appear often will tend to be similar to many others. This can be a strong limitation since it might produce less stable clusters.

Two-Steps Markov Chain: co-T. Consider now the representation obtained by walking two steps on the bipartite graph (i.e., tag→resource→tag). Similar to *co-R* we can compute the *co-T* representation by $\mathbf{T}' = \langle \mathbf{T}_{\mathcal{T} \to \mathcal{R}}, \mathbf{T}_{\mathcal{R} \to \mathcal{T}} \rangle$, where $\langle \cdot, \cdot \rangle$ indicates the dot-product. Entries (i, j) of $\mathbf{T}'$ corresponds to the probability of a Markov chain that starts in tag t_i ends in tag t_j. The value $T'_{i,j}$ corresponds to the probability of observing a tag, given that the other is present.

From a semantic perspective, the rationale behind this representation consists in representing in a similar way tags that appear with the same tags. We consider this representation relevant due to the fact that, while users do not tag resources with synonyms and identical concepts, they will likely use similar keywords in similar contexts, therefore with similar tags.

Although this reasoning can be extended to an arbitrary number of steps by multiplying alternatively $\mathbf{T}_{\mathcal{T} \to \mathcal{R}}$ and $\mathbf{T}_{\mathcal{R} \to \mathcal{T}}$, we argue that more than two steps will lead to representations that cannot be directly interpreted.

3.2 Domain Agnostic Representations

One of the emerging research challenges in the Natural Language Processing (NLP) community is the development of models able to define a suitable and compact representation (or embedding) of words. Several algorithms have been recently introduced in the literature for this purpose [5,13,15]. The two main categories of these methods are known as word- and character-level embeddings. The former aims at learning a feature vector which relates a given word with its ordinary context and its semantic relations with other words. The latter learns a representation which depends on the inner structure of the input word, i.e. the sequence of characters which compose it.

On one hand, in the case of word-level representations, i.e. word embeddings, two words are similar if they have a semantic relation. On the other hand, two words have similar character-level representations if they are visibly similar. For instance, the word-level similarity between the words *king* and *queen* is high, whereas their character-level similarity is low. As a counter-example, the words *weather* and *whether* have low word-level similarity but high character-level similarity. See [5,9,16,20] for some recent applications of word- and character-embeddings.

Despite this introductory distinction, character-level embeddings have not been considered in this work due to the nature of movie tags. Indeed, the semantic connection between tags is more important than their writing similarity. In this work, two word-embeddings have been used, which are the popular word-2-vec[13] algorithm, and GloVe [15].

word-2-vec. Word-2-vec is an unsupervised algorithm used to learn a general-purpose semantic representation of words and n-grams. The algorithm relies on a shallow single hidden layer neural architecture. Two models derive from the algorithm, which are Continuous Bag-Of-Word (CBOW) and Skip-gram. The objective of Skip-gram is to predict the context given a word, whereas the CBOW model predicts a word given its context. The algorithm scans large corpora to acquire probabilities related to the co-occurrences of words, and n-grams in general, in the same sliding window. However, this kind of model suffers from the disadvantage that it do not operate directly on the co-occurrence statistics of the corpus [15].

Global Vectors (GloVe). The GloVe algorithm overcomes the aforementioned issue by pre-computing statistics of co-occurrences for a given window dimension. Consequently, the training is performed on the basis of these statistics. As is the case of word-2-vec, GloVe is a general-purpose unsupervised method. It combines the advantages of the two major model families in the literature, which are the global matrix factorization and local context window methods.

4   Experimental Assessment

To evaluate different representations and test how users interact with different approaches, we applied tag clustering in the specific domain of movies. Information about the association between tags and movies have been gathered from The Movie Database (TMDb[1]). TMDb is a crowed sourced movie and TV database. It represents one of the richest sources of metadata about movies with details of over 26778 movies. Among metadata that can be mined from TMDb, tags are the most relevant to our task. Tags are generated and applied to movies by users.

---

[1] https://www.themoviedb.org/.

17057 tags are available on TMDb. Since many tags are applied to a very small number of movies, we ignored all tags that appear less than 5 times throughout the entire dataset. This left us with 4543 unique tags. Table 2 shows examples of associations between movies and tags in TMDb.

In our user study we used a pre-trained version of word-2-vec based on *Wikipedia* and *PubMed* corpora (publicly available[2]). This model represents words in a 200 dimensional latent space. Similarly, for GloVe we used a pre-trained model on *Wikipedia* corpus, and has embeddings of size 250 [10]. Since we used pre-trained embeddings, tags available in the TMDb dataset needed to be pre-processed in order to have a correspondent matches in these representations. In particular, tag pre-processing includes the following steps:

1. split the tag in words and remove punctuation;
2. for each word check whether it is included into the pre-trained model;
3. capitalize the first letter and go to (2);
4. considering again the original word, substitute punctuation with spaces;
5. for each new word apply (2) and (3).

Pre-trained models have been trained to represent either single words and highly popular n-grams. Because of that, a special case regards those tags formed by more than one word, that corresponds to roughly the 43% of the selected ones. In this case, a particular representation is needed, since, in general, they are not available in the model. Multi-words tags are represented as the average representation of each contained word. This problem doesn't subsists in *ad-hoc* bipartite-graph based representation, since multi-words tags are considered as a whole. At the end of the pre-processing, misspelled tags and few non-English words were not available in the pre-trained models. In particular, 9 words in word-2-vec model and 8 words in GloVe have not been found. For these words, we assigned a random representation. The chosen clustering algorithm has been *hierarchical agglomerative clustering*. The agglomerative clustering is the most natural clustering approach due to the fact that tags can be easily sorted according to their specificity. General tags can include more specific ones, producing a hierarchical structure that can be captured by the chosen algorithm. Additionally, this decision is supported by a rich literature [1,8,17]. A second aspect that needs to be addressed in developing a hierarchical clustering is the linkage method. Table 1 shows the comparison among clusters' dimensions statistics according to different representations. It is possible to observe that, while single linkage produces always clusters with the highest variance in size, *complete linkage* is the strategy that produces the most evenly sized clusters. The only exception is the one step Markov chain (*co-R*) representation, with *average linkage* that produces lowest variance. In our experiments we used the configuration which achieve lowest variance. The number of clusters has been set to 1000.

---

[2] http://evexdb.org/pmresources/vec-space-models/.

Table 1. Variation in clusters' features using different representations and linkage strategies. Median, maximum and standard deviation refer to clusters' size in number of tags. Minimum size is 1 for all representations and linkage.

Representation	Linkage	Median	Max	Std	Number of singletons
co-R	Single	1	3463	109.42	931
	Average	4	**20**	**2.58**	14
	Complete	3	1328	41.90	**9**
co-T	Single	1	3249	102.65	815
	Average	4	48	4.38	37
	Complete	4	**32**	**3.50**	**7**
w2v	Single	1	3343	105.63	888
	Average	2	195	10.55	406
	Complete	3	**45**	**4.80**	**197**
GloVe	Single	1	3454	109.14	941
	Average	2	702	23.8	432
	Complete	3	**90**	**6.28**	**133**

Table 2. Examples of movies and associated tags in TMDb.

Movie title	Tags
Gladiator	parent child relationship, chariot, philosopher, slavery, fictionalized history, barbarian horde, 2nd century, successor, senate, gladiator, roman empire, emperor, battlefield, combat, rome italy, ancient world, arena
Inception	dream, hotel, spy, danger, rescue, mission, philosophy, imagination, fistfight, dream world, virtual reality, los angeles, zero gravity, car crash, allegory, subconscious, paris france, idea
Titanic	love, inspired by true events, tragic love, panic, ship, class differences, imax, disaster, star crossed lovers, iceberg, titanic, ocean liner, steerage, rich snob, rich woman poor man, epic

5   Data Collection

The aforementioned representations have been evaluated by considering users' preferences regarding the output of the clustering system.

Humans' preferences have been collected through the Amazon Mechanical Turk (AMT) [2,18] service. In short, the AMT is a crowdsourcing service which allows a requester to use human intelligence to perform tasks that computers are currently unable to do. These tasks could be, for instance, annotations of documents, labelling images, and so on. The annotations are conducted by participants (or workers) through ad-hoc Human Intelligence Tasks (HITs), which are described in the following. Given an input tag $t \in \mathcal{T}$, the clusters containing the tag t obtained by using two representations have been proposed to the user. The users have been asked to accurately inspect the two clusters, and to select the one which best fits the input tag. Alternatively, in case of indecision, the user had the possibility to not express its preference by selecting a third option. Some examples of HITs are reported in Table 3.

The first obstacle with the data collection process is that the user could abstract from the context of movies, and his selection could not consider the relation between tags and associated movies. To reduce this issue, users have been minutely instructed with a clear explanation of the objective, the goals, and the relation between tags and movies. Moreover, some examples have been shown to clarify ambiguities (see Table 2).

Table 3. Examples of HITs.

HIT id	Input tag	Cluster 1	Cluster 2
1	Motel	mall, street, restaurant, apartment, tenement, inn, farmhouse, cafe, casino, diner	highway, lover, murderer, nevada, hitchhike
2	Cycling	race, horse racing, bicycle, horse, street race, horse race, horse track, bike, racing	doping, bicycle
3	Christmas eve	precocious child, christmas party, new year's eve, home alone, reindeer, christmas tree, scrooge, christmas, bishop, little boy	thanksgiving, weekend, save the day, valentine's day, stag night, halloween, christmas, dinner, night watchman, night shift

In order to guarantee a high quality of the collected data, HITs have been restricted to workers with an approval rating on AMT equal or greater than 85%. The time allotted for each task has been set to 4 min. Only a single submission per participant was allowed.

The total amount of available HITs is a combinatorial expression consisting of the number of input tags times the possible pairwise representations. In this case, 4543 tags have been used, with resulting 54516 different HITs. A random sub-sample of 1500 HITs has been considered to alleviate the effort on the data acquisition process. However, 57 of the selected HITs contain at least one cluster with a single element, which corresponds to the input tag. These trivial HITs have not been administered to workers, and the comparison of their associated representations has been automatically solved in favour of the non-trivial cluster. One reason why trivial HITs have been automatically solved is that singletons do not provide valuable information, and they are useless in a recommendation context. Hence, we consider singletons as bad clusters a-priori.

In order to limit the cognitive load on workers, in case of large clusters, at most 10 tags have been shown. These tags have been randomly selected.

6   Discussion

The quality of a representation has been defined as the ratio between the number of HITs with a positive vote for the representation, and the number of HITs which involve the representation. Draws, i.e. the HITs where the workers do not express a preference, are considered apart.

The results of the human judgement are reported in Fig. 3. The results have been analyzed with and without the effect of the trivial HITs. What is striking in the figure is the similar trend of preferences for each representation. The best result is achieved by the word-2-vec algorithm, with only 52.8% of winning HITs. However, word-2-vec is the method which suffers the most from the trivial HITs ratings (55.7% of winning HITs when trivial results have not considered). In other words, one of the main problems of this method is that it produces several singletons, as it is also shown in Table 1.

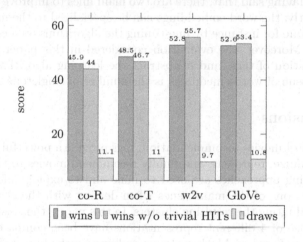

Fig. 3. Preferences expressed (in percentage) in the HITs.

A pairwise comparison between representations is presented in Table 4. Here, the quality score of pairwise representations has been considered, i.e. the ratio between the winning HITs against the opponent representation, and the total number of HITs which involve the pair. The pairwise comparison provides us further meaningful food for thought. It is possible to summarize these results in three points:

– classical represenations achieve comparable results, i.e. 50.7% of winning co-R against co-T. There is not a significant improvement;
– word embeddings also achieve similar results among them, especially when considering trivial HITs in the evaluation (50.4% in favour of word-2-vec);
– word embeddings achieve always better results against classical representations. The two highest peaks are obtained by word-2-vec against co-R (62.1%), and GloVe against co-T (57.3%).

Table 4. Quality pairwise comparison. Left: the ij entry contains the percentage of winning HITs achieved by the representation i (row) while compared against the representation j (column). Right: the same approach when including the trivial HITs.

	co-R	co-T	w2v	GloVe		co-R	co-T	w2v	GloVe
co-R	–	49.3	37.9	44.7	co-R	–	49.3	42.1	46.4
co-T	50.7	–	47.1	42.7	co-T	50.7	–	49.8	45.3
w2v	62.1	52.9	–	52.5	w2v	57.9	50.2	–	50.4
GloVe	55.3	57.3	47.5	–	GloVe	53.6	54.7	49.6	–

However, the results of evaluation do not clearly highlight which representation is better in this specific context. Simple domain aware and sophisticated domain agnostic methods achieve comparable results even if they produce different clusters. Having said that, there are two main lines to improve the users' satisfaction. Firstly, the word embeddings can be specialized to the movies domain. This can be done for instance by fine-tuning the algorithms on a corpus containing synopsis. Moreover, the evaluation considered in this paper concerns only the representation of tags, and it abstract the learning algorithm. Indeed, one of the main issue of word-emeddings is the number of singletons.

7   Conclusions

In the context of movies recommendation, tags provide a powerful mechanism to annotate, organize, represent, and finally recommend movies to the users. Generally, clustering approaches are used to mine and to exploit information from tags. However, one of the main issues when dealing with these methods is the difficult of evaluating the resulting clusters. In this paper, the clusters developed through the use of 4 different representations have been compared. These representations are based on Markov chains and word embeddings. The evaluation has been carried out by considering personal opinions and users' preferences on the resulting clusters. Results show that general-purpose word-embeddings best fit the users' preferences. In the near future, we plan to intensively exploit further humans' opinions, aiming at creating a dataset of preferences which can be used to develop semi-supervised clustering algorithms able to produce results that are more satisfying for the users.

References

1. Brooks, C.H., Montanez, N.: Improved annotation of the blogosphere via autotagging and hierarchical clustering. In: Proceedings of the 15th International Conference on World Wide Web, pp. 625–632. ACM (2006). https://doi.org/10.1145/1135777.1135869
2. Buhrmester, M., Kwang, T., Gosling, S.D.: Amazon's mechanical turk: a new source of inexpensive, yet high-quality, data? Perspect. Psychol. Sci. **6**(1), 3–5 (2011). https://doi.org/10.1177/1745691610393980

3. Cui, J., Li, P., Liu, H., He, J., Du, X.: A neighborhood search method for link-based tag clustering. In: Huang, R., Yang, Q., Pei, J., Gama, J., Meng, X., Li, X. (eds.) ADMA 2009. LNCS (LNAI), vol. 5678, pp. 91–103. Springer, Heidelberg (2009). https://doi.org/10.1007/978-3-642-03348-3_12

4. Cui, J., Liu, H., He, J., Li, P., Du, X., Wang, P.: TagClus: a random walk-based method for tag clustering. Knowl. Inf. Syst. **27**(2), 193–225 (2011). https://doi.org/10.1007/s10115-010-0307-y

5. Devlin, J., Chang, M.W., Lee, K., Toutanova, K.: BERT: pre-training of deep bidirectional transformers for language understanding. arXiv preprint arXiv:1810.04805 (2018)

6. Faggioli, G., Polato, M., Aiolli, F.: Tag-based user profiling: a game theoretic approach. In: Proceedings of the 27th ACM Conference on User Modelling, Adaptation And Personalization (UMAP 2019) (2019). https://doi.org/10.1145/3314183.3323462

7. Garcia-Plaza, A.P., Zubiaga, A., Fresno, V., Martinez, R.: Reorganizing clouds: a study on tag clustering and evaluation. Expert Syst. Appl. **39**(10), 9483–9493 (2012). https://doi.org/10.1016/j.eswa.2012.02.108

8. Gemmell, J., Shepitsen, A., Mobasher, B., Burke, R.D.: Personalization in folksonomies based on tag clustering. In: Association for the Advancement of Artificial Intelligence (AAAI 2008) (2008). https://doi.org/10.1007/978-3-540-85836-2_19

9. Levy, O., Goldberg, Y.: Neural word embedding as implicit matrix factorization. In: Advances in Neural Information Processing Systems, pp. 2177–2185 (2014)

10. Li, B., et al.: Investigating different syntactic context types and context representations for learning word embeddings. In: Proceedings of the 2017 Conference on Empirical Methods in Natural Language Processing, pp. 2411–2421 (2017). https://doi.org/10.18653/v1/D17-1257

11. Li, X., et al.: Inducing taxonomy from tags: an agglomerative hierarchical clustering framework. In: Zhou, S., Zhang, S., Karypis, G. (eds.) ADMA 2012. LNCS (LNAI), vol. 7713, pp. 64–77. Springer, Heidelberg (2012). https://doi.org/10.1007/978-3-642-35527-1_6

12. Liu, Z., Bao, J., Ding, F.: An improved k-means clustering algorithm based on semantic model, pp. 30:1–30:5 (2018). https://doi.org/10.1145/3148453.3306269

13. Mikolov, T., Sutskever, I., Chen, K., Corrado, G.S., Dean, J.: Distributed representations of words and phrases and their compositionality. In: Advances in Neural Information Processing Systems, pp. 3111–3119 (2013)

14. Papadopoulos, S., Kompatsiaris, Y., Vakali, A.: A graph-based clustering scheme for identifying related tags in folksonomies. In: Bach Pedersen, T., Mohania, M.K., Tjoa, A.M. (eds.) DaWaK 2010. LNCS, vol. 6263, pp. 65–76. Springer, Heidelberg (2010). https://doi.org/10.1007/978-3-642-15105-7_6

15. Pennington, J., Socher, R., Manning, C.: Glove: global vectors for word representation. In: Proceedings of the 2014 Conference on Empirical Methods in Natural Language Processing (EMNLP), pp. 1532–1543 (2014). https://doi.org/10.3115/v1/D14-1162

16. Santos, C.D., Zadrozny, B.: Learning character-level representations for part-of-speech tagging. In: Proceedings of the 31st International Conference on Machine Learning (ICML 2014), pp. 1818–1826 (2014)

17. Shepitsen, A., Gemmell, J., Mobasher, B., Burke, R.: Personalized recommendation in social tagging systems using hierarchical clustering. In: Proceedings of the 2008 ACM Conference on Recommender Systems, RecSys 2008, pp. 259–266. ACM, New York (2008). https://doi.org/10.1145/1454008.1454048

18. Sorokin, A., Forsyth, D.: Utility data annotation with Amazon mechanical turk. In: 2008 IEEE Computer Society Conference on Computer Vision and Pattern Recognition Workshops, pp. 1–8. IEEE (2008). https://doi.org/10.1109/CVPRW. 2008.4562953

19. Tang, J.: Improved k-means clustering algorithm based on user tag. J. Converg. Inf. Technol. **5**, 124–130 (2010). https://doi.org/10.4156/jcit.vol5.issue10.16

20. Turian, J., Ratinov, L., Bengio, Y.: Word representations: a simple and general method for semi-supervised learning. In: Proceedings of the 48th Annual Meeting of the Association for Computational Linguistics, pp. 384–394. Association for Computational Linguistics (2010)

21. Yang, J., Wang, J.: Tag clustering algorithm LMMSK: improved k-means algorithm based on latent semantic analysis. J. Syst. Eng. Electron. **28**(2), 374–384 (2017). https://doi.org/10.21629/JSEE.2017.02.18

Special Session: Deep Learning in Image Reconstruction

Special Session: Deep Learning in Image Reconstruction

Deep learning (DL) has recently gained a lot of attention as superior method in solving challenging problems in machine learning. DL algorithms have successfully tackled many computer vision problems. Recently, researchers have started to apply DL to image reconstruction for various modalities. A bottleneck in this approach arises from the lack of labeled data for training large-scale networks. To overcome this, simulations of the forward model for a given imaging modality have been used to create sufficient data for training. The reconstruction problem can then be formulated as a supervised learning problem. Taking the forward computation into account, such approaches typically result in a kind of autoencoder for solving the inverse problem in image reconstruction, thus incorporating additional knowledge in form of the forward model into the data-driven approach.

This special session aimed at bringing together researchers working in this field. By discussing the trade-offs between the data-driven approach of DL and the traditional model-driven reconstruction algorithms, we hoped to improve our understanding of strengths and weaknesses of DL algorithms in image reconstruction. Overall, making better use of the underlying model of the imaging modality at hand is expected to lead to smarter algorithms for data-driven image reconstruction.

Organization

Werner Dubitzky	Helmholtz Zentrum München (GmbH), Germany
Keiichi Ito	Helmholtz Zentrum München (GmbH), Germany
Carlos Garcia Perez	Helmholtz Zentrum München (GmbH), Germany
Wolfgang zu Castell	Helmholtz Zentrum München (GmbH), Germany

A Sparse Filtering-Based Approach for Non-blind Deep Image Denoising

Rafael G. Pires[1] , Daniel S. Santos[2] , Gustavo B. Souza[1] ,
Alexandre L. M. Levada[1] , and João Paulo Papa[2](✉)

[1] UFSCar - Federal University of São Carlos, São Carlos, Brazil
rafapires@gmail.com, gustavo.botelho@gmail.com,
alexandre.levada@gmail.com
[2] UNESP - São Paulo State University,
Av. Eng. Luiz Edmundo Carrijo Coube, 14-01, Bauru, SP 17033-360, Brazil
danielfssantos1@gmail.com, joao.papa@unesp.br

Abstract. During the image acquisition process, some level of noise is usually added to the data mainly due to physical limitations of the sensor, and also regarding imprecisions during the data transmission and manipulation. Therefore, the resultant image needs to be further processed for noise attenuation without losing details. In this work, we attempt to denoise images using the advantage of sparse-based encoding and deep networks. Experiments on public images corrupted by different levels of Gaussian noise support the effectiveness of the proposed approach concerning some state-of-the-art image denoising approaches.

1 Introduction

Noise is an undesirable artifact usually present in images acquired by onboard cameras in satellites, and it is often generated by physical limitations of the acquisition sensor or even by unsuitable environmental conditions. These issues, however, are often unavoidable in practical situations, which turns out the presence of noise in images a prevalent problem. Such degradation deliberates a bad aspect to images; meanwhile, it can also impose difficulties to computational tasks such as edge detection, segmentation, and image classification. Denoising images is challenging since the noise is related to the high-frequency content, i.e., the details of the image [5]. The goal, therefore, is to find a compromise between suppressing the noise and not losing too much important information.

In the past decades, filter-based image denoising techniques figured as the most relevant ones, such as the well-known Inverse and Median filters, as well as the Wiener Filter [2]. Nowadays, non-linear approaches based on deep models have been successfully used in many real-world problems. Regarding the task of image restoration, it is worth noting the use of Multi-Layer Perceptron networks

The authors are grateful to Capes, CNPq grant #306166/2014-3, and FAPESP grants #2014/16250-9, #2014/12236-1, 2016/19403-6, and #2018/21934-5.

© Springer Nature Switzerland AG 2019
I. V. Tetko et al. (Eds.): ICANN 2019, LNCS 11731, pp. 471–482, 2019.
https://doi.org/10.1007/978-3-030-30493-5_46

(MLP) [1,9], Convolutional Neural Networks (CNN) [6,13], Denoising Autoencoders (DA), and Stacked Denoising Autoencoders (SDA) [11,12].

Another approach that has received attention recently concerns the sparse coding, which is the study of algorithms that attempt to learn sparse representations of the data. Such techniques have been extensively employed in several fields of image processing, such as compressed sensing, image classification, face recognition, super-resolution, and image restoration. Sparse coding methods reconstruct the signal by performing simple linear combinations of an overcomplete dictionary with reasonable results in several applications. Also, the dictionary can be learned from the data itself, as proposed by Elad and Aharon [3].

Although sparse coding and deep networks have been widely used in a number of applications in the literature, only a few works have explored their synergy in the context of image denoising. Xie et al. [12], for instance, proposed to combine sparse coding with Stacked Denoising Autoencoders (SDA) by considering the Kullback-Leibler divergence in the loss function to induce sparsity. However, the proposed approach requires fine-tuning parameters, which is usually time-consuming and data-dependent.

In this work, we propose an approach that makes use of both sparse coding theory and deep networks for robust image denoising. We used the Sparse Filtering (SF) technique [8] to learn sparse representations in a Denoising Autoencoder (DA) that is employed as a pre-trained model for further feeding an SDA network. In a nutshell, the Sparse Filter is used to induce sparsity in the hidden layers of DA, besides having one parameter only. To the best of our knowledge, SF has never been used together with DA for image denoising.

Experiments on public images corrupted by different levels of Gaussian noise support the effectiveness of the proposed approach overcoming state-of-the-art deep learning restoration techniques. The remainder of this paper is organized as follows. Section 2 presents a background theory concerning Denoising Autoencoders, Stacked Denoising Autoencoders, and Sparse Filters. Section 3 describes the proposed approach, and Sect. 4 presents the methodology adopted in this work. Finally, Sects. 5 and 6, state the experiments and conclusions, respectively.

2 Theoretical Background

2.1 Denoising Autoencoder

The process of corrupting images is usually formulated as follows:

$$g = f + \eta, \tag{1}$$

where $g \in \Re^{m \times n}$ and $f \in \Re^{m \times n}$ denote the corrupted and clean images, respectively. Additionally, the term $\eta \in \Re^{m \times n}$ stands for the additive noise (usually a Gaussian noise).

A Denoising Autoencoder stands for a two-layered neural network (i.e., a visible and a hidden unit $h \in \Re^z$) that aims to generate clean images from their noisy versions. As depicted in Fig. 1, given a corrupted image g, the network model attempts to recover the original image f by means of a learning step that minimizes a loss function $L(\hat{f}, f)$ (e.g., the Mean Square Error - MSE), where $\hat{f}$ denotes the denoised (i.e., reconstructed) image. The learning procedure aims at computing a set of encoding (Θ_{enc}) and decoding parameters (Θ_{dec}) that are further used to obtain the smoothed image, such that $\Theta_{enc} = \{W_{enc}, b_{enc}\}$ and $\Theta_{dec} = \{W_{dec}, b_{dec}\}$. In this case, $W_{enc} \in \Re^{mn \times z}$ and $W_{dec} \in \Re^{z \times mn}$ denote the weight matrices concerning the encoder and decoder, respectively. Similarly, $b_{enc} \in \Re^z$ and $b_{dec} \in \Re^z$ stand for biases regarding the encoder and decoder, respectively.

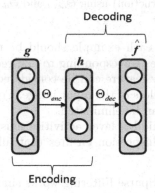

Fig. 1. DA architecture: a corrupted image g is used to feed the model that aims at learning the encoding and decoding weights that minimize a loss function that considers the similarity between the reconstructed image and its clean version.

2.2 Stacked Denoising Autoencoder

An SDA is composed of two or more stacked DAs, as shown in Fig. 2. In this example, the neural architecture comprises three hidden layers (i.e., h_1, h_2, and h_3) and a set of encoding, i.e., $\Theta_{enc1} = \{W_{enc1}, b_{enc1}\}$ and $\Theta_{enc2} = \{W_{enc2}, b_{enc2}\}$, and decoding, i.e., $\Theta_{dec1} = \{W_{dec1}, b_{dec1}\}$ and $\Theta_{dec2} = \{W_{dec2}, b_{dec2}\}$, weights.

Once again, the corrupted image g is used as input to the model, which employs the hidden layers h_1 and h_2 for the encoding step, and further layers h_2 and h_3 for the decoding process. Similarly, the model uses a loss function to minimize the difference between the reconstructed image and its clean version.

2.3 Sparse Filtering

According to Ngiam et al. [8], SF is an efficient technique that is based on three fundamental principles of sparsity:

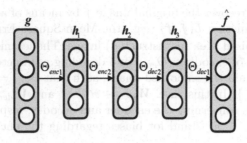

Fig. 2. SDA architecture: firstly, a corrupted image g is encoded by Θ_{enc1} and Θ_{enc2}, for further decoding (reconstruction) using Θ_{dec2} and Θ_{dec1}.

- **Population sparsity:** each example should be represented by only a few active (non-zero) features corresponding to non-redundant image areas.
- **Lifetime sparsity:** any feature can be considered meaningful if it is active most of the time, regardless of the input image. Constant inactivations stand for redundancies that can be eliminated.
- **High Dispersal:** the hidden layer activities must be roughly the same for different images. High hispersion ensures that all (non-zero) features have equal contribution.

Roughly speaking, the Sparse Filtering works similarly to a neural network, i.e., it comprises a hidden layer that aims at computing a set of weights to learn an input-output model. The main difference is that SF involves first normalizing the data with respect to the feature distribution and then to the data itself.

3   Proposed Approach

In this section, we describe the proposed approach for image denoising, which consists of three main steps: the first two aim at training two different DAs, while the last step attempts to learn a final SDA. The Sparse Filtering is employed during the DA learning procedure only (i.e., the first two steps), whose weights are transferred to the SDA model for a fine-tuning step.

For the sake of explanation, we denote DA$_1$ and DA$_2$ as the denoisining autoencoders used in the first two steps, as described earlier. In this case, $\Theta_1 = (\Theta_{enc1}, \Theta_{dec1})$ and $\Theta_2 = (\Theta_{enc2}, \Theta_{dec2})$ stand for the set of parameters concerning DA$_1$ and DA$_2$, respectively.

Since a denoising autoencoder comprises two phases, i.e., an encoding and a decoding, we train the encoder part of DA$_1$ using the Sparse Filtering background. The idea is to learn the set of parameters Θ_1 that minimizes the following loss function:

$$J^1(\mathcal{G}, \mathcal{F}; \Theta_{enc1}, \Theta_{dec1}) = J^1_{SF}(\mathcal{G}; \Theta_{enc1}) + J^1_{MSE}(\mathcal{G}, \mathcal{F}; \Theta_{enc1}, \Theta_{dec1}), \quad (2)$$

where J^1_{SF} and J^1_{MSE} stand for the loss functions that induce sparsity and reduces the reconstruction error, respectively.

The loss function J_{SF}^1, which basically minimizes the activation values of the hidden layer, can be computed as follows:

$$J_{SF}^1(\mathcal{G}; \Theta_{enc1}) = \frac{1}{M} \sum_{i=1}^{M} \hat{h}^{(i)}, \tag{3}$$

where $\mathcal{G} = (g^{(1)}, g^{(2)}, \ldots, g^{(M)})$ stands for the set of input images, $g^{(i)}$ corresponds to i^{th} corrupted image, M denotes the number of images that are available for training purposes, and $\hat{h}^{(i)} = \tilde{h}^{(i)}/||\tilde{h}^{(i)}||$ such that $\tilde{h}^{(i)}$ represents the normalized version of $h^{(i)}$. In this case, $\tilde{h}^{(i)}$ is obtained by dividing each term $h_j^{(i)}$ by β_j, where $\beta_j = \sum_{i=1}^{M} h_j^{(i)}$, $j = 1, 2, \ldots, z$.

To compute the activation of each hidden unit, we employed a Rectified Linear Unit [7] activation function (τ), which is computed as follows:

$$h^{(i)} = \tau(g^{(i)} W_{enc1} + b_{enc1}) = \max(0, g^{(i)} W_{enc1} + b_{enc1}). \tag{4}$$

After learning the set of parameters Θ_{enc1}, we can now proceed to the decoding step of DA$_1$, which aims at learning Θ_{dec1}. In this case, we used another loss function than that one described in Eq. 3, since we not interested in inducing sparsity, but minimizing the reconstruction error instead. The loss function used in the decoding step is given as follows:

$$J_{MSE}^1(\mathcal{G}, \mathcal{F}; \Theta_{enc1}, \Theta_{dec1}) = \frac{1}{2M} \sum_{i=1}^{M} ||f^{(i)} - \hat{f}^{(i)})||^2, \tag{5}$$

where $f^{(i)}$ and $\hat{f}^{(i)}$ stand for the ith input (clear) image and its reconstructed version such that $\mathcal{F} = \{f^{(1)}, f^{(2)}, \ldots, f^{(M)}\}$. Notice that $\hat{f}^{(i)} = \Psi_1(g^{(i)})$, where $\Psi_1(\cdot)$ stands for the function that describes the entire DA$_1$:

$$\Psi_1(g^{(i)}) = h^{(i)} W_{dec1} + b_{dec1}. \tag{6}$$

Algorithm 1 implements the learning process described earlier concerning the encoding and decoding parts of DA$_1$. Lines 1–2 initialize the set of parameter for the encoding and decoding steps. The main loop in Lines 3–7 is in charge of iterating the learning procedure until the criterium of the maximum number of epochs is met. The inner loop in Lines 4–7 considers all training samples for updating the set of parameters Θ_{enc1} and Θ_{dec1}. Line 5 considers the loss function presented in Eq. 3, while Lines 6–7 make use of the loss function described in Eq. 5.

The next step is to train DA$_2$, i.e., to learn the set of parameters Θ_2. This procedure works similarly to the one described for DA$_1$, but now using the following loss function:

$$J^2(\mathcal{H}, \bar{\mathcal{H}}; \Theta_{enc2}, \Theta_{dec2}) = J_{SF}^2(\mathcal{H}; \Theta_{enc2}) + J_{MSE}^2(\mathcal{H}, \bar{\mathcal{H}}; \Theta_{enc2}, \Theta_{dec2}), \tag{7}$$

where $\mathcal{H} = \{h^{(1)}, h^{(2)}, \ldots, h^{(z)}\}$ stands for the set of hidden layer activations of DA$_1$, which are now used as an input for the DA$_2$ model. Roughly speaking, the

---

Algorithm 1. Learning process of DA_1.

Input : set $\mathcal{F}$ of clean images, set $\mathcal{G}$ of corrupted images, α_{MSE} (learning rate of J^1_{MSE}), α_{SF} (learning rate of J^1_{SF}), E (number of epochs), B (number of batches), and M (size of each training batch)

Output: Θ_{enc1} and Θ_{dec1}

1   $\Theta_{enc1} \leftarrow U(0,1)$;
2   $\Theta_{dec1} \leftarrow U(0,1)$;
3   **for** $e \in \{1,2,\ldots,E\}$ **do**
4   **for** $m \in \{1,2,\ldots,B\}$ **do**
5         $\Theta_{enc1} \leftarrow \Theta_{enc1} + \frac{\alpha_{SF}}{M} \frac{\partial J^1_{SF}(\mathcal{G};\Theta_{enc1})}{\partial \Theta_{enc1}}$;
6         $\Theta_{enc1} \leftarrow \Theta_{enc1} + \frac{\alpha_{MSE}}{M} \frac{\partial J^1_{MSE}(\mathcal{F},\mathcal{G};\Theta_{enc1},\Theta_{dec1})}{\partial \Theta_{enc1}}$;
7         $\Theta_{dec1} \leftarrow \Theta_{dec1} + \frac{\alpha_{MSE}}{M} \frac{\partial J^1_{MSE}(\mathcal{F},\mathcal{G};\Theta_{enc1},\Theta_{dec1})}{\partial \Theta_{dec1}}$;

---

training process of DA_2 is pretty much similar to the one of DA_1, but now we replace $\mathcal{G}$ by $\mathcal{H}$ and $\mathcal{F}$ by $\bar{\mathcal{H}}$. In this case, $\bar{\mathcal{H}} = \{\bar{\boldsymbol{h}}^{(1)}, \bar{\boldsymbol{h}}^{(2)}, \ldots, \bar{\boldsymbol{h}}^{(z)}\}$ denotes the output of DA_2, where $\bar{\boldsymbol{h}}^{(i)} \in \Re^{z'}$ can be computed as follows:

$$\bar{\boldsymbol{h}}^{(i)} = \tau(\boldsymbol{h}^{(i)}\boldsymbol{W}_{enc2} + \boldsymbol{b}_{enc2}).\tag{8}$$

Notice that DA_2 also figures a sparsity-based learning in the encoding step, similarly to DA_1. In this case, the loss function $J^2_{SF}(\mathcal{H};\Theta_{enc2})$ is in charge of such process, and it can be formulated as follows:

$$J^2_{SF}(\mathcal{H};\Theta_{enc2}) = \frac{1}{M} \sum_{i=1}^{M} \hat{\bar{\boldsymbol{h}}}^{(i)},\tag{9}$$

$\hat{\bar{\boldsymbol{h}}}^{(i)} = \tilde{\bar{\boldsymbol{h}}}^{(i)}/\|\tilde{\bar{\boldsymbol{h}}}^{(i)}\|$ such that $\tilde{\bar{\boldsymbol{h}}}^{(i)}$ represents the normalized version of $\bar{\boldsymbol{h}}^{(i)}$. In this case, $\tilde{\bar{\boldsymbol{h}}}^{(i)}$ is obtained by dividing each term $\bar{h}^{(i)}_j$ by $\bar{\beta}_j$, where $\bar{\beta}_j = \sum_{i=1}^{M} \bar{h}^{(i)}_j$, $j = 1, 2, \ldots, z'$.

Further, the decoding step of DA_2 takes place, but now using the loss function $J^2_{MSE}(\mathcal{H}, \bar{\mathcal{H}}; \Theta_{enc2}, \Theta_{dec2})$, which is given as follows:

$$J^2_{MSE}(\mathcal{H}, \bar{\mathcal{H}}; \Theta_{enc2}, \Theta_{dec2}) = \frac{1}{2M} \sum_{i=1}^{M} \|\boldsymbol{h}^{(i)} - \hat{\boldsymbol{h}}^{(i)}_R)\|^2,\tag{10}$$

where $\hat{\boldsymbol{h}}^{(i)}_R$ stands for the reconstructed version if $\boldsymbol{h}^{(i)}$. Notice that $\hat{\boldsymbol{h}}^{(i)}_R = \Psi_2(\boldsymbol{h}^{(i)})$, where $\Psi_2(\cdot)$ stands for the function that describes the entire DA_2:

$$\Psi_2(\boldsymbol{h}^{(i)}) = \hat{\boldsymbol{h}}^{(i)}_R \boldsymbol{W}_{dec2} + \boldsymbol{b}_{dec2}.\tag{11}$$

Finally, the last step concerns using an SDA to combine both DA_1 and DA_2 into a single deep model for image denoising. Given that we have Θ_1 and Θ_2

learned in the previous steps, we can then build a model like the one depicted in Fig. 2. Therefore, we go for an extra round of learning process by feeding up the SDA model with the very same training data used before and use another loss function to fine-tune the model:

$$J_{MSE}^3(\mathcal{G},\mathcal{F};\Theta_1,\Theta_2) = \frac{1}{2M}\sum_{i=1}^{M}||\boldsymbol{f}^{(i)} - \hat{\boldsymbol{f}}^{(i)})||^2,\qquad(12)$$

where $\hat{\boldsymbol{f}}^{(i)} = \Psi_4(\Psi_3(\boldsymbol{h}^{(i)}))\boldsymbol{W}_{dec1}+\boldsymbol{b}_{dec1}$, $\Psi_4(\Psi_3(\boldsymbol{h}^{(i)})) = \tau(\Psi_3(\boldsymbol{h}^{(i)})\boldsymbol{W}_{dec2}+\boldsymbol{b}_{dec2})$, and $\Psi_3(\boldsymbol{h}^{(i)}) = \tau(\boldsymbol{h}^{(i)}\boldsymbol{W}_{enc2} + \boldsymbol{b}_{enc2})$. Algorithm 2 implements the learning process of the proposed approach described earlier, which is called Stacked Denoising Autoencoder Sparse Filtering (SDASF).

Algorithm 2. Learning process of SDASF.

Input : $\Omega_{enc} = \{\Theta_{enc1}, \Theta_{enc2}\}$, $\Omega_{dec} = \{\Theta_{dec1}, \Theta_{dec2}\}$, set of clean images $\mathcal{F}$, set of corrupted images $\mathcal{G}$, α_{MST} (learning rate of J_{MSE}^3), E (number of epochs), B (number of batches), and M (size of each training batch)

Output: Ω_{enc} and Ω_{dec}

1 **for** $e \in \{1, 2, \ldots, E\}$ **do**
2   **for** $m \in \{1, 2, \ldots, B\}$ **do**
3         $\Omega_{enc} \leftarrow \Omega_{enc} + \frac{\alpha_{MSE}}{M}\frac{\partial J_{MSE}^3(f,g;\Omega_{enc},\Omega_{dec})}{\partial \Omega_{enc}}$;
4         $\Omega_{dec} \leftarrow \Omega_{dec} + \frac{\alpha_{MSE}}{M}\frac{\partial J_{MSE}^3(f,g;\Omega_{enc},\Omega_{dec})}{\partial \Omega_{dec}}$;

---

4 Methodology

In this section, we present the methodology used to evaluate the proposed approach. The learning procedure presented in the previous section was defined over the entire image, but we can also use patches to turn the whole process faster. Therefore, we build a training dataset composed of 180,000 gray-scale patches of size 7×7 from images obtained from *Caltech* dataset [4][1]. Notice that all training patches were normalized[2] accordingly to Burger et al. [1], and their sizes were chosen empirically.

The learning step requires both the clean image and its corrupted version. Therefore, we added Gaussian noise with different severities to all images of the dataset, thus ending up in three distinct experiments. We considered Gaussian noise with standard deviations within the range $\sigma = \{15, 25, 35\}$. The trained models were further used to denoise eight satellite images obtained from the Internet, which are displayed in Fig. 3.

---

[1] Pictures of objects belonging to 101 categories, and about 40 to 800 images per category.

[2] The normalization consists in subtracting all pixel values by 0.5 and dividing them by 0.2.

Fig. 3. Remote sensing images used in this work.

The proposed approach was compared against three state-of-the-art machine learning denoising techniques:

- **MLP**: a plain Multi-Layer Perceptron applied for denoising tasks by Burger et al. [1];
- **CNN**: a bidimensional Convolutional Neural Network applied for denoising tasks by Jain and Seung [6];
- **SSDA**: a Stacked Sparse Denoising Autoencoder with sparsity constraints proposed by Xie et al. [12]; and
- **ND**: a simple comparison between a noise-contaminated patch and its noiseless version to be used as baseline, i.e., no denoising (ND) is applied in this case.

As mentioned earlier, the proposed approach SDASF comprises three phases:

- DA_1: a denoising autoencoder with architecture as of 49-245-49 is firstly trained using the 180,000 patches extracted from *Caltech* dataset. For learning purposes, we used the set of corrupted patches as input and the set of clean patches as output. The input and output layers contain $7 \times 7 = 49$ neurons each, while the hidden layer contains 245 neurons.
- DA_2: a denoising autoencoder with architecture as of 245-1,225-245 is trained using the 180,000 patches extracted from *Caltech* dataset as well. The input layer contains 245 neurons since the output of the encoding part of DA_1 is used as input to the encoding part of DA_2 in the next phase (SDA). The hidden layer contains 1,225 neurons, and the output layer figures 245 neurons (decoding part) that are used to feed the decoding part of DA_1 in the final model.
- SDA: the full approach comprises an SDA with architecture as of 49-245-1,225-245-49, which is trained once more using the very same patches employed in the previous two phases. In this step, the set of parameters learned in the previous phases are transferred to the SDA for a fine-tuning procedure.

Table 1. Average PSNR values (dB) considering the remote sensing images and the Standard dataset.

Techniques	$\sigma = 15$	$\sigma = 25$	$\sigma = 35$
MLP [1]	29.42 (29.73)	26.91 (27.15)	25.40 (24.44)
CNN [6]	28.60 (29.62)	25.99 (27.12)	24.45 (25.44)
SSDA [12]	29.65 (30.02)	27.09 (27.33)	25.52 (25.57)
SDASF	**30.08 (30.53)**	**27.23 (27.56)**	**25.56 (25.68)**
ND	24.60 (24.61)	20.16 (20.17)	17.24 (17.25)

Different learning rates (i.e., $\alpha_{SF} = 0.0002$ and $\alpha_{MSE} = 0.0001$) were used for each loss function to ensure a good trade-off between sparsity and good reconstructions. Also, all three phases make use of Adam optimizer with batches of size 128 and 30 epochs for convergence purposes. Notice that the neural architectures considered in this work are the same used by Xie et al. [12] to allow a fair comparison.

The denoising phase consists first of splitting the noisy image into 7×7 patches for feeding SDASF further. The smoothed counterparts (i.e., denoised patches) are assembled to compose the output image. The denoised patches are overlapped using a stride of 1 to avoid undesirable artifacts and the overlapped region is averaged to produce the final image.

To provide a deeper insight about the proposed approach, we also considered a dataset (hereinafter called "Standard") with some images well-known to the scientific community: "Cameraman", "Lena", "Barbara", "Boat", "Couple", "Fingerprint", "Hill", "House", "Man", "Montage", and "Peppers" [10].

5 Experimental Section

We assessed the effectiveness of SDASF using the well-known Peak Signal-to-Noise Ratio (PSNR), i.e., the larger the values, the better denoised the images are. Table 2 presents the mean PSNR values considering the satellite images and the Standard dataset (values in parenthesis), where the best average results are in bold.

Table 2. Training and denoising times in seconds.

Techniques	Train	Denoising
MLP	142.71	4,29
CNN	575,36	0.10
SSDA	171,04	3,44
SDASF	244,68	3,36

As one can observe, the proposed approach obtained the best results considering both datasets and all levels of noise. Deep neural networks pre-trained with denoising auto-encoders using SF figure the advantage of having only one hyperparameter, i.e., the patch size. Additionally, based on empirical tests, the use of non-overlapping training patches provided better results (Table 1).

The proposed approach presents good quantitative and qualitative results, as one can observe in Figs. 4 and 5. Even under the presence of higher noise levels ($\sigma = 35$), the results are still promising, with PSNR values of around 55.16%

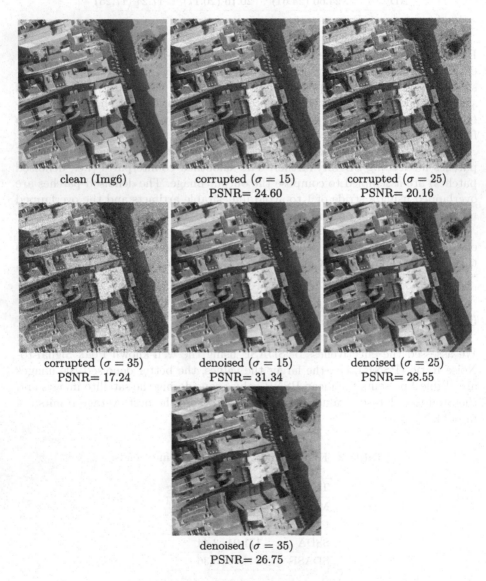

Fig. 4. Qualitative and quantitative results considering different levels of noise and image "Img6".

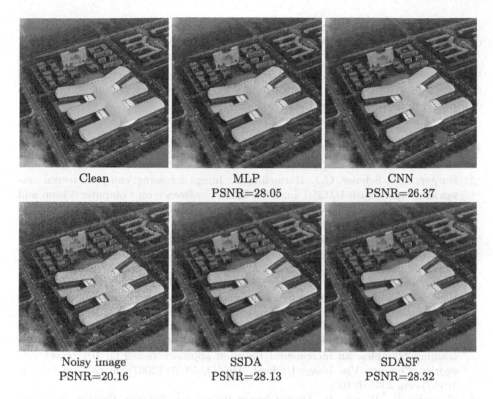

Clean

MLP
PSNR=28.05

CNN
PSNR=26.37

Noisy image
PSNR=20.16

SSDA
PSNR=28.13

SDASF
PSNR=28.32

Fig. 5. Qualitative and quantitative results considering "Img2" corrupted by $\sigma = 25$.

better when compared to the corrupted image. We believe the results can be further improved by increasing the number of training samples, as performed by Burger et al. [1] and Schuler et al. [9], as well as using more images related to the main application of this work, i.e., satellite imagery.

Table 2 presents the training and denoising times. Considering the training phase, we can observe that CNN was slower than all techniques. On the other hand, with respect to the denoising time, we observed that the techniques that need to process the image patch by patch such as MLP, SSDA, and the proposed approach SDASF, were more time consuming than CNN.

6 Conclusions

In this work, we present a novel approach called SDASF to image denoising that combines Sparse Filtering and a Stacked Denoising Autoencoder pre-trained with two distinct Denoising Autoencoders. The main contributions of this paper are two-fold: (i) to design an SDA composed of two different DAs trained separately, and to (ii) use SF background in the encoding step.

The proposed method induces a sparse-based regulation during the encoding process, which is further used to fine-tune a deep network. Experiments on public images corrupted with different levels of Gaussian noise support the effectiveness of the proposed approach when compared to other techniques.

Regarding future works, we intend to evaluate the proposed approach under both blur and noise effects, as well as to induce sparsity in the SDA fine-tuning step as well. Additionally, we also intend to evaluate deeper models, i.e., SDA composed of more than two DAs.

References

1. Burger, H.C., Schuler, C.J., Harmeling, S.: Image denoising: can plain neural networks compete with BM3D? In: 2012 IEEE Conference on Computer Vision and Pattern Recognition (CVPR), pp. 2392–2399. IEEE (2012). https://doi.org/10.1109/CVPR.2012.6247952
2. Dabov, K., Foi, A., Katkovnik, V., Egiazarian, K.: Image denoising with block-matching and 3D filtering. In: Image Processing: Algorithms and Systems, Neural Networks, and Machine Learning, vol. 6064, p. 606414. International Society for Optics and Photonics (2006). https://doi.org/10.1117/12.643267
3. Elad, M., Aharon, M.: Image denoising via sparse and redundant representations over learned dictionaries. IEEE Trans. Image Process. **15**(12), 3736–3745 (2006). https://doi.org/10.1109/TIP.2006.881969
4. Fei-Fei, L., Fergus, R., Perona, P.: Learning generative visual models from few training examples: an incremental Bayesian approach tested on 101 object categories. Comput. Vis. Image Underst. **106**(1), 59–70 (2007). https://doi.org/10.1016/j.cviu.2005.09.012
5. Gonzalez, R., Woods, R.: Digital Image Processing, 3rd edn. Prentice-Hall Inc., Upper Saddle River (2006)
6. Jain, V., Seung, S.: Natural image denoising with convolutional networks. In: Advances in Neural Information Processing Systems, pp. 769–776 (2009)
7. Nair, V., Hinton, G.E.: Rectified linear units improve restricted Boltzmann machines. In: Proceedings of the 27th International Conference on Machine Learning (ICML 2010), pp. 807–814 (2010)
8. Ngiam, J., Chen, Z., Bhaskar, S.A., Koh, P.W., Ng, A.Y.: Sparse filtering. In: Advances in Neural Information Processing Systems, pp. 1125–1133 (2011)
9. Schuler, C.J., Burger, H.C., Harmeling, S., Schölkopf, B.: A machine learning approach for non-blind image deconvolution. In: 2013 IEEE Conference on Computer Vision and Pattern Recognition (CVPR), pp. 1067–1074. IEEE (2013). https://doi.org/10.1109/CVPR.2013.142
10. Tampere University of Technology: Image and video denoising by sparse 3D transform-domain collaborative filtering. http://www.cs.tut.fi/~foi/GCF-BM3D/. Accessed May 2018
11. Vincent, P., Larochelle, H., Lajoie, I., Bengio, Y., Manzagol, P.A.: Stacked denoising autoencoders: learning useful representations in a deep network with a local denoising criterion. J. Mach. Learn. Res. **11**(Dec), 3371–3408 (2010)
12. Xie, J., Xu, L., Chen, E.: Image denoising and inpainting with deep neural networks. In: Advances in Neural Information Processing Systems, pp. 341–349 (2012)
13. Xu, L., Ren, J.S., Liu, C., Jia, J.: Deep convolutional neural network for image deconvolution. In: Advances in Neural Information Processing Systems, pp. 1790–1798 (2014)

Hybrid Attention Driven Text-to-Image Synthesis via Generative Adversarial Networks

Qingrong Cheng(iD) and Xiaodong Gu(✉)(iD)

Department of Electronic Engineering, Fudan University,
Shanghai 200433, China
xdgu@fudan.edu.cn

Abstract. With the development of generative models, image synthesis conditioned on the specific variable becomes an important research theme gradually. This paper presents a novel spectral normalization based *Hybrid Attentional Generative Adversarial Networks* (HAGAN) for text to image synthesis. The hybrid attentional mechanism is composed of text-image crossmodal attention and self-attention of image sub regions. Cross-modal attention mechanism contributes to synthesize more fine-grained and text-related image by introducing word-level semantic information in generative model. The self-attention solves the long distance reliance of image local-region features when generate image. With spectral normalization, the training of GANs become more stable than traditional GANs, which conduces to avoid model collapse and gradient vanishing or explosion. We conduct experiments on widely used Oxford-102 flower dataset and CUB bird dataset to validate our proposed method. During quantitative and non-quantitative experimental comparison, the results indicate that the proposed method achieves the best performance on Inception score (IS), Fréchet Inception Distance (FID) and visual effect.

Keywords: Text to image synthesis · Spectral normalization · Self-attention · Cross-modal attention · Generative Adversarial Networks

1 Introduction

Recent years have witnessed the great progress of Deep Neural Networks (DNNs), especially various kinds of generative tasks and discriminative tasks. Particularly, Convolutional Neural Networks (CNNs) have shown excellent performance on the challenging multi-category classification [1]. Besides, another branch of research focus is generative task, which is inverse mapping of discriminative task. In particular, generative tasks based on Generative Adversarial Networks (GANs) have achieved promising results [2] in image synthesis. Recently, photo-realistic image synthesis gradually becomes an important research direction with many potential applications, such as, computer graphics and photo retouching. To be specific, methods for text-to-image synthesis need generate image that are highly similar to meanings embedded in texts. However, image synthesis, conditioned on the given text descriptions, is also a knotty problem because of the great gap between text modality and image modality.

© Springer Nature Switzerland AG 2019
I. V. Tetko et al. (Eds.): ICANN 2019, LNCS 11731, pp. 483–495, 2019.
https://doi.org/10.1007/978-3-030-30493-5_47

Almost all existing text-to-image synthesis methods are based on GANs and some of them achieve remarkable performance. Generative Adversarial Networks (GANs) is proposed by Goodfellow in 2014 [3], which has made impressive performance in generative tasks. It is composed of two sub-networks, generator and discriminator, trained with a competing goal in an adversarial manner. From them on, GANs related work become a focused research direction. Meanwhile, adversarial learning mechanisms have shown great progress in many complex simulating problems [4].

Although excellent performance in many tasks, GANs are well known for difficulty in training and mode collapse. Many research works indicate that the instability in training is due to the disjoint of the generated data distribution and the real data distribution [5]. Besides, the mode collapse in GANs shows that the model will synthesize similar samples with uniform color and single texture. For addressing the knotty problem, many methods were proposed until now, such as WGAN [6], WGAN-GP [7] and SNGAN [8]. Some of those methods achieve excellent performance in stabilizing the training process and avoiding mode collapse.

Text-to-image synthesis is more challengeable than simply generate image from random noise or category condition. Text description contains more abundant and detailed image features, which should be drawn in synthesized image. Aiming at synthesizing photo-realistic image, there are two main branches of methods, VAE-based methods [9–11] and GAN-based methods [12–16]. Cai et al. [9] propose an image synthesis framework for fine-grained image in a multi-stage variational auto-encoder manner. Gulrajani et al. [10] present an improved PixelCNN-based model named PixelVAE, which introduces an autoregressive decoder for natural image synthesis. Deep Recurrent Attentive Writer (DRAW) [11] combines spatial attention mechanism with sequential VAE framework for constructing complex images.

Apart from VAE-based methods, GAN-based approaches also show great effectiveness in text to image synthesis. Specifically, Reed et al. [12] firstly introduce the traditional GAN into text to image synthesis in 2016. Following on the previous work, they propose a Generative Adversarial What-Where Network (GAWWN) [17] by using position box as additional supervision, which achieves better performance. However, the images synthesized by the first model are blurry and unclear. Inspired by the drawing step of human beings, multi-stage strategy is introduced into image synthesis in recent years, such as StackGAN [13, 14], AttnGAN [16] and CWPGGAN [15]. To be specific, StackGAN has two versions, StackGAN-v1 [13] and StackGAN-v2 [14]. StackGAN-v1 is based on two-stage GANs, while the StackGAN-v2 is an advanced three-sage model. Therefore, the images synthesized by the second model are more realistic and richly-textured than the first method. Progressive growing mechanism [18] is adopted in CWPGGAN [15], which can gradually improve the resolution and quality.

Attention mechanism shows effectiveness in many applications, especially in natural language process and computer vision. More specifically, self-attention mechanism is introduced in image generation [19]. Besides, attention mechanism is also adopted in text to image generative task, such as alignDRAW [20] and AttnGAN [16]. The alignDRAW [20] based on the mentioned DRAW introduces soft attention mechanism for attending to the relevant words of image feature. Xu et al. [16] propose a multi-stage Attentional Generative Adversarial Network (AttnGAN) for fine-grained image synthesis from text. Their methods not only use generator to generate high-resolution realistic image but also add word-level feature into generator, while others' methods only adopt sentence feature.

Inspired by previous work, we propose a spectral normalization based *Hybrid Attentional Generative Adversarial Networks* (HAGAN) that combines the image self-attention and text-image cross-modal attention mechanism for fine-grained image synthesis in this paper. Firstly, the features are extracted by the pretrained model name DAMSM [16], which contains both text and image feature embedding. Then, we feed the encoded text feature into three-stage hybrid attentional generative adversarial networks for image synthesis. The self-attention mechanism is introduced in the first-stage network and cross-modal attention is adopted in second and third stage generators. We mainly use the publicly available Oxford-102 flowers dataset and the Caltech CUB-200 birds dataset to conduct the experimental analysis. During the evaluation metric and side-by-side comparison with the state-of-the-art methods, the results indicate that our proposed method can get better visual effect and competitive evaluation value.

Compared to existing works, the main contributions of our work are as follows.

(1) By developing a hybrid attention mechanism for text to image synthesis, self-attention of image generation can solve long distance reliance between local features and cross-modal attention can add word-level features in generator for fine-grained image details.

(2) Due to spectral normalization, the training of the model becomes more stable than traditional GANs. Therefore, the generator can synthesize more realistic image due to discriminator satisfied with K-Lipschitz constraint can provide useful and effective gradient information for model optimizing.

The rest of this paper is organized as follows. The second section presents our proposed HAGAN approach. The third section shows the experimental results and comparison, and the last section concludes this paper.

2   The Proposed Method

2.1   Background

A. Generative Adversarial Networks

The GANs consists of two sub-networks, a discriminator D and a generator G, that cooperate and compete in a minimax game until the game achieves zero-sum game. Such minimax game can be described as the following object function $V(G, D)$.

$$\min_{G} \max_{D} V(G, D) = E_{x \sim p_{data}(x)}[\log(D(x))] \\ + E_{x \sim p_z(z)}[\log(1 - D(G(z)))], \tag{1}$$

where x is the real image and z is random noise. In the training process, the discriminator tries to maximize V, however, the generator wants to minimize the object function. In the last, the game of the two networks achieves the Nash Equilibrium that both can obtain the best performance.

B. Conditional Generative Adversarial Networks

Conditional GANs add conditional variable y to control the features of output image. The object function of conditional GAN can be described as follows.

$$\min_G \max_D V(G, D) = E_{x \sim p_{data}(x)}[\log(D(x|y))] \\ + E_{x \sim p_z(z)}[\log(1 - D(G(z|y)))], \tag{2}$$

where y is a conditional variable. The function of generator $G(z|y)$ allows the generator G to generate images conditioned on the given conditioning variable. The discriminator $D(x|y)$ evaluates whether the generated image is matched with conditioning variable y or not.

2.2 The Framework of Hybrid Attentional Generative Adversarial Networks

The HAGAN enables the generator to draw different sub-regions conditioned on related words and other long distance related image sub-regions. Meanwhile, the spectral normalization stabilize the training of the discriminator, which will contribute the optimization of the generator. The framework of the HAGAN, as shown in Fig. 1.

A. Hybrid Attentional Generative Adversarial Networks

Suppose the texts and images are stored in a N-pair document corpus (X^T, X^I). Here, X^T is text data and X^I is image data. The text feature and image feature are extracted by the well-trained embedding model DAMSM [16], which is based on the bi-directional Long Short-Term Memory (LSTM) and Convolutional Neural Network (CNN).

$$(\bar{\varphi}, \varphi; \bar{\phi}, \Phi) = F_{DAMSM}(X^T, X^I), \tag{3}$$

where φ indicates word feature matrix, $\bar{\varphi}$ denotes sentence feature, $\bar{\phi}$ is global image feature and Φ presents the sub-region feature matrix.

The encoded sentence feature $\bar{\varphi}$ will be pretreated before input into the multi-stage generative networks. As following,

$$\tilde{\varphi} = F_{cat}(z, F^{ca}(\bar{\varphi})), \tag{4}$$

where z is random noise vector, F^{ca} denotes the Conditioning Augmentation [14] which converts the sentence feature vector $\bar{\varphi}$ to the conditioning vector, and F_{cat} is concatenate function. After several upsample operation, the hidden feature gradually denotes the image features. The self-attention mechanism acts on the hidden feature maps $(\hat{h}_0, \hat{h}_1)$. As following,

$$\hat{h}_i = \hat{F}_i(h_{i-1}, F_i^{self-attn}(\hat{h}_{i-1})), \text{ where } i = 1, 2. \tag{5}$$

Here, $F_i^{self-attn}$ is the self-attention mechanism. The first-stage generator synthesizes image conditioned on the output of self-attention block directly. The generative networks consists of three generators (G_0, G_1, G_2), which use the previous hidden feature (h_0, h_1, h_2) to generate different-scale images $(\hat{x}_0, \hat{x}_1, \hat{x}_2)$. To be specific, the process of multi-stage generator is defined as following.

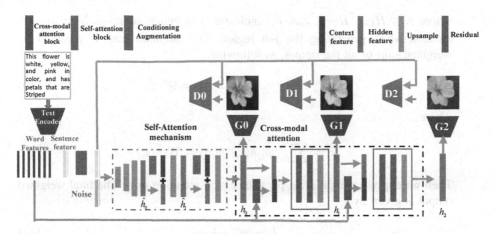

Fig. 1. The overall pipeline of the hybrid attentional generative adversarial networks.

$$\hat{x}_i = G_i(h_i), \text{ where } i = 1, 2. \tag{6}$$

The cross-modal mechanism is introduced in the second and the third networks, which can add more detailed attribute informations in the feature matrix. Specifically, the operation of cross-modal attention is defined as following.

$$h_i = F_i(h_{i-1}, F_i^{cro\_attn}(\varphi, h_{i-1})), \text{ where } i = 1, 2. \tag{7}$$

Here, φ is the feature matrix of word features, and $F_i^{cro\_attn}$ is the cross-attention model of the i-th stage generator. All of these functions are modeled as neural networks.

(1) *Self-Attention mechanism for the first stage generator*

The generator G and discriminator D of GAN models usually consist of convolutional neural networks. However, the convolutional filter only process the information in a local neighborhood, such as window size 3×3. Hence, long-range dependencies cannot be considered in the convolutional process. By introducing the self-attention mechanism into the GANs model, the generator can use the long-distance relationships between widely separated sub-regions.

In the deep model, the feature map $\hat{h} \in \mathbb{R}^{C \times N}$ of previous layer presents the hidden features of an image. We use two 1×1 convolutional layer to convert the feature map into two space $\bar{H}, \hat{H}$, and then calculate the attention of the two feature maps.

$$\beta_{j,i} = \frac{\exp(s_{ij})}{\sum_{i=1}^{N} \exp(s_{ij})}, \tag{8}$$

where $s_{ij} = \bar{H}(x_i)^T \hat{H}(x_j)$, and $\beta_{j,i}$ indicates how much attention from the i-th location when generating the j-th region. The attention map is obtained by weighted sum of all the output, as following.

$$\hat{C}_i = (\hat{c}_1, \cdots, \hat{c}_j, \cdots, \hat{c}_N) \in \mathbb{R}^{C \times N}, \tag{9}$$

where,

$$\hat{c}_j = \sum_{i=1}^{N} \beta_{j,i} h(x_i), h(x_i) = \mathbf{W}_h \hat{h}(x_i). \tag{10}$$

Then we apply a weight scale parameter γ on attention map. The final weighted output is given by,

$$\hat{h}_i = \gamma \hat{C}_i + \hat{h}_{i-1}, \tag{11}$$

where γ is initialized as 0.
In short, the self-attention mechanism can be denoted as

$$\hat{h}_i = \hat{F}_i(h_{i-1}, F_i^{self\_attn}(\varphi, \hat{h}_{i-1})). \tag{12}$$

(2) *Cross-modal Attention mechanism for the second and third stage generators*
Cross-modal attention mechanism is adopted to add relevant word-level information to networks for producing fine-grained image. The input of the cross-modal attention mechanism is the previous hidden feature $h \in \mathbb{R}^{\hat{D} \times N}$ of image and the word-level features $\varphi \in \mathbb{R}^{D \times T}$, which is encoded by the optimized model. Then, the word features are converted to a common space by adding a perceptron layer. Specifically, word feature $\hat{\varphi} \in \mathbb{R}^{D \times T}$ is converted by $\hat{\varphi} = \mathbf{U}\varphi$, where $\mathbf{U} \in \mathbb{R}^{\hat{D} \times D}$. Then, we calculate the word-context vector of the j-th sub-region by attention mechanism. Hidden feature h denotes the query, and the converted word features are the value. In detail, the word-context of the j-th sub-region is calculated as follows.

$$c_j = \sum_{i=0}^{T-1} \beta_{j,i} \hat{\varphi}_i, \tag{13}$$

where

$$\beta_{j,i} = \frac{\exp(s'_{j,i})}{\sum_{k=0}^{T-1} \exp(s'_{j,k})}. \tag{14}$$

Here, the similarity is computed by dot-product similarity

$$s'_{j,i} = h_j^T \hat{\varphi}_i. \tag{15}$$

In short, the word-context can be denotes as

$$\mathbf{C} = F^{cro\_attn}(\varphi, h) = (c_0, c_1, \ldots, c_{N-1}) \in \mathbb{R}^{\hat{D} \times N} \tag{16}$$

Then, the word-context and original image hidden feature is concatenated and feed in next layer.

(3) *Objective function of multi-stage GANs*

In our work, we adopt three generators and three discriminators in text-image translation. Each stage of generator $G_i (i = 0, 1, 2)$ has a corresponding discriminator D_i. The same with the conditional GANs, the objective function of the i-th generator is defined as follows.

$$L_{G_i} = -\frac{1}{2} E_{\hat{x}_i \sim p_{G_i}}[\log(D_i(\hat{x}_i))] - \frac{1}{2} E_{\hat{x}_i \sim p_{G_i}}[\log(D_i(\hat{x}_i, \bar{\varphi}))], \tag{17}$$

where the first part is unconditional loss and the second term is conditional loss. Meanwhile, in order to ensure the generated image is match with the text description, we introduce the DAMSM loss [16] into the objective function of the last-stage generator. As following,

$$L = L_{G_2} + \lambda_2 L_{DAMSM}, \tag{18}$$

where λ_2 is a balance factor.

In the adversarial learning, the discriminators evaluate whether the synthesized image is realistic and matched with the text or not. The objective function of each stage discriminator is defined as follows.

$$L_{D_i} = -\frac{1}{2} E_{x_i \sim p_{data}}[\log(D_i(x_i))] - \frac{1}{2} E_{\hat{x}_i \sim p_{G_i}}[\log(1 - D_i(\hat{x}_i))] + \\ -\frac{1}{2} E_{x_i \sim p_{data_i}}[\log(D_i(x_i, \bar{e}))] - \frac{1}{2} E_{\hat{x}_i \sim p_{G_i}}[\log(1 - D_i(\hat{x}_i, \bar{\varphi}))], \tag{19}$$

where x_i is from the real i-th scale image and $\hat{x}_i$ the generated image from the i-th stage generator. By optimizing the discriminator and generator alternately, the network will achieve zero-sum game that the generators and discriminators obtain the best performance.

B. Spectral Normalization for Stabilizing Training

Model Collapse, gradient vanishing and gradient explosion are very popular phenomena in the training of GANs. Besides, the balance of training between generator and discriminator is hard to control, which leads to converge difficultly. In order to solve the problem, many methods were proposed to improve the stability of model, such as WGAN [6] and WGAN-GP [7]. The original WGAN introduces Wasserstein distance to measure the distance between the real data and the generated data and minimize it. The Wasserstein distance is calculated as follows,

$$W(P_r, P_g) = \sup_{\|f\|_{Lip} \leq K} E_{x \sim P_r}[f(x)] - E_{x \sim P_g}[f(x)]. \tag{20}$$

Here, the formula $\|f\|_{Lip} \leq K$ indicates that the function $f(\bullet)$ is satisfied with K-Lipschitz constraint. The original WGAN presents a way of clipping the weights of discriminator in $[-c, c]$, which drops the fitting capacity of deep neural network. WGAN-GP adopts Gradient Penalty in discriminator to satisfy K-Lipschitz constraint, which increases computational effort. Therefore, those methods could not solve the problem absolutely. For stabilizing GAN-based model, the discriminator D should follow the Lipschitz continuity hypothesis. In other words, we need constrain the function of discriminator to satisfy the K-Lipschitz constraint.

$$\arg \max_{\|f\|_{Lip} \leq K} V(G, D), \tag{21}$$

where the $\|f\|_{Lip}$ is the smallest value of K such that $|f(x_1) - f(x_2)| \leq K|x_1 - x_2|$ for any x_1, x_2. Miyato et al. [8] propose a novel weight normalization named spectral normalization, which stabilize the training of discriminator by forcing the network to satisfy the Lipschitz constraint. Therefore, normalizing the weight parameters W of each layer can ensure the $\|f\|_{Lip}$ is bounded from above by 1. As following

$$\|\nabla_x(f(x))\|_2 = \|D_N \frac{W_N}{\sigma(W_N)} \cdots D_1 \frac{W_1}{\sigma(W_1)}\|_2 \leq \prod_{i=1}^{N} \frac{\sigma(W_i)}{\sigma(W_i)} = 1. \tag{22}$$

where $\sigma(W)$ is spectral normalization and D_N is nonlinear activation function of the N-th layer. With spectral normalization, the discriminator provides useful gradient to generator for optimization so that the network optimize better and generate images that are more realistic.

3   Experimental Results and Evaluation

3.1   Datasets and Evaluation Metric

We conduct experiments for text to image synthesis on the widely used CUB dataset [21] and Oxford-102 dataset [22]. The statistics of each datasets as shown in Table 1. In order to verify the effectiveness fairly, Inception Score (IS) [23] and Fréchet Inception Distance [24] are adopted for quantitative evaluation of generative model.

Table 1. Statistics of the datasets.

Datasets	CUB		Oxford-102	
	Train	Test	Train	Test
Number of samples	8,855	2,933	7,034	1,155
Captions/image	10	10	10	10
Categories	200		102	

Inception Score. The Inception Score (IS) is current well-known metric for evaluating the generative performance of GANs. The motivation of Inception Score is that excellent generative models should generate realistic, various and meaningful images. The calculation of IS score as follows.

$$IS = \exp(E_{X \sim P_G}[KL(P_{Y|X}(y|x))||P_Y(y)]), \tag{23}$$

where x denotes sample of generated image, and y is image label predicted by the inception model. The Eq. (22) indicates that classes of generated image should be as diverse as possible and the label prediction probability should be as accurate as possible. Therefore, the higher KL divergence shows excellent generative ability of model.

Fréchet Inception Distance. Assuming that both the real data and the generated data distribution following Gaussian distribution, so they have two major parameters, mean and covariance (m, C). The distance between the two data distribution is measured by Fréchet distance. The calculation is as following.

$$FID = ||m - m_r||_2^2 + Tr(C + C_r - 2(CC_r)^{\frac{1}{2}}), \tag{24}$$

where (m, C) are mean and covariance of generated data, and (m_r, C_r) are mean and covariance of real data. The lower distance of the mentioned two distributions presents that the synthesized image are more similar to the original data.

3.2 Experimental Results and Comparison

(1) Evaluation Metric Comparison

In experiment, we make quantitative and non-quantitative comparison with many state-of-art methods. Tables 2 and 3 show the quantitative comparison details of IS and FID score on Oxford-102 dataset and CUB dataset. For fair comparison, we choose some IS and FID value from the published paper [14, 15]. On the Oxford-102 dataset, the proposed method achieves 3.95 of inception score and 47.32 of Fréchet Inception Distance, which outperforms the previous methods. Likewise, the proposed method obtains the highest IS value (from 4.36 to 4.43) and competitive FID value (44.64). Comparing to the Oxford dataset, the CUB dataset is more difficult for text to image generation. The bird dataset can better reflect the performance of different methods. Significantly, the results show that the proposed method is able to achieve better performance than other state-of-art text to image synthesis methods.

Table 2. Fréchet Inception Distance and Inception Score for the Oxford-102 dataset.

Model	Resolution	FID	IS
GAN-INT-CLS	64 × 64	79.55	2.66 ± 0.03
WGAN-CLS	64 × 64	–	3.11 ± 0.02
WGAN-CLS with TTUR	64 × 64	–	3.20 ± 0.01
StackGAN-v1	256 × 256	55.28	3.71 ± 0.04
StackGAN-v2	256 × 256	48.68	3.82 ± 0.06
CLSPGGAN	256 × 256	–	3.76 ± 0.03
CWPGGAN	256 × 256	–	3.86 ± 0.02
AttnGAN	265 × 256	50.24	3.89 ± 0.02
HAGAN	256 × 256	47.32	3.95 ± 0.03

Table 3. Fréchet Inception Distance and Inception Score for the CUB dataset.

Model	Resolution	FID	IS
GAN-INT-CLS	64 × 64	68.79	2.88 ± 0.04
StackGAN-v1	256 × 256	51.89	3.70 ± 0.04
StackGAN-v2	256 × 256	–	3.82 ± 0.06
GAWWN	256 × 256	67.22	3.62 ± 0.07
CWPGGAN	256 × 256	–	4.09 ± 0.03
AttnGAN	256 × 256	46.43	4.36 ± 0.04
HAGAN	256 × 256	44.64	4.43 ± 0.03

(2) Visual Effect Comparison

The comparisons of state-of-art text-to-image generative methods by side-by-side comparison are shown in Fig. 2. Life part of Fig. 2 is various images generated by different methods, which are conditioned on the same text description of the Oxford-102 dataset. By scrutinizing the image details and text description roughly, the results show that all images generated by different methods matches with the text, and all those images are realistic and natural. However, the detailed comparison indicates that the image generated by our method are more realistic. On the challengeable CUB dataset, we can find that some previous methods have difficulty in generating highly real and clear image conditioned on the given text, such as GAN-CLS, GAWWN and

Fig. 2. Side-by-side comparison on the Oxford-102 dataset and CUB bird dataset.

StackGAN_v1. On the contrary, our proposed method can generate photo-realistic and fine-grained image, especially the bird of the third column. Therefore, in conclusion, our proposed method generates more realistic, more fine-grained and more natural images than other methods in visual evaluation.

(3) Word-Level Attention Visualization

For better evaluating the performance of attention mechanism, we visualize the word-level attention results as shown in Fig. 3. The attention visualization are shown below the red box. The words belong to the paired text description, and the bright region is the corresponding attention area of the words. However, some words do not give attention to right area, such as articles and verbs, which make less contribution to image synthesis. The words describing object attributes, such as colours, shape, and parts of objects, can give attention to correct regions. With adding word-level semantic information in the latter two generators, the generators can redraw the word's information in the corresponding region, which can saliently enhance the significant details of generated image as well as make it be suitable for the human system (Fig. 4).

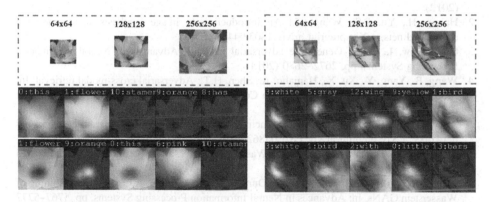

Fig. 3. Word-level attention visualization of the Oxford-102 flower dataset and CUB bird dataset.

Fig. 4. More examples synthesized by our proposed method.

4   Conclusion

This paper presents a hybrid attentional model to fulfill text-to-image synthesis. The hybrid attentional mechanism contributes to improve performance of generating fine-grained and realistic image. Meanwhile, the training of network become more stable by introducing spectral normalization in discriminator network. The conducted experiments show that our proposed method synthesizes realistic images in visual comparison, and outperforms the state-of-the-art approaches in FID and IS metric.

Acknowledgments. This work is supported by National Natural Science Foundation of China under grant 61771145 and grant 61371148.

References

1. Krizhevsky, A., Sutskever, I., Hinton, G.E.: ImageNet classification with deep convolutional neural networks. In: Advances in Neural Information Processing Systems, pp. 1097–1105 (2012)
2. Huang, H., Yu, P.S., Wang, C.: An introduction to image synthesis with generative adversarial nets. arXiv preprint arXiv:1803.04469 (2018)
3. Goodfellow, I., et al.: Generative adversarial nets. In: Advances in Neural Information Processing Systems, pp. 2672–2680 (2014)
4. Wang, B., Yang, Y., Xu, X., Hanjalic, A., Shen, H.T.: Adversarial cross-modal retrieval. In: Proceedings of the 2017 on Multimedia Conference, pp. 154–162. ACM Press, California (2017)
5. Arjovsky, M., Bottou, L.: Towards principled methods for training generative adversarial networks. arXiv preprint arXiv:1701.04862 (2017)
6. Arjovsky, M., Chintala, S., Bottou, L.: Wasserstein GAN. arXiv preprint arXiv:1701.07875 (2017)
7. Gulrajani, I., Ahmed, F., Arjovsky, M., Dumoulin, V., Courville, A.C.: Improved training of Wasserstein GANs. In: Advances in Neural Information Processing Systems, pp. 5767–5777 (2017)
8. Miyato, T., Kataoka, T., Koyama, M., Yoshida, Y.: Spectral normalization for generative adversarial networks. arXiv preprint arXiv:1802.05957 (2018)
9. Cai, L., Gao, H., Ji, S.: Multi-stage variational auto-encoders for coarse-to-fine image generation. arXiv preprint arXiv:1705.07202 (2017)
10. Gulrajani, I., et al.: PixeLVAE: a latent variable model for natural images. arXiv preprint arXiv:1611.05013 (2016)
11. Gregor, K., Danihelka, I., Graves, A., Rezende, D.J., Wierstra, D.: Draw: a recurrent neural network for image generation. In: International Conference on Machine Learning, Lille, pp. 1462–1471 (2015)
12. Reed, S., Akata, Z., Yan, X., Logeswaran, L., Schiele, B., Lee, H.: Generative adversarial text to image synthesis. arXiv preprint arXiv:1605.05396 (2016)
13. Zhang, H., et al.: StackGAN: text to photo-realistic image synthesis with stacked generative adversarial networks. In: Proceedings of the IEEE International Conference on Computer Vision, pp. 5907–5915. IEEE Press, Venice (2017)
14. Han, Z., Tao, X., Hongsheng, L., Shaoting, Z., Xiaogang, W., Xiaolei, H.: StackGAN++: realistic image synthesis with stacked generative adversarial networks. IEEE Trans. Pattern Anal. Mach. Intell. **41**, 1947–1962 (2018)

15. Bodnar, C.: Text to image synthesis using generative adversarial networks. arXiv preprint arXiv:1805.00676 (2018)
16. Xu, T., et al.: AttnGAN: fine-grained text to image generation with attentional generative adversarial networks. arXiv preprint arXiv:1711.10485 (2017)
17. Reed, S.E., Akata, Z., Mohan, S., Tenka, S., Schiele, B., Lee, H.: Learning what and where to draw. In: Advances in Neural Information Processing Systems, pp. 217–225 (2016)
18. Karras, T., Aila, T., Laine, S., Lehtinen, J.: Progressive growing of GANs for improved quality, stability, and variation. arXiv preprint arXiv:1710.10196 (2017)
19. Zhang, H., Goodfellow, I., Metaxas, D., Odena, A.: Self-attention generative adversarial networks. arXiv preprint arXiv:1805.08318 (2018)
20. Mansimov, E., Parisotto, E., Ba, J.L., Salakhutdinov, R.: Generating images from captions with attention. In: International Conference on Learning Representations, San Juan (2016)
21. Wah, C., Branson, S., Welinder, P., Perona, P., Belongie, S.: The Caltech-UCSD Birds-200-2011 dataset (2011)
22. Nilsback, M.E., Zisserman, A.: Automated flower classification over a large number of classes. In: IEEE Sixth Indian Conference on Computer Vision, Graphics & Image Processing, pp. 722–729. IEEE Press, Bhubaneswar (2008)
23. Salimans, T., Goodfellow, I., Zaremba, W., Cheung, V., Radford, A., Chen, X.: Improved techniques for training GANs. In: Advances in Neural Information Processing Systems, pp. 2234–2242 (2016)
24. Heusel, M., Ramsauer, H., Unterthiner, T., Nessler, B., Klambauer, G., Hochreiter, S.: GANs trained by a two time-scale update rule converge to a nash equilibrium. arXiv preprint arXiv:1706.08500 (2017)

Hypernetwork Functional Image Representation

Sylwester Klocek[iD], Łukasz Maziarka[iD], Maciej Wołczyk[iD], Jacek Tabor[iD], Jakub Nowak[iD], and Marek Śmieja[✉][iD]

Faculty of Mathematics and Computer Science, Jagiellonian University,
Łojasiewicza 6, 30-348 Kraków, Poland
sylwekklocek@gmail.com, l.maziarka@gmail.com, maciej.wolczyk@gmail.com,
jacek.tabor@uj.edu.pl, pl.jakub.nowak@gmail.com, marek.smieja@ii.uj.edu.pl

Abstract. Motivated by the human way of memorizing images we introduce their functional representation, where an image is represented by a neural network. For this purpose, we construct a hypernetwork which takes an image and returns weights to the target network, which maps point from the plane (representing positions of the pixel) into its corresponding color in the image. Since the obtained representation is continuous, one can easily inspect the image at various resolutions and perform on it arbitrary continuous operations. Moreover, by inspecting interpolations we show that such representation has some properties characteristic to generative models. To evaluate the proposed mechanism experimentally, we apply it to image super-resolution problem. Despite using a single model for various scaling factors, we obtained results comparable to existing super-resolution methods.

Keywords: Hypernetwork · Image representation · Deep learning

1 Introduction

Classical machine learning approaches are based on optimizing a predefined class of functions, such as linear or kernel classifier [31], Gaussian clustering [3], least squares regression [13], etc. With the emergence of deep learning, we learned how to employ general nonlinear functions given by complex neural networks. Replacing shallow methods by deep neural networks opened new opportunities in machine learning because arbitrarily complex functions can be approximated [14].

However, when one considers a typical representation of the data, we still use a somehow shallow and restrictive vector approach. Although real data, such as sound or image, have analog character, we represent them in an artificial vector form. In consequence, one cannot easily access an arbitrary position of the image, rescale the image or perform (even linear) operations like the rotation [9,11,25] without the additional use of interpolation [12]. Concluding, it is impossible to satisfactory utilize advanced deep learning methods without creating natural mechanisms for data representation.

© Springer Nature Switzerland AG 2019
I. V. Tetko et al. (Eds.): ICANN 2019, LNCS 11731, pp. 496–510, 2019.
https://doi.org/10.1007/978-3-030-30493-5_48

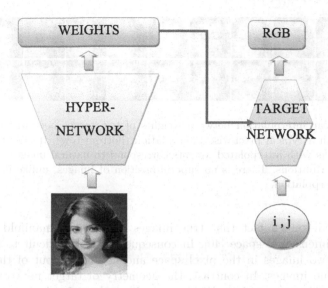

Fig. 1. The scheme of our approach. The hypernetwork takes an image and produces the weights to target network, which is responsible for approximating an image at every real-valued coordinate pair $(i, j) \in [0, 1]^2$.

The aim of this paper is a proof of concept that one can effectively construct and train *functional representations* of images. By a *functional (or deep) representation of an image* we understand a function (neural network) $\mathcal{I} : \mathbb{R}^2 \to \mathbb{R}^3$ which given a point (with arbitrary coordinates) (x, y) in the plane returns the point in $[0, 1]^3$ representing the RGB values of the color of the image at (x, y). Observe that given a functional image representation, we can easily obtain a corresponding vector representation by taking $(\mathcal{I}(i, j))_{i=1..k, j=1..n}$. In contrast to the vector form, the functional image representation in its idea can be compared by the way a human represents the image[1].

The main achievement of the paper is the functional image representation constructed with the use of hypernetwork [15]. The hypernetwork takes an image and produces a target neural network, which is an approximation of the input image, see Fig. 1 for illustration. Instead of creating the whole architecture of the target network from scratch, the hypernetwork returns only the weights to its fixed, predefined architecture. This allows us to effectively train both hypernetwork and target networks at the same time, using stochastic gradient descent.

We summarize the advantages and applications of our representation:

1. We use a single hypernetwork, which gives a recipe of how to represent images by target networks. In consequence, we return the functional representation of an input image at test time without additional training.

---

[1] We can reasonably hypothesize that a human representation of an image in the memory is given by some neural network.

Fig. 2. A linear interpolation between weights of two target networks (upper row) compared with a typical pixel-wise interpolation (bottom row). Images produced by target networks with interpolated weights correspond to natural images coming from true data distributions. There is no superimposition of images, unlike in the case of pixel-wise interpolation.

2. It is a well-known fact that true images represent a manifold embedded in high-dimensional space [40]. In consequence, it is difficult to interpolate between two images in the pixel space and not to fall out of the distribution of true images. In contrast, the geometry of target networks' weights is less complex (Sect. 4.1). We have verified that simple linear interpolation between weights of target networks representing images leads to natural images (Fig. 2).
3. Due to the continuity of this representation, we are not limited to a fixed resolution but can operate on a continuous range of coordinates. To confirm this property, we applied our approach to a super-resolution task (Fig. 3). In contrast to typical super-resolution methods [11,43], which are trained for a single scaling factor, our model is able to upscale the image to any size. While its effects are competitive compared to existing approaches (Sect. 4.2), we can also use non-standard scales at test time (Fig. 4).

2   Related Work

We briefly outline related approaches. We start by describing the hypernetwork model. Next, we discuss typical methods for image representation. Finally, we move on to super-resolution techniques.

Hypernetworks were introduced in [15] to refer to a network generating weights for a target network solving a specific task. The authors aimed to reduce the number of trainable parameters by designing a hypernetwork with a smaller number of parameters than the target network being generated. Making an analogy between hypernetworks and generative models, the authors of [32], used this mechanism to generate a diverse set of target networks approximating the same function. This is slightly similar to our technique, but instead of creating multiple networks for the same task, we aim at generating an individual network for each task (image). In [32], the hypernetwork was used to directly maximize the conditional likelihood of target variables given a certain input. Moreover, hypernetworks were also applied in Bayesian context [21,29]. Finally, the authors of [5,28,42] used the hypernetwork mechanism to create or improve the search of the whole network architecture.

Fig. 3. Higher resolution images (scaling factor ×4) obtained with a use of the bicubic interpolation (left) and our method (middle). A low-resolution input image is on the right side.

Fig. 4. Resizing the image to non-standard 2.5 × 1 resolutions.

Images are commonly represented as two-dimensional matrices with a fixed size (resolution). Due to the high redundancy in image features, one can use auto-encoders to create compressed representation in lower-dimensional space [2]. By adding controlled noise to input data at the training stage, we can obtain a representation, which is less sensitive to the image perturbations at the test stage [37]. Auto-encoders can also be used as a basis for generative models, such as VAE [20] or WAE [35], which can learn even more compact representations and generate new images using a decoder network [26]. Although auto-encoders can be used to transfer knowledge to less explored domains [38], the representations they learn are in a vector space. To represent an image as a function, one can approximate pixel intensities by a regression function, e.g. polynomial or kernel regression [34], or interpolate between neighbor pixels, e.g. bicubic or B-spline interpolation [12,17,36]. More advanced approaches rely on using wavelets [23] or ridgelets [10], which play a key role in JPEG2000 compressor [6]. Nevertheless, the aforementioned methods require manual selection of the regression model, which makes them difficult to use in practice. To overcome these problems we follow a deep learning approach and allow the hypernetwork to select the optimal function based on the data set.

The super-resolution area has been dominated by deep learning models. One of the earliest approaches applied a lightweight convolutional neural network to directly map a low-resolution image to its high-resolution counterpart [11]. In contrast, the authors of [19] used a very deep convolutional network inspired

by the VGG Net to predict residuals instead of the output image itself. The
paper [22] introduced a discriminator network as in the case of GANs to make
upscaled images more realistic. In [24] the authors reused residual networks but
removed unnecessary modules, which resulted in increased speed and stabilization of the model. The authors of [43] also used residual networks, but exploited
the hierarchical features from all the convolutional layers instead of the last one.
Despite huge progress in the super-resolution area, most of the existing models are trained for a single scale factor. In contrast, the proposed hypernetwork
technique allows us to upscale the image to multiple sizes using only a single
hypernetwork.

3   Functional Image Representation

In this section, we introduce our functional image representation. First, we
describe our learning model and define its cost function. Next, we discuss the
architectures of the hypernetwork and the target network.

3.1   Hypernetwork Model

Let $f : [0,1]^2 \to [0,255]^3$ be a function describing the image. In practice, we only
observe pixel intensities in a fixed grid. To improve this discrete representation,
we aim at creating a function:

$$T_\theta(i,j) = T((i,j),\theta) : [0,1]^2 \times \Theta \to [0,255]^3,$$

which approximates RGB values of each coordinate pair $(i,j) \in [0,1]^2$. Our
objective is to find an optimal weight vector $\theta \in \Theta$ for every image.

In the simplest case, T_θ can be obtained by linear or quadratic interpolation
[12], which however may not be sufficient for approximating complex image
structures. To achieve higher flexibility, one could model every image with a
specific neural network T_θ. Nevertheless, training separate networks for each
image using backpropagation may be computationally inefficient.

We approach this task by introducing a hypernetwork

$$H_\varphi : X \ni x \to \theta \in \Theta,$$

which for an image $x \in X$ returns weights θ to the corresponding target network
T_θ. Thus, an image x is represented by a function $T((i,j); H_\varphi(x))$, which for any
coordinates $(i,j) \in [0,1]^2$ returns corresponding RGB intensities of the image
$x \in X$.

To use the above model, we need to train the weights φ of the hypernetwork.
For this purpose, we minimize classical mean squared error (MSE) over training
images. More precisely, we take an input image $x \in X$ and pass it to H_φ. The
hypernetwork returns weights θ to target network T_θ. Next, the input image x

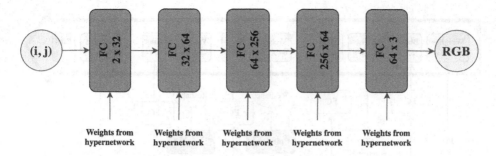

Fig. 5. The architecture of the target network.

is compared with the output from target network T_θ over known pixels. In other words, we minimize the expected MSE over the training set of images:

$$MSE = \sum_{x \in X} \sum_{(i,j)} [x[i,j] - T((i,j); H_\varphi(x))]^2.$$

Observe that we only train a single neural model (hypernetwork), which allows us to produce a great variety of functions at test time. In consequence, we might expect that target networks for similar images will be similar (see experimental section). In contrast, if we created a single network for every image, such identification would be misleading.

3.2 Architecture

In this part, we present the architectures used for creating functional image representation.

Target Network. An architecture of the target network is supposed to be simple and small. This allows us to keep the performance of training phase at the highest possible level as the target network is not directly trained. Moreover, small networks can be easily reused for other applications.

Target network maps a pair of two coordinates to corresponding three dimensional RGB intensities. The target network consists of five fully-connected layers, see Fig. 5 with biases added in each of them. The layers' dimensions are being gradually increased. This is happening up to the middle layer. Later on, they are being decreased. This is because steep transitions of layers' dimensions negatively affect the learning ability of neural network. The matrices used in the target network have the following dimensions: 2×32, 32×64, 64×256, 256×64, 64×3. Additionally, batch normalization is used after each layer [18]. We have chosen cosine to be the activation function between two consecutive layers. This choice is motivated by a typical approach used in mathematical image transformation, which is based on discrete cosine transform (DCT), see JPEG compression. For our purposes bounded cosine function worked much better than ReLU, which

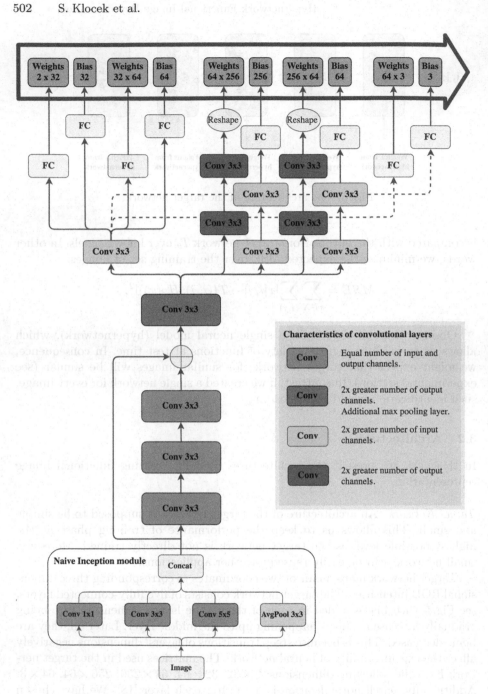

Fig. 6. The architecture of the hypernetwork.

can give arbitrary high outputs[2]. We use sigmoid as the activation function for the last layer. No convolutions were used, because the input of the target network is too simple.

Hypernetwork. The hypernetwork used in our model is a convolutional neural network with one residual connection, see Fig. 6. There are two main parts of the hypernetwork's architecture. The first one is common and takes part in generating weights for all of the target network's layers. Second part, on the other hand, contains several branches. Each branch calculates weights for a different layer of target network. This approach enabled faster training, compared to creating a separate hypernetwork for each layer of the target network.

The task of common layers is to extract meaningful features. This extraction is performed using Naive Inception Module [33] followed by four convolution layers. Inception module leverages three different convolutions and average pooling. It improves network accuracy and does not negatively influence training time. At the very end of common part, we introduced max pooling to decrease the number of features.

Next, there are multiple branches responsible for converting extracted features into actual weights of target network. Dimensions of the target network's layers vary. Therefore, convolutions with different number of input and output channels are used in the branches. Their role is to adjust the sizes of tensors. Shapes of tensors also need to be modified, hence convolutions are followed by either fully-connected layers or simple reshapes. Fully-connected layers are used in branches generating weights for smaller layers of target network. Batch normalization is used after each layer of hypernetwork and ReLU is chosen as the activation function.

Finally, to accelerate training phase even more, we reduced number of trainable parameters by adapting an approach from inception network [7]. More precisely, we replaced each $n \times n$ convolution with $1 \times n$ convolution followed by a $n \times 1$ convolution.

4   Experiments

In this section, we examine the proposed functional image representation. First, we analyze the space of target networks' weights and show some interesting geometrical properties. Next, we use the continuity of our representation and apply it to super-resolution.

4.1   Target Networks Geometry

It is believed that high dimensional data, e.g. images, is embedded in low-dimensional manifolds [14]. In consequence, direct linear interpolation between images does not produce pictures from true data distributions.

---

[2] Other experimental studies report that there are not much difference between using cosine and ReLU as activity function [14].

In this experiment, we would like to inspect the space of target network weights. In particular, we verify whether linear interpolation between the weights of two target networks produces true images. We use the CelebA data set [27] with trivial preprocessing of cropping central 128×128 pixels from the image and resizing it to 64×64 pixels.

In the test phase, we generate target networks for two images and linearly interpolate between weights of these networks. Figure 7a presents exemplary images returned by interpolated target networks. In most cases, interpolation produces natural images that come from the true data distribution (first four rows). It means that we transformed a manifold of images into a more compact structure, where linear interpolation can be applied. It allows us to suspect that similar images have similar weights in their corresponding target networks. Occasionally, interpolated weights produce images from outside of images manifold (last row). This negative effect sometimes occurred when interpolated images were significantly different, e.g. in this examples, we have a pair of images of people of different gender, hairstyle, skin color and photographed in different poses. These rare cases are accepted because we did not use any additional constraints. For a comparison, we generate classical pixel-wise interpolation between analogical examples. As can be seen in Fig. 7b, the results are much worse, resulting in a superimposition of images.

Going further, we verify a layer-wise interpolation. Namely, we take weights to one target network and gradually change weights of the first i layers in the direction of corresponding weights in the second target network. As can be seen in Fig. 8, each layer may be understood as having different functionality, i.e. third layer is responsible for the general shape, while the last layer corrects the colors in the image.

4.2   Super-Resolution

The hypernetwork allows us to describe every image as a continuous function (target network) defined on a unit 2D square. As a result, we can evaluate this function on every grid and upscale the image to any size. In this experiment, we compare the effects of this process with super-resolution approaches.

To make this approach successful we feed the hypernetwork with low resolution images and evaluate MSE loss on high resolution ones. More precisely, we take the original image of the size $m \times n$, downscale it to $k \times l$ using bicubic interpolation and input it to the hypernetwork. The hypernetwork produces the weights for the target network, which defines the functional representation of the input image. To evaluate its quality, we take a grid of the size $m \times n$ on the image returned by the target network and compare the values of this grid with pixels intensities of the original image.

Since input images can have different resolutions, we split them into overlapping parts of fixed sizes. In consequence, the value at each coordinate is described by multiple target networks. To produce a single output for every coordinate in the test phase, we take the (weighted) average of values returned by all target networks covering this coordinate. This also allows us to smooth the output

(a) Interpolation between target networks weights.

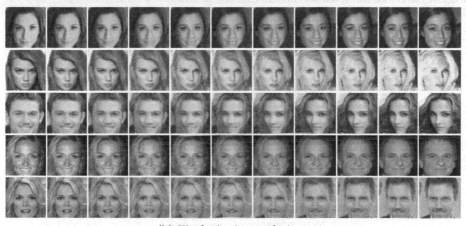

(b) Pixel-wise interpolation.

Fig. 7. Comparison of the interpolation between weights of two target networks (a) and linear pixel-wise interpolation (b). We also include an example of unsuccessful interpolation (last row), where interpolation went beyond the manifold of true images.

function. Moreover, we supplied a target network with an additional parameter α, which indicated the scaling factor.

To test our approach, we train the model on 800 examples from DIV2K data set [1]. Its performance is evaluated on Set5 [4], Set14 [41], B100 [30], and Urban100 [16]. As a quality measure we use PSNR and SSIM [39], which are common score functions applied in super-resolution. Their high values indicate better performance of a model. We consider scale factors $\times 2$, $\times 3$, $\times 4$.

As a baseline, we use bicubic interpolation. Moreover, we compare our approach with SRCNN [11], which was a state-of-the-art method in 2016. We also

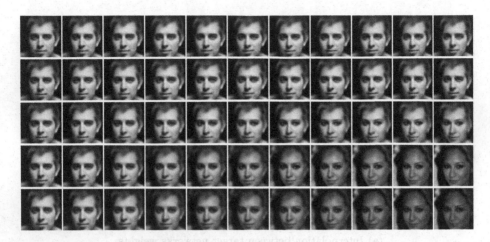

Fig. 8. Layerwise interpolation on the CelebA. In i-th row we interpolate only over the first i layers of the network.

Table 1. The average PSNR values obtained for a super-resolution task.

	Scale	Bicubic	SRCNN [11]	RDN [43]	Hypernetwork
Set5	2×	33.64	36.66	38.30	36.09
	3×	30.41	32.75	34.78	32.85
	4×	28.42	30.49	32.61	30.69
Set14	2×	30.33	32.45	34.10	32.30
	3×	27.63	29.30	30.67	29.37
	4×	26.08	27.50	28.92	27.61
B100	2×	29.48	31.36	32.40	31.11
	3×	27.12	28.41	29.33	28.31
	4×	25.87	26.90	27.80	26.86
Urban100	2×	26.85	29.50	33.09	29.43
	3×	24.43	26.24	29.00	26.26
	4×	23.11	24.52	26.82	24.56

include a recent state-of-the-art method – RDN [43]. Our goal is to train a single hypernetwork model to generate images at various scales. This is a more general solution than typical super-resolution approaches, where every model is responsible for upscaling the image to only one resolution. Therefore, it is expected that both SRCNN and RDN will perform better than our method.

The results presented in Tables 1 and 2 demonstrate that our model gives significantly better performance than baseline bicubic interpolation (see also Fig. 3 for sample result). Surprisingly, a single hypernetwork trained on all scales achieves a comparable performance to SRCNN, which created a separate model for each scale factor. It shows high potential of our model. Nevertheless, our

Table 2. The average SSIM values obtained for a super-resolution task.

	Scale	Bicubic	SRCNN [11]	RDN [43]	Hypernetwork
Set5	2×	0.930	0.9542	0.9616	0.9505
	3×	0.869	0.909	0.930	0.910
	4×	0.812	0.863	0.900	0.869
Set14	2×	0.869	0.907	0.922	0.900
	3×	0.775	0.822	0.848	0.816
	4×	0.703	0.751	0.789	0.751
B100	2×	0.843	0.888	0.902	0.879
	3×	0.738	0.786	0.811	0.778
	4×	0.666	0.710	0.743	0.706
Urban100	2×	0.839	0.895	0.937	0.891
	3×	0.733	0.799	0.868	0.798
	4×	0.656	0.722	0.807	0.723

model was not able to obtain scores as high as recent state-of-the-art super-resolution methods. It might be caused by the insufficient architecture of hyper-network. In our opinion, designing similar architecture of hypernetwork to RDN should lead to comparable performance. The main advantage of our approach is its generality – we train a single model for various scale factors.

5   Conclusion

In this work, we have presented an extension of hypernetworks mechanism, which makes it possible to create functional representations of images. Due to the continuity of the representation, we were able to upscale images to any resolution, while obtaining results comparable to those achieved by specialized super-resolution models. We also observed that the hypernetwork transforms a manifold of images to a more compact space with a convenient topology. In particular, we were able to traverse linearly from one image to another (using the weights of their target networks) without falling out from the true data distribution.

In future work, we will investigate other applications of the proposed hypernetwork-based functional image representation. One natural direction of further research is testing other image restoration tasks, such as deblurring, denoising and deblocking. We will also investigate the potential of hypernetworks in image inpainting. Furthermore, we are going to ensure better continuity of the learned representation, by using derivative-based methods such as Sobolev Training [8], and based on the improved representation, we will work on implementing continuous filters for convolutional neural networks.

Another research direction is to further explore the geometry of the target network space. For example, we will try to use the distance in the weights space between two target network-based representations as an alternative to the typical mean squared error loss used in many deep learning models, e.g. autoencoders or generative models.

Acknowledgements. This work was partially supported by the National Science Centre (Poland) grant no. 2018/31/B/ST6/00993 and by the Foundation for Polish Science grant no. POIR.04.04.00-00-14DE/18-00.

References

1. Agustsson, E., Timofte, R.: Ntire 2017 challenge on single image super-resolution: dataset and study. In: Proceedings of the IEEE Conference on Computer Vision and Pattern Recognition Workshops, pp. 126–135 (2017). https://doi.org/10.1109/CVPRW.2017.150
2. Baldi, P.: Autoencoders, unsupervised learning and deep architectures. In: Proceedings of the 2011 International Conference on Unsupervised and Transfer Learning Workshop, UTLW 2011, vol. 27, pp. 37–50. JMLR.org (2011). http://dl.acm.org/citation.cfm?id=3045796.3045801
3. Banfield, J.D., Raftery, A.E.: Model-based gaussian and non-gaussian clustering. Biometrics **49**(3), 803–821 (1993). https://doi.org/10.2307/2532201. http://www.jstor.org/stable/2532201
4. Bevilacqua, M., Roumy, A., Guillemot, C., Alberi-Morel, M.L.: Low-complexity single-image super-resolution based on nonnegative neighbor embedding (2012). https://doi.org/10.5244/C.26.135
5. Brock, A., Lim, T., Ritchie, J.M., Weston, N.: SMASH: one-shot model architecture search through hypernetworks. CoRR abs/1708.05344 (2017). arXiv:abs/1708.05344
6. Christopoulos, C., Skodras, A., Ebrahimi, T.: The JPEG2000 still image coding system: an overview. IEEE Trans. Consum. Electron. **46**(4), 1103–1127 (2000). https://doi.org/10.1109/30.920468
7. Czarnecki, W.M., Osindero, S., Jaderberg, M., Swirszcz, G., Pascanu, R.: Rethinking the inception architecture for computer vision. In: Advances in Neural Information Processing Systems, pp. 4278–4287 (2017). https://doi.org/10.1109/CVPR.2016.308
8. Czarnecki, W.M., Osindero, S., Jaderberg, M., Swirszcz, G., Pascanu, R.: Sobolev training for neural networks. In: Advances in Neural Information Processing Systems, pp. 4278–4287 (2017)
9. Danelljan, M., Robinson, A., Shahbaz Khan, F., Felsberg, M.: Beyond correlation filters: learning continuous convolution operators for visual tracking. In: Leibe, B., Matas, J., Sebe, N., Welling, M. (eds.) ECCV 2016. LNCS, vol. 9909, pp. 472–488. Springer, Cham (2016). https://doi.org/10.1007/978-3-319-46454-1_29
10. Do, M.N., Vetterli, M.: The finite ridgelet transform for image representation. IEEE Trans. Image Process. **12**(1), 16–28 (2003). https://doi.org/10.1109/TIP.2002.806252
11. Dong, C., Loy, C.C., He, K., Tang, X.: Image super-resolution using deep convolutional networks. IEEE Trans. Pattern Anal. Mach. Intell. **38**(2), 295–307 (2016). https://doi.org/10.1109/TPAMI.2015.2439281
12. Gao, S., Gruev, V.: Bilinear and bicubic interpolation methods for division of focal plane polarimeters. Opt. Express **19**(27), 26161–26173 (2011). https://doi.org/10.1364/OE.19.026161

13. Geladi, P., Kowalski, B.R.: Partial least-squares regression: a tutorial. Analytica chimica acta **185**, 1–17 (1986). https://doi.org/10.1016/0003-2670(86)80028-9
14. Goodfellow, I., Bengio, Y., Courville, A.: Deep Learning. The MIT Press, Cambridge (2016)
15. Ha, D., Dai, A., Le, Q.V.: Hypernetworks. arXiv preprint arXiv:1609.09106 (2016)
16. Huang, J.B., Singh, A., Ahuja, N.: Single image super-resolution from transformed self-exemplars. In: Proceedings of the IEEE Conference on Computer Vision and Pattern Recognition, pp. 5197–5206 (2015). https://doi.org/10.1109/CVPR.2015.7299156
17. Hwang, J.W., Lee, H.S.: Adaptive image interpolation based on local gradient features. IEEE Signal Process. Lett. **11**(3), 359–362 (2004). https://doi.org/10.1109/LSP.2003.821718
18. Ioffe, S., Szegedy, C.: Batch normalization: accelerating deep network training by reducing internal covariate shift. arXiv preprint arXiv:1502.03167 (2015)
19. Kim, J., Kwon Lee, J., Mu Lee, K.: Accurate image super-resolution using very deep convolutional networks. In: Proceedings of the IEEE Conference on Computer Vision and Pattern Recognition, pp. 1646–1654 (2016). https://doi.org/10.1109/CVPR.2016.182
20. Kingma, D.P., Welling, M.: Auto-encoding variational bayes. arXiv preprint arXiv:1312.6114 (2013)
21. Krueger, D., Huang, C.W., Islam, R., Turner, R., Lacoste, A., Courville, A.: Bayesian hypernetworks. arXiv preprint arXiv:1710.04759 (2017)
22. Ledig, C., et al.: Photo-realistic single image super-resolution using a generative adversarial network. In: Proceedings of the IEEE Conference on Computer Vision and Pattern Recognition, pp. 4681–4690 (2017). https://doi.org/10.1109/CVPR.2017.19
23. Lee, T.S.: Image representation using 2D gabor wavelets. IEEE Transactions on pattern analysis and machine intelligence **18**(10), 959–971 (1996). https://doi.org/10.1109/34.541406
24. Lim, B., Son, S., Kim, H., Nah, S., Mu Lee, K.: Enhanced deep residual networks for single image super-resolution. In: Proceedings of the IEEE Conference on Computer Vision and Pattern Recognition Workshops, pp. 136–144 (2017). https://doi.org/10.1109/CVPRW.2017.151
25. Lin, T.Y., Dollár, P., Girshick, R., He, K., Hariharan, B., Belongie, S.: Feature pyramid networks for object detection. In: Proceedings of the IEEE Conference on Computer Vision and Pattern Recognition, pp. 2117–2125 (2017). https://doi.org/10.1109/CVPR.2017.106
26. Liu, M.Y., Breuel, T., Kautz, J.: Unsupervised image-to-image translation networks. In: Advances in Neural Information Processing Systems, pp. 700–708 (2017)
27. Liu, Z., Luo, P., Wang, X., Tang, X.: Deep learning face attributes in the wild. In: Proceedings of International Conference on Computer Vision (ICCV) (2015). https://doi.org/10.1109/ICCV.2015.425
28. Lorraine, J., Duvenaud, D.: Stochastic hyperparameter optimization through hypernetworks. CoRR abs/1802.09419 (2018). arXiv:abs/1802.09419
29. Louizos, C., Welling, M.: Multiplicative normalizing flows for variational bayesian neural networks. In: Proceedings of the 34th International Conference on Machine Learning, vol. 70, pp. 2218–2227. JMLR. org (2017)
30. Martin, D., Fowlkes, C., Tal, D., Malik, J.: A database of human segmented natural images and its application to evaluating segmentation algorithms and measuring ecological statistics. In: null, p. 416. IEEE (2001). https://doi.org/10.1109/ICCV.2001.937655

31. Scholkopf, B., Smola, A.J.: Learning with Kernels: Support Vector Machines, Regularization, Optimization, and Beyond. MIT press (2001). https://doi.org/10.1109/TNN.2005.848998

32. Sheikh, A.S., Rasul, K., Merentitis, A., Bergmann, U.: Stochastic maximum likelihood optimization via hypernetworks. arXiv preprint arXiv:1712.01141 (2017)

33. Szegedy, C., et al.: Going deeper with convolutions. In: Proceedings of the IEEE Conference on Computer Vision and Pattern Recognition, pp. 1–9 (2015). https://doi.org/10.1109/CVPR.2015.7298594

34. Takeda, H., Farsiu, S., Milanfar, P., et al.: Kernel regression for image processing and reconstruction. Ph.D. thesis, Citeseer (2006). https://doi.org/10.1109/TIP.2006.888330

35. Tolstikhin, I., Bousquet, O., Gelly, S., Schoelkopf, B.: Wasserstein auto-encoders. arXiv preprint arXiv:1711.01558 (2017)

36. Unser, M., Aldroubi, A., Eden, M.: Fast B-spline transforms for continuous image representation and interpolation. IEEE Trans. Pattern Anal. Mach. Intell. 3, 277–285 (1991). https://doi.org/10.1109/34.75515

37. Vincent, P., Larochelle, H., Bengio, Y., Manzagol, P.A.: Extracting and composing robust features with denoising autoencoders. In: Proceedings of the 25th International Conference on Machine Learning, pp. 1096–1103. ACM (2008). https://doi.org/10.1145/1390156.1390294

38. Wang, N., Yeung, D.Y.: Learning a deep compact image representation for visual tracking. In: Advances in Neural Information Processing Systems, pp. 809–817 (2013)

39. Wang, Z., Bovik, A.C., Sheikh, H.R., Simoncelli, E.P., et al.: Image quality assessment: from error visibility to structural similarity. IEEE Trans. Image Process. 13(4), 600–612 (2004). https://doi.org/10.1109/TIP.2003.819861

40. Yeh, R.A., Chen, C., Yian Lim, T., Schwing, A.G., Hasegawa-Johnson, M., Do, M.N.: Semantic image inpainting with deep generative models. In: Proceedings of the IEEE Conference on Computer Vision and Pattern Recognition, pp. 5485–5493 (2017). https://doi.org/10.1109/CVPR.2017.728

41. Zeyde, R., Elad, M., Protter, M.: On single image scale-up using sparse-representations. In: Boissonnat, J.-D., et al. (eds.) Curves and Surfaces 2010. LNCS, vol. 6920, pp. 711–730. Springer, Heidelberg (2012). https://doi.org/10.1007/978-3-642-27413-8_47

42. Zhang, C., Ren, M., Urtasun, R.: Graph hypernetworks for neural architecture search. CoRR abs/1810.05749 (2018). arXiv:abs/1810.05749

43. Zhang, Y., Tian, Y., Kong, Y., Zhong, B., Fu, Y.: Residual dense network for image super-resolution. In: Proceedings of the IEEE Conference on Computer Vision and Pattern Recognition, pp. 2472–2481 (2018)

Instance-Based Segmentation for Boundary Detection of Neuropathic Ulcers Through Mask-RCNN

H. V. L. C. Gamage[✉][iD], W. O. K. I. S. Wijesinghe[iD], and Indika Perera

Department of Computer Science and Engineering,
University of Moratuwa, Moratuwa, Sri Lanka
{chathuri.12,isuru.10,indika}@cse.mrt.ac.lk

Abstract. Neuropathic ulcers form and proliferate because of peripheral neuropathy, usually in diabetic patients. The existing ulcer assessment process which relies on visual examination, potentially be imprecise and inefficient. Therefore this indicates the necessity of a more quantitative and cost-effective solution that enables ulcer diagnosing process much faster. In the current literature, different deep learning approaches are available for diagnosing illnesses through medical imagery. When diagnosing diabetic patients who are suffering from neuropathic ulcers through imagery, the locating and segmenting of ulcer boundaries is of great importance. In this study, we propose an approach to automate the process of locating and segmenting ulcers through Mask-RCNN model. We use a dataset of 400 ulcer imagery and corresponding annotations of ulcers for this task. This approach achieves an overall ulcer detection average precision (AP) at Intersection over union (IoU) threshold 0.5 of 0.8632 and mean average precision (mAP) at Intersection over union (IoU) threshold 0.5 to 0.95 by steps of size 0.05 of 0.5084 for ResNet-101 backbone.

Keywords: Neuropathic Ulcers · Diabetics · Instance Segmentation · Mask-RCNN · Intersection over Union (IoU) · Mean Average Precision (mAP) · Convolutional Neural Network (CNN)

1 Introduction

Diabetes mellitus, syndrome of carbohydrate metabolism characterized by the impaired ability of the body to produce (Type 2) or respond to insulin and thereby maintain (Type 1) proper levels of glucose in the blood [1]. Major life-threatening complications like blindness, kidney failure, cardiovascular and Neuropathic Ulcer are the results of this disorder. According to the statistics of International Diabetes Federation, over 425 million people are suffering from Diabetes Mellitus worldwide. Every year, more than 1 million patients are suffering from diabetes amputation since the failure of early diagnosis and treatments of wounds [17].

© Springer Nature Switzerland AG 2019
I. V. Tetko et al. (Eds.): ICANN 2019, LNCS 11731, pp. 511–522, 2019.
https://doi.org/10.1007/978-3-030-30493-5_49

In current clinical practices, the wound assessment process is completely based on visual examination by the experts and estimate wound area with manual measurements directly on wound or on high resolution wound imagery. Furthermore, in most cases, high-resolution imagery use to keep track the growth of wound, early diagnose, keep patient history etc. In addition to that CT scans, MRI, X-Ray tests also use to evaluate the status of the wound. The accuracy of manual inspection process possibly be inaccurate because failures in wound area determination due to irregular patterns and uncertain boundaries of the wounds. A consistent wound assessment system can only be accomplished by regularly performing accurate measurements of wound area, analyzing its colours, and the relative sizes of different wound tissues. Due to the lack of consistency, current measurements, which captured through the existing tools, cannot be guaranteed. Therefore, these facts imply the need for a computer-aided system to quantitate neuropathic ulcers through computer vision techniques to improve the accuracy of wound assessment process.

When considering an autonomous wound assessment system, the initial step is to identify the wound boundary using image segmentation algorithms. Since the wounds consist of complex visual characteristics, which make precise automatic segmentation a challenging task, where traditional methods usually fail. The main advantage of autonomous boundary detection of wounds is to monitor the different features of the wound such as area, wound healing status etc. Moreover, this automated process helps to reduce the workload of clinicians considerably and may be an effective way to wound care. The electronic records for wound healing process may help clinicians to track the state of a wound for further treatments.

In this paper, we mainly focus on the core process of wound assessment, determination of wound boundaries and propose a deep learning approach to locate and segment wound areas automatically through Mask-RCNN model.

2   Literature Review

In recent years, various schemes for detect and segment neuropathic ulcers appeared in the literature. Researchers have developed many schemes and techniques for detection and segmentation of ulcer imagery. Perez et al. 2001 [16] propose a leg ulcer segmentation algorithm which based on the region growing. The segmentation has obtained through an automatic analysis made in the RGB and SI channels by changing the RGB colour space to the HIS colour space. The authors have tried to figure out the characteristics of each of the five channels that will make the segmentation more efficient. Then the selected best channel had segmented and used as a mask over the original image. Jones et al. 2000 [9] propose an active contour based models and level set- based methods to adaptively regularize the segmentation contour upon existing manual delineation process. However, these approaches have limitations such as dependent on the initial contour settings, sensitivity to illumination conditions and instability when dealing with the poorly defined wound boundaries or false edges. Wang et al. [25] present an approach to determine the boundaries of the foot ulcer images which were captured through an image capture box using cascaded two-stage Support Vector Machine (SVM) based classification. A set of k binary SVM classifiers had trained and applied to a different subset of entire training dataset and incorrectly classified instances had collected in the first stage. Another SVM classifier had trained on misclassified instances as the second stage. The authors had used several image processing algorithms to extract features such as colour, texture etc. Then Principal Component Analysis (PCA) was applied to extracted features. This approach has achieved 73.3% of average sensitivity and 94.6% of average specificity. Veredas et al. [23] had done a research on binary tissue classification on wound images using neural networks and Bayesian classifiers. The authors of the research had used a clustering based approach, a hybrid model which is based on mean shift procedure and a region growing strategy for the implementation of region segmentation.

There are several software systems have built for wound image analysis and monitoring. Filko et al. [5] has implemented a system called WITA to analyze and management of the wounds. PictZar Digital Planimetry [15] is a Commercial Software to segmentation and measurement of chronic wound images. Another software called MOWA – Mobile Wound Analyzer [14] was implemented by team of developers and it analyses the wound images and calculate the features of the wounds. Silhouette TM [21] is also another wound assessment software implemented by Aranz Medical. This system is able to capture 3D image of the wound and measure the area, depth, and volume of the wound and its healing progress. Although these systems facilitate to analyze the wound features, do not support for an automatic boundary detection process. Therefore, the clinicians have to draw the boundaries of the wounds manually.

Even though the wound segmentation models which had developed using traditional machine learning techniques have achieved good performance, those models have limitations since they have the dependencies with the low level hand crafted features and poor performance on complex scenarios due to the limited learning ability. In recent studies, the algorithms, which based on Convolutional Neural Network (CNN), had shown an enormous success in computer vision tasks such as object detection, localization, semantic segmentation, and instance segmentation [11]. Goyal et al. 2017 [6] propose a deep learning based approach to automatically segment the diabetic wounds. The proposed architecture of this research is two-tier transfer learning to train the fully convolutional networks. In first tire transfer learning, CNN models that make to FCNNs have trained on ImageNet dataset [20]. In second tire of transfer learning, trained FCN models on the Pascal VOC segmentation dataset [4]. The authors had used different transfer learning models like FCN-AlexNet, FCN-32s, FCN-16s, and FCN-8s. Out of these models, FCN-16s was the best performer. Furthermore, 5-fold cross validation was used to train the model and it has been achieved Dice Similarity Coefficient of 0.794 for ulcer region, 0.851 for surrounding skin region, and 0.899 for the combination of both regions. The limitation of this system is the segmented output appears without the background of the image. It does not visualize a clear distribution of the wound throughout the limb. Wang et al. [24] has proposed a unified framework for wound segmentation and analysis using deep convolutional neural networks. The authors have been used deep CNN encoder decoder network in an end-to-end style. The model had achieved 95% of pixel accuracy and 47.3% of mean IoU. However, the traditional FCN are not instance aware which means they cannot separate multiple objects with same type because the method had developed to differentiate similar pixels in to different instances. Le et al. [12] proposed a solution to adhere these problems by creating the Fully Convolutional Instance Segmentation architecture. This also has the limitations with discerning instance of the same class and systematic errors on overlapping instances. Kaiming He et al. 2017 [7] had introduced Mask-RCNN architecture which is an extension of Faster – RCNN [18]. This architecture has been used in different instance segmentation approaches [8,10].

3   DataSet

The dataset used for the experiments of this research consists 400 imagery of neuropathic ulcers which are collected from diabetic clinics. The annotation of the imagery is done with the help of expert diabetologists. We split the dataset as 90% of the data as training and 10% of the data as our test set. We annotated all three sets of data separately by using VGG Image Annotator (VIA) tool [22]. Then we exported the annotated details into a JSON file. These are a set of high-resolution imagery captured from different cameras in a variety of conditions such as colours, lighting, different orientation etc.

4   Methodology

4.1   Preprocessing

We have a quiet small dataset for our training imagery. Therefore we perform data augmentation to increase the size of our training dataset. This is one of the solutions to reduce the overfitting problem. We use off-line data augmentation technique, which indicates dataset was augmented and annotated before the training process. We used following strategies to do the augmentation.

- Horizontal image flips.
- 90°, 180°, 270° image rotations.
- Gaussian blur with standard deviation 2.5.
- Luminosity scaling in the range of [0.8,1.5].

Moreover, none of the above transformations change the essential characteristics of the neuropathic ulcers, hence all are valid for use in augmentation.

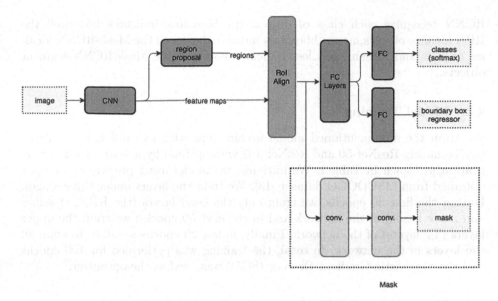

Mask

Fig. 1. Mask-RCNN architecture

4.2   Model Architecture

The Authors in [7] introduce mask-RCNN model (see Fig. 1) as simple flexible and general framework for object instance segmentation. Mask RCNN is an extended version of the Faster RCNN [18], which is a RPN (Region Proposal Network) based model that consists of two outputs for each object namely a class label and a anchor box for the object region. Mask-RCNN outputs an object mask in addition to the other two outputs of Faster RCNN and this mask supports to do the object segmentation more accurately. During the first phase of the model, it scans the image and generates regional proposals. At the second phase, it classifies the proposals and generates bounding boxes and masks. It uses a backbone network to extract the features, where early layers detect low level features (edge, corners etc.) and later layers detect high level features (object instances). The object detection process achieves using regional proposal network (RPN) which scans the images in a sliding–window fashion. The RPN generate two outputs; Region of Interest (ROIs) and Bounding Boxes. Then ROIs are classified by ROI classifier and bounding boxes are refined through the bounding box regressor. Finally, it uses the regions from the classified ROIs and generates mask for them. The model uses multi task loss function given by $L = L_{cls} + L_{bbox} + L_{mask}$.

$L_{cls} = \text{rpn\_class\_loss} + \text{mrcnn\_class\_loss}$
$L_{bbox} = \text{rpn\_bbox\_loss} + \text{mrcnn\_bbox\_loss}$
$L_{mask} = \text{mrcnn\_mask\_loss}$

Where rpn_class_loss indicates how well the Region Proposal Network separates background with objects,mrcnn_class_loss indicates how well the Mask-

RCNN recognize each class of object, rpn_bbox_loss indicates how well the RPN localize objects,mrcnn_bbox_loss indicates how well the Mask-RCNN localize objects and mrcnn_mask_loss indicates how well the Mask-RCNN segment objects.

4.3   Model Training

We train the aforementioned model architecture with two different backbone CNNs namely ResNet-50 and ResNet-101 with optimal hyper-parameter configuration as shown in Table 1. We initialize the model using pre-trained weights obtained from MSCOOCO dataset [13]. We train the layers under three stages. During the first 30 epochs, we train only the head layers (the RPN, classifier and mask heads) of the network and in the next 50 epochs; we train the upper layers (4+ layers) of the network. Finally, in last 20 epochs we allow to train all the layers in the network. In total, the training was performed for 100 epochs and the Stochastic Gradient Descent (SGD) was used as the optimizer.

4.4   Hyperparameter Tuning

We set the number of classes to 2 as shown in the above table because we need to perform the segmentation between ulcer and background region. We evaluate our model under two backbone architectures: ResNet-50 and ResNet-101. ResNet 101 gave the best results in terms of accuracy and performance. We test with learning rate ranging from 0.01 to 0.0001. We got an optimal performance at learning rate 0.001 and we do not observe any noticeable improvement for other values. We use 0.9 learning momentum for SGD and set L2 regularization (weight decay) parameter to 0.0001. The rest of the parameters choose according to our memory of the machine.

Table 1. Configurations of Mask-RCNN.

Number of classes	2
Backbone	ResNet-101
Images per GPU	2
RPN anchor scales	32, 64, 128, 256, 512
Train ROIs per image	128
Mask shape	28×28
Anchors per image	256
Learning rate	0.001
Learning momentum	0.9
Weight decay	0.0001
Batch size	2
Image dimensions	256×256

4.5   Baseline

In order to compare our proposed approach, we use U-Net [19] as the baseline model. U-Net is a popular deep CNN architecture which can be used for small medical datasets with fairly good performance in medical image segmentation task. The architecture of this network consists of two paths: encoding path and decoding path as shown in Fig. 2. In the training process we could achieve best performance for Adam optimizer with 1×10^{-3} learning rate. We train the model up to 100 epochs. All weights initialize using Xavier initialization scheme.

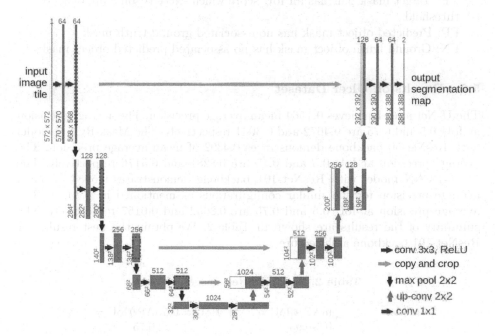

Fig. 2. U-Net architecture

5   Evaluations and Results

5.1   Evaluation Metrics

We used Average Precision (AP) at different intersection over union (IoU) thresholds as our evaluation metrics for this study. The IoU of predicted set of object pixels and a set of ground truth object pixels is calculated as:

$$IoU(target, predicted) = \frac{(target \cap prediction)}{(target \cup prediction)}$$

We considered a range of IoU threshold values I.e. range from 0.5 to 0.95 with a step size of 0.05 and at each point we calculate an average precision value.

At each threshold value t, a precision value calculated by considering the true positives (TP), false negatives (FN) and false positives (FP) of comparing the ground truth and the predicted masks. The mean average precision of a single image is then calculated as the mean of the above average precision values at each IoU threshold.

$$mAP = \frac{1}{|thresholds|} \sum_t \frac{TP(t)}{(TP(t) + FP(t) + FN(t))}$$

TP: Target mask pair has an IoU score which exceeds some pre-defined threshold

FP: Predicted object mask has no associated ground truth mask

FN: Ground truth object mask has no associated predicted object masks

5.2   Results on Ulcer Dataset

The U-Net model achieves 0.1564 mean average precision. The average precision at IoU 0.5 and 0.75 are 0.4672 and 0.3041 respectively. The Mask-RCNN model with ResNet-50 backbone demonstrates 0.4392 of mean average precision. The average precision at IoU 0.5 and 0.75 are 0.8281 and 0.5176 respectively. The Mask-RCNN model with ResNet-101 backbone demonstrates 0.5084 of mean average precision for the similar configurations as mentioned in Table 1. The average precision at IoU 0.5 and 0.75 are 0.8632 and 0.6157 respectively. The summary of the results are shown in Table 2. We obtain the best results for ResNet-101 backbone architecture.

Table 2. Results for test imagery dataset

Model	mAP@[IoU = 0.5-0.95]	AP@IoU = 0.5	AP@IoU = 0.75
U-Net	0.1564	0.4672	0.3041
Mask-RCNN (Backbone = ResNet-50)	0.4392	0.8281	0.5176
Mask-RCNN (Backbone = ResNet-101)	**0.5084**	**0.8632**	**0.6157**

Figure 3 demonstrates the detailed results of boundary detection on few images from our test dataset: (a) Input image (b) Ground truth mask (c) Predicted Wound Mask on ulcer (d) Predicted Binary Mask (e) Predicted contour of ulcer. Figure 4 illustrates few more segmentation results. Green contour indicates the ground truth boundary and the red contour indicates the predicted boundary.

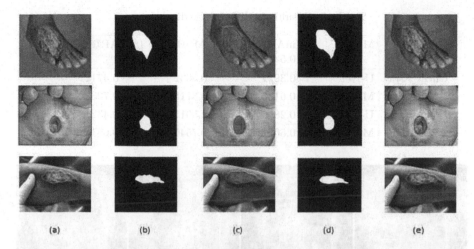

(a)   (b)   (c)   (d)   (e)

Fig. 3. Visualization of qualitative results for several neuropathic ulcers.

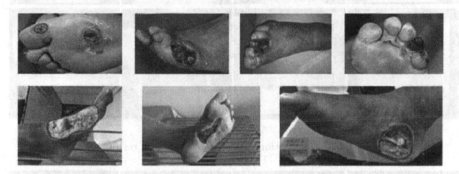

Fig. 4. Visualization of ulcer boundary detection results.

5.3   Results on Different Medical Datasets

In addition to the foot ulcer dataset we consider two different medical datasets namely chest X-ray dataset [2] and polyps dataset [3]. Dataset [2,3] consists of 662 number of X-ray chest imagery and 300 number of polyps imagery captured from colonoscopy videos respectively. Both datasets provide the ground truth masks. We train the Mask-RCNN model per each dataset with the configurations as mentioned in Table 1. These two datasets also split using the same proportion as mentioned in Sect. 3.

Table 3 summarize the obtained results for dataset [2,3]. Furthermore, Figs. 5 and 6 visualize the segmentation results for selected sample imagery from test set of both datasets. Green contour indicates the ground truth boundary and the red contour indicates the predicted boundary.

Table 3. Results on different medical datasets

Dataset	Model	mAP@[IoU = 0.5-0.95]	AP@IoU = 0.5	AP@IoU = 0.75
Chest X-ray	U-Net	0.3021	0.5632	0.4712
	Mask-RCNN	0.6135	0.8715	0.7230
Polyps	U-Net	0.2896	0.5017	0.4321
	Mask-RCNN	0.5031	0.7546	0.6013

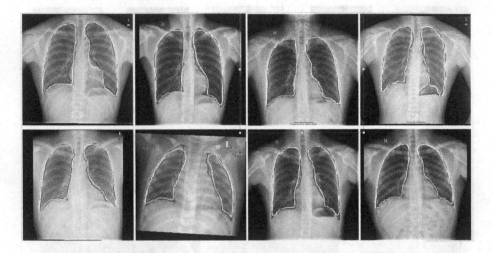

Fig. 5. Visualization of lung segmentation results.

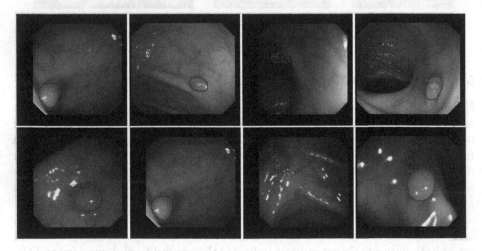

Fig. 6. Visualization of polyps segmentation results.

6   Conclusions and Future Work

In this paper, we demonstrate an automated solution for ulcer detection and segmentation. The proposed solution based on Mask-RCNN model, which had designed for object detection, localization and instance segmentation of imagery. The proposed approach gave a promising results in terms of accuracy and performance compared to U-Net model. Overall, this solution can be considered as a promising approach to replace the manual measurements of ulcers, which can benefit both patients and clinicians. In order to improve the model in terms of accuracy and performance, we can add more data points to the existing dataset or apply different augmentation techniques such as jittering, distortions etc. Furthermore, this automated process will help in further implementations such as determine the various pathology of neuropathic ulcers as multi-class classification, implementation of automatic annotating tool which can automatically detect and classify the ulcers and related pathology, and implementation of different systems such as mobile applications and computer vision based remote telemedicine systems for the detection of ulcers and provide feedbacks. According to the experiments mentioned in Sect. 5.3, the Mask-RCNN model provides significant improvement in medical image segmentation task with respective to the baseline model. Moreover, this segmentation approach can spread on other medical imagery segmentation tasks such as brain tumor segmentation, skin lesion segmentation etc.

References

1. Britannica, T.E.o.E.: Diabetes mellitus, May 2019. https://www.britannica.com/science/diabetes-mellitus
2. Shenzhen hospital chest x-ray set. https://www.kaggle.com/yoctoman/shcxr-lung-mask
3. CVC colon DB. http://mv.cvc.uab.es/projects/colon-qa/cvccolondb
4. Everingham, M., Van Gool, L., Williams, C.K., Winn, J., Zisserman, A.: The pascal visual object classes (VOC) challenge. Int. J. Comput. Vis. **88**(2), 303–338 (2010). https://doi.org/10.1007/s11263-009-0275-4
5. Filko, D., Antonic, D., Huljev, D.: Wita—application for wound analysis and management. In: The 12th IEEE International Conference on e-Health Networking, Applications and Services, pp. 68–73. IEEE (2010). https://doi.org/10.1109/HEALTH.2010.5556533
6. Goyal, M., Yap, M.H., Reeves, N.D., Rajbhandari, S., Spragg, J.: Fully convolutional networks for diabetic foot ulcer segmentation. In: 2017 IEEE International Conference on Systems, Man, and Cybernetics (SMC), pp. 618–623. IEEE (2017). https://doi.org/10.1109/SMC.2017.8122675
7. He, K., Gkioxari, G., Dollár, P., Girshick, R.: Mask R-CNN. In: Proceedings of the IEEE International Conference on Computer Vision, pp. 2961–2969 (2017). https://doi.org/10.1109/TPAMI.2018.2844175
8. Johnson, J.W.: Adapting Mask-RCNN for automatic nucleus segmentation. arXiv preprint arXiv:1805.00500 (2018). https://doi.org/10.1007/978-3-030-17798-0

9. Jones, T.D., Plassmann, P.: An active contour model for measuring the area of leg ulcers. IEEE Trans. Med. Imaging **19**(12), 1202–1210 (2000). https://doi.org/10. 1109/42.897812
10. Kaggle RSNA pneumonia detection challenge explained. https://medium. com/@sebastiannorena/kaggle-rsna-pneumonia-detection-challenge-explained- c140b19bf903
11. LeCun, Y., Bengio, Y., Hinton, G.: Deep learning. Nature **521**(7553), 436 (2015). https://doi.org/10.1038/nature14539
12. Li, Y., Qi, H., Dai, J., Ji, X., Wei, Y.: Fully convolutional instance-aware semantic segmentation. In: Proceedings of the IEEE Conference on Computer Vision and Pattern Recognition, pp. 2359–2367 (2017). https://doi.org/10.1109/CVPR.2017. 472
13. Lin, T.-Y., et al.: Microsoft COCO: common objects in context. In: Fleet, D., Pajdla, T., Schiele, B., Tuytelaars, T. (eds.) ECCV 2014. LNCS, vol. 8693, pp. 740–755. Springer, Cham (2014). https://doi.org/10.1007/978-3-319-10602-1_48
14. Mowa-mobile wound analyzer. http://www.diabetesincontrol.com/mowa-mobile- wound-analyzer-wound-care-solutions-ulcer-management/
15. Pictzar digital planimetry software. http://www.pictzar.com/
16. Perez, A.A., Gonzaga, A., Alves, J.M.: Segmentation and analysis of leg ulcers color images. In: Proceedings International Workshop on Medical Imaging and Augmented Reality, pp. 262–266. IEEE (2001). https://doi.org/10.1109/MIAR. 2001.930300
17. Piemonte, L.: Type 2 diabetes. https://idf.org/52-about-diabetes.html
18. Ren, S., He, K., Girshick, R., Sun, J.: Faster R-CNN: towards real-time object detection with region proposal networks. In: Advances in Neural Information Processing Systems, pp. 91–99 (2015). https://doi.org/10.1109/TPAMI.2016.2577031
19. Ronneberger, O., Fischer, P., Brox, T.: U-net: convolutional networks for biomedical image segmentation. In: Navab, N., Hornegger, J., Wells, W.M., Frangi, A.F. (eds.) MICCAI 2015. LNCS, vol. 9351, pp. 234–241. Springer, Cham (2015). https://doi.org/10.1007/978-3-319-24574-4_28
20. Russakovsky, O., et al.: ImageNet large scale visual recognition challenge. Int. J. Comput. Vision **115**(3), 211–252 (2015). https://doi.org/10.1007/s11263-015- 0816-y
21. Silhouette. https://www.aranzmedical.com/
22. Vgg image annotator (via). http://www.robots.ox.ac.uk/~vgg/software/via/
23. Veredas, F., Mesa, H., Morente, L.: Binary tissue classification on wound images with neural networks and bayesian classifiers. IEEE Trans. Med. Imaging **29**(2), 410–427 (2009). https://doi.org/10.1109/TMI.2009.2033595
24. Wang, C., et al.: An unified framework for automatic wound segmentation and analysis with deep convolutional neural networks. In: 2015 37th Annual International Conference of the IEEE Engineering in Medicine and Biology Society (EMBC), pp. 2415–2418. IEEE (2015). https://doi.org/10.1109/EMBC.2015.7318881
25. Wang, L., Pedersen, P.C., Agu, E., Strong, D.M., Tulu, B.: Area determination of diabetic foot ulcer images using a cascaded two-stage SVM-based classification. IEEE Trans. Biomed. Eng. **64**(9), 2098–2109 (2016). https://doi.org/10.1109/ TBME.2016.2632522

Capsule Networks for Attention Under Occlusion

Antonio Rodríguez-Sánchez[✉] and Tobias Dick

Department of Computer Science, University of Innsbruck, Innsbruck, Austria
antonio.rodriguez-sanchez@uibk.ac.at, tobias.dick@student.uibk.ac.at
https://iis.uibk.ac.at

Abstract. Capsule Neural Networks (CapsNet) serve as an attempt to model the neural organization in biological neural networks. Through the routing-by-agreement algorithm, the attention mechanism is implemented as individual capsules that focus on specific upstream capsules while ignoring the rest. By using the routing algorithm, CapsNets are able to attend overlapping digits from the MNIST dataset. In this work, we evaluate the attention capabilities of Capsule Networks using the routing-by-agreement with occluded shape stimuli as presented in neurophysiology. We do so by implementing a more compact type of capsule network. Our results in classifying the non-occluded as well as the occluded shapes show that indeed, CapsNets are able to differentiate occlusions from near-occlusion situations as in real biological neurons. In our experiments, performing the reconstruction of the occluded stimuli also shows promising results.

Keywords: Capsule Networks · Overlapping datasets · Deep learning

1 Introduction

In recent years, biological inspiration - and even biological realism - has become of great interest in the machine learning community, especially in deep learning. The visual cortex seems to be organized into areas where neurons perform similar tasks. The first researchers to shed some light on how visual information may be processed in the primate visual cortex were Hubel and Wiesel [1]. In this (more or less) hierarchical organization of the visual cortex, the object recognition pathway starts in Visual Area 1 (V1) and ends in the inferotemporal cortex (IT). In an intermediate area, Visual Area 4 (V4), cells are selective to local curvatures in such a way that groups of those cells can be considered to encode shapes [2]. Fukushima's Neocognitron [3] was the first model inspired on neurons of the visual cortex, the focus was V1 neurons, namely the simple and complex cells that Hubel and Wiesel found. Before the advent of deep artificial neural networks, biologically plausible neural networks were hand designed and first compared to real neuronal responses evaluated on stimuli used in neurophysiology [4–6], after which they were evaluated on real computer vision images [7,8].

© Springer Nature Switzerland AG 2019
I. V. Tetko et al. (Eds.): ICANN 2019, LNCS 11731, pp. 523–534, 2019.
https://doi.org/10.1007/978-3-030-30493-5_50

Convolutional neural networks (CNNs), first introduced in 1989 [9], were inspired by the Neocognitron and have been extensively used in almost every computer vision application during the present decade. Many variations of CNNs have been developed with different goals in mind, such as object recognition [10], mobile vision [11], large-scale image recognition [12] or medical image analysis [13], just to name a few. Differently to the earlier biologically inspired models, CNNs achieve much better results on computer vision tasks but work more like a black box. Thus the recent interest in evaluating these algorithms with psychological or neurological stimuli in order to better analyze what is encoded inside these networks and evaluate their biological plausibility or improve them including biological inspiration. Examples of this include evaluating the classical Gestalt principles, like similarity [14,15], symmetry [16] or closure [17].

Similarly to CNNs, Capsule Networks (or CapsNets) cover regions of the image and use convolutional layers. But they overcome one important limitation of CNNs, that of using scalar feature values for neuronal activations, CapsNets instead use capsules [18] as outputs, which are vector valued groups of neurons. Additionally, CapsNets do not use max-pooling, thus keeping important information about items location inside a region (e.g. the part of an object). Training and inference is usually performed using a dynamic routing algorithm between capsules of different layers [19]. In that work, Sabour and colleagues were able to show excellent results on a dataset consisting of two MNIST [20] overlapping digits with an average overlap of 80%, called MultiMNIST. CapsNet classification accuracy on this dataset was 95% using just a 3 layer network. This is the same accuracy Ba et al. [21] achieved with a much deeper more traditional CNN model on a dataset overlapping MNIST digits just like MultiMNIST but with less than 4% overlap. During training CapsNets, the routing-by-agreement algorithm has the ability to select upstream capsules that are useful for the final result while disregarding others without relying exclusively on the training of weights, which would be a way of implementing attention in an artificial neural network. CapsNets were able to extract the instantiation parameters of both overlapped digits and reconstruct them using a comparatively simple fully connected network.

In the seminal paper by Bushnell et al. [22], they created different images for studying shape occlusion responses from biological neurons (located in area V4). These images resulted from a combination of the so called primary shapes and contextual stimuli. Primary shapes correspond to shapes that are obtained from a partially occluded full circle, citing Bushnell: *The primary shapes were designed such that, when presented adjoining the corresponding contextual stimulus, the percept is that of a partially occluded circle.* On the other hand, contextual stimuli would correspond to the occluding shapes. Bushnell et al. showed how neurons encoded occluded stimuli in area V4 of the monkey. V4 neurons seem to suppress accidental contours (those present in occlusion) and that the spatial proximity of contextual stimulus alone is not sufficient for suppression. Neurons in area V4 are selective to specific shapes, be occluded or not. In this work we want to test the ability of CapsNets to detect occlusion patterns. We selected the 2D

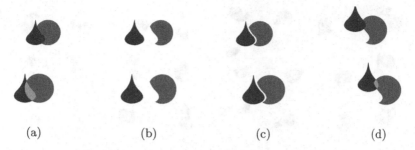

(a)   (b)   (c)   (d)

Fig. 1. Some sample images generated following Bushnell et al. [22]. The top row shows the overlap as occlusion, the bottom row highlights the overlapping regions in light blue. (a) A tear shape is partially occluding a full circle. (b) A primary shape (moon-like) next to a tear shape. (c) Same case as b, but shapes much closer (almost occluding) (d) This example shows the tear shape overlapping a primary shape (Color figure online)

shapes from Bushnell et al. [22] in order to investigate if the neurons in CapsNets perform a similar suppression as V4 neurons in the presence of occlusions.

2   Methods

2.1   Stimuli

We created more than one million images similar to the ones in the influential work of Bushnell et al. [22]. In that work, primary shapes were combined with context stimuli (the rest of shapes). Figure 1 shows an example where a tear shape is present with a primary shape. In Fig. 1a the primary shape is the result of a partially occluded full circle, while in the cases of Fig. 1b-d, it does not correspond to an occluded full circle. We will then consider two conditions:

1. Occlusion: In the case of occlusion, a shape is occluding a circle or a context shape. These correspond to the cases in Fig. 1a and d. In Fig. 1a, a tear shape is next to a primary shape, i.e. occluding a circle, in Fig. 1d, the shape occluded is not a circle (but a primary shape).
2. Non-occlusion: In this case, there is no occlusion. These correspond to the cases in Fig. 1b and c. In the former, there is large gap between the two shapes (primary and context shapes), in the latter, there is still no overlap in the image even though the shapes are much closer.

Note that the shape classes present in Fig. 1b and c are the same as in Fig. 1d. While the shape classes in Fig. 1a contains a tear shape as in the other three cases, but the other stimulus is different, i.e. a full circle (partially occluded), as opposed to a primary shape (moon-like) as in the other three.

In order to create the more than one million images, we considered much more variability than in Bushnell et al. [22]. Both shapes are placed in random

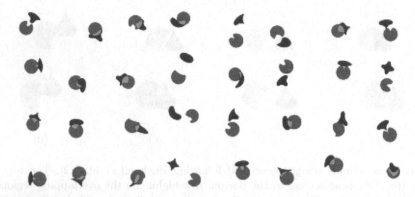

Fig. 2. Some examples from the used dataset are shown. These are examples of the actual images used for training. Different colors are shown only for illustration purposes. (Color figure online)

locations with a small random distance between them in any direction. Further variation was achieved by adding random rotation. Stimuli were normalized to have the diameter of the circle. Examples used in the experimental evaluation for both cases are shown in Fig. 2 and include combinations of a primary shape or a full circle with a context shape (the rest). Note that our CapsNet is only fed a single channel image containing both shapes in white over a black background (as in the insets of Fig. 4), thus making it non-trivial to detect the overlapping patterns.

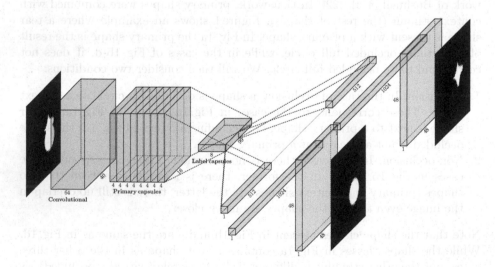

Fig. 3. A visual representation of the CapsNet used in this work.

2.2 CapsNet Architecture and Training

CapsNets were introduced by Hinton et al. [18]. To summarize, capsules in Capsule Networks are groups of neurons that represent an object or a part of an object in a parse tree. The output of a capsule is a so called activation or instantiation vector representing properties of this object. The length of a vector encodes the probability whether an object is present or not. CapsNets are usually trained with the routing-by-agreement algorithm [19] which routes instantiation vectors from lower level capsules to upper level capsules. In this architecture the output of a convolutional layer with multiple filters is vectorized by the primary capsules layer to create the input to the first capsule layer. The routing-by-agreement algorithm then routes the vectors from this first capsule layer to the output capsules.

The first task is to classify which shapes are in an image independently if there is occlusion or not. For example, the output neurons active for the cases in Fig. 1b, c and d should be the same, namely, a moon-like and a tear shape. In the first two cases, there is no occlusion, but in the third one, the moon-like shape is partially occluded. The shape classes in Fig. 1a contains also a tear shape, but the other class is different (a circle).

Figure 3 shows our CapsNet architecture where a context shape overlapping a primary shape is used as an example input along with its ideal respective reconstruction output. Each block represents one layer with subdivisions indicating capsule channel boundaries. The gray block represents the operation of calculating the length of the label capsule outputs. The three layers belonging to the classification part are shown in light blue while the reconstruction layers are shown in light red. The first and second reconstruction layers share their weights. Even though, this network architecture shares similarities to the CapsNet implementation used for Multi-MNIST [19], we implemented the following modifications for the task at hand:

- The size of the input images is larger (three tests: 36×36, 48×48 and 60×60 pixels), meaning a larger number of input neurons.
- The task consists of many more classes (99), thus, a larger number of output neurons. These 99 classes correspond to the different shapes used: 1 full circle, 49 primary shapes (representing incomplete circles) and 49 contextual stimuli (occluding shapes).
- Less instantiation parameters during reconstruction (8-D). This is because in our case the stimuli is also rotated and shifted, but unlike in Multi-MNIST left undistorted.
- For the first layer (a convolutional layer), we used a kernel side length of 9 pixels and 64 filters. Activation function remains ReLU.
- For the primary capsules layer we use 8 channels of 4-D capsules with a kernel side length of 9 pixels and a stride of 2.

These modifications on the CapsNet [19] has several effects over the network. The first one is that the total number of capsule outputs is higher than the amount of channels because of the weight sharing achieved through the convolution. In total, hidden output capsules are $16 \times 16 \times 8$ 4-D vectors for the 48×48 case. The hidden capsule outputs are activated using the following squashing function for normalizing the length of the vectors to be in the range $[0, 1]$ [19]:

$$\text{squash}(\mathbf{x}) := \frac{||\mathbf{x}||^2}{1 + ||\mathbf{x}||^2} \frac{\mathbf{x}}{||\mathbf{x}||} \tag{1}$$

The label capsules layer will encode the predicted class in their vector length as well as the instantiation values of the input images. We use 8-D capsules for representing the learned instantiation parameters in each class. This layer uses the trained weights for each capsule in order to project each of the primary capsules layer outputs to the output space, which in our case corresponds to the labels. These values are then used in a weighted sum using weights determined by the dynamic routing algorithm. During each pass, these weights are initialized to $1/n_c$, where $n_c = 99$ in our case (99 classes). The algorithm iteratively compares the orientation of the squashed result (Eq. 1) of the weighted sum with the orientation of each weighted primary capsules layer. We use 3 iterations of the routing algorithm.

The second task is the reconstruction of the occluded or non-occluded shape and its occluding or non occluding counterpart. For the reconstruction, the values of the actual label capsule activation vectors (and not just their length, used for classification) are used. The output layer is trained separately for the different context shapes as well as the primary shapes while the rest of the network is shared (Fig. 3). The weights of two fully connected layers consisting of 512 and 1024 hidden units respectively are shared. Both of these layers are activated using the following leaky ReLU function [23].

$$\text{leaky\_relu}(x) := \begin{cases} x & \text{if } x \geq 0 \\ 0.01 \cdot x & \text{if } x < 0 \end{cases} \tag{2}$$

The leakiness helps at the beginning of training where the inputs to the reconstruction network are still largely random. Without leakiness some parts of the network might be inadvertently disabled during training [23]. The output layer is activated using a sigmoid function and is reshaped and interpreted as a image of the size of the input image. The weights corresponding to the primary shapes or full circle and context stimuli are trained separately.

For this task we define a loss function consisting of three terms:

$$L = \sum_{k=1}^{99} L_{c_k} + \frac{L_o}{100} + \frac{L_r}{1000} \tag{3}$$

where L is the total loss term minimized during training and L_{c_k} is the margin loss for existence of class k as given by [19]:

$$L_{c_k} = \mathbf{y}_k \max\{0, 0.9 - ||\mathbf{v}_k||\}^2 + 0.5 (1 - \mathbf{y}_k) \max\{0, ||\mathbf{v}_k|| - 0.1|\}^2 \tag{4}$$

where $\mathbf{y}_k$ is 1 if a shape from class k is in the image and 0 otherwise. $\mathbf{v}_k$ is the label capsule output vector for class k. L_o is an additional squared error loss term specific for the circle class, as it is the most relevant shape class to encourage higher activation margins:

$$L_o = (\mathbf{y}_1 - ||\mathbf{v}_1||)^2 \tag{5}$$

$\mathbf{y}_1$ is the label for the full circle class and $\mathbf{v}_1$ is the label capsule output for the full circle class. The reconstruction loss term L_r is an average of the sum of squared errors between the original input shapes and their reconstruction as given by:

$$L_r = \frac{1}{2} \left(\sum_{i=1}^{l} \sum_{j=1}^{l} \left(\mathbf{r}_{ij}^{(c)} - \mathbf{x}_{ij}^{(c)} \right)^2 + \sum_{i=1}^{l} \sum_{j=1}^{l} \left(\mathbf{r}_{ij}^{(p)} - \mathbf{x}_{ij}^{(p)} \right)^2 \right) \tag{6}$$

where $\mathbf{r}^{(c)}$ is the reconstructed context shape, $\mathbf{x}^{(c)}$ is the original input context shape, $\mathbf{r}^{(p)}$ is the reconstructed primary shape and $\mathbf{x}^{(p)}$ is the input primary shape.

The training set consists of one million images. In order to avoid overfitting, during each epoch, 40000 additional dynamically generated training samples were used and not reused again. The validation and test sets contain 100000 samples. The test set was used to experimentally determine good model parameters. The validation set was used to determine the final accuracy and error numbers. Context stimuli were equally represented, which then were combined with either a primary shape (incomplete circle) or a full circle with equal probability. In total, for training we generated 37500 images per input stimulus.

We use the default Adam optimizer [24] as implemented in TensorFlow [25] keeping the default parameter values. Additionally we use an exponential learning rate decay as follows:

$$lr = 0.001 \cdot 0.95^{\left\lfloor \frac{i_{\text{batch}}}{10000} \right\rfloor} \tag{7}$$

where lr is the resulting learning rate for the respective batch and i_{batch} is a simple counter that is incremented after each batch [26]. The batch size used is 32.

3   Experimental Evaluation

We did our evaluation on accuracy, precision, recall and $F1$-score. Accuracy is calculated by determining if the correct class is one of the two longest label capsule activations of the network:

$$A_c = \frac{1}{2} \cdot \sum_{k=1}^{99} \mathbf{y}_k \cdot f(\mathbf{v}, k) \tag{8}$$

$$f(\mathbf{v}, k) := \begin{cases} 1 & \text{if } ||\mathbf{v}_k|| = \max ||\mathbf{v}|| \\ 1 & \text{if } ||\mathbf{v}_k|| = \max\{x | x \in ||\mathbf{v}|| \wedge x \neq \max ||\mathbf{v}||\} \\ 0 & \text{otherwise} \end{cases} \tag{9}$$

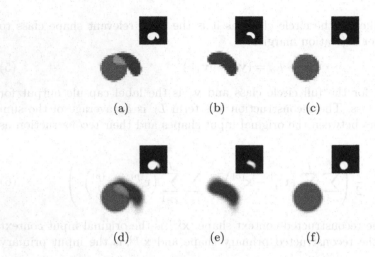

(a)   (b)   (c)

(d)   (e)   (f)

Fig. 4. Occlusion sample and reconstructed stimuli. The stimuli are shown in different colors only for illustration purposes. The real input to CapsNet and its reconstruction are shown in the inset (white shapes over black background). (a-c) is the input while (d-f) is the reconstruction. (Color figure online)

where v_k is the label capsule output vector for class k, f checks if v_k is one of the two longest vectors in v, which indicates the presence of class k in the input image. If both image stimuli are classified correctly, $A_c = 1$, if only one is correct, $A_c = 0.5$, otherwise (both are wrong), it is 0.

For the first task, namely the classification of the two stimuli present in every image under the presence or absence of occlusion, we achieved overall (99 classes) classification accuracies of 83.3% ($stdev = 0.004\%$), 90.8% ($stdev = 0.003\%$) and 91.8% ($stdev = 0.003\%$) on the test set, for input size images 36×36, 48×48, and 60×60 pixels respectively. If we consider the case of detecting the overlap condition where a full circle is in the image partially occluded (as in Fig. 1a), i.e. detecting that it is a full circle vs. one of the primary shapes (as in Fig. 1b-d), accuracy rises to 96.3%, 97.7% and 97.1% respectively. For this latter case, we can evaluate precision, recall and $F1$-score as a combination of the true positives T_p, false positives F_p and false negatives F_n:

$$P = \frac{T_p}{T_p + F_p}, \ R = \frac{T_p}{T_p + F_n}, \ F1 = \frac{2PR}{P + R} \tag{10}$$

where P is the precision and R is the recall. The $F1$ measure combines both values into one so that false positives and false negatives are taken into account in this one value. $F1$-score were 0.964, 0.978 and 0.972 for the corresponding three image sizes (36×36, 48×48, and 60×60 pixels). See Table 1 for the quantitative results.

Table 1. CapsNet evaluation on the validation set after 20 epochs of training on three image sizes.

Metric	36×36	48×48	60×60
Classification accuracy (overall)	83.3%	90.8%	91.8%
Classification accuracy (circle)	96.3%	97.7%	97.1%
Precision (circle)	93.1%	95.6%	94.5%
Recall (circle)	100%	100%	100%
F1-score (circle)	0.964	0.978	0.972
Reconstruction error (overall)	0.19	0.19	0.20

For the case of the reconstruction task, we define the reconstruction error as:

$$E_r = \sqrt{\sum_{i=1}^{l}\sum_{j=1}^{l}\left(\mathbf{r}_{ij}^{(c)} - \mathbf{x}_{ij}^{(c)}\right)^2 + \sum_{i=1}^{l}\sum_{j=1}^{l}\left(\mathbf{r}_{ij}^{(p)} - \mathbf{x}_{ij}^{(p)}\right)^2} \tag{11}$$

where $\mathbf{r}^{(c)}$ is the reconstructed context shape, $\mathbf{x}^{(c)}$ is the original input context shape, $\mathbf{r}^{(p)}$ is the reconstructed primary shape and $\mathbf{x}^{(p)}$ is the input primary shape.

We obtained reconstruction errors of 0.19 to 0.20 (Table 1) depending on the images input size. Figure 4 shows an example of stimuli reconstruction. Figure 4a is the only input being supplied to the network (without colors, as shown in the black background inset), while Fig. 4b, c are the two stimuli that were combined for that input image. Figure 4d is generated by overlapping the two individual reconstructed images (Fig. 4e and f).

We can see that even though the reconstruction is not perfect, it is very close to the original stimuli before occlusion. The network was able to do so by correctly determining the instantiation parameters of the shapes although the reconstruction of the fine details was missing. This is somewhat expected as the reconstruction network would have to store all possible stimuli in its parameters.

4   Discussion and Conclusions

Deep artificial neural networks provide exceptional performance in almost every field of computer science that requires machine learning, such as computer vision, natural language processing, speech and audio recognition or bioinformatics. Probably, the most prominent example of deep learning are Convolutional Neural Networks (CNNs), first introduced in 1989 [9]. CNNs have achieved the best results on most computer vision tasks but work like a black box. Thus the recent interest in evaluating these algorithms with psychological or neurological stimuli in order to better analyze what is encoded inside these networks and evaluate their biological plausibility or improve them including biological inspiration.

Several works have proposed that CNNs are similar to how the primate visual cortex [27,28] process visual information.

On the other hand, CNNs suffer from several limitations (e.g. limited capability to encode position and orientation) due to the use of max-pooling, scalar feature values and learning based purely on the backpropagation algorithm, whose biological plausibility has been and still is a subject of discussion [29]. Even though we cannot solve all at once, there are recent steps on the right direction to overcoming them. Our belief is that Capsule Networks is one of those steps, instead of just computing an error in the output layer and backpropagating it through the hidden layers, CapsNets combined with the routing-by-agreement algorithm provide an additional feedback in form of agreement which is generated on each layer. The routing-by-agreement is more consistent with biological models in terms of how to use feedback connections, which is thought to be driven by attention mechanisms. The Dynamic Routing algorithm [30] was such a biologically-plausible model, where control neurons would dynamically modify the intracortical connections in order to route regions from lower areas to higher areas, in striking resemblance to CapsNets and routing-by-agreement. In the primate visual cortex, one such higher areas include Visual Area 4 (V4) which represent shapes in terms of groups of neurons encoding their contours and curvature features.

Rich natural scenes include partially occluded objects. The work of Bushnell et al. [22] studies how neurons in area V4 respond to occluded stimuli. The hypothesis is that those responses are modulated through feedback attention, thus active neurons would be those that represent the correct full unoccluded shape. In our study we have tested if this modulation can be simulated through the use of CapsNets as a deep learning architecture that uses capsules as containing groups of neurons to represent objects based on their properties. More importantly, to investigate if the routing-by-agreement algorithm can serve to simulate such an attentional mechanism, thus modulating the responses of those neurons. We have shown that CapsNet can detect partially-occluded objects even with a high degree of occlusion. Table 1 shows that this is the case independently of the resolution of the input image, although higher resolution is linked to slightly higher accuracy. Shape classification when any two shapes are in the image is an outstanding 91.8%. This number was even higher (97.7%) when the task was to differentiate among a primary shape (unoccluded) or an occluded full circle, thus proving that CapsNets can attend and reconstruct an occluded shape. The reconstruction of the stimuli was achieved also to a high degree of accuracy even though some fine details were missing. Both these experiments show the capabilities of the CapsNet networks at attending one part of the image and neglect the occluding stimulus. In light to the results obtained in this work, we can consider for future work a more detailed analysis on the capabilities of CapsNet. As the network is able to roughly reconstruct the original input stimuli using the learned instantiation parameters produced by the classification network, we could additionally use Shahroudnejad et al. [31] methodology to analyze the effect of the learned instantiation parameters in order to obtain more insight into what is exactly what the network attends to inside the shapes. Additionally, a comparison with the performance of classical CNNs like AlexNet [10], MobileNets [11], VGG [12] and any of the variants that include attention gates [32] could be useful to further show the capabilities of Capsule Neural Networks.

Acknowledgments. We would like to thank Sebastian Stabinger for his useful comments and Prof. Anitha Pasupathy for providing the program to create the single shape stimuli.

References

1. Hubel, D.H., Wiesel, T.N.: Receptive fields, binocular interaction and functional architecture in the Cat's visual cortex. J. Physiol. **160**(1), 106–154 (1962)
2. Pasupathy, A., Connor, C.E.: Shape representation in area V4: position-specific tuning for boundary conformation. J. Neurophysiol. **86**(5), 2505–2519 (2001)
3. Fukushima, K., Wake, N.: Handwritten alphanumeric character recognition by the neocognitron. IEEE Trans. Neural Netw. **2**(3), 355–365 (1991)
4. Riesenhuber, M., Poggio, T.: Hierarchical models of object recognition in cortex. Nat. Neurosci. **2**(11), 1019 (1999)
5. Rodríguez-Sánchez, A., Tsotsos, J.: The roles of endstopped and curvature tuned computations in a hierarchical representation of 2D shape. PLoS ONE **7**(8), e42058 (2012)
6. Rodríguez-Sánchez, A., Oberleiter, S., Xiong, H., Piater, J.: Learning V4 curvature cell populations from sparse endstopped cells. In: Villa, A.E.P., Masulli, P., Pons Rivero, A.J. (eds.) ICANN 2016. LNCS, vol. 9887, pp. 463–471. Springer, Cham (2016). https://doi.org/10.1007/978-3-319-44781-0_55
7. Serre, T., Wolf, L., Bileschi, S., Riesenhuber, M., Poggio, T.: Robust object recognition with cortex-like mechanisms. IEEE Trans. Pattern Anal. Mach. Intell. **3**, 411–426 (2007)
8. Rodríguez-Sánchez, A.J., Tsotsos, J.K.: The importance of intermediate representations for the modeling of 2D shape detection: endstopping and curvature tuned computations. In: CVPR 2011, June 2011, pp. 4321–4326 (2011)
9. LeCun, Y., et al.: Backpropagation applied to handwritten zip code recognition. Neural Comput. **1**(4), 541–551 (1989)
10. Krizhevsky, A., Sutskever, I., Hinton, G.E.: ImageNet classification with deep convolutional neural networks. In: Advances in Neural Information Processing Systems, pp. 1097–1105 (2012)
11. Howard, A.G., et al.: MobileNets: efficient convolutional neural networks for mobile vision applications. arXiv preprint arXiv:1704.04861 (2017)
12. Simonyan, K., Zisserman, A.: Very deep convolutional networks for large-scale image recognition. arXiv preprint arXiv:1409.1556 (2014)
13. Ronneberger, O., Fischer, P., Brox, T.: U-Net: convolutional networks for biomedical image segmentation. In: Navab, N., Hornegger, J., Wells, W.M., Frangi, A.F. (eds.) MICCAI 2015. LNCS, vol. 9351, pp. 234–241. Springer, Cham (2015). https://doi.org/10.1007/978-3-319-24574-4_28
14. Stabinger, S., Rodríguez-Sánchez, A.: Evaluation of deep learning on an abstract image classification dataset. In: Proceedings of the IEEE Conference on Computer Vision and Pattern Recognition, pp. 2767–2772 (2017). Workshop on Mutual Benefits of Cognitive and Computer Vision (MBCC)

15. Kim, J., Ricci, M., Serre, T.: Not-so-CLEVR: visual relations strain feedforward neural networks (2018)
16. Stabinger, S., Rodríguez-Sánchez, A., Piater, J.: 25 years of CNNs: can we compare to human abstraction capabilities? In: Villa, A.E.P., Masulli, P., Pons Rivero, A.J. (eds.) ICANN 2016. LNCS, vol. 9887, pp. 380–387. Springer, Cham (2016). https://doi.org/10.1007/978-3-319-44781-0_45
17. Kim, B., Reif, E., Wattenberg, M., Bengio, S.: Do neural networks show Gestalt phenomena? An exploration of the law of closure. arXiv preprint arXiv:1903.01069 (2019)
18. Hinton, G.E., Krizhevsky, A., Wang, S.D.: Transforming auto-encoders. In: Honkela, T., Duch, W., Girolami, M., Kaski, S. (eds.) ICANN 2011. LNCS, vol. 6791, pp. 44–51. Springer, Heidelberg (2011). https://doi.org/10.1007/978-3-642-21735-7_6
19. Sabour, S., Frosst, N., Hinton, G.E.: Dynamic routing between capsules. In: Advances in Neural Information Processing Systems, pp. 3859–3869 (2017)
20. LeCun, Y.: MNIST handwritten digit database. http://yann.lecun.com/exdb/mnist/. Accessed 05 Mar 2019
21. Ba, J., Mnih, V., Kavukcuoglu, K.: Multiple object recognition with visual attention. arXiv preprint arXiv:1412.7755 (2014)
22. Bushnell, B.N., Harding, P.J., Kosai, Y., Pasupathy, A.: Partial occlusion modulates contour-based shape encoding in primate area V4. J. Neurosci. 31(11), 4012–4024 (2011)
23. Maas, A.L., Hannun, A.Y., Ng, A.Y.: Rectifier nonlinearities improve neural network acoustic models. In: Proceedings of ICML, vol. 30, no. 1, p. 3 (2013)
24. Kingma, D.P., Ba, J.: Adam: a method for stochastic optimization. arXiv preprint arXiv:1412.6980 (2014)
25. Abadi, M., et al.: TensorFlow: a system for large-scale machine learning. In: 12th USENIX OSDI 2016, pp. 265–283 (2016)
26. Tensorflow Contributors: tf.train.exponential_decay. https://www.tensorflow.org/api_docs/python/tf/train/exponential_decay. Accessed 05 Mar 2019
27. Cadieu, C.F., et al.: Deep neural networks rival the representation of primate it cortex for core visual object recognition. PLoS Comput. Biol. 10(12), e1003963 (2014)
28. LeCun, Y., Bengio, Y., Hinton, G.E.: Deep learning. Nature 521(7553), 436 (2015)
29. Crick, F.: The recent excitement about neural networks. Nature 337(6203), 129–132 (1989)
30. Olshausen, B.A., Anderson, C.H., Van Essen, D.C.: A neurobiological model of visual attention and invariant pattern recognition based on dynamic routing of information. J. Neurosci. 13(11), 4700–4719 (1993)
31. Shahroudnejad, A., Afshar, P., Plataniotis, K.N., Mohammadi, A.: Improved explainability of capsule networks: relevance path by agreement. In: 2018 IEEE Global Conference on Signal and Information Processing (GlobalSIP). IEEE, pp. 549–553 (2018)
32. Fu, J., Zheng, H., Mei, T.: Look closer to see better: recurrent attention convolutional neural network for fine-grained image recognition. In: Proceedings of the IEEE Conference on Computer Vision and Pattern Recognition, pp. 4438–4446 (2017)

IP-GAN: Learning Identity and Pose Disentanglement in Generative Adversarial Networks

Bassel Zeno[1]($\boxtimes$) (ID), Ilya Kalinovskiy[2] (ID), and Yuri Matveev[1] (ID)

[1] ITMO University, Saint Petersburg, Russia
basilzeno@gmail.com, matveev@mail.ifmo.ru
[2] STC-innovations Ltd., Saint Petersburg, Russia
kalinovskiy@speechpro.com

Abstract. Synthesizing realistic multi-view face images from a single-view input is an effective and cheap way for data augmentation. In addition it is promising for more efficiently training deep pose-invariant models for large-scale unconstrained face recognition. It is a challenging generative learning problem due to the large pose discrepancy between the synthetic and real face images, and the need to preserve identity after generation. We propose IP-GAN, a framework based on Generative Adversarial Networks to disentangle the identity and pose of faces, such that we can generate face images of a specific person with a variety of poses, or images of different identities with a particular pose. To rotate a face, our framework requires one input image of that person to produce an identity vector, and any other input face image to extract a pose embedding vector. Then we recombine the identity vector and the pose vector to synthesize a new face of the person with the extracted pose. Two learning pathways are introduced, the generation and the transformation, where the generation path focuses on learning complete representation in the latent embedding space. While the transformation path focuses on synthesis of new face images with target poses. They collaborate and compete in a parameter-sharing manner, and in an unsupervised settings. The experimental results demonstrate the effectiveness of the proposed framework.

Keywords: Generative Adversarial Networks · Disentangled Representation · Latent embedding space

1 Introduction

Benefiting from the convolutional neural networks trained on large-scale face databases [1, 2], the performance of face recognition systems has been significantly improved over the past few years. However, pose variations are still the bottleneck for many real-world face recognition scenarios. Furthermore, they are relatively underrepresented in training datasets of these face recognition systems.

Existing methods that address pose variations can be categorized into two classes. The first one tries to obtain pose-invariant features [3], while the other resorts to synthesis approaches to generate face images of a specific person in the frontal pose

© Springer Nature Switzerland AG 2019
I. V. Tetko et al. (Eds.): ICANN 2019, LNCS 11731, pp. 535–547, 2019.
https://doi.org/10.1007/978-3-030-30493-5_51

which can be directly used by face recognition systems without retraining the recognition models.

For the first class, metric learning [4] is a common way to obtain pose-invariant features. Moreover, feature mapping method [5] transforms a profile face representation to a frontal pose in order to simplify the recognition.

The second class is often called face rotation, which resorts to computer graphics or deep learning methods to rotate profile faces to the frontal viewpoints. In [6] 3D Morphable Model (3DMM) derives a morphable face model by transforming the facial shape and texture to match an input image as closely as possible. Consequently, the recovered shape and texture can be used to generate a target face image under novel poses. These methods are good at generating small pose faces, but their performance decreases under large face poses due to severe loss of texture.

Recently, deep learning architectures and algorithms have already shown impressive capability to recover a frontal face in a data-driven way. The framework TP-GAN [7] adopts a two-pathway generator networks to learn photorealistic frontal view synthesis, the first one focuses on the inference of global structure, while the other focuses on the transformation of local texture. In FF-GAN [8] the 3DMM conditioned GAN attempts to recover a frontal view image from a large pose face image by following the single-pathway framework architecture. DR-GAN [9] can generate an image with a specific pose. More recent work, such as CR-GAN [10] creates view-specific face images using a two-pathway learning scheme.

However, we argue that there are several drawbacks for the above class of approaches. First, TP-GAN has a complex architecture to preserve global and local texture information. It contains a global network and four landmark located patch networks whose training and inference are very time-consuming. Second, in FF-GAN, the 3DMM conditioned GAN can retain the visual quality under occlusions during frontalization. Third, their architecture is specific for face frontalization and is not applicable to the synthesis of arbitrary poses. Fourth, they can only manipulate limited face pose variations. In addition they train GAN in a supervised fashion, or in its conditional setting, since they require full annotation of poses for training the models. Last, all these methods do not learn a pose-invariant feature representation, since factors of variation such as identity and pose might still be entangled.

To address the above issues, in this paper we propose a novel GAN framework for face rotation by learning disentangled identity and pose codes in unsupervised fashion. Factoring the pose and identity enhances the interpretability of the learned representations, it improves the performances of learning tasks which are reliant on the sole identity or pose representation, and enables the user to generate face images of a specific person with a variety of poses, or several images of a particular pose.

Our framework consists of a single generator G, in contrast to work Huang et al. [7], whose input is the concatenation of identity and pose codes. The generator has several upsampling, convolutional layers, and a set of residual blocks. In addition to the generator G, two encoders E_I and E_P are used to map the data x to its latent identity and pose representation.

To summarize, the main contributions are as follows:

- We present the Identity and Pose Disentangled GAN (IP-GAN) that learns to disentangle identity and pose representations of the data in an unsupervised fashion.
- To the best of our knowledge, we are the first to propose a complete representations of GAN models in unsupervised fashion, in order to rotate face to any arbitrary pose, by transferring pose between data samples.
- We propose the two path-way learning scheme to train the proposed IP-GAN framework.
- A combinations of different loss functions are proposed to work together within the proposed learning scheme.

2 Related Work

2.1 Generative Adversarial Network

As one of the most significant improvements on the research of deep generative models, GAN framework was introduced by Goodfellow et al. [11] for generative modeling of data through learning a transformation from points belonging to the simple prior distribution $(z \sim P_z)$ to those from the data distribution $(x \sim P_{data})$. A typical GAN model consists of two modules that play an adversarial game: a discriminator and a generator. While the generator learns to generate fake samples $G(z)$ that are indistinguishable from real samples, the discriminator learns to distinguish between these fake samples $(G(z) \sim P_G)$ and real ones $(x \sim P_{data})$, thus giving a scalar output $(y \in \{0, 1\})$. The goal of the generator is to fool the discriminator by generating photorealistic samples that resemble those from the real data while that of the discriminator is to accurately distinguish between real and generated data. The two models, typically designed as neural networks, play a min max game with the objective function as shown in Eq. (1):

$$\min_{G} \max_{D} V(D, G) = E_{x \sim P_{data}}[logD(x)] + E_{z \sim P_z}[\log(1 - D(G(z)))]. \quad (1)$$

where P_{data} is the real data distribution and P_z is the prior distribution.

2.2 Face Frontalization

Synthesizing a frontal view of a face from a single image with large pose variations is very challenging due to self-occlusion and recovering the 3D information from 2D projections is ambiguous. Many approaches have been proposed to frontalize profile faces, such as deep learning methods [7, 8]. Since the proposed method in [8] relies on any 3D knowledge for geometry shape estimation, while the method in [7] infers it through data-driven learning. Huang et al. [7] proposed a Two-Pathway Generative Adversarial Network (TP-GAN) for synthesizing photorealistic frontal view human faces from profile images by simultaneously preserving identity and perceiving global structures and local details. Yin et al. [8] proposed Face Frontalization Generative

Adversarial Network framework (FF-GAN), which incorporates elements from both deep 3DMM and face recognition CNNs to achieve high-quality and identity-preserving frontalization using a single input image under various head poses including extreme profile views up to $\pm 90°$. Both TP-GAN and FF-GAN obtained impressive results on face frontalization, but they need an explicit annotations of the frontal faces.

2.3 Multi-view Synthesis

Tran et al. [9] proposed (DR-GAN) Disentangled Representation Learning-Generative Adversarial Network. In DR-GAN, the encoder-decoder structure of the generator enables this model to learn an identity representation that is both discriminative and generative. The input of the encoder is a face image of any pose, the output of the decoder is a synthetic face, frontal or rotated with a target pose, even the extreme profile $\pm 90°$. The discriminator in DR-GAN is trained to not only distinguish real vs. synthetic (or fake) images, but also predict the identity and pose of a face. Very recently, Tian et al. [10] proposed the Complete Representation GAN-based method, follows a single-pathway design: an encoder-decoder network followed by a discriminator network. While it uses a two-pathway learning scheme to learn the "complete" representations. Both DR-GAN and CR-GAN can generate realistic, identity-preserved, view-specific images from a single-view input, but they need to explicitly label the faces whether in supervised or self-supervised learning settings.

3 Proposed Approach

In this section, we present our proposed GAN framework for face rotation. To generate a face image of any specific identity with an arbitrary target pose, our framework incorporates the pose information in the synthesis process. Different from the recent work [8] that uses a 3D morphable face simulator to generate pose information and the works [7, 9, 10] that encode this information in an one-hot vector, our proposed framework can learn such information by explicitly disentangling identity and pose representation from a single face image.

3.1 Identity and Pose Disentangled GAN Model

Let $x_1^s \in X$ be input image of a certain subject identity, and $x_2^p \in X$ be another input image to extract the target pose features. Our goal is to synthesize a new face image x' of the subject of x_1^s with the extracted pose of x_2^p. To achieve this goal, we assume that each image $x \in X$ is generated from an identity latent code $s \in S$ and a pose latent code $p \in P$. In other words, x_1^s, x_2^p are generated by the pair (s_1, p_1) and the pair (s_2, p_2) respectively. As the result, the new face image x' is synthesized by the pair (s_1, p_2).

The overall framework of our proposed Identity and Pose Disentangled GAN (IP-GAN) is depicted in Fig. 1, which contains five parts: (1) the identity encoder network E_I; (2) the head pose encoder network E_P; (3) the generative network G; (4) the identity classification network C; and (5) the discriminative network D. The function of the network E_I is to extract the identity latent code s_1 from the subject image x_1^s,

$s_1 = E_I(x_1^s)$. While the function of the network E_P is to extract the pose latent code p_2 of the target pose image x_2^p, $p_2 = E_P(x_2^p)$. We then use the generator network G to produce the final output image x' using the combined identity latent code s_1 and the extracted pose latent code p_2, $x' = G(E_I(x_1^s), E_P(x_2^p))$. All networks are included in the training phase, while in the testing phase, the network C and the network D are not included. The network C is used to preserve the identity by measuring the posterior probability $P(c|x_1^s)$, where c is the subject identity of x_1^s. The discriminative network D distinguishes between real and generated images.

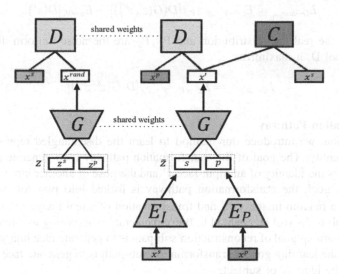

Fig. 1. Our IP-GAN.

3.2 Learning Disentangled Representations of Identity and Pose

Since we usually have a limited number of training samples, the identity encoder E_I, the pose encoder E_P and the generator G can map only a subspace of Z^s and Z^P latent spaces. This would lead to a severe problem in testing phase when using "unseen" data as the input. It is highly probable that the encoders may map the novel input data out of these learned subspaces, which necessarily leads to poor syntheses since G has never "seen" the embedding.

This fact motivates us to train both encoders E_P, E_I and G that can "cover" the whole Z^S and Z^P spaces, so we can learn complete representations. To achieve this goal, inspired by [10], we propose new method for training the model networks. We introduce two learning pathways, generation and transformation, where the generation pathway focuses on mapping the entire Z^S and Z^P spaces to high-quality images. While the transformation pathway focuses on synthesis new face images with target poses.

Generation Pathway

The generation pathway trains generator G and discriminator D. Here the encoders E_P and E_I are not involved since G tries to generate from random noises. Given two random noises from noise uniform distribution $z = (z^s, z^p); z^s \in Z^S$ and $z^p \in Z^P$, G aims to synthesis a realistic image $x^{rand} = G(z^s, z^p)$ under the random identity z^s and the random pose z^p. Similar to vanilla GAN [11], the generative network G competes in a two-player minimax game with the discriminative network D. Network D tries to distinguish real data from G's output, while network G tries to fool the network D. Concretely, network D tries to minimize the loss function:

$$L_{D-adv_{gen}} = E_{z^s \sim P_{z^s}, z^p \sim P_{z^p}}[D(G(z^s, z^p))] - E_{x \sim P_x}[D(x^s)] \qquad (2)$$

where P_x is the real data distribution and P_{z^s}, P_{z^p} are the noise uniform distributions. G tries to fool D, it maximizes:

$$L_{G-adv_{gen}} = E_{z^s \sim P_{z^s}, z^p \sim P_{z^p}}[D(G(z^s, z^p))] \qquad (3)$$

Transformation Pathway

In this section, we introduce our method to learn the disentangled representation of pose and identity. The goal of the transformation pathway is to generate an image x' that combines the identity of an input face x^s and the pose of another input face x^p. To achieve this goal, the transformation pathway is forked into two sub-paths, reconstruction of a random image x^{rand} and transformation of a real image x^s. Only one of these two sub-paths will be enabled in the same iteration according to random boolean value. The learning goal of reconstruction sub-path is to generate face image with target pose, while the learning goal of transformation sub-path is to generate face image with preserving the identity of subject.

Reconstruction Sub-path. The reconstruction sub-path trains the networks E_P, D and G but keeping the networks E_I and C fixed. The encoders learn the embedding vectors z^s, z^p by trying to reconstruct the random face image x^{rand} under the pose $E_P(x^p)$. More specifically, After extracting the identity vector $E_I(x^{rand})$ and pose vector $E_P(x^p)$, we concatenate them in the latent space and feed the combined vector, $\bar{z} = [E_I(x^{rand}), E_P(x^p)]$ into the network G to reconstruct the random face image under the pose of x^p. The generator G should produce $\hat{x}^{rand}$, the reconstruction of x^{rand}. D is trained to distinguish the fake image $\hat{x}^{rand}$ from the real one x^p. Thus D minimizes:

$$L_{D-adv_{recon}} = E_{x \sim P_x, x^{rand} \sim P_{x^{rand}}}[D(G(E_I(x^{rand}), E_P(x^p)))] - E_{x^{rand} \sim P_{x^{rand}}}[D(x^{rand})] \quad (4)$$

where, $\hat{x}^{rand} = G(E_I(x^{rand}), E_P(x^p))$. G tries to fool D, it maximizes:

$$L_{G-adv_{recon}} = E_{x \sim P_x, x^{rand} \sim P_{x^{rand}}}[D(G(E_I(x^{rand}), E_P(x^p)))] \qquad (5)$$

In order to insure that the generator learns correctly synthesizing the random face image under the pose of x^p, we use the following loss function after reconstructing the image x^p:

$$L_{GR_{recon}} = \|G(E_I(x^p), E_P(x^p)) - x^p\|_1 \tag{6}$$

$$L_{GP} = \|E_P(\hat{x}^{rand}) - E_P(x^p)\|_2^2 \tag{7}$$

Transformation Sub-path. The transformation sub-path trains the networks E_I, C, D and G but keeping the network E_P fixed. The output of E_I should be identity-preserved so the multiview images will present the same identity. We propose a cross reconstruction task to make E_P and E_I disentangle the viewpoint from the identity. More specifically, we sample a real image pair (x_i^s, x_j^p) that share the same identity but different poses. The goal is to transform x_i^s to x_j^p. To achieve this, E_I takes x_i^s as input and outputs a pose-invariant identity representation s_i, while E_P takes x_j^p as input and outputs a pose representation p_j. We concatenate these latent vectors and feed the combined vector, $\bar{z} = (s_i, p_j) = [\left(E_I(x_i^s), E_P(x_j^p)\right)]$ into the network G. The generator G should produce $\hat{x}$, the transformation of x_i^s. D is trained to distinguish the fake image $\hat{x}$ from the real one x_j^p. Thus D minimizes:

$$L_{D-adv_{trans}} = E_{x \sim P_x}\left[D\left(G\left(E_I(x_i^s), E_P(x_j^p)\right)\right)\right] - E_{x \sim P_x}\left[D\left(x_j^p\right)\right] \tag{8}$$

G tries to fool D, it maximizes:

$$L_{G-adv_{trans}} = E_{x_i^s, x_j^p \sim P_x}\left[D\left(G\left(E_I(x_i^s), E_P(x_j^p)\right)\right)\right] \tag{9}$$

In order to train the networks E_P and E_I in a fully unsupervised fashion to extract the embedding vectors, we use the following loss function:

$$L_{GT} = \left\|\hat{x}, x_j^p\right\|_1 \tag{10}$$

where L1-norm loss function is utilized to enforce that $\hat{x}$ is the reconstruction of x_j^p.

The encoders E_P, E_I help G to generate high quality image with pose of x_j^p and content of x_i^s, so to achieve that we make use of a VGG-19 network, Φ [12], pretrained as a fixed loss function that measure differences in high-level content between images. Let $\Phi_l(x)$ denote the feature maps of lth layer of the loss network for the input image x. Then the content-consistency loss function between images $\hat{x}$ and x_j^p is defined as:

$$L_{GC_{trans}} = \left\| \Phi_l(\hat{x}) - \Phi_l\left(x_j^p\right) \right\|_2^2 \qquad (11)$$

Similar to $L_{GR_{recon}}$, we use the following loss function after reconstructing the image of target identity x_i^s:

$$L_{GR_{trans}} = \left\| G\left(E_I\left(x_i^s\right), E_P\left(x_i^s\right)\right) - x_i^s \right\|_1 \qquad (12)$$

With regard to training the identity encoder E_I, given a set of face images with the identity labels $\{x_i^s, c_i\}$, we use the softmax loss for training it to perform face classification task. Therefore, the same individuals have approximately the same feature which can be used as the identity vector. Formally, the loss function of the network E_I is:

$$L_I = -E_{x \sim P_x}[log(P(c|x^s))] \qquad (13)$$

Meanwhile, classification network C tries to classify the generated faces of different identities, meaning it tries to minimize the loss function:

$$L_C = -E_{\hat{x} \sim P_{\hat{x}}}[log(P(c|\hat{x}))] \qquad (14)$$

where, $\hat{x}$ the transformation of x_i^s, $P_{\hat{x}}$ is the generated data distribution. The network C and network I share the parameters.

3.3   Overall Objective Function

The final synthesis loss function is the weighted sum of all losses defined in Eqs. 2–14. Each network relates to only part of the loss function according to learning path-way and sub-path, as shown in Table 1.

$$\min_{E_I, E_P, C, G} \max_D L(E_I, E_P, C, G, D)$$
$$= \lambda_D(L_{D-adv_{gen}} + L_{D-adv_{recon}} + L_{D-adv_{trans}})$$
$$+ \lambda_G\left(L_{G-adv_{gen}} + L_{G-adv_{recon}} + L_{G-adv_{trans}}\right) + \lambda_I(L_I + L_C) \qquad (15)$$
$$+ \lambda_C L_{GC_{trans}} + \lambda_T L_{GT} + \lambda_P L_{GP} + \lambda_R(L_{GR_{recon}} + L_{GR_{trans}})$$

where $\lambda_D, \lambda_G, \lambda_I, \lambda_C, \lambda_T, \lambda_P, \lambda_R$ are weights that control the importance of reconstruction terms.

Table 1. Networks and their related loss functions.

Generation pathway		Transformation path-way					
D	G	Reconstruction sub-path		Transformation sub-path			
$L_{D-adv_{gen}}$	$L_{G-adv_{gen}}$	D	G, E_P	E_I	D	G	C
		$L_{D-adv_{recon}}$	$L_{G-adv_{recon}}$ $L_{GR_{recon}}$ L_{GP}	$L_I, L_{GC_{trans}}$	$L_{D-adv_{trans}}$	$L_{G-adv_{trans}}$ $L_{GR_{trans}}$ $L_{GC_{trans}}$ L_{GT}	L_C

4   Experiments

4.1   Database

We train and evaluate IP-GAN on datasets without pose labels. Multi-PIE is a labeled database [13] for evaluating face recognition under pose, illumination, and expression variations in controlled settings. It contains 337 identities and provides annotations of the head pose vertical rotation angles. We use 250 subjects from the first session with different illuminations and expressions. The first 170 subjects are chosen for training and the rest 80 for testing.

4.2   Implementation Details

Our implementations of the identity encoder, the generator and the discriminator are modified from the residual networks in work Tian et al. [10], where E_I shares a similar network structure with D. For networks E_I and C, we use the same structure. Meanwhile, E_I and C share parameters in the training stage to speed up convergence. For network E_P, we use the same network structure of style encoder as in work Huang et al. [14]. Values of the two random noises z^s, z^p in range $[-1, +1]$. The model is implemented using the deep learning toolbox Torch. The batch size is 8. Adam optimizer is used with the learning rate of 0.0005 and momentum of [0, 0.9].

4.3   Features Visualization

In this section we perform an experiment to show the performance of our model in disentangling the representation of head view from the identity. We visualize two groups of the 128-dim deep features on a two dimensional space. 3000 images with 10 subjects, 9 poses and different illuminations are selected from Multi-PIE dataset. The first group of features are extracted by Light-CNN [15], while we use our pose encoder to extract the second group of embedding vectors. We use t-SNE [16] to visualize both groups. The left side of Fig. 2 illustrates the deep feature space of first group, while on the right side pose embedding vectors are visualized. It's clear that images can be easily separated into different groups according to their poses.

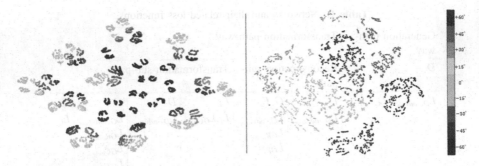

Fig. 2. Feature space using Light-CNN (left) and our pose encoder (right). Each color represents a different pose. (Color figure online)

We conduct another experiment to show the performance of our model in learning the identity representation with different poses. 1400 face images are selected with 4 identities and 9 different head poses ranging from -60° to +60°. The left side of Fig. 3 shows the deep feature space of Light-CNN embedding vectors, while the right side visualizes the identity embedding vectors which are extracted by our identity encoder. It is noticeable from the results in the left side that the Light-CNN failed to learn the identity representation under large head poses ($-60°$, $+60°$), while our model is able to learn that despite the small training dataset and only after 20 epochs.

Fig. 3. Feature space using Light-CNN (left) and our identity encoder (right). Each color represents a different identity. (Color figure online)

4.4 Face Pose Transformation

This section presents the results of face pose transformation. The goal of face pose transformation is to generate an image $\hat{x}$ that combines the identity of an input face x^s and the pose of another input face x^p. Figure 4 presents the face synthesis results of the identities that appear in the training dataset. Our method performs well in face synthesis, preserving both identity and pose. In addition, it can recover photorealistic frontal views from very large poses. Figure 5 shows IP-GAN ability to frontalize profile faces with compelling identity-preserving.

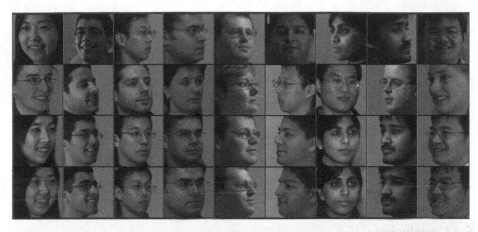

Fig. 4. Face Synthesis results by IP-GAN under different poses. From up to down, identities, poses, transformation results. The ground truth images are provided at the last row.

Fig. 5. Frontal view synthesis by IP-GAN. From up to down, identities, poses, frontalization results. The ground truth images are provided at the last row.

4.5 Random Face Generation

Another important feature of our framework is that it can synthesize unseen faces from the training set using the generation path. Given random identity latent vector and random pose latent vector, the generator can synthesize new face images. Figure 6 shows random face generation results.

Fig. 6. Random face generation results.

5 Conclusion

In this paper, we propose the Identity and Pose Disentangled GAN (IP-GAN) that learns to disentangle identity and pose representations of the data in an unsupervised fashion. The framework learns the complete representations through the proposed learning schema which consists of two-path ways; the generation path-way which focuses on learning the entire pose and identity latent spaces, and the transformation path-way that focuses on synthesis new face images of specific subject identity with target poses. Our GAN framework demonstrates the possibility to synthesize photo-realistic and identity-preserving faces with arbitrary poses and achieves state-of-the-art of generating real and identity preserving face images under large pose variations. We hope to explore whether our framework can be applied to other datasets larger than ours, and to conduct quantitative experiments to show the effectiveness of using our identity encoder to extract embedding vectors for pose-invariant face recognition.

Acknowledgements. This work was financially supported by the Government of the Russian Federation (Grant 08-08).

References

1. Cao, Q., Shen, L., Xie, W., Parkhi, O.M., Zisserman, A.: VGGFace2: a dataset for recognising faces across pose and age. In: IEEE FG (2018). https://doi.org/10.1109/fg.2018.00020
2. Guo, Y., Zhang, L., Hu, Y., He, X., Gao, J.: MS-Celeb-1M: a dataset and benchmark for large-scale face recognition. In: Leibe, B., Matas, J., Sebe, N., Welling, M. (eds.) ECCV 2016. LNCS, vol. 9907, pp. 87–102. Springer, Cham (2016). https://doi.org/10.1007/978-3-319-46487-9_6
3. Chen, D., Cao, X., Wen, F., Sun, J.: Blessing of dimensionality: high-dimensional feature and its efficient compression for face verification. https://doi.org/10.1109/cvpr.2013.389
4. Schroff, F., Kalenichenko, D., Philbin, J.: FaceNet: a unified embedding for face recognition and clustering. In: CVPR (2015). https://doi.org/10.1109/cvpr.2015.7298682

5. Cao, K., Rong, Y., Li, C., Tang, X., Loy, C.C.: Pose-Robust face recognition via deep residual equivariant mapping. In: CVPR (2018). https://doi.org/10.1109/cvpr.2018.00544
6. Blanz, V., Vetter, T.: A morphable model for the synthesis of 3D faces. In: SIGGRAPH 1999. ACM Press (1999). https://doi.org/10.1145/311535.311556
7. Huang, R., Zhang, S., Li, T., He, R.: Beyond face rotation: global and local perception GAN for Photorealistic and identity preserving frontal view synthesis. In: ICCV (2017). https://doi.org/10.1109/iccv.2017.267
8. Yin, X., Yu, X., Sohn, K., Liu, X., Chandraker, M.: Towards Large-pose face frontalization in the wild. In: ICCV (2017). https://doi.org/10.1109/iccv.2017.430
9. Tran, L., Yin, X., Liu, X.: Disentangled representation learning GAN for pose-invariant face recognition. In: CVPR (2017). https://doi.org/10.1109/cvpr.2017.141
10. Tian, Y., Peng, X., Zhao, L., Zhang, S., Metaxas, D.N.: CR-GAN: learning complete representations for multi-view generation (2018). https://doi.org/10.24963/ijcai.2018/131
11. Goodfellow, I.J., et al.: Generative adversarial nets. In: NIPS (2014). https://arxiv.org/abs/1406.2661
12. Simonyan, K., Zisserman, A.: Very deep convolutional networks for large-scale image recognition. In: ICLR (2015). https://arxiv.org/abs/1409.1556
13. Gross, R., Matthews, I., Cohn, J., Kanade, T., Baker, S.: Multi-PIE. Image Vision Comput. **28**, 807–813 (2010). https://doi.org/10.1016/j.imavis.2009.08.002
14. Huang, X., Liu, M.-Y., Belongie, S., Kautz, J.: Multimodal unsupervised image-to-image translation. In: Ferrari, V., Hebert, M., Sminchisescu, C., Weiss, Y. (eds.) ECCV 2018. LNCS, vol. 11207, pp. 179–196. Springer, Cham (2018). https://doi.org/10.1007/978-3-030-01219-9_11
15. Wu, X., He, R., Sun, Z., Tan, T.: A light CNN for deep face representation with noisy labels (2018). https://doi.org/10.1109/tifs.2018.2833032
16. Van der Maaten, L., Hinton, G.: Visualizing data using t-SNE (2008). http://citeseerx.ist.psu.edu/viewdoc/summary?doi=10.1.1.457.7213

Special Session: Machine Learning with Graphs: Algorithms and Applications

Special Session: Machine Learning with Graphs: Algorithms and Applications

Networks have become ubiquitous as data from many diverse disciplines can naturally be modeled as graph structures. Characteristic examples include social and information networks, technological networks, web graphs, as well as networks from the domain of biology and neuroscience. Developing machine learning algorithms for graph data is a crucial task with a plethora of cross-disciplinary applications. The special session aimed to bring together researchers from both academia and industry that are interested in state-of-the-art algorithmic techniques and methodologies in machine learning for graphs and their related applications.

The topics covered in the special session included:

- Algorithms and Methods:
  - Representation Learning on Graphs
  - Deep Learning and Graph Neural Networks
  - Learning Rich Network Structures (e.g., signed graphs)

- Application Domains:
  - Social Media and Social Network Analysis
  - Knowledge Graphs and Semantic Networks
  - Neuroscience and Network Biology

Organization

Danai Koutra   University of Michigan Ann Arbor, USA
Fragkiskos D. Malliaros   CentraleSupélec, Inria Saclay, France
Evangelos Papalexakis   University of California Riverside, USA

Program Committee

Nesreen Ahmed   Intel Research Labs, USA
Narges Ahmidi   Johns Hopkins University, USA
Abdulkadir Çelikkanat   CentraleSupélec, Inria Saclay, France
Stergios Christodoulidis   University of Bern, Switzerland
Maximilien Danisch   Sorbonne University, France
Tyler Derr   Michigan State University, USA
Christos Giatsidis   Amadeus, France
Mark Heimann   University of Michigan Ann Arbor, USA
Meng Jiang   University of Notre Dame, USA
Argyris Kalogeratos   ENS Paris-Saclay, France
Yao Ma   Michigan State University, USA
Duong Nguyen   University of California San Diego, USA
Giannis Nikolentzos   École Polytechnique, France
George Panagopoulos   École Polytechnique, France
Ryan Rossi   Adobe Research, USA
Tara Savafi   University of Michigan Ann Arbor, USA
Kai Shu   Arizona State University, USA
Konstantinos Skianis   École Polytechnique, France
Jian Tang   HEC, MILA, Canada
Anton Tsitsulin   University of Bonn, Germany
Maria Vakalopoulou   CentraleSupélec, Inria Saclay, France
Sagar Verma   CentraleSupélec, France

Hypernetwork Knowledge Graph Embeddings

Ivana Balažević[1]([✉]), Carl Allen[1], and Timothy M. Hospedales[1,2]

[1] School of Informatics, University of Edinburgh, Edinburgh, UK
{ivana.balazevic,carl.allen,t.hospedales}@ed.ac.uk
[2] Samsung AI Centre, Cambridge, UK

Abstract. Knowledge graphs are graphical representations of large databases of facts, which typically suffer from incompleteness. Inferring missing relations (links) between entities (nodes) is the task of link prediction. A recent state-of-the-art approach to link prediction, ConvE, implements a convolutional neural network to extract features from concatenated subject and relation vectors. Whilst results are impressive, the method is unintuitive and poorly understood. We propose a hypernetwork architecture that generates simplified relation-specific convolutional filters that (i) outperforms ConvE and all previous approaches across standard datasets; and (ii) can be framed as tensor factorization and thus set within a well established family of factorization models for link prediction. We thus demonstrate that convolution simply offers a convenient computational means of introducing sparsity and parameter tying to find an effective trade-off between non-linear expressiveness and the number of parameters to learn.

1 Introduction

Knowledge graphs, such as WordNet, Freebase, and Google Knowledge Graph, are large graph-structured databases of facts, containing information in the form of triples (e_1, r, e_2), with e_1 and e_2 representing subject and object entities and r a relation between them. They are considered important information resources, used for a wide variety of tasks ranging from question answering to information retrieval and text summarization. One of the main challenges with existing knowledge graphs is their incompleteness: many of the links between entities in the graph are missing. This has inspired substantial work in the field of *link prediction*, i.e. the task of inferring missing links in knowledge graphs.

Until recently, many approaches to link prediction have been based on different factorizations of a 3-moded binary tensor representation of the training triples [12,17,22,23]. Such approaches are shallow and linear, with limited expressiveness. However, attempts to increase expressiveness with additional fully connected layers and non-linearities often lead to overfitting [12,17]. For this reason, Dettmers et al. introduce ConvE, a model that uses 2D convolutions over reshaped and concatenated entity and relation embeddings [3]. They motivate the use of convolutions by being parameter efficient and fast to compute

© Springer Nature Switzerland AG 2019
I. V. Tetko et al. (Eds.): ICANN 2019, LNCS 11731, pp. 553–565, 2019.
https://doi.org/10.1007/978-3-030-30493-5_52

on a GPU, as well as having various robust methods from computer vision to prevent overfitting. Even though results achieved by ConvE are impressive, it is highly unintuitive that convolution – particularly 2D convolution – should be effective for extracting information from 1D entity and relation embeddings.

In this paper, we introduce HypER, a model that uses a *hypernetwork* [5] to generate convolutional filter weights for each relation. A hypernetwork is an approach by which one network generates weights for another network, that can be used to enable weight-sharing across layers and to dynamically synthesize weights given an input. In our context, we generate relation-specific filter weights to process input entities, and also achieve *multi-task knowledge sharing* across relations in the knowledge graph. Our proposed HypER model uses a hypernetwork to generate a set of 1D relation-specific filters to process the subject entity embeddings. This simplifies the interaction between subject entity and relation embeddings compared to ConvE, in which a global set of *2D* filters are convolved over *reshaped and concatenated* subject entity and relation embeddings, which is unintuitive as it suggests the presence of 2D structure in word embeddings. Moreover, interaction between subject and relation in ConvE depends on an arbitrary choice about how they are reshaped and concatenated. In contrast, HypER's hypernetwork generates relation-specific filters, and thus extracts *relation-specific features* from the subject entity embedding. This necessitates no 2D reshaping, and allows entity and relation to interact more completely, rather than only around the concatenation boundary. We show that this simplified approach, in addition to improving link prediction performance, can be understood in terms of tensor factorization, thus placing HypER within a well established family of factorization models. The apparent obscurity of using convolution within word embeddings is thereby explained as simply a convenient computational means of introducing *sparsity* and *parameter tying*.

We evaluate HypER against several previously proposed link prediction models using standard datasets (FB15k-237, WN18RR, FB15k, WN18, YAGO3-10), across which it consistently achieves state-of-the-art performance. In summary, our key contributions are:

- proposing a new model for link prediction (HypER) which achieves state-of-the-art performance across all standard datasets;
- showing that the benefit of using convolutional instead of fully connected layers is due to restricting the number of dimensions that interact (i.e. explicit regularization), rather than finding higher dimensional structure in the embeddings (as implied by ConvE); and
- showing that HypER in fact falls within a broad class of tensor factorization models despite the use of convolution, which serves to provide a good trade-off between expressiveness and number of parameters to learn.

2   Related Work

Numerous matrix factorization approaches to link prediction have been proposed. An early model, RESCAL [12], tackles the link prediction task by optimizing a scoring function containing a bilinear product between vectors for each of the subject and object entities and a full rank matrix for each relation. Dist-Mult [23] can be viewed as a special case of RESCAL with a diagonal matrix per relation type, which limits the linear transformation performed on entity vectors to a stretch. ComplEx [22] extends DistMult to the complex domain. TransE [1] is an affine model that represents a relation as a translation operation between subject and object entity vectors.

A somewhat separate line of link prediction research introduces Relational Graph Convolutional Networks (R-GCNs) [15]. R-GCNs use a convolution operator to capture locality information in graphs. The model closest to our own and which we draw inspiration from, is ConvE [3], where a convolution operation is performed on the subject entity vector and the relation vector, after they are each reshaped to a matrix and lengthwise concatenated. The obtained feature maps are flattened, put through a fully connected layer, and the inner product is taken with all object entity vectors to generate a score for each triple. Advantages of ConvE over previous approaches include its expressiveness, achieved by using multiple layers of non-linear features, its scalability to large knowledge graphs, and its robustness to overfitting. However, it is not intuitive why convolving across concatenated and reshaped subject entity and relation vectors should be effective.

The proposed HypER model does no such reshaping or concatenation and thus avoids both implying any inherent 2D structure in the embeddings and restricting interaction to the concatenation boundary. Instead, HypER convolves *every* dimension of the subject entity embedding with *relation-specific* convolutional filters generated by the hypernetwork. This way, entity and relation embeddings are combined in a non-linear (quadratic) manner, unlike the linear combination (weighted sum) in ConvE. This gives HypER more expressive power, while also reducing parameters.

Interestingly, we find that the differences in moving from ConvE to HypER in fact bring the factorization and convolutional approaches together, since the 1D convolution process is equivalent to multiplication by a highly sparse tensor with tied weights (see Fig. 2). The multiplication of this "convolutional tensor" (defined by the relation embedding and hypernetwork) and other weights gives an implicit relation matrix, corresponding to those in e.g. RESCAL, DistMult and ComplEx. Other than the method of deriving these relation matrices, the key difference to existing factorization approaches is the ReLU non-linearity applied prior to interaction with the object embedding.

3   Link Prediction

In link prediction, the aim is to learn a scoring function ϕ that assigns a score $s = \phi(e_1, r, e_2) \in \mathbb{R}$ to each input triple (e_1, r, e_2), where $e_1, e_2 \in \mathcal{E}$ are subject and object entities and $r \in \mathcal{R}$ a relation. The score indicates the strength of prediction that the given triple corresponds to a true fact, with positive scores meaning true and negative scores, false. Link prediction models typically map entity pair e_1, e_2 to their corresponding distributed embedding representations $\mathbf{e}_1, \mathbf{e}_2 \in \mathbb{R}^{d_e}$ and a score is assigned using a *relation-specific* function,

Table 1. Scoring functions of state-of-the-art link prediction models, the dimensionality of their relation parameters, and their space complexity. d_e and d_r are the dimensions of entity and relation embeddings respectively, $\overline{\mathbf{e}}_2 \in \mathbb{C}^{d_e}$ denotes the complex conjugate of $\mathbf{e}_2$, and $\underline{\mathbf{e}}_1, \underline{\mathbf{w}}_r \in \mathbb{R}^{d_w \times d_h}$ denote a 2D reshaping of $\mathbf{e}_1$ and $\mathbf{w}_r$ respectively. $*$ is the convolution operator, $\mathbf{F}_r = \text{vec}^{-1}(\mathbf{w}_r \mathbf{H})$ the matrix of relation specific convolutional filters, vec is a vectorization of a matrix and vec^{-1} its inverse, f is a non-linear function, and n_e and n_r respectively denote the number of entities and relations.

Model	Scoring function	Relation parameters	Space complexity
RESCAL [12]	$\mathbf{e}_1^\top \mathbf{W}_r \mathbf{e}_2$	$\mathbf{W}_r \in \mathbb{R}^{d_e^2}$	$\mathcal{O}(n_e d_e + n_r d_e^2)$
TransE [1]	$\|\mathbf{e}_1 + \mathbf{w}_r - \mathbf{e}_2\|$	$\mathbf{w}_r \in \mathbb{R}^{d_e}$	$\mathcal{O}(n_e d_e + n_r d_e)$
NTN [17]	$\mathbf{u}_r^\top f(\mathbf{e}_1 \mathbf{W}_r^{[1..k]} \mathbf{e}_2 + \mathbf{V}_r \begin{bmatrix} \mathbf{e}_1 \\ \mathbf{e}_2 \end{bmatrix} + \mathbf{b}_r)$	$\mathbf{W}_r \in \mathbb{R}^{d_e^2 k}, \mathbf{V}_r \in \mathbb{R}^{2 d_e k},$ $\mathbf{u}_r \in \mathbb{R}^k, \mathbf{b}_r \in \mathbb{R}^k$	$\mathcal{O}(n_e d_e + n_r d_e^2 k)$
DistMult [23]	$\langle \mathbf{e}_1, \mathbf{w}_r, \mathbf{e}_2 \rangle$	$\mathbf{w}_r \in \mathbb{R}^{d_e}$	$\mathcal{O}(n_e d_e + n_r d_e)$
ComplEx [22]	$\text{Re}(\langle \mathbf{e}_1, \mathbf{w}_r, \overline{\mathbf{e}}_2 \rangle)$	$\mathbf{w}_r \in \mathbb{C}^{d_e}$	$\mathcal{O}(n_e d_e + n_r d_e)$
ConvE [3]	$f(\text{vec}(f([\underline{\mathbf{e}}_1 ; \underline{\mathbf{w}}_r] * w)) \mathbf{W}) \mathbf{e}_2$	$\mathbf{w}_r \in \mathbb{R}^{d_r}$	$\mathcal{O}(n_e d_e + n_r d_r)$
HypER (ours)	$f(\text{vec}(\mathbf{e}_1 * \text{vec}^{-1}(\mathbf{w}_r \mathbf{H})) \mathbf{W}) \mathbf{e}_2$	$\mathbf{w}_r \in \mathbb{R}^{d_r}$	$\mathcal{O}(n_e d_e + n_r d_r)$

$s = \phi_r(\mathbf{e}_1, \mathbf{e}_2)$. The majority of link prediction models apply the logistic sigmoid function $\sigma(\cdot)$ to the score to give a probabilistically interpretable prediction $p = \sigma(s) \in [0, 1]$ as to whether the queried fact is true. The scoring functions for models from across the literature and HypER are summarized in Table 1, together with the dimensionality of their relation parameters and the significant terms of their space complexity.

4 Hypernetwork Knowledge Graph Embeddings

In this work, we propose a novel hypernetwork model for link prediction in knowledge graphs. In summary, the hypernetwork projects a vector embedding of each relation via a fully connected layer, the result of which is reshaped to give a set of convolutional filter weight vectors for each relation. We explain this process in more detail below. The idea of using convolutions on entity and relation embeddings stems from computer vision, where feature maps reflect patterns in the image such as lines or edges. Their role in the text domain is harder to interpret, since little is known of the meaning of a single dimension in a word embedding. We believe convolutional filters have a regularizing effect when applied to word embeddings (compared to the corresponding full tensor), as the filter size restricts which dimensions of embeddings can interact. This allows nonlinear expressiveness while limiting overfitting by using few parameters. A visualization of HypER is given in Fig. 1.

4.1 Scoring Function and Model Architecture

The relation-specific scoring function for the HypER model is:

$$\begin{aligned} \phi_r(\mathbf{e}_1, \mathbf{e}_2) &= f(\text{vec}(\mathbf{e}_1 * \mathbf{F}_r) \mathbf{W}) \mathbf{e}_2 \\ &= f(\text{vec}(\mathbf{e}_1 * \text{vec}^{-1}(\mathbf{w}_r \mathbf{H})) \mathbf{W}) \mathbf{e}_2, \end{aligned} \tag{1}$$

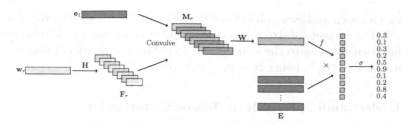

Fig. 1. Visualization of the HypER model architecture. Subject entity embedding $\mathbf{e}_1$ is convolved with filters $\mathbf{F}_r$, created by the hypernetwork $\mathbf{H}$ from relation embedding $\mathbf{w}_r$. The obtained feature maps $\mathbf{M}_r$ are mapped to d_e-dimensional space via $\mathbf{W}$ and the non-linearity f applied before being combined with all object vectors $\mathbf{e}_2 \in \mathbf{E}$ through an inner product to give a score for each triple. Predictions are obtained by applying the logistic sigmoid function to each score.

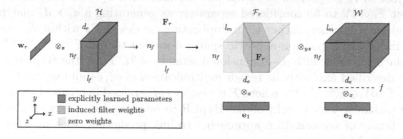

Fig. 2. Interpretation of the HypER model in terms of tensor operations. Each relation embedding $\mathbf{w}_r$ generates a set of filters $\mathbf{F}_r$ via the hypernetwork $\mathcal{H}$. The act of convolving $\mathbf{F}_r$ over $\mathbf{e}_1$ is equivalent to multiplication of $\mathbf{e}_1$ by a tensor $\mathcal{F}_r$ (in which $\mathbf{F}_r$ is diagonally duplicated and zero elsewhere). The tensor product $\mathcal{F}_r \otimes_{yz} \mathcal{W}$ gives a $d_e \times d_e$ matrix specific to each relation. Axes labels indicate the modes of tensor interaction (via inner product).

where the vec^{-1} operator reshapes a vector to a matrix, and non-linearity f is chosen to be a rectified linear unit (ReLU).

In the feed-forward pass, the model obtains embeddings for the input triple from the entity and relation embedding matrices $\mathbf{E} \in \mathbb{R}^{n_e \times d_e}$ and $\mathbf{R} \in \mathbb{R}^{n_r \times d_r}$. The hypernetwork is a fully connected layer $\mathbf{H} \in \mathbb{R}^{d_r \times l_f n_f}$ (l_f denotes filter length and n_f the number of filters per relation, i.e. *output channels* of the convolution) that is applied to the relation embedding $\mathbf{w}_r \in \mathbb{R}^{d_r}$. The result is reshaped to generate a matrix of convolutional filters $\mathbf{F}_r = \mathrm{vec}^{-1}(\mathbf{w}_r \mathbf{H}) \in \mathbb{R}^{l_f \times n_f}$. Whilst the overall dimensionality of the filter set is $l_f n_f$, the rank is restricted to d_r to encourage parameter sharing between relations.

The subject entity embedding $\mathbf{e}_1$ is convolved with the set of relation-specific filters $\mathbf{F}_r$ to give a 2D feature map $\mathbf{M}_r \in \mathbb{R}^{l_m \times n_f}$, where $l_m = d_e - l_f + 1$ is the feature map length. The feature map is vectorized to $\mathrm{vec}(\mathbf{M}_r) \in \mathbb{R}^{l_m n_f}$, and projected to d_e-dimensional space by the weight matrix $\mathbf{W} \in \mathbb{R}^{l_m n_f \times d_e}$. After applying a ReLU activation function, the result is combined by way of inner product with each and every object entity embedding $\mathbf{e}_2{}^{(i)}$, where i varies over all entities in the dataset (of size n_e), to give a vector of scores. The logistic sigmoid is applied element-wise to the score vector to obtain the predicted probability of each prospective triple being true $\mathbf{p}_i = \sigma(\phi_r(\mathbf{e}_1, \mathbf{e}_2{}^{(i)}))$.

4.2 Understanding HypER as Tensor Factorization

Having described the HypER architecture, we can view it as a series of tensor operations by considering the hypernetwork $\mathbf{H}$ and weight matrix $\mathbf{W}$ as tensors $\mathcal{H} \in \mathbb{R}^{d_r \times l_f \times n_f}$ and $\mathcal{W} \in \mathbb{R}^{l_m \times n_f \times d_e}$ respectively. The act of convolving $\mathbf{F}_r = \mathbf{w}_r \otimes \mathcal{H}$ over the subject entity embedding $\mathbf{e}_1$ is equivalent to the multiplication of $\mathbf{e}_1$ by a sparse tensor $\mathcal{F}_r$ within which $\mathbf{F}_r$ is diagonally duplicated with zeros elsewhere (see Fig. 2). The result is multiplied by $\mathcal{W}$ to give a vector, which is subject to ReLU before the final dot product with $\mathbf{e}_2$. Linearity allows the product $\mathcal{F}_r \otimes \mathcal{W}$ to be considered separately as generating a $d_e \times d_e$ matrix for each relation. Further, rather than duplicating entries of $\mathbf{F}_r$ within $\mathcal{F}_r$, we can generalize $\mathcal{F}_r$ to a relation-agnostic sparse 4 moded tensor $\mathcal{F} \in \mathbb{R}^{d_r \times d_e \times n_f \times l_m}$ by replacing entries with d_r-dimensional strands of $\mathcal{H}$. Thus, the HypER model can be described explicitly as tensor multiplication of $\mathbf{e}_1, \mathbf{e}_2$ and $\mathbf{w}_r$ with a core tensor $\mathcal{F} \otimes \mathcal{W} \in \mathbb{R}^{d_e \times d_e \times d_r}$, where $\mathcal{F}$ is heavily constrained in terms of its number of free variables. This insight allows HypER to be viewed in a very similar light to the family of factorization approaches to link prediction, such as RESCAL, DistMult and ComplEx.

4.3 Training Procedure

Following the training procedure introduced by [3], we use *1-N scoring* with the Adam optimizer [8] to minimize the binary cross-entropy loss:

$$\mathcal{L}(\mathbf{p}, \mathbf{y}) = -\frac{1}{n_e} \sum_i (\mathbf{y}_i \log(\mathbf{p}_i) + (1 - \mathbf{y}_i)\log(1 - \mathbf{p}_i)), \tag{2}$$

where $\mathbf{y} \in \mathbb{R}^{n_e}$ is the label vector containing ones for true triples and zeros otherwise, subject to *label smoothing*. **Label smoothing** is a widely used technique shown to improve generalization [14,20]. Label smoothing changes the ground-truth label distribution by adding a uniform prior to encourage the model to be less confident, achieving a regularizing effect. **1-N scoring** refers to simultaneously scoring $(e_1, r, \mathcal{E})$, i.e. for all entities $e_2 \in \mathcal{E}$, in contrast to *1-1 scoring*, the practice of training individual triples (e_1, r, e_2) one at a time. As shown by [3], 1-N scoring offers a significant speedup (3x on train and 300x on test time) and improved accuracy compared to 1-1 scoring. A potential extension of the HypER model described above would be to apply convolutional filters to *both* subject and object entity embeddings. However, since this is not trivially implementable with 1-N scoring and wanting to keep its benefits, we leave this to future work.

4.4   Number of Parameters

Table 2 compares the number of parameters of ConvE and HypER (for the FB15k-237 dataset, which determines n_e and n_r). It can be seen that, overall, HypER has fewer parameters (4.3M) than ConvE (5.1M) due to the way HypER directly transforms relations to convolutional filters.

Table 2. Comparison of number of parameters for ConvE and HypER on FB15k-237. h_m and w_m are height and width of the ConvE feature maps respectively.

Model	E	R	Filters	W
ConvE	$n_e \times d_e$	$n_r \times d_r$	$l_f n_f$	$h_m w_m n_f \times d_e$
	$2.9M$	$0.1M$	$0.0M$	$2.1M$
HypER	$n_e \times d_e$	$n_r \times d_r$	$d_r \times l_f n_f$	$l_m n_f \times d_c$
	$2.9M$	$0.1M$	$0.1M$	$1.2M$

5   Experiments

5.1   Datasets

We evaluate our HypER model on the standard link prediction task using the following datasets (see Table 3):

FB15k. [1] a subset of Freebase, a large database of facts about the real world.

WN18. [1] a subset of WordNet, containing lexical relations between words.

FB15k-237. created by [21], noting that the validation and test sets of FB15k and WN18 contain the inverse of many relations present in the training set, making it easy for simple models to do well. FB15k-237 is a subset of FB15k with the inverse relations removed.

WN18RR. [3] a subset of WN18, created by removing the inverse relations.

YAGO3-10. [3] a subset of YAGO3 [10], containing entities which have a minimum of 10 relations each.

Table 3. Summary of dataset statistics.

Dataset	Entities (n_e)	Relations (n_r)
FB15k	$14,951$	$1,345$
WN18	$40,943$	18
FB15k-237	$14,541$	237
WN18RR	$40,943$	11
YAGO3-10	$123,182$	37

5.2   Experimental Setup

We implement HypER in PyTorch [13] and make our code publicly available.[1]

Implementation Details. We train our model with 200 dimension entity and relation embeddings ($d_e = d_r = 200$) and 1-N scoring. Whilst the relation embedding dimension does not have to equal the entity embedding dimension, we set $d_r = 200$ to match ConvE for fairness of comparison.

To accelerate training and prevent overfitting, we use batch normalization [6] and dropout [18] on the input embeddings, feature maps and the hidden layer. We perform a hyperparameter search and select the best performing model by mean reciprocal rank (MRR) on the validation set. Having tested the values $\{0., 0.1, 0.2, 0.3\}$, we find that the following combination of parameters works well across all datasets: input dropout 0.2, feature map dropout 0.2, and hidden dropout 0.3, apart from FB15k-237, where we set input dropout to 0.3. We select the learning rate from $\{0.01, 0.005, 0.003, 0.001, 0.0005, 0.0001\}$ and exponential learning rate decay from $\{1., 0.99, 0.995\}$ for each dataset and find the best performing learning rate and learning rate decay to be dataset-specific. We set the convolution stride to 1, number of feature maps to 32 with the filter size 3×3 for ConvE and 1×9 for HypER, after testing different numbers of feature maps $n_f \in \{16, 32, 64\}$ and filter sizes $l_f \in \{1 \times 1, 1 \times 2, 1 \times 3, 1 \times 6, 1 \times 9, 1 \times 12\}$ (see Table 9). We train all models using the Adam optimizer with batch size 128. One epoch on FB15k-237 takes approximately 12 seconds on a single GPU compared to 1 min for e.g. RESCAL, largely due to 1-N scoring.

Evaluation. Results are obtained by iterating over all triples in the test set. A particular triple is evaluated by replacing the object entity e_2 with all entities $\mathcal{E}$ while keeping the subject entity e_1 fixed and vice versa, obtaining scores for each combination. These scores are then ranked using the "filtered" setting only, i.e. we remove all true cases other than the current test triple [1].

We evaluate HypER on five different metrics found throughout the link prediction literature: mean rank (MR), mean reciprocal rank (MRR), hits@10, hits@3, and hits@1. Mean rank is the average rank assigned to the true triple, over all test triples. Mean reciprocal rank takes the average of the reciprocal rank assigned to the true triple. Hits@k measures the percentage of cases in which the true triple appears in the top k ranked triples. Overall, the aim is to achieve high mean reciprocal rank and hits@k and low mean rank. For a more extensive description of how each of these metrics is calculated, we refer to [3].

5.3   Results

Link prediction results for all models across the five datasets are shown in Tables 4, 5 and 6. Our key findings are:

---

[1] https://github.com/ibalazevic/HypER.

– whilst having fewer parameters than the closest comparator ConvE, HypER consistently outperforms all other models across all datasets, thereby achieving state-of-the-art results on the link prediction task; and
– our filter dimension study suggests that no benefit is gained by convolving over reshaped 2D entity embeddings in comparison with 1D entity embedding vectors and that most information can be extracted with very small convolutional filters (Table 9).

Overall, HypER outperforms all other models on all metrics apart from mean reciprocal rank on WN18 and mean rank on WN18RR, FB15k-237, WN18, and YAGO3-10. Given that mean rank is known to be highly sensitive to outliers [11], this suggests that HypER correctly ranks many true triples in the top 10, but makes larger ranking errors elsewhere.

Given that most models in the literature, with the exception of ConvE, were trained with 100 dimension embeddings and 1-1 scoring, we reimplement previous models (DistMult, ComplEx and ConvE) with 200 dimension embeddings and 1-N scoring for fair comparison and report the obtained results on WN18RR in Table 7. We perform the same hyperparameter search for every model and present the mean and standard deviation of each result across five runs (different random seeds). This improves most previously published results, except for ConvE where we fail to replicate some values. Notwithstanding, HypER remains the best performing model overall despite better tuning of the competitors.

Table 4. Link prediction results on WN18RR and FB15k-237. The RotatE [19] results are reported without their self-adversarial negative sampling (see Appendix H in the original paper) for fair comparison, given that it is not specific to that model only.

	WN18RR					FB15k-237				
	MR	MRR	H@10	H@3	H@1	MR	MRR	H@10	H@3	H@1
DistMult [23]	5110	.430	.490	.440	.390	254	.241	.419	.263	.155
ComplEx [22]	5261	.440	.510	.460	.410	339	.247	.428	.275	.158
Neural LP [24]	–	–	–	–	–	–	.250	.408	–	–
R-GCN [15]	–	–	–	–	–	–	.248	.417	.264	.151
MINERVA [2]	–	–	–	–	–	–	–	.456	–	–
ConvE [3]	**4187**	.430	.520	.440	.400	244	.325	.501	.356	.237
M-Walk [16]	–	.437	–	.445	.414	–	–	–	–	–
RotatE [19]	–	–	–	–	–	**185**	.297	.480	.328	.205
HypER (ours)	5798	**.465**	**.522**	**.477**	**.436**	250	**.341**	**.520**	**.376**	**.252**

To ensure that the difference between reported results for HypER and ConvE is not simply due to HypER having a reduced number of parameters (implicit regularization), we trained ConvE reducing the number of feature maps to 16

Table 5. Link prediction results on WN18 and FB15k.

	WN18					FB15k				
	MR	MRR	H@10	H@3	H@1	MR	MRR	H@10	H@3	H@1
TransE [1]	**251**	–	.892	–	–	125	–	.471	–	–
DistMult [23]	902	.822	.936	.914	.728	97	.654	.824	.733	.546
ComplEx [22]	–	.941	.947	.936	.936	–	.692	.840	.759	.599
ANALOGY [9]	–	.942	.947	.944	.939	–	.725	.854	.785	.646
Neural LP [24]	–	.940	.945	–	–	–	.760	.837	–	–
R-GCN [15]	–	.819	**.964**	.929	.697	–	.696	.842	.760	.601
TorusE [4]	–	.947	.954	.950	.943	–	.733	.832	.771	.674
ConvE [3]	374	.943	.956	.946	.935	51	.657	.831	.723	.558
SimplE [7]	–	.942	.947	.944	.939	–	.727	.838	.773	.660
HypER (ours)	431	**.951**	.958	**.955**	**.947**	**44**	**.790**	**.885**	**.829**	**.734**

Table 6. Link prediction results on YAGO3-10.

	YAGO3-10				
	MR	MRR	H@10	H@3	H@1
DistMult [23]	5926	.340	.540	.380	.240
ComplEx [22]	6351	.360	.550	.400	.260
ConvE [3]	**1676**	.440	.620	.490	.350
HypER (ours)	2529	**.533**	**.678**	**.580**	**.455**

instead of 32 to have a comparable number of parameters to HypER (explicit regularization). This showed no improvement in ConvE results, indicating HypER's architecture does more than merely reducing the number of parameters.

Hypernetwork Influence. To test the influence of the hypernetwork and, thereby, knowledge sharing between relations, we compare HypER results on WN18RR and FB15k-237 with the hypernetwork component removed, i.e. without the first fully connected layer and with the relation embeddings directly corresponding to a set of convolutional filters. Results presented in Table 8 show that the hypernetwork component improves performance, demonstrating the value of multi-task learning across different relations.

Filter Dimension Study. Table 9 shows results of our study investigating the influence of different convolutional filter sizes on the performance of HypER. The lower part of the table shows results for 2D filters convolved over reshaped (10 × 20) 2D subject entity embeddings. It can be seen that reshaping the embeddings is of no benefit, especially on WN18RR. These results indicate that the purpose of convolution on word embeddings is not to find patterns in a 2D embedding (as with images), but perhaps to limit the number of dimensions that can interact with each other, thereby avoiding overfitting. In the upper part of the table, we

Table 7. Link prediction results on WN18RR; all models trained with 200 dimension embeddings and 1-N scoring.

	WN18RR				
	MR	MRR	H@10	H@3	H@1
DistMult [23]	**4911 ± 109**	.434 ± .002	.508 ± .002	.447 ± .001	.399 ± .002
ComplEx [22]	5930 ± 125	.446 ± .001	**.523 ± .002**	.462 ± .001	.409 ± .001
ConvE [3]	**4997 ± 99**	.431 ± .001	.504 ± .002	.443 ± .002	.396 ± .001
HypER (ours)	5798 ± 124	**.465 ± .002**	.522 ± .003	**.477 ± .002**	**.436 ± .003**

Table 8. Results with and without hypernetwork on WN18RR and FB15k-237.

	WN18RR		FB15k-237	
	MRR	H@10	MRR	H@10
HypER	**.465 ± .002**	**.522 ± .003**	**.341 ± .001**	**.520 ± .002**
HypER (no **H**)	.459 ± .002	.511 ± .002	.338 ± .001	.515 ± .001

vary the length of 1D filters, showing that comparable results can be achieved with filter sizes 1×6 and 1×9, with diminishing results for smaller (e.g. 1×1) and larger (e.g. 1×12) filters.

Label Smoothing. Contrary to the ablation study of [3], showing the influence of hyperparameters on mean reciprocal rank for FB15k-237, from which they deem label smoothing unimportant, we find label smoothing to give a significant improvement in prediction scores for WN18RR. However, we find it does have a negative influence on the FB15k scores and as such, exclude label smoothing from our experiments on that dataset. We therefore recommend evaluating the

Table 9. Influence of different filter dimension choices on prediction results.

	WN18RR		FB15k-237	
Filter size	MRR	H@1	MRR	H@1
1×1	.455	.422	.337	.248
1×2	.458	.428	.337	.248
1×3	.457	.427	.339	.250
1×6	.459	.429	.340	.251
1×9	**.465**	**.436**	**.341**	**.252**
1×12	.457	.428	**.341**	**.252**
2×2	.456	.429	.340	.250
3×3	.458	.430	.339	.250
5×5	.452	.423	.340	**.252**

influence of label smoothing on a per dataset basis and leave to future work analysis of the utility of label smoothing in the general case.

6   Conclusion

In this work, we introduce HypER, a hypernetwork model for link prediction on knowledge graphs. HypER generates relation-specific convolutional filters and applies them to subject entity embeddings. The hypernetwork component allows information to be shared between relation vectors, enabling multi-task learning across relations. To our knowledge, HypER is the first link prediction model that creates non-linear interaction between entity and relation embeddings by convolving relation-specific filters over the entity embeddings.

We show that no benefit is gained from 2D convolutional filters over 1D, dispelling the suggestion that 2D structure exists in entity embeddings implied by ConvE. We also recast HypER in terms of tensor operations showing that, despite the convolution operation, it is closely related to the established family of tensor factorization models. Our results suggest that convolution provides a good trade-off between expressiveness and parameter number compared to a dense network. HypER is fast, robust to overfitting, has relatively few parameters, and achieves state-of-the-art results across almost all metrics on multiple link prediction datasets.

Future work might include expanding the current architecture by applying convolutional filters to both subject and object entity embeddings. We may also analyze the influence of label smoothing and explore the interpretability of convolutional feature maps to gain insight and potentially improve the model.

Acknowledgements. We thank Ivan Titov for helpful discussions on this work. Ivana Balažević and Carl Allen were supported by the Centre for Doctoral Training in Data Science, funded by EPSRC (grant EP/L016427/1) and the University of Edinburgh.

References

1. Bordes, A., Usunier, N., Garcia-Duran, A., Weston, J., Yakhnenko, O.: Translating embeddings for modeling multi-relational data. In: Advances in Neural Information Processing Systems (2013)
2. Das, R., et al.: Go for a walk and arrive at the answer: reasoning over paths in knowledge bases using reinforcement learning. In: International Conference on Learning Representations (2018)
3. Dettmers, T., Minervini, P., Stenetorp, P., Riedel, S.: Convolutional 2D knowledge graph embeddings. In: Association for the Advancement of Artificial Intelligence (2018)
4. Ebisu, T., Ichise, R.: TorusE: knowledge graph embedding on a Lie group. In: Association for the Advancement of Artificial Intelligence (2018)
5. Ha, D., Dai, A., Le, Q.V.: Hypernetworks. In: International Conference on Learning Representations (2017)

6. Ioffe, S., Szegedy, C.: Batch normalization: accelerating deep network training by reducing internal covariate shift. In: International Conference on Machine Learning (2015)
7. Kazemi, S.M., Poole, D.: Simple embedding for link prediction in knowledge graphs. In: Advances in Neural Information Processing Systems (2018)
8. Kingma, D.P., Ba, J.: Adam: a method for stochastic optimization. In: International Conference on Learning Representations (2015)
9. Liu, H., Wu, Y., Yang, Y.: Analogical inference for multi-relational embeddings. In: International Conference on Machine Learning (2017)
10. Mahdisoltani, F., Biega, J., Suchanek, F.M.: Yago3: a knowledge base from multilingual Wikipedias. In: Conference on Innovative Data Systems Research (2013)
11. Nickel, M., Rosasco, L., Poggio, T.A.: Holographic embeddings of knowledge graphs. In: Association for the Advancement of Artificial Intelligence (2016)
12. Nickel, M., Tresp, V., Kriegel, H.P.: A three-way model for collective learning on multi-relational data. In: International Conference on Machine Learning (2011)
13. Paszke, A., et al.: Automatic differentiation in PyTorch. In: NIPS-W (2017)
14. Pereyra, G., Tucker, G., Chorowski, J., Kaiser, L., Hinton, G.: Regularizing neural networks by penalizing confident output distributions. arXiv preprint arXiv:1701.06548 (2017)
15. Schlichtkrull, M., Kipf, T.N., Bloem, P., van den Berg, R., Titov, I., Welling, M.: Modeling relational data with graph convolutional networks. In: Gangemi, A., et al. (eds.) ESWC 2018. LNCS, vol. 10843, pp. 593–607. Springer, Cham (2018). https://doi.org/10.1007/978-3-319-93417-4_38
16. Shen, Y., Chen, J., Huang, P.S., Guo, Y., Gao, J.: M-Walk: learning to walk over graphs using Monte Carlo tree search. In: Advances in Neural Information Processing Systems (2018)
17. Socher, R., Chen, D., Manning, C.D., Ng, A.: Reasoning with neural tensor networks for knowledge base completion. In: Advances in Neural Information Processing Systems (2013)
18. Srivastava, N., Hinton, G., Krizhevsky, A., Sutskever, I., Salakhutdinov, R.: Dropout: a simple way to prevent neural networks from overfitting. J. Mach. Learn. Res. 15(1), 1929–1958 (2014)
19. Sun, Z., Deng, Z.H., Nie, J.Y., Tang, J.: RotatE: knowledge graph embedding by relational rotation in complex space. In: International Conference on Learning Representations (2019)
20. Szegedy, C., Vanhoucke, V., Ioffe, S., Shlens, J., Wojna, Z.: Rethinking the inception architecture for computer vision. In: Computer Vision and Pattern Recognition (2016)
21. Toutanova, K., Chen, D., Pantel, P., Poon, H., Choudhury, P., Gamon, M.: Representing text for joint embedding of text and knowledge bases. In: Empirical Methods in Natural Language Processing (2015)
22. Trouillon, T., Welbl, J., Riedel, S., Gaussier, É., Bouchard, G.: Complex embeddings for simple link prediction. In: International Conference on Machine Learning (2016)
23. Yang, B., Yih, W.t., He, X., Gao, J., Deng, L.: Embedding entities and relations for learning and inference in knowledge bases. In: International Conference on Learning Representations (2015)
24. Yang, F., Yang, Z., Cohen, W.W.: Differentiable learning of logical rules for knowledge base reasoning. In: Advances in Neural Information Processing Systems (2017)

Signed Graph Attention Networks

Junjie Huang[1,2]([✉]), Huawei Shen[1], Liang Hou[1,2], and Xueqi Cheng[1]

[1] CAS Key Laboratory of Network Data Science and Technology,
Institute of Computing Technology, Chinese Academy of Sciences, Beijing, China
{huangjunjie17s,shenhuawei,houliang17z,cxq}@ict.ac.cn
[2] University of Chinese Academy of Sciences, Beijing, China

Abstract. Graph or network data is ubiquitous in the real world, including social networks, information networks, traffic networks, biological networks and various technical networks. The non-Euclidean nature of graph data poses the challenge for modeling and analyzing graph data. Recently, Graph Neural Network (GNN) is proposed as a general and powerful framework to handle tasks on graph data, e.g., node embedding, link prediction and node classification. As a representative implementation of GNNs, Graph Attention Networks (GAT) is successfully applied in a variety of tasks on real datasets. However, GAT is designed to networks with only positive links and fails to handle signed networks which contain both positive and negative links. In this paper, we propose Signed Graph Attention Networks (SiGAT), generalizing GAT to signed networks. SiGAT incorporates graph motifs into GAT to capture two well-known theories in signed network research, i.e., balance theory and status theory. In SiGAT, motifs offer us the flexible structural pattern to aggregate and propagate messages on the signed network to generate node embeddings. We evaluate the proposed SiGAT method by applying it to the signed link prediction task. Experimental results on three real datasets demonstrate that SiGAT outperforms feature-based method, network embedding method and state-of-the-art GNN-based methods like signed graph convolutional networks (SGCN).

Keywords: Signed network · Network embedding ·
Graph Neural Network

1 Introduction

Graph Neural Networks (GNNs) have been receiving more and more research attentions, achieving good results in many machine learning tasks e.g., semi-supervised node classification [11], network embedding [12] and link prediction. GNNs introduce neural networks into graph data by defining convolution [11], attention [22] and other mechanisms. However, previous GNNs methods mainly focus on undirected and unsigned networks (i.e., networks consisting of only positive edges). How to apply GNNs to signed, directed, weighted and other complex networks is an important research direction.

© Springer Nature Switzerland AG 2019
I. V. Tetko et al. (Eds.): ICANN 2019, LNCS 11731, pp. 566–577, 2019.
https://doi.org/10.1007/978-3-030-30493-5_53

With the growing popularity of online social media, many interactions are generated by people on the Web. Most of these interactions are positive relationships, such as friendship, following, support and approval. Meanwhile, there are also some negative links that indicate disapproval, disagreement or distrust. Some recent researches about signed social networks suggest that negative links bring more valuable information over positive links in various tasks [14]. In the early years, the methods for modeling signed networks were mainly based on feature extraction [14], matrix factorization [8] and network embedding [10,23]. They have achieved good performances in tasks such as signed link prediction and node classification. The core of modeling signed network is two important sociological theories, i.e., balance theory and status theory. Derr et al. [2] proposed a signed graph convolution networks (SGCN) method to model signed network. It utilizes balance theory to aggregate and propagate the information. However, it only applies to the undirected signed networks.

In this paper, we try to model the directed signed network into graph attention network (GAT). To the best of our knowledge, it's the first time to introduce GAT to directed signed networks. Attention mechanism can characterize the different effects of different nodes on the target node, e.g., A is more kind to me than B.

The main contributions of this paper are as follows:

- We analyze the key elements of the signed link prediction problem. Triads motifs are used to describe two important sociological theories, i.e., balance theory and status theory.
- We introduce the GAT to model the signed network and design a new motif-based GNN model for signed networks named SiGAT.
- We conduct experiments on some real signed social network data sets to demonstrate the effectiveness of our proposed framework SiGAT.

The rest of paper is organized as follows. In Sect. 2, we review some related work. We propose our model SiGAT in Sect. 3, which consists of introducing related social theory, incorporating these theories into motifs and finally presenting our Signed Graph Attention Network (SiGAT). In Sect. 4, experiments are performed to empirically evaluate the effectiveness of our framework for learning node embedding. We discuss the results of some signed network embedding and SGCN, analyze the impact of hyper-parameters in our model. Finally, we conclude and discuss future work in Sect. 5.

2 Related Work

2.1 Signed Network Embedding

Signed social networks are such social networks in which social ties can have two signs: positive and negative [15]. To mine signed networks, many algorithms have been developed for tasks such as community detection, node classification, link prediction and spectral graph analysis. In particular, [15] proposed a feature engineering-based methods to do the signed link prediction tasks. Recently, with the development of network representation learning [5,17,20], researchers begin to learn low-dimensional vector representations for a social network. For signed network embedding methods, SNE [25] adopts the log-bilinear model and incorporates two signed-type vectors to capture the positive or negative relationship of each edge along the path. SiNE [23] designs a new objective function guided by social theories to learn signed network embedding, it proposed to add a virtual node to enhance the training process. SIDE [10] provides a linearly scalable method that leverages balance theory along with random walks to obtain the low-dimensional vector for the directed signed network. SIGNet [9] combines balance theory with a specialized random and new sampling techniques in directed signed networks. These methods are devoted to defining an objective function that incorporates sociological theory and then using some machine learning techniques e.g., sampling and random walks to optimize look-up embedding.

2.2 Graph Neural Networks

GNNs have achieved a lot of success in the semi-supervised node classification task [11], graph classification [4] and other graph analysis problems. The concept of graph neural network (GNN) was first proposed in [18], which extended neural networks for processing the data represented in graph domains. Recently, researchers tried to apply convolution [11], attention [22], auto-encoder [12] and other mechanisms into graphs.

GCN in [11] is designed with the focus to learning representation at the node level in the semi-supervised node classification task. GraphSAGE [6] extends it to the large inductive graphs. GAE [12] firstly integrates GCN into a graph auto-encoder framework. This approach is aimed at representing network vertices into a low-dimensional vector space by using neural network architectures in unsupervised learning tasks. GAT [22] applies attention mechanisms to GNNs. Compared to the convolution, the attention mechanism can describe the influence of different nodes on the target node. On the semi-supervised task of the node, it shows better results than GCN. SGCN [1] introduced GNNs to a signed network for the first time. It designed a new information aggregation and propagation mechanism for the undirected signed network according to the balance theory. However, the majority of current GNNs tackle with static homogeneous graphs [24], how to apply GNNs into other complex networks (e.g., directed signed networks) is still a challenge.

3   Methods

3.1   Signed Network Theory

Balance theory and status theory play an important role in analyzing and modeling signed networks. These theories can enable us to characterize the differences between the observed and predicted configurations of positive and negative links in online social networks [14].

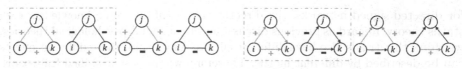

(a) Illustration of structural balance theory   (b) Illustration of status theory

Fig. 1. The balance theory and status theory

Balance Theory. Balance theory originated in social psychology in the mid-20th-century [7]. It was initially intended as a model for undirected signed networks. All signed triads with an even number of negative edges are defined as balanced. In other words, for the four triangles in Fig. 1(a), the triangles which all three of these users are friends or only one pair of them are friends are defined as balanced i.e., the first two triads are balanced. Balance theory suggests that people in a social network tend to form into a balanced network structure. It can be described as "the friend of my friend is my friend" and "the enemy of my enemy is my friend". This theory has a wide range of applications in the field of sociology. [14] found that balance theory has widely existed in social networks. It was proven that if a connected network is balanced, it can be split into two opposing groups and vice versa [3].

Status Theory. Status theory is another key social psychological theory for signed network. It is based on directed signed networks. It supposes directed relationship labeled by a positive sign "+" or negative sign "−" means target node has a higher or lower status than source node [21]. For example, a positive link from A to B, it does not only mean "B is my friend" but also "I think B has a higher status than I do". For the triangles in Fig. 1(b), the first two triangles satisfy the status ordering and the latter two do not satisfy it. For the first triangle, when Status(j) > Status(i) and Status(k) > Status(j), we have Status(k) > Status(i). Based on status theory, it can be found that "opinion leaders" have higher social status (manager or advisor) than ordinary users [21].

Based on the analysis of the above two theories on social network datasets, [14] extracts a total of 23 features of two given nodes u and v and then uses logistic regression to train and infer the positive and negative of edges. Experiments show that these features can help to achieve good results in machine learning tasks. These two theories can be described in triangles. These directed edges, signed edges and triangles can all be defined as motifs of a network [16]. In signed social networks, nodes not only spread the message/information/influence by just directed positive or negative relationship but also triad motifs.

3.2 Signed Motifs

Since positive neighbors and negative neighbors should have different effects on the target node, they should obviously be treated separately. Furthermore, for directed signed networks, the direction also contains some knowledge, e.g., status theory. Regard as previous discussions about social-psychological theories, we think that triads also have different effects. As we said, these different effects can be described by different motifs. Therefore, we propose a unified framework to effectively model signed directed networks based on motifs.

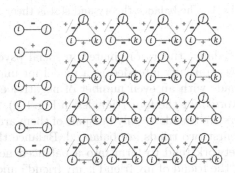

Fig. 2. 38 different motifs in SiGAT, mean different influences from node j to node i

In our framework, we define positive/negative neighbors (2 motifs), positive/negative with direction neighbors (4 motifs) and 32 different triangle motifs in Fig. 2. We first extract these motifs and then apply them to our model. The method of effectively extracting these motifs can be found in [19].

3.3   Signed Graph Attention Networks

GAT [22] introduces attention mechanism into the graph. It uses a weight matrix to characterize the different effects of different nodes on the target node. GAT firstly computes the α_{ij} for node i and node j by the attention mechanism $\boldsymbol{a}$ and LeakyReLU nonlinearity (with negative input slope $\alpha = 0.2$) as:

$$\alpha_{ij} = \frac{\exp\left(\text{LeakyReLU}\left(\mathbf{a}^T[\mathbf{W}h_i \| \mathbf{W}h_j]\right)\right)}{\sum_{k \in \mathcal{N}_i} \exp\left(\text{LeakyReLU}\left(\mathbf{a}^T[\mathbf{W}h_i \| \mathbf{W}h_k]\right)\right)}, \tag{1}$$

where $\cdot^T$ represents transposition and $\|$ is the concatenation operation, $\mathcal{N}_i$ are the neighborhoods of node i, $\mathbf{W}$ is the weight matrix parameter, h_i is the node feature of node i.

After computing the normalized attention coefficients, a linear combination of the features is used and served to the final output features for every node:

$$h_i' = \sigma\left(\sum_{j \in \mathcal{N}_i} \alpha_{ij} \mathbf{W} h_j\right). \tag{2}$$

Based on previous discussions, we proposed our SiGAT model in Fig. 3. For a node u, in different motifs, we get different neighborhoods. For example, in

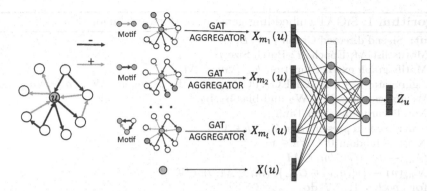

Fig. 3. For a signed network, the target u is the orange node. Red/Green links mean positive/negative links, the arrows indicate the directions. We define different motifs and obtain the corresponding neighborhoods in them. Under a motif definition, the gray color nodes are the neighborhoods, we use GAT to describe the influences from these neighborhoods. After aggregating the information from different motifs, we use these $\mathbf{X}_{m_i}(u)$ and $\mathbf{X}(u)$ to get the representation $\mathbf{Z}_u$. (Color figure online)

$v \rightarrow^+ u$ motifs definition, the gray color nodes v are the neighborhoods, we use GAT to describe the influence from these neighborhoods. After aggregating the information from different motifs, we concatenate these $\mathbf{X}_{m_i}(u)$ and $\mathbf{X}(u)$ to a two-layer fully connected neural network to generate the representation $\mathbf{Z}_u$ for node u, m_i is the i-th motif. We also take the scalability into considerations and propose a mini batch algorithm like GraphSAGE [6].

The details are described in Algorithm 1. First, for each node u, we initialize embeddings $\mathbf{X}(u)$ as features. We extracted new graphs under different motif definitions using the motif graph extract function F_{m_i}, based on motif list which is given before. In other word, for each target node u, we get its neighbor nodes $\mathcal{N}_{m_i}(u)$ under motif m_i definition. Then, for every motif m_i, we use a GAT model GAT-AGGREGATOR$_{m_i}$ with the parameter $\mathbf{W}_{m_i}, \boldsymbol{a}_{m_i}$ as follows to get the message $\mathbf{X}_{m_i}(u)$ in this motif:

$$\alpha_{uv}^{m_i} = \frac{\exp\left(\text{LeakyReLU}\left(\mathbf{a}_{m_i}^T[\mathbf{W}_{m_i}\mathbf{X}(u)\|\mathbf{W}_{m_i}\mathbf{X}(v)]\right)\right)}{\sum_{k \in \mathcal{N}_{m_i}(u)} \exp\left(\text{LeakyReLU}\left(\mathbf{a}_{m_i}^T[\mathbf{W}_{m_i}\mathbf{X}(u)\|\mathbf{W}_{m_i}\mathbf{X}(k)]\right)\right)}, \quad (3)$$

$$\mathbf{X}_{m_i}(u) = \sum_{v \in \mathcal{N}_{m_i}(u)} \alpha_{uv}^{m_i}\mathbf{W}_{m_i}\mathbf{X}(v). \quad (4)$$

Finally, we concatenate all the messages from different motifs with $\mathbf{X}(u)$ to a two-layer fully connected neural network to obtain the final embedding $\mathbf{Z}_u$ for the target node u. Here, $\mathbf{W_1}, \mathbf{W_2}, \mathbf{b_1}, \mathbf{b_2}$ are the parameters for the 2-layer neural network.

After computing the loss, we propagate the gradient back and update the parameters. In our models, we define an unsupervised loss function in Eq. 5 which reflects friend embeddings are similar, and the enemy embeddings are dissimilar.

Algorithm 1. SiGAT embedding generation (forward) algorithm

Input: Sigend directed Graph $G(V, E, s)$;
    Motifs list $\mathcal{M}$; Epochs T; Batch Size B;
    Motifs graph extract function $F_{m_i}, \forall m_i \in \mathcal{M}$;
    Aggregator GAT-AGGREGATOR$_{m_i}$ with the parameter $\mathbf{W}_{m_i}, \boldsymbol{a}_{m_i}, \forall m_i \in \mathcal{M}$;
    Weight matrices $\mathbf{W_1}, \mathbf{W_2}$ and bias $\mathbf{b_1}, \mathbf{b_2}$;
    Non-linearity function Tanh ;
Output: Node representation $\mathbf{Z}_u, \forall u \in V$
1: $\mathbf{X}(u) \leftarrow$ random$(0, 1), \forall u \in V$
2: $G_{m_i} \leftarrow F_{m_i}(G), \forall m_i \in \mathcal{M}$
3: $\mathcal{N}_{m_i}(u) \leftarrow \{v | (u, v) \in G_{m_i}\}, \forall m_i \in \mathcal{M}, \forall u \in V$
4: **for** $epoch = 1, ..., T$ **do**
5:    **for** $batch = 1, ..., |V|/B$ **do**
6:       $\mathcal{B} \leftarrow V_{(batch-1) \times B+1:batch \times B}$
7:       **for** $u \in \mathcal{B}$ **do**
8:          **for** $m_i \in \mathcal{M}$ **do**
9:            $\mathbf{X}_{m_i}(u) \leftarrow$ GAT-AGGREGATOR$_{m_i}(\{\mathbf{X}_v, \forall v \in \mathcal{N}_{m_i}(v)\})$
10:         **end for**
11:         $\mathbf{X}'(u) \leftarrow$ CONCAT$(\mathbf{X}(u), \mathbf{X}_{m_1}(u), ..., \mathbf{X}_{m_{|\mathcal{M}|}}(u))$
12:         $\mathbf{Z}_u \leftarrow \mathbf{W_2} \cdot$ Tanh$(\mathbf{W_1} \cdot \mathbf{X}'(u) + \mathbf{b_1}) + \mathbf{b_2}$
13:       **end for**
14:    **end for**
15: **end for**
16: **return** $\mathbf{Z}_u$

$$J_{\mathcal{G}}(\mathbf{Z}_u) = - \sum_{v^+ \in \mathcal{N}(u)^+} \log\left(\sigma(\mathbf{Z}_u^\top \mathbf{Z}_{v+})\right) - Q \sum_{v^- \in \mathcal{N}(u)^-} \log\left(\sigma(-\mathbf{Z}_u^\top \mathbf{Z}_{v-})\right), \quad (5)$$

where σ is the sigmoid function, $\mathcal{N}(u)^+$ is the set of positive neighborhoods of node u, $\mathcal{N}(u)^-$ is the set of negative neighborhoods of node u and Q is the balanced parameter for the unbalanced positive and negative neighborhoods.

4 Experiments

4.1 Link Sign Prediction and Dataset Description

Link sign prediction is the task of predicting unobserved signs of existing edges in the test dataset given train dataset. It's the most fundamental signed network analysis task [14].

We do experiments on three real-world signed social network datasets. e.g., Bitcoin-Alpha[1], Slashdot[2] and Epinions[3].

Bitcoin-Alpha [13] is from a website where users can trade with Bitcoins. Because the Bitcoin accounts are anonymous, the users need to build online

---

[1] http://snap.stanford.edu/data/soc-sign-bitcoin-alpha.html.
[2] http://snap.stanford.edu/data/soc-sign-Slashdot090221.html.
[3] http://snap.stanford.edu/data/soc-sign-epinions.html.

trust networks for their safety. Members of Bitcoin-Alpha rate other members in a scale of -10 (total distrust) to $+10$ (total trust) in steps of 1, which can help to prevent transactions with fraudulent and risky users. In this paper, we treat the scores greater than 0 as positive and others as negative. Slashdot [15] is a technology-related news website where users are allowed to tag each other as friends or foes. In other words, it is a common signed social network with friends and enemies labels. Epinions [15] was a product review site. Members of the site can indicate their trust or distrust of the reviews of others. The network reflects people's opinions to others. In these networks, the edges are inherently directed [15].

The statistics of three datasets are summarized in Table 1. We can see that positive and negative links are imbalanced in these datasets.

Table 1. Statistics of three datasets

Dataset	# nodes	# pos links	# neg links	% pos ratio
Bitcoin-Alpha	3,783	22,650	1,536	93.65
Slashdot	82,140	425,072	124,130	77.40
Epinions	131,828	717,667	123,705	85.30

4.2 Baselines

To validate the effectiveness of SiGAT, we compare it with some state-of-the-art baselines and one simple version $\text{SiGAT}_{+/-}$. The baselines are as follows:

- Random: It generates d dimensional random values, $Z_j = (z_1, z_2, ..., z_d), z_i \in [0.0, 1.0)$. It can be used to show the logistic regression's ability in this task.
- Unsigned network embedding: We use some classical unsigned network embedding methods (e.g., DeepWalk [17], Node2vec [5], LINE [20]) to validate the importance of signed edges.
- Signed network embedding: We use some signed network embedding methods(e.g., SIDE [10], SiNE [23], SIGNet [9]) to show the effectiveness of modeling signed edges.
- SGCN [2]: It makes a dedicated and principled effort that utilizes balance theory to correctly aggregate and propagate the information across layers of a signed GCN model. It is the latest attempt to introduce GNNs to signed networks.
- FeExtra [14]: This method extracts two parts, a total of 23 features from the signed social network.
- $\text{SiGAT}_{+/-}$: We made some simplifications, considering only the positive and negative neighbors, the model can verify if the attention mechanism works.

For a fair comparison, the embedding dimension d of all methods is set to 20 which is closed to FeExtra [14]. We use the authors released code for DeepWalk,

Node2vec, LINE, SIDE, SiNE and SIGNet. For SGCN, we use the code from github[4]. We follow the authors recommended settings in Slashdot and Epinions and try to fine-tune parameters to achieve the best results in Bitcoin-Alpha. Like previous works [2,10,23], we first use these methods to get node representations. For edge e_{ij}, we concatenate these two learned representation z_i and z_j to compose an edge representation z_{ij}. After that, we train a logistic regression classifier on the training set and use it to predict the edge sign in the test set. We perform 5-fold cross-validation experiments to get the average scores as the same as [10]. Each training set is used to train both embedding vectors and logistic regression model. Our models[5] were implemented by PyTorch with the Adam optimizer (Learning Rate = 0.0005, Weight Decay = 0.0001, Batch Size = 500). The epoch was selected from $\{10, 20, 50, 100\}$ by 5-fold cross-validation on the training folds.

4.3 Results

We report the average Accuracy, F1, macro-F1 and AUC in Table 2. We have bolded the highest value of each row. From Table 2, we can find that:

Table 2. The results of Signed Link Prediction on three datasets

Dataset	Metric	Random	Unsigned Network Embedding			Signed Network Embedding			Feature Engineering	Graph Neural Network		
		Random	Deepwalk	Node2vec	LINE	SiNE	SIDE	SIGNet	FExtra	SGCN	SiGAT$_{+/-}$	SiGAT
Bitcoin-Alpha	Accuracy	0.9365	0.9365	0.9274	0.9350	0.9424	0.9369	0.9443	0.9477	0.9351	0.9427	**0.9480**
	F1	0.9672	0.9672	0.9623	0.9662	0.9699	0.9673	0.9706	0.9725	0.9658	0.9700	**0.9727**
	Macro-F1	0.4836	0.4836	0.5004	0.5431	0.6683	0.5432	0.7099	0.7069	0.6689	0.6570	**0.7138**
	AUC	0.6395	0.6435	0.7666	0.7878	0.8788	0.7832	**0.8972**	0.8887	0.8530	0.8699	0.8942
Slashdot	Accuracy	0.7740	0.7740	0.7664	0.7638	0.8269	0.7776	0.8391	0.8457	0.8200	0.8331	**0.8482**
	F1	0.8726	0.8726	0.8590	0.8655	0.8921	0.8702	0.8984	**0.9061**	0.8860	0.8959	0.9047
	Macro-F1	0.4363	0.4363	0.5887	0.4463	0.7277	0.5469	0.7559	0.7371	0.7294	0.7380	**0.7660**
	AUC	0.5415	0.5467	0.7622	0.5343	0.8423	0.7627	0.8749	0.8859	0.8440	0.8639	**0.8864**
Epinions	Accuracy	0.8530	0.8518	0.8600	0.8262	0.9131	0.9186	0.9116	0.9206	0.9092	0.9124	**0.9293**
	F1	0.9207	0.9198	0.9212	0.9040	0.9502	0.9533	0.9491	0.9551	0.9472	0.9498	**0.9593**
	Macro-F1	0.4603	0.4714	0.6476	0.4897	0.8054	0.8184	0.8065	0.8075	0.8102	0.8020	**0.8449**
	AUC	0.5569	0.6232	0.8033	0.5540	0.8882	0.8893	0.9091	**0.9421**	0.8818	0.9079	0.9333

- For signed networks, itself is the positive and negative imbalance. Given random embedding, logistic regression can learn a part of the linear features to a certain extent.
- After using unsigned network embedding methods, all metrics have been increased. Node2Vec has made an obvious improvement, and DeepWalk and LINE have different performances on different datasets.
- SiNE, SIDE, and SIGNet are designed for the signed network; the result is significantly higher than other unsigned network embedding methods. These algorithms modeled some related sociological theories and had achieved good results for signed network analysis. SIGNet showed more stable and better

---

[4] https://github.com/benedekrozemberczki/SGCN.
[5] https://github.com/huangjunjie95/SiGAT.

results in our experiments. Besides, The performance of SIDE is not as good as that in [10]. This can be due to the different pre-processing processes and low dimension parameter.

- Although FeExtra was proposed earlier, it has performed well because of its successful use of relevant sociological theory. However, it should be pointed out that this method relies on feature engineering to extract features manually and models only edge features without node representations. Its generalization ability to other tasks is weak.
- SGCN shows a performance close to SiNE as they said in [2], but it cannot effectively model the signed directed networks because the algorithm did not consider the direction. Its mechanism is the simple average of neighbors' hidden state, which can not describe the influence of different nodes to the target node.
- SiGAT$_{only+/-}$ shows a performance close to or slightly better than SGCN, indicating that the attention mechanism can better model the problem. Compared with the signed graph convolution network method, attention mechanism considers the influence of different nodes on the target node, which is reasonable.
- Our SiGAT model achieves almost the best performance on three datasets. It shows that the relevant theories of sociology have also been successfully modeled by designing different motifs and attention mechanism.

4.4 Parameter Analysis

In this subsection, we analyze the some hyper-parameters. We randomly select 80% training edges and 20% testing edges. When analyzing epoch, we set $d = 20$, and record the AUC performance of different epoch generated representations and the total loss value during training. When discussing dimension, we set the corresponding epoch number to 100 (Bitcoin-Alpha), 20 (Slashdot), 10 (Epinions) to discuss the robustness of different dimensions.

As shown in Fig. 4, we can see that the value of loss decreases and the value of AUC increases at the earlier epochs and then gradually converges. In different datasets, the best epochs are different. This can be due to the different scales and network structure. Many nodes are updated multiple times quickly due to the mini-batch algorithm. In Fig. 4(d), the performance increases quickly and becomes almost stable with the increasing number of dimension d. In conclusion, our model shows relatively robust results on hyper-parameters.

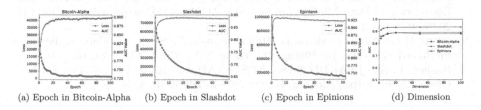

(a) Epoch in Bitcoin-Alpha   (b) Epoch in Slashdot   (c) Epoch in Epinions   (d) Dimension

Fig. 4. Parameter Analysis for the epoch and dimension in three datasets

5   Conclusion

In this paper, we try to introduce GNNs into signed networks. GNNs have recently received widespread attention and achieved great improvements in many tasks. However, most GNNs have been on unsigned networks. We analyzed some key points in signed network analysis and redesigned the GNNs architectures. Based on the basic assumptions of social psychology, we leverage the attention mechanism and different motifs to model the signed network. Experimental results on three real-world signed networks show that our proposed SiGAT outperforms different state-of-the-art signed network embedding methods. Moreover, we analyze the hyper-parameters and show the robustness of our methods. In future work, we will further investigate the performance of SiGAT on more signed network tasks, such as node classification and clustering. Further, we will consider more effective methods to model signed networks.

Acknowledgement. This work is funded by the National Natural Science Foundation of China under grant numbers 61425016, 61433014, and 91746301. Huawei Shen is also funded by K.C. Wong Education Foundation and the Youth Innovation Promotion Association of the Chinese Academy of Sciences.

References

1. Arinik, N., Figueiredo, R., Labatut, V.: Signed graph analysis for the interpretation of voting behavior. arXiv preprint arXiv:1712.10157 (2017)
2. Derr, T., Ma, Y., Tang, J.: Signed graph convolutional network. arXiv preprint arXiv:1808.06354 (2018). https://doi.org/10.1109/ICDM.2018.00113
3. Easley, D., Kleinberg, J.: Networks, Crowds, and Markets: Reasoning About a Highly Connected World. Cambridge University Press, Cambridge (2010)
4. Gilmer, J., Schoenholz, S.S., Riley, P.F., Vinyals, O., Dahl, G.E.: Neural message passing for quantum chemistry. arXiv preprint arXiv:1704.01212 (2017)
5. Grover, A., Leskovec, J.: node2vec: scalable feature learning for networks. In: Proceedings of the 22nd ACM SIGKDD International Conference on Knowledge Discovery and Data Mining, pp. 855–864. ACM (2016). https://doi.org/10.1145/2939672.2939754
6. Hamilton, W., Ying, Z., Leskovec, J.: Inductive representation learning on large graphs. In: Advances in Neural Information Processing Systems, pp. 1025–1035 (2017)
7. Heider, F.: Attitudes and cognitive organization. J. Psychol. **21**(1), 107–112 (1946)
8. Hsieh, C.J., Chiang, K.Y., Dhillon, I.S.: Low rank modeling of signed networks. In: Proceedings of the 18th ACM SIGKDD International Conference on Knowledge Discovery and Data Mining, pp. 507–515. ACM (2012). https://doi.org/10.1145/2339530.2339612
9. Islam, M.R., Aditya Prakash, B., Ramakrishnan, N.: SIGNet: scalable embeddings for signed networks. In: Phung, D., Tseng, V.S., Webb, G.I., Ho, B., Ganji, M., Rashidi, L. (eds.) PAKDD 2018. LNCS (LNAI), vol. 10938, pp. 157–169. Springer, Cham (2018). https://doi.org/10.1007/978-3-319-93037-4_13

10. Kim, J., Park, H., Lee, J.E., Kang, U.: Side: representation learning in signed directed networks. In: Proceedings of the 2018 World Wide Web Conference on World Wide Web, pp. 509–518. International World Wide Web Conferences Steering Committee (2018). https://doi.org/10.1145/3178876.3186117
11. Kipf, T.N., Welling, M.: Semi-supervised classification with graph convolutional networks. arXiv preprint arXiv:1609.02907 (2016)
12. Kipf, T.N., Welling, M.: Variational graph auto-encoders. arXiv preprint arXiv:1611.07308 (2016)
13. Kumar, S., Spezzano, F., Subrahmanian, V., Faloutsos, C.: Edge weight prediction in weighted signed networks. In: 2016 IEEE 16th International Conference on Data Mining (ICDM), pp. 221–230. IEEE (2016). https://doi.org/10.1109/ICDM.2016.0033
14. Leskovec, J., Huttenlocher, D., Kleinberg, J.: Predicting positive and negative links in online social networks. In: Proceedings of the 19th International Conference on World Wide Web, pp. 641–650. ACM (2010). https://doi.org/10.1145/1772690.1772756
15. Leskovec, J., Huttenlocher, D., Kleinberg, J.: Signed networks in social media. In: Proceedings of the SIGCHI Conference on Human Factors in Computing Systems, pp. 1361–1370. ACM (2010). https://doi.org/10.1145/1753326.1753532
16. Milo, R., Shen-Orr, S., Itzkovitz, S., Kashtan, N., Chklovskii, D., Alon, U.: Network motifs: simple building blocks of complex networks. Science 298(5594), 824–827 (2002). https://doi.org/10.1515/9781400841356.217
17. Perozzi, B., Al-Rfou, R., Skiena, S.: Deepwalk: online learning of social representations. In: Proceedings of the 20th ACM SIGKDD International Conference on Knowledge Discovery and Data Mining, pp. 701–710. ACM (2014). https://doi.org/10.1145/2623330.2623732
18. Scarselli, F., Gori, M., Tsoi, A.C., Hagenbuchner, M., Monfardini, G.: The graph neural network model. IEEE Trans. Neural Netw. 20(1), 61–80 (2009). https://doi.org/10.1109/TNN.2008.2005605
19. Schank, T., Wagner, D.: Finding, counting and listing all triangles in large graphs, an experimental study. In: Nikoletseas, S.E. (ed.) WEA 2005. LNCS, vol. 3503, pp. 606–609. Springer, Heidelberg (2005). https://doi.org/10.1007/11427186_54
20. Tang, J., Qu, M., Wang, M., Zhang, M., Yan, J., Mei, Q.: Line: large-scale information network embedding. In: Proceedings of the 24th International Conference on World Wide Web, pp. 1067–1077. International World Wide Web Conferences Steering Committee (2015). https://doi.org/10.1145/2736277.2741093
21. Tang, J., Lou, T., Kleinberg, J.: Inferring social ties across heterogenous networks. In: Proceedings of the Fifth ACM International Conference on Web Search and Data Mining, pp. 743–752. ACM (2012). https://doi.org/10.1145/2124295.2124382
22. Velickovic, P., Cucurull, G., Casanova, A., Romero, A., Lio, P., Bengio, Y.: Graph attention networks. arXiv preprint arXiv:1710.10903 1(2) (2017)
23. Wang, S., Tang, J., Aggarwal, C., Chang, Y., Liu, H.: Signed network embedding in social media. In: Proceedings of the 2017 SIAM International Conference on Data Mining, pp. 327–335. SIAM (2017). https://doi.org/10.1137/1.9781611974973.37
24. Wu, Z., Pan, S., Chen, F., Long, G., Zhang, C., Yu, P.S.: A comprehensive survey on graph neural networks. arXiv preprint arXiv:1901.00596 (2019)
25. Yuan, S., Wu, X., Xiang, Y.: SNE: signed network embedding. In: Kim, J., Shim, K., Cao, L., Lee, J.-G., Lin, X., Moon, Y.-S. (eds.) PAKDD 2017. LNCS (LNAI), vol. 10235, pp. 183–195. Springer, Cham (2017). https://doi.org/10.1007/978-3-319-57529-2_15

Graph Classification with 2D Convolutional Neural Networks

Antoine J.-P. Tixier[1]([✉]), Giannis Nikolentzos[1], Polykarpos Meladianos[2],
and Michalis Vazirgiannis[1,2]

[1] École Polytechnique, Palaiseau, France
{anti5662,nikolentzos,mvazirg}@lix.polytechnique.fr
[2] Athens University of Economics and Business, Athens, Greece
pmeladianos@aueb.gr

Abstract. Graph learning is currently dominated by graph kernels, which, while powerful, suffer some significant limitations. Convolutional Neural Networks (CNNs) offer a very appealing alternative, but processing graphs with CNNs is not trivial. To address this challenge, many sophisticated extensions of CNNs have recently been introduced. In this paper, we reverse the problem: rather than proposing yet another graph CNN model, *we introduce a novel way to represent graphs as multi-channel image-like structures that allows them to be handled by vanilla 2D CNNs*. Experiments reveal that our method is more accurate than state-of-the-art graph kernels and graph CNNs on 4 out of 6 real-world datasets (with and without continuous node attributes), and close elsewhere. Our approach is also preferable to graph kernels in terms of time complexity. Code and data are publicly available (https://github.com/Tixierae/graph_2D_CNN).

1 Introduction

Graph Classification. Graphs, or networks, are rich, flexible, and universal structures that can accurately represent the interaction among the components of many natural and human-made complex systems. A central graph mining task is that of *graph* classification (not to be mistaken with *node* classification). The instances are full graphs and the goal is to predict the category they belong to. The applications of graph classification are numerous and range from determining whether a protein is an enzyme or not in bioinformatics, to categorizing documents in NLP, and social network analysis. Graph classification is the task of interest in this study.

Limitations of Graph Kernels. The state-of-the-art in graph classification has traditionally been dominated by *graph kernels*. Graph kernels compute the similarity between two graphs as the sum of the pairwise similarities between some of their substructures, and then pass the similarity matrix computed on the entire dataset to a kernel-based supervised algorithm such as the Support Vector Machine [7] to learn soft classification rules. Graph kernels mainly vary

© Springer Nature Switzerland AG 2019
I. V. Tetko et al. (Eds.): ICANN 2019, LNCS 11731, pp. 578–593, 2019.
https://doi.org/10.1007/978-3-030-30493-5_54

based on the substructures they use, which include random walks [11], shortest paths [2], and subgraphs [30], to cite only a few. While graph kernels have been very successful, they suffer significant limitations:

L1: High time complexity. This problem is threefold: first, populating the kernel matrix requires computing the similarity between every two graphs in the training set (say of size N), which amounts to $N(N-1)/2$ operations. The cost of training therefore increases much more rapidly than the size of the dataset. Second, computing the similarity between a pair of graphs (i.e., performing a single operation) is itself polynomial in the number of nodes. For instance, the time complexity of the shortest path graph kernel is $\mathcal{O}(|V_1|^2|V_2|^2)$ for two graphs (V_1, V_2), where $|V_i|$ is the number of nodes in graph V_i. Processing large graphs can thus become prohibitive, which is a serious limitation as big networks abound in practice. Finally, finding the support vectors is $\mathcal{O}(N^2)$ when the C parameter of the SVM is small and $\mathcal{O}(N^3)$ when it gets large [4], which can again pose a problem on big datasets.

L2: Disjoint feature and rule learning. With graph kernels, the computation of the similarity matrix and the learning of the classification rules are two independent steps. In other words, the features are fixed and not optimized for the task.

L3: Graph comparison is based on small independent substructures. As a result, graph kernels focus on local properties of graphs, ignoring their global structure [24]. They also underestimate the similarity between graphs and suffer unnecessarily high complexity (due to the explosion of the feature space), as substructures are considered to be orthogonal dimensions [38].

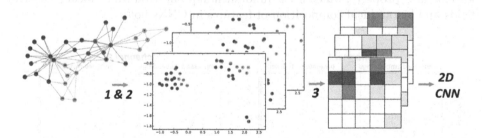

Fig. 1. Our 3-step approach represents graphs as "images" suitable for vanilla 2D CNNs. Continuous node attribute vectors can be passed as extra channels. Steps 1 & 2: graph node embeddings and compression with PCA. Step 3: computation and stacking of the 2D histograms.

2   Proposed Method

Overview. We propose a simple approach to turn a graph into a multi-channel image-like structure suitable to be processed by a traditional 2D CNN. The process (summarized in Fig. 1) can be broken down into 3 steps: (1) graph node embedding, (2) embedding space compression, (3) repeated extraction of 2D slices from the compressed space and computation of a 2D histogram for each slice.

The "image" representation of the graph is finally given by the stack of its 2D histograms (each histogram making for a channel). Note that the dimensionality of the final representation of a graph does not depend on its number of nodes or edges. Big and small graphs are represented by images of the same size.

Our method addresses the limitations of graph kernels in the following ways:

L1. By converting all graphs in a given dataset to representations of the same dimensionality, and by using a classical 2D CNN architecture for processing those graph representations, our method offers *constant time* complexity at the instance level, and *linear time* complexity at the dataset level. Moreover, state-of-the-art node embeddings can be obtained for a given graph in *linear time* w.r.t. the number of nodes in the graph, for instance with node2vec [13].

L2. Thanks to the 2D CNN classifier, features are learned directly from the raw data during training such that classification accuracy is maximized.

L3. Our approach capitalizes on state-of-the-art graph node embedding techniques that capture both local and global properties of graphs. In addition, we remove the need for handcrafted features.

How to Represent Graphs as Structures that Verify the Spatial Dependence Property? Convolutional Neural Networks (CNNs) are feed-forward neural networks specifically designed to work on regular grids [19]. A regular grid is the d-dimensional Euclidean space discretized by parallelotopes (rectangles for $d = 2$, cuboids for $d = 3$, etc.). Regular grids satisfy the *spatial dependence*[1] property, which is the fundamental premise on which local receptive fields and hierarchical composition of features in CNNs hold.

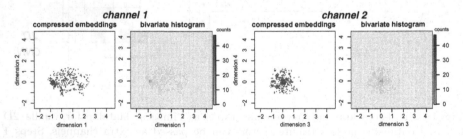

Fig. 2. Node embeddings and image representation of graph ID #10001 (577 nodes, 1320 edges) from the REDDIT-12K dataset.

Traditionally, a graph $G(V, E)$ is encoded as its adjacency matrix A or Laplacian matrix L. A is a square matrix of dimensionality $|V| \times |V|$, symmetric in

---

[1] the concept of spatial dependence is well summarized by: "everything is related to everything else, but near things are more related than distant things" [35]. For instance in images, close pixels are more related than distant pixels.

the case of undirected graphs, whose $(i, j)^{th}$ entry $A_{i,j}$ is equal to the weight of the edge $e_{i,j}$ between nodes v_i and v_j, if such an edge exists, or to 0 otherwise. On the other hand, the Laplacian matrix L is equal to $D - A$, where D is the diagonal degree matrix. One could initially consider passing one of those structures as input to a 2D CNN. However, unlike in images, where close pixels are more strongly correlated than distant pixels, adjacency and Laplacian matrices are not associated with spatial dimensions and the notion of Euclidean distance, and thus do not satisfy the spatial dependence property. As will be detailed next, we capitalize on *graph node embeddings* to address this issue.

Step 1: Graph node embeddings. There is local correlation in the node embedding space. In that space, the Euclidean distance between two points is meaningful: it is inversely proportional to the similarity of the two nodes they represent. For instance, two neighboring points in the embedding space might be associated with two nodes playing the same structural role (e.g., of flow control), belonging to the same community, or sharing some other common property.

Step 2: Alignment and compression with PCA. As state-of-the-art node embedding techniques (such as `node2vec`) are neural, they are stochastic. Dimensions are recycled from run to run, which means that a given dimension will not be associated with the same latent concepts across graphs, or across several runs on the same graph. Therefore, to ensure that the embeddings of all the graphs in the collection are comparable, we apply PCA and retain the first $d \ll D$ principal components (where D is the dimensionality of the original node embedding space). PCA also serves an information maximization (compression) purpose. Compression is desirable in terms of complexity, as it greatly reduces the shape of the tensors fed to the CNN (and thus the number of channels, for reasons that will become clear in what follows), at the expense of a negligible information loss.

Step 3: Computing and stacking 2D histograms. We finally repeatedly extract 2D slices from the d-dimensional PCA node embedding space, and turn those planes into regular grids by discretizing them into a finite, fixed number of equally-sized bins, where the value associated with each bin is the count of the number of nodes falling into that bin. In other words, we represent a graph as a stack of $d/2$ 2D histograms of its (compressed) node embeddings[2]. As illustrated in Fig. 2, the first histogram is computed from the coordinates of the nodes in the plane made of the first two principal directions, the second histogram from directions 3 and 4, and so forth. Note that using adjacent and following PCA dimensions is an arbitrary choice. It ensures at least that channels are sorted by decreasing order of informativeness.

Using computer vision vocabulary, bins can be viewed as *pixels*, and the 2D slices of the embedding space as *channels*. However, in our case, instead of having 3 channels (R, G, B) like with color images, we have $d/2$ of them. That is, each pixel (each bin) is associated with a vector of size $d/2$, whose entries are the counts of the nodes falling into that bin in the corresponding 2D slice of the embedding space. Finally, the *resolution* of the image is determined by the

---

[2] our representation is unrelated to the widespread *color histogram* encoding of images.

number of bins of the histograms, which is constant for a given dataset across all channels.

3 Experiments

2D CNN Architecture. We implemented a variant of LeNet-5 [19] with which we reached 99.45% accuracy on the MNIST handwritten digit classification dataset. As illustrated in Fig. 3 for an input of shape (5, 28, 28), this simple architecture deploys four convolutional-pooling layers (each repeated twice) in parallel, with respective region sizes of 3, 4, 5 and 6, followed by two fully-connected layers. Dropout [33] is employed for regularization at every hidden layer. The activations are ReLU functions (in that, our model differs from LeNet-5), except for the ultimate layer, which uses a softmax to output a probability distribution over classes. For the convolution-pooling block, we employ 64 filters at the first level, and as the signal is halved through the (2, 2) max pooling layer, the number of filters in the subsequent convolutional layer is increased to 96 to compensate for the loss in resolution.

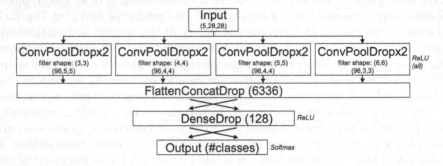

Fig. 3. 2D CNN architecture used in our experiments. The number within parentheses refer to the *output* dimensions of the tensors.

Random Graphs. To quickly test the viability of our pipeline, we created a synthetic 5-class dataset containing 2000 undirected and unweighted networks in each category. For the first and second classes, we generated graphs with the Stochastic Block Model, featuring respectively 2 and 3 communities of equal sizes. The in-block and cross-block probabilities were respectively set to 0.1 and 0.7, and the size $|V|$ of each graph was randomly sampled from the Normal distribution with mean 150 and standard deviation 30, i.e., $\mathcal{N}(150, 30)$. The third category was populated with scale-free networks, that is, graphs whose degree distributions follow a power law, using the Barabási-Albert model. The number of incident edges per node was sampled for each graph from $\mathcal{N}(5, 2)$, and the size of each graph was drawn from $\mathcal{N}(150, 30)$ like for the first two classes.

Finally, the fourth and fifth classes were filled with Erdös-Rényi graphs whose sizes were respectively sampled from $\mathcal{N}(300, 30)$ and $\mathcal{N}(150, 30)$, and whose edge probabilities were drawn from $\mathcal{N}(0.3, 0.15)$. This overall, gave us a large variety of graphs.

Spectral Embeddings. We started with the most naive way to embed the nodes of a graph, that is, using the eigenvectors of its adjacency matrix (we also experimented with the Laplacian but observed no difference). Here, no PCA-based compression was necessary. We retained the eigenvectors associated with the 10 largest eigenvalues in magnitude, thus embedding the nodes into a 10-dimensional space.

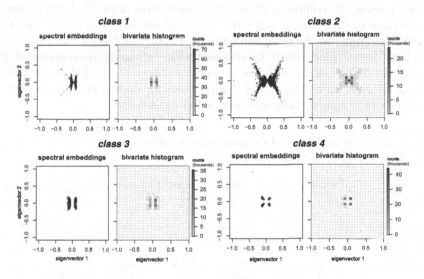

Fig. 4. Overlayed node embeddings in the space made of the first two eigenvectors of the adjacency matrices (first channel) and associated overlayed bivariate histograms for all graphs in the first four classes of the *random* dataset

Image Resolution and Channels. As we computed the unit-normed eigenvectors, the coordinates of any node in any dimension belonged to the $[-1, 1]$ range. Furthermore, inspired by the MNIST images which are 28×28 in size, and on which we initially tested our 2D CNN architecture, we decided to learn 2D histograms featuring 28 bins in each direction. This gave us a resolution of $28/(1-(-1))$, that is, 14 pixels per unit, which we write 14:1 for brevity in the remainder of this paper. Finally, we decided to make use of the information carried out by all 10 eigenvectors, and thus kept 5 channels. Any graph in the dataset was thus represented as a tensor of shape (5, 28, 28).

Results. In a 10-fold cross validation setting where each fold was repeated 3 times, we reached a mean classification accuracy of 99.08% (± 3.21), usually within 3 epochs, which is a very good performance. Even though we injected some variance, categories are associated with specific patterns (see Fig. 4) which are easily captured by the CNN.

Real-World Graphs. We conducted experiments on 6 real-world datasets whose statistics are summarized in Table 1. In all datasets, graphs are unweighted, undirected, with unlabeled nodes, and the task is to predict the class they belong to. Classes are mutually exclusive. The first five datasets, REDDIT-B, REDDIT-5K, REDDIT-12K, COLLAB, and IMDB-B, come from [38] and contain social networks. The sixth dataset, PROTEINS_full, is a bioinformatics one and comes from [3,14]. In this dataset, each node is associated with a 29-dimensional continuous vector. To incorporate this extra information into our images, we compressed the attribute vectors with PCA, retaining the same number of dimensions as for the node embeddings, and normalized them to have same range as the node embeddings. Finally, each node was represented by a single vector made of its compressed node embedding concatenated with its compressed and normalized continuous attribute vector. That is, we had $d/2$ channels for node embeddings and $d/2$ channels for node attributes, where d is the number of principal components retained in each case.

Table 1. Statistics of the social network datasets (first 5 columns) and the bioinformatics dataset used in our experiments.

	IMDB-B	COLLAB	REDDIT-B	REDDIT-5K	REDDIT-12K	PROTEINS_full
Max # vertices	136	492	3782	3648	3782	620
Min # vertices	12	32	6	22	2	4
Average # vertices	19.77	74.49	429.61	508.50	391.40	39.05
Max # edges	1249	40120	4071	4783	5171	1049
Min # edges	26	60	4	21	1	5
Average # edges	96.53	2457.78	497.75	594.87	456.89	72.82
# graphs	1000	5000	2000	4999	11929	1113
Average diameter	1.861	1.864	9.72	11.96	10.91	11.57
Average density (%)	52.06	50.92	2.18	0.90	1.79	21.21
# classes	2	3	2	5	11	2
Max class imbalance	1:1	1:3.4	1:1	1:1	1:5	1:1.5

Neural Embeddings. We used the `node2vec` algorithm [13] as our node embedding method for the real-world networks, as it was giving much better results than the basic spectral method. `Node2vec` applies the very fast Skip-Gram language model [20] to truncated biased random walks performed on the graph. The algorithm scales linearly with the number of nodes in the network. We used the high performance, publicly available C++ implementation of the authors[3].

Image Resolution. We stuck to similar values as in our initial experiments involving spectral embeddings, i.e., 14 pixels per unit (14:1). We also tried 9 pixels per unit (9:1).

---

[3] https://github.com/snap-stanford/snap/tree/master/examples/node2vec.

Image Size. On a given dataset, image size is calculated as the range $|max(\text{coordinates}) - min(\text{coordinates})| \times$ resolution, where `coordinates` are the flattened node embeddings. For instance, on COLLAB with a resolution of 9:1, image size is equal to 37×37, since $|2.78 - (-1.33)| \times 9 \approx 37$.

Number of Channels. With the p and q parameters of `node2vec` held constant and equal to 1, we conducted a search on the coarse grid $\{(14,9); (2,5)\}$ to get more insights about the impact of resolution and number of channels (respectively). When using 5 channels, the graphs with less than 10 nodes were removed, because for these graphs, we cannot get a 10-dimensional node embedding space (there cannot be more dimensions than data points). This represented only a few graphs overall.
On the PROTEINS_full dataset, we only experimented with 2 node embeddings channels and 2 node attribute channels.

p, q, c, and d_{n2v} `node2vec` Parameters. With the best resolution and number of channels, we then tuned the return and in-out parameters p and q of `node2vec`. Those parameters respectively bias the random walks towards exploring larger areas of the graph or staying in local neighborhoods, allowing the embeddings to encode a similarity that interpolates between structural equivalence (two nodes acting as, e.g., flow controllers, are close to each other) and homophily (two nodes belonging to the same community are close to each other). Following the `node2vec` paper [13], we tried 5 combinations of values for (p, q): $\{(1,1); (0.25,4); (4,0.25); (0.5,2); (2,0.5)\}$. Note that $p = q = 1$ is equivalent to `DeepWalk` [27].

Since the graphs in the COLLAB and the IMDB-B datasets are very dense (>50%, average diameter of 1.86), we also tried to set the context size c to smaller values of 1 and 2 (the default value is $c = 10$). The context size is used when generating the training examples (`context`, `target`) to pass to the skip-gram model: in a given random walk, it determines how many nodes *before* and *after* the target node should be considered part of the context.

Finally, d_{n2v} is the dimension of the node embeddings learned by `node2vec`. The default value is 128. Since the graphs in COLLAB, IMDB-B, and PRO-TEINS_full are small (20, 74, and 39 nodes on average), we experimented with lower values than the default one on this dataset, namely 12 and 4. Note that with $d_{n2v} = 4$, we can get at most 2 channels. The final values of p, q, c, and d_{n2v} for each dataset are summarized in Table 2.

Table 2. Final resolution, number of channels, and p, q, c, and d_{n2v} `node2vec` parameters for each dataset. *number of channels for node embeddings and continuous node attributes. - means default value(s).

	REDDIT-B	REDDIT-5K	REDDIT-12K	COLLAB	IMDB-B	PROTEINS_full
Res	9:1	9:1	9:1	9:1	14:1	9:1
#Chann	5	2	5	5	5	2/2*
p,q	2, 0.5	4, 0.25	1, 1	0.5, 2	1, 1	0.5, 2
c,d_{n2v}	-	-	-	2, 12	-	-, 4

Baselines. On the social network datasets (on which there are no continuous node attributes), we re-implemented two state-of-the-art graph kernels, the graphlet kernel [30] (we sampled 2000 graphlets of size up to 6 from each graph) and the Weisfeiler-Lehman (WL) kernel framework [31] with the subtree graph kernel [11] (we used node degrees as labels).

We also report for comparison purposes the performance of multiple state-of-the-art baselines that all share with us the same experimental setting: Deep Graph Kernels (DGK) [38], the best graph CNN from [23], namely PSCN $k = 10$, and Deep Graph CNN (DGCNN) [39]. Note that to be fair, we excluded DGK from the comparison on the bioinformatics dataset since it doesn't make use of node attributes.

On the PROTEINS_full dataset, we compared with the following baselines, which all take into account node attribute vectors: Hash Graph Kernel (HGK-SP) with shortest-path base kernel [21], Hash Graph Kernel with Weisfeiler-Lehman subtree base kernel (HGK-WL) [21], the best performing variant of Graph invariant kernels (GIK) [25], GraphHopper [10] and baselines within, propagation kernel with diffusion scheme (PROP-diff) [22], and propagation kernel with hashing-based label discretization and WL kernel update (PROP-WL) [22].

Configuration. Following best practice, we used 10-fold cross validation and repeated each fold 3 times in all our experiments. For the graphlet and WL kernels, we used a C-SVM classifier[4] [26]. The C parameter of the SVM and the number of iterations in WL were jointly optimized on a $90-10$ % partition of the training set of each fold by searching the grid $\left\{ (10^{-4}, 10^4, \text{len} = 10); (2, 7, \text{step} = 1) \right\}$.

For our 2D CNN, we used Xavier initialization [12], a batch size of 32, and for regularization, a dropout rate of 0.3 and early stopping with a patience of 5 epochs (null delta). The categorical cross-entropy loss was optimized with Adam [15] (default settings). We implemented our model in `Keras` [6] version $1.2.2^5$ with `tensorflow` [1] backend. The hardware used consisted in an NVidia Titan X Pascal GPU with an 8-thread Intel Xeon 2.40 GHz CPU and 16 GB of RAM. The graph kernel baselines were run on an 8-thread Intel i7 3.4 GHz CPU, with 16 GB of RAM.

4   Results

Results are reported in Table 3 for the social network datasets and Table 4 for the bioinformatics dataset.

---

[4] http://scikit-learn.org/stable/modules/generated/sklearn.svm.SVC.html.

[5] https://faroit.github.io/keras-docs/1.2.2/.

Table 3. 10-fold CV average test set classification accuracy of our proposed method compared to state-of-the-art graph kernels and graph CNNs, on the social network datasets. $\pm$ is standard deviation. Best performance per column in **bold**. *indicates stat. sign. at the $p < 0.05$ level (our 2D CNN vs. WL) using the Mann-Whitney U test.

Method	Dataset				
	REDDIT-B (size = 2,000; nclasses = 2)	REDDIT-5K (4,999; 5)	REDDIT-12K (11,929; 11)	COLLAB (5,000; 3)	IMDB-B (1,000; 2)
Graphlet Shervashidze2009	77.26 ($\pm$2.34)	39.75 ($\pm$1.36)	25.98 ($\pm$1.29)	73.42 ($\pm$2.43)	65.40 ($\pm$5.95)
WL Shervashidze2011	78.52 ($\pm$2.01)	50.77 ($\pm$2.02)	34.57 ($\pm$1.32)	**77.82*** ($\pm$1.45)	**71.60** ($\pm$5.16)
Deep GK Yanardag2015	78.04 ($\pm$0.39)	41.27 ($\pm$0.18)	32.22 ($\pm$ 0.10)	73.09 ($\pm$0.25)	66.96 ($\pm$0.56)
PSCN $k = 10$ Niepert2016	86.30 ($\pm$1.58)	49.10 ($\pm$0.70)	41.32 ($\pm$0.42)	72.60 ($\pm$2.15)	71.00 ($\pm$2.29)
DGCNN Zhang2018	-	-	-	73.76 ($\pm$0.49)	70.03 ($\pm$0.86)
2D CNN (our method)	**89.12*** ($\pm$1.70)	**52.11** ($\pm$2.24)	**48.13*** ($\pm$1.47)	71.33 ($\pm$1.96)	70.40 ($\pm$3.85)

Table 4. 10-fold CV average test set classification accuracy of our proposed method compared to state-of-the-art graph kernels and graph CNNs on the bioinformatics dataset (PROTEINS_full).

	2D CNN our method	DGCNN Zhang18	PSCN $k = 10$ Niepert16	HGK-SP Morris16	HGK-WL Morris16	GIK Orsini15	GraphHopper Feragen13	PROP-diff Neumann12	PROP-WL Neumann12
Acc.	**77.12**	75.54	75.00	75.14	74.88	76.6	74.1	73.3	73.1
Std. dev.	2.79	0.94	2.51	0.47	0.64	0.6	0.5	0.4	0.8

Our approach shows statistically significantly better than all baselines on the REDDIT-12K and REDDIT-B datasets, with large absolute improvements of **6.81** and **2.82** in accuracy over the best performing competitor, respectively. We also reach best performance on the REDDIT-5K and PROTEINS_full datasets, with respective improvements in accuracy of **1.34** and **0.52** over the best performing baselines. In particular, the fact that we reach best performance on PROTEINS_full shows that our approach is flexible enough to leverage not only the topology of the network, but also continuous node attributes, in a very simple and unified way (we simply concatenate node embeddings with node attribute vectors). Note that when not using node attributes (only the first 2 node embeddings channels), the performance of our model decreases from 77.12 to 73.43, showing that our approach is capable of using both topological and attribute information.

Finally, on the IMDB-B dataset, we get third place, very close ($\leq$1.2) to the top performers (no statistically significant difference). The only dataset on which a baseline proved significantly better than our approach is actually COLLAB (WL graph kernel). On this dataset though, the WL kernel beats *all* models by a wide margin, and we are relatively close to the other Deep Learning approaches ($\leq$2.43 difference).

Runtimes. Even if not directly comparable, we report in Table 5 kernel matrix computation time for the two graph kernel baselines, along with the time per epoch of our 2D CNN model.

Table 5. Runtimes in seconds, rounded to the nearest integer. *without using node attributes

	REDDIT-B	REDDIT-5K	REDDIT-12K	COLLAB	IMDB-B	PROTEINS_full
Size, average (# nodes, # edges)	2000, (430,498)	4999, (509,595)	11929, (391,457)	5000, (74,2458)	1000, (20,97)	1113, (39,73)
Input shapes (for our approach)	(5,62,62)	(2,65,65)	(5,73,73)	(5,36,36)	(5,37,37)	(4,70,70) (2,70,70)*
Graphlet Shervashidze2009	551	5046	12208	3238	275	-
WL Shervashidze2011	645	5087	20392	1579	23	-
2D CNN (our approach)	6	16	52	5	1	1

Discussion. Replacing the raw counts by the empirical joint probability density function, either by normalizing the histograms, or with a Kernel Density Estimate, significantly deteriorated performance. This suggests that keeping the absolute counts is important, which makes sense, as some categories might be associated with larger or smaller graphs, on average. We also observed that increasing the number of channels to more than 5 does not yield better results. This was expected, as higher order channels contain less information than lower order channels. However, reducing the number of channels improves performance in some cases, probably because it plays a regularization role.

Data augmentation. Generating more training examples by altering the input images is known to improve performance in image classification [17]. However, since we had direct access to the underlying data that were used to generate the images of the graphs, which is typically not the case in computer vision, we thought it would be more sensible to implement a synthetic data generation scheme at the *node embeddings* level rather than at the image level. More precisely, we used a simple but powerful nonparametric technique known as the *smoothed bootstrap with variance correction* [32], detailed in Algorithm 1. This generator is used in hydroclimatology to improve modeling of precipitation [18] or streamflow [29]. Unlike the traditional bootstrap [9] which simply draws with replacement from the initial set of observations, the smoothed bootstrap can generate values outside of the original range while still being faithful to the structure of the underlying data [28,34], as shown in Fig. 5.

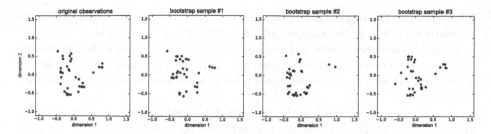

Fig. 5. Example of data augmentation with the smoothed bootstrap generator, for graph ID #3 of IMDB-B dataset (35 nodes, 133 edges).

Algorithm 1. Smoothed bootstrap with variance correction

Input : list L of M arrays of original d-dimensional node embeddings, $p_b \in \mathbb{R}^+$
Output: list L' of $M' = \text{int}(p_b \times M)$ arrays of synthetic node embeddings
1 $L'' \leftarrow$ select M' elements from L at random without replacement
2 **for** $A \in \mathbb{R}^{|V_A| \times d}$ *in* L'' **do**
3 $\quad$ $\mu_A \leftarrow$ apply(mean, A)
4 $\quad$ $\sigma_A^2 \leftarrow$ apply(var, A)
5 $\quad$ $h_A \leftarrow$ apply(compute_kde_bandwidth, A)
6 $\quad$ $A' \leftarrow$ empty array
7 $\quad$ **for** $j \leftarrow 1$ **to** $\text{int}\left(\mathcal{N}\left(|V_A|, \frac{|V_A|}{5}\right)\right)$ **do**
8 $\quad\quad$ sample i from $U(1, |V_A|)$
9 $\quad\quad$ $\epsilon \leftarrow$ apply($\mathcal{N}(0, \sqrt{x})$, h$_A$)
10 $\quad\quad$ **for** $k \leftarrow 1$ **to** d **do**
11 $\quad\quad\quad$ $A'[j,k] \leftarrow \mu_A[k] + (A[i,k] - \mu_A[k] + \epsilon[k]) / \sqrt{1 + h_A[k] / \sigma_A^2[k]}$
12 $\quad\quad$ **end**
13 $\quad$ **end**
14 $\quad$ store A' in L'
15 **end**
16 **return** L'

$|V_A|$ is the number of nodes in the graph whose node embeddings are contained in A. Note that $\mu_A, \sigma_A^2, h_A, \epsilon$ in lines 3, 4, 5 and 9 are all d-dimensional vectors (i.e., the functions are applied column-wise). $\mathcal{N}$ and U in lines 7 and 8 are the normal and discrete uniform distributions. The function $\mathcal{N}(0, \sqrt{x})$ in line 9 is applied to each element of h_A. It generates Gaussian noise in each direction. Line 5 computes the bandwidths of the Kernel Density Estimate (KDE) over each dimension. The bandwidth controls the degree of smoothing. While there are many different ways to evaluate the bandwidth, we used the classical and widespread Silverman's rule of thumb [32], as shown in Eq. 1 below for a sample X of N scalars:

$$\text{silverman}(X) = \frac{0.9 \min\left(\sigma_X, \frac{Q_3(X) - Q_1(X)}{1.34}\right)}{N^{1/5}} \tag{1}$$

Where Q_3 and Q_1 represent respectively the third and first quartiles, σ_X is the standard deviation of the sample, and N is the size of the sample.

Line 11 simulates a new embedding vector. Using the smoothed bootstrap scheme can be viewed as sampling from the KDE. This is consistent with our way of representing graphs as images, since a KDE is nothing more than a smoothed histogram.

In our experiments, using the smoothed bootstrap improved performance only on REDDIT-B (+0.33% in accuracy), for $p_b = 0.2$ (i.e., augmenting the dataset with 20% of synthetic graphs). Other values of p_b (0.05, 0.1, 0.5) were not successful. Further research is thus necessary to understand how to make the proposed data augmentation strategy more effective.

Limitations. Even though results are very good out-of-the-box in most cases, finding an embedding algorithm that works well, or the right combination of parameters for a given dataset, can require some efforts. For instance, on COLLAB and IMDB-B (the only two datasets on which we do not reach best performance), we hypothesize that our results are inferior because the default parameter values of node2vec may not be well-suited to very dense graphs such as the ones found in COLLAB and IMDB-B (diameter < 2, density > 50). Optimizing the node2vec parameters on these datasets probably requires more than a coarse grid search.

5   Related Work

Motivated by the outstanding performance recently reached by CNNs in computer vision, e.g. [17,37], much research has been devoted to generalizing CNNs to graphs. Solutions fall into two broad categories: *spatial* and *spectral* techniques [5]. Spectral approaches [8,16] invoke the convolution theorem from signal processing theory to perform graph convolutions as pointwise multiplications in the Fourier domain of the graph. The basis used to send the graph to the Fourier domain is given by the SVD decomposition of the Laplacian matrix of the graph, whose eigenvalues can be viewed as frequencies. By contrast, spatial methods [23,36,39] operate directly on the graph structure. For instance, in [23], the algorithm first determines the sequence of nodes for which neighborhood graphs (of equal size) are created. To serve as receptive fields, the neighborhood graphs are then normalized, i.e., mapped to a vector space with a linear order, in which nodes with similar structural roles in the neighborhood graphs are close to each other. Normalization is the central step, and is performed via a labeling procedure. A 1D CNN architecture is finally applied to the receptive fields. While the aforementioned sophisticated frameworks have made great strides, we showed in this paper that graphs can also be processed by vanilla 2D CNN architectures.

6   Conclusion

We showed that CNN for images can be used for learning graphs in a completely off-the-shelf manner. Our approach is flexible and can take continuous node attributes into account. We reach better results than state-of-the-art graph kernels and graph CNN models on 4 real-world datasets out of 6. Furthermore, these good results were obtained with limited parameter tuning and by using a basic 2D CNN model. From a time complexity perspective, our approach is preferable to graph kernels too, allowing to process bigger datasets featuring larger graphs.

Acknowledgments. We thank the anonymous reviewers for their helpful comments. The GPU used in this project was donated by NVidia as part of their GPU grant program.

References

1. Abadi, M., et al.: TensorFlow: large-scale machine learning on heterogeneous distributed systems. arXiv preprint arXiv:1603.04467 (2016)
2. Borgwardt, K.M., Kriegel, H.: Shortest-path kernels on graphs. In: Proceedings of the 5th International Conference on Data Mining, pp. 74–81 (2005)
3. Borgwardt, K.M., Ong, C.S., Schönauer, S., Vishwanathan, S., Smola, A.J., Kriegel, H.P.: Protein function prediction via graph kernels. Bioinformatics **21**(Suppl. 1), i47–i56 (2005)
4. Bottou, L., Lin, C.J.: Support vector machine solvers. Large Scale Kernel Mach. **3**(1), 301–320 (2007)
5. Bruna, J., Zaremba, W., Szlam, A., LeCun, Y.: Spectral networks and locally connected networks on graphs. arXiv preprint arXiv:1312.6203 (2013)
6. Chollet, F., et al.: Keras (2015). https://github.com/fchollet/keras
7. Cortes, C., Vapnik, V.: Support-vector networks. Mach. Learn. **20**(3), 273–297 (1995)
8. Defferrard, M., Bresson, X., Vandergheynst, P.: Convolutional neural networks on graphs with fast localized spectral filtering. In: Advances in Neural Information Processing Systems, pp. 3837–3845 (2016)
9. Efron, B.: Bootstrap methods: another look at the jackknife. In: Kotz, S., Johnson, N.L. (eds.) Breakthroughs in Statistics. Springer Series in Statistics (Perspectives in Statistics), pp. 569–593. Springer, New York (1992). https://doi.org/10.1007/978-1-4612-4380-9_41
10. Feragen, A., Kasenburg, N., Petersen, J., de Bruijne, M., Borgwardt, K.: Scalable kernels for graphs with continuous attributes. In: Advances in Neural Information Processing Systems, pp. 216–224 (2013)
11. Gärtner, T., Flach, P., Wrobel, S.: On graph kernels: hardness results and efficient alternatives. In: Schölkopf, B., Warmuth, M.K. (eds.) COLT-Kernel 2003. LNCS (LNAI), vol. 2777, pp. 129–143. Springer, Heidelberg (2003). https://doi.org/10.1007/978-3-540-45167-9_11
12. Glorot, X., Bengio, Y.: Understanding the difficulty of training deep feedforward neural networks. In: AISTATS, vol. 9, pp. 249–256 (2010)
13. Grover, A., Leskovec, J.: node2vec: scalable feature learning for networks. In: Proceedings of the 22nd ACM SIGKDD International Conference on Knowledge Discovery and Data Mining, pp. 855–864. ACM (2016)
14. Kersting, K., Kriege, N.M., Morris, C., Mutzel, P., Neumann, M.: Benchmark data sets for graph kernels (2016). http://graphkernels.cs.tu-dortmund.de

15. Kingma, D., Ba, J.: Adam: a method for stochastic optimization. arXiv preprint arXiv:1412.6980 (2014)
16. Kipf, T.N., Welling, M.: Semi-supervised classification with graph convolutional networks. arXiv preprint arXiv:1609.02907 (2016)
17. Krizhevsky, A., Sutskever, I., Hinton, G.E.: ImageNet classification with deep convolutional neural networks. In: Advances in Neural Information Processing Systems, pp. 1097–1105 (2012)
18. Lall, U., Rajagopalan, B., Tarboton, D.G.: A nonparametric wet/dry spell model for resampling daily precipitation. Water Resour. Res. **32**(9), 2803–2823 (1996)
19. LeCun, Y., Bottou, L., Bengio, Y., Haffner, P.: Gradient-based learning applied to document recognition. Proc. IEEE **86**(11), 2278–2324 (1998)
20. Mikolov, T., Chen, K., Corrado, G., Dean, J.: Efficient estimation of word representations in vector space. arXiv preprint arXiv:1301.3781 (2013)
21. Morris, C., Kriege, N.M., Kersting, K., Mutzel, P.: Faster kernels for graphs with continuous attributes via hashing. In: 2016 IEEE 16th International Conference on Data Mining (ICDM), pp. 1095–1100. IEEE (2016)
22. Neumann, M., Patricia, N., Garnett, R., Kersting, K.: Efficient graph kernels by randomization. In: Flach, P.A., De Bie, T., Cristianini, N. (eds.) ECML PKDD 2012. LNCS (LNAI), vol. 7523, pp. 378–393. Springer, Heidelberg (2012). https://doi.org/10.1007/978-3-642-33460-3_30
23. Niepert, M., Ahmed, M., Kutzkov, K.: Learning convolutional neural networks for graphs. In: Proceedings of the 33rd Annual International Conference on Machine Learning. ACM (2016)
24. Nikolentzos, G., Meladianos, P., Vazirgiannis, M.: Matching node embeddings for graph similarity. In: AAAI, pp. 2429–2435 (2017)
25. Orsini, F., Frasconi, P., De Raedt, L.: Graph invariant kernels. In: Proceedings of the Twenty-fourth International Joint Conference on Artificial Intelligence, pp. 3756–3762 (2015)
26. Pedregosa, F., et al.: Scikit-learn: machine learning in Python. J. Mach. Learn. Res. **12**, 2825–2830 (2011)
27. Perozzi, B., Al-Rfou, R., Skiena, S.: DeepWalk: online learning of social representations. In: Proceedings of the 20th ACM SIGKDD International Conference on Knowledge Discovery and Data Mining, pp. 701–710. ACM (2014)
28. Rajagopalan, B., Lall, U., Tarboton, D.G., Bowles, D.: Multivariate nonparametric resampling scheme for generation of daily weather variables. Stoch. Hydrol. Hydraulics **11**(1), 65–93 (1997)
29. Sharma, A., Tarboton, D.G., Lall, U.: Streamflow simulation: a nonparametric approach. Water Res. Res. **33**(2), 291–308 (1997)
30. Shervashidze, N., Petri, T., Mehlhorn, K., Borgwardt, K.M., Vishwanathan, S.: Efficient graphlet kernels for large graph comparison. In: Proceedings of the International Conference on Artificial Intelligence and Statistics, pp. 488–495 (2009)
31. Shervashidze, N., Schweitzer, P., Van Leeuwen, E.J., Mehlhorn, K., Borgwardt, K.M.: Weisfeiler-Lehman graph kernels. J. Mach. Learn. Res. **12**, 2539–2561 (2011)
32. Silverman, B.W.: Density Estimation for Statistics and Data Analysis, vol. 26. CRC Press, Boca Raton (1986)
33. Srivastava, N., Hinton, G.E., Krizhevsky, A., Sutskever, I., Salakhutdinov, R.: Dropout: a simple way to prevent neural networks from overfitting. J. Mach. Learn. Res. **15**(1), 1929–1958 (2014)
34. Tixier, A.J.P., Hallowell, M.R., Rajagopalan, B.: Construction safety risk modeling and simulation. Risk Anal. **37**, 1917–1935 (2017)

35. Tobler, W.R.: A computer movie simulating urban growth in the Detroit region. Econ. Geogr. **46**(sup1), 234–240 (1970)
36. Vialatte, J.C., Gripon, V., Mercier, G.: Generalizing the convolution operator to extend CNNs to irregular domains. arXiv preprint arXiv:1606.01166 (2016)
37. Vinyals, O., Toshev, A., Bengio, S., Erhan, D.: Show and tell: a neural image caption generator. In: Proceedings of the IEEE Conference on Computer Vision and Pattern Recognition, pp. 3156–3164 (2015)
38. Yanardag, P., Vishwanathan, S.: Deep graph kernels. In: Proceedings of the 21th ACM SIGKDD International Conference on Knowledge Discovery and Data Mining, pp. 1365–1374. ACM (2015)
39. Zhang, M., Cui, Z., Neumann, M., Chen, Y.: An end-to-end deep learning architecture for graph classification (2018)

Community Detection via Joint Graph Convolutional Network Embedding in Attribute Network

Di Jin[1], Bingyi Li[1]([✉]), Pengfei Jiao[1,2], Dongxiao He[1], and Hongyu Shan[1]

[1] College of Intelligence and Computing, Tianjin University, Tianjin 300350, China
{jindi,libingyi,pjiao,hedongxiao,shhy}@tju.edu.cn
[2] Center of Biosafety Research and Strategy, Tianjin University,
Tianjin 300350, China

Abstract. Community detection is a foundational task in network analysis. Besides the topology information, in recent years, there have been many methods utilizing network attribute information for community detection. The key of introducing the attribute information is how to integrate these two sources of information for better community detection. Graph Convolutional Networks (GCN) is an effective way to integrate network topologies and node attributes. However, when GCN is used in community detection, it often suffers from two problems: (1) the embedding derived from GCN is not community-oriented, and (2) this model is semi-supervised rather than unsupervised. To address these problems, we propose an unsupervised model for community detection via joint GCN embedding, i.e. JGE-CD. We employ GCN as the basic structure of encoder to match the above two sources of information, and use a dual encoder to derive two different embeddings by using an attribute network and its variant with random transformation. We further introduce a community detection module considering the community properties into the joint learning process. It derives two community detection results for a relative-entropy minimization which work together with a topology reconstruction module in order to make the model discover community structure in an unsupervised way. Extensive experiments on seven real-world networks show a superior performance of our model over some state-of-the-art methods.

Keywords: Community detection ·
Graph Convolutional Networks (GCN) ·
Network embedding

1 Introduction

Many complex systems in the real world can be represented in the form of networks, such as the Internet, Protein-Protein Interaction networks and so on. The community structure of a network is defined as the dense relationship between

© Springer Nature Switzerland AG 2019
I. V. Tetko et al. (Eds.): ICANN 2019, LNCS 11731, pp. 594–606, 2019.
https://doi.org/10.1007/978-3-030-30493-5_55

the nodes within the same community and the sparse relationship across different communities. Community detection is very helpful for analyzing the complex networks, which has attracted the attention of many researchers.

Several different community detection methods have been proposed, most of which utilize network topological information [5,12]. However, besides the network topologies, the attribute information of the network also plays an important role in community detection, which can help to better explain the community semantically. Many community detection methods use the attribute information of the network. They adopt techniques from nonnegative matrix factorization (NMF) [17], topic model-based [19] to deep learning [2]. In particular, the deep learning methods have attracted a lot of attention since they always own a satisfactory performance in terms of both efficiency and scalability, when dealing with the attribute network which is always large-scale and high-dimensional.

The key of introducing the attribute information into the community detection is that how to better integrate attribute information with the network topologies. However, existing deep learning methods typically use some simple ways to integrate these two sources of information, e.g. concatenating them directly [8]. As a result, these methods fail to match well these two sources for community detection.

Graph Convolution Network (GCN) [6] is an effective way to integrate the above two sources of information and extract more feature of attribute networks, which is widely used in the task of network analysis [7,11,13]. Some related works also apply GCN to community detection [3]. However, GCN has at least two drawbacks when it is used in community detection. First, the embedding derived by the GCN is often a general representation. As a result it is often not directly suitable for community detection. That is, such embedding does not consider community properties which makes it often ineffective for the problem of discovering communities. Second, existing GCN models are typically semi-supervised, while, in real networks, community detection are often unsupervised. Though there have been many related extensions of GCN, while they have not solved the above two problems.

To solve these problems, we propose an unsupervised model that joints GCN embedding for community detection within an autoencoder framework, named as JGE-CD. The model contains three main parts. The first is a dual encoder structure, which uses different attribute networks (i.e., clean network data and noisy network data) as the input instances, to derive two different embeddings. In specific, we use the GCN as the shared structure of these two encoders to effectively integrate the attributes with topological information. The second is a community detection module which stacks on top of the dual encoder. In this module, we employ modularity as the constraint of the layer of community detection. It incorporates the community properties into the joint learning process of both community detection and network embedding. In addition, through the community detection layer, the different embeddings are finally converted to two community detection results. We then perform KL divergence minimization between these two results, which can be taken as an unsupervised adversarial process of updating parameters. Finally, besides the KL divergence minimization, we also employ a topology reconstruction module as the third module to form a unified unsupervised model. These modules work together to make our model discover precise community structure in an unsupervised way. Meanwhile, the joint process also makes the embedding incorporate high-level information (i.e., community structure). So, the embedding not only consider the microscopic information (i.e., the compression process of GCN) but also consider the macrostructure which is produced as a by-product.

Overall, the contributions of our this work are summarized as follows:

- We propose an unsupervised model for community detection via joint GCN embedding, which can not only effectively integrate the network topology with the network semantics, but also serve well for community detection.
- Technically, we design a community module considering the community properties, employ an unsupervised relative-entropy minimization process and give a topology reconstruction module, in order to make the model discover community structure effectively in an unsupervised way.
- We comprehensively conduct experiments using JGE-CD on several real-world networks to show its superior performance over some state-of-art methods.

2 Related Works

A number of community detection methods have been proposed. Some of them utilize the network topological information alone based on different techniques, such as modularity [12], stochastic block model [5]. Recently, many methods use attribute information besides the topological information. These methods include nonnegative matrix factorization (NMF) [17], topic model-based [19], deep learning [2] and so on. For example, the method in [17] combines the information of network topologies with information of node attributes using a framework of NMF. [19] distinguishes the topics into two levels and proposes a Bayesian probabilistic framework with two-level community model to find semantic community structures. The method in [2] proposes an autoencoder framework which constructs the modularity matrix and Markov matrix as the input and then obtains an embedding for community detection. Some works also combine the community detection with network embedding [9]. [9] joints the network structure embedding with the node attributes and then formulates the problem of community detection as an NMF problem.

Another type of works related to our model are the Graph Convolutional Network (GCN). As a way to efficiently integrate node attributes and network topology, GCN has been widely applied to various network analysis tasks. For example, [13] employs a GCN to encode the attribute network and use an adversarial training strategy to regularize the embedding. [11] proposes a multi-dimensional GCN to learn embedding which considers the relations between nodes within- and across-dimensions. The most relevant to our work are two works [3,7]. [7] proposes a VAE-based latent representations learning framework, which employs

GCN as the encoder. However, this model is not designed for community detection, and the learned embedding is general rather than community oriented. Another related work integrates the GCN and MRF for semi-supervised community detection [3]. However, this model is semi-supervised, while in real networks, community detection problems are often unsupervised. Compared with these two works, the JGE-CD is an unsupervised model via joint GCN embedding for community detection, which considers the community-oriented constraints in the learning process.

3   The Method

Given an undirected and attribute network $G = (V, E)$, where $V = \{v_1, v_2, ..., v_n\}$ denotes a set of n nodes and $E = \{e_{ij}\}$ denotes a set of edges. The network topology of G is represented by an adjacent matrix $A \in \mathrm{R}^{n \times n}$ and node semantics are denoted by an attribute matrix $X \in \mathrm{R}^{n \times m}$, where m is the maximal number of the attribute. The community detection in G is to divide the nodes into c communities.

3.1   An Overview of the Method

The JGE-CD method for community detection has three modules, namely, a dual encoder, a community detection module and a topology reconstruction module (Fig. 1). The dual encoder (the blue dashed border), including a clean encoder and an augmented encoder, uses two Graph Convolutional Networks (GCN) as encoders to compress the attribute network and its variant, respectively, and then derives the two corresponding embedding. A modularity-based community detection module (the red dashed border) stacks on top of the above dual encoders to discover the communities. Followed by the augmented encoder, a topology reconstruction module (the yellow dashed border) is used to reconstruct the network topology to form an reconstruction loss.

In the encoder module, in order to integrate network topology with node attributes effectively and then extract more valid feature, we use GCN as the structure of the encoder to compress the attribute networks. Specifically, we use a clean encoder to compress the clean attribute network and an augmented encoder to compress an augmented attribute network, in which we shuffle the orders of some nodes. After that, the two encoders derive their respective embedding for the community detection module.

In the community detection module, we first use the multinomial logistic regression (soft-max) function to predict which community the node belongs to, using the above two embedding and produce two results, i.e., a clean result and an augmented result. Then we introduce modularity as a constraint on the clean result, which is can be seen as knowledge of the community detection to make the training process community-oriented. After that, we minimize the relative entropy (KL divergence) between the two results in the iterative training process, which is used as an unsupervised loss and helps to obtain a precise community detection result.

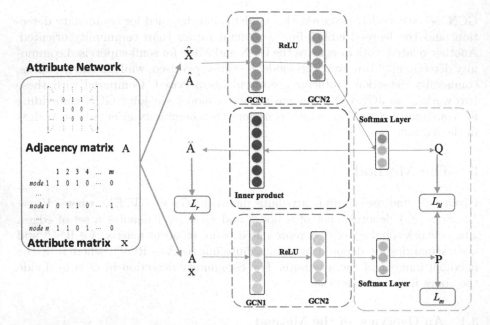

Fig. 1. The JGE-CD architecture. The model contains three modules, a dual encoder (the blue dashed border), a community detection module (the green dashed border) and a topology reconstruction module (the red dashed border). (Color figure online)

In the topology reconstruction module, we employ an inner product operation to reconstruct the network topology based on the embedding derived by the augmented encoder. We then define an reconstruction loss as a regularization item, which can work together with the KL divergence minimization to make this model discover the communities in an unsupervised way.

3.2 The Dual Encoder

The dual encoder contains a clean encoder to generate clean embedding, and an augmented encoder to generate an augmented embedding. We will give a detailed introduction in the Sect. 3.3 about why we design an augmented encoder to derive the augmented embedding. In specific, we employ GCN as the structure of these two encoders since GCN is believed to an effective way to integrate network topologies with node attributes.

A Clean Encoder. We use the original attribute network (i.e., adjacency matrix A and attribute matrix X) as input to the clean encoder. Then we compress the input data using a GCN with two layers. The output, denoted as clean embedding Z, is defined as:

$$Z = \tilde{D}^{-\frac{1}{2}}\tilde{A}\tilde{D}^{-\frac{1}{2}}\text{ReLU}\left(\tilde{D}^{-\frac{1}{2}}\tilde{A}\tilde{D}^{-\frac{1}{2}}XW^{(0)}\right)W^{(1)} \tag{1}$$

where $W^{(0)}$ (and $W^{(1)}$) are the parameters in the first (and second) convolutional layer and ReLU($\cdot$) is the activation function. Here $\tilde{A} = A + I_n$ where I_n is the identity matrix and $\tilde{D}$ is the diagonal matrix where $\tilde{D}_{ii} = \sum_j A_{ij}$.

An Augmented Encoder. We use a variant of the original attribute network via some random transformations as the input data of the augmented encoder. In specific, we shuffle the order of some nodes and obtain another adjacency matrix $\hat{A}$ and attribute matrix $\hat{X}$. The structure of the augmented encoder is also a GCN with two-layers. The output, denoted as augmented embedding $\hat{Z}$, is defined as:

$$\hat{Z} = \bar{D}^{-\frac{1}{2}} \bar{A} \bar{D}^{-\frac{1}{2}} \text{ReLU} \left(\bar{D}^{-\frac{1}{2}} \bar{A} \bar{D}^{-\frac{1}{2}} \hat{X} W^{(0)} \right) W^{(1)} \qquad (2)$$

This encoder shares the same set of neural network parameters with the clean encoder, i.e, $W^{(0)}$ and $W^{(1)}$. Here, $\bar{A} = \hat{A} + I_n$ and $\bar{D}_{ii} = \sum_j \hat{A}_{ij}$.

The dual GCN structure helps to integrate the network topology with node attribute and derive two embedding, which can further contribute to the joint unsupervised community detection.

3.3 The Community Detection Module

This is core of our model. We design a community detection module that joints GCN embedding. We further define community-oriented objective function based on modularity and defined an unsupervised regularization term.

After obtain the embedding, we use a multinomial logistic regression (soft-max) function to predict the probability that the node belongs to each community (we denote it as $P \in R^{n \times c}$) as follows.

$$p_{ik} = P(y_i = k | z_i, \Theta) = \frac{\exp\left(\theta_k^T z_i\right)}{\sum\limits_{k'=1}^{c} \exp\left(\theta_{k'}^T z_i\right)} \qquad (3)$$

where z_i is an instance of Z, which represents the embedding of the node v_i . And $\Theta = [\theta_1, ..., \theta_c] \in R^{d_z \times c}$ are parameters of the soft-max function where d_z is the dimension of Z and p_{ik} is an element of P which indicates the probability of the i-th node belonging to the k-th community. Similarly, we obtain another probability prediction Q for community detection using the augmented embedding $\hat{Z}$ via (3).

In order to make this module suitable for community detection, we introduce a modularity as a constraint on P.

$$M = tr\left(P^T B P\right) \qquad (4)$$

where $B \in R^{n \times n}$ whose element $B_{ij} = A_{ij} - \frac{d_i d_j}{2e}$. d_i denotes the degree of the node v_i and the e is the number of the edges. $tr(X)$ is the trace of matrix X. We can incorporate this constraint into the model as a loss by adding a minus to (4). In this way, we can maximize the modularly as the training process. We further take the logarithm for M to make it the same magnitude as other losses. So, we define the modularity-based loss as follow.

$$L_m = -\ln M \tag{5}$$

Furthermore, we design an unsupervised regularization term via the Relative Entropy Minimization (i.e., KL divergence) between the above two probabilistic community predictions.

$$L_{kl} = KL\,(\mathrm{P}||\mathrm{Q}) = \frac{1}{N}\sum_{i=1}^{n}\sum_{k=1}^{c} q_{ik}\log\frac{q_{ik}}{p_{ik}} \tag{6}$$

This regularization term forces the model to have invariant features with respect to augmented data [4]. It generally can be seen as an adversarial operation and make the model discovery the community structure in an unsupervised way. This is why we design an augmented encoder to derive the augmented embedding.

With incorporating the knowledge in community detection (i.e., modularly) into this module and employ an adversarial-like process via an KL divergence minimization, our model can produce a precise community detection result (we use the P as the final community detection result) in an unsupervised way.

3.4 The Topology Reconstruction Module

To unify the network embedding and community detection as an unsupervised model, we further employ a topology reconstruction module to reconstruct network topology using the embedding. In specific, followed by the augmented encoder, the decoder pathway reconstructs the topological structure, denoted by $\ddot{\mathrm{A}}$, using embedding through an inner product operation as follow.

$$\ddot{\mathrm{A}} = \hat{\mathrm{Z}}\hat{\mathrm{Z}}^{\mathrm{T}} \tag{7}$$

The loss of this topology reconstruction module is defined as

$$L_r = \ell(\ddot{\mathrm{A}}, \mathrm{A}) \tag{8}$$

where the $\ell\,(\cdot)$ is the mean square error. A is the adjacency matrix of clean attribute network. This module further promotes accuracy of the community detection results via constraining the parameter of GCN.

3.5 The Unified Model and Its Optimization

To sum up, the loss function of JGE-CD is defined as:

$$L = L_m + L_{kl} + L_r \tag{9}$$

The first item, L_m in (5), is the community-modularity loss, which incorporates domain knowledge of community detection into the model. The second item, L_{kl} in (6), is the adversarial-like loss, i.e., KL divergence between the community prediction of the clean embedding and the augmented embedding, which improves the accuracy of the prediction by iteratively training. The last item, L_r in (8), is the topology reconstruction loss, which improves embedding by reducing reconstruction loss and improves the accuracy of community detection. The joint unsupervised model JGE-CD above can then be optimized by using stochastic gradient descent in back-propagation.

4   Experiments

We first introduce the experiment settings and then compare our approach with several state-of-the-art methods for evaluation. We also give validation of our method on embedding learning. We finally give an analysis of why our method works.

4.1   Experiment Setup

Datasets Description. We used seven real-world of with varying sizes (Table 1). WebKB is a webpage dataset from four different universities, which contains four sub-datasets, namely Cornell, Texas, Washington, Wisconsin. Facebook is the largest sub-network in Facebook data. Citeseer is a citation network of scientific publications. Pubmed is a scientific publications from Pubmed database.

Table 1. Statistics of the datasets used, where n is the number of nodes, e the number of edges, m the number of attributes and c the number of categories.

Datasets	Cornell	Texas	Washington	Wisconsin	Facebook	Citeseer	Pubmed
n	195	183	217	262	1,045	3,312	1,9717
e	304	328	446	530	2,6749	4,732	44,338
m	1,588	1,498	1,578	1623	576	3,698	500
c	5	5	5	5	6	6	3

Baseline Methods. We compared JGE-CD with two types of state-of-the-art methods. To ensure fairness, these two type method both have topology-based methods and methods using both topological and attribute information. (1) Community detection methods. Topology-based: DCSBM [5]; Both: Block-LDA [1], PCL-DC [18], SCI [17], ASCD [16], TLSC [19]. (2) Network embedding methods. Since JGE-CD is a joint network embedding for community detection, we also compared it with several embedding methods for community detection. Topology-based: Deepwalk [15]. Both: TriDNR [14], ARGA [13], GAE [7]. ARGA and GAE are both GCN-based methods. We employed k-means to cluster the node embedding derived from these network embedding methods.

Parameter Settings. All community detection methods we used as the comparison methods need to be pre-specified the number of communities, as well as our approach. As for the network embedding methods, the k-mean method to cluster the embedding needs to specified the number of communities. So this comparison is fair in our experiment. We set these number of the communities to the same value that in the ground truth.

We used Graph Convolutional Network (GCN) with two layers for all datasets and the dimension of the GCN is 128 and 64 for the first layers and second layers, respectively (The dimension of the second layers is also the dimension of embedding). As for the parameters in the comparison methods, we used the default values which used in their respective paper. We implemented our approach using TensorFlow at a learning rate of 0.001.

4.2 Community Detection

Compare with the Community Detection Methods. After getting the community membership P, we treated the community number with the greatest probability as the community label of the node. We used Accuracy (AC) and normalized mutual information (NMI) [10] as evaluation metrics, which are widely used in community detection experiments. From the result (Table 2), we can see that JGE-CD performed the best on 5 of the 7 datasets. In specific, our method improved upon the best baseline (i.e., TLSC) by 6.00% in AC and 3.48% in NMI.

Table 2. Comparison of our JGE-CD with community detection methods in terms of AC and NMI. N/A denotes run time > 100 h

Metrics	Methods	Datasets						
		Cornell	Texas	Washington	Wisconsin	Facebook	Citeseer	Pubmed
AC (%)	DCSBM	37.95	48.09	31.81	32.82	31.33	38.48	53.64
	Block-LDA	46.15	54.11	39.17	49.62	37.66	25.52	49.01
	PCL-DC	30.26	38.81	29.95	30.15	40.38	34.08	**63.55**
	SCI	36.92	49.73	46.09	46.42	51.04	29.53	N/A
	ASCD	48.21	60.66	50.69	**53.05**	47.45	32.63	50.38
	TLSC	47.69	65.02	51.61	49.23	47.25	35.74	61.38
	JGE-CD	**48.72**	**66.67**	**61.29**	45.04	**65.42**	**50.25**	62.56
NMI (%)	DCSBM	9.69	16.56	9.87	3.14	29.41	4.13	12.28
	Block-LDA	6.81	4.21	3.69	10.09	9.28	1.41	6.58
	PCL-DC	7.23	10.37	5.66	5.01	38.63	17.54	26.84
	SCI	6.81	12.49	6.83	13.28	20.81	7.17	N/A
	ASCD	16.68	22.05	14.31	**20.56**	58.29	9.66	14.85
	TLSC	13.61	23.92	17.63	16.65	53.01	**23.16**	19.63
	JGE-CD	**17.93**	**28.24**	**18.31**	12.21	**69.61**	22.24	**27.25**

Compare with the Network Embedding Methods. We used the network embedding methods to derive the embedding and then cluster the nodes with the k-means. From the result (Table 3), we can see that our method beats 5 of the 7 datasets in AC and 6 of the 7 datasets in NMI. On average, JGE-CD improved upon the best baseline by 10.10% in AC and 10.06% in NMI.

Table 3. Comparison of our JGE-CD with network embedding methods in terms of AC and NMI.

Metrics	Methods	Datasets						
		Cornell	Texas	Washington	Wisconsin	Fackbook	Citeseer	Pubmed
AC (%)	DeepWalk + k-mean	36.05	46.72	40.76	38.76	38.94	36.21	64.84
	TriDNR + k-mean	38.21	47.54	43.59	43.70	40.05	34.44	59.29
	ARGA + k-means	37.33	41.48	43.66	42.81	35.50	43.50	58.76
	GAE + k-means	37.74	48.49	43.91	**46.85**	37.63	48.83	**65.83**
	JGE-CD	**48.72**	**66.67**	**61.29**	45.04	**65.42**	**50.25**	62.56
NMI (%)	DeepWalk + k-mean	7.06	6.16	5.66	7.65	47.88	10.58	26.55
	TriDNR + k-mean	7.20	4.32	8.10	6.60	51.02	**32.37**	19.28
	ARGA + k-means	10.26	7.28	12.60	11.92	67.55	22.72	18.40
	GAE + k-means	9.83	10.16	9.75	7.19	30.31	25.85	25.29
	JGE-CD	**17.93**	**28.24**	**18.31**	**12.21**	**69.61**	22.24	**27.25**

4.3   Network Embedding

In the process of the joint training, the network embedding performance is also improved since the network embedding learning process takes into account the community structure of the network. We conducted the classic network embedding experiment, node classification, to validate this effect. In specific, we used the output of the second GCN as the embedding of our method. We classified these nodes with ground-truth using the LibSVM and LibLINEAR software packages in Weka. 10-fold cross-validation was employed in each network and the accuracy (AC) [10] was used as the evaluation metrics (Since the NMI is not suitable the classification task, we discarded it). From the result (Table 4), we can see that our method performs the best on 6 of the 7 networks in term of AC and 5 of the 7 networks in terms of NMI. It validated that JGE-CD promoted the accuracy of the embedding.

Table 4. Comparison on node classification using the embedding in terms of AC.

Packages	Methods	Datasets						
		Cornell	Texas	Washington	Wisconsin	Facebook	Citeseer	Pubmed
LibSVM /AC (%)	DeepWalk	38.97	49.18	55.30	49.24	66.89	52.52	78.79
	TriDNR	37.95	48.09	47.01	40.46	65.23	54.47	79.07
	ARGA	42.56	**56.28**	58.99	49.26	66.69	65.10	80.64
	GAE	45.13	55.01	54.38	53.82	69.64	68.97	85.42
	JGE-CD	**55.89**	55.73	**59.45**	**54.96**	**70.33**	**70.92**	**85.55**
LibLINEAR /AC (%)	DeepWalk	38.46	48.09	53.92	49.62	66.50	48.42	78.36
	TriDNR	34.87	42.08	43.32	41.60	68.09	52.91	78.40
	ARGA	41.54	59.02	**60.37**	56.11	71.11	66.71	80.59
	GAE	45.64	51.91	54.84	54.49	71.70	69.25	**87.81**
	JGE-CD	**52.82**	**59.56**	59.44	**61.06**	**72.88**	**71.90**	85.31

4.4 Why Our Approach Works

To explain why our JGE-CD works, we examined three factors and conducted some analytical experiments. Due to space limit, we only used three different scale datasets (i.e., Texas (183), Facebook (1045), Pubmed (19717)) and one metrics (i.e., AC) as example.

A Joint GCN Embedding Architecture. Recall the architecture of our methods in the Fig. 1, which is a community detection framework via joint GCN embedding. To evaluate the effectiveness of this architecture, we compared JGE-CD with its two variants. The first variant is JFE-CD, which we used a fully connected neural network instead of the GCN as the way to integrate the topological information with the attribute information. This variant aims to evaluate the effectiveness of the GCN. The second variant is SGE-CD, which we separated the network embedding with the community detection to evaluate the effectiveness of the "joint". From the result (Fig. 2(a)), we can see that the JGE-CD has a much better performance than the other two variants. It validated two aspects about our approach. First, the GCN integrates the network topology well with the node attribute than some simple way, which is help for the community detection. Second, the joint framework can further make the GCN embedding suitable for the community detection.

Two Unsupervised Mechanisms. We tested the effectiveness of the two mechanisms (i.e., relative entropy minimization and a topology reconstruction module). We trained the model that discard the KL loss (denoted by JGE-DKL) and the topology reconstruction loss (JGE-DR), respectively. From the results (Fig. 2(b)), the JGE-CD always can beat the remaining two variants, which highlight the performance of the unsupervised mechanisms in the JGE-CD.

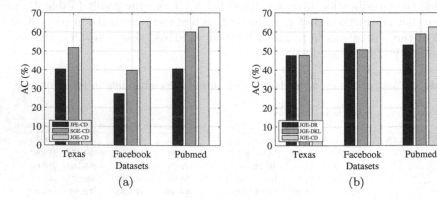

Fig. 2. (a) Comparison of JFE-CD, SGE-CD and JGE-CD on community detection in term of AC. (b) Comparison of JGE-DR, JGE-DKL and JGE-CD on community detection in term of AC.

Community-Considering Module. To analyze the modularity-constraint we integrate into the community detection module, we trained a variant model that discard the modularity loss (denoted by JGE-DM). By evaluating the experimental results, the AC of JGE-DM is 55.19 (Texas), 53.05 (Facebook) and 42.03 (Pubmed), which is much lower than our result (66.67 for Texas, 65.42 for Facebook and 62.56 for Pubmed). This result revealed that by considering the community properties (i.e., modularity), which can be seen as domain knowledge, we can obtain a better result of the community detection.

5   Conclusion

In this paper, we propose an unsupervised model for community detection via joint GCN embedding. To effectively integrate the network topology and network semantics, we employ GCN as the structure of the encoder and incorporate a community detection module into it. We design this community detection module considering network modularity as a constraint term, which make the joint model is a finally community-oriented. In addition, we introduce a dual encoder architecture to derive two adversarial-like community detection results via an unsupervised relative-entropy (KL divergence) minimization process. This dual encoder finally works together with a topology reconstruction module to make the model discover community structure in an unsupervised way. Extensive experiments on some real-world networks demonstrate the superior performance of our new approach over existing methods.

References

1. Balasubramanyan, R., Cohen, W.W.: Block-LDA: jointly modeling entity-annotated text and entity-entity links. In: Proceedings of the 2011 SIAM International Conference on Data Mining, pp. 450–461. SIAM (2011). https://doi.org/10.1137/1.9781611972818.39
2. Cao, J., Jin, D., Yang, L., Dang, J.: Incorporating network structure with node contents for community detection on large networks using deep learning. Neurocomputing **297**, 71–81 (2018). https://doi.org/10.1016/j.neucom.2018.01.065
3. Di, J., Ziyang, Liu, W.L., Dongxiao, H., Weixiong, Z.: Graph convolutional networks meet Markov random fields: semi-supervised community detection in attribute networks. In: Thirty-Third AAAI Conference on Artificial Intelligence
4. Guo, X., Zhu, E., Liu, X., Yin, J.: Deep embedded clustering with data augmentation. In: Asian Conference on Machine Learning, pp. 550–565 (2018). https://doi.org/10.1109/ACCESS.2018.2855437
5. Karrer, B., Newman, M.E.: Stochastic blockmodels and community structure in networks. Phys. Rev. E **83**(1), 016107 (2011)
6. Kipf, T.N., Welling, M.: Semi-supervised classification with graph convolutional networks. arXiv preprint arXiv:1609.02907 (2016)
7. Kipf, T.N., Welling, M.: Variational graph auto-encoders. stat **1050**, 21 (2016)
8. Li, H., Wang, H., Yang, Z., Liu, H.: Effective representing of information network by variational autoencoder. In: IJCAI, pp. 2103–2109. Morgan Kaufmann (2017). https://doi.org/10.24963/ijcai.2017/292

9. Li, Y., Sha, C., Huang, X., Zhang, Y.: Community detection in attributed graphs: an embedding approach. In: Thirty-Second AAAI Conference on Artificial Intelligence. AAAI (2018)
10. Liu, H., Wu, Z., Li, X., Cai, D., Huang, T.S.: Constrained nonnegative matrix factorization for image representation. IEEE Trans. Pattern Anal. Mach. Intell. **34**(7), 1299–1311 (2012). https://doi.org/10.1109/TPAMI.2011.217
11. Ma, Y., Wang, S., Aggarwal, C.C., Yin, D., Tang, J.: Multi-dimensional graph convolutional networks. arXiv preprint arXiv:1808.06099 (2018)
12. Pan, G., Zhang, W., Wu, Z., Li, S.: Online community detection for large complex networks. PLoS One **9**(7), e102799 (2014). https://doi.org/10.1371/journal.pone.0102799
13. Pan, S., Hu, R., Long, G., Jiang, J., Yao, L., Zhang, C.: Adversarially regularized graph autoencoder for graph embedding. In: Proceedings of the 27th International Joint Conference on Artificial Intelligence, pp. 2609–2615. AAAI (2018)
14. Pan, S., Wu, J., Zhu, X., Zhang, C., Wang, Y.: Tri-party deep network representation. Network **11**(9), 12 (2016). https://doi.org/10.1145/2623330.2623732
15. Perozzi, B., Al-Rfou, R., Skiena, S.: DeepWalk: online learning of social representations. In: Proceedings of the 20th ACM SIGKDD International Conference on Knowledge Discovery and Data Mining, pp. 701–710. ACM (2014)
16. Qin, M., Jin, D., Lei, K., Gabrys, B., Musial-Gabrys, K.: Adaptive community detection incorporating topology and content in social networks. Knowl.-Based Syst. **161**, 342–356 (2018). https://doi.org/10.1016/j.knosys.2018.07.037
17. Wang, X., Jin, D., Cao, X., Yang, L., Zhang, W.: Semantic community identification in large attribute networks. In: Thirtieth AAAI Conference on Artificial Intelligence. AAAI (2016)
18. Yang, T., Jin, R., Chi, Y., Zhu, S.: Combining link and content for community detection: a discriminative approach. In: Proceedings of the 15th ACM SIGKDD International Conference on Knowledge Discovery and Data Mining, pp. 927–936. ACM (2009). https://doi.org/10.1145/1557019.1557120
19. Zhang, G., Jin, D., Gao, J., Jiao, P., Fogelman-Soulié, F., Huang, X.: Finding communities with hierarchical semantics by distinguishing general and specialized topics. In: Proceedings of the 27th International Joint Conference on Artificial Intelligence, pp. 3648–3654. AAAI (2018)

Temporal Coding of Neural Stimuli

Adrian Horzyk[1]([✉]) [iD], Krzysztof Gołdon[1] [iD],
and Janusz A. Starzyk[2,3] [iD]

[1] AGH University of Science and Technology, Krakow, Poland
horzyk@agh.edu.pl, krzysztofgoldon@gmail.com
[2] University of Information Technology and Management in Rzeszow,
Rzeszow, Poland
starzykj@gmail.com
[3] School of EECS, Ohio University, Athens, USA

Abstract. Contemporary artificial neural networks use various metrics to code input data and usually do not use temporal coding, unlike biological neural systems. Real neural systems operate in time and use the time to code external stimuli of various kinds to produce a uniform internal data representation that can be used for further neural computations. This paper shows how it can be done using special receptors and neurons which use the time to code external data as well as internal results of computations. If neural processes take different time, the activation time of neurons can be used to code the results of computations. Such neurons can automatically find data associated with the given inputs. In this way, we can find the most similar objects represented by the network and use them for recognition or classification tasks. Conducted research and results prove that time space, temporal coding, and temporal neurons can be used instead of data feature space, direct use of input data, and classic artificial neurons. Time and temporal coding might be an important branch for the development of future artificial neural networks inspired by biological neurons.

Keywords: Temporal coding · Temporal neurons ·
Feature representation in the time space ·
Stimuli receptor transformation into the time space ·
Associative temporal neural networks · Associative graph data structure

1 Introduction

Different types of artificial neural networks either directly use external data as inputs, or use normalization, standardization, various transformations of input data space (e.g. PCA, ICA) [1, 7], or code them through preprocessing operations. Most of the artificial neural networks use discrete time iterations where all or a part of neurons are evaluated and calculate outputs in the same iteration (discrete time step) making computation totally synchronous (e.g. Hebb's rule, Oja's rule, McCulloch-Pitts model, Perceptron) [7, 21]. It can be perceived as positive because some processes in the brain seem to be synchronized (due to brain waves of various frequencies). Nonetheless, the majority of neuron activations are asynchronous [1, 3] and difficult to implement because they

© Springer Nature Switzerland AG 2019
I. V. Tetko et al. (Eds.): ICANN 2019, LNCS 11731, pp. 607–621, 2019.
https://doi.org/10.1007/978-3-030-30493-5_56

require ordering and simulation in time on contemporary computers. Time is used in spiking models of neurons [6, 14, 15, 19] that handle various internal processes in time, and time dependencies influence the stimulation processes in their networks [7, 8, 10, 14].

On the other hand, synchronous processes are more likely used because discrete time iterations can be easily parallelized using GPUs. Modern deep neural network architectures [7, 8] are also updated in discrete time steps. The synchronicity is an important factor to achieve efficient computational models on contemporary computers thanks to the use of vectorization and parallelization of computations of the same operations on multiple data. Nevertheless, we should not ignore the nature where asynchronous processes take place and have a great part in brain processes [1, 18]. Such processes have significance in the creation of new biology inspired strategies and algorithms.

The major difference between asynchronous and synchronous computations is that the synchronously fired neurons lose the possibility to code and differentiate external stimuli and internal results using time because the time is the same for all synchronously triggered neurons (e.g. in one iterative epoch or training cycle). Biological neural systems do not work in this way, and each neuron can be activated independently of other neurons if only it is charged to its activation threshold level [16, 18]. The synchronicity between the biological neurons is secondary and depends on the synchronic stimulations from outside. Asynchronous activations of neurons in continuous time allow neurons to compete and produce time-coded results based on such competitions. The first activated neurons in time identify objects or classes the most associated with the input stimuli, can influence other processes, inhibit competing neurons, and produce memories. In nature, the memories are usually created for the strongest and the most frequent stimuli associated with the earliest activated neurons representing memorized objects or their classes, temporal sequences, or spatial neighborhood of objects.

This paper defines a new model of neural networks which are based on temporal coding, temporal neurons, and special receptors that transform external stimuli (input data) into an internal temporal form (time space). Temporal neurons work asynchronously and must be simulated using a global queue mechanism responsible for ordering processes in time and allowing for competitions between such neurons. In our paper, this mechanism will be described and used in the conducted experiments. Temporal neurons are connected using associative rules that link features and objects in the same way as in the associative graph data structures (AGDS) [12, 13] where each unique value of each attribute is represented by a single node that aggregates the representation of all duplicates of this value in a dataset. Unlike many other approaches which use many-to-many connections between neurons organized in layers or matrices, the presented approach does not use layers but links similar features and connects them to the objects defined by these features. In nature and various biologically inspired models of neural networks, there is a wide variety of connections that only seldom form a multilayer structure where neurons are connected to other consecutive layers [1, 16, 18, 23]. Hence, the major part of the contemporary used artificial neural network models is based on unrealistic and highly restrictive fundamentals. We use these models in computational intelligence because they produce valuable results, but they do not work like real neuronal systems that can exhibit the ability to self-develop appropriate cognitive and intelligent architectures [4, 5, 17, 22]. This paper partially removes the established limitations of contemporary neuronal models and shows new abilities of neural systems that can be used and exploited in future research. These abilities will be achieved using the time space, temporal coding, and time-based processing in the neural network.

2 Temporal Coding and Receptors

Temporal coding means the transformation of values from the input feature space into the time space, i.e. to the appropriate charging periods of the neurons connected to receptors. Each receptor is most sensitive to the given feature value v_k^n and less sensitive to other close feature values. The sensitiveness of the n-th receptor of the k-th attribute and simultaneously its strength with which it stimulates the connected neuron can be expressed by the following condition:

$$s_k^n = 1 - \frac{|v_k^n - v_k|}{R_k} \qquad (1)$$

where $R_k = v_k^{max} - v_k^{min}$ is a range of values of the attribute k. Each receptor is connected to a neuron that is charged by this receptor for the period p_n according to the similarity of the input stimulus v_k to the value v_k^n represented by this receptor:

$$p_n = 1 - s_k^n = \frac{|v_k^n - v_k|}{R_k} \qquad (2)$$

This provides the firing of neurons at a different time according to the presented input value. Using the continuous time of activations, input stimuli can be precisely coded in time space, i.e. without rounding. Thus, input data (external stimuli) are transformed into different charging periods and activation times of the connected neurons.

Figure 1 illustrates possible stimulation flow as a reaction to the input value v_k sensed by the receptors that charge the V_k^n and V_k^{n+1} neurons, their neighbors V_k^{n-1} and V_k^{n+2}, the connected temporal object neurons O^{j_2} and O^{j_4} of the strongest stimulated temporal value neuron V_k^n, the subsequent class neuron C^{l_1}, and finally, the stop neuron S which activation stops the stimulation process of the neural network for the presented input.

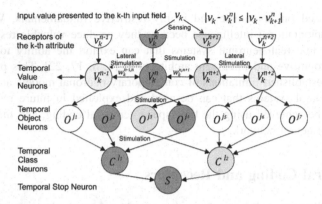

Fig. 1. Stimulation schema of temporal processes in the described neural network. The temporal process is started by the input stimulation of value v_k that is sensed by the receptors representing the closest values with the different strengths (1) that influence charging periods (2) of connected temporal value neurons which start stimulation of connected neighboring temporal value neurons (through lateral connections) and temporal object neurons when activated. Activated temporal object neurons subsequently activate temporal class neurons, and finally, the temporal stop neuron.

Always one or two receptors react to the input value v_k, computing the charging periods p_n of the connected neurons. If the input value v_k exactly equals to the value v_k^n represented by one of the receptors, then only a single receptor reacts to such an input stimulus and the activation period p_n of the connected neuron V_k^n is equal to 0 according to (2), i.e. it activates immediately after receiving the input stimulus. If the input v_k differs from all values v_k^n represented by the receptors of the attribute k, then two receptors representing the closest smaller and the closest bigger values react to such an input stimulus (except when the value v_k is minimal or maximal). In this case, the activation periods p_n of two connected neurons are computed after (2), and those periods are greater than 0.

3 Associative Temporal Neurons

Associative Temporal Neuron (ATN) is a new proposed model of neurons that differentiate charging periods according to the stimulation strengths and create associative connections representing various relations between input features and objects. ATN neurons work in time, but their way of working is different from classic spiking neuron models [6, 14, 15, 19]. The different charging periods determine the activation moments of ATN neurons and the charging processes of other connected neurons in the network. The order of the activations of neurons is used to produce and interpret results. The earliest activated neurons represent the most strongly associated objects to the input stimuli, where the associations can represent similarity, the order in sequence or spatial neighborhood. Thus, the earliest activated neurons can point out the most similar objects represented by the neural network. In conscious thinking, we usually take into account the first thoughts in response to the asked question or external stimuli. The ATN model also defines a minimum number of charge stimuli for each neuron, which allows it to be activated. This number is defined as a number of features which define the object represented by the temporal object neuron. It is equal to one for all value neurons representing simple one-value input features. This number increases the confidence of the classification when the neuron represents a class. Summarizing, ATN neuron n can be activated when two following conditions are simultaneously true:

1. The number of input stimuli s_n is equal to or greater than a threshold number θ_n of the required stimuli.
2. Current simulation time T of the network for the current run operation can be established from the f function value (4) of the sum S_n (3) of the g function values of the earliest s_n charging periods $p_1, \ldots, p_{s_n}$, where functions f and g must be positive:

$$S_n = \sum_{i=1}^{s_n} g(p_i) \tag{3}$$

$$p_n = f(S_n) \tag{4}$$

$$p_i = T_i - T_o \tag{5}$$

where p_i is a period (5) computed as a difference between the activation time T_i of the neuron sending the stimulus and the starting time T_o of the currently running operation in the network, where $p_1 \leq \cdots \leq p_{s_n}$. Each ATN neuron counts incoming input stimuli (s_n) and sums the charging periods transformed by the function g according to (3). When this count achieves the threshold number θ_n, then the function f is evaluated, and the final activation period p_n of the charged neuron n is calculated (4). This means that after the period p_n of charging elapses, the neuron will be activated, its activation time T_n during simulation will be established according to this period and the simulation time T_o when the charging operation was started:

$$T_n = T_o + p_n \tag{6}$$

The activation time T_n together with the information about the neuron is added to the global event queue (GEQ) that sorts events in the incremental order to control the sequence of updating neurons. The GEQ is responsible for activating the neurons in the appropriate time (6) allowing for their competitions.

The activated ATN neuron switches to the refraction period that prevents cyclic retrograde stimulations of mutually stimulating neurons. During the refraction period, ATN neurons are insensitive to any input stimuli. After the refraction period, the neuron returns to its resting (initial) state waiting for the next stimuli. To switch the neuron from its refraction state to its resting state, its parameters must be initialized. To perform this efficiently, each ATN neuron stores a special counter of the network operations (OPCntr). It points to the number of operation during which it has been updated the last time. The OPCntr counter is also stored in the network and incremented by one each time when the network is stimulated by new inputs (e.g. starts a new classification process). If the OPCntr counter of the neuron is less than the OPCntr counter of the network, the neuron variables are outdated and must be updated (e.g. initialized) before the next operation on this neuron. After each operation, the neuron

OPCntr is updated to be equal to the network OPCntr of to avoid the repetitive updating (e.g. initialization) before the subsequent updates of this neuron during the currently running operation. Owing to this mechanism, neurons are initialized exactly once before they are used in the next operation. This mechanism avoids the necessity to initialize variables of all neurons of the neural network before starting a new operation, which saves a lot of time.

When the neuron is activated, it sends stimuli to the connected neurons. If the connected neuron represents another numerical value v_k^m of the same attribute k, then the charging period p_m of this neuron is calculated from:

$$p_m = 1 - w_k^{m,n} \tag{7}$$

where $w_k^{m,n}$ is a connection weight between neurons V_k^n and V_k^m representing neighbor values v_k^n and v_k^m. It is defined as the absolute difference between these values v_k^n and v_k^m normalized by the range R_k of values of the k-th attribute:

$$w_k^{m,n} = 1 - \frac{|v_k^m - v_k^n|}{R_k} \tag{8}$$

The weights between value neurons V_k^m and V_k^n are symmetrical, i.e. $w_k^{n,m} = w_k^{m,n}$. Value neurons representing symbolic data are not connected and weighted.

In summary, each temporal value neurons V_k^n representing an input feature v_k^n is connected to at least one object neuron that represents a training sample defined by a set of input features. The activated neuron representing a feature v_k^n stimulates all connected object neurons. The stimulus contains the time when this neuron has been activated. Object neurons count up the incoming stimuli and compute the sums as defined by (3).

4   Associative Temporal Neural Networks

Associative Temporal Neural Networks (ATNN) are built from associative temporal neurons (ATN) and special receptors which transform input data into the time space and implement temporal processes between associatively connected neurons. ATN neurons connect to reproduce relations between elements represented by these neurons, where elements can be single-value features or various objects defined by features or other objects. Some objects can define classes or clusters or have a special meaning in the network. Associative connections between neurons are created for the known relationships which can be deduced from the input (training) data. All unique single-value data (like numbers, symbols, or strings) are represented by separate neurons connected to the receptors which are sensitive for these values. Neurons representing orderable values are additionally connected to the neurons representing neighbor values, and the connections are weighted (8). The associative representation means that all duplicates of the features of each attribute separately are aggregated and represented by the same neurons [9, 13]. Therefore, objects defined by a subset of the same features are indirectly connected by a subset value neurons representing these features. Objects defined by similar (not the same) features are also indirectly connected through connected value neurons representing similar features. Such associative organization of the network automatically groups objects according to their similarities and facilitates the inference processes [11]. For classic training data used for classification tasks composed from objects defined by a given number of features, the created network of elements can be described by the associative graph data structure (AGDS) [12, 13] that is used as a backbone structure of a proposed ATNN structure. Both these structures can be further accelerated using AVB+trees [13], which allow typically for less than the logarithmic time of the search of appropriate receptors for the input data. Any training dataset can be easily transformed into this graph structure as described in [12, 13].

Assume that we have a dataset $\mathbb{S} = \{P^1, \ldots, P^N\}$ consisting of N samples $P^n = [v_1^n, \ldots, v_k^n, \ldots, v_K^n]$ where v_k^n is a k-th attribute value defining the n-th sample (object) P^n, and K is the number of attributes. Each unique sample P^n is represented in the ATNN network by the **temporal object neuron** O^j where the number of object neurons J is less or equal to the number of training samples ($J \leq N$) because of possible duplicates of samples in the dataset $\mathbb{S}$ and their aggregated representation in the ATNN network.

The training samples represent L classes $\{c^1, \ldots, c^L\}$ represented by L **temporal class neurons** $\{C^1, \ldots, C^L\}$, and each training sample can belong to one or more classes dependently on the considered classification type (single-label classification or multi-label classification) [11, 20, 24]. In this paper, we consider only single-label classification problems to compare ATNN networks to KNN classifiers. Each unique k-th attribute value (feature) v_k^n is represented by a so-called **temporal value neuron** V_k^m in the ATNN network. Because the same feature can define many objects in the training data set, the number M_k of the temporal value neurons $\{V_k^1, \ldots, V_k^{M_k}\}$ representing the k-th attribute unique values is typically smaller than the number N of all training samples, i.e. $M_k \leq N$ (usually $M_k \ll N$ for real datasets due to the big number of duplicated attribute values defining various objects).

Temporal value neurons are also connected to object neurons, which represent training samples defined by the values represented by these connected value neurons. Connection weights between value and object neurons are equal to one. A group of object neurons can also define classes represented by temporal class neurons. The connections between object and class neurons are also equal to one. The activation threshold number of stimuli of temporal class neurons is set to one, so they immediately react to activation of any object neuron that defines a sample of the given class.

The ATNN network can be used as a k nearest neighbor (KNN) classifier, where we will also use a single **temporal stop neuron** that is connected to all class neurons. This neuron is responsible for stopping the network calculations when the demanded goal of computations (e.g. recognition, classification, or sorting) is finished. We defined the threshold of this neuron accordingly to the task (e.g. 1 for recognition, N for sorting) where this network is used. The weights between class and stop neurons are also equal to one, reacting to all classified samples immediately, but the threshold of this neuron can be set to the required number of searched samples. The activated ATN neurons of all kinds will stimulate other connected neurons, and each stimulus includes information about the time T_n (6) of the activation to compute appropriate periods (4) for computing sums (3) and charging periods (5) for the connected neurons.

5 Applying ATNN for Ordering Objects

The main idea of ATNN networks is to allow neurons to be activated over time in a succession that comes from the associative (relationship) strengths, e.g. similarity, between input data and the object. Therefore, the object neurons representing objects that are similar to input data (external stimuli) should be activated faster than neurons representing less similar objects. In fact, the ATNN can quickly order all represented objects (training samples) according to their similarities to the input data (Fig. 2), starting from the most similar one(s) (activated earlier, darker red in Fig. 2) and finishing with the least similar one(s) (activated later, lighter red in Fig. 2). Here, the similarities between represented objects by object neurons and given inputs will be defined by the Euclidean measure in the time space of charging periods of temporal value neurons representing transformed features into the charging periods.

Associative Distance Sorting Algorithm using an ATNN network:

```
Input: T: training data, x: classified sample,
Output: sortedObjects: objects sorted in descending order

OrderingObjects(T, x)
   atnn = CreateATNN(T)
   GEQ.InitializeSimulationTime()
   foreach attribute of atnn do
       attribute.FindAndStartStimulate(x[numberOfAttribute]) // using AVB+trees
   while (stopNeuron.State < threshold)
       currentEvent = GEQ.RemoveFirst
       simulationTime=currentEvent.UpdateTime
       currentEvent.Simulate() // add activated object neurons to sortedObjects
   return sortedObjects
```

In the first step of this algorithm, the structure of the ATNN network based on AGDS [13] is created. Values of each attribute are represented via an AVB+tree [13] that supports efficient access, and all operations are processed in at most logarithmic time. The stop condition of this algorithm is provided by the temporal stop neuron which is connected to all object neurons and which threshold is equal to the number of all pattern neurons to stop stimulation when all object neurons are activated. To aggregate representation of all duplicates during the construction of the ATNN structure, a special insertion operation using an AVB+tree [13] is used to increase counter if the inserted value already exists in this structure or to insert it and rebalance the tree (if necessary) when the inserted value is new and must be added:

```
SearchOrInsertNeuron(value)
   foreach neuron in node
       if (neuron.Value == value)
           neuron.IncrementCounterOfDuplicates() // aggregate their representation
           return true
   if (childList.Count==0) // this is a leaf of AVB+tree
       CreateNewNeuronAndInsertItIntoAVBTreeNode() // rebalance a tree if necessary
   else
       if (Neurons.First.Value < value)
           return Neurons.First.SearchOrInserNeuron(value)
       elseif (Neurons.Last.Value > value)
           return Neurons.Last.SearchOrInserNeuron(value)
       else return Neurons.Middle.SearchOrInsertNeuron(value)
```

In the second step, the ATNN is stimulated by the sample as long as the stop neuron is activated. During this process, a list of activated object neurons is created in the order coming from the activation moments of these neurons. Thus, the samples (represented by object neurons) are ordered according to the similarity (e.g. defined by the Euclidean distance) to the sample presented on the input of the network. To start the stimulation process of the ATNN network quickly, the previously created AVB+trees for all attributes are used to search for temporal value neurons which will be stimulated appropriately to the presented input values. Subsequently, temporal value neurons compute their activation moments and put them to the GEQ that handles their ordering and triggering in the right simulation time. In each simulation step, always the first event from the GEQ is run until the temporal stop neuron is activated. During the simulation process, all activated neurons stimulate connected neurons which insert new events to the GEQ according to the computed activation periods as described in the previous sections.

The similarity to input data in Fig. 2 is represented by the activation times (marked with t in the lowest raw of circles of neurons that represents training samples) – the smaller times are, the more similar the training samples are. Such ordering is used by various methods, e.g. KNN classifiers, which base their classification decisions on the given number of nearest neighbors, various sorting routines, recognition, clustering, and classification algorithms. This ordering of ATNN networks is also quite fast because the algorithm needs to stimulate the neurons only once for a given input (training) dataset to get results.

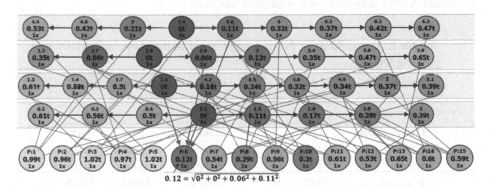

$$0.12 = \sqrt{0^2 + 0^2 + 0.06^2 + 0.11^2}$$

Fig. 2. Example of sorting objects by the ATNN according to the similarity to the given input sample [5.4, 2.8, 3.6, 1.3] to 15 Iris samples [25]. The temporal value neurons representing features of four attributes are stimulated laterally by the previously activated value neurons as marked by blue arrows. The sorted training samples are represented by temporal object neurons in the last row which compute their activation times using an Euclidean distance of activation times of the stimulating value neurons as presented on a sample object neuron activated in time 0.12. Each neuron presents the value or ID in the upper part, the activation time in the middle, and the number of its activations in the bottom part of its circle. The presented activation times of object neurons reflect the similarity of the input sample to all training samples from this dataset. The similarities expressed by the activation times produced the following sequence of training samples: P:6, P:10, P:8, P:12, P:7, P:9, P:15, P:14, P:11, P:13, P:2, P:4, P:1, P:3, and P:5. (Color figure online)

The computational complexity of ordering all objects according to their similarities to the given input sample is linear $O\left(N + \sum_{k=1}^{K} h_k(N)\right) \leq O((K+1) \cdot N)$, where N is the number of objects, and K is the number of attributes, and the function h_k defines the number of unique values for the k-th attribute. If we do the same operation on a tabular data structure, we need to compute similarity factors for each object and sort final results according to these factors. This operation would cost log-linear time $O(K \cdot N \log N)$ for quicksort, heapsort, or merge sort algorithms [2] or $O(K \cdot N)$ for counting sort or radix sort [2]. Thus ATNN networks based on the AGDS structure can theoretically have the same or better complexity than the best classic sorting algorithms based on a tabular structure especially when the sorted data consist of many duplicated values defining objects in the dataset. Unfortunately, the simulation of temporal neurons on a sequential machine (which are most common today) takes extra time because events (representing activation times of neurons) must be ordered, so the presented sorting on ATNN is not always faster than classic approaches, however, on asynchronously parallel machines, it would run much faster. Therefore, the presented sorting approach using ATNN networks can only explain how biological neurons can deal with ordering tasks when working on a different computational platform than contemporary computers.

The ATNN sorts objects starting from the most similar, so the efficiency of ATNN will be better when searching for a limited subset of the most similar objects because the classic approaches usually require to compute all similarities and sort all objects before selecting the most similar ones. The results presented in Table 1 show the sorting times of the most similar objects to the given input sample computed using the classic quicksort approach on data stored in tables compared to the sorting made by the simulated ATNN networks on a sequential machine.

Table 1. Comparisons of the recognition speed of the classic and ATNN sorting routines.

Training datasets*	No of instances	No of attributes	Unique features	Sorting time [timer ticks]	
				Classic	ATNN
Immunotherapy	90	7	27,46%	5456	487
Iris	150	4	20,50%	21825	596
Wine	178	12	59,74%	35466	2565
Banknote	1372	4	91,40%	43651	16136
Wine quality red	1599	11	8,26%	35466	10924
Eye	14980	14	2,58%	504719	428724
HTRU2	17898	8	52,81%	373765	1150166
Telescope	19020	10	77,34%	428329	1591250
Credit card	30000	23	25,30%	2515411	2272482
Shuttle	43500	9	0,26%	774812	2600721
Drive	58509	48	59,75%	18036210	27835709
Skin	245057	3	0,10%	1355921	556790

*Datasets were taken from UCI ML Repository [25]

In most of the cases, ATNN network is faster than quicksort, but the major goal of this paper is to prove that data can be transformed into the uniform time space and temporal spikes, not a better processing speed on the contemporary computer. Thus a higher efficiency of our approach is an extra bonus point.

6   Applying ATNN for Recognition Tasks

The first studies in comparisons of classic implementation of KNN classifiers based on the associative graph data structures (AGDS) were described in [12]. The results demonstrated that the associative organization of data with additional relations between values of the same attribute could greatly accelerate KNN classifiers. In this work, we show that KNN classifiers can also be modeled by ATNN networks and that the classification process can be fully automatic.

If $g(p_i) = p_i$ and $f(S_j) = S_j$ are defined as the identity functions, then the computed activation periods reflects the normalized Manhattan distance between input data and the features defining an object represented by an object neuron:

$$P_n^k = \sum\nolimits_{i=1}^{n_j} p_i = \sum\nolimits_{i=1}^{n_j} \frac{\left|v_k^m - v_k^n\right|}{R_k} \tag{9}$$

If $g(p_i) = p_i^2$ and $f(S_j) = \sqrt{S_j}$, then the computed activation time reflects the Euclidean distance between input data and the features defining an object represented by an object neuron:

$$P_n^k = \sqrt{\sum\nolimits_{i=1}^{n_j} p_i^2} = \sqrt{\sum\nolimits_{i=1}^{n_j} \left(\frac{\left|v_k^m - v_k^n\right|}{R_k}\right)^2} \tag{10}$$

Hence, ATN neurons can be used to compute the nearest neighbors for KNN classifiers because the Manhattan or Euclidean distances of training samples to the classified sample can be transformed into the activation periods of object neurons representing these samples. The charging periods linearly change as these distances, so they model original training data space in the internal time space of ATNN networks. Therefore, the first k activated ATN temporal object neurons represent the same k nearest neighbors (objects) as computed by the KNN classifier.

When k = 1 the classification process points out the most similar object to the presented sample on the input, so this process can also be viewed as a recognition process as shown in Fig. 3. The comparisons of the speed of the classic KNN classifier with k = 1 and ATNN used in recognition tasks are presented in Table 2 where ATNN networks win in recognition speed.

In most of the cases, the proposed approach was more efficient than the classical one. The results were obtained on a sequential machine and were from 2% slower in case of Credit Card data set to 250 times faster in case of Skin data set than the classic KNN approach. Notice, that our temporal coding network would be much faster if implemented on a network of neurons, since neurons pass information to its neighbors

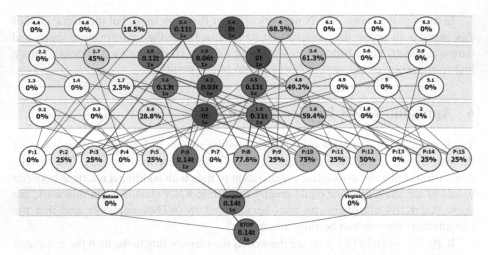

Fig. 3. Example of recognition (for k = 1) of the most similar object (here P:6) to the presented inputs [5.6, 3.0, 4.1, 1.3] made by the ATNN network created for the training data consisting of 15 Iris samples [25]. The temporal value neurons for each attribute separately are presented in the four top rows. In the fifth row, there are temporal object neurons defined by the appropriate combinations of temporal value neurons (representing training samples). The sixth row consists of temporal class neurons. The temporal STOP neuron in the last row stops the simulation when the most similar object is found. Neurons painted in red symbolized the activated neurons at different times. Only activated neurons stimulate other connected neurons which do not always achieve activations (neurons painted in green). White colored neurons do not take part in the presented recognition process, which saves time in comparison to the KNN classification process, where all training samples take part in the classification process. (Color figure online)

Table 2. Comparisons of the recognition speed of the classic and ATNN approaches.

Training datasets*	No of instances	No of attributes	Unique features	Recognition [timer ticks]	
				Classic	ATNN (k = 1)
Immunotherapy	90	7	27,46%	301	38
Iris	150	4	20,50%	282	25
Wine	178	12	59,74%	309	84
Banknote	1372	4	91,40%	476	33
Wine quality red	1599	11	8,26%	550	81
Eye	14980	14	2,58%	2814	363
HTRU2	17898	8	52,81%	2863	84
Telescope	19020	10	77,34%	3233	85
Credit card	30000	23	25,30%	7319	7472
Shuttle	43500	9	0,26%	5629	4803
Drive	58509	48	59,75%	25948	1429
Skin	245057	3	0,10%	19741	79

*Datasets were taken from UCI ML Repository [25]

concurrently, while classic algorithm would have no such additional efficiency increase since finding the nearest neighbor would require sequential comparisons. However, once again, the high efficiency of our approach was its extra bonus. We aimed to demonstrate that both ordering and recognition of the input data was possible using temporal coding of input stimuli and associative temporal neurons that may represent sensory values, objects, and classes. A network made of such associative temporal neurons may be a useful addition to neural network structures that work with numeric as well as symbolic data. Such networks may be good to obtain grounded semantic memories needed for efficient learning and motor control functions in robots.

7   Conclusions and Remarks

This paper presented temporal coding of the input data using associative temporal neurons and the resulting associative structures named associative temporal neural networks. The contribution of this paper was to show that it is possible to code features of various attributes using time, namely, various periods of charging and activations of neurons. The temporal coding also requires an appropriate associative neural structure which reproduces relations between features and objects. That is why we used the associative graph data structure AGDS to reproduce the structure of features and samples by value neurons and object neurons which were appropriately connected, and the connections weighted to emphasize the relationships between them. On this basis, neurons and their connections were set up to be charged in different periods according to the similarity of the features and objects to the presented input samples. As a result of such coding, temporal neurons were activated over time after different periods, which reproduced the Euclidean or Manhattan distance between training samples and the input samples. Using such a network, we could define a sorting routine which allowed to designate the most similar samples and sort them due to the computed similarities. The computed similarities were used to designate k nearest neighbors, order samples, or recognize the most similar sample. The created networks can be used for classification, sorting, and recognition tasks. Thanks to the use of the associative structure of the created associative temporal neural networks, not all objects need to be compared in ATNN networks, so we can usually compute results faster, especially when running them in parallel. In spite of the experimentally proven efficiency, the major goal of this paper was to show how original data space can be transformed into the time space. Experiments proved that the time-based computations using associative temporal neurons are possible, and classic coding of input data can be replaced by the presented temporal one using the time space instead of the original one.

Acknowledgement. This work was supported by the grant from the National Science Centre, Poland DEC-2016/21/B/ST7/02220 and AGH 11.11.120.612.

References

1. Carpenter, G.A., Grossberg, S.: Adaptive resonance theory. In: Arbib, M. (ed.) The Handbook of Brain Theory and Neural Networks, pp. 87–90. MIT Press, Cambridge (2003)
2. Cormen, T., Leiserson, Ch., Rivest, R., Stein, C.: Introduction to Algorithms. 3nd edn., pp. 484–504. MIT Press and McGraw-Hill (2009)
3. Deuker, L., et al.: Memory consolidation by replay of stimulus-specific neural activity. J. Neurosci. **33**(49), 19373–19383 (2013)
4. Duch, W.: Brain-inspired conscious computing architecture. J. Mind Behav. **26**, 1–22 (2005)
5. Franklin, S., Madl, T., D'Mello, S., Snaider, J.: LIDA: a systems-level architecture for cognition, emotion, and learning. IEEE Trans. Auton. Ment. Dev. **6**(1), 19–41 (2014). https://doi.org/10.1109/TAMD.2013.2277589
6. Gerstner, W., Kistler, W.: Spiking Neuron Models. Cambridge University Press, Cambridge (2002). https://doi.org/10.1017/cbo9780511815706
7. Goodfellow, I., Bengio, Y., Courville, A.: Deep Learning. MIT Press, Cambridge (2016)
8. Graupe, D.: Deep Learning Neural Networks. World Scientific, Singapore (2016). https://doi.org/10.1142/10190
9. Horzyk, A.: Neurons can sort data efficiently. In: Rutkowski, L., Korytkowski, M., Scherer, R., Tadeusiewicz, R., Zadeh, L.A., Zurada, J.M. (eds.) ICAISC 2017. LNCS (LNAI), vol. 10245, pp. 64–74. Springer, Cham (2017). https://doi.org/10.1007/978-3-319-59063-9_6
10. Horzyk, A.: Deep associative semantic neural graphs for knowledge representation and fast data exploration. In: Proceedings of KEOD 2017, pp. 67–79. Scitepress Digital Lib. (2017). https://doi.org/10.5220/0006504100670079
11. Horzyk, A., Starzyk, J.A.: Fast neural network adaptation with associative pulsing neurons. In: 2017 IEEE Symposium Series on Computational Intelligence, pp. 339–346. IEEE Xplore (2017). https://doi.org/10.1109/ssci.2017.8285369
12. Horzyk, A., Gołdon, K.: Associative graph data structures used for acceleration of K nearest neighbor classifiers. In: Kůrková, V., Manolopoulos, Y., Hammer, B., Iliadis, L., Maglogiannis, I. (eds.) ICANN 2018. LNCS, vol. 11139, pp. 648–658. Springer, Cham (2018). https://doi.org/10.1007/978-3-030-01418-6_64
13. Horzyk, A.: Associative graph data structures with an efficient access via AVB+trees. In: 2018 11th International Conference on Human System Interaction (HSI), pp. 169–175. IEEE Xplore (2018). https://doi.org/10.1109/hsi.2018.8430973
14. Izhikevich, E.M.: Neural excitability, spiking, and bursting. Int. J. Bifurcat. Chaos **10**, 1171–1266 (2000). https://doi.org/10.1142/S0218127400000840
15. Izhikevich, E.M.: Simple model of spiking neurons. IEEE Trans. Neural Netw. **14**(6), 1569–1572 (2003). https://doi.org/10.1109/TNN.2003.820440
16. Kalat, J.W.: Biological Grounds of Psychology, 10th edn. Wadsworth Publishing, Belmont (2008)
17. Laird, J.E.: Extending the soar cognitive architecture. In: Proceedings of the First Conference on AGI, Memphis, Tennessee, pp. 224–235 (2008)
18. Longstaff, A.: Neurobiology. PWN, Warsaw (2006)
19. Maass, W.: Networks of spiking neurons: the third generation of neural network models. Neural Netw. **10**(9), 1659–1671 (1997)
20. Read, J., Pfahringer, B., Holmes, G., Frank, E.: Classifier chains for multi-label classification. Mach. Learn. J. **85**(3), 333 (2011). https://doi.org/10.1007/978-3-642-38067-9_13
21. Rutkowski, L.: Techniques and Methods of Artificial Intelligence. PWN, Warsaw (2012)

22. Tadeusiewicz, R.: Introduction to Intelligent Systems, Fault Diagnosis. Models, Artificial Intelligence, Applications. CRC Press, Boca Raton (2011)
23. Tyukin, I., Gorban, A.N., Calvo, C., Makarova, J., Makarov, V.A.: High-dimensional brain: a tool for encoding and rapid learning of memories by single neurons. Bull. Math. Biol. 1–33 (2018). https://doi.org/10.1007/s11538-018-0415-5. Special Issue: Modelling Biological Evolution: Developing Novel Approaches
24. Zhang, M.L., Zhou, Z.H.: Multi-label neural networks with applications to functional genomics and text categorization. IEEE Trans. Knowl. Data Eng. **18**, 1338–1351 (2006). https://doi.org/10.1007/978-3-319-00563-8_22
25. UCI ML Repository. http://archive.ics.uci.edu/ml/datasets/Iris. Accessed 14 Apr 2018

Heterogeneous Information Network Embedding with Meta-path Based Graph Attention Networks

Meng Cao⬤, Xiying Ma⬤, Ming Xu, and Chongjun Wang(✉)⬤

National Key Laboratory for Novel Software Technology,
Department of Computer Science and Technology, Nanjing University,
Nanjing 210023, China
{caomeng,mf1833043}@smail.nju.edu.cn, xuming0830@gmail.com,
chjwang@nju.edu.cn

Abstract. Network embedding is an emerging research field which aims at projecting network elements into lower dimensional spaces. However, most network embedding algorithms focus on homogeneous networks, thus cannot be directly applied to the Heterogeneous Information Networks (HINs) which are prevalent in real world systems. Therefore, how to effectively preserve both the structural and semantic information, as well as higher-level proximity of large-scale HINs still remains an open problem. In this paper, we propose a novel heterogeneous network embedding model called **Meta**-path based **G**raph **AT**tention n**E**twork (Meta-GATE). To tackle the problem of preserving both structural and semantic information for HINs, we adopt a multiple meta-paths based random walk scheme with Skip-Gram for sampling and pre-training. In addition, the model involves graph attention networks to collect and aggregate heterogeneous information, which may reveal higher-level implicit semantics. Besides, to ensure stability, multi-head attention mechanism is employed in the graph attention model by concatenating the embedding of multiple independent self-attention processes. Experiments on two real-world bibliography networks are conducted. Compared to state-of-the-art baselines, our method achieves the best performance on the node classification task, which shows the effectiveness of the proposed model.

Keywords: Heterogeneous information networks ·
Network embedding · Graph attention networks

1 Introduction

Heterogeneous information networks (HINs) are ubiquitous in human society, which consist of multi-typed components and interactions as well as rich side information like texts, images, etc. An example of a bibliography network schema is shown in Fig. 1(a), and the corresponding network instance is displayed in

© Springer Nature Switzerland AG 2019
I. V. Tetko et al. (Eds.): ICANN 2019, LNCS 11731, pp. 622–634, 2019.
https://doi.org/10.1007/978-3-030-30493-5_57

Fig. 1(b). The rich semantics and multi-relations in HINs have attracted a lot of attention. Therefore, how to efficiently discover latent patterns and hidden knowledge in HINs has become a hot research topic over the recent years, and this brings great challenges to network representation learning.

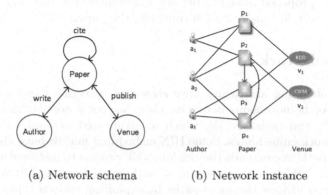

(a) Network schema   (b) Network instance

Fig. 1. A heterogeneous information network example.

Network representation learning is also known as network embedding, which aims at projecting network elements, i.e. nodes, edges, to a lower dimensional vector space, meanwhile preserving the properties of the original network. In this way, the high computational complexity brought by high-dimensional matrix operations can be avoided to some extent, and the vectors can be further used as features for downstream machine learning tasks, such as classification, clustering, and so on.

However, existing network embedding methods mainly focus on homogeneous networks, while real-life systems are mostly heterogeneous information networks (HINs). The embedding learning on HINs is challenging due to the following reasons: (1) the embedding of network elements should preserve not only the topological structures, but also the heterogeneity of nodes and links; (2) the rich semantics and high-order proximity in HINs should also be considered. Therefore, efficient methods for analyzing and extracting the latent knowledge in HINs are highly desired.

In this paper, we propose a novel heterogeneous network embedding model called **Meta**-path based **G**raph **AT**tention n**E**twork (MetaGATE). To effectively preserve the structural and semantic heterogeneity of HINs, multiple meta-paths based random walks are adopted for information sampling and pre-training. To tackle the problem of preserving higher-level proximity of HINs, and considering the fact that different neighbors have different extent of influence on the central node, hereby we employ the Graph Attention Network (GAT) for effective node information aggregation and embedding learning. Different from contemporary GAT model [11], we apply GAT on HINs with the pre-trained network embeddings as input node features, which contain the semantic information related to different node types. In this way, the three major goals in the study of HIN embedding, i.e., to preserve structural/semantic/higher-level information in HINs, are solved under the unified framework.

The remainder of this paper is organized as follows. We first review the related work in Sect. 2, and provide the preliminary concepts in Sect. 3. In Sect. 4, we introduce the proposed model in details. Experimental results and analysis are presented in Sect. 5. Finally Sect. 6 concludes the paper.

2   Related Work

Networks embedding algorithms have shown an emerging trend over the past few years. For homogeneous networks, there are some popular models such as DeepWalk [9] and node2vec [5], which use truncated random walk sequences to learn network embeddings. Some HIN embedding models have also been proposed. Metapath2vec extends the random walk process to meta-path based sampling schemes, which can preserve meta-path related semantic information [3]. HIN2VEC also utilizes the meta-paths by exploiting various types of relationships [4]. However, these methods only focus on utilizing meta-path to preserving the semantics in HINs, without considering the higher level proximities in HINs.

To better analyze graph-shaped data, the Graph Neural Network model is proposed, and soon becomes a hot research topic due to its outstanding performance in graph-related tasks. State-of-the-art works include spectral methods like Graph Convolutional Network (GCN) [7] which looks at the complete 1-hop neighborhood around the node for aggregation and utilize multiple depths of the model to capture higher-order information, and non-spectral methods such as GraphSAGE [6] which preserves partially sampled neighborhood around the node. FastGCN [2] further improves the performance of GraphSAGE with importance sampling to make the model more efficient. Although these methods have achieved remarkable results in network analytical tasks, they are designed for processing homogeneous networks. Therefore, how to apply GNN models to learn the latent knowledge in HINs remains an open question, and an effective model for the embedding of HINs is highly demanded.

3   Preliminaries

Definition 1. *Heterogeneous Information Network (HIN).* *A heterogeneous information network is defined as a graph $\mathcal{G} = (\mathcal{V}, \mathcal{E})$ with a node mapping function $\phi : \mathcal{V} \to \mathcal{A}$, and a link mapping function $\psi : \mathcal{E} \to \mathcal{R}$. $\mathcal{A}$ and $\mathcal{R}$ denote the sets of node types and link types, and $|\mathcal{A}| + |\mathcal{R}| > 2$.*

Definition 2. *Meta-Path.* *In a HIN, Meta-path ρ is defined as a specific sequence of node types $a_1, a_{2,...,a_n}$ and/or edge types $r_1, r_{2,...,r_{n-1}}$:*

$$\rho = a_1 \xrightarrow{r_1} \dots a_i \xrightarrow{r_i} \dots \xrightarrow{r_{n-1}} a_n$$

*In Fig. 1 we present an illustration of a bibliography network consists of node types as Author(A), Paper(P), and Venue(V). Take the meta-path Author-Paper-Author(APA) for example, it denotes two authors collaborating on the same paper, and $a_1 - p_1 - a_2$ is a **meta-path instance** of the meta-path APA.*

Definition 3. HIN embedding. *Given a heterogeneous network $\mathcal{G} = (\mathcal{V}, \mathcal{E})$, HIN embedding aims at learning a function $f : \mathcal{V} \to \mathbb{R}^d$ that projects each node $v \in \mathcal{V}$ to a vector in a d-dimensional space $\mathbb{R}^d$, where $d \ll |\mathcal{V}|$.*

4   The Proposed Model

In this section, we will introduce the proposed model. The overall framework of the model is displayed in Fig. 2. For a given HIN, the model adopts multiple meta-paths as sampling schemes. Then a Skip-Gram-based network embedding layer is used for heterogeneity embedding training. Finally a graph attention network is trained for specific tasks. The model is designed for un-directed and un-weighted HINs, but can be easily extended to directed and weighted circumstances. We first introduce each part of the model, and then formulate the corresponding algorithm hereinafter.

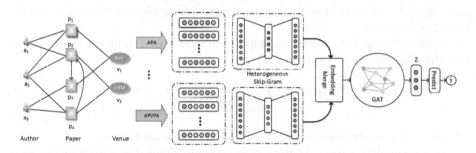

Fig. 2. The overall framework of the proposed MetaGATE model.

4.1   Meta-path Sampling Scheme

Given a HIN $\mathcal{G} = (\mathcal{V}, \mathcal{E})$, we can define multiple meta-path-based random walk schemes to capture both semantic information and structural correlations between different types of nodes.

Specifically, a meta-path scheme ρ can be defined as $v_1^{T_1} \to v_2^{T_2} \to \cdots \to v_i^{T_k} \to v_{i+1}^{T_{k+1}} \to \cdots \to v_n^{T_K}$, where T_k denotes node type k.

In this way, the transition probability at step i is defined as follows:

$$p\left(v_{i+1}|v_i^{T_k}\right) = \begin{cases} \frac{1}{\left|\mathcal{N}_{k+1}\left(v_i^{T_k}\right)\right|} & \left(v_{i+1}, v_i^{T_k}\right) \in \mathcal{E}, \phi\left(v_{i+1}\right) = T_{k+1} \\ 0 & \left(v_{i+1}, v_i^{T_k}\right) \in \mathcal{E}, \phi\left(v_{i+1}\right) \neq T_{k+1} \\ 0 & \left(v_{i+1}, v_i^{T_k}\right) \notin \mathcal{E} \end{cases} \quad (1)$$

where $v_i^{T_k} \in \mathcal{V}_i$ denotes node i with type k, $\mathcal{N}_{k+1}\left(v_i^{T_k}\right)$ represents the neighborhood of node $v_i^{T_k}$ with node type T_{k+1}, and $\phi\left(\cdot\right)$ represents the node type mapping function. In this way, the walk will be constrained to follow the predefined meta-path ρ when deciding the next move. Practically, the meta-paths are usually set to be symmetric so that the walk can be recursively guided for sufficient sampling in long distances [3].

4.2   Heterogeneity Embedding Training

For each type of meta-path scheme ρ_i, we can obtain the meta-path-based walk sequences of $\mathcal{G}$, similar to word2vec [8], a Skip-Gram model which aims to learn the embedding vectors of nodes by predicting the node context is adopted. Generally, the objective is to maximize the probabilities of the context appearance with given central nodes, that is:

$$\arg\max_{\theta} \prod_{v \in \mathcal{V}} \prod_{c \in \mathcal{N}(v)} p\left(c|v; \theta\right) \tag{2}$$

where $\mathcal{N}\left(v\right)$ is the neighborhood of node v, $p\left(c|v; \theta\right)$ represents the conditional probability of the appearance of context node c for a given node v, and θ is the model parameters.

Inspired by the heterogeneous Skip-Gram model used in [3], we can reformulate the objective function in Eq. (2) as:

$$\arg\max_{\theta} \sum_{v \in \mathcal{V}} \sum_{t \in T} \sum_{c_T \in \mathcal{N}_T(v)} \log p\left(c_T|v; \theta\right) \tag{3}$$

where $\mathcal{N}_T\left(v\right)$ represents the neighborhood of node v with type T.

According to literature [1], $p\left(c_T|v; \theta\right)$ is usually defined as a softmax function, that is,

$$p\left(c_T|v; \theta\right) = \frac{e^{X_{c_T} \cdot X_v}}{\sum_{u_T \in \mathcal{V}_T} e^{X_{u_T} \cdot X_v}} \tag{4}$$

where X_v is the embedding vector of node v, $\mathcal{V}_T$ is the node set of type T in $\mathcal{G}$.

In order to calculate Eq. (4) efficiently, negative sampling [8] is introduced. Given a negative sample size M, the objective function in Eq. (3) is updated as:

$$\mathcal{O}\left(X\right) = \log \sigma\left(X_{c_T} \cdot X_v\right) + \sum_{m=1}^{M} \mathbb{E}_{u_T^m \sim P_T(u_T)} \left[\log \sigma\left(-X_{u_T^m} \cdot X_v\right)\right] \tag{5}$$

where $\sigma\left(x\right) = \frac{1}{1+e^{-x}}$, and $P_T(u_T)$ is the distribution of negative samples, from which a negative node u_T^m of type T is sampled for M times. The network embedding X_i of meta-path ρ_i is then optimized by using stochastic gradient descent algorithm.

For meta-path set $\rho = \{\rho_1, \rho_2, ..., \rho_k\}$, we perform the same heterogeneous embedding training process, then apply a merge operation on the generated embeddings to reserve different types of semantics, i.e.:

$$\mathbf{X} = \text{Merge}(X_1, X_2, ..., X_k) \tag{6}$$

To evaluate the influence of the merge operation on the model performance, three types of merge operations are compared in the experiments, i.e. max-pooling, mean-pooling, and concatenation, along with detailed analysis presented in Sect. 5.

4.3   Message Passing and Aggregation with GAT

In our model, a graph attention module is employed, which consists of two graph attention layers, each containing an input layer, a hidden layer, and an output layer. Firstly, the input layer performs a linear transformation on the heterogeneous embedding of node i, i.e.

$$\mathbf{z}_i = W\mathbf{x}_i \tag{7}$$

where $\mathbf{x}_i \in \mathbf{X}$ is the node embedding, and W is the weight matrix of the input layer.

After the linear transformation, a pair-wise unnormalized attention score between each pair of neighbors is calculated:

$$e_{ij} = \text{LeakyReLU}\left(\boldsymbol{a}^T\left(\mathbf{z}_i\|\mathbf{z}_j\right)\right) \tag{8}$$

where $\|$ denotes concatenation, and $\boldsymbol{a}$ is a learnable weight vector.

To normalize Eq. (8), a softmax function is applied on each node's in-coming edges:

$$\alpha_{ij} = softmax_j\left(e_{ij}\right) = \frac{\exp\left(e_{ij}\right)}{\sum_{k \in \mathcal{N}_i} \exp\left(e_{ik}\right)} \tag{9}$$

where $\mathcal{N}_i$ is the set of node i's one-hop neighbors, including itself.

Finally, in the output layer, an aggregation function is performed to generate the feature embedding of the next layer:

$$h_i = \sigma\left(\sum_{j \in \mathcal{N}(i)} \alpha_{ij}\mathbf{z}_j\right) \tag{10}$$

For model stabilization, a multi-head attention is adopted by concatenating the embedding of K independent self-attention processes:

$$h_i' = \|_{k=1}^{K}\sigma\left(\sum_{j \in \mathcal{N}(i)} \alpha_{ij}^k\mathbf{z}_j^k\right) \tag{11}$$

Finally, the output results are trained on specific tasks to obtain graph attention parameters. For example, in node classification, we try to minimize the cross entropy loss between the predictions and the ground truth, the loss function is:

$$L = -\sum_{i=1}^{N} y_i \log \left(\frac{e^{h'_i}}{\sum_j e^{h'_j}} \right) \tag{12}$$

where y_i is the label of node i.

The algorithm of MetaGATE is presented in Algorithm 1 below. It is notable that the model we propose is a unified framework, where the feature acquisition can be seen as a pre-processing step, and this can also be replaced with other state-of-the-art embedding algorithms.

Algorithm 1. MetaGATE

Input: Hererogeneous network $\mathcal{G} = (\mathcal{V}, \mathcal{E})$, number of walks w, walk length l, meta-path set ρ, number of negative samples M, window size q, label set L, number of train epoch k

Output: Network embedding $\mathbf{X}$

1: Initialize X
2: **for** $i \leftarrow 1$ to $|\rho|$ **do**
3:     **for** $v \in V$ **do**
4:         $MP(v) \leftarrow$ GenerateMetaPathWalk$(\mathcal{G}, \rho_i, v, l, w)$ (Eq. (1))
5:     **end for**
6:     $X_i \leftarrow$ HererogeneousEmbeddingTraining (X, M, MP, q) (Eq. (5))
7: **end for**
8: $\mathbf{X} \leftarrow$ Merge$(X_1, X_2, ..., X_{|\rho|})$
9: **for** epoch$\leftarrow 0$ to k **do**
10:     $\mathbf{X} \leftarrow$ GAT$(\mathcal{G}, \mathbf{X}, L)$ (Eq. (7)~(12))
11: **end for**
12: **return** $\mathbf{X}$

---

5   Experiments

5.1   Datasets

To verify the effectiveness of the proposed MetaGATE model, we conduct experiments on two real-world heterogeneous information networks. The detailed description of the datasets are shown in Table 1.

- **DBLP**[1]. A computer science bibliography network which contains 14375 papers (P), 14475 authors (A), 20 conferences (C), and 8811 terms (T). The authors, papers, conferences are divided into four areas: *database, data mining, machine learning, and information retrieval*. Two meta-paths, APVPA and APTPA, are employed for the experiments.

---

[1] https://dblp.uni-trier.de/.

Table 1. Statistics of the datasets.

Dataset	Relations (A-B)	Number of A	Number of B	Number of A-B	Ave.Degrees of A	Ave.Degrees of B	Meta-paths
DBLP	Paper-Author	14375	14475	41794	2.91	2.89	APVPA
	Paper-Venue	14375	20	14375	1.00	718.75	APTPA
	Paper-Term	14375	8811	88683	6.17	10.07	
AMiner	Paper-Author	127623	164473	355072	2.78	2.16	APA
	Paper-Venue	127623	101	127623	1.00	1263.59	APVPA

- **AMiner**[2], a bibliography network whose schema is the same as shown in Fig. 1(a). The network consists of 127,623 papers (P), 164,473 authors (A), and 101 venues (V), for each venue we match it with one of 10 labels according to the *CCF International conference catalogue*[3], and we also match the papers and authors with the same set of labels, where papers are labeled the same as the venue in which they are published, and authors are labeled as the most frequently appeared label of their papers. Two meta-paths, APA and APVPA, are employed for the experiments.

5.2   Baseline Algorithms

We experimented on the following network embedding methods:

- **DeepWalk** [9]: A random walk based network embedding method for homogeneous networks. Here we run DeepWalk on the whole network ignoring the different node types.
- **LINE** [10]: An edge modeling based method for homogeneous network embedding which directly learn vertex representations from vertex-vertex connections. Here we use LINE with both 1st and 2nd order of proximity as comparison, and the difference of node types is also ignored.
- **metapath2vec/metapath2vec++** [3]: A heterogeneous network embedding method which adopts metapaths and heterogeneous Skip-Gram to learn network semantic features. metapath2vec adopts full-size node set as the output of Skip-Gram model, and metapath2vec++ extended the output by distinguishing different node types.
- **HIN2VEC** [4]: A meta-path parameterized embedding method for HINs. It consists of a single-layer neural network with the meta-path types as optimization objective.
- **GAT** [11]: A supervised deep neural network model for graphs with attention mechanism. The Graph Attention Network is designed for homogeneous networks, and requires node features as input. Here we implement GAT on the whole network ignoring node types, and we design two types of node fea-

---

[2] https://www.aminer.cn/aminernetwork.
[3] https://www.ccf.org.cn/xspj/gyml/.

tures[4], denoted as GAT_{encode} and $GAT_{partial}$, respectively. The experiment is conducted based on the Deep Graph Library (DGL)[5].

- **MetaGATE**: Our proposed method. For the merge of embeddings from different meta-paths, we evaluated three types of operation, i.e. max-pooling, mean-pooling, and concatenation. The corresponding methods are abbreviated as $MetaGATE_{max}$, $MetaGATE_{mean}$, and $MetaGATE_{concat}$, respectively.

5.3   Parameter Settings

For random walk involved methods, including DeepWalk, metapath2vec(++), and HIN2VEC, the walk length is set to 48, and the random walks are repeated 50 times per node, the window size and number of negative samples per node are set to 7 and 5, respectively. The embedding size for all baselines are set to 128. For LINE, second-order proximity is used, and the starting value of learning rate is 0.025; for metapath2vec/metapath2vec++, the metapath is set as "APVPA". For GAT, we initialize the node features in two above-mentioned ways, and the other parameters are set the same as our proposed model.

For the proposed model, the embedding size of the heterogeneity embedding training module is set to 128. For the GAT module, the output dimensions of the first and second graph attention network layers are set to 64 and 10, respectively, and the number of hidden attention head is set to 8, for the output layer the number of heads is set to 1. We train the dataset for 200 epochs and use Adam to optimize MetaGATE. The learning rate is 0.005.

5.4   Node Classification

We conducted node classification experiments on DBLP and AMiner to evaluate the effectiveness of proposed model. The datasets are randomly divided into train set and test set, then we use logistic regression to predict the label of the test samples. We set the train sample percentage from 5% to 90%, and the training of each group are repeated for 10 times. The average values of Macro-F1 and Micro-F1 of the two datasets are reported in Tables 2 and 3, respectively.

The results indicate that, the proposed MetaGATE model achieves the best performance among the baselines. From Table 2 it can be observed that, for DBLP, heterogeneous network embedding methods, i.e., metapath2vec and HIN2VEC, performed slightly better than homogeneous network embedding methods, i.e., DeepWalk and LINE. For the supervised baselines, GAT_{encode} and $GAT_{partial}$ fail to outperform all the other baselines, and the difference between the two methods is also significant, these observations imply that GAT

---

[4] The first method is to randomly-initialize feature vectors with node type encoded as binaries (e.g., for node type "P", we encode [1 0 0] as the first three bits of the feature vector). The second method is to assign random values only to part of the feature vector, the rest are set to zeros (e.g., for node type "P", we only assign non-zero values to the first 32 bits of feature vector).

[5] https://github.com/dmlc/dgl.

has a strong dependency on node features. Of all methods, MetaGATE with concatenation achieves the best performance.

In Table 3, the evaluation results show that homogeneous network embedding methods such as DeepWalk and LINE fail to perform well on the AMiner dataset, while metapath2vec and HIN2VEC achieve better results than homogeneous network embedding methods, but are still inferior to MetaGATE. Both GAT_{encode} and $GAT_{partial}$ perform poorly, which can further reveal the fact that lacking of correct node features will deteriorate the performance of GAT. When comparing different merge operations for MetaGATE, it demonstrates that concatenation gives the best performance. By integrating the graph attention network model, features learned from heterogeneous Skip-Gram are further aggregated with multi-head attention mechanism, then optimal attention weights are learned, therefore generating the best performance in node classification task.

Table 2. Node classification results in DBLP.

Metric	Method	5%	10%	20%	30%	40%	50%	60%	70%	80%	90%
Macro-F1	DeepWalk	0.8939	0.9131	0.9242	0.9279	0.9308	0.9328	0.9346	0.9364	0.9377	0.9371
	LINE	0.8730	0.8921	0.9001	0.9023	0.9050	0.9036	0.9051	0.9045	0.9041	0.9061
	metapath2vec	0.9270	0.9296	0.9302	0.9302	0.9307	0.9316	0.9317	0.9327	0.9324	0.9321
	metapath2vec++	0.9249	0.9284	0.9299	0.9320	0.9323	0.9322	0.9337	0.9322	0.9330	0.9357
	HIN2VEC	0.8999	0.9206	0.9297	0.9343	0.9358	0.9373	0.9394	0.9395	0.9418	0.9409
	GAT_{encode}	0.2956	0.4236	0.5043	0.6027	0.6689	0.6339	0.7360	0.7133	0.6492	0.6429
	$GAT_{partial}$	0.6243	0.7301	0.8501	0.8307	0.8366	0.8210	0.8173	0.8446	0.7933	0.8225
	$MetaGATE_{max}$	0.9154	0.9295	0.9329	0.9403	0.9476	0.9418	0.9462	0.9464	0.9344	0.9378
	$MetaGATE_{mean}$	0.9295	0.9328	0.9315	0.9380	0.9473	0.9440	0.9511	0.9492	0.9418	0.9408
	$MetaGATE_{concat}$	**0.9307**	**0.9436**	**0.9445**	**0.9519**	**0.9560**	**0.9544**	**0.9555**	**0.9602**	**0.9457**	**0.9439**
Micro-F1	DeepWalk	0.9015	0.9191	0.9293	0.9328	0.9355	0.9373	0.9388	0.9406	0.9421	0.9416
	LINE	0.8835	0.8998	0.9068	0.9087	0.9111	0.9097	0.9111	0.9106	0.9106	0.9120
	metapath2vec	0.9317	0.9340	0.9345	0.9345	0.9349	0.9357	0.9360	0.9368	0.9363	0.9365
	metapath2vec++	0.9296	0.9330	0.9344	0.9363	0.9366	0.9366	0.9380	0.9366	0.9374	0.9404
	HIN2VEC	0.9065	0.9253	0.9339	0.9381	0.9395	0.9410	0.9429	0.9435	0.9454	0.9446
	GAT_{encode}	0.1141	0.2797	0.3667	0.5091	0.5783	0.7458	0.6293	0.6105	0.5839	0.5747
	$GAT_{partial}$	0.7382	0.8029	0.8697	0.8540	0.8560	0.8463	0.8461	0.8698	0.8186	0.8476
	$MetaGATE_{max}$	0.9229	0.9342	0.9374	0.9438	0.9502	0.9450	0.9498	0.9498	0.9356	0.9429
	$MetaGATE_{mean}$	0.9344	0.9375	0.9361	0.9418	0.9498	0.9468	0.9537	0.9534	0.9427	0.9429
	$MetaGATE_{concat}$	**0.9364**	**0.9477**	**0.9484**	**0.9549**	**0.9589**	**0.9569**	**0.9585**	**0.9630**	**0.9475**	**0.9476**

5.5 Parameter Sensitivity Analysis

To evaluate the sensitivity of different parameters including meta-path length, number of walks per node, and number of heads in multi-head attention, we conducted experiments on AMiner data with 50% of nodes as train set and 10% as validation set, and the remaining 40% as test set. The classification accuracy, F1-macro and F1-micro are evaluated.

First we evaluated the the performance of MetaGATE with varying number of meta-path length. As shown in Fig. 3(a), the performance does not have evident

Table 3. Node classification results in AMiner.

Metric	Method	5%	10%	20%	30%	40%	50%	60%	70%	80%	90%
Macro-F1	DeepWalk	0.2784	0.2915	0.3011	0.3034	0.3048	0.3067	0.3077	0.3085	0.3097	0.3095
	LINE	0.2046	0.2074	0.2107	0.2093	0.2110	0.2108	0.2109	0.2111	0.2100	0.2101
	metapath2vec	0.4161	0.4209	0.4241	0.4249	0.4259	0.4263	0.4260	0.4268	0.4261	0.4261
	metapath2vec++	0.4212	0.4231	0.4246	0.4247	0.4249	0.4248	0.4249	0.4252	0.4249	0.4259
	HIN2VEC	0.5396	0.5463	0.5501	0.5512	0.5523	0.5524	0.5529	0.5504	0.5526	0.5528
	GAT_{encode}	0.0496	0.0399	0.0395	0.0394	0.0512	0.0572	0.0811	0.1014	0.0410	0.0395
	$GAT_{partial}$	0.1674	0.1439	0.1719	0.1498	0.1580	0.1543	0.1640	0.1439	0.1464	0.1794
	$MetaGATE_{max}$	0.8872	0.8942	0.8891	0.8924	0.9036	0.8991	0.8990	0.8909	0.9034	0.8947
	$MetaGATE_{mean}$	0.9090	0.9019	0.9201	0.9140	0.9059	0.9116	0.9114	0.9154	0.9134	0.9084
	$MetaGATE_{concat}$	**0.9542**	**0.9543**	**0.9568**	**0.9565**	**0.9557**	**0.9565**	**0.9549**	**0.954**	**0.9525**	**0.9553**
Micro-F1	DeepWalk	0.4051	0.4101	0.4133	0.4142	0.4142	0.4151	0.4153	0.4155	0.4163	0.4145
	LINE	0.3540	0.3576	0.3596	0.3607	0.3609	0.3609	0.3615	0.3612	0.3605	0.3614
	metapath2vec	0.5281	0.5324	0.5349	0.5357	0.5361	0.5365	0.5365	0.5369	0.5364	0.5366
	metapath2vec++	0.5549	0.5569	0.5587	0.5591	0.5593	0.5595	0.5595	0.5597	0.5596	0.5602
	HIN2VEC	0.6343	0.6394	0.6430	0.6427	0.6443	0.6438	0.6442	0.6436	0.6440	0.6442
	GAT_{encode}	0.2563	0.2457	0.2461	0.2456	0.2586	0.2659	0.3006	0.3346	0.2446	0.2465
	$GAT_{partial}$	0.4219	0.4102	0.4437	0.4202	0.4240	0.4306	0.4345	0.4091	0.4169	0.4374
	$MetaGATE_{max}$	0.9182	0.9242	0.9210	0.9230	0.9294	0.9278	0.9270	0.9200	0.9275	0.9234
	$MetaGATE_{mean}$	0.9323	0.9273	0.9380	0.9346	0.9301	0.9349	0.9349	0.9354	0.9355	0.9325
	$MetaGATE_{concat}$	**0.9667**	**0.9663**	**0.9679**	**0.9677**	**0.9671**	**0.9682**	**0.9664**	**0.9654**	**0.9661**	**0.9677**

changes when meta-path length increases, which indicates that a relatively short
meta-path length can already provide a good result. Next we evaluated the
number of walks per node, which is depicted in Fig. 3(b). It can be observed
that the performance remained relatively stable when number of walks grows.
Finally, the number of heads in multi-head attention is assessed in Fig. 3(c). It
is clear to see that the performance first grows then remains stable with number
of heads increasing, from this result we can choose appropriate number of heads
to achieve optimum model performance.

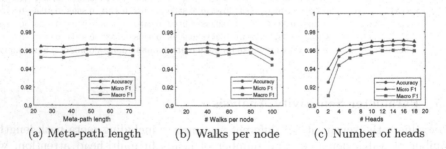

(a) Meta-path length   (b) Walks per node   (c) Number of heads

Fig. 3. Parameter sensitivity analysis.

6   Conclusion

In this paper, we proposed the MetaGATE model for HIN embedding. To tackle the heterogeneity property of HINs, a meta-path based sampling scheme is adopted and a Skip-Gram model is used for pre-training. By employing the graph attention network for message passing and aggregation in HINs, higher-level proximity and semantic information are preserved, which provides a new perspective for the study of heterogeneous network embedding. Experiments of task-specific learning on real-world dataset verified the effectiveness of the proposed model. In the future, we hope to include other side information of HINs, such as node attributes, to further discover latent knowledge in HINs.

Acknowledgments. This paper is supported by the National Key Research and Development Program of China (Grant No. 2018YFB1403400), the National Natural Science Foundation of China (Grant No. 61876080), the Collaborative Innovation Center of Novel Software Technology and Industrialization at Nanjing University.

References

1. Bengio, Y., Courville, A., Vincent, P.: Representation learning: a review and new perspectives, vol. 35, pp. 1798–1828. IEEE (2013). https://doi.org/10.1109/TPAMI.2013.50
2. Chen, J., Ma, T., Xiao, C.: FastGCN: fast learning with graph convolutional networks via importance sampling. arXiv preprint arXiv:1801.10247 (2018)
3. Dong, Y., Chawla, N.V., Swami, A.: metapath2vec: scalable representation learning for heterogeneous networks. In: Proceedings of the 23rd ACM SIGKDD International Conference on Knowledge Discovery and Data Mining, pp. 135–144. ACM (2017). https://doi.org/10.1145/3097983.3098036
4. Fu, T.Y., Lee, W.C., Lei, Z.: Hin2Vec: explore meta-paths in heterogeneous information networks for representation learning. In: Proceedings of the 2017 ACM on Conference on Information and Knowledge Management, pp. 1797–1806. ACM (2017). https://doi.org/10.1145/3132847.3132953
5. Grover, A., Leskovec, J.: node2vec: scalable feature learning for networks. In: Proceedings of the 22nd ACM SIGKDD International Conference on Knowledge Discovery and Data Mining, pp. 855–864. ACM (2016). https://doi.org/10.1145/2939672.2939754
6. Hamilton, W., Ying, Z., Leskovec, J.: Inductive representation learning on large graphs. In: Advances in Neural Information Processing Systems, pp. 1024–1034 (2017)
7. Kipf, T.N., Welling, M.: Semi-supervised classification with graph convolutional networks. arXiv preprint arXiv:1609.02907 (2016)
8. Mikolov, T., Sutskever, I., Chen, K., Corrado, G.S., Dean, J.: Distributed representations of words and phrases and their compositionality. In: Advances in Neural Information Processing Systems, pp. 3111–3119 (2013)
9. Perozzi, B., Al-Rfou, R., Skiena, S.: DeepWalk: online learning of social representations. In: Proceedings of the 20th ACM SIGKDD International Conference on Knowledge Discovery and Data Mining, pp. 701–710. ACM (2014). https://doi.org/10.1145/2623330.2623732

10. Tang, J., Qu, M., Wang, M., Zhang, M., Yan, J., Mei, Q.: Line: large-scale information network embedding. In: Proceedings of the 24th International Conference on World Wide Web, pp. 1067–1077. International World Wide Web Conferences Steering Committee (2015). https://doi.org/10.1145/2736277.2741093
11. Veličković, P., Cucurull, G., Casanova, A., Romero, A., Lio, P., Bengio, Y.: Graph attention networks. In: ICLR 2018 (2018)

Dual-FOFE-net Neural Models for Entity Linking with PageRank

Feng Wei(✉), Uyen Trang Nguyen, and Hui Jiang

Department of Electrical Engineering and Computer Science, York University,
4700 Keele St, Toronto, ON M3J 1P3, Canada
{fwei,utn,hj}@cse.yorku.ca

Abstract. This paper presents a simple and computationally efficient approach for entity linking (EL), compared with recurrent neural networks (RNNs) or convolutional neural networks (CNNs), by making use of feedforward neural networks (FFNNs) and the recent dual fixed-size ordinally forgetting encoding (dual-FOFE) method to fully encode the sentence fragment and its left/right contexts into a fixed-size representation. Furthermore, in this work, we propose to incorporate PageRank based distillation in our candidate generation module. Our neural linking models consist of three parts: a PageRank based candidate generation module, a dual-FOFE-net neural ranking model and a simple NIL entity clustering system. Experimental results have shown that our proposed neural linking models achieved higher EL accuracy than state-of-the-art models on the TAC2016 task dataset over the baseline system, without requiring any in-house data or complicated handcrafted features. Moreover, it achieves a competitive accuracy on the TAC2017 task dataset.

Keywords: Neural network · Entity linking · Knowledge base

1 Introduction

Named entities (NEs) have received much attention over the last two decades [17], mostly focused on recognizing the boundaries of textual NE mentions and classifying them as, e.g., Person (PER), Organization (ORG), Facility (FAC), Geo-political Entity (GPE) or Location (LOC). In 2009, NIST proposed the shared task challenge of EL [15]. EL is a similar but broader task than named entity disambiguation (NED). NED is concerned with disambiguating a textual NE mention where the correct entity is known to be one of the Knowledge Base (KB) entries, while EL also requires systems to deal with the case where there is no entry for the NE in the reference KB.

In [10], the authors group and summarise different approaches to EL taken by participating systems. There is a vast body of research on NED, highlighted by [9]. The problem has been studied extensively by employing a variety of machine learning, and inference methods, including a pipeline of deterministic modules

© Springer Nature Switzerland AG 2019
I. V. Tetko et al. (Eds.): ICANN 2019, LNCS 11731, pp. 635–645, 2019.
https://doi.org/10.1007/978-3-030-30493-5_58

[13], simple classifiers [22], graphical models [3], classifiers augmented with ILP inference [1], and more recently, neural approaches [14,23,24,29].

In this paper, we propose to use the recent dual-FOFE method [25] to fully encode the left/right contexts for each target mention, and then a simple FFNNs can be trained to make a precise linking for each target mention based on the fixed-size presentation of the contextual information. Moreover, we propose to incorporate PageRank based distillation in our candidate generation system. Compared with [14,24,29], our proposed neural linking models, without requiring any in-house data or complicated handcrafted features, can efficiently achieve higher EL accuracy in terms of computing than the baseline system, and achieves a competitive accuracy on both TAC2016 and TAC2017 task datasets.

The remainder of this paper is organized as follows. Section 2 describes our proposed neural linking models. In Sect. 3, we discuss experimental results and compare the performance of our proposed models with that of existing state-of-the-art systems. Finally, Sect. 4 draws the conclusions and outlines our future work.

2   Our Proposed Neural Linking Models

In this section, we discuss our proposed neural linking models, which consist of three parts: PageRank based candidate generation module, dual-FOFE-net neural ranking model and NIL entity clustering system.

2.1   PageRank Based Candidate Generation

Inspired by [19], we propose to extend the previous work in [14] to incorporate a PageRank based distillation in our candidate generation module to generate candidates for each detected mention. Candidates are generated based on KBs, including *Freebase* and *Wikipedia* [30]. Lucene fuzzy search strategy is applied in the implementation. The input to this module is a detected mention, and the output is a candidate list, which consists of a group of *Freebase* nodes potentially matching this mention, as shown in Fig. 1.

Following are five types of mention extensions implemented in the candidate generation module:

- *Substring Extension*: For each mention, all the recognized named entities in its original context document containing that mention will be selected. For instance, given the mention "Trump" in document *d*, "Donald Trump" will be selected as its substring extension if the named entity "Donald Trump" is found in *d*.
- *Translation Extension*: If a mention is in Chinese or Spanish, we invoke Google Translation to obtain its English translation as Translation Extension.
- *Country Extension*: The abbreviation of a country name can be extended to a more concrete one. For example, the mention of the geo-political entity "UK" will be extended to "United Kingdom".

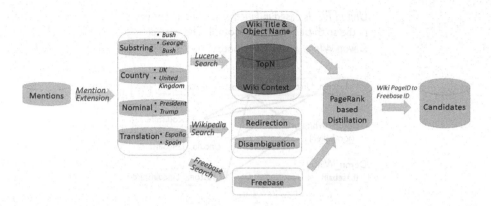

Fig. 1. The diagram of the PageRank based candidate generation module.

- *Nominal Extension*: The nearest recognized entity with the same entity type as its nominal extension will be selected to be added to the query list.
- *Traditional Chinese Extension*: If a mention is in the form of simplified Chinese, its traditional Chinese version will be obtained.

After the mention extensions, three parallel strategies are applied respectively:

- To invoke Lucene fuzzy search on *Wikipedia* titles, first paragraphs, document context and *Freebase* object names.
- To query a database with *Wikipedia* redirection and disambiguation information.
- To query a database with all *Freebase* entities.

PageRank Based Distillation. As discussed above, although in general the mention extension step helps to enhance the candidate coverage, it also products too many candidates for a single mention. This behavior leads to much noise, and slows down the whole system.

In this work, inspired by [19], we propose to incorporate PageRank based distillation at the last step. As depicted in Fig. 2, a toy document graph includes three entity mentions and seven candidates: three candidates generated for Lincolnshire, and two candidates generated for United F.C. and Devon White each. Each graph node $e(m, c)$ is a pair of an entity mention m and a candidate c. An edge is drawn between two candidates of different entities whenever there is a link from the *Wikipedia* page for one candidate to the *Wikipedia* page for the other. There is no edge between candidates competing for the same entity.

It is worth noting that edges in our graph model represent relations between candidates. We insert an edge between two candidates if the *Wikipedia* entry corresponding to either of the two candidates contains a link to the other candidate. We assume that this relation is bidirectional and thus this edge is undirected.

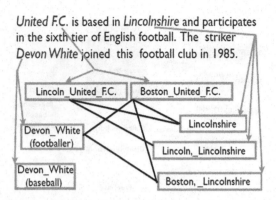

Fig. 2. A toy document graph for three entity mentions: United F.C., Lincolnshire, Devon White. Source: Adapted from Pershina et al. [21]

We rank the candidates of each mention based on their outbound link counts to all the recognized mentions in the same document, and keep the top τ candidates for each mention, where τ is a distillation factor. We name this step "distillation". The score is calculated by the *Wikipedia*'s anchors as follows:

$$score(c) = \sum_{m'} count(c, m') \tag{1}$$

where c is the candidate entity, m' is the linked page (a mention identified in the same document) and $count(c, m')$ is the total co-occurrence count of c and m'.

Finally, for each detected mention m, the candidate generation module generates a list of K candidates as $C = \{c_1, \cdots, c_K\}$, with $K \in [0, \tau]$.

2.2   Dual-FOFE-net Neural Ranking Models

Dual Fixed-Size Ordinally Forgetting Encoding (Dual-FOFE). FOFE [31] was proposed as an alternative to commonly used sequence embedding representations, and achieved competitive results in language modeling. There is a nice theoretical property to guarantee that FOFE codes can almost uniquely encode any variable-length sequence of words into a fixed-size representation without losing any information.

Given a vocabulary V, where each word can be represented by a 1-of-$|V|$ one-hot vector. Let $S = \{w_1, \cdots, w_N\}$ denote a sequence of N words from V, and e_n denote the one-hot vector of the n-th word in S, where $1 \leq n \leq N$. Assuming $z_0 = 0$, the FOFE code z_n of the sequence from word w_1 to w_n is shown as follows:

$$z_n = \alpha \cdot z_{n-1} + e_n \tag{2}$$

where α is a constant forgetting factor. Thus, z_n can be viewed as a fixed-size representation of the subsequence $\{w_1, \cdots, w_n\}$. We can see that, according to the theoretical properties presented in [31], any sequence of variable length can be uniquely and losslessly encoded into a fixed-size representation by FOFE.

This simple ordinally-forgetting mechanism has been applied to some NLP tasks, e.g., [26–28] and have achieved very competitive results.

The main idea of dual-FOFE is to generate augmented FOFE encoding codes by concatenating two FOFE codes using two different forgetting factors. Each of these FOFE codes is still computed in the same way as the mathematical formulation shown in Eq. (2). The difference between them is that we may select to use two different values for the forgetting factor for additional modeling benefits.

Our Proposed Dual-FOFE-net. As described in 2.1, we generate a candidate list C for each detected mention m. This list contains a special NIL candidate and some *Freebase* node IDs that match the mention in the candidate generation process. In this work, we propose to use a FFNNs probability ranking model to assign probabilities to all candidates in the list. The candidate with the highest probability is chosen as the final linking result. Each time, the FFNNs probability ranking model takes a mention m and a candidate c_k from the list C to compute a matching score, e_k. In order to do this, we make use of dual-FOFE to encode mention context features for the neural network.

As shown in Fig. 3, the input feature vector to the FFNNs probability ranking model is a concatenation of all the following features:

- *Mention string embedding*: Each detected mention is represented as a bag-of-words vector. This bag-of-words vector is projected into a 128-dimension dense vector.
- *Document context*: The left and right contexts of each mention are encoded by dual-FOFE, and projected into a 256-dimension dense vector.
- *Knowledge base description*: The corresponding KB, *Freebase*, description of each candidate and target mention is individually represented as one bag-of-words vectors (weighted using the TFIDF schema), which is mapped to a 128-dimension dense vector. As for Chinese and Spanish, since the languages have fewer resources than English in *Freebase*, we invoke *Google* APIs, which extract the translation to expand their Chinese and Spanish descriptions separately.

In this paper, we use the rectified linear activation function, i.e., $f(x) = \max(0, x)$, to compute from activations to outputs in each hidden layer, which are in turn fed to the next layer as inputs. For the output layer, we make use of the softmax function to compute posterior probabilities between two nodes, standing for correct links or incorrect links, shown as follows:

$$P_r(c_k|m) = \frac{exp(e_k)}{\sum_{k=1}^{K} exp(e_k)}. \tag{3}$$

2.3   NIL Entity Clustering

For all mentions identified as NIL by the above dual-FOFE-net neural ranking model, we perform a simple rule-based algorithm to cluster them: Different

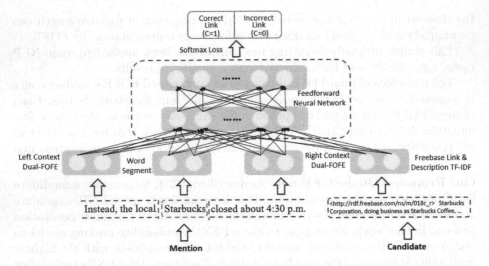

Fig. 3. Our proposed neural ranking model using dual-FOFE codes as input and a feed-forward neural network.

named NIL mentions are grouped into one cluster only if their mention strings are the same (case-insensitive).

3   Experiments and Results

In this section, we evaluate the effectiveness of our proposed methods on the benchmark datasets: TAC2016 and TAC2017 Trilingual Entity Discovery and Linking (EDL) tasks, and compare the performance of our proposed neural linking model with state-of-the-art models [14,29] on both TAC2016 and TAC2017 task datasets.

In [14], the authors use some in-house data annotated by themselves, which consists of about 10,000 Chinese and English documents acquired through their web crawler. These documents are internally labelled using some annotation rules similar to the KBP guidelines. In [29], many complicated handcrafted features are created, including the mention level feature, entity level feature, mention-to-entity feature and entity-to-entity feature. It is worth noting that we have not used any in-house data or handcrafted features in our models, and all used features (either word or character level) are automatically derived from the data based on the simple FOFE formula.

3.1   Dataset

Given a document collection in three languages (English, Chinese and Spanish), the TAC trilingual EDL task [11,12] automatically identifies entities from a source collection of textual documents in multiple languages, as shown in Table 1,

Table 1. Number of Documents in TAC2015-2017

	English	Chinese	Spanish	ALL
15 Train	168	147	129	444
15 Eval	167	166	167	500
16 Eval	168	167	168	503
17 Eval	167	167	166	500

classifies them into one of the following pre-defined five types: Person (PER), Geo-political Entity (GPE), Organization (ORG), Location (LOC) and Facility (FAC), links them to an existing KB (BaseKB)[1], and clusters mentions for those NIL entities that do not have corresponding BaseKB entries. The corpus consists of news articles and discussion forum posts published in recent years, related to but non-parallel across languages.

3.2 Neural Model Setup

Three models are trained and evaluated independently. Three sets of word embeddings of 128 dimensions are derived from English Gigaword [20], Chinese Gigaword [5] and Spanish Gigaword [16], respectively. Since Chinese segmentation is not reliable, based on a predefined set of all possible characters, we view the focus token as a character sequence and encode it using dual-FOFE. We then project the character encodings using a trainable character embedding matrix.

Relying on the development experiments, the set of hyper-parameters used in our experiments is summarized as follows: (i) Learning rate: All models are trained using the stochastic gradient descent (SGD) algorithm while the learning rate is set to be 0.1; (ii) Network structure: Three hidden layers and ReLUs [18] as the nonlinear activation function, randomly initialized based on a uniform distribution between $-\sqrt{\frac{6}{N_i+N_o}}$ and $\sqrt{\frac{6}{N_i+N_o}}$ [4]; (iii) Dropout [6] is adopted during training; (iv) Number of epochs: 30; (v) Chinese character embeddings: 64 dimensions, randomly initialized; (vi) Forgetting factor: $\alpha = (0.5, 0.9)^2$; (vii) Distillation factor: $\tau = 20$.

3.3 Evaluation Metrics

To evaluate the effectiveness of our proposed models, we use the standard NERLC and CEAFmC metrics, which are combined measures of linking and clustering performance.[3]

---

[1] http://basekb.com/.

[2] The choice of the forgetting factors α is empirical. We've evaluated on a development set in some early experiments. It turns out that $\alpha = (0.5, 0.9)$ is the best. As a result, (0.5, 0.9) is used for all EL tasks throughout this paper.

[3] More details regarding data format and scoring metric can be found in http://nlp.cs.rpi.edu/kbp/.

Table 2. Performance comparison with the best system on the TAC2016 datasets (in terms of NERLC F_1 and CEAFmC F_1).

	[14] TAC Rank 1		Our proposed models	
	NERLC	CEAFmC	NERLC	CEAFmC
Trilingual	64.7	66.0	**65.9**	**67.1**
English	66.6	67.6	**67.7**	**69.0**
Chinese	65.0	70.2	**66.4**	**70.7**
Spanish	61.6	63.5	**62.5**	**64.4**

3.4   Results and Discussion

Table 2 shows the performance of our proposed model on the TAC2016 task dataset along with the TAC Rank 1 system [14]. Our model outperforms the other system by 1.2% in terms of NERLC, and by 1.1% in terms of CEAFmC with the overall trilingual EL performance. Furthermore, each of the three individual models is better than its counterpart in the TAC Rank 1 system in terms of both NERLC and CEAFmC. As for the TAC2017 task dataset, shown in Table 3, encouragingly, the NERLC performance of English, Chinese and trilingual overall, outperforms the best system [29], by 0.4%, 0.6% and 0.2% separately, and the CEAFmC performance of English, Spanish and trilingual overall, is slightly better than the best system [29], by 0.5%, 0.4% and 0.4% separately.

Following are the advantages of our proposed models over the state-of-the-art on both TAC2016 and TAC2017 task datasets. First, unlike the systems in [14,29], our models do not rely on any in-house data or complicated handcrafted features. It is very time consuming and labour intensive to prepare clean annotated in-house data, or to collect and select good handcrafted features. More importantly, we have not used any handcrafted features in our models, and all used features (either word or character level) are automatically derived from the data based on the simple FOFE formula. Secondly, we present a simple and computationally efficient approach compared with recurrent neural networks (RNNs) or convolutional neural networks (CNNs), by making use of feedforward neural networks (FFNNs) and the recent dual fixed-size ordinally forgetting encoding (dual-FOFE) method to fully encode the sentence fragment and its left/right contexts into a fixed-size representation. Feedforward neural networks (FFNNs) use rather simple structures consisting of several fully-connected layers. These neural networks are known to be powerful as universal approximators [8], and they are simpler and faster to train and inference than the more recent variants such as long short-term memory (LSTM) [7], Gated Recurrent Unit (GRU) [2] or CNNs. Thus, our proposed dual-FOFE-net neural model is light and highly efficient compared with RNNs or CNNs. Last but not least, our proposed PageRank based distillation not only enhances the candidate coverage, but also speeds up the whole models.

Table 3. Performance comparison with the best system on the TAC2017 datasets (in terms of NERLC F_1 and CEAFmC F_1).

	[29] TAC Rank 1		Our proposed models	
	NERLC	CEAFmC	NERLC	CEAFmC
Trilingual	67.8	70.5	**68.0**	**70.9**
English	66.8	68.8	**67.2**	**69.3**
Chinese	71.0	73.2	**71.6**	72.4
Spanish	65.0	68.9	64.8	**69.3**

4 Conclusion

This paper presents a simple and computationally efficient approach to EL by applying FFNNs on top of dual-FOFE features. Furthermore, we propose to incorporate PageRank based distillation in our candidate generation module. Our experiments have shown that, without requiring any in-house data or complicated handcrafted features, it achieves higher EL accuracy than state-of-the-art systems on the TAC2016 task dataset, and offers a competitive accuracy on the TAC2017 task dataset.

In our future work, we will evaluate our neural linking models on more datasets and conduct more experiments to measure the sensitivity of the system to the values of some hyperparameters (e.g., number of hidden layers). In addition, we will explore more architectures (e.g., convolutional layers), to quantify the contribution of some modules.

References

1. Cheng, X., Roth, D.: Relational inference for Wikification. In: Proceedings of the 2013 Conference on Empirical Methods in Natural Language Processing, pp. 1787–1796 (2013)
2. Cho, K., et al.: Learning phrase representations using RNN encoder-decoder for statistical machine translation. arXiv preprint arXiv:1406.1078 (2014)
3. Durrett, G., Klein, D.: A joint model for entity analysis: coreference, typing, and linking. Trans. Assoc. Comput. Linguist. **2**, 477–490 (2014)
4. Glorot, X., Bordes, A., Bengio, Y.: Deep sparse rectifier neural networks. In: Proceedings of the Fourteenth International Conference on Artificial Intelligence and Statistics, pp. 315–323 (2011)
5. Graff, D., Chen, K.: Chinese gigaword. LDC Catalog No.: LDC2003T09 (2005). ISBN 1:58563-58230
6. Hinton, G.E., Srivastava, N., Krizhevsky, A., Sutskever, I., Salakhutdinov, R.R.: Improving neural networks by preventing co-adaptation of feature detectors. arXiv preprint arXiv:1207.0580 (2012)
7. Hochreiter, S., Schmidhuber, J.: Long short-term memory. Neural Comput. **9**(8), 1735–1780 (1997)
8. Hornik, K.: Approximation capabilities of multilayer feedforward networks. Neural Netw. **4**(2), 251–257 (1991)

9. Ji, H.: Entity discovery and linking reading list (2016). http://nlp.cs.rpi.edu/kbp/2014/elreading.html
10. Ji, H., Grishman, R.: Knowledge base population: successful approaches and challenges. In: Proceedings of the 49th Annual Meeting of the Association for Computational Linguistics: Human Language Technologies, vol. 1, pp. 1148–1158. Association for Computational Linguistics (2011)
11. Ji, H., Nothman, J., Dang, H.T., Hub, S.I.: Overview of TAC-KBP2016 tri-lingual EDL and its impact on end-to-end cold-start KBP. Proc. TAC (2016)
12. Ji, H., et al.: Overview of TAC-KBP2017 13 languages entity discovery and linking. In: Proceedings of the Tenth Text Analysis Conference (TAC2017) (2017)
13. Ling, X., Singh, S., Weld, D.S.: Design challenges for entity linking. Trans. Assoc. Comput. Linguist. **3**, 315–328 (2015)
14. Liu, D., Lin, W., Wei, S., Zhang, S., Jiang, H.: The USTC NELSLIP systems for trilingual entity detection and linking tasks at TAC KBP 2016. In: TAC (2016)
15. McNamee, P., Dang, H.T.: Overview of the TAC 2009 knowledge base population track. In: Text Analysis Conference (TAC), vol. 17, pp. 111–113 (2009)
16. Mendonca, A., Graff, D.A., DiPersio, D.: Spanish Gigaword, 2nd edn. Linguistic Data Consortium, Philadelphia (2009)
17. Nadeau, D., Sekine, S.: A survey of named entity recognition and classification. Lingvisticae Investigationes **30**(1), 3–26 (2007)
18. Nair, V., Hinton, G.E.: Rectified linear units improve restricted Boltzmann machines. In: Proceedings of the 27th International Conference on Machine Learning (ICML-2010), pp. 807–814 (2010)
19. Page, L., Brin, S., Motwani, R., Winograd, T.: The PageRank citation ranking: bringing order to the web. Technical report, Stanford InfoLab (1999)
20. Parker, R., Graff, D., Kong, J., Chen, K., Maeda, K.: English Gigaword. Linguistic Data Consortium, Philadelphia (2011)
21. Pershina, M., He, Y., Grishman, R.: Personalized page rank for named entity disambiguation. In: Proceedings of the 2015 Conference of the North American Chapter of the Association for Computational Linguistics: Human Language Technologies, pp. 238–243 (2015)
22. Ratinov, L., Roth, D., Downey, D., Anderson, M.: Local and global algorithms for disambiguation to Wikipedia. In: Proceedings of the 49th Annual Meeting of the Association for Computational Linguistics: Human Language Technologies, vol. 1, pp. 1375–1384. Association for Computational Linguistics (2011)
23. Sil, A., Dinu, G., Kundu, G., Florian, R.: The IBM systems for entity discovery and linking at TAC 2017. In: Proceedings of TAC2017 (2017)
24. Tang, S., et al.: The ZHI-EDL system for entity discovery and linking at TAC KBP 2017. In: TAC (2017)
25. Watcharawittayakul, S., Xu, M., Jiang, H.: Dual fixed-size ordinally forgetting encoding (FOFE) for competitive neural language models. In: Proceedings of the 2018 Conference on Empirical Methods in Natural Language Processing, pp. 4725–4730 (2018)
26. Xu, M., Jiang, H., Watcharawittayakul, S.: A local detection approach for named entity recognition and mention detection. In: Proceedings of the 55th Annual Meeting of the Association for Computational Linguistics (Volume 1: Long Papers), vol. 1, pp. 1237–1247 (2017)
27. Xu, M., Nosirova, N., Jiang, K., Wei, F., Jiang, H.: FOFE-based deep neural networks for entity discovery and linking. In: TAC (2017)
28. Xu, M., Wei, F., Watcharawittayakul, S., Kang, Y., Jiang, H.: The yorkNRM systems for trilingual EDL tasks at TAC KBP 2016. In: TAC (2016)

29. Yang, T., Du, D., Zhang, F.: The TAI system for trilingual entity discovery and linking track in TAC KBP 2017. In: TAC (2017)
30. Zesch, T., Müller, C., Gurevych, I.: Extracting lexical semantic knowledge from Wikipedia and wiktionary. LREC **8**, 1646–1652 (2008)
31. Zhang, S., Jiang, H., Xu, M., Hou, J., Dai, L.: The fixed-size ordinally-forgetting encoding method for neural network language models. In: Proceedings of ACL (2015)

Spatial-Temporal Graph Convolutional Networks for Sign Language Recognition

Cleison Correia de Amorim[⊠] [iD], David Macêdo[iD], and Cleber Zanchettin[iD]

Centro de Informática, Universidade Federal de Pernambuco,
Recife, PE 50740-560, Brazil
{cca5,dlm,cz}@cin.ufpe.br
http://www.cin.ufpe.br

Abstract. The recognition of sign language is a challenging task with an important role in society to facilitate the communication of deaf persons. We propose a new approach of Spatial-Temporal Graph Convolutional Network for sign language recognition based on the human skeletal movements. The method uses graphs to capture the dynamics of the signs in two dimensions, spatial and temporal, considering the complex aspects of the language. Additionally, we present a new dataset of human skeletons for sign language based on ASLLVD to contribute to future related studies.

Keywords: Sign language · Convolutional Neural Network · Spatial Temporal Graph

1 Introduction

Sign language is a visual communication skill that enables individuals with different types of hearing impairment to communicate in society. It is the language used by most deaf people in daily lives and, moreover, the symbol of identification between the members of that community and the main force that unites them [11].

According to the World Health Organization, the number of people with disabling hearing loss is about 360 million, which is equivalent to a forecast of 1 in 10 individuals around the world [9]. This data, in turn, highlights the breadth and importance of sign language in the communication of many of these people.

Despite that, there are only a small number of hearing people able to communicate through sign language. This ends up characterizing an invisible barrier that interferes with the communication between deaf and hearing persons, making a more effective integration among these people impossible [12]. In this context, it is essential to develop tools that can fill this gap by promoting the integration among the population.

Research related to the sign recognition have been developed since the 1990s, and it is possible to verify significant results [7,21]. The main challenges are primarily related to considering the dynamic aspects of the language, such as move-

© Springer Nature Switzerland AG 2019
I. V. Tetko et al. (Eds.): ICANN 2019, LNCS 11731, pp. 646–657, 2019.
https://doi.org/10.1007/978-3-030-30493-5_59

ments, articulations between body parts and non-manual expressions, rather than merely recognizing static signs or isolated hand positions.

In this work, we adopt a new approach to performing the recognition of human actions based on spatial-temporal graphs called Spatial-Temporal Graph Convolutional Networks (ST-GCN) [20]. Using graph representations of the human skeleton, it focuses on body movement and the interactions between its parts, disregarding the interference of the environment around them. Besides, it addresses the movements under the spatial and temporal dimensions, and this allows us to capture the dynamic aspects of the actions over time. These characteristics are consistent to dealing with the challenges and peculiarities of sign language recognition.

The main contributions of the paper are: (1) the proposition of a new technique to sign recognition based on human movement, which considers different aspects of its dynamics and contributes to overcoming some of the main field challenges; and (2) the creation of a new dataset of human skeletons for sign language, currently non-existent, which aims to support the development of studies in this area.

Sections 2 and 3 present the related work and the details of ST-GCN. In Sect. 4, the creation of the new database is discussed. Section 5 addresses the adjustments in ST-GCN and Sect. 6 the conduction of the experiments. Finally, Sects. 7 and 8 contain the results and final remarks, respectively.

2   Related Work

The recognition of sign languages obtained significant progress in recent years, which is motivated mainly by the advent of advanced sensors, new machine learning techniques, and more powerful hardware [21]. Besides, approaches considered intrusive and requiring the use of sensors such as gloves, accelerometers, and markers coupled to the body of the interlocutor have been gradually abandoned and replaced by new approaches using conventional cameras and computer vision techniques.

Due to this movement, it is also notable the increase in the adoption of techniques for feature extraction such as SIFT, HOG, HOF and STIP to preprocess images obtained from cameras and provide more information to machine learning algorithms [6,7].

Convolutional Neural Networks (CNN), as in many computer vision applications, obtained remarkable results in this field with accuracy reached 90% depending on the dataset [15,18]. There are still some variations as 3D CNNs, the combination with other models such as Inception or the Regions of Interest applications [3,4,16]. Recurrent Neural Networks and Temporal Residual Networks also obtained interesting results in the same purpose [5,13].

Despite the above advances, a large portion of these studies addresses static signs or single-letter images, from the dactylology[1] [3,18]. The problem is the

---

[1] Dactylology - digital or manual alphabet, generally used by the deaf to introduce a word that does not yet have an equivalent sign [11].

negative effect on the intrinsic dynamics of the language, such as its movements, non-manual expressions, and articulations between parts of the body [14]. In this sense, it is extremely relevant that new studies observe such important characteristics.

With this purpose, we present an approach based on skeletal body movement to perform sign recognition. This technique is known as Spatial-Temporal Graph Convolutional Network (ST-GCN)[2] and was introduced in [20]. The approach aims for methods capable of autonomously capturing the patterns contained in the spatial configuration of the body joints as well as their temporal dynamics.

3   Spatio-Temporal Graph Convolutional Networks

The ST-GCN uses as a base of its formulation a sequence of skeleton graphs representing the human body obtained from a series of action frames of the individuals. Figure 1a shows this structure, where each node corresponds to an articulation point. The intra-body vertices are defined based on the body's natural connections. The inter-frame vertices, in turn, connect the same joints between consecutive frames to denote their trajectory over time [20].

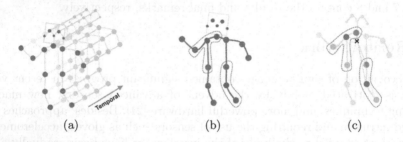

(a)   (b)   (c)

Fig. 1. (a) Sequence of skeleton graphs, denoting human movement in space and time. (b) Sampling strategy in a convolution layer for a single frame. (c) Spatial Configuration Partitioning strategy [20]. (Color figure online)

Figure 2 gives an overview of this technique. First, the estimation of individuals' skeletons in the input videos, as well as the construction of space-time graphs based on them. Then, multiple ST-GCN convolution layers are applied, gradually generating higher and higher levels of feature maps for the presented graphs. Finally, they are submitted to a classifier to identify the corresponding action.

To understand the operation of the ST-GCN, it is necessary first to introduce its sampling and partitioning strategies. When we are dealing with convolutions over 2D images, it is easy to imagine the existence of a rigid grid (or rectangle) around a central point that represents the sampling area of the convolutional

---

[2] Available at https://github.com/yysijie/st-gcn.

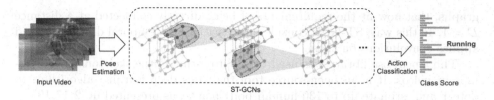

Fig. 2. Overview of the ST-GCN approach [20, p. 3].

filter, which delimits the neighborhood. In the case of graphs, however, it is necessary to look beyond this definition and consider the neighborhood of the center point as points that are directly connected by a vertex. Figure 1b represents this definition for a single frame. Note that for the red center points, the dashed edges represent the sampling area of the convolutional filter. Note also that although there are other points physically close to the central points (such as the points of the feet, knees, and waist), the method does not consider these points unless there are vertices connecting them to the red ones. This sequence of steps is the **sampling strategy** of ST-GCN.

Figure 1b shows how the convolutional filter considers only points immediately connected to the central points. In other words, the delimitation of the filter area considers the neighbors with distance $D = 1$. The authors defined this distance in ST-GCN proposition paper [20].

The **partitioning strategy** is based on the location of the joints and the characteristics of the movement of the human body, as shown in Fig. 1c. According to the authors, concentric or eccentric movements categorize the body parts, and the points in the sampling region are partitioned into three subsets:

- The root node (or center point, marked green in Fig. 1c);
- The centripetal group (blue dots in the Fig. 1c), which are the neighborhood nodes that are closest to the center of gravity of the skeleton (black cross);
- The centrifugal group (yellow dots in the Fig. 1c), which are the nodes farther from the center of gravity.

The center of gravity is taken to be the average coordinate of all joints of the skeleton in one frame. During convolution, each point of the body is labeled according to one of the above partitions named Spatial Configuration Partitioning [20]. It is through this method that the authors also establish the weights of the model, making each partition to receive a different weight to be learned.

To learn the temporal dimension, the ST-GCN extends the concept of graph convolution shown above, considering this dimension as a sequence of skeletons graphs stacked consecutively, as in Fig. 1a. With this, we have in hands a set of graphs that are neighbors to each other. Let us now assume that each articulation of the body in a graph must be connected using a vertex to itself in the graph of the previous neighbor frame and also in the next neighbor frame. Given that, if we return to the definition of sampling introduced above, we verify that the convolutional filter contemplates those points belonging to the neighboring graphs, that now fit the requirements of being directly connected at a distance $D = 1$. In this way, ST-GCN considers the spatial and temporal dimensions and applies convolutions on them.

The OpenPose library was used by [20] to estimate the skeletons of individuals in videos. It is an open-source tool that uses deep learning algorithms to detect and estimate up to 130 human body points, as presented in [2,17,19].

4   New Dataset of Human Skeletons for Sign Language

We introduce a new dataset of human skeletons for sign language based on the American Sign Language Lexicon Video Dataset (ASLLVD). The ASLLVD is a broad public dataset[3] containing video sequences of thousands of American Sign Language (ASL) signs, as well as their annotations, respective start point, and end frame markings, and class labels for each sample [1,8].

According to the authors, each sign in ASLLVD is articulated by native individuals in ASL, and video sequences are collected using a four-camera system that simultaneously captures two frontal views, one side view and one enlarged view of the face of these individuals. Figure 3 exemplifies capturing three of these views for the "MERRY-GO-ROUND" sign.

Fig. 3. Example of the sign "MERRY-GO-ROUND" from three different perspectives, in the ASLLVD [1, p. 2].

There is a total of 2,745 signs, represented in approximately 10 thousand samples. Each sign contains between 1 and 18 samples articulated by different individuals (the average number of samples per sign is 4).

Figure 4 shows a series of preprocessing steps to make the ASLLVD samples compatible with the ST-GCN model input. These steps, in turn, gave rise to a new dataset consisting of the skeletal estimates for all the signs contained therein, which was named **ASLLVD-Skeleton**[4].

The first step consists of **obtaining the videos** based on the associated metadata file used to guide this process. We consider only the videos captured by the frontal camera, once they simultaneously contemplate movements of the trunk, hands, and face of the individuals.

---

[3] Available at http://csr.bu.edu/asl/asllvd/annotate/index.html.
[4] Available at http://www.cin.ufpe.br/~cca5/asllvd-skeleton.

Fig. 4. Preprocessing steps for creating the ASLLVD-Skeleton dataset.

The next step is to **segment the videos** to generate a video sample for each sign. Once every file in the original dataset corresponds to multiple signs per individual, it is necessary to change this organization in such a way that each sign is arranged individually with its label. The output of this step consists of small videos with a few seconds.

The third step consists of **estimating the skeletons** of the individuals in segmented videos. In other words, the coordinates of the individuals' joints are estimated for all frames, composing the skeletons that can be used to generate the graphs of the ST-GCN method. As in [20], we used the OpenPose library in this process, and a total of 130 key points were estimated. Figure 5 illustrates the reconstruction in a 2D image of the estimated coordinates for the "EXAGGERATE" sign.

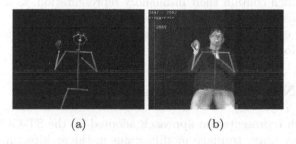

(a)   (b)

Fig. 5. Reconstruction of the skeleton from the coordinates estimated by OpenPose for the sign "EXAGGERATE" (a); and the overlapping this skeleton in the original video (b).

The fourth step involves **filtering the key points**. We use only 27 of the 130 estimated key points, which 5 refer to the shoulders and arms, and 11 refer to each hand, as illustrated in Fig. 6.

The fifth step concerns the **division of the dataset** into smaller subsets for training and test. For this procedure, a cross-validation dataset tool called *"train_test_split"*, available by the Scikit-Learn [10] library was used. In this division, we assign a proportion of 80% of the samples for training (corresponding to 7,798 samples) and 20% for tests (corresponding to 1,950 samples). This division is a commonly adopted proportion, and we understand that the number of samples in these groups is enough to validate the performance of most machine learning models.

Finally, the sixth step is to **normalize and serialize** the samples to make them compatible with the ST-GCN input. Normalization aims to make the

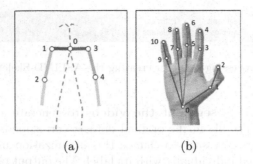

Fig. 6. Representation of the 27 used key points, which 5 refer to the shoulders and arms (a) and 11 refer to each hand (b).

length of all samples uniform by applying the repetition of their frames sequentially to the complete filling of an established fixed number of frames. The number of fixed frames adopted here is 63 (using a rate of 30 FPS, it is equivalent to a video with a duration of approximately 2 s). Serialization, in turn, consists of preloading the normalized samples and translating them into physical Python files which contain their in-memory representation, as in ST-GCN [20]. We adopted this format to optimize the data loading process.

The source code that performs these steps is also available in the ASLLVD-Skeleton download area[5].

5   ST-GCN for Sign Language Recognition

Since the graph representation approach adopted by the ST-GCN is very flexible, it is not necessary to make modifications in the architecture of the model. Instead, only specific adaptations to consider the new coordinates of the sign language domain were needed.

The first one is to modify the algorithm to make it able to use new customized layouts of graphs beyond those defined initially in [20]. Thus, a new type of layout called "custom" was defined in the configuration file of the model. Besides, a new parameter called "custom_layout" has been created within the "graph_args" attribute to allow the number of nodes, the central node, and the edges of new graphs to be informed.

With this configuration, it is possible to map the topology of the graph used in this paper, which is composed of the 27 joints and vertices shown in Fig. 6. Finally, small adjustments were required in the dimensions of the matrices used by the data feeders and in the file that defines the graph domain of the model so that they can now support layouts with dynamic dimensions.

The source code containing these adaptations is public available (see footnote 5). It consists of a *forked* repository created from the one developed by the authors in [20].

---

[5] Available at http://www.cin.ufpe.br/~cca5/st-gcn-sl.

6   Experiments

We used as reference the experiments proposed in [7], which evaluate the performance of models as the Block-Based Histogram of Optical Flow (BHOF) (see Sect. 2) and popular techniques such as Motion Energy Image (MEI), Motion History Image (MHI), Principal Component Analysis (PCA), and Histogram of Optical Flow (HOF) in the ASLLVD dataset.

The authors used a subset containing 20 signs selected from the ASLLVD, as presented in Table 1. To reproduce this configuration, we selected the estimated skeletons for these signs from the ASLLVD-Skeleton dataset.

Table 1. Selected signs for the experiments in [7].

Dataset	Selected signs
ASLLVD	adopt, again, all, awkward, baseball, behavior, can, chat, cheap, cheat, church, coat, conflict, court, deposit, depressed, doctor, don't want, dress, enough

We also identified that among the selected signs, there were different articulations for the signs AGAIN, BASEBALL, CAN, CHAT, CHEAP, CHEAT, CONFLICT, DEPRESS, DOCTOR, and DRESS. To solve this, we chose to keep only the articulation that contained the most significant number of samples for their signs.

Since the number of samples resulting from this process was small, totaling 131 items, it was necessary to apply a new division making 77% of this subset for training and 33% for tests. With this strategy, we improved the balance in the number of samples for an adequate evaluation of the model. Finally, as we have a new sample division, it was also necessary to normalize and serialize the samples to be compatible with the ST-GCN. The resulting subset was named **ASLLVD-Skeleton-20**[6].

Due to the characteristics of this small dataset and based on preliminary experiments, the size of the used batch is of 8 samples. In the same way, as in the original implementation of the ST-GCN training algorithm, we used as optimizer the Stochastic Gradient Descent (SGD) with Nesterov Momentum.

For the learning rate, a decay strategy was adopted, which consists of initializing it with a higher value and gradually reducing it in the later epochs of the learning process to allow for more and more refined adjustments of the weights, as in [20]. Thus, in the experiment with the 20 selected signs, the total number of epochs was 200, adopting an initial rate of 0.01, which was decreased to the values of 0.001, 0.0001 and 0.00001 after the end of the epochs 50, 100 and 150, respectively.

Besides, an experiment with the complete ASLLVD dataset was also conducted to establish a reference value. In this experiment, we set the batch size to

---

[6] Available at http://www.cin.ufpe.br/~cca5/asllvd-skeleton-20.

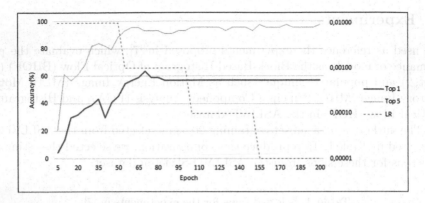

Fig. 7. Accuracy obtained by the presented approach in recognition of the 20 signs selected from the ASLLVD. (Color figure online)

24. The learning rate had similar behavior to the experiment above. Next section presents the obtained results in both scenarios, and the pre-trained models are available along with the adaptations made in the ST-GCN, according to Sect. 5.

7   Results

Figure 7 presents the first experiment using the approach presented in [7], which considers the selection of 20 specific signs of the ASLLVD. The red line presents the accuracy of the model (*top-1*) and its evolution throughout training epochs. The gray line represents the *top-5* accuracy, which corresponds to the accuracy based on the five most likely responses presented by the model. Finally, the blue dashed line represents the evolution of the learning rate used in the respective epochs and its decay behavior.

From the image, we can observe that the model was able to achieve an accuracy of 56.82% from the epoch 80 in sign recognition. The *top-5* accuracy, in turn, was able to reach 95.45%. This performance was superior to the results presented by traditional techniques such as MEI, MHI, and PCA, but was not able to overcome that obtained by the HOF and BHOF techniques [7]. Table 2 presents the comparison of these results.

We performed a second experiment using 2,745 signs to establish a reference to the complete ASLLVD dataset. In this scenario, an accuracy (*top-1*) of 20.85% and a *top-5* accuracy of approximately 40.15% was obtained. Of course, this is a much more challenging task than the one proposed in [7], and these results reflect this complexity.

From the table, we can see that the approach presented in this paper, based on graphs of the coordinates of human articulations, has not yet been able to provide such remarkable results as that based on the description of the individual movement of the hands through histograms adopted by BHOF. Indeed, the application of consecutive steps for optical flow extraction, color map creation,

Table 2. Sign recognition accuracy using different approaches as proposed in [7].

	Accuracy (%)
MHI	10.00
MEI	25.00
PCA	45.00
ST-GCN SL	**56.82**
HOF	70.00
BHOF	85.00

block segmentation and generation of histograms from them were able to ensure that more enhanced features about the hand movements were extracted favoring its sign recognition performance. This technique is derived from HOF and differs only by the approach of focusing on the hands of individuals while calculating the optical flow histogram.

Methods such as MEI and MHI, however, present more primitive approaches, which mainly detect the movements and their intensity from the difference between the consecutive frames of actions. They are not able to differentiate individuals or to focus on specific parts of their body, causing movements of any nature considered equivalently. The PCA, in turn, adds the ability to reduce the dimensionality of the components based on the identification of those with more significant variance and that, consequently, is more relevant for movement detection in the frames.

The ASLLVD-Skeleton database, in turn, presented high relevance and was able to comply with its purpose of making feasible and supporting the development of this work. Through its approach and format, it allowed the benefits of adopting a robust and assertive technique to read human skeleton without restrictions or performance penalty during experiments. Considering that such techniques usually have a high computational cost, especially when combined with complex deep learning models, this is a factor that commonly restricts or impedes the progress of researches that seeks to evolve in this direction, based on the coordinates of the body. Thus, this paper contributes and extends to the academic community the benefits found here, encouraging and enabling other advances in sign language recognition.

8   Final Remarks

The results obtained with the presented approach did not reach such expressive performance as those obtained by some of the related techniques. However, its contributions are significant in guiding the next steps to be taken by future studies in this field.

Based on our observations, it is relevant to seek approaches capable of enriching the information about the movements of the estimated coordinates, especially those of the hands and fingers, which, although very subtle, play a central role in the articulation and meaning of the signs.

This assumption may be related, for example, to the definition of a new partitioning strategy that would allow more emphasis on the subtle traces of hands and fingers to the detriment of the other parts of the body. Also, the definition of specific weights for these parts would enable the model to learn more about its dynamics; today, the model does not distinguish the type of learned joint. Finally, including the depth information of coordinates may provide to the model more details about the trajectory of these parts, enabling the observation through the three-dimensional plane.

References

1. Athitsos, V., et al.: The American sign language lexicon video dataset. In: 2008 IEEE Computer Society Conference on Computer Vision and Pattern Recognition Workshops, pp. 1–8, June 2008. https://doi.org/10.1109/CVPRW.2008.4563181
2. Cao, Z., Simon, T., Wei, S.E., Sheikh, Y.: Realtime multi-person 2D pose estimation using part affinity fields. In: CVPR (2017). https://doi.org/10.1109/CVPR.2017.143
3. Das, A., Gawde, S., Suratwala, K., Kalbande, D.: Sign language recognition using deep learning on custom processed static gesture images. In: 2018 International Conference on Smart City and Emerging Technology (ICSCET), pp. 1–6, January 2018. https://doi.org/10.1109/ICSCET.2018.8537248
4. ElBadawy, M., Elons, A.S., Shedeed, H.A., Tolba, M.F.: Arabic sign language recognition with 3D convolutional neural networks. In: 2017 Eighth International Conference on Intelligent Computing and Information Systems (ICICIS), pp. 66–71, December 2017. https://doi.org/10.1109/INTELCIS.2017.8260028
5. Konstantinidis, D., Dimitropoulos, K., Daras, P.: Sign language recognition based on hand and body skeletal data. In: 2018–3DTV-Conference: The True Vision - Capture, Transmission and Display of 3D Video (3DTV-CON), pp. 1–4, June 2018. https://doi.org/10.1109/3DTV.2018.8478467
6. Laptev, I., Marszalek, M., Schmid, C., Rozenfeld, B.: Learning realistic human actions from movies. In: 2008 IEEE Conference on Computer Vision and Pattern Recognition, pp. 1–8, June 2008. https://doi.org/10.1109/CVPR.2008.4587756
7. Lim, K.M., Tan, A.W., Tan, S.C.: Block-based histogram of optical flow for isolated sign language recognition. J. Vis. Commun. Image Represent. **40**, 538–545 (2016). https://doi.org/10.1016/j.jvcir.2016.07.020
8. Neidle, C., Thangali, A., Sclaroff, S.: Challenges in development of the American sign language Lexicon video dataset (ASLLVD) corpus. In: 5th Workshop on the Representation and Processing of Sign Languages: Interactions between Corpus and Lexicon, LREC 2012, Istanbul, Turkey, May 2012. https://open.bu.edu/handle/2144/31899
9. World Health Organization: Deafness and hearing loss, March 2018. https://www.who.int/news-room/fact-sheets/detail/deafness-and-hearing-loss. Online
10. Pedregosa, F., et al.: Scikit-learn: machine learning in Python. J. Mach. Learn. Res. **12**, 2825–2830 (2011). http://www.ee.columbia.edu/~ronw/pubs/jmlr2011-scikit-learn.pdf
11. Pereira, M.C.d.C., Choi, D., Vieira, M.I., Gaspar, P., Nakasato, R.: Libras - Conhecimento Além Dos Sinais, 1st edn. Pearson, São Paulo (2011)

12. Peres, S.M., Flores, F.C., Veronez, D., Olguin, C.J.M.: Libras signals recognition: a study with learning vector quantization and bit signature. In: 2006 Ninth Brazilian Symposium on Neural Networks (SBRN 2006), pp. 119–124, October 2006. https://doi.org/10.1109/SBRN.2006.26

13. Pigou, L., Herreweghe, M.V., Dambre, J.: Gesture and sign language recognition with temporal residual networks. In: 2017 IEEE International Conference on Computer Vision Workshops (ICCVW), pp. 3086–3093, October 2017. https://doi.org/10.1109/ICCVW.2017.365

14. de Quadros, R.M., Karnopp, L.B.: Língua de sinais brasileira: estudos linguísticos, vol. 1. Artmed, Porto Alegre (2004)

15. Rao, G.A., Syamala, K., Kishore, P.V.V., Sastry, A.S.C.S.: Deep convolutional neural networks for sign language recognition. In: 2018 Conference on Signal Processing And Communication Engineering Systems (SPACES), pp. 194–197, January 2018. https://doi.org/10.1109/SPACES.2018.8316344

16. Sajanraj, T.D., Beena, M.: Indian sign language numeral recognition using region of interest convolutional neural network. In: 2018 Second International Conference on Inventive Communication and Computational Technologies (ICICCT), pp. 636–640, April 2018. https://doi.org/10.1109/ICICCT.2018.8473141

17. Simon, T., Joo, H., Matthews, I.A., Sheikh, Y.: Hand keypoint detection in single images using multiview bootstrapping (2017). http://arxiv.org/abs/1704.07809

18. Taskiran, M., Killioglu, M., Kahraman, N.: A real-time system for recognition of American sign language by using deep learning. In: 2018 41st International Conference on Telecommunications and Signal Processing (TSP), pp. 1–5 (2018). https://doi.org/10.1109/TSP.2018.8441304

19. Wei, S., Ramakrishna, V., Kanade, T., Sheikh, Y.: Convolutional pose machines. In: 2016 IEEE Conference on Computer Vision and Pattern Recognition (CVPR), pp. 4724–4732, June 2016. https://doi.org/10.1109/CVPR.2016.511

20. Yan, S., Xiong, Y., Lin, D.: Spatial temporal graph convolutional networks for skeleton-based action recognition. CoRR abs/1801.07455 (2018). http://arxiv.org/abs/abs/1801.07455

21. Zheng, L., Liang, B., Jiang, A.: Recent advances of deep learning for sign language recognition. In: 2017 International Conference on Digital Image Computing: Techniques and Applications (DICTA), pp. 1–7, November 2017. https://doi.org/10.1109/DICTA.2017.8227483

Graph Convolutional Networks Improve the Prediction of Cancer Driver Genes

Roman Schulte-Sasse[1]([⊠]) [iD], Stefan Budach[1], Denes Hnisz[1] [iD],
and Annalisa Marsico[1,2]([⊠]) [iD]

[1] Max Planck Institute for Molecular Genetics,
Ihnestrasse 63-73, 14195 Berlin, Germany
{sasse,marsico}@molgen.mpg.de
[2] Institute for Computational Biology, Helmholtz Zentrum Munich,
Ingolstädter Landstr. 1, 85764 Neuherberg, Germany

Abstract. Despite the vast increase of high-throughput molecular data, the prediction of important disease genes and the underlying molecular mechanisms of multi-factorial diseases remains a challenging task. In this work we use a powerful deep learning classifier, based on Graph Convolutional Networks (GCNs) to tackle the task of cancer gene prediction across different cancer types. Compared to previous cancer gene prediction methods, our GCN-based model is able to combine several heterogeneous omics data types with a graph representation of the data into a single predictive model and learn abstract features from both data types. The graph formalizes relations between genes which work together in regulatory cellular pathways. GCNs outperform other state-of-the-art methods, such as network propagation algorithms and graph attention networks in the prediction of cancer genes. Furthermore, they demonstrate that including the interaction network topology greatly helps to characterize novel cancer genes, as well as entire disease modules. In this work, we go one step forward and enable the interpretation of our deep learning model to answer the following question: what is the molecular cause underlying the prediction of a disease genes and are there differences across samples?

Keywords: Graph Convolutional Networks (GCNs) ·
Cancer gene prediction · Multi-omics ·
Layer-wise Relevance Propagation (LRP)

1 Introduction

Sequencing studies on a large scale from consortia such as ENCODE, GTEx and the Cancer genome Atlas (TCGA) have provided the scientific community with an ever-increasing amount of genomic, transcriptomic and epigenetic data sets for over 10,000 cancer patients. This has led to the development of bioinformatics tools for the prediction of cancer genes from genomic data [1] and to annotated sets of driver cancer genes through initiatives such as the Network of

© Springer Nature Switzerland AG 2019
I. V. Tetko et al. (Eds.): ICANN 2019, LNCS 11731, pp. 658–668, 2019.
https://doi.org/10.1007/978-3-030-30493-5_60

Cancer Genes (NCG) [2] and the COSMIC cancer gene census (CGC) [3]. But despite this vast increase of data, accurate prediction of cancer driving genes remains a challenging task. This is mostly due to the complexity of the disease, where heterogeneity, long recognized as an important clinical determinant of patient outcomes, is still poorly understood at a molecular level. Heterogeneity in cancer is not limited to differences between patients, but occurs also within a tumor. The highly heterogeneous tumor micro-environment partly comes from the organization of genes in regulatory pathways where the disruption of two different genes might lead to the same phenotype. Therefore, including the topology of interactions between genes greatly helps to characterize novel disease genes or modules. Many network models have been developed to make use of known gene-gene interactions for prediction, relying on the assumption that interacting genes tend to produce similar phenotypes [4]. Such methods most often use random walks to propagate scores from known disease genes or highly mutated ones through the network [4]. Among recent network methods for cancer gene prediction, HotNet2 and hierarchical HotNet [5,6] identify highly mutated subnetworks across 12 and 33 cancer types. By using a directed heat diffusion model they simultaneously take into account cancer mutations in individual genes and the local topology of protein-protein interactions.

Cancer has traditionally been viewed as a set of diseases driven by the accumulation of genetic mutations. Classical approaches to predict cancer genes rely on the identification of significantly hyper-mutated genes compared to background, i.e. genes with a cancer mutation frequency significantly higher than the average background [1]. Recent genomic technologies have shown, however, that alteration of the epi-genome - mainly disruption of DNA methylation patterns at oncogenes or tumor suppressor genes - contributes to cancer biogenesis and progression, and offers new possibilities for therapy. Therefore, integrating different layers of genomic, epigenetic and transcriptomic data is crucial for the accurate prediction of cancer driver genes [7].

Current advances in machine learning, mainly in deep learning algorithms, have led to unprecedented results in molecular biology [7–9] due to the vast increase in genomic data set sizes and the development of scalable and interpretable models. And while they have been notoriously difficult to interpret in the past, advances in feature interpretation strategies for deep neural networks make it possible to investigate why decisions were made, leveraging deep understanding of the underlying data [10]. Recently, graph deep learning tools have emerged to incorporate graph structures into a deep learning setting [11–13]. In particular, deepWalk [14] is regarded as one of the first unsupervised graph embedding methods based on latent representation learning . It was used to learn social representations of a graph's vertices by treating short random walks on the graph as equivalent to sentences. Graph convolutional networks (GCNs) [15] follow a different approach by directly trying to classify the nodes of a network based on node features and network architecture. They aggregate information from neighboring nodes in a hierarchical fashion, similar to convolutional neural networks (CNNs) and can work in a semi-supervised setting where labeled nodes are scarce. Likewise graph attention networks (GATs) are a recent supervised node-classification method where the graph convolutions are replaced by graph attention layers, thus offering high performance and interpretability out of the box.

In this work we set out to classify and predict novel pan-cancer genes using multiple sources of omics data and gene-gene interaction networks with a GCN classifier. Compared to network propagation and graph embedding algorithms, our GCN can handle high-dimensional node feature vectors, therefore exploiting the predictive power coming from different complementary data types and the underlying network simultaneously. In addition, their classification results can be interpreted with LRP (or similar methods) by mapping the network output back to the input space. Our method outperforms classical feature-based machine learning approaches, network-based approaches such as PageRank and HotNet2 diffusion processes and graph attention networks for cancer gene prediction. It is comparable to the deep representation learning method DeepWalk, whose results are, however, hard to interpret due to the latent feature embedding and therefore less useful for the task of understanding individual contributions of molecular features to cancer gene predictions.

To dissect the mechanisms underlying cancer, we employ layer-wise relevance propagation (LRP) [10] to identify the important components of both the gene input signal and network topology of the learned model in the neighborhood of a gene. Although GCNs have been successfully applied to other domains, to our knowledge this is the first time that such an approach was adapted to predict novel cancer driver genes from several omics data sets and network properties, while also being capable of disentangling the underlying molecular mechanisms driving the etiology of cancer.

2   Methods

2.1   Data Collection and Preprocessing

We collected mutations, DNA methylation and gene expression data of more than 8000 samples from the TCGA, covering 16 different cancer types. We limited our analysis to those cancer types for which DNA methylation information in tumor and normal tissue was available and for which already preprocessed gene expression data existed [16].

Gene Mutation Rates. We processed Mutation Annotation Format (MAF) files from TCGA, adapting the preprocessing pipeline from HotNet2 [5]. The cancer mutation rate for each gene in each cancer type was defined as the sum of both single nucleotide variations (SNVs) and copy number aberrations (CNAs) averaged across all samples of that cancer type.

Gene DNA Methylation. We collected DNA methylation data from 450k Illumina bead arrays deposited in TCGA for both tumor and adjacent normal tissue. For each gene i we define a measure of differential DNA methylation at its promoter in cancer type c, dm_i^c, as the difference in methylation signal between cancer

β_i^t, and matched normal sample β_i^n, averaged across all samples S_c available for cancer type c:

$$dm_i^c = \frac{1}{|S_c|} \sum_{s \in S_c} \left(\beta_i^t - \beta_i^n \right) \tag{1}$$

Gene Expression. To measure the expression level of each gene in each sample we use the data set from Wang et al. [16], where RNA-seq data of both tumor and control samples from TCGA, along with expression data from the GTEx consortium have been already quantile-normalized and batch-corrected using ComBat [17]. For each gene differential expression was computed as log2 fold change between expression in cancer versus a matched normal sample and then averaged across samples.

Each gene is represented by a 16×3-dimensional vector, containing 16 mutation rates, 16 methylation values and 16 differential gene expression values for each of the 16 analyzed cancer types. Values from different omics data on different scales are subjected to min-max normalization before being concatenated and fed as input to the GCN model (see Fig. 1). We collected protein-protein interactions from Consensus Path DB (CPDB) [18] and discarded interactions scored below 0.5.

Collection of Positive and Negative Labels. We obtain known cancer genes from the network of cancer genes (NCG) [2]. The expert-curated list from NCG is largely overlapping with the COSMIC cancer gene census (CGC) [3] and offers a list of genes known to play a significant role in multiple cancers. To derive a list of negatives, genes most likely not being associated with cancer, we started from the set of all genes and recursively removed genes part of the NCG, COSMIC or OMIM databases, as well genes part of KEGG cancer pathways. At the end our final set comprises 654 positives and 1424 negatives.

2.2 Graph Convolutional Networks

Graph convolutional networks (GCNs) [15] extend convolutional neural network (CNN) frameworks to non-euclidean data. As opposed to images that are organized as a regular grid, nodes can have variable number of neighbors and local topologies that are important for the classification result. Similarly to CNNs, GCNs scan a filter over a signal $x \in \mathrm{R}^N$ (N being the number of nodes in the graph), trying to recognize patterns in a local neighborhood of a node.

In spectral graph theory, we can define a graph convolution by decomposing the graph signal in its spectral domain and then applying a filter on the components of the signal x [19]. However, such a decomposition requires computation of the eigenvectors of the graph laplacian L which is defined as $L = \widetilde{D}^{-\frac{1}{2}} A \widetilde{D}^{-\frac{1}{2}}$ where A is the adjacency matrix and D the degree matrix of the graph. Unfortunately, computing eigenvectors of such large matrices is often unfeasible [12].

By simplifying the notation of spectral graph convolution, GCNs are able to average neighboring information around a node by multiplication of the adja-

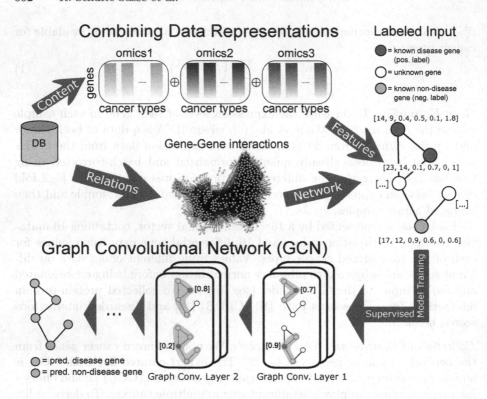

Fig. 1. Overview of our GCN model. Different omics data matrices are collected and concatenated to one large feature matrix. Together with a gene-gene interaction network and few gene labels we obtain a partially labeled graph that is fed to a GCN classifier. At each layer, node and graph features are aggregated and transformed to form input for the next layer. The GCN outputs is a fully labeled graph where each gene is assigned a probability to be a disease gene.

cency matrix with its feature matrix $X \in \mathrm{R}^{N \times F}$. We can define a simple propagation rule for each layer in a graph convolutional network:

$$H^{(l+1)} = \sigma\left(LH^{(l)}\Theta^{(l)}\right) \qquad (2)$$

where the adjacency matrix $\widetilde{A} = A + I$ has added self connections and σ denotes a non-linearity, such as the ReLU activation function. The first layer receives X as input, so $H^{(0)} = X$.

The multiplication of L and X can be regarded as a 1-step random walk, smoothing the features across the local neighborhood of every node. Due to the added self connections, the original node signal is also preserved and incorporated into the smoothing. These smoothed values are then transformed by the weight matrix Θ and the non-linearity σ, similar to a standard neural network definition.

Interestingly, stacking of multiple layers of graph convolutions increases the flow of information and therefore the degree of smoothing for the features, alleviating the need for pooling operations [20].

2.3   Model Training

We randomly split the labeled data (see Sect. 3.1) into training (75%) and test (25%) sets, stratifying them such that the ratio of positives to negatives is the same in both sets. The input to the algorithm consists of a network represented by its normalized adjacency matrix L, a feature matrix X, containing one row per node and labels y for some of the nodes.

To train the model, we compute the cross-entropy loss $\mathcal{L}$ for our training nodes as:

$$\mathcal{L} = -\big(y\log(p) + (1-y)\log(1-p)\big) \tag{3}$$

where p is the output of the network after softmax activation and y the original node label (0 or 1). Since we are dealing with class imbalance (more negatives than positives), we scale the loss for positives by a factor optimized during model training. We use Tensorflow for computation of the gradients, and the ADAM optimizer for training for a fixed number of epochs. To select the best hyperparameters for the model, we run a grid search by using internal 10-fold cross validation (CV). The best combination of hyper-parameters resulted in a dropout rate of 0.5, a learning rate of 0.001, multiplication of the loss for positives by 60, a weight decay of 0.05 and two graph convolutional layers with 50 and 100 filters, respectively.

2.4   Interpretation of GCN Predictions

To fully take advantage of the data integration combining different omics data and the PPI network, we also examine why a gene of interest was labeled as a cancer gene or not. To this end we use layer-wise relevance propagation (LRP), a general interpretation method for non-linear classification architectures [10]. The DeepExplain python package[1] is used to apply LRP to the GCN after the model training. Conceptually, the LRP algorithm performs a backward pass through a network to redistribute the network output $f(x)$ (also called *relevance* in this context) over the input by computing

$$R_i^{(l)} = \sum_j \frac{a_i w_{ij}}{\sum_i a_i w_{ij}} R_j^{(l+1)} \tag{4}$$

where R represents the relevance of node i and j in layer l and $l+1$, respectively, and where a is the output of node i and w_{ij} the weight connecting node i and j. For all input features of a gene of interest we thereby obtain relevance values that illustrate the importance of individual features, such as PPI network neighbors and omics type, for the GCN classification. Relevance values can be both positive or negative, indicating whether the presence of a feature has a positive or negative impact on the classification result.

---

[1] https://github.com/marcoancona/DeepExplain.

3 Results and Discussion

3.1 Ability of the GCN Model to Predict Cancer Genes

We applied our GCN model to predict pan-cancer genes from mutations, DNA methylation and gene expression data (Sect. 2.1) and compared it to state-of-the-art methods on our held-out test set. In detail, we compared GCNs to two network-only and one feature-only baseline, namely PageRank [21] and Deep-Walk [14] for the network and a random forest for the features. This helps us to investigate how informative network relationships and node features are for the classification task. For DeepWalk, we used the learned embeddings as input to a non-linear SVM classifier with radial basis function for node classification. To enable a fair comparison, we conducted a grid search over possible hyper-parameters on our validation set, in order to find an optimal combination of the length and number of random walks, size of embedding and context window used to learn the embedding.

We further included in the comparison analysis two cancer-specific baselines, which represent the state-of-the-art prediction methods in the cancer bioinformatics community: HotNet2 [5] and MutSigCV [1]. These two methods are based on different principles. While HotNet2 uses a diffusion process to smooth mutation information across the network and relies on the assumption of label homophily (neighboring nodes are likely to share the same mutation score), MutSigCV prioritizes highly mutated genes, accounting for biological biases such as sequence composition and gene length. For HotNet2 we used the steady-state distribution for performance evaluation.

Finally, we also compared the performance of our method to graph attention models (GAT) [22]. Those are a more recent improvement of GCNs on citation networks and can profit from the features and the network as well. We tried the method with 4 and 8 attention heads only, as suggested in the original paper. The GCN outperforms all of the other methods in terms of area under the precision-recall curve (AUPR) and area under the ROC-curve (AUROC) apart from DeepWalk, whose performance is similar or slightly lower than the GCN performance.

The high performance of the network-only methods PageRank and DeepWalk most probably comes from a large study bias in the field of cancer genomics where genes that have been reported to be involved with cancer are over-represented in protein-protein interaction studies and databases [23]. Hence, we observe that known cancer genes have a more than 5-fold higher median node degree than non-cancer genes in our PPI network. Furthermore, the genes most prominently involved in cancer are lying in central pathways, such as those responsible for DNA repair or cell growth [23]. This leads us to believe that the extreme difference in node degree between well-known cancer genes and non-cancer genes leads to a highly informative network structure and ultimately to the high performance of network-based methods.

To investigate this issue further, we tested how sensitive DeepWalk and our GCN model are on an independent set of candidate cancer genes with the fol-

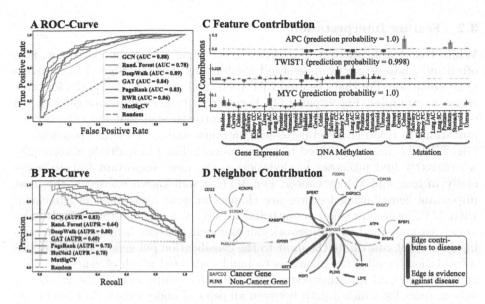

Fig. 2. Results of GCN on TCGA data. A & B Performance comparison. **C** LRP feature importance for three genes: *APC*, *TWIST1* and *MYC*. **D** Strongly connected component containing the *SAPCD2* cancer gene and relative LRP importance of its neighborhood for classification. Red nodes correspond to known cancer genes. Edge thickness denotes the strength of the LRP contribution and the arrow's color indicates whether an edge is evidence for (red) or against (blue) the classification of the gene as cancer gene. (Color figure online)

lowing characteristics: first, this set comes from a manually curated collection of genes for which associations to cancer have been reported while a definitive proof is still lacking; second, this set contains genes which have been less studied according to the scientific literature and therefore not over-represented in PPI databases [2]. While the GCN finds 53% of the candidate cancer genes, Deep-Walk only recovers 40% of them at a similar number of predictions. This clearly points to the fact that GCNs, compared to network-only methods, represent a more sensitive approach to predict general properties of cancer genes, beyond PPI network hubs.

Overall, our method is powerful in recovering high confidence cancer genes and less sensitive - while still being better than network-only methods - with genes where the cancer association is more uncertain, as expected. False negatives mainly correspond to known genes associated with blood cancers. For those cancer types, we lack complete data sets in the TCGA database and therefore did not include them into our analysis. Among the top model predictions we confidently recover well-known cancer genes such as *KRAS*, *TP53*, *MYC*, *BRCA1* and *BRCA2*.

3.2   Feature Interpretation

While the identification of new driver genes is an important ingredient, it is often more interesting to understand the features that contributed to a certain gene prediction. For that, we used LRP (Sect. 2.4) and show the results for three genes as examples (Fig. 2, panel C). For *APC*, a tumor suppressor gene known to be mutated in colorectal cancer [24], we correctly find that the relevance for mutations in colon and rectal tissues contribute the most to the classification. For *TWIST1*, a gene reported to be hyper-methylated in multiple cancers [25], we correctly find promoter methylation as the most important feature, especially in lung and kidney cancer. For *MYC*, a well-known oncogene, the most important contributing features are changes in gene expression, in agreement with *MYC* over-expression in epithelial tissues, pancreas and lung [26]. Besides omics features, the LRP framework can also identify those neighboring interacting genes that contribute the most to the classification per gene. Combining this information for all genes allows us to build a directed graph of gene-gene LRP contributions and investigate strongly connected components of the graph, i.e. sub-modules for which a path between all pairs of nodes exists. As example we show the module containing the *SAPCD2* cancer gene together with its important neighbors. The component is enriched for other cancer genes, highlighting the importance of interactions with other cancer genes inside the same pathway for classification.

4   Outlook and Conclusions

Our results show that graph convolutional networks are highly suitable to handle multi-dimensional omics data sets and gene-gene interaction networks for the purpose of cancer gene prediction, and provide an elegant integration of graph data into a conventional deep learning framework. Combining them with feature interpretation techniques such as LRP, we can gain meaningful mechanistic insights into how a gene might contribute to specific cancer types. Due to the averaging of features across patients for a given cancer type, the GCN is blind to distinct sub-populations within a cancer tissue. We are currently adapting our method to work directly at sample/patient level to perform patient stratification based on the learned classification features. In addition, a potential disadvantage of the current setting is the rather general notion of cancer genes. While there are some genes with a strong influence across several cancer types, a multi-class, multi-label scenario would be even more suitable and allow the prediction of those cancer genes which are highly specific to a certain cancer type, paving the way to precision oncology.

Although in this work we focus on cancer gene prediction from different patients and cancer types, the method can be easily generalized to predict disease genes in potentially any multi-factorial disease where multi-omics data are available.

References

1. Lawrence, M.S., et al.: Mutational heterogeneity in cancer and the search for new cancer-associated genes. Nature **499**(7457), 214–218 (2013). https://doi.org/10.1038/nature12213
2. Repana, D., et al.: The network of cancer genes (NCG): a comprehensive catalogue of known and candidate cancer genes from cancer sequencing screens. Genome Biol. **20**(1), 1–12 (2019). https://doi.org/10.1186/s13059-018-1612-0
3. Sondka, Z., et al.: The COSMIC cancer gene census: describing genetic dysfunction across all human cancers. Nat. Rev. Cancer **18**(11), 696–705 (2018). https://doi.org/10.1038/s41568-018-0060-1
4. Cowen, L., et al.: Network propagation: a universal amplifier of genetic associations. Nat. Rev. Genet. **18**(9), 551–562 (2017). https://doi.org/10.1038/nrg.2017.38
5. Leiserson, M.D., et al.: Pan-cancer network analysis identifies combinations of rare somatic mutations across pathways and protein complexes. Nat. Genet. **47**(2), 106–114 (2015). https://doi.org/10.1038/ng.3168
6. Reyna, M.A., Leiserson, M.D., Raphael, B.J.: Hierarchical HotNet: identifying hierarchies of altered subnetworks. Bioinformatics **34**(17), i972–i980 (2018). https://doi.org/10.1093/bioinformatics/bty613
7. Camacho, D.M., et al.: Next-generation machine learning for biological networks. Cell **173**(7), 1581–1592 (2018). https://doi.org/10.1016/j.cell.2018.05.015
8. Alipanahi, B., et al.: Predicting the sequence specificities of DNA- and RNA binding proteins by deep learning. Nat. Biotechnol. **33**(8), 831–838 (2015). https://doi.org/10.1038/nbt.3300
9. Zhou, J., Troyanskaya, O.G.: Predicting effects of noncoding variants with deep learning-based sequence model. Nat. Methods **12**(10), 931–934 (2015). https://doi.org/10.1038/nmeth.3547
10. Bach, S., et al.: On pixel-wise explanations for non-linear classifier decisions by layer-wise relevance propagation. PLoS ONE **10**(7), 1–46 (2015). https://doi.org/10.1371/journal.pone.0130140
11. Zhou, J., et al.: Graph neural networks: a review of methods and applications, arXiv, pp. 1–20, December 2018
12. Defferrard, M., Bresson, X., Vandergheynst, P.: Convolutional neural networks on graphs with fast localized spectral filtering. In: NIPS, pp. 1–14 (2016)
13. Rossi, R.A., Zhou, R., Ahmed, N.K.: Deep inductive network representation learning. In: Companion of the Web Conference 2018 - WWW 2018, New York, New York, USA. ACM Press, pp. 953–960 (2018). https://doi.org/10.1145/3184558.3191524
14. Perozzi, B., Al-Rfou, R., Skiena, S.: DeepWalk: Online Learning of Social Representations, arXiv, pp. 1–10, March 2014. https://doi.org/10.1145/2623330.2623732
15. Kipf, T.N., Welling, M.: Semi-supervised classification with graph convolutional networks. In: ICLR 2017, pp. 1–10 (2016)
16. Wang, Q., et al.: Data descriptor: unifying cancer and normal rna sequencing data from different sources. Sci. Data 1–8 (2018). https://doi.org/10.1038/sdata.2018.61
17. Johnson, W.E., Li, C., Rabinovic, A.: Adjusting batch effects in microarray expression data using empirical Bayes methods. Biostatistics **8**(1), 118–127 (2007). https://doi.org/10.1093/biostatistics/kxj037
18. Kamburov, A., et al.: ConsensusPathDB: toward a more complete picture of cell biology. Nucleic Acids Res. **39**(SUPPL.1), D712–D717 (2011). https://doi.org/10.1093/nar/gkq1156

19. Bruna, J., et al.: Spectral networks and locally connected networks on graphs. In: ICLR, p. 14 (2013). https://doi.org/10.1088/1464-4258/4/5/302
20. Li, Q., Han, Z., Wu, X.-M.: Deeper insights into graph convolutional networks for semi-supervised learning. arXiv, pp. 1–9 (2018)
21. Brin, S., Page, L.: The anatomy of a large-scale hypertextual web search engine. Comput. Netw. ISDN Syst. **30**(1–7), 107–117 (1998). https://doi.org/10.1016/S0169-7552(98)00110-X. ISSN 0169-7552
22. Veličković, P., et al.: Graph Attention Networks. arXiv, pp. 1–8 (2017)
23. Hakes, L., et al.: Protein-protein interaction networks and biology–what's the connection? Nat. Biotechnol. **26**(1), 69–72 (2008). https://doi.org/10.1038/nbt0108-69
24. Fodde, R.: The APC gene in colorectal cancer. Eur. J. Cancer **38**(7), 867–871 (2002). https://doi.org/10.1016/S0959-8049(02)00040-0
25. Zhao, Z., et al.: Multiple biological functions of Twist1 in various cancers. Oncotarget **8**(12), 20380–20393 (2017). 10.18632/oncotarget.14608
26. Dang, C.V.: MYC on the path to cancer. Cell **149**(1), 22–35 (2012). https://doi.org/10.1016/j.cell.2012.03.003

CNN-Based Semantic Change Detection in Satellite Imagery

Ananya Gupta(✉) ⓘD, Elisabeth Welburn ⓘD, Simon Watson ⓘD, and Hujun Yin ⓘD

The University of Manchester, Manchester, UK
{ananya.gupta,elisabeth.welburn,simon.watson,hujun.yin}@manchester.ac.uk

Abstract. Timely disaster risk management requires accurate road maps and prompt damage assessment. Currently, this is done by volunteers manually marking satellite imagery of affected areas but this process is slow and often error-prone. Segmentation algorithms can be applied to satellite images to detect road networks. However, existing methods are unsuitable for disaster-struck areas as they make assumptions about the road network topology which may no longer be valid in these scenarios. Herein, we propose a CNN-based framework for identifying accessible roads in post-disaster imagery by detecting changes from pre-disaster imagery. Graph theory is combined with the CNN output for detecting semantic changes in road networks with OpenStreetMap data. Our results are validated with data of a tsunami-affected region in Palu, Indonesia acquired from DigitalGlobe.

Keywords: Convolutional Neural Networks · Semantic segmentation · Graph theory · Satellite imagery

1 Introduction

Hundreds of natural disasters strike every year across the globe[1]. Timely damage assessment and mapping of disaster-struck areas are extremely important to disaster relief efforts. They are especially important in developing countries where the affected areas may not have been mapped. Furthermore, routes into affected areas can be blocked due to the effects of the disaster, rendering pre-existing maps ineffectual. An example of a disaster-struck area is shown in Fig. 1.

Volunteer initiatives around the world make use of publicly available satellite imagery to map out such areas following natural disasters to help provide prompt assistance [5]. However, due to inconsistency across different initiatives and inexperience of the volunteers, this process is often error-prone and time-consuming [24].

---

[1] https://ourworldindata.org/natural-disasters.

A. Gupta is funded by the School of Electrical and Electronic Engineering, The University of Manchester and the ACM SIGHPC/Intel Computational and Data Science Fellowship. E. Welburn is funded by the EPSRC HOME Offshore project grant EP/P009743/1.

© Springer Nature Switzerland AG 2019
I. V. Tetko et al. (Eds.): ICANN 2019, LNCS 11731, pp. 669–684, 2019.
https://doi.org/10.1007/978-3-030-30493-5_61

Fig. 1. Extracted images from satellite imagery of Palu, Indonesia showing the devastation due to the tsunami and earthquake in September, 2018 [9]. *Left*: Before the tsunami. *Right*: The day after the tsunami.

There is increasing demand for automating the process of road extraction from satellite imagery since up-to-date road maps are important for location and navigation services [21]. The current research in road identification treats it as a semantic segmentation task where satellite images are used to predict the probability of a pixel being a road [20]. However, due to variations, shadows and occlusions caused by buildings and trees, a number of road segments are often misidentified by the segmentation algorithms. Heuristics based methods are often used to help alleviate this problem by reasoning about missing connections between broken roads. However, these assumptions do not work well for post-disaster scenarios as there are broken connections due to blockages and damages caused by the disaster.

Comparing pixel values of satellite imagery from before and after a disaster is a potentially useful approach for identifying the effect of the disaster. However, comparing pixels directly is implausible due to various effects such as illuminations and seasons, which can cause significant changes in image statistics, shadows and changes in vegetation. The use of Convolutional Neural Networks (CNNs) has been proposed for damage assessment in buildings to order to cope with these challenges [2,13,26]. However, these approaches require a large amount of manually annotated training data for each location, which is expensive, time-consuming and unscalable.

Hence, instead of comparing pixel values or requiring manually annotated data we propose to use data from OpenStreetMap (OSM) [23] to train a CNN to identify semantic features such as roads in satellite imagery. This is used with pre-disaster and post-disaster imagery to help (a) identify the changes due to the disaster and identify the impacted areas; and (b) map out the road networks in the post-disaster landscape to aid disaster relief efforts. We further combine the OSM data with graph theory to obtain a more robust estimate of road networks.

The framework allows for assessing the level of damage so as to identify high impact areas in a timely manner. A costing function is used to express the usability of affected roads for accessibility.

2 Related Work

2.1 Road Segmentation

There are several existing approaches for extracting road maps from satellite imagery. A number of these approaches are based on probabilistic models. Geometric probabilistic models have been developed for road image generation followed by MAP estimation over image windows for road network identification [3]. Wenger et al. [30] proposed a probabilistic network structure to minimise a high order Conditional Random Field model to determine road connectivity. Another approach is based on manually identifying road points to define a road segment followed by matching further connected segments using a Kalman Filter [29]. Heuristic methods based on radiometric, geometric, and topological characteristics have been used to define road models, which are further refined based on contextual knowledge about objects such as buildings [17].

The drawback with both the heuristics and probabilistic approaches is that they work under an inherent assumption that road networks are connected and patches of roads do not exist in isolation. However, in post-disaster scenarios, this assumption is often invalid. Major points that we are interested in are actually the paths that are missing and/or broken connections left in the aftermath of the disaster.

More recently, CNN-based methods have been used to segment road pixels from non-road pixels in satellite imagery. Some methods have approached this as a segmentation problem and reported pixel based metrics [1,8,22]. However, since this approach does not take overall network topology into account, small gaps in the resultant network are not penalised, though they cause lengthy detours in practice. Another approach extracted topological networks from the segmented output and used smart heuristics in post processing to connect missing paths and remove small stubs which were seen as noise [20,27]. However, similar to the heuristic approach, it is not suitable for disaster-struck areas because the assumption of roads being connected is no longer valid.

2.2 Disaster Analysis

Satellite imagery is becoming an increasingly popular resource for disaster response management [28]. Recent research has mostly focused on identifying buildings affected by floods and hurricanes. For instance, a CNN-based fusion of multi-resolution, multi-temporal and multi-sensor images was used to extract spatial and temporal characteristics for finding flooded buildings [26]. CNNs have also been used to classify the probability of washed-away buildings by using clips of pre-disaster and post-disaster imagery [2,13].

Automated road identification in post-disaster scenarios is a nascent topic with concurrent research [14]. Some work is based on identifying road obstacles such as fallen trees and standing water using vehicle trajectory data [7]. Estimating road registration errors following earthquakes using post-event images has also been studied [19]. However, it could only correct for ground shifts due to earthquakes but could not address the problem of missing roads. A crowd-sourced pedestrian map builder was proposed [4] but it would require people walking around in potentially inaccessible disaster-struck areas.

Similar to our work, Doshi *et al.* [10] proposed a framework for change detection using satellite images in conjunction with CNNs. They identified buildings and roads in images before and after a disaster and used the per-pixel differences to quantify the disaster impact. By contrast, we focus on identifying road networks and correlating the changes with data obtained from OSM. We further propose a cost-based routing approach to take the affected areas into account when identifying possible routes for first responses.

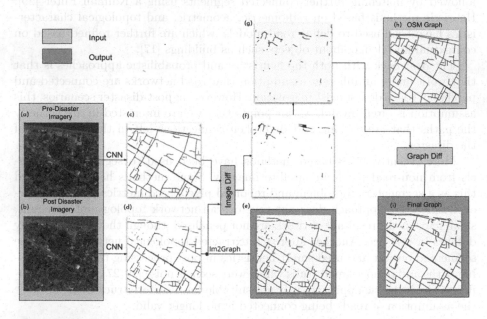

Fig. 2. Block diagram of proposed methodology: pre-disaster satellite image (a) and post-disaster satellite image (b) are converted to road masks, (c) and (d), respectively using a CNN-based segmentation model. The post-disaster road mask is converted to a road graph (e) using the process described in Sect. 3.2. The difference in mask images (f) is converted to a network graph (g), which is then subtracted from road network graph taken from OpenStreetMap (h), to produce the final post-disaster road network graph (i).

3   Methodology

The proposed framework is based on four distinct steps: (1) Road segmentation from satellite images using a CNN; (2) Creation of a road network graph from the segmented images; (3) Comparing pre-disaster and post-disaster road segments to identify possible changes; (4) Registering the changed segments with the OSM data to get a more realistic road network map. This pipeline is visualised in Fig. 2 and further described in the following sections.

3.1   Segmentation

We have developed a LinkNet [6] based network for the task of semantic road segmentation. It belongs to the family of encoder-decoder segmentation models [25] and the architecture is shown in Fig. 3. A ResNet34 [16] model pretrained on the ImageNet is used as the encoder since it has been found to yield good performances for the task without excessive computational costs. The encoding part starts with a convolutional block of 64 3×3 filters, followed by a MaxPool layer with a kernel size of 3×3. This is followed by 4 encoding blocks as shown in Fig. 3. Each encoder layer consists of a number of residual blocks as shown in Fig. 4a. Each convolutional layer in this case is followed by a batch normalisation and a ReLU layer. The output of each encoder layer feeds into the corresponding decoder layer to help recover the fine details lost in the downsampling by the convolution and pooling layers.

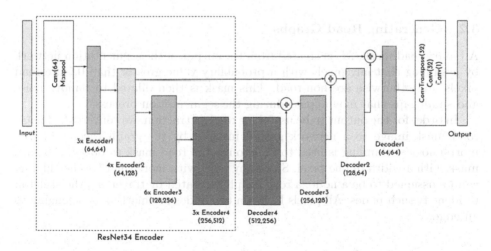

Fig. 3. Segmentation model: input is a satellite image and output is a single channel mask with each pixel representing probability being a road.

The architecture of a decoder block is shown in Fig. 4b, where the transposed convolution layer acts as an upsampling layer. The final layers in the decoding section include a transposed convolution layer, followed by a convolution layer with 32 channels as input and output and a final convolution layer with one

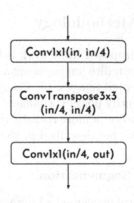

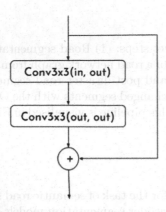

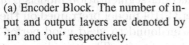

(a) Encoder Block. The number of input and output layers are denoted by
'in' and 'out' respectively.

(b) Decoder Block. The number of input and output layers are denoted by
'in' and 'out' respectively.

Fig. 4. Encoder and Decoder Blocks.

output channel corresponding to the class label, either 'road' or 'no road' in
this case. The network is trained with a binary cross-entropy loss function using
satellite images as input and a probability mask as output, which is compared
to the binary mask of the ground truth in order to compute the loss.

3.2   Generating Road Graphs

A binary road mask, M, is created from the output of the segmentation network
by assuming that any pixels with a probability value greater than 0.5 are road
pixels and otherwise are non-road. This mask is then dilated to remove noise
and small gaps that may appear during the segmentation process.

In order for the output to be useful for route extraction, we convert the binary
road mask image to a network graph, inspired by the graph theory. Firstly,
morphological thinning is used to skeletonise the road mask to obtain a binary
mask with a width of one pixel. Since any pixel with more than two neighbours
can be assumed to be a node, a road graph is created by traversing the skeleton
to identify such nodes. All pixels between two nodes are marked as belonging to
an edge.

3.3   Comparing Graphs

Graph theory offers a number of ways to compare graph similarity. However,
these methods are typically based on logical topology whereas in the case of
roads, we are also interested in the physical topology. Comparing road network
graphs is a non-trivial task since corresponding nodes in the two graphs may have
an offset and do not necessarily coincide in spatial coordinates. Furthermore, the
edges are not uniform and can have a complex topology.

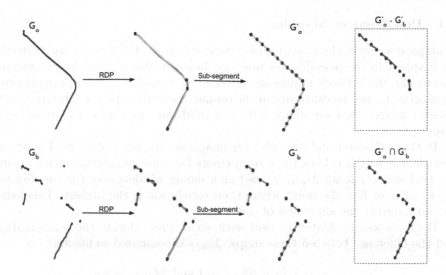

Fig. 5. Comparing graphs to find corresponding edges. Initial graphs (G_a and G_b) are approximated as combinations of linear segments using RDP [11]. They are further divided into sub-segments of length, l, in graphs G'_a and G'_b which can then be compared on a per-segment basis giving the difference between the graphs in the dashed box on the top right and the intersection in the box on the bottom right of the figure.

We simplify the graphs, G_a and G_b, in order to find correspondences as shown in Fig. 5. The edges of a graph are approximated with piece-wise linear segments using the Ramer-Douglas-Pecker algorithm [11]. The networks then contain edges that are linear, nodes that are incident to two linear edges and junctions that are incident to three or more edges. The weight of an edge is calculated as the euclidean distance between the vertices of the edge.

Each linear segment is then sliced into smaller sub-segments of a fixed length, l, to create the simplified graphs G'_a and G'_b, respectively. The sub-segments are compared to find which sub-segments in the two graphs are corresponding as follows:

$$\forall e_a, e_b; e_a \in G'_a, e_b \in G'_b$$
$$e_a = \{v_{a1}, v_{a2}\}; e_b = \{v_{b1}, v_{b2}\} \tag{1}$$
$$e_a = e_b, \quad \text{iff} \quad |a1 - b1| < l/2 \quad \text{and} \quad |a2 - b2| < l/2$$

where the sub-segments in graphs G'_a and G'_b are given as e_a and e_b, which are defined in terms of their two vertices, v_{a1} and v_{a2}, and v_{b1} and v_{b2}, respectively. The euclidean distance between two vertices is given by $|a1 - b1|$ where $a1$ and $b1$ are the coordinates of the first vertices of v_{a1} and v_{b1} respectively. Essentially, two segments, e_a and e_b, are assumed to be corresponding if both the vertices of e_a are within a certain distance of both vertices of e_b. The corresponding segments can be used to find the intersection of the two graphs as shown in Fig. 5.

3.4   Post Disaster Mapping

In an ideal scenario, the road graphs generated from Sect. 3.2 can be used directly for mapping in the post-disaster imagery. However, due to non-ideal segmentation masks, the network graphs are often missing available connections. Herein, we propose to use the difference in the output masks from pre-disaster and post-disaster imagery in conjunction with the OSM data to obtain a more realistic map.

Both pre-disaster and post-disaster images are used to obtain road masks as described in Sect. 3.1. In order to compensate for some image registration errors, the post-disaster mask M_{post} is used as a sliding window over the pre-disaster mask, M_{pre} to find the point where their correlation is the highest. This helps partially correct the alignment of the images.

Both masks are dilated to deal with small errors during the segmentation and the difference between these maps, M_{diff} is computed as follows:

$$M_{diff_p} = \begin{cases} 1 \text{ if } M_{pre_p} = 1 \text{ and } M_{post_p} = 0 \\ 0 \qquad\qquad \text{otherwise} \end{cases} \tag{2}$$

where M_{pre_p} is the value of pixel p in the pre-disaster mask and M_{post_p} is the value of the corresponding pixel in the post-disaster mask. This function computes the change where a road existed in the pre-disaster mask but is absent in the post-disaster mask.

All output lines in M_{diff} that are thinner than a certain threshold can be assumed as noise due to registration error and are removed using erosion. Morphological opening is carried out to remove further noise from the image. The final image provides the routes that changed due to the disaster. A road graph G_{diff} is generated from this image following the process described in Sect. 3.2. An ideal pre-disaster road network $G_{pre-ideal}$ for the region is obtained from the publicly available OSM dataset. Although the OSM data is not completely accurate [20], our experiments have found that using the prior knowledge from the OSM is useful for creating a more robust output.

For each edge in G_{diff}, the closest edge in $G_{pre-ideal}$ is found as explained in Sect. 3.3 and the cost of the corresponding edge, C_e, is updated according to the following equation:

$$C_e = \alpha \times \frac{s_{e,diff}}{d^2} \tag{3}$$

where $s_{e,diff}$ is the size of the missing segment in G_{diff}, d is the distance between the missing point in G_{diff}, and the corresponding edge in $G_{pre-ideal}$ and α is an impact factor based on the scale of disaster. The value of α can be varied from 1, implying no effect from disaster, to ∞ in areas where the roads are completely disconnected or missing such as in areas that might have been washed away by floods.

4   Experiments

4.1   Datasets

DigitalGlobe provides high resolution satellite imagery in wake of natural disasters as part of its Open Data Initiative [9] to support disaster recovery. We have identified a dataset from DigitalGlobe that has both pre-event and post-event imagery and visible damages to human settlements due to disasters. It is based on the earthquake and tsunami that devastated Sulawesi Island, Indonesia on 28th September, 2018. We extracted an area of approximately $45\,\mathrm{km}^2$ around Palu city. An area of $14\,\mathrm{km}^2$ with noticeable disaster damage was used for testing.

Following the standard practice of using separate areas for training and testing, the first experiment used imagery over an area of $31\,\mathrm{km}^2$ for training and validation while the remaining area of $14\,\mathrm{km}^2$ for testing. The split is shown in Fig. 6, where the area showing noticeable disaster damage was used as the test area. This dataset has been referred to as '*splitDataset*' in the following sections.

Fig. 6. Road map showing the extent of the dataset used. The section outlined in the green box was used as the test dataset and the section in orange was used for training in the case of '*splitDataset*'. (Color figure online)

Since the training dataset only used pre-disaster imagery but the inference was done over both the pre-disaster and post-disaster imagery, a second set of experiments used the entire 45 km^2 pre-disaster imagery for training. This dataset has been called '*wholeDataset*' in the experimental section. In both cases, a random subset of 10% of the training data was used for validation.

We also downloaded the publicly available data from the OSM for the entire region and extracted all the features marked as roads from this dataset, including all highways, lanes and bicycle paths. The available data from the OSM only includes labels from the pre-disaster imagery in a vector form.

In order to form suitable data for training the network, we converted the vector road labels to a raster format by converting the lat-long coordinates to pixel coordinates and using a 2 m buffer around the lines identifying roads for the mask. The roads in the post-disaster imagery were manually labelled for testing purposes.

4.2   Metrics

There are two primary types of metrics for road extraction: the first is a pixel-wise metric to quantify the performance of the segmentation algorithm, and the second is based on the structure and completeness of the graph.

When defining road networks, a pixel-wise metric such as intersection-over-union (IoU) is not suitable since smalls gaps in the output road mask may cause a small error in IoU but, in reality, can lead to large detours if used in a graph for navigation purposes. Graph-based metrics are more difficult to define and quantify since determining how similar two topological graphs are is a non-trivial problem.

In this case, we first compare graphs by using their sub-segments as described in Sect. 3.3 and report the standard metrics of precision, recall and F-score defined as follows:

$$p = \frac{TP}{TP + FP}$$
$$r = \frac{TP}{TP + FN} \tag{4}$$
$$F_{score} = 2 \times \frac{p \times r}{p + r}$$

where TP is true positive rate of the segments, FN false negative rate, FP false positive rate, and F_{score} is a measure of the overall accuracy.

These metrics provide a measure of similarity of the road networks. However, they do not take road connectivity into account, which is of particular importance when calculating routes. Hence, we also report the metrics proposed in [30] by generating a large set of source-destination pairs and finding the shortest paths for those pairs in the ground truth graph, $G_{post-ideal}$, and the predicted graph.

Based on the length of the extracted paths, it is possible to measure if the two graphs are identical as the path lengths should be similar. If the extracted path length is too short compared to the original graph, there will be incorrect shortcuts predicted in the network. Conversely, if the length of the output path is too long or there are no connections, there will be gaps in the graph where there should be roads.

4.3 Baseline

We have compared our method to the basic version of DeepRoadMapper (DRM) [20] that uses a ResNet55 based encoder-decoder network and a soft-jaccard loss for training. The output is similar to the post-disaster mask in our proposed framework shown in Fig. 2(e).

4.4 Training Details

The satellite images and corresponding masks were clipped to 416×416 pixels. Only the pre-disaster imagery was used as training images with the corresponding OSM information as labels. The post-disaster imagery was used purely for inference. The network was trained using the Adam optimiser [18] with a learning rate of 0.0001 for 100 epochs and a batch size of 12. A pretrained ResNet34 was used to initialise the encoder and the He initialisation [15] was used for the decoder.

5 Results

5.1 Quantitative Results

The performance in terms of precision, recall and accuracy is shown in Table 1, where the results of DRM [20] are also compared with that of the proposed approach, termed as *OSM Diff*. In this case, the value of α was set to ∞ so the changed road segments were completely removed from the road network. This allowed for a fair comparison since all broken roads have been marked as disconnected in the ground truth of the dataset. As can be seen from the results, our method outperformed the baseline by a large margin for both splits of the dataset. This was because that a number of the road segments were missed by the segmentation approach in [20]. Note that our method benefited from the prior knowledge of OSM and had better connectivity than methods that assumed no prior knowledge other than a training dataset.

Table 1. Quality, completeness, correctness of sub-segments

Method	Dataset Split	TP	FP	FN	Precision	Recall	F-score
DeepRoadMapper [20]	splitDataset	5899	1011	597	0.85	0.90	0.88
DeepRoadMapper [20]	wholeDataset	6073	995	423	0.86	0.93	0.89
OSM Diff (**Ours**)	splitDataset	6453	395	43	0.94	0.99	0.96
OSM Diff (**Ours**)	wholeDataset	6451	387	45	0.94	0.99	0.96

Another point worth noting is that the test results were similar across the datasets, regardless of the method used. This is possibly because the training

Fig. 7. Visualisation of results with extracted roads from post-disaster imagery shown in blue. *Left*: GT (Manually Labelled). *Middle*: Baseline DRM [20]. *Right*: OSM Diff (Ours) (Color figure online)

was always done on the pre-disaster imagery whereas the test dataset included only the post-disaster imagery. Hence, even though the areas overlapped, the training and testing dataset were disparate and sufficiently large to allow for generalisation.

We report the connectivity results as described in Sect. 4.2 in Table 2. Our method outperformed DRM by a large margin. This was again due to a number of missing connections. As can be seen from the table, the path planner was

Table 2. Connectivity Results. Correct implies that the shortest paths are similar in length to the ground truth, *No Connections* is where there was no possible path, *Too Short* is where the paths were too short compared to the ground truth and *Too Long* is where the path was too long. All values are given as percentages.

Graph type	Dataset split	Correct	No connections	Too short	Too long
DeepRoadMapper [20]	splitDataset	25.81	53.11	2.95	18.08
DeepRoadMapper [20]	wholeDataset	41.31	40.17	6.21	12.26
OSM Diff(**Ours**)	splitDataset	68.97	20.05	**1.53**	9.36
OSM Diff(**Ours**)	wholeDataset	73.38	16.76	2.03	7.72
OSM Weighted Diff (**Ours**)	wholeDataset	**86.59**	0	8.18	**5.13**

unable to find any paths for about 16% of the pairs in the output from *OSM Diff* whereas DRM had over twice the number of missing connections.

The ablation study across different dataset splits for connectivity results showed that both methods performed measurably better when trained with a larger dataset. This was contrary to the precision and recall metrics, which were found similar across the datasets. These results show that the segmentation network performed better at identifying connected segments when provided with more training data.

We also report the results for *OSM Weighted Diff* where the value of α was set to 5. This allowed for a higher number of correct paths with no missing connections since the network graph was similar to the OSM graph but with a higher cost on affected roads. However, this method gave a larger number of 'too short' paths since it allowed paths which might be impossible to traverse in the post-disaster scenario due to roads that might have been flooded or washed away. Figure 7 shows the outputs of the proposed method along with the ground truth and the results from [20].

5.2   Qualitative Results

The difference in the segmentation masks, M_{diff}, can be used for identifying the most impacted areas. An area under consideration can be divided into small grids of a fixed size and all pixels in a grid summed to get an estimate of how affected the area is. This was done over the results from our experiments on the Palu imagery and the output has been plotted as a heat map shown in Fig. 8.

The results matched the conclusions of the European Commission report [12] on the impact of the disaster. The earthquake caused soil liquefaction in the south-west region of Palu, which can be seen as an area of major impact. The coast was also mostly impacted due to the tsunami, which again can be seen in the figure.

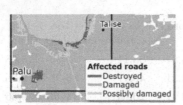

Fig. 8. *Left*: Map of affected roads extracted from [12]. *Right*: Heatmap of our results indicating the severity of the impact overlaid onto the satellite image. Yellow (Extremely severe) > Red (Less severe) (Color figure online)

6   Conclusions

This paper outlines a framework for identifying road networks in post-disaster scenarios using both satellite imagery and OSM data. It is based on the use of CNNs for road segmentation and graph theory for comparing the changes detected from pre-disaster and post-disaster satellite imagery to leverage knowledge from OSM. This mapping process is currently done manually, while the proposed method can reduce the annotation time down from days to minutes, enabling provision of timely assistance to subsequent relief and rescue work.

The proposed method has been tested on a dataset of Palu, Indonesia from 2018 around the time it was struck by a tsunami and an earthquake. Both quantitative and qualitative results were promising in identifying accessible routes in the region, and the method also successfully identified the highly affected areas in the city.

The work can be further improved by identifying the reasons for the broken roads and updating the cost function accordingly. For example, standing water or road debris are identified as obstacles and can have a lower cost in the network map but areas that have had landslides should have a much higher cost.

Acknowledgment. The authors would like to thank Drs. Andrew West and Thomas Wright for their valuable insights.

References

1. Aich, S., Van Der Kamp, W., Stavness, I.: Semantic binary segmentation using convolutional networks without decoders. In: IEEE Computer Society Conference on Computer Vision and Pattern Recognition Workshops, pp. 182–186, 2018 June (2018). https://doi.org/10.1109/CVPRW.2018.00032
2. Amit, S.N.K.B., Aoki, Y.: Disaster detection from aerial imagery with convolutional neural network. In: International Electronics Symposium on Knowledge Creation and Intelligent Computing, pp. 239–245. IEEE, september 2017. https://doi.org/10.1109/KCIC.2017.8228593
3. Barzohar, M., Cooper, D.B.: Automatic finding of main roads in aerial images by using geometricstochastic models and estimation. IEEE Trans. Pattern Anal. Mach. Intell. **18**(7), 707–721 (1996). https://doi.org/10.1109/34.506793
4. Bhattacharjee, S., Roy, S., Das Bit, S.: Post-disaster map builder: crowdsensed digital pedestrian map construction of the disaster affected areas through smartphone based DTN. Comput. Commun. **134**, 96–113 (2019). https://doi.org/10.1016/j.comcom.2018.11.010
5. Boccardo, P., Giulio Tonolo, F.: Remote sensing role in emergency mapping for disaster response. In: Lollino, G., Manconi, A., Guzzetti, F., Luino, F., Culshaw, M., Bobrowsky, P. (eds.) Engineering Geology for Society and Territory - Volume 5, pp. 17–24. Springer, Cham (2015). https://doi.org/10.1007/978-3-319-09048-1_3
6. Chaurasia, A., Culurciello, E.: LinkNet: exploiting encoder representations for efficient semantic segmentation. In: 2017 IEEE Visual Communications and Image Processing, VCIP Description 2017, January 2018, pp. 1–4, June 2018. https://doi.org/10.1109/VCIP.2017.8305148
7. Chen, L., et al.: RADAR. Proc. ACM Interact. Mob Wearable Ubiquit. Technol. **1**(4), 1–23 (2018). https://doi.org/10.1145/3161159
8. Demir, I., et al.: DeepGlobe 2018: a challenge to parse the earth through satellite images. In: IEEE Computer Society Conference on Computer Vision and Pattern Recognition Workshops, June 2018, pp. 172–181, May 2018. https://doi.org/10.1109/CVPRW.2018.00031
9. DigitalGlobe: Open Data Initiative
10. Doshi, J., Basu, S., Pang, G.: From Satellite Imagery to Disaster Insights (2018)
11. Douglas, D.H., Peucker, T.K.: Algotrithms for the reduction of the number of points required to represent a digitized line or its caricature. Cartographica: Int. J. Geograph. Inf. Geovisualization **10**(2), 112–122 (1973). https://doi.org/10.3138/FM57-6770-U75U-7727
12. European Commision Joint Research Centre: Mw 7.5 Earthquake in Indonesia 28, Emergency Report. Technical Report (2018)
13. Fujita, A., Sakurada, K., Imaizumi, T., Ito, R., Hikosaka, S., Nakamura, R.: Damage detection from aerial images via convolutional neural networks. In: 5th IAPR International Conference on Machine Vision Applications, pp. 5–8. IEEE, May 2017. https://doi.org/10.23919/MVA.2017.7986759
14. Gupta, A., Welburn, E., Watson, S., Yin, H.: Post disaster mapping with semantic change detection in satellite imagery. In: Computer Vision and Pattern Recognition Workshops, pp. 0–0 (2019)
15. He, K., Zhang, X., Ren, S., Sun, J.: Delving deep into rectifiers: surpassing human-level performance on ImageNet classification. In: Proceedings of the IEEE International Conference on Computer Vision, 2015, pp. 1026–1034, February 2015. https://doi.org/10.1109/ICCV.2015.123

16. He, K., Zhang, X., Ren, S., Sun, J.: Deep residual learning for image recognition. In: IEEE Conference on Computer Vision and Pattern Recognition (2016). https://doi.org/10.1109/CVPR.2016.90

17. Hinz, S., Baumgartner, A.: Automatic extraction of urban road networks from multi-view aerial imagery. ISPRS J. Photogram. Remote Sens. **58**(1–2), 83–98 (2003). https://doi.org/10.1016/S0924-2716(03)00019-4

18. Kingma, D.P., Ba, J.: Adam: a method for stochastic optimization. In: International Conference on Learning Representations (2015). https://doi.org/10.1063/1.4902458

19. Liu, Z., Zhang, J., Li, X.: An automatic method for road centerline extraction from post-earthquake aerial images, January 2019. https://doi.org/10.1016/j.geog.2018.11.008

20. Mattyus, G., Luo, W., Urtasun, R.: DeepRoadMapper: extracting road topology from aerial images. In: IEEE International Conference on Computer Vision, October 2017, pp. 3458–3466. IEEE, October 2017. https://doi.org/10.1109/ICCV.2017.372

21. Miller, G.: The Huge, Unseen Operation Behind the Accuracy of Google Maps (2014)

22. Mnih, V., Hinton, G.E.: Learning to detect roads in high-resolution aerial images. In: European Conference on Computer Vision, pp. 1–14 (2010). https://doi.org/10.1007/978-3-642-15567-3_16

23. OpenStreetMap Contributors: Planet dump (2017). https://planet.osm.org

24. Poiani, T.H., Rocha, R.D.S., Degrossi, L.C., de Albuquerque, J.P.: Potential of collaborative mapping for disaster relief: a case study of OpenStreetMap in the Nepal earthquake 2015. In: Proceedings of the Annual Hawaii International Conference on System Sciences, March 2016, pp. 188–197. IEEE, January 2016. https://doi.org/10.1109/HICSS.2016.31

25. Ronneberger, O., Fischer, P., Brox, T.: U-Net: convolutional networks for biomedical image segmentation. In: Navab, N., Hornegger, J., Wells, W.M., Frangi, A.F. (eds.) MICCAI 2015. LNCS, vol. 9351, pp. 234–241. Springer, Cham (2015). https://doi.org/10.1007/978-3-319-24574-4_28

26. Rudner, T.G.J., et al.: Multi3Net: segmenting flooded buildings via fusion of multiresolution, multisensor, and multitemporal satellite imagery. In: Thirty-Third AAAI Conference on Artificial Intelligence (2019)

27. Van Etten, A., Lindenbaum, D., Bacastow, T.M.: SpaceNet: a remote sensing dataset and challenge series. Arxiv (2018)

28. Voigt, S., Kemper, T., Riedlinger, T., Kiefl, R., Scholte, K., Mehl, H.: Satellite image analysis for disaster and crisis-management support. IEEE Trans. Geosci. Remote Sens. **45**(6), 1520–1528 (2007). https://doi.org/10.1109/TGRS.2007.895830

29. Vosselman, G., de Knecht, J.: Road tracing by profile matching and Kaiman filtering. In: Automatic Extraction of Man-Made Objects from Aerial and Space Images, pp. 265–274 (2011). https://doi.org/10.1007/978-3-0348-9242-1_25

30. Wegner, J.D., Montoya-Zegarra, J.A., Schindler, K.: Road networks as collections of minimum cost paths. ISPRS J. Photogram. Remote Sens. **108**, 128–137 (2015). https://doi.org/10.1016/j.isprsjprs.2015.07.002

Axiomatic Kernels on Graphs for Support Vector Machines

Marcin Orchel[1,2]([⊠]) [iD] and Johan A. K. Suykens[1] [iD]

[1] ESAT-STADIUS, KU Leuven, 3001 Leuven (Heverlee), Belgium
{marcin.orchel,johan.suykens}@esat.kuleuven.be
[2] Department of Computer Science, AGH University of Science and Technology,
Kraków, Poland

Abstract. We solve the problem of classification on graphs by generating a similarity matrix from a graph with virtual edges created using predefined rules. The rules are defined based on axioms for similarity spaces. Virtual edges are generated by solving the problem of computing paths with maximal fixed length. We perform experiments by using the similarity matrix as a kernel matrix in support vector machines (SVM). We consider two versions of SVM: for inductive and transductive learning. The experiments show that virtual edges reduce the number of support vectors. When comparing to kernels on graphs, the SVM method with virtual edges is faster while preserving similar generalization performance.

Keywords: Support vector machines · Graph kernels

We solve a problem of collective classification [11] also known as iterative classification or link-based classification where the goal is to determine correct label assignments of all objects in the network. One of the approaches is to use attributes of neighbors' examples, which is called relational classification. Another approach is to use class labels assigned to neighbor instances. This approach is called iterative collective classification. The iterative classification algorithm uses a local classifier that takes class labels of neighbors and return a label value and repeat the process. The collective classification has been applied to a number of real-world problems [11], for example document classification. Specifically, we solve a problem of classification without features. It is called similarity-based classification [4]. The example for such problem is fraud detection for anonymized cell phone network [6]. It is also known as graph-based semi-supervised learning when a graph has unlabeled nodes and weighted edges [9]. The problem of collective classification for partially labeled data is also known as within-network classification [6].

We focus on using SVM for solving the collective classification problem. The requirements for using SVM is to define a kernel matrix, which is a similarity matrix with an additional property of positive semi-definiteness. For networks, we usually have similarities only between connected objects, which leads to a

© Springer Nature Switzerland AG 2019
I. V. Tetko et al. (Eds.): ICANN 2019, LNCS 11731, pp. 685–700, 2019.
https://doi.org/10.1007/978-3-030-30493-5_62

problem of generating a kernel matrix for a graph. The technique of generating kernels for networks is called "kernels on graphs". This technique has been already applied for recommendation tasks [5,20] and for semi-supervised classification [5]. The general issue of kernels on graphs is high computational cost (usually $O(n^3)$). The alternative approach is to map the data to the Euclidean space using spectral embedding. The disadvantage is high computational cost and forcing data to be in a specific space. So discrete kernels may be preferable [7]. We focus on another workaround using a similarity matrix with SVM which may not be positive semi-definite. Recently, an efficient solver for SVM has been proposed [10], which does not use a regularization term. It solves a convex optimization problem, regardless of the positive semi-definiteness of a kernel matrix. The straightforward approach to define a similarity matrix is to use value 0 for disconnected vertices meaning no similarity at all. However, this assumption may not be met in collective classification when relations are supposed to exist also for disconnected vertices. Another problem with the straightforward approach specific to SVM is the unknown classification when the example is not connected with a support vector. One of the approaches to solve these problems is to generate additional edges in a graph with provided similarities. This idea has been proposed in [6]. The authors use "ghost edges" with proximities generated by a Random Walk method using specific measure based on a Laplacian matrix. The potential problems with this measure are related to nonstability of the randomness process. The measure is used in a supervised manner by adding ghost edges only between labeled and unlabeled pair of nodes. The supervised setting is prone to cascading errors for iterative methods [13]. Moreover, the measure for ghost edges is defined for a classification on graphs, while it is unclear how to define the measure on weighted graphs which is a more general problem considered in this paper. The weighted graphs has been mentioned in [9] with the example of a weight being the number of hyperlinks between websites.

One of the approaches to define a kernel on graphs is to compute the shortest path distance between nodes [14]. The potential problems have been mentioned like constructing positive definite function and sensitiveness to the insertion/deletion of individual edges. Computing the shortest path between all examples is also expensive, the Floyd-Warshall algorithm has complexity $O(n^3)$. The potential problem with the idea of using the shortest path is that it operates on distances instead of similarities between examples. So the similarities need to be converted to distances and vice versa. The idea in this paper is to generate "virtual edges" by finding paths with maximal similarity and a fixed size. Another idea is to compute similarities for "virtual edges" by using axioms defined for similarity spaces. Such an approach has the advantage of simple interpretation of particular added weights and can be regarded as prior knowledge for similarities. It uses local information instead of global, so it can be used for streaming graphs.

Related Work. There are two approaches for using a Non-Positive Semidefinite Similarity Matrix for kernel machines [19]: algorithmic and spectrum-transformation. In the algorithmic approach, one uses the NPSD similarity matrix as a kernel. We need a special formulation of SVM, or a heuristic to find a local solution. In the spectrum-transformation methods the kernel matrix is generated. The representative of the second approach is a diffusion kernel [7] based on a diffusion equation, that considers the data distribution when computing pairwise similarity. We use a novel type of solver for SVM without regularization term which solves a convex optimization problem with any similarity matrix [10].

The approach of using generated PSD kernels for graph data has been investigated in [5]. The authors investigated nine kernels on graphs: the exponential diffusion kernel, the Laplacian exponential diffusion kernel, the von Neumann diffusion kernel, the regularized Laplacian kernel, the commute-time kernel, the random-walk-with-restart similarity matrix, the regularized commute-time kernel, the Markov diffusion kernel and the relative-entropy diffusion matrix. The graph kernels based on normalized Laplacian, mainly Regularized Laplacian, Diffusion Process, p-step Random Walk and Inverse Cosine have been investigated in [7,16]. In [14], the authors mentioned a random walk kernel, where multiple paths are created with the size T. Potential problems include choosing suitable T and inability to reach a vertex due to cycles. Recently, a random walk method has been used for learning representations in the deep learning framework for classification of graph data [13]. In [1], authors proposed shortest path kernels in a different problem of comparing two graphs with each other. Some kernels have been proposed to improve the computational performance for comparing graphs like Weisfeiler-Lehman Graph Kernels [15]. However, it is not clear whether they can be used to classification on graphs. The diffusion kernel requires diagonalizing the Laplacian which is of order n^3. Moreover, it may have problems with accuracy [3].

The outline of the paper is as follows. First, we define a problem, then the methods and rules, then we derive rules from axioms. After that, we show experiments on real world data sets.

1   Problem Definition

We consider the following problem.

Definition 1 (Classification space $\mathfrak{C}$). For a universe X of objects, we have a set C of classes and a set of mappings $M_T : X_T \subset X \to C$ called a *training set* T, $X_{Te} \subseteq X$, where X_{Te} is a set called a *test set*.

For example in the Fig. 1, $\{x_1, x_2, x_6\}$ is a training set, $\{x_3, x_4, x_5, x_7, x_8\}$ is a test set.

Definition 2 (Classification space on a graph $\mathfrak{G}$). We define a classification space on a graph, as a classification space with a graph $G = (X_T \cup X_{Te}, E)$ with weights for each $e \in E$.

We need to know X_{Te} for a graph. We interpret the weights as similarities between examples. The graph without weights can be represented by a weighted graph with binary weights. The example graph is depicted in the Fig. 1.

Problem 1 (Semi-supervised classification problem). A *semi-supervised classification problem* is to find a class for each element of X_{Te} given M_T and X_{Te} on $\mathfrak{C}$.

Problem 2 (Semi-supervised classification problem on a graph). A *semi-supervised classification problem on a graph* is to find a class for each element of $x \in X_{Te}$ given M_T and X_{Te} on $(\mathfrak{C}, \mathfrak{G})$.

For example in the Fig. 1, we need to find a class for all test examples x_3, x_4, x_5, x_7, x_8.

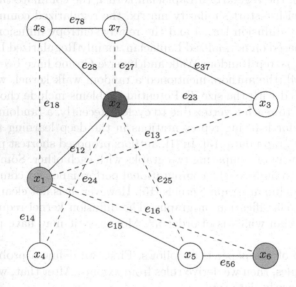

Fig. 1. Problem definition. Filled circles – training data examples, unfilled circles – test data examples, solid lines – edges in a given graph, dashed lines – virtual edges added to a graph.

2   Main Contribution

The idea of a method is to generate "virtual edges" between some disconnected examples. In the Fig. 1, the virtual edges are depicted with a dashed line, and these are $e_{16}, e_{13}, e_{45}, e_{25}, e_{24}, e_{18}, e_{27}$.

Another idea in the proposed approach is to generate virtual edges for all examples for which there exists a path of a length 2. For example in the Fig. 1, there is a path from x_1 to x_3 of length 2 through the example x_2, so we generate the edge e_{13}. There is no path of a length 2 between x_5 and x_3, so there is no virtual edge between them. We generate paths in a semi-supervised way based on both training and test examples when both are available as in Problem 2. For example in order to generate a path between x_1 and x_6, we use connections with a test example x_5. We do not need to generate edges between test examples, like e_{45} or e_{38}. The generation of virtual edges can be regarded as an unsupervised setting in the sense, that we do not use labels. We do not consider as a path, a sequence of vertices with possible duplicates, which is called a walk, for example x_1 to x_1.

The next idea is to compute the maximal similarity for the virtual edge. Sometimes, there are two paths of a length 2 between examples. In the Fig. 1, there are two paths between x_2 and x_7, that are $x_2 \to x_3 \to x_7$ and $x_2 \to x_8 \to x_7$. We compute the similarity for the virtual edge e_{27} based on each path and we choose the maximal similarity.

We propose three rules for generating weights for virtual edges being similarities between examples. These are

$$\tilde{s}(x_1, x_3) \leftarrow \frac{s(x_1, x_2)\, s(x_2, x_3)}{s(x_1, x_2) + s(x_2, x_3)}, \tag{1}$$

where s is a similarity measure between two examples. For example, in the Fig. 1, when $s(x_1, x_5) = 0.6$, $s(x_5, x_6) = 0.9$, we induce similarity $\tilde{s}(x_1, x_6) = 0.36$. The next rule is

$$\tilde{s}(x_1, x_3) \leftarrow \exp\left(-\left(\sqrt{-\log s(x_2, x_3)} + \sqrt{-\log s(x_1, x_2)}\right)^2\right). \tag{2}$$

For example, in the Fig. 1, when $s(x_1, x_5) = 0.6$, $s(x_5, x_6) = 0.9$, we induce similarity $\tilde{s}(x_1, x_6) \approx 0.34$. The third rule is

$$\tilde{s}(x_1, x_3) \leftarrow s_{\max} - \frac{1}{2}\left(\sqrt{2s_{\max} - 2s(x_2, x_3)} + \sqrt{2s_{\max} - 2s(x_1, x_2)}\right)^2, \tag{3}$$

where $s_{\max}$ is a maximal possible value of a similarity. For example, in the Fig. 1, assuming $s_{\max} = 1.0$, when $s(x_1, x_5) = 0.6$, $s(x_5, x_6) = 0.9$, we induce similarity $\tilde{s}(x_1, x_6) \approx 0.1$.

We propose three methods for solving a semi-supervised classification on graphs by using SVM.

Method 1 (Reference Graph Support Vector Machines (RGSVM)). *Train a model with SVM with all training examples with a similarity matrix defined as* $s(x_i, x_j) = w(x_i, x_j)$ *for* $(x_i, x_j) \in E$, *otherwise* $s(x_i, x_j) = 0$.

For the example in the Fig. 1 the training set for SVM is x_1, x_2, x_6. The similarity matrix is

$$\begin{bmatrix} s(x_1, x_1) & s(x_1, x_2) & 0 \\ s(x_2, x_1) & s(x_2, x_2) & 0 \\ 0 & 0 & s(x_6, x_6) \end{bmatrix}. \tag{4}$$

The decision boundary is

$$y_1 \alpha_1 s(x_1, x) + y_2 \alpha_2 s(x_2, x) + y_6 \alpha_6 s(x_6, x) + b = 0, \tag{5}$$

where y_i is a label of the ith example, α_i and b are parameters computed by SVM. The next method uses virtual edges.

Method 2 (Axiomatic Kernel Graph Support Vector Machines (AKGSVM)). *Train a model with SVM with all training examples with a similarity matrix defined as $s(x_i, x_j) = w(x_i, x_j)$ for $(x_i, x_j) \in E$, otherwise for each two not connected vertices x_1, x_2 find a path of length 2 between them with maximal induced similarity. So when $(x_1, x_3) \notin E$, find all x_i, such as $(x_1, x_i) \in E$ and $(x_i, x_3) \in E$, and*

$$\max_{x_i} \tilde{s}(x_1, x_3) \tag{6}$$

and then set $s(x_1, x_3) = \tilde{s}(x_1, x_3)$ for a path through the optimal x_i^*, where the induced similarity is computed based on (1) or (2) or (3), otherwise $s(x_i, x_j) = 0$. For the (1) the method is called AKGSVM1, for (2) AKGSVM2, for (3) AKGSVM3.

For the AKGSVM3, we may have a negative bound (3), so we compute the maximum of the bound (3) and (2). For the example in the Fig. 1, the training set for SVM is x_1, x_2, x_6. The similarity matrix is

$$\begin{bmatrix} s(x_1, x_1) & s(x_1, x_2) & \max(\tilde{s}_b(x_1, x_6), \tilde{s}(x_1, x_6)) \\ s(x_2, x_1) & s(x_2, x_2) & 0 \\ \max(\tilde{s}_b(x_6, x_1), \tilde{s}(x_6, x_1)) & 0 & s(x_6, x_6) \end{bmatrix}, \tag{7}$$

where $\tilde{s}_b$ is the similarity induced by (2). The decision boundary is

$$y_1\alpha_1 s(x_1, x) + y_2\alpha_2 s(x_2, x) + y_6\alpha_6 s(x_6, x) + b = 0, \tag{8}$$

where s is the original similarity when exist or the induced similarity otherwise. The third method is based on creating a local model for each test example.

Method 3 (Reduced Transductive Graph Support Vector Machines (RTGSVM)). *Perform AKGSVM for each test example x_p separately given a subset of training data with the r most similar examples connected with x_p, where r is some parameter.*

The property of creating local models to particular test examples which are not designed to generalize to other test examples is called transductive learning. When there is no enough connected examples to match the r value, the subset is smaller than r. For the example in the Fig. 1, for $r = 2$, the training set for $x_p = x_5$ is a subset of training data with two nearest training examples to x_5 computed based on similarities. It may be $\{x_1, x_6\}$ or $\{x_2, x_1\}$ or $\{x_2, x_6\}$. For the first case the similarity matrix is

$$\begin{bmatrix} s(x_1, x_1) & \tilde{s}(x_1, x_6) \\ \tilde{s}(x_6, x_1) & s(x_2, x_2) \end{bmatrix}. \tag{9}$$

The decision boundary is

$$y_1\alpha_1 \tilde{s}(x_1, x) + y_6\alpha_6 \tilde{s}(x_6, x) + b = 0, \tag{10}$$

where $\tilde{s}$ is the original similarity when exist, otherwise it is the induced similarity. Overall, we have 5 separate models one per each test example. In the last two methods, virtual edges are used also when classifying test data. The RTGSVM method can be used with any type of virtual edges either based on (1) or (2) or (3).

3 Analysis of Rules

The rules are generated based on defined similarity spaces. First, we define a similarity space corresponding to (1).

Definition 3 (Similarity space $\mathfrak{S}$). For a binary relation R on X, we define a *similarity measure* $s_R : X \times X \to \mathbb{R}$ where s_R is a restriction of the function s to a binary relation R, which is a subset of the Cartesian product that is $R \subseteq X \times X$, shortly we note s_R as s. Similarity measure fulfills the similarity axioms: $0 < s(x_1, x_2) \leq s_{max}$, where $s_{max} \geq 0$, $s(x_1, x_2) = s_{max} \iff x_1 = x_2$, $s(x_1, x_2) = s(x_2, x_1)$ and

Axiom 1.

$$s(x_1, x_3) \geq \frac{s(x_1, x_2)\, s(x_2, x_3)}{s(x_1, x_2) + s(x_2, x_3)}. \tag{11}$$

We call the assumptions axioms following [2]. The Axiom 1 gives a lower bound for similarity. We define additionally a *pseudosimilarity*, when the second axiom is replaced with $s(x_1, x_1) = s_{max}$. The similarity concept is related to the distance. The distance is also called a metric and is part of a definition of a metric space. That is the metric or distance is defined as a function $d : X \times X \to \mathbb{R}_+$ satisfying the following properties: $d(x_1, x_2) \geq 0$ and $d(x_1, x_2) = 0 \iff x_1 = x_2$ (non-negativity), $d(x_1, x_2) = d(x_2, x_1)$ (symmetry) and

Axiom 2.

$$d(x_1, x_3) \leq d(x_2, x_3) + d(x_1, x_2). \tag{12}$$

One difference between the definition of similarity and the distance is that the distance is defined for all $x \in X$, while the similarity we define only on some subset of $X \times X$, that is the binary relation R. We derive axioms for similarity from axioms for the distance. In particular, when we replace similarity by the inverse of a distance assuming $d(x, y) > 0$ in (11), so we substitute

$$s(x_1, x_2) = 1/d(x_1, x_2), \tag{13}$$

we get the distance axiom Axiom 2. The kernel matrix for some kernel functions can be interpreted as similarities between all examples. One notable example is the radial basis function (RBF) kernel, for which the kernel value is defined in terms of distance. By using this property, we can get alternative triangle inequality for the similarity (1). We can convert it to similarity by using $\log s(x_1, x_2) = -d(x_1, x_2)^2$, so $d(x_1, x_2) = \pm\sqrt{-\log s(x_1, x_2)}$, d is nonnegative so

$$d(x_1, x_2) = \sqrt{-\log s(x_1, x_2)}. \tag{14}$$

We additionally assume that $s_{max} = 1$. Then after substituting (14) to (12), we get

Axiom 3.

$$s(x_1, x_3) \geq \exp\left(-\left(\sqrt{-\log s(x_2, x_3)} + \sqrt{-\log s(x_1, x_2)}\right)^2\right). \tag{15}$$

Based on this axiom, we created a rule (2). It gives another possible lower bound on the similarity. It holds that

$$s\left(x_1, x_2\right) > \frac{s\left(x_1, x_2\right) s\left(x_2, x_3\right)}{s\left(x_1, x_2\right) + s\left(x_2, x_3\right)}. \tag{16}$$

The same holds for the left side being $s(x_2, x_3)$. So the lower bound is smaller than any similarity involved. The maximal possible induced similarity for $s_{max} = 1$ is 0.5, and it is achieved for $s_1 = 1$ and $s_2 = 1$. For any s_{max} the maximal possible similarity is $s_{max}/2$ and is achieved for $s_1 = s_{max}$ and $s_2 = s_{max}$.

When generating similarities using the RBF kernel for some given particular distances, the question is about satisfying the triangle inequality for similarity Axiom 1. So after substitution $s(x_1, x_2) = \exp\left(-d(x_1, x_2)^2\right)$ to Axiom 1, we get

$$d^2\left(x_1, x_3\right) \leq$$
$$\log\left(\exp\left(-d^2\left(x_1, x_2\right)\right) + \exp\left(-d^2\left(x_2, x_3\right)\right)\right) + d^2\left(x_1, x_2\right) + d^2\left(x_2, x_3\right). \tag{17}$$

This is alternative triangle inequality for a distance which can be used for defining alternative metric space.

We can also substitute (12) to Axiom 3 and we get

$$d\left(x_1, x_3\right) \leq \frac{1}{\exp\left(-\left(\sqrt{\log d\left(x_2, x_3\right)} + \sqrt{\log d\left(x_1, x_2\right)}\right)^2\right)}. \tag{18}$$

This is another way of defining triangle inequality for a metric space. For vector spaces, the feature map for kernel machines exists only when the kernel is positive-definite. The similarity is a broader concept and this condition may not be met. The relation between metric and the positive-definite kernels extended to the concept of similarity is as follows

$$\tilde{d}\left(x_1, x_2\right) = \sqrt{2s_{max} - 2s\left(x_1, x_2\right)}, \tag{19}$$

where

$$2s_{max} - 2s\left(x_1, x_2\right) \geq 0, \tag{20}$$

$\tilde{d}$ is a pseudometric. The disadvantage of such definition is that a distance is bounded by a value $\sqrt{2s_{max}}$. For a pseudometric the second axiom is replaced by $d(x, x) = 0$, so the distance between different examples can be 0. The pseudometric is related to pseudosimilarity. The condition (20) is the existence of s_{max}. The relation can be reformulated as

$$s\left(x_1, x_2\right) = s_{max} - \frac{1}{2}\tilde{d}\left(x_1, x_2\right)^2. \tag{21}$$

So positive definiteness is related to a specific assumption about connection between similarity and metric. For such definition, we can see how the triangle inequality axiom for similarities Axiom 1 relates to distances by substituting (21) to Axiom 1 and we get

$$\tilde{d}(x_1, x_3)^2 \leq \frac{1}{s_{\max}} \frac{1}{4} \tilde{d}(x_1, x_2)^2 \tilde{d}(x_2, x_3)^2 - \frac{1}{s_{\max}} \frac{1}{4} \tilde{d}(x_1, x_2)^2 \tilde{d}(x_1, x_3)^2$$
$$- \frac{1}{4} \tilde{d}(x_2, x_3)^2 d(x_1, x_3)^2 - s_{\max}. \tag{22}$$

We derive the axiom for similarity by substituting distance from (19) to Axiom 2 and we get

Axiom 4.

$$s(x_1, x_3) \geq s_{\max} - \frac{1}{2} \left(\sqrt{2s_{\max} - 2s(x_2, x_3)} + \sqrt{2s_{\max} - 2s(x_1, x_2)} \right)^2. \tag{23}$$

The problem is that sometimes we get loose negative bound for similarity, which is useless for substitution. Based on this axiom, we created a rule (3).

4   Analysis of Methods

In addition to the proposed methods, we analyze the following methods: knearest neighbors (KNN), SVM with a diffusion kernel based on a normalized Laplacian and SVM with a shortest path kernel. The shortest path kernel minimizes distances, so given a graph with similarities, we convert each similarity on a path to a distance and vice versa by using (13).

Complexity Analysis. For the SVM methods trained on the whole kernel matrix, that are SVM with a diffusion kernel and SVM with a shortest path kernel the computational complexity for the stochastic gradient solver with the worst violators [10] is $O(sn)$, where s is the number of support vectors, n is the number of examples. The maximal number of iterations is s. In each iteration, we need to update a functional margin value for each remaining parameter and find the worst violator. Computing a kernel matrix for a diffusion kernel requires eigendecomposition of a normalized Laplacian matrix which has complexity $O(n^3)$. Computing a kernel matrix for a shortest path kernel leads to all-pairs shortest paths problem. We need to compute paths between all vertices, except those already connected. The Floyd-Warshall algorithm has complexity $O(n^3)$, where n is the number of vertices. For sparse graphs with nonnegative weights which is the case here, Dijkstra's algorithm can be used which has complexity $O(|E|n + n^2 \log n)$, where $|E|$ is the number of edges. Both kernels are impractical in solving large-scale machine learning problems. Regarding the method AKGSVM1, AKGSVM2, AKGSVM3 the complexity of adding virtual edges is $O(|E||E_c|)$, where $|E_c|$ is the number of edges connected with each vertex for the algorithms which iterates over all edges and all connected edges to both vertices of the edge. The connected edges can be found in a constant time using hash structures. This algorithm is suitable for sparse graphs. The question about solving the longest path problem for a fixed size which has linear complexity $O(l!2^l n)$, where l is the length of the path for AKGSVM1, AKGSVM2, AKGSVM3, when we do not sum weights remains open. This algorithm performs deep-first search, path decomposition and apply dynamic programming. For the KNN, there is no training phase. During testing, the nearest neighbors must be found for each test example which can done in $O(tn)$, where t is the number of test examples. For the RGSVM, the complexity is as for SVM. For the RTGSVM, the complexity is the same as for AKGSVM however computed only for nearest neighbors.

Memory Complexity. For graphs, especially sparse graphs, we can store information about edges in a sparse matrix structure. The problem with the methods with global kernels that are SVM with a diffusion kernel and SVM with a shortest path kernel is that they create a dense matrix with n^2 elements (including the existing edges). So we lose sparsity of a representation for a problem. For AKGSVM1, AKGSVM2, AKGSVM3, we add only a limited number of edges. For RTGSVM, it is enough to add a virtual edge only for a pair of edges, when one of them is connected with a test example.

Generalization Performance. The SVM has roots in statistical learning theory. They have been developed using generalization bounds based on Vapnik--Chervonenkis (VC) dimension. Vapnik derived generalization bounds for the transductive learning in [17], which are better than for inductive learning. The proposed method RTGSVM uses transductive learning approach. We use classification on graphs, where the assumption about independent and identically distributed (i.i.d.) data may not be met. Moreover, we deal with discrete spaces, for which we need to use combinatorial bounds. Such bounds competitive to the VC dimension bounds have been derived [18].

For the RGSVM, the kernel with 0s can be indefinite (the quadratic form is neither convex nor concave). The consequence is that the objective function of the dual problem of SVM can be non-convex. The problem that is related to indefinite kernels is that the feature map may not exist for such kernel. However, for graphs, we do not have feature representation for data examples. We have similarities that are defined a priori. Recently, in [10] the stochastic gradient method has been proposed with extreme early stopping. The method does not use a regularization term, thus the kernel values are only related to the linear term.

When we compute similarities according to (11) as in AKGSVM1, then the maximal similarity path will be equivalent to the minimal distance path (the shortest path) for a path length of 2.

For the decision boundary of SVM

$$\sum_{i=1}^{n} y_i \alpha_i K\left(x_i, x\right) + b = 0, \tag{24}$$

where y_i is a class of the ith example, K is a kernel function for RGSVM, when the testing example is not connected with any support vector, then its classification depends on the sign of b. We expect that during tuning of hyperparameters

Table 1. Generalization performance. The numbers in descriptions of the columns mean the methods: 1 - KNN, 2 - RGSVM, 3 - AKGSVM1, 4 - AKGSVM2, 5 - AKGSVM3, 6 - RTGSVM, 7 - diffusion kernel SVM, 8 - the shortest path kernel SVM. Column descriptions: *no* – experiment name, *data* – data set, k_g – the number of the nearest neighbors for generating a structure, k/r – the k value for KNN and r value for RTGSVM, *train* – maximal training set size, *test* – maximal testing set size, *ra* – the average rank for the mean misclassification error; the best method is in bold.

no	data	k_g	k/r	train	test	ra1	ra2	ra3	ra4	ra5	ra6	ra7	ra8
Ex1	All	5	5	100	20	4.55	4.6	4.31	4.35	4.62	4.48	4.88	**4.2**
DBLPEx1	dblp		5	500	50	6.05	6.4	3.65	3.65	3.65	**3.1**	6.4	**3.1**

of SVM, the solutions with support vectors close to validation examples will be promoted. Because AKGSVM increases the number of connections, we expect that the requirement for the number of support vectors will be lower. In particular, AKGSVM increases the number of test examples connected with at least one training example. In Fig. 1, x_7 after generating a virtual edge e_{27} becomes connected to a training example x_2. In the case, when the number of test examples is much bigger than the number of training examples additional procedure might be needed for AKGSVM and KNN in order to improve classification of test examples not connected with any training examples, for example iterative labeling.

5   Experiments

We perform two types of experiments. The first type is on generated weighted graphs with distances from standard classification data sets. We generated graphs by connecting each example with k_g nearest neighbors from a training data set found using the Euclidean distance. The weights in graphs are distances, which are later converted to similarities based on the RBF kernel as a function of a distance by each of the method. Local methods that are KNN and RTGSVM use the similarity to find nearest neighbors. The second type of experiment is on

Table 2. The number of support vectors. The numbers in descriptions of the columns mean the methods: 1 - KNN, 2 - RGSVM, 3 - AKGSVM1, 4 - AKGSVM2, 5 - AKGSVM3, 6 - RTGSVM, 7 - diffusion kernel SVM, 8 - the shortest path kernel SVM. Column descriptions: *no* – experiment name, *sv* – the number of support vectors reported for inductive methods, the best method is in bold, *svB* – Bayesian signed rank probability for the number of support vectors.

no	sv2	sv3	sv4	sv5	sv7	sv8	svB23	svB38
Ex1	81	71	73	76	68	**57**	0.86	**0.95**
DBLPEx1	497	430	500	500	385	**293**	0.63	0.75

Table 3. Computational performance. The numbers in descriptions of the columns mean the methods: 1 - KNN, 2 - RGSVM, 3 - AKGSVM1, 4 - AKGSVM2, 5 - AKGSVM3, 6 - RTGSVM, 7 - diffusion kernel SVM, 8 - the shortest path kernel SVM. Column descriptions: no – experiment name, trT – cumulative training time in seconds, teT – cumulative testing time in seconds.

no	trT2	trT3	trT4	trT5	trT7	trT8	teT1	teT6
Ex1	**0.78**	1.9	1.93	1.86	4.66	4.61	**13.0**	65.0
DBLPEx1	**0.11**	0.18	0.2	0.2	5.97	3.31	2.0	**0.0**

real world graph with similarities for a citation network DBLP. The nodes are articles indexed in the DBLP data set. We use specifically the DBLP data set v1 [12]. The edges are citations between articles. The weights for this graph are binary similarities. When there is a citation the weight is 1, otherwise it is 0. It is a special case of a weighted graph. The graph is undirected.

We use our implementation of the stochastic gradient descent (SGD) solver from [10] for all variants of SVM. For the first type of experiment, we compare all methods on data sets listed in Table 4 for binary classification. More details about data sets are on the LibSVM site [8]. We selected all data sets from this site for binary classification, except those that was too large to store and process them using dense structures due to memory limitation. We plan to implement sparse representation of data in our framework in the future. For all data sets, every feature is scaled linearly to $[0, 1]$. For SVM based methods the number of hyperparameters to tune is 2, σ and C, for KNN only σ for the experiment 1. In the second experiment, we do not tune σ, because we have already similarities. For all hyperparameters, we use a double grid search method for finding the best values – first a coarse grid search is performed, then a finer grid search. The range for σ values is from 2^{-9} to 2^9, for C it is from 2^{-9} to 2^{14} on the first level. The β parameter for a diffusion kernel is set to 0.3. We use the procedure similar to repeated double (nested) cross validation for performance comparison. For the outer loop, we run a modified k-fold cross validation for $k = 10$, with the training set size set to 80% of all examples. When it is not possible to create the next fold, we shuffle data and start from the beginning. For the DBLP data set, we shuffle data with balancing classes only during preprocessing graph data to the internal format. There is no need to shuffle data for our experiments due to limited size of a training data set. We use additional automated standardization of a training matrix after dividing data to folds. We use the 5-fold cross validation for the inner loop for finding optimal values of the hyperparameters. In the first experiment, we generate a graph for each iteration of the inner loop. After that, we run the method on a training set, and we report results on a test set. Here, we also generate a graph in the first experiment. We additionally limit the size of a training set to speed up the experiments. The limitation for a training set is listed in Table 1 for each experiment. We also limit the number of test examples in experiments with the RTGSVM method to speed up the experiments, because for this method, a new model must be trained for each test example. In the future, we plan to implement aggregation of test examples to limit the number of models. We also use Bayesian statistical tests which are preferred over null hypothesis significance testing (NHST). In particular, we use Bayesian signed rank test implemented in R. For the RTGSVM method, we generate virtual edges as in AKGSVM1.

The overall results for generalization performance are in Table 1. The results for the number of support vectors are in Table 2. The results for computational performance are in Table 3. The example of results for particular data sets are in Table 4. The observations are as follows.

- The generalization performance of local models like KNN and RTGSVM are competitive to inductive SVM. For standard classification SVM has small advantage in terms of accuracy over KNN [10]. For graph data, this difference is even less noticeable. However, the requirement for local models is to get enough connections to the nearest neighbors to match the expected number of neighbors k for KNN, and r for RTGSVM. For the increased number of connections to other test examples, local models may start to degrade. Moreover, in the Ex1, the number of the nearest neighbors matches the number of neighbors used during graph generation. In real world graphs the number of neighbors in a graph varies, thus choosing the optimal value of k/r may require some tuning, and even then the optimal value can be different for different test examples. The generalization performance of RGSVM is slightly worse than AKGSVM which is noticeable for the DBLP data set, but without statistical significance. Thus, reducing sparsity of a similarity matrix is beneficial for SVM. However, generating the full kernel matrix does little noticeable improvement in the generalization performance over AKGSVM, for the shortest path kernel. Specifically, for the DBLP data set, AKGSVM is little better than KNN and RGSVM. The most competitive for the shortest path kernel is RTGSVM with almost the same performance.
- The number of support vectors is much better for the AKGSVM than for RGSVM which is almost statistically significant for combined results (column svB23 in Table 2). It is almost statistically significant for particular data sets (column svB23 in Table 4). The global kernels have still advantage in terms of the number of support vectors compared to other methods (column svB39 in Table 2). The potential reason for this is that introducing sparsity in the kernel matrix by putting 0s leads to non-smoothness, which requires more complex functions. For the DBLP data sets, the number of support vectors is rather high due to the limited number of edges, which is affected by the size of a training data set and connections with other test examples.
- Training time for SVM with global kernels that are SVM with a diffusion kernel, and with the shortest path kernel are considerably higher than for other methods due to computational complexity. We also lose advantage of sparsity of a kernel matrix, so memory consumption may be an issue. Testing time for local models can be greater than training time for inductive models, which depends on the number of test examples.

Overall, from the practical point of view, the local models like KNN can be used for data sets when there are enough neighbor connections and we have a graph with connections between similar examples and relatively small number of test examples. If the weighted connections are between nonsimilar examples, the local classifiers may not work properly. For bigger number of connections, SVM methods may have advantage of taking into account structure of the data. Due to the computational performance and memory requirements, the proposed axiomatic kernels are a better choice than global kernels, especially for big data sets.

Table 4. Results per data set. The numbers in descriptions of the columns mean the methods: 1 - KNN, 2 - RGSVM, 3 - AKGSVM1, 4 - AKGSVM2, 5 - AKGSVM3. Column descriptions: no – experiment name, dn – the name of a data set, s – the number of all examples, d – the dimension of a problem, ce – the mean misclassification error; the best method is in bold, sv – the number of support vectors reported for inductive methods, svB – Bayesian signed rank probability for the number of support vectors.

no	dn	s	d	err1	err2	err3	err4	err5	sv2	sv3	svB23
Ex1	a1a	24947	123	**0.34**	0.39	0.385	0.39	0.36	81	**53**	0.79
Ex1	australian	690	14	0.205	0.205	0.21	0.195	**0.17**	92	**89**	0.41
Ex1	breast-cancer	675	10	0.065	0.06	0.07	0.07	**0.05**	98	**58**	0.83
Ex1	cod-rna	100000	8	0.24	0.195	**0.17**	0.18	0.21	79	**55**	0.75
Ex1	colon-cancer	62	2000	0.25	0.26	**0.22**	0.245	0.255	31	**27**	0.53
Ex1	covtype	100000	54	**0.305**	0.33	0.355	0.34	0.355	**69**	76	0.24
Ex1	diabetes	768	8	**0.215**	0.235	0.27	0.245	0.26	82	**78**	0.54
Ex1	fourclass	862	2	0.05	**0.03**	0.035	**0.03**	0.035	74	**53**	0.63
Ex1	german_numer	1000	24	**0.3**	0.365	0.35	0.335	0.33	79	**74**	0.53
Ex1	heart	270	13	0.19	0.2	0.16	**0.155**	0.18	99	**92**	0.53
Ex1	HIGGS	100000	28	**0.345**	0.355	0.43	0.395	0.4	91	**87**	0.36
Ex1	ijcnn1	100000	22	0.015	**0.0**	**0.0**	0.01	0.01	94	100	0.0
Ex1	ionosphere_sc	350	33	0.26	0.195	0.1	**0.075**	0.145	57	**55**	0.5
Ex1	liver-disorders	341	5	0.5	**0.48**	0.485	0.485	0.49	**88**	96	0.13
Ex1	madelon	2600	500	**0.415**	0.43	0.44	0.47	0.455	86	**77**	0.6
Ex1	mushrooms	8124	111	0.05	**0.025**	0.05	0.045	0.035	79	**44**	0.85
Ex1	phishing	5785	68	**0.185**	0.225	0.2	0.21	0.205	98	**87**	0.61
Ex1	skin_nonskin	51432	3	0.07	**0.035**	0.04	0.04	**0.035**	85	**82**	0.36
Ex1	splice	2990	60	0.345	0.355	0.375	**0.34**	0.365	85	**70**	0.73
Ex1	sonar_scale	208	60	0.29	0.23	0.215	**0.205**	0.25	69	**66**	0.52
Ex1	SUSY	100000	18	0.365	**0.355**	0.38	**0.355**	0.41	90	**85**	0.61
Ex1	svmguide1	6910	4	**0.08**	0.1	0.09	0.09	0.085	80	**63**	0.57
Ex1	svmguide3	1243	21	**0.25**	0.285	0.26	0.295	0.275	83	**70**	0.66
Ex1	w1a	34703	300	**0.0**	**0.0**	**0.0**	**0.0**	**0.0**	91	**75**	0.42
Ex1	websam_unigr	100000	134	0.155	0.175	**0.1**	0.135	0.185	71	**56**	0.72
DBLPEx1	dblp	7967		0.162	0.164	**0.144**	**0.144**	**0.144**	497	**430**	0.63

6 Summary

We proposed a novel idea of generating rules based on axioms for generating virtual edges, which are used in SVM for classification on graphs. Potentially, virtual edges can be used with any other method for classifying graphs. The results are promising in terms of computational performance when compared to kernels on graphs. The proposed methods may be preferable in some scenarios over simple local models like KNN. The framework of axiomatic rules can be extended by introducing combination of rules and tuning of rules by incorporating uncertainty in the form of hyperparameters.

Acknowledgments. The theoretical analysis and the method design are financed by the National Science Centre in Poland, project id 289884, UMO-2015/17/D/ST6/04010, titled "Development of Models and Methods for Incorporating Knowledge to Support Vector Machines" and the data driven method is supported by the European Research Council under the European Union's Seventh Framework Programme. Johan Suykens acknowledges support by ERC Advanced Grant E-DUALITY (787960), KU Leuven C1, FWO G0A4917N. This paper reflects only the authors' views, the Union is not liable for any use that may be made of the contained information.

References

1. Borgwardt, K.M., Kriegel, H.: Shortest-path kernels on graphs. In: Proceedings of the 5th IEEE International Conference on Data Mining (ICDM 2005), Houston, Texas, USA, 27–30 November 2005, pp. 74–81 (2005). https://doi.org/10.1109/ICDM.2005.132
2. Bronshtein, I.N., Semendyayev, K., Musiol, G., Muehlig, H.: Handbook of Mathematics. In: Functional Analysis, pp. 596–641. Springer, Heidelberg (2007). https://doi.org/10.1007/978-3-540-72122-2_12
3. Can, T., Çamoglu, O., Singh, A.K.: Analysis of protein-protein interaction networks using random walks. In: Proceedings of the 5th International Workshop on Bioinformatics, BIOKDD 2005, Chicago, Illinois, USA, 21 August 2005, pp. 61–68 (2005). https://doi.org/10.1145/1134030.1134042
4. Chen, Y., Garcia, E.K., Gupta, M.R., Rahimi, A., Cazzanti, L.: Similarity-based classification: concepts and algorithms. J. Mach. Learn. Res. **10**, 747–776 (2009)
5. Fouss, F., Françoisse, K., Yen, L., Pirotte, A., Saerens, M.: An experimental investigation of kernels on graphs for collaborative recommendation and semisupervised classification. Neural Netw. **31**, 53–72 (2012). https://doi.org/10.1016/j.neunet.2012.03.001
6. Gallagher, B., Tong, H., Eliassi-Rad, T., Faloutsos, C.: Using ghost edges for classification in sparsely labeled networks. In: Proceedings of the 14th ACM SIGKDD International Conference on Knowledge Discovery and Data Mining, Las Vegas, Nevada, USA, 24–27 August 2008, pp. 256–264 (2008). https://doi.org/10.1145/1401890.1401925
7. Kondor, R., Lafferty, J.D.: Diffusion kernels on graphs and other discrete input spaces. In: Machine Learning, Proceedings of the Nineteenth International Conference (ICML 2002), Sydney, Australia, 8–12 July 2002, pp. 315–322, University of New South Wales (2002)

8. LIBSVM Data Sets, July 2011. www.csie.ntu.edu.tw/~cjlin/libsvmtools/datasets/
9. Lin, F., Cohen, W.W.: Semi-supervised classification of network data using very few labels. In: International Conference on Advances in Social Networks Analysis and Mining, ASONAM 2010, Odense, Denmark, 9–11 August 2010, pp. 192–199 (2010). https://doi.org/10.1109/ASONAM.2010.19
10. Melki, G., Kecman, V., Ventura, S., Cano, A.: OLLAWV: online learning algorithm using worst-violators. Appl. Soft Comput. **66**, 384–393 (2018)
11. Namata, G., Sen, P., Bilgic, M., Getoor, L.: Collective classification. In: Sammut, C., Webb, G.I. (eds.) Encyclopedia of Machine Learning and Data Mining, pp. 238–242. Springer, Boston (2017). https://doi.org/10.1007/978-1-4899-7687-1_44
12. Pan, S., Zhu, X., Zhang, C., Yu, P.S.: Graph stream classification using labeled and unlabeled graphs. In: 29th IEEE International Conference on Data Engineering, ICDE 2013, Brisbane, Australia, 8–12 April 2013, pp. 398–409 (2013). https://doi.org/10.1109/ICDE.2013.6544842
13. Perozzi, B., Al-Rfou, R., Skiena, S.: DeepWalk: online learning of social representations. In: The 20th ACM SIGKDD International Conference on Knowledge Discovery and Data Mining, KDD 2014, New York, NY, USA, 24–27 August 2014, pp. 701–710 (2014). https://doi.org/10.1145/2623330.2623732
14. Schölkopf, B., Tsuda, K., Vert, J.P.: Kernel methods in computational biology. In: Diffusion Kernels, pp. 171–192, 16 July 2004. The MIT Press, January 2003
15. Shervashidze, N., Schweitzer, P., van Leeuwen, E.J., Mehlhorn, K., Borgwardt, K.M.: Weisfeiler-Lehman graph kernels. J. Mach. Learn. Res. **12**, 2539–2561 (2011)
16. Smola, A.J., Kondor, R.: Kernels and regularization on graphs. In: Computational Learning Theory and Kernel Machines, Proceedings of the 16th Annual Conference on Computational Learning Theory and 7th Kernel Workshop, COLT/Kernel 2003, Washington, DC, USA, 24–27 August 2003, pp. 144–158 (2003)
17. Vapnik, V.N.: Statistical Learning Theory. Wiley-Interscience (1998)
18. Vorontsov, K., Ivahnenko, A.: Tight combinatorial generalization bounds for threshold conjunction rules. In: Kuznetsov, S.O., Mandal, D.P., Kundu, M.K., Pal, S.K. (eds.) PReMI 2011. LNCS, vol. 6744, pp. 66–73. Springer, Heidelberg (2011). https://doi.org/10.1007/978-3-642-21786-9_13
19. Wu, G., Chang, E.Y., Zhang, Z.: An analysis of transformation on non-positive semidefinite similarity matrix for kernel machines. In: Proceedings of the 22nd International Conference on Machine Learning (2005)
20. Yajima, Y., Kuo, T.: Efficient formulations for 1-SVM and their application to recommendation tasks. JCP **1**(3), 27–34 (2006). https://doi.org/10.4304/jcp.1.3.27-34

Multitask Learning on Graph Neural Networks: Learning Multiple Graph Centrality Measures with a Unified Network

Pedro Avelar[✉] [ID], Henrique Lemos[ID], Marcelo Prates[ID], and Luis Lamb[ID]

Institute of Informatics, UFRGS, Porto Alegre, Brazil
{phcavelar,hlsantos,morprates,lamb}@inf.ufrgs.br

Abstract. The application of deep learning to symbolic domains remains an active research endeavour. Graph neural networks (GNN), consisting of trained neural modules which can be arranged in different topologies at run time, are sound alternatives to tackle relational problems which lend themselves to graph representations. In this paper, we show that GNNs are capable of multitask learning, which can be naturally enforced by training the model to refine a single set of multidimensional embeddings $\in \mathbb{R}^d$ and decode them into multiple outputs by connecting MLPs at the end of the pipeline. We demonstrate the multitask learning capability of the model in the relevant relational problem of estimating network centrality measures, focusing primarily on producing rankings based on these measures, i.e. is vertex v_1 more central than vertex v_2 given centrality c?. We then show that a GNN can be trained to develop a *lingua franca* of vertex embeddings from which all relevant information about any of the trained centrality measures can be decoded. The proposed model achieves 89% accuracy on a test dataset of random instances with up to 128 vertices and is shown to generalise to larger problem sizes. The model is also shown to obtain reasonable accuracy on a dataset of real world instances with up to 4k vertices, vastly surpassing the sizes of the largest instances with which the model was trained ($n = 128$). Finally, we believe that our contributions attest to the potential of GNNs in symbolic domains in general and in relational learning in particular.

Keywords: Graph neural networks · Graph networks · Centrality measures · Network centrality

1 Introduction

A promising technique for building neural networks on symbolic domains is to enforce permutation invariance by connecting adjacent elements of the domain of discourse through neural modules with shared weights which are themselves subject to training. By assembling these modules in different configurations one can

© Springer Nature Switzerland AG 2019
I. V. Tetko et al. (Eds.): ICANN 2019, LNCS 11731, pp. 701–715, 2019.
https://doi.org/10.1007/978-3-030-30493-5_63

reproduce each graph's structure, in effect training neural components to compute the appropriate messages to send between elements. The resulting architecture can be seen as a message-passing algorithm where the messages and state updates are computed by trained neural networks. This model and its variants are the basis for several architectures such as message-passing neural networks [11], recurrent relational networks [28], graph networks [3], graph convolutional networks [22] and graph neural networks (GNN) [13,31,32] whose terminology we adopt.

GNNs have been successfully employed on both theoretical and practical combinatorial domains, with [28] showing how they can tackle Sudoku puzzles, [34] developing a GNN which is able to predict the satisfiability of CNF boolean formulas, [11] using them to predict quantum properties about molecules and [30] using it outside of the context of graphs for relational question answering about pictures. Some of these approaches can even extend their computation to more iterations of message-passing, showing that GNNs can not only learn from examples, but may learn to reason in an iterative fashion.

With this in mind, we turn to the relevant problem of approximating centrality measures on graphs, a combinatorial problem with very relevant applications in our highly connected world, including the detection of power grid vulnerabilities [27], influence inside interorganisational and collaboration networks [8], social network analysis [21], pattern recognition on biological networks [9] among others. This work concerns itself with whether a neural network can approximate centrality measures solely from a network's structure.

The remainder of the paper is structured as follows. First we provide a survey of related work on approximating centrality measures with neural networks. Then, we present the basic concepts of centrality measures used in this paper, introduce some GNN-based models for approximating and learning the relations between centralities in graphs, describe our experimental evaluation, and verify the models' generalisation and interpretability. Finally, we conclude with the contributions and shortcomings of our work, and point out direction for further research.

2   Related Work

With regard to our problem at hand, related work can be divided into two fronts. One is that of neural networks approximating centralities on graphs, and the other is the related work on modelling node-level embeddings with graph neural networks. On this first aspect, there are works such as [14–17] and [23], all of which uses neural networks to estimate centrality measures. However, in [15,16] and [14] they use a priori knowledge of other centralities to approximate a different one, on [17] they also produce a ranking of the centrality measures, but do so using the degree and eigenvector centralities as input, and in [16] local features such as number of vertices in a network, number of edges in a network, degree and the sum of the degrees of all of the vertex's neighbours are used. These contributions differ from ours in that we feed our neural network solely with the network structure – that is, they use a simple MLP which receives numeric information about a specific node and outputs an approximation of one desired centrality measure, while our builds a message-passing procedure with only the network structure and no numeric information whatsoever. [32] fits both fronts, since it also uses GNNs to compute rankings for the PageRank centrality measure for a single graph, and does not focus on other centralities nor analyses the transfer between centralities.

In [29] latent representations of networks are learned, akin to our models, but their work does not focus on predicting centrality measures per-se. The works in [10,19,37] also concern themselves with creating node embeddings, but again their work does not focus on centrality measures and use features other than the network structure as inputs, specifically [37] trains its predictor for each specific graph, while also using the node's degree as information. Most of these, however, either train their models using the same network distributions which they use for evaluation, or learn embeddings specific to each graph, while we learn an algorithmic procedure from synthetic data distributions and try to extrapolate to other distributions, which makes direct comparison somewhat difficult. On the other hand, works such as [25,39,40] work on learning generative models for graph generation, the first and the latter applying their work to the generation of chemical compounds, this can be seen as a step forward in modelling parameters for graphs, but their analyses do not study the relationships between the latent space and the characteristics of the networks itself. For a survey of the area of Graph Neural Networks in general, one can look at [3,11,41].

Given this, our contributions are: (1) The experimental analyses of different graph neural network approaches to ranking nodes and how well they scale in a multitask environment – in contrast with non-GNN models and those which were not used for multitasking. (2) The proposal of using a learned native comparison method – instead of approximating the metrics themselves to produce rankings. (3) The training of a model which predicts values solely from the network structure, which is trained in graphs altogether different from those where the model is tested – instead of optimising the model for a specific graph or training it in graphs from the same distribution as the target ones. (4) We briefly analyse the relationship behind the learned computed embeddings and their target functions.

3   On Centrality Measures

In general, node-level centralities attempt to summarise a node's contribution to the network cohesion. Several centralities have been proposed and many models and interpretations have been suggested, namely: autonomy, control, risk, exposure, influence, etc. [6]. Despite their myriad of applications and interpretations, in order to calculate some of these centralities one may face both high time and space complexity, thus making it costly to compute them on large networks. Although some studies pointed out a high degree of correlation between some of the most common centralities [24], it is also stated that these correlations are attached to the underlying network structure and thus may vary across different network distributions [33]. Therefore, techniques to allow faster centrality computation are topics of active research [14].

Here, however, we are not concerned as much with the time complexity of computing the centrality measure per-se, but with whether a neural network can infer a node's centrality solely from the network structure – that is, without any numeric information about the node, its neighbourhood, or the network itself – even if the complexity of the methods presented here is similar to that of matrix operations, as the underlying procedures are based on these, and is polynomial with the size of the input. With this in mind, we selected four well-known node centralities to investigate in our study: **degree** – first proposed by [36], it simply calculates to how many neighbours a node is connected; **betweenness** – it calculates the number of shortest paths which cross by the given node. High betweenness nodes are more important to the graph's cohesion, i.e., their removal may disconnect the graph. A fast algorithm version was introduced by [7]; **closeness** – as defined by [4], it is a distance-based centrality which measures the average geodesic distance between a given node and all other reachable nodes; **eigenvector** – this centrality uses the largest eigenvalue of the adjacency matrix to compute its eigenvector [5] and assigns to each node a score based upon the score of the nodes to whom it is connected. It is usually computed via a power iteration method with no convergence guaranteed.

4   A GNN Model for Learning Relations Between Centrality Measures

On a conceptual level, the GNN application considered here assigns multidimensional embeddings $\in \mathbb{R}^d$ to each vertex in the input graph. These embeddings are refined through t_{max} iterations of message-passing. At each iteration, each vertex adds up all the messages received along its edges and adds up all the messages received along its outcoming edges, obtaining two $\mathbb{R}^d$ tensors. These two tensors are concatenated to obtain a $\mathbb{R}^{2d}$ tensor, which is fed to a Recurrent Neural Network (RNN) which updates the embedding of the vertex in question. Note that a "message" sent by a vertex embedding in this sense is the output of a Multilayer Perceptron (MLP) which is fed with the embedding of the vertex in question.

In summary, these models can be seen as a message-passing algorithm in which the update (V_u) and message-computing ($src_{msg} : \mathbb{R}^d \rightarrow \mathbb{R}^d$, $tgt_{msg} : \mathbb{R}^d \rightarrow \mathbb{R}^d$) modules are trained neural networks. With this setup we tested two different methods of extracting the centrality measures from the propagated embeddings. Our baseline method is the straightforward application of the methods commonly proposed in the GNN literature, where a MLP $approx_c : \mathbb{R}^d \rightarrow \mathbb{R}$, which tries to approximate each centrality measure c in question directly, is trained. In the method we propose here, however, we train a MLP $cmp_c : \mathbb{R}^{2d} \rightarrow \mathbb{R}$, which is assigned with computing the probability that $v_i >_c v_j$ given their embeddings, where $>_c$ here denotes the total ordering imposed by the centrality measure c, that is, the node v_i whose embedding is on the first d dimensions has a strictly higher c-centrality than v_j whose embedding is on the last d dimensions of the input to cmp_c).

The first method computes the centralities directly, minimising the mean squared error (MSE) between the prediction and the true value of the centrality, and a ranking can be extracted from these directly. For such method we considered three different setups, in one we learn the normalised centrality measures directly (named AN, for "Approximate the Normalised centrality"), in the second one we learn the unnormalised version of the centrality measures (called AU, for "Approximate the Unnormalised centrality"), and a third approach was to learn from the normalised centrality values, but perform a normalisation of the model's approximated value before using it as its final output (dubbed AM, for "Approximate the normalised centrality, with normalisation on the Model"). We perform Stochastic Gradient Descent (SGD), more specifically TensorFlow's Adam implementation, on the MSE loss, and evaluate these models using both absolute and relative errors, as well as computing the Kendall-τ correlation coefficient for the set of predictions in each graph.

In our proposed method, however, we only have one model (that we name as RN, for "Rank centralities Natively by comparison") in which for each pair of vertices $(v_i, v_j) \in \mathcal{V} \times \mathcal{V}$ and for each centrality $c \in \mathcal{C}$, our network guesses the probability that $v_i >_c v_j$. To train such a network we perform Stochastic Gradient Descent (SGD) on the binary cross entropy loss between the probabilities computed by the network and the binary "labels" obtained from the total ordering provided by c. This process can be made simple by organising the n^2 network outputs for each centrality, as well as the corresponding labels, into $n \times n$ matrices, as Fig. 1 exemplifies. In such a matrix, the binary cross entropy is computed as $H(\mathbf{M}_{\gtrsim_c}, \mathbf{T}) = -\sum_{i,j} P(v_i >_c v_j) \log T_{ij}$. Arranging the labels and cross entropy in such a way allows one to compute an accuracy on these binary predictions, which can be seen as a more strict version of the Kendall-τ correlation coefficient, since this accuracy metric penalises ties in the ranking as much as discordant rankings. With this in mind, we focus on this accuracy metric when comparing the baseline models (AN,AU,AM) with the RN model. We also provide other common metrics, calculated in the same fashion, such as Precision, Recall and True Negative rates, for the sake of completeness.

$$
\begin{pmatrix}
P(v_1 >_c v_1) & P(v_2 >_c v_1) & P(v_3 >_c v_1) \\
P(v_1 >_c v_2) & P(v_2 >_c v_2) & P(v_3 >_c v_2) \\
P(v_1 >_c v_3) & P(v_2 >_c v_3) & P(v_3 >_c v_3)
\end{pmatrix}
\begin{pmatrix}
0 & 1 & 1 \\
0 & 0 & 1 \\
0 & 0 & 0
\end{pmatrix}
$$

Fig. 1. Example of a predicted fuzzy comparison matrix $\mathbf{M}_{\gtrsim_c}$ at the left and the training label given by an upper triangular matrix $\mathbf{T}$ at the right, for a graph with three vertices $\mathcal{V} = \{v_1, v_2, v_3\}$ sorted in ascending centrality order as given by the centrality measure c. The binary cross entropy is computed as $H(\mathbf{M}_{\gtrsim_c}, \mathbf{T}) = -\sum_{i,j} P(v_i >_c v_j) \log T_{ij}$

In summary, the models used here are the three baseline models, which are identical in structure, and whose only difference is the function they optimise: the AN model optimises $MSE(pred_c, norm(c))$, the AM model optimises $MSE(norm(pred_c), norm(c))$, while the AU model optimises $MSE(pred_c, c)$. In contrast, our proposed model learns a different function and computes the rankings of the centralities through a learned native comparison on the internal embedding representation of nodes, using the cross entropy on the binary comparison between all pairs of embeddings as a loss. A complete description of our RN algorithm is presented in Algorithm 1.

Algorithm 1. Graph Neural Network Centrality Predictor

1: **procedure** GNN-CENTRALITY($\mathcal{G} = (\mathcal{V}, \mathcal{E}), \mathcal{C}$)
2:
3:     // Compute adj. matrix
4:     $\mathbf{M}[i,j] \leftarrow 1$ if $(v_i, v_j) \in \mathcal{E}$ else 0
5:     // Initialise all vertex embeddings with the initial embedding V_{init} (this initial embedding is a parameter learned by the model)
6:     $V^1[i,:] \leftarrow V_{init} \mid \forall v_i \in \mathcal{V}$
7:     // Run t_{max} message-passing iterations
8:     **for** $t = 1 \dots t_{max}$ **do**
9:         // Refine each vertex embedding with messages received from incoming edges
10:        // either as a source or a target vertex
11:        $\mathbf{V}^{t+1}, \mathbf{V}_h^{t+1} \leftarrow V_u(\mathbf{V}^t, \mathbf{M} \times src_{msg}(\mathbf{V}^t), \mathbf{M}^T \times tgt_{msg}(\mathbf{V}^t))$
12:
13:    **for** $c \in \mathcal{C}$ **do**
14:        // Compute a fuzzy comparison matrix $M_{\gtrsim_c} \in \mathbb{R}^{|\mathcal{V}| \times |\mathcal{V}|}$
15:        $\mathbf{M}_{\gtrsim_c}[i,j] \leftarrow cmp_c(\mathbf{V}^{t_{max}}[i,:], \mathbf{V}^{t_{max}}[j,:]) \mid \forall\ v_i, v_j \in \mathcal{V}$
16:        // Compute a strict comparison matrix $M_{<_c} \in \{\top, \bot\}^{|\mathcal{V}| \times |\mathcal{V}|}$
17:        $\mathbf{M}_{>_c} \leftarrow M_{\gtrsim_c} > \frac{1}{2}$

---

5   Experimental Setup

We generated a training dataset by producing 4096 graphs between 32 and 128 vertices for each of the four following random graph distributions (total 16384): (1) Erdő-Rényi [2], (2) Random power law tree[1], (3) Connected Watts-Strogatz small-world model [38], (4) Holme-Kim model [20]. Further details are reported in Table 1. We use 5 different datasets for evaluating our models, called "test", "large", "different", "sizes" and "real". The "test" dataset is composed of new instances sampled from the same distributions as the training dataset, and has

---

[1] This refers to a tree with a power law degree distribution specified by the parameter γ.

the same number of instances. The "large" dataset has fewer instances, sampled from the same distributions but the number of nodes in the networks are counted between 128 and 512 vertices. The "different" dataset has the same number of nodes as in the training data, but these are sampled from two different distributions, namely the Barabási-Albert model [1] with and shell graphs[2] [35]. The "sizes" dataset again uses the same distributions seen during training, but with the specific sizes on a range from 32 to 256 with strides of 16 to allow us to analyse the generalisation to larger instance sizes and how the performance of the models vary with the graph's size, beginning in the range with which the model was trained up until the size of the largest instances of the "large" dataset. All these graphs were generated with the Python NetworkX package[3] [18].

Table 1. Training instance generation parameters

Graph distribution	Parameters
Erdős-Rényi	$p = 0.25$
Random power law tree	$\gamma = 3$
Watts-Strogatz	$k = 4, p = 0.25$
Holme-Kim	$m = 4, p = 0.1$

The "real" dataset consists of real instances obtained from either the Network Repository[4] or from the Stanford Large Network Dataset Collection[5], which were *power-eris1176*, a power grid network, *econ-mahindas*, an economic network, *socfb-haverford76* and *ego-Facebook*, Facebook networks, *bio-SC-GT*, a biological network, and *ca-GrQc*, a scientific collaboration network. The size range of these networks significantly surpass that of the training instances, overestimating from ×9 to ×31 the size of the largest ($n = 128$) networks which will be seen during training, while also pertaining to entirely different graph distributions than those described in Table 1.

In the results presented here we instantiate our models with size $d = 64$ vertex embeddings, using a simple layer-norm LSTM with d-dimensional output and hidden state, as well as three-layered (d, d, d) MLPs src_{msg}, tgt_{msg}, $(d, d, 1)$ and $(2d, 2d, 1)$ MLPs for the $approx_c$ and cmp_c functions $\forall c \in C^6$. Rectified Linear Units (ReLU) were used for all hidden layers as non-linearities and the output layer was linear, the LSTM used ReLUs instead of the traditional hyperbolic tangent. The embedding dimensionality of 64 was chosen since it improved

---

[2] The shell graphs used here were generated with the number of points on each shell proportional to the "radius" of that shell. I.E., $n_i \approx \pi \times i$ with n_i being the number of nodes in the i-th shell.

[3] https://networkx.github.io/.

[4] http://networkrepository.com/.

[5] https://snap.stanford.edu/data/.

[6] C here denotes the set of centrality measures.

performance over lower dimensional embeddings, no other hyperparameters were optimised.

The message-passing kernel weights are initialised with TensorFlow's Xavier initialisation method described in [12] and the biases are initialised to zeroes. The LSTM assigned with updating embeddings has both its kernel weights and biases initialised with TensorFlow's Glorot Uniform Initialiser [12], with the addition that the forget gate bias had 1 added to it. The number of message-passing timesteps is set at $t_{max} = 32$.

Each training epoch is composed by 32 SGD operations on batches of size 32, randomly sampled from the training dataset (The sampling inhibited duplicates in an epoch, but duplicates are allowed across epochs). Instances were batched together by performing a disjoint union on the graphs, producing a single graph with every graph being a disjoint subgraph of the batch-graph, in this way the messages from one graph were not be passed to another, effectively separating them.

All models were tested both by training a single model for each centrality as well as building a model which shared the computed node embedding over different centralities, learning only a different $approx_c$ or cmp_c function as necessary for each centrality. This allows us to measure whether such multitask learning of different centrality measures has any impact, positive or negative, on the model's performance.

6   Experimental Analysis

After 32 training epochs, the models were able to compute centrality predictions and comparisons. The approximation models had high errors for some of the centralities, but the Kendall-τ correlation coefficient on the AN and AM models was satisfactory when calculated on the generated datasets, results similar to those reported in [14–17]. The AU model had a low Kendall-τ correlation coefficient for the multitasking model, even for the "test" dataset, which mirrors how the normalisation for each centrality allows the network to perform better due to a common range of possible values.

Thus, we turned ourselves to comparing the three remaining models which presented a reasonable accuracy, often achieving over 80% training accuracy (averaged over all centralities). However the AM model had difficulties performing as good consistently on the "test" dataset, where the AN and RN models did not have any problem whatsoever. The accuracy of the multitasking AN model was consistently and significantly worse than the non-multitasking models, while in the RN model the multitasking is outperformed by the basic model in many cases, but the overall accuracy is not significantly different (see Table 2 for further results and comparisons). In the big picture, however, the RN model consistently outperformed the others, lagging only on the Recall metric.

In this context, recall that the multitask learning model is required to develop a "lingua franca" of vertex embeddings from which information about any of the studied centralities can be easily extracted, so in a sense it is solving a harder

Table 2. Performance metrics (Precision, Recall, True Negative rate, Accuracy) computed for the trained models on the "test" dataset. All values are presented as percentages and the best value for each metric and centrality combination is highlighted.

Centrality	AN				AM				RN			
	P	R	TN	Acc	P	R	TN	Acc	P	R	TN	Acc
Without multitasking												
Betweenness	82.4	**88.6**	84.6	86.0	39.9	45.6	43.5	44.1	**90.3**	88.5	**91.0**	**89.8**
Closeness	83.1	**85.4**	84.3	84.8	83.1	**85.4**	84.4	84.9	**88.4**	84.5	**89.8**	87.3
Degree	75.3	**98.4**	81.6	87.5	63.3	85.8	70.4	75.6	**99.3**	94.9	**99.4**	**97.6**
Eigenvector	87.0	87.0	87.6	87.3	86.5	86.5	87.0	86.8	86.2	**90.2**	82.3	86.3
Average	82.0	**89.8**	84.5	86.4	68.2	75.8	71.3	72.8	**91.0**	89.5	90.6	**90.3**
With multitasking												
Betweenness	73.6	79.6	76.1	77.3	41.9	46.7	46.3	46.2	87.2	87.2	88.9	87.9
Closeness	71.4	73.0	73.6	73.3	80.2	82.5	81.7	82.1	86.9	82.0	88.7	85.5
Degree	73.7	96.0	80.4	85.9	72.4	94.3	79.3	84.7	98.3	92.4	99.0	96.4
Eigenvector	83.5	83.5	84.2	83.9	85.5	85.5	86.2	85.9	**89.8**	88.3	**90.4**	**89.4**
Average	75.5	83.0	78.6	80.1	70.0	77.3	73.4	74.7	90.5	87.5	**91.8**	89.8

problem. We also computed performance metrics for the "large" dataset where the overall accuarcy of the AM model falls to about 65%, while it surprisingly was able to maintain performance for the eigenvector centrality. The accuracy of the AN model falls a few percent points to 81% and 78% (without/with multitasking), while the RN model also had its performance lowered to about 80%, a sharper drop but still performing on par with the AN model. This result shows that our proposed model is able to generalise to larger problem sizes than those it was trained on, with a expected decrease accuracy, while also being able of handling multitasking without interference on the performance of other centralities.

6.1   Generalising to Other Distributions

Having obtained good performance for the distributions the network was trained on, we wanted to assess the possibility of accurately predicting centrality comparisons for graph distributions it has never seen. That was done by computing performance metrics on the "different" dataset for which the results are reported in Table 3. Although its accuracy is reduced in comparison (81.9% vs 90.3% overall), the model can still predict centrality comparisons with high performance, obtaining its worst result at 82.0% recall for the closeness centrality. The model without multitasking outperforms the multitasking one only by a narrow margin (0.5% at the overall accuracy), while the multitasking model has a better accuracy than even the best baseline model. The comparison model was only outperformed in the recall metric and only by 0.3% in average.

Table 3. Performance metrics (Precision, Recall, True Negative rate, Accuracy) computed for the trained models on the "different" dataset. All values are presented as percentages and the best value for each metric and centrality combination is highlighted.

Centrality	AN				AM				RN			
	P	R	TN	Acc	P	R	TN	Acc	P	R	TN	Acc
Without multitasking												
Betweenness	80.2	**80.3**	80.8	**80.5**	43.1	43.2	44.8	44.0	**81.2**	77.5	**81.8**	79.7
Closeness	80.8	82.4	81.6	82.0	80.8	82.4	81.6	82.0	81.7	75.3	**84.2**	79.9
Degree	82.7	94.6	85.2	89.0	75.3	86.8	78.3	81.7	86.4	72.5	89.0	82.1
Eigenvector	70.4	70.4	71.2	70.8	69.9	69.9	70.7	70.3	**84.9**	**87.9**	**83.8**	**85.8**
Average	78.5	**81.9**	79.7	80.6	67.3	70.6	68.9	69.5	**83.6**	78.3	**84.7**	**81.9**
With multitasking												
Betweenness	73.4	73.6	74.3	73.9	46.6	46.8	48.2	47.5	77.9	77.0	78.5	77.8
Closeness	81.3	82.9	82.2	82.5	**81.9**	**83.5**	82.7	**83.1**	79.6	77.5	81.4	79.5
Degree	85.0	**97.0**	87.4	**91.3**	84.1	96.1	86.5	90.3	**87.4**	74.9	**91.0**	84.0
Eigenvector	67.5	67.5	68.4	68.0	70.4	70.4	71.1	70.7	79.6	80.5	79.9	80.2
Average	76.8	80.3	78.1	78.9	70.8	74.2	72.1	72.9	81.1	77.5	82.7	80.4

Testing the models on the "sizes" dataset provides insight on how their performance decays on larger instances. Our multitask model accuracy presents an expected decay with increasing problem sizes (see Fig. 2), while the non-multitasking models performed similarly. However, this decay is not a free fall towards 50% for instances with almost twice the size of the ones used to train the model, in fact the overall accuracy remains around 80% when $n = 240$ which implies that some level of generalisation to even larger problem sizes is achievable.

The approximation models, however, had different behaviours for each centrality. The RM model was able to maintain its performance relatively stable for the eigenvector centrality, while with degree and closeness it presented sharp drops inside and outside the range of sizes it was trained upon (the betweenness centrality did not reach 50% accuracy and was not taken into consideration). The RN model was the most stable overall, only having a significant drop in performance with the closeness centrality when tested on instances with 176 or more nodes.

We also wanted to assess the model's performance on real world instances. We analysed the performace of the model on the "real" dataset. The RN model was able to obtain up to 86.04% accuracy (on degree) and 81.79% average accuracy on the best case (*bio-SC-GT*), and 57.85% accuracy (closeness) and 65.47% average accuracy on the worst case (*ego-Facebook*). Overall the accuracy for the models without/with multitask were 74.6% and 73.8%, which are only slightly worse than the non-multitasking AN model, which has a 76.8%, owing much of this to its high performance on the degree centrality, which it predicts with

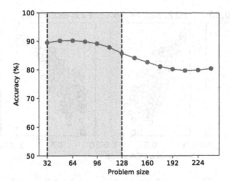

Fig. 2. The overall accuracy decays with increasing problem sizes, although it still does not approach 50% (equivalent to randomly guessing each vertex-to-vertex centrality comparison) for the largest instances tested here. The dotted lines delimit the range of problem sizes used to train the network ($n = 32 \ldots 128$).

almost 90% accuracy over all graphs. Due to the extreme overestimation of the number of nodes and the different distributions present in these graphs, we found it impressive that these two models can predict with such high accuracies. And although our model was outclassed in the non-multitasking environment, the multitasking RN model outperforms the multitasking AN model (with 68.34% accuracy) by a large margin, while being faster by computing all four centralities at once.

6.2 Interpretability

Machine learning has achieved impressive feats in the recent years, but the interpretability of the computation that takes place within trained models is still limited [26]. [34] have shown that it is possible to extract useful information from the embeddings of CNF literals, which they manipulate to obtain satisfying assignments from the model (trained only as a classifier). This allowed them to deduce that Neurosat works by guessing UNSAT as a default and changing its prediction to SAT only upon finding a satisfying assignment.

In our case, we can obtain insights about the learned algorithm by projecting the refined set of embeddings $V^{t_{max}} \in \mathbb{R}^{64}$ onto one-dimensional space by the means of Principal Component Analysis (PCA) and plotting the projections against the centrality values of the corresponding vertices. We do this only for the RN model, since the comparison framework might suggest that the learned embeddings need not represent the centrality values directly. Figure 3 shows the evolution of this plot through message-passing iterations, from which we can infer some aspects of the learned algorithm. First of all, the zeroth step is suppressed due to space limitations, but because all embeddings start out the same way, it corresponds to a single vertical line. One can see how the PCA of the embeddings seem to sort the vertices along their eigenvector centrality while the model improves its performance. The cause for the most central vertex shifting

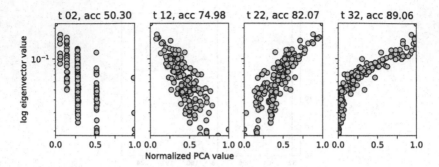

Fig. 3. Evolution of the 1D projection of vertex embeddings of a non-multitask model plotted against the corresponding eigenvector centralities (plotted on a log scale) through the message-passing iterations for a graph sampled from the Watts-Strogatz small world distribution of the "large" dataset.

from being on the left to the right is most likely due to the normalisation of the PCA projection. As the solution process progresses, the network progressively manipulates each individual embedding to produce a correlation between the centrality values and the $\mathbb{R}^d$ vector, which can be visualised here as reordering data points along the horizontal axis.

The case shown here, however, is not universally true, and vary somewhat depending on the distribution from which the graph was drawn. Graphs sampled from the power law tree distribution, for example, seem to be more exponential in nature when comparing the log-centrality value and the normalised 1-dimensional PCA value. But most of the distributions trained on had a similar behaviour of making a line between the logarithm of the centrality and the normalised PCA values. However, even in the cases where the centrality model did not achieve a high accuracy, we can still look at the PCA values and see whether they yield a somewhat sensible answer to the problem. Thus, the embeddings generated by the network can be seen as the GNN trying to create a centrality measure of its own with parts of, or the whole embedding being correlated with those centralities with which the network was trained.

7   Conclusions

In this paper, we demonstrated how to train a neural network to predict graph centrality measures, while feeding it with only the raw network structure. In order to do so, we enforced permutation invariance among graph elements by engineering a message-passing algorithm composed of neural modules with shared weights. These modules can be assembled in different configurations to reflect the network structure of each problem instances. We show that our proposed model which is trained to predict centrality comparisons (i.e. is vertex v_1 more central than vertex v_2 given the centrality measure c?) performs slightly better than the baseline model of predicting the centrality measure directly. We draw parallels between the accuracy defined in this setup and the Kendall-τ correlation coefficient, and show that our model performs reasonably well within this metric and that this performance generalises reasonably well to other problem distributions and larger problem sizes. We also show that the model shows promising performance for very large real world instances, which overestimate the largest instances known at training time from $\times 9$ to $\times 31$ (4,000 as opposed to 128 vertices). We also show that although our model can be instantiated separately for each centrality measure, it can also be trained to predict all centralities simultaneously, with minimal effect to the overall accuracy.

The model presented here however, isn't without its failings. Some models in the literature are able to provide meaningful node embeddings with lower dimensionality values than the ones we present here, however their work does not focus on centrality measures and some use information about the graph other than its topology, as well as learn the embeddings for similar graph distributions or even the same graph used for training. Although our model is presented here with four sample centralities, some of which are easier to compute precisely rather than approximating with our method, we believe that the joint prediction of multiple centralities is still useful, and that the time complexity of our model, which is dependant mostly on matrix multiplications, still remains polynomial even if one increases the amount of centralities being predicted.

In summary, this work presents, to the best of our knowledge, the first application of Graph Neural Networks to multiple centrality measures and the proposal of a comparison framework for processing node embeddings. We yield an effective model and provide ways to have such a model work with various centralities at once, in a more memory-efficient way than having a different model for every centrality – with minimal loss in its performance. Finally, our work attests to the power of relational inductive bias in neural networks, allowing them to tackle graph-based problems, and also showing how the proposed model can be used to provide a network that condenses multiple information about a graph in a single embedding.

Acknowledgements. This study was financed in part by the Coordenação de Aperfeiçoamento de Pessoal de Nível Superior - Brasil (CAPES) - Finance Code 001, and by the Brazilian Research Council CNPq. We gratefully acknowledge the support of NVIDIA Corporation with the donation of the Quadro P6000 GPU used for this research.

References

1. Albert, R., Barabási, A.L.: Statistical mechanics of complex networks. Rev. Mod. Phys. **74**(1), 47 (2002). https://doi.org/10.1103/RevModPhys.74.47
2. Batagelj, V., Brandes, U.: Efficient generation of large random networks. Phys. Rev. E **71**(3), 036113 (2005). https://doi.org/10.1103/PhysRevE.71.036113
3. Battaglia, P.W., et al.: Relational inductive biases, deep learning, and graph networks. arXiv preprint arXiv:abs/1806.01261 (2018)
4. Beauchamp, M.: An improved index of centrality. Behav. Sci. **10**(2), 161–163 (1965). https://doi.org/10.1002/bs.3830100205

5. Bonacich, P.: Power and centrality: a family of measures. Am. J. Sociol. **92**(5), 1170–1182 (1987). https://doi.org/10.1086/228631
6. Borgatti, S.P., Everett, M.G.: A graph-theoretic perspective on centrality. Soc. Netw. **28**(4), 466–484 (2006). https://doi.org/10.1016/j.socnet.2005.11.005
7. Brandes, U.: A faster algorithm for betweenness centrality. J. Math. Sociol. **25**(2), 163–177 (2001). https://doi.org/10.1080/0022250X.2001.9990249
8. Chen, K., Zhang, Y., Zhu, G., Mu, R.: Do research institutes benefit from their network positions in research collaboration networks with industries or/and universities? Technovation (2017). https://doi.org/10.1016/j.technovation.2017.10.005
9. Estrada, E., Ross, G.J.: Centralities in simplicial complexes applications to protein interaction networks. J. Theor. Biol. **438**, 46–60 (2018). https://doi.org/10.1016/j.jtbi.2017.11.003
10. García-Durán, A., Niepert, M.: Learning graph representations with embedding propagation. In: NIPS, pp. 5125–5136 (2017)
11. Gilmer, J., Schoenholz, S.S., Riley, P.F., Vinyals, O., Dahl, G.E.: Neural message passing for quantum chemistry. In: ICML, Proceedings of Machine Learning Research, PMLR, vol. 70, pp. 1263–1272 (2017)
12. Glorot, X., Bengio, Y.: Understanding the difficulty of training deep feedforward neural networks. In: AISTATS, JMLR Proceedings, vol. 9, pp. 249–256. JMLR.org (2010)
13. Gori, M., Monfardini, G., Scarselli, F.: A new model for learning in graph domains. In: IJCNN, vol. 2, pp. 729–734. IEEE (2005). https://doi.org/10.1109/IJCNN.2005.1555942
14. Grando, F., Granville, L.Z., Lamb, L.C.: Machine learning in network centrality measures: tutorial and outlook. ACM Comput. Surv. **51**(5), 102:1–102:32 (2019). https://doi.org/10.1145/3237192
15. Grando, F., Lamb, L.C.: Estimating complex networks centrality via neural networks and machine learning. In: IJCNN, pp. 1–8. IEEE (2015). https://doi.org/10.1109/IJCNN.2015.7280334
16. Grando, F., Lamb, L.C.: On approximating networks centrality measures via neural learning algorithms. In: IJCNN, pp. 551–557. IEEE (2016). https://doi.org/10.1109/IJCNN.2016.7727248
17. Grando, F., Lamb, L.C.: Computing vertex centrality measures in massive real networks with a neural learning model. In: IJCNN, pp. 1–8. IEEE (2018). https://doi.org/10.1109/IJCNN.2018.8489690
18. Hagberg, A., Swart, P., Chult, D.S.: Exploring network structure, dynamics, and function using networkx, Technical report, Los Alamos National Laboratory (LANL) (2008)
19. Hamilton, W.L., Ying, Z., Leskovec, J.: Inductive representation learning on large graphs. In: NIPS, pp. 1025–1035 (2017)
20. Holme, P., Kim, B.J.: Growing scale-free networks with tunable clustering. Phys. Rev. E **65**(2), 026107 (2002). https://doi.org/10.1103/PhysRevE.65.026107
21. Kim, J., Hastak, M.: Social network analysis: characteristics of online social networks after a disaster. Int J. Inf. Manag. **38**(1), 86–96 (2018). https://doi.org/10.1016/j.ijinfomgt.2017.08.003
22. Kipf, T.N., Welling, M.: Semi-supervised classification with graph convolutional networks. In: ICLR, OpenReview.net (2017)

23. Kumar, A., Mehrotra, K.G., Mohan, C.K.: Neural networks for fast estimation of social network centrality measures. In: Ravi, V., Panigrahi, B.K., Das, S., Suganthan, P.N. (eds.) Proceedings of the Fifth International Conference on Fuzzy and Neuro Computing (FANCCO - 2015). AISC, vol. 415, pp. 175–184. Springer, Cham (2015). https://doi.org/10.1007/978-3-319-27212-2_14

24. Lee, C.Y.: Correlations among centrality measures in complex networks. arXiv preprint arXiv: physics/0605220 (2006)

25. Li, Y., Vinyals, O., Dyer, C., Pascanu, R., Battaglia, P.: Learning deep generative models of graphs. arXiv preprint arXiv:abs/1803.03324 (2018)

26. Lipton, Z.C.: The mythos of model interpretability. arXiv preprint arXiv:abs/1606.03490 (2016)

27. Liu, B., Li, Z., Chen, X., Huang, Y., Liu, X.: Recognition and vulnerability analysis of key nodes in power grid based on complex network centrality. IEEE Trans. Circ. Syst. II 65(3), 346–350 (2018). https://doi.org/10.1109/TCSII.2017.2705482

28. Palm, R.B., Paquet, U., Winther, O.: Recurrent relational networks for complex relational reasoning. arXiv preprint arXiv: abs/1711.08028 (2017)

29. Perozzi, B., Al-Rfou, R., Skiena, S.: DeepWalk: online learning of social representations. In: KDD, pp. 701–710. ACM (2014). https://doi.org/10.1145/2623330.2623732

30. Santoro, A., et al.: A simple neural network module for relational reasoning. In: NIPS, pp. 4974–4983 (2017)

31. Scarselli, F., Gori, M., Tsoi, A.C., Hagenbuchner, M., Monfardini, G.: The graph neural network model. IEEE Trans. Neural Netw. 20(1), 61–80 (2009). https://doi.org/10.1109/TNN.2008.2005605

32. Scarselli, F., Yong, S.L., Gori, M., Hagenbuchner, M., Tsoi, A.C., Maggini, M.: Graph neural networks for ranking web pages. In: Web Intelligence, pp. 666–672. IEEE Computer Society (2005). https://doi.org/10.1109/WI.2005.67

33. Schoch, D., Valente, T.W., Brandes, U.: Correlations among centrality indices and a class of uniquely ranked graphs. Soc. Netw. 50, 46–54 (2017). https://doi.org/10.1016/j.socnet.2017.03.010

34. Selsam, D., Lamm, M., Bünz, B., Liang, P., de Moura, L., Dill, D.L.: Learning a SAT solver from single-bit supervision. arXiv preprint arXiv:abs/1802.03685 (2018)

35. Sethuraman, G., Dhavamani, R.: Graceful numbering of an edge-gluing of shell graphs. Discrete Math. 218(1–3), 283–287 (2000). https://doi.org/10.1016/S0012-365X(99)00360-X

36. Shaw, M.: Group structure and the behavior of individuals in small groups. J. Psychol. 38(1), 139–149 (1954). https://doi.org/10.1080/00223980.1954.9712925

37. Tang, J., Qu, M., Wang, M., Zhang, M., Yan, J., Mei, Q.: LINE: large-scale information network embedding. In: WWW, pp. 1067–1077. ACM (2015). https://doi.org/10.1145/2736277.2741093

38. Watts, D.J., Strogatz, S.H.: Collective dynamics of 'small-world' networks. Nature 393(6684), 440 (1998). https://doi.org/10.1038/30918

39. You, J., Liu, B., Ying, Z., Pande, V.S., Leskovec, J.: Graph convolutional policy network for goal-directed molecular graph generation. In: NeurIPS, pp. 6412–6422 (2018)

40. You, J., Ying, R., Ren, X., Hamilton, W.L., Leskovec, J.: GraphRNN: generating realistic graphs with deep auto-regressive models. In: ICML, Proceedings of Machine Learning Research, PMLR, vol. 80, pp. 5694–5703 (2018)

41. Zhang, Z., Cui, P., Zhu, W.: Deep learning on graphs: a survey. arXiv preprint arXiv:abs/1812.04202 (2018)

Special Session: BIGCHEM: Big Data and AI in Chemistry

Special Session: BIGCHEM:
Big Data and AI in Chemistry

Artificial Intelligence and machine learning are increasingly used in the chemical industry, in particular with respect to Big Data. These developments have the potential to automate, facilitate, and speed-up the key steps in drug research. However, their applications are still at an early stage. In particular, this is due to the need to develop "chemistry-aware" methods and/or to adapt the existing methods to work with chemical data. The goal of this session was to show progress and exemplify the current needs, trends, and requirements for AI and machine learning for chemical data analysis. In particular, it focused on the use of chemical informatics and machine learning methodologies to analyze chemical Big Data, e.g., to predict biological activities and physico-chemical properties, facilitate property-oriented data mining, predict biological targets for compounds on a large scale, design new chemicals, and analyze large virtual chemical spaces. The covered topics included:

- Big Data Analysis in Chemistry
- Machine Learning in Drug Discovery
- Machine Learning Methodologies for Mining Very Large Compound Data Sets
- Big Data Visualization and Modeling
- Virtual Screening Methods to Exploit Large Virtual Chemical Spaces
- Machine Learning and the Use of HTS Data for Compound Activity Predictions
- Analysis of Compound Promiscuity and Frequent Hitters
- Reaction-Driven *De Novo* Design to Explore the Chemical Space
- Accessing New Chemical Space Based on Predictive Models
- Reaction Informatics and Synthetic Route Prediction
- Molecular Dynamics and Quantum Chemistry Calculations Using Neural Networks

The authors of articles/abstracts of the BIGCHEM Special Session were invited to submit their articles to the special issue of *J. Cheminformatics*.

Organization and PC Members

Igor V. Tetko   Helmholtz Zentrum München (GmbH), Germany
Pavel Karpov   Helmholtz Zentrum München (GmbH), Germany
Monica Campillos   Helmholtz Zentrum München (GmbH), Germany
Igor Baskin   Moscow State University, Russia
Christian Bauckhage   Fraunhofer IAIS, Germany
Hongming Chen   AstraZeneca, Sweden
Artem Cherkasov   University of British Columbia, Canada
Vladimir Chupakhin   Janssen Pharmaceutical Companies, USA
Djork-Arné Clevert   Bayer, Germany
Ola Engkvist   AstraZeneca, Sweden
Peter Ertl   Novartis Institutes for BioMedical Research,
                           Switzerland
Olexandr Isayev   University of North Carolina at Chapel Hill, USA
Dmitry Karlov   Skoltech, Russia
Uwe Koch   LDC GmbH, Germany
Gilles Marcou   University of Strasbourg, France
Florian Nigsch   Novartis Institutes for BioMedical Research,
                           Switzerland
Dmitry Osolodkin   Chumakov FSC R&D IBP RAS, Russia
Vladimir Palyulin   Moscow State University, Russia
Mike Preuss   Leiden University, The Netherlands
Eugene Radchenko   Moscow State University, Russia
Jean-Louis Reymond   University of Bern, Switzerland
Andrei Tolstikov   Merck Group, Germany
Antoni Valencia   Independent Consultant, Spain
Bruno Villoutreix   Molecular Informatics for Health, France

Neural Network Guided Tree-Search Policies for Synthesis Planning

Amol Thakkar[1,2](✉), Esben Jannik Bjerrum[1](✉), Ola Engkvist[1],
and Jean-Louis Reymond[2]

[1] Hit Discovery, Discovery Sciences, IMED Biotech Unit, AstraZeneca,
Gothenburg, Sweden
{amol.thakkar, esben.bjerrum}@astrazeneca.com,
av.mol.th@gmail.com
[2] Department of Chemistry and Biochemistry,
University of Bern, Bern, Switzerland

Abstract. Developments and accessibility of computational methods within machine learning and deep learning have led to the resurgence of methods for computer assisted synthesis planning (CASP). In this paper we introduce our viewpoints on the analysis of reaction data, model building and evaluation. We show how the models' performance is affected by the specificity of the extracted reaction rules (templates) and outline the direction of research within our group.

Keywords: Reaction informatics · Neural networks · Classification · Synthesis prediction

1 Introduction

With the increasing availability of reaction data, developments and accessibility of computational methods, and a drive to further automate design, make, test, analyze (DMTA) cycles within drug discovery [1], computer assisted synthesis planning (CASP) has seen renewed interest as of late [2]. This has been spurred by recent achievements in the application of neural networks combined with search algorithms [3, 4], learning from breakthroughs in their application to games such as chess and Go [5].

CASP or retrosynthetic analysis refers to the strategy used by chemists to deconstruct a compound into its simpler precursors. In likeness with games, both are fundamentally decision-making tasks, in intractable search spaces, and with complex optimal solutions, where evaluation of the position and available moves at each step is difficult (Table 1). A knowledge base of reactions a chemist has learned throughout their career, coupled with extensive literature searching, form the basis of an initial pattern recognition step, from which applicable reactions can be identified and prioritized.

Recent studies have shown that neural network policies framed as multi-class classification problems can identify likely reactions through the noisy knowledge base [3, 4]. However, we have found they are heavily weighted towards frequently occurring reactions, owing to imbalanced datasets. Thus, miss out on less frequent yet feasible alternatives. In the present study, we explore and tune neural network architectures

© The Author(s) 2019
I. V. Tetko et al. (Eds.): ICANN 2019, LNCS 11731, pp. 721–724, 2019.
https://doi.org/10.1007/978-3-030-30493-5_64

Table 1. Comparison of search spaces in games vs retrosynthetic analysis.

	Chess	Go	Retrosynthesis
Search breadth	35	250	>500,000
Search depth	80	150	ca. 10

with the aim of maximizing the number of synthetically feasible options at each step. This is supplemented by curation and analysis of the underlying knowledge base, extracted from available reaction datasets. The number of which is limited when publicly available data is considered.

2   Methods

The US patent office extracts are a set of text mined reactions from the patent literature [6]. Given the reaction SMILES, an extension of the SMILES notation used to represent molecular structures [7], we used a modified version of Coley and coworkers algorithm to extract reaction templates [8]. That is the transformation required to convert the reactants into the products. These form the core of our knowledge base from which we can train a policy to enumerate retrosynthetic pathways in the form of a tree. To evaluate performance on a state-of-the-art model, we have opted to reimplement a variant of the policy used by Segler and Waller [4].

3   Results

We found that the applicability of the predicted templates from our trained policies varied with the template specificity (Fig. 1), the size of the template library, and the network architecture. Additionally, we developed a method for determining the selectivity of our templates during extraction and validation, the effect of which we are in the process of investigating on the trained model.

Preliminary evaluation of the models was performed on a random selection of 10,000 compounds from each ChEMBL [9] and FDB17 [10]. This enabled assessment of both the model's predictive ability, and the validity of templates across a range of druglike and novel scaffolds. Using this assessment criteria, we aim to maximize the number of options available to our policy at each step in the subsequent tree search. Thereby, enabling the later prediction of full synthetic pathways, which is a necessity in accelerating automated DMTA cycles.

Whilst our viewpoint on the performance of the models has shed new light on the way in which a model may be evaluated, there is still much detail to investigate. This paper introduces preliminary results for a template-based synthesis planning methodology, and highlights that a more rigorous study is currently underway. This will encompass larger datasets, template design, data curation, the network architecture, and the implementation of appropriate metrics. These results will follow in a more rigorous study of the problems faced in computer assisted synthesis planning.

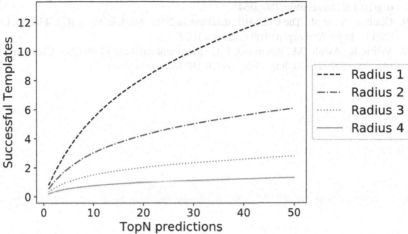

Fig. 1. Assessment and comparison of the model's ability to predict templates that can be successfully applied for a subset of 10,000 randomly sampled compounds from each ChEMBL and FDB17. The radius refers to the number of bonds from the reaction center that are considered.

Funding. Amol Thakkar is supported financially by the European Union's Horizon 2020 research and innovation program under the Marie Skłodowska-Curie Grant Agreement No. 676434, "Big Data in Chemistry" ("BIGCHEM," http://bigchem.eu).

References

1. Nicolaou, C.A., et al.: Idea2Data: toward a new paradigm for drug discovery. ACS Med. Chem. Lett. (2019). https://doi.org/10.1021/acsmedchemlett.8b00488
2. Boström, J., Brown, D.G., Young, R.J., Keserü, G.M.: Expanding the medicinal chemistry synthetic toolbox. Nat. Rev. Drug Discov. **17**, 709 (2018). https://doi.org/10.1038/nrd.2018.116
3. Segler, M.H.S., Preuss, M., Waller, M.P.: Planning chemical syntheses with deep neural networks and symbolic AI. Nature **555**, 604 (2018). https://doi.org/10.1038/nature25978
4. Segler, M.H.S., Waller, M.P.: Neural-symbolic machine learning for retrosynthesis and reaction prediction. Chem. Eur. J. **23**(25), 5966–5971 (2017). https://doi.org/10.1002/chem.201605499
5. Silver, D., et al.: Mastering the game of Go with deep neural networks and tree search. Nature **529**, 484 (2016). https://doi.org/10.1038/nature16961
6. Daniel, L.: Extraction of chemical structures and reactions from the literature, Doctoral thesis, University of Cambridge (2012)
7. Weininger, D.: SMILES, a chemical language and information system. 1. introduction to methodology and encoding rules. J. Chem. Inf. Comput. Sci. **28**(1), 31–36 (1988). https://doi.org/10.1021/ci00057a005

724   A. Thakkar et al.

8. Coley, C.W., Barzilay, R., Jaakkola, T.S., Green, W.H., Jensen, K.F.: Prediction of organic reaction outcomes using machine learning. ACS Cent. Sci. **3**(5), 434–443 (2017). https://doi.org/10.1021/acscentsci.7b00064

9. Gaulton, A., et al.: The ChEMBL database in 2017. Nucleic Acids Res. **45**(D1), D945–D954 (2017). https://doi.org/10.1093/nar/gkw1074

10. Visini, R., Awale, M., Reymond, J.-L.: Fragment database FDB-17. J. Chem. Inf. Model. **57** (4), 700–709 (2017). https://doi.org/10.1021/acs.jcim.7b00020

Open Access This chapter is licensed under the terms of the Creative Commons Attribution 4.0 International License (http://creativecommons.org/licenses/by/4.0/), which permits use, sharing, adaptation, distribution and reproduction in any medium or format, as long as you give appropriate credit to the original author(s) and the source, provide a link to the Creative Commons license and indicate if changes were made.

The images or other third party material in this chapter are included in the chapter's Creative Commons license, unless indicated otherwise in a credit line to the material. If material is not included in the chapter's Creative Commons license and your intended use is not permitted by statutory regulation or exceeds the permitted use, you will need to obtain permission directly from the copyright holder.

LSTM and 1-D Convolutional Neural Networks for Predictive Monitoring of the Anaerobic Digestion Process

Mark McCormick[✉] and Alessandro E. P. Villa

NeuroHeuristic Research Group, University of Lausanne, Quartier UNIL-Dorigny,
1015 Lausanne, Switzerland
mark.mccormick@unil.ch,
http://www.neuroheuristic.org

Abstract. Anaerobic digestion is a natural process that transforms organic substrates to methane and other products. Under controlled conditions the process has been widely applied to manage organic wastes. Improvements in process control are expected to lead to improvements in the technical and economic efficiency of the process. This paper presents and compares 3 different neural network model architectures for use as anaerobic digestion process predictive models. The models predict the future biogas production trend from measured physical and chemical parameters. The first model features an LSTM layer, the second model features a 1-D convolutional layer and the third model combines 2 separate inputs and parallel treatment using LSTM and 1-D convolutional layers followed by merging to produce a single prediction. The predictions can be used to adaptively adjust the substrate feeding rate in accordance with the transient state of the digestion process as defined by liquid feeding rate, the organic acid and ammonium ion concentrations and the pH of the digester liquid phase. The training and testing data were obtained during 1 year of continuous operation of a pilot-plant treating restaurant wastes. PLS regression and ICA were used to select the most relevant process parameters from the data. The 1-D Convolutional based model comprising 272 trainable parameters predicted the future biogas flow rate changes with accuracy as high as 89% and an average accuracy of 58% . The work-flow can be applied to optimize the control of the study digester and to control bioreactors in general.

Keywords: LSTM, 1-D Convolutional Neural Network · Bioprocess

1 Introduction

Restaurants, hotels, markets, fisheries and other small to medium size agro-food industries generate 88 million tonnes of organic waste per year in Europe of which 47 million tonnes per year are household food wastes and 17 million tonnes per year are food processing wastes [1]. This waste stream contains valuable components such as water (approximately 80% by mass) and valuable energy and

© The Author(s) 2019
I. V. Tetko et al. (Eds.): ICANN 2019, LNCS 11731, pp. 725–736, 2019.
https://doi.org/10.1007/978-3-030-30493-5_65

substances such as proteins and organic acids. In the context of the circular economy and current objectives to improve energy efficiency, recycle water and valuable substances, produce energy from renewable sources, reduce greenhouse gas emissions and close nutrient cycles, the waste management sector offers many opportunities to implement innovative technical solutions. Anaerobic digestion has been proposed as a technical solution to manage organic wastes. This study aims to contribute to the strategy used to control a high-performance anaerobic digester supplied by Digesto Sarl. The strength of the Digesto®. digestion module is the ability to treat small to medium quantities of waste on their site of production, without any transport and in an automated mode (programmable logic controller for distance monitoring of the digester performances and for breakdown prevention). The elimination of transport costs and a high degree of automation make it the appropriate treatment solution for many end-users [2].

Currently, anaerobic digestion is successfully applied to reduce the organic matter content and to improve the dewatering properties of excess wastewater treatment sludge. Anaerobic digestion is also used to produce energy from municipal solid wastes, garden wastes and from energy crops. These processes are characterized by large bioreactors, long hydraulic retention times between 20 and 60 days and by regular feeding of well characterized and homogeneous substrates. In contrast, small-scale, on-site digestion of restaurant wastes would require an intensified bioprocess featuring a small digester, a short hydraulic retention time of less than 10 days and the capacity to tolerate wide variations in the feeding rate and the substrate composition.

Anaerobic digestion is a complex process characterized by non-linear functions and interactions between many different biochemical processes. A high-performance control strategy is thus required to intensify the anaerobic digestion process and to achieve the goals of small size, a short hydraulic retention time and long-duration autonomous operation. Artificial Neural Networks, and especially Deep Neural Networks, have demonstrated their capacity to map complex and non-linear relations between data. The purpose of this work is to construct models based on some popular Neural Network architectures and to assess their capacity to predict biogas production.

2   Materials and Methods

2.1   Machine Learning

Multivariate statistical analysis approaches have been used to detect faults and abnormal operation of anaerobic digestors treating waste activated sludge (WAS). Using Principal Component Analysis (PCA), Hotelling's T-squared and Shewhart control charts, the transitions to unstable periods were associated with accumulation of volatile fatty acids [3]. Computational self-adapting methods (Support Vector Machines, SVM) were compared with an analytical method to predict the total ammonia nitrogen (TAN) concentration in the effluent from a two-stage anaerobic digestion (AD) process treating poultry wastes. The SVM-based model outperformed the analytical method for the TAN prediction, achieving a relative average error of 15.2% against 43% for the analytical method. Moreover, SVM showed higher prediction accuracy in comparison with Artificial Neural Networks [4]. Machine learning methods have been used to estimate the state of biogas plants, as defined by the ADM-1 model of the anaerobic digestion process, using on-line measurements of parameters such as biogas production, CH_4 and CO_2 content in the biogas, pH value and substrate feed volume [5].

In contrast to the conventional approach of using recurrent networks, especially LSTM layers, for sequence processing, convolutional architectures have demonstrated longer effective memory and have outperformed recurrent networks across a diverse range of tasks [6]. In the field of aircraft control, a "state-image" approach to capture the inflight state variables produced a feature map that contained the values of the in-flight parameters at a given moment. The control strategy also included historical data. A Convolutional Neural Network (CNN) and a Recursive Neural Network with Long Short-Term Memory (RNN (LTSM)) layers were implemented in parallel to output vectors that were subsequently merged and processed in fully connected layers. Inputting data to different branches to allow separate extraction of the time dependent and the current parameter values was expected to lead to improved accuracy when compared to sequential treatment by CNN and LSTM layers [7].

2.2 Experimental Set-Up

The study digester was a machine comprising tanks, pumps, heating elements, sensors and command and data acquisition capacities. The total liquid working volume was 630 liters. Shredded restaurant wastes were fed to the digester at regular intervals. The study data was acquired from on-line measurements and by analysis of samples collected at intervals between 1 and 7 days during 366 days of continuous operation. The study data included 22 parameters that characterize the input waste stream. Characterization of the digestion process included manual measurement of 44 parameters that describe the liquid and gas phases. The on-line measurements included redundant measurements of liquid and gas phase physical-chemical properties and the measurement of 7 mechanical parameters such as internal pumping and agitation. A total of 83 parameters were considered for the study. The digestion process was controlled manually using only the expert knowledge of the operators. The operating conditions were adjusted *ad hoc* during the study to optimize the rates of biogas production and the removal of the organic fraction of the waste. In this study, fixed duration feeding intervals were defined. At the start of each time interval a decision was made to feed or to not feed the substrate to the digestion machine.

2.3 Problem Solving Approach

Autonomous operation of the digester would require automated transfer of raw waste from a buffer tank to the digestion machine. The problem is to decide when the raw waste should be transferred (fed) to the digestion machine. Overfeeding would upset the bioprocess and can lead to ceased waste degradation and biogas production. Underfeeding would lead to inefficient use of the digester. A useful model should make predictions that agree with the actual values that were measured after the prediction. The digester should be fed only when the biological state is conducive to waste degradation and biogas production. Therefore, the purpose of the neural network model is to automate decision making. To solve the problem, a supervised approached based on mapping measured process predictor variables to a single measured response variable was chosen. The models were trained and tested using simultaneously measured prediction and response values. The models were tested by comparing predicted to measured values of the response variable.

The proposed anaerobic digester control strategy is to compare the predicted future biogas flow rate trend to the actual biogas flow rate TREND. Increased biogas production rate is associated with improved bioprocess quality. Decreased or stable biogas production rate is associated with deteriorating bioprocess quality. If the predicted future biogas flow rate is higher than the actual measured biogas flow rate, then the decision is to feed the digestion machine. If the predicted biogas flow rate was the same as or lower than the actual measured biogas flow rate, then the decision was to not feed the digestion machine.

2.4   Data Analysis and Preprocessing

Analysis and data augmentation techniques were used to prepare a single dataset for use in developing all the Neural Network models. Raw data analysis aimed to identify the parameters to use to train and test the Neural Network model. Principal Component Analysis (PCA), Independent Component Analysis (FastICA) and Partial Least Squares (PLS) regression algorithms from Scikit-learn [8] were used for dimensionality reduction and to identify the parameters that make the greatest contribution to process stability. A plot of the weights of the first principal components shows that the waste loading rate, total solids, carbon and nitrogen content of the feed, and the total solids concentration in the of the digester make the greatest contributions to overall variance of the system. However, these parameters can be measured only manually and manipulation by the operators is difficult or impossible. Consequently, they are not practically useful in a neural network model for process control.

ICA is recommended when the data has a high degree of kurtosis. Since sharp peeks were observed for biogas H_2 content, and liquid phase ammonium and volatile fatty acid content, the raw data were analyzed using ICA. ICA analysis showed that the H_2 content of the biogas and the volatile fatty acid content of the digester were the most important independent components of the data. The raw data was used in the DNN models without pre-treatment to remove kurtosis because spiking of both H_2 and volatile fatty acids is not unusual in the anaerobic digestion process. The H_2 concentration in the biogas can be measured on-line. However, no sensors for volatile fatty acids are currently available at an affordable price. The method of Partial Least Squares Regression was used to map the matrix of predictor variables to the biogas production rate. The results of PLS analysis showed that regression equation coefficients for the nitrogen content of the feed and the digester loading rate had the greatest magnitudes.

Considering the results of PCA, ICA and PLS regression analysis, knowledge of the anaerobic digestion process and the practical aspects of data acquisition, the following parameters were selected for inclusion in the dataset used to build the Neural Network models with 4 input features and 1 target variable.

- *Predictor variables (input features)*
  - Total mass of feed (water + waste + co-substrate) [kg/day]
  - pH
  - Ammonium concentration [mg N/liter]
  - Volatile fatty acid concentration [mg Acetic Acid/liter]
- *Response variable (target)*
  - Biogas flow rate [liter/hour]

Pre-processing included augmentation of the number of data points. In particular, the frequency of values obtained from off-line measurements of physical-chemical parameters was increased from approximately 1 per week to 1 per hour. The number of data points was increased by resampling to create one-hour time intervals followed by linear interpolation of the off-line data to fill the new sampling times. Interpolation was assumed to be valid because the changes in biological systems occur slowly over several days. The augmented dataset included 8733 rows of data representing the selected input features and the target. Using the Scikit-learn MinMaxScaler function, the previously standardized data were then normalized to values between 0 and 1. Figure 1 shows the pre-processed dataset that was used to build the Neural Network models. Five series of training and test data were obtained using the Scikit-learn TimeSeriesSplit function which respects the sequential order of the original data.

2.5 Neural Network Architectures

The study aimed to compare the recurrent, convolutional and combined neural network modeling approaches to solving a time series analysis problem having at 4 features. The models were built using Python 3.6.8 and the Keras 2.2.4 API running on top of TensorFlow 1.3.0 library for machine learning. The 3 architectures selected for development and the rationale for selection are summarized in Table 1 and the models are summarized in Fig. 2.

Long Short-Term Memory (LSTM). The sequential model was constructed using a single LSTM layer followed by a Dense layer. The input was a 3D tensor with shape (`batch_size`, `timesteps`, `input_dim`) where `batch_size` is the number of sample batches, `timestep` was set to four hours and `input_dim` was equal to the number of features. The LSTM layer had 160 units, used the `tanh` activation and `hard_sigmoid` recurrent activation functions with zero dropout. `Unit_forget_bias` was set to false since this setting was found to give a more

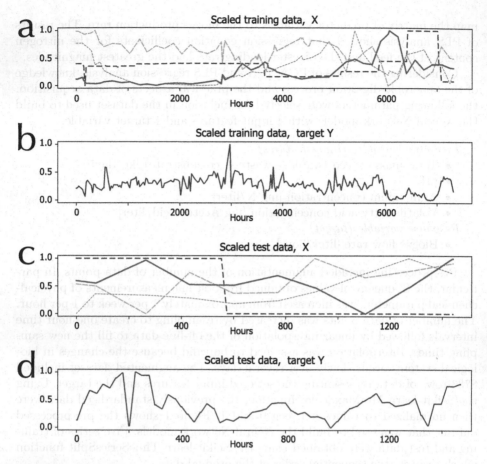

Fig. 1. Training and testing data after preprocessing. **(a)** Normalized training data of predictor variables. blue line: total volume of feed; red line: pH; green curve: ammonium concentration; orange line: volatile fatty acid concentration. **(b)** Normalized training data of the response variable, i.e. biogas flow rate. **(c)** Normalized test data of predictor variables. Same color code of panel (a). **(d)** Normalized test data of the response variable. (Color figure online)

accurate prediction. The model output was from a single dense layer having 1 unit, `relu` activation and `use_bias` set to False. The model had 105,760 trainable parameters.

1-Dimensional Convolutional Layer (Conv1-D). The sequential model was constructed using a single Conv1D layer followed by a MaxPooling1D layer followed by a Dense layer. The input was a 3D tensor with shape (`batch_size`, `timesteps`, `input_dim`), where batch size is the number of sample batches, `timesteps` was set to four hours and `input_dim` was equal to the number of features. The Conv1D layer had 16 filters, a kernel size of 4, stride of 1, padding

Table 1. Evaluated Neural Network architectures

Neural network architecture	Rationale for selection
LSTM, biogas flow rate predictions	Expert knowledge that bioprocesses have long and variable lag-times between measured predictor values and responses. Historical information
1-D CNN, biogas flow rate predictions	Potential to extract useful information from detailed analysis of features at given moments. State image concept
LSTM/1-D CNN hybrid, biogas flow rate predictions	Potential to improve accuracy by separate processing of historical and state image information followed by merging the information

set to same, and used the `relu` activation function. The pool size of the Max-Pooling layer was set to 4. The model output was from a single dense layer having 1 unit, `relu` activation and and `use_bias` set to False. The model had 272 trainable parameters.

LSTM/Conv1-D Hybrid. The Keras Model Class API was used to construct a model having 2 separate input and processing branches and a single output. The first branch included an LSTM layer followed 2 dense layers. The LSTM layer had the same configuration as the layer described above. The first dense layer had 48 nodes and used the `relu` activation function. The final dense layer had 1 node and used the `relu` activation function. The second branch included a Conv1D layer followed by MaxPooling1D Dense and Flatten layers. The Conv1D layer had the same configuration as the layer described above. The MaxPooling1D layer strides were set to 1 and padding was set to same. The two branches were merged using the Keras concatenate layer. The merged output was further passed to a dense layer with 8 hidden nodes. Model output was from a dense layer having 1 node and using the `relu` activation. The model had 106,158 trainable parameters.

3   Results

The augmented dataset was split into 5 different pairs of training and test sets using the TimeSeriesSplit function. Each pair comprised separate sets of consecutive 1-h intervals and a total of between 5000 and 8500 timestamps. The LSTM, Conv1D and hybrid LSTM/Conv1D models were trained for 1000 epochs with a batch size of 4 h. Loss during training was assessed using mean square error. The loss during training of the 3 models is shown in Fig. 3.

The models were tested on data obtained previously using the TimeSeriesSplit function. The prediction accuracy was assessed visually by plotting the predicted values and the measured values at 4-h intervals corresponding to the

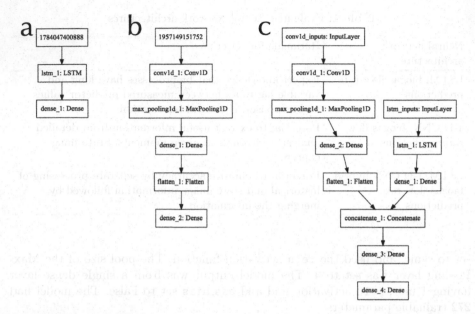

Fig. 2. Artificial Neural Network architectures: **(a)** LSTM, **(b)** Conv1D and **(c)** Hybrid LSTM/Conv1D architectures

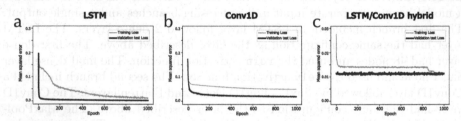

Fig. 3. Comparison of training loss for the 3 ANN architectures.

batch-size of 4 times 1-h timestamps. The RMSE for the individual sample pairs was calculated at the same 4-h interval (batch).

The predicted and the observed biogas flow rates for the 3 different regression models are shown in Fig. 4abc. The RMSE of each batch is shown in Fig. 4def. To assess the accuracy of the 3 models over a long duration, the overall RMSE was also calculated for the entire population of batches.

Table 2 shows the overall RMSE of the 3 models.

Since the purpose of the model is to decide if the digester should be fed at the beginning of a time interval, a special evaluation procedure was developed to rate the models in terms of their capacity to make the correct decision regarding digester feeding. To rate the models, the consequences of the model's decision were compared to the actual change in measured biogas flow rate 4 h after the decision. Any decision to feed the digester, followed by an observed

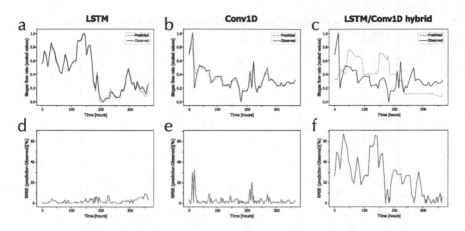

Fig. 4. Predicted (dashed red curve) and observed (solid blue curve) biogas production for LSTM (panel a), 1-D CNN (panel b), and LSTM/1-D CNN hybrid (panel c). The corresponding errors (RMSE) are presented in panels d,e,f for LSTM, 1-D CNN and LSTM/1-D CNN hybrid ANN architectures, respectively. (Color figure online)

Table 2. RMSE of the entire population

Model (ANN architecture)	RMSE (%)
LSTM	1.7
1-D CNN	2.6
LSTM/1-D CNN hybrid	19.0

decrease in biogas production, was an incorrect decision. Any decision to feed the digester, followed by an observed stable or increased biogas production, was a correct decision. The model accuracy was simply the ratio of correct decisions to requested decisions expressed as a percentage. To validate the 1-D convolutional DNN model, 12 new data sets generated by splitting the original data were used to predict the biogas production trend 4-h in the future. The average accuracy of the prediction was 58% with a standard deviation of 24% and a range between 28% and 100%.

To investigate the response time between changes in the measured predictor and response variables, the error of the LSTM model predictions was evaluated at different lag-times after the feeding decision. In this case, error was defined as the RMSE of the measurements and predictions made at 1-h intervals during the time sequence. Population RMSE was calculated for lag-times from 1 to 100 h after a feeding decision. The results show that the model is most accurate when the lag-time between the feeding decision and the time off comparison of the predicted to the observed values is much greater than 4 h. This result suggests that the biological response time of biogas production to changes in the 4 measured process parameters used this study is longer than 4 h (Fig. 5) .

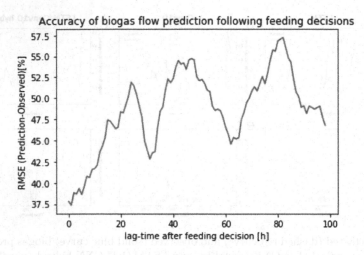

Fig. 5. RMSE of the predicted and observed biogas production at different times after measurement of predictor variables (LSTM model).

4   Discussion

The treatment of restaurant wastes is a particularly challenging application of anaerobic digestion because the loading rate and composition of the wastes vary over a relatively large range when compared to the treatment conditions for anaerobic digestion of waste activated sludge, manure and energy crops. Consequently, the feed controller must have the capacity to map a wide range of values of the relevant process parameters and account for long-duration trends. The results of this work show that a one-dimensional convolutional Deep Neural Network model trained using state image data obtained by measuring 4 parameters could be implemented as a control model to regulate substrate feeding to an anaerobic digester treating restaurant wastes. This results suggests that the convolutional network and state image approach leads to a more effective model because this approach is less influenced by the long and variable lag-times obeserved in a real bioprocess. In contrast, the LSTM model takes long range trends into account. Both the 1-D convolutional model and the LSTM models achieved low RMSEs between predicted and observed values of the first data set. However, model testing with new data showed that the LSTM did not have the ability to generalise. The hybrid LSTM/1-D convoltional network had the highest RMSE and the least ability to generalise.

The study demonstrated that the prediction accuracy of the neural network model is similar to the prediction accuracy of simply ajusting the feeding rate according to the observed trend in biogas production.

The average prediction accuracy of 58% should be compared to the accuracy of 63% obtained by simply following the trend in biogas production. The trend in biogas production was the result of *ad hoc* control of the digester by skilled operators working in a laboratory. It seems unlikely that a prediction accuracy as high as 63% could have been achieved in an industrial setting without control by skilled operators. Importantly, the measured parameters of input liquid flow, output biogas flow, digester pH and digester ammonium content can be measured on-line using commercially available sensors. However, no sensor to measure volatile fatty acids is available commercially.

Considering that approximately 80% of the data points for pH, ammonium and volatile fatty acids concentrations were obtained by interpolation between values measured only 1 or 2 times per week, the accuracy of the model could be improved by increasing the sampling frequency thereby obtaining real data. Higher data resolution and consequently higher model accuracy could be achieved using on-line sensors.

5   Conclusion

This study evaluated 3 Neural Network architectures for use as supervised learning models to predict biogas production in an anaerobic digester. The models were constructed, trained and tested on the same time series dataset obtained from long-duration operation of an anaerobic digester treating restaurant wastes. Of the 3 models evaluated, the 1-D convolutional model was best able to accurately predict biogas production trends on new data as evaluated in terms of the ratio of correct predictions to the total number of requested predictions. The results suggest that the feeding rate of the study digester can be controlled using an 1-D Convolutional based DNN controller. Future work should aim to improve the resolution of the training data set by implementing on-line sensing of the relevant process parameters. During this study, the digester feeding rate was the only manipulated parameter. Future work should also aim to control additional manipulatable parameters and the mechanical parameters of the digestion process. Considering the wide range of variation observed in the input and target variables, the collection of time series data over a very long duration is required to make an accurate model.

Acknowledgement. This work was supported by Swiss Innovation cheque #32609 and by F.R. Mahrer of Digesto Sarl, Puplinge(GE), Switzerland.

References

1. Stenmarck, Å., et al.: FUSIONS: estimates of European food waste levels, Stockholm (2016). ISBN 978-91-88319-01-2
2. Capaccioli, S., et al.: Socio-economic benefits of small anaerobic digestion systems. In: Proceeding of the 24th European Biomass Conference and Exhibition, pp. 1530–1534, Amsterdam (2016)
3. Leite, W.R.M., Belli Filho, P., Gottardo, M., Pavan, P., Bolzonella, D.: Monitoring and control improvement of single and two stage thermophilic sludge digestion through multivariate analysis. Waste Biomass Valorization **9**(6), 985–994 (2016)

4. Alejo, L., Atkinson, J., Guzmán-Fierro, V., Roeckel, M.: Effluent composition prediction of a two-stage anaerobic digestion process: machine learning and stoichiometry techniques. Environ. Sci. Pollut. Res. **25**(21), 21149–21163 (2018)
5. Gaida, D., et al.: State estimation for anaerobic digesters using the ADM1. Water Sci. Technol. **66**(5), 1088–1095 (2012)
6. Bai, S., Kolter, J.Z., Koltun, V.: An empirical evaluation of generic convolutional and recurrent networks for sequence modeling. arXiv: e1803.01271 (2018)
7. Dong, Y.: An application of deep neural networks to the in-flight parameter identification for detection and characterization of aircraft icing. Aerosp. Sci. Technol. **77**, 34–49 (2018)
8. Pedregosa, F., et al.: Scikit-learn: machine learning in Python. J. Mach. Learn. Res. **12**(Oct), 2825–2830 (2011)

Open Access This chapter is licensed under the terms of the Creative Commons Attribution 4.0 International License (http://creativecommons.org/licenses/by/4.0/), which permits use, sharing, adaptation, distribution and reproduction in any medium or format, as long as you give appropriate credit to the original author(s) and the source, provide a link to the Creative Commons license and indicate if changes were made.

The images or other third party material in this chapter are included in the chapter's Creative Commons license, unless indicated otherwise in a credit line to the material. If material is not included in the chapter's Creative Commons license and your intended use is not permitted by statutory regulation or exceeds the permitted use, you will need to obtain permission directly from the copyright holder.

Progressive Docking - Deep Learning Based Approach for Accelerated Virtual Screening

Vibudh Agrawal[1,2] , Francesco Gentile[1] , Michael Hsing[1] ,
Fuqiang Ban[1] , and Artem Cherkasov[1(✉)]

[1] Vancouver Prostate Centre, Department of Urologic Sciences,
Faculty of Medicine, University of British Columbia, 2660 Oak Street,
Vancouver, BC V6H 3Z6, Canada
{vagrawal,fgentile,mhsing,fban,
acherkasov}@prostatecentre.com
[2] The Bioinformatics Graduate Program, University of British Columbia,
Vancouver, Canada

Abstract. We have developed a novel, hybrid QSAR-docking approach (called 'progressive docking') that can speed up the process of virtual screening by enhancing it with Deep Learning models trained on-the-go on produced docking scores. The developed method can, therefore, predict docking outcome for yet unprocessed molecular entries and hence to progressively remove unfavorable chemical structures from the remaining docking base. This approach provides 50–100X speed increase for the standard docking procedures while retaining >90% of qualified molecules. We demonstrate that the use of PD allows processing of about 360 million molecules just in 2 weeks using a standard 200 CPU setup.

Keywords: Computer aided drug discovery · QSAR · Progressive docking ·
Deep Learning · Artificial intelligence · Cheminformatics

Drug screening is an extensive and expensive process [1], where computational (docking) models can significantly increase both the speed and hit rate of the discovery [2]. On the other hand, even the most advanced and elaborate docking algorithms [3–5], coupled with significant computational resources, are not fast enough to fully scan currently available molecular database (such as ZINC [6]), which has reached billions of entries in size.

The current solution to the problem is to apply filters based on simplistic physicochemical properties such as molecular weight and number of H-bond donors/acceptors [7], and various drug-likeness filters to reduce the size of the docking base to typically 10 s of millions of entries. With such a reduction, the vast majority of available chemical entities in sources like ZINC remain unprocessed.

In this work, we propose a new pipeline (Fig. 1) integrated with the machine learning framework which can speed the process of virtual screening (molecular docking) of the entire ZINC database by above 50X while retaining >90% of qualified molecules and using a fairly modest 200 CPUs setup (Fig. 2). The different statistics (AUC, enrichment, precision, recall) for every model at each iteration is shown in Table 2. Our method is based on the idea of 'progressive docking' pipeline which

© The Author(s) 2019
I. V. Tetko et al. (Eds.): ICANN 2019, LNCS 11731, pp. 737–740, 2019.
https://doi.org/10.1007/978-3-030-30493-5_66

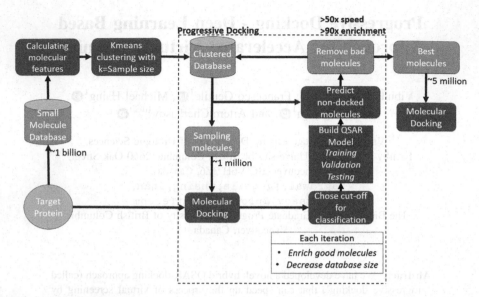

Fig. 1. Flowchart of progressive docking. Features are extracted for all the molecules from the small molecule database and are clustered. About 1 million molecules are then sampled from this clustered dataset. These 1 million molecules are docked and a cut-off value is chosen to convert this problem of regression to classification. A QSAR model (DNN) is built and is used to predict the classes for all the molecules in the clustered database. All the molecules classified as 'bad' are rejected. This process is repeated multiple times to bring down the total molecules in the database to a manageable size.

utilizes docking scores generated by the slower but more accurate docking method (such as GLIDE [3]) on a small subset of molecules, to quickly predict the docking score of the remaining undocked molecules in the database [8]. For our machine learning model, we use DNN (Deep Neural Networks) with the number of hidden layers ranging from 1–4 (hyperparameter) and same number of neurons (hyperparameter) in each of the layers. Since the dataset is very unbalanced, we also use oversampling along with class weights (both oversampling ratios, as well as weights, are hyperparameters). Each model is trained on a single Tesla P100 NVIDIA GPU. We used Morgan fingerprints with radius 2 and 1024 bits for all the molecules.

Docking and model training are two major time-consuming steps in our method. Out of the two, docking is the rate determining step for our case (docking one million takes ~4 days while hyperparameter search takes <1 day). Hence, we can conclude that a more accurate model will also result in a faster progressive docking implementation (since fewer molecules will be left in the end for docking). We further compared the performance of Deep Neural Networks (DNN) model with other machine learning approaches like Random Forest and Logistic Regression to demonstrate that the use of DNN provides the most accurate results (Table 1) and hence a faster PD implementation. All generated Enrichment Factor (EF) and AUC_ROC values are significant and consistent.

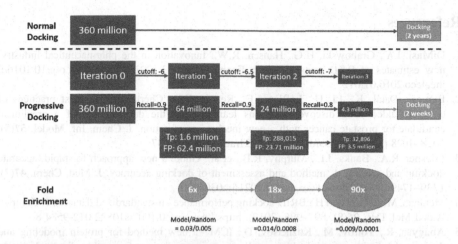

Fig. 2. Flowchart of progressive docking for ERG protein. The time required to dock 360 M molecules without PD is about ~2years based on available resources. After each iteration, the size of the remaining dataset decreases with only losing 10% good molecules (not compounded after each iteration). The definition of a good molecule gets better after each iteration (cutoff). Enrichment of good molecules compared to random also increases exponentially. The overall time for PD to dock 360 M million molecules is about 2–3 weeks

Table 1. Performance of different QSAR models on the same dataset. A decrease in precision value, as well as AUC_ROC, is observed from DNN to logistic regression. DNN is trained on 1 GPU while random forest and logistic regression are trained on 1 CPU. Training time is very quick for logistic regression while almost similar for DNN and random forest. Predicting time increases from logistic regression to random forest.

Model	Enrichment factor	AUC_ROC	CPU/GPU	Training time	Predicting time
DNN	2.5	0.88	GPU	30 min	0.5 min
Random forest	2	0.85	CPU	40 min	4 min
Logistic regression	1.88	0.82	CPU	1 min	1 s

Table 2. Performance statistics of different QSAR models (DNN) at each iteration steps. The model is selected to have a recall of 90% (capturing 90% of good molecules). The enrichment factor, as well as AUC, gets better after each iteration.

Iteration	Enrichment factor	AUC_ROC	Docking score cut-off
1	6	0.81	−6.0
2	18	0.90	−6.5
3	90	0.91	−7.0

We anticipate that the progressive docking framework can be integrated with any choice of docking software that the user trusts and can be used with any machine learning algorithm (we recommend random forest if no GPUs are available or else DNN).

References

1. DiMasi, J.A., Grabowski, H.G., Hansen, R.W.: Innovation in the pharmaceutical industry: new estimates of R&D costs. J Health Econ. **47**, 20–33 (2016). https://doi.org/10.1016/j.jhealeco.2016.01.012
2. Ban, F., Dalal, K., Li, H., LeBlanc, E., Rennie, P.S., Cherkasov, A.: Best practices of computer-aided drug discovery: lessons learned from the development of a preclinical candidate for prostate cancer with a new mechanism of action. J. Chem. Inf. Model. **57**(5), 1018–1028 (2017). https://doi.org/10.1021/acs.jcim.7b00137
3. Friesner, R.A., Banks, J.L., Murphy, R.B., et al.: Glide: a new approach for rapid, accurate docking and scoring. 1. method and assessment of docking accuracy. J. Med. Chem. **47**(7), 1739–1749 (2004). https://doi.org/10.1021/jm0306430
4. McGann, M.: FRED and HYBRID docking performance on standardized datasets. J. Comput. Aided Mol. Des. **26**(8), 897–906 (2012). https://doi.org/10.1007/s10822-012-9584-8
5. Abagyan, R., Totrov, M., Kuznetsov, D.: ICM—a new method for protein modeling and design: Applications to docking and structure prediction from the distorted native conformation. J. Comput. Chem. **15**(5), 488–506 (1994). https://doi.org/10.1002/jcc.540150503
6. Sterling, T., Irwin, J.J.: ZINC 15–ligand discovery for everyone. J. Chem. Inf. Model. **55**(11), 2324–2337 (2015). https://doi.org/10.1021/acs.jcim.5b00559
7. Lipinski, C.A.: Lead- and drug-like compounds: the rule-of-five revolution. Drug Discov. Today Technol. **1**(4), 337–341 (2004). https://doi.org/10.1016/j.ddtec.2004.11.007
8. Cherkasov, A., Ban, F., Li, Y., Fallahi, M., Hammond, G.L.: Progressive docking: a hybrid QSAR/docking approach for accelerating in silico high throughput screening. J. Med. Chem. **49**(25), 7466–7478 (2006). https://doi.org/10.1021/jm060961+

Open Access This chapter is licensed under the terms of the Creative Commons Attribution 4.0 International License (http://creativecommons.org/licenses/by/4.0/), which permits use, sharing, adaptation, distribution and reproduction in any medium or format, as long as you give appropriate credit to the original author(s) and the source, provide a link to the Creative Commons license and indicate if changes were made.

The images or other third party material in this chapter are included in the chapter's Creative Commons license, unless indicated otherwise in a credit line to the material. If material is not included in the chapter's Creative Commons license and your intended use is not permitted by statutory regulation or exceeds the permitted use, you will need to obtain permission directly from the copyright holder.

Predictive Power of Time-Series Based Machine Learning Models for DMPK Measurements in Drug Discovery

Modest von Korff[(⊠)] ⓘ, Olivier Corminboeuf ⓘ, John Gatfield,
Sébastien Jeay ⓘ, Isabelle Reymond ⓘ, and Thomas Sander ⓘ

Idorsia Pharmaceuticals Ltd., 4123 Allschwil, Switzerland
{modest.korff,olivier.corminboeuf,john.gatfield,
sebastien.jeay,isabelle.reymond,
thomas.sander}@idorsia.com

Abstract. Four datasets measuring DMPK (drug metabolism and pharma-cokinetics) parameters, and one target protein-specific dataset were analyzed by machine learning methods. Parameters measured for the five compound sets were biological activity data, plasma protein binding, permeability in MDCK I cell layers, intrinsic clearance by human liver microsomes, and plasma exposure in orally dosed rats. The measured data were sorted chronologically, reflecting the order in which they had been obtained in the discovery project. Subsets of the chronologically sorted data that appeared early in the project were used as training datasets to build predictive models for subsequent compounds based on kNN, partial least squares regression (PLSR), nonlinear PLSR, random forest regression, and support vector regression. A median model was used as a baseline to assess the machine learning model prediction quality. Data sets sorted in order of increasing test set prediction error: intrinsic clearance, plasma protein binding, cell layer permeability, biological activity on target protein, and bioavailability as AUC in rats. Our results give a first estimation of the power of machine learning to predict DMPK properties of compounds in an ongoing drug discovery project.

Keywords: Machine learning · DMPK · Drug discovery · Biological processes

1 Introduction

In drug discovery, new molecules undergo clinical trials with human subjects only after passing numerous checks for safety and potency in biological test systems. Often desired is a drug suitable for oral administration, i.e., a molecule that can cross cellular membranes separating the gastrointestinal system from blood vessels. Cell assays with MDCK I cells are used to assess membrane penetration. After absorption, blood vessels distribute the molecule across the organism and bring it to its site of action. Blood contains many proteins that bind a substantial fraction of the compound. This is measured as plasma protein binding (PPB). On its way, the molecule passes through the liver that contains enzymes able to metabolize many types of chemical substances, thus reducing the active drug's concentration (clearance). In drug discovery, a

© The Author(s) 2019
I. V. Tetko et al. (Eds.): ICANN 2019, LNCS 11731, pp. 741–746, 2019.
https://doi.org/10.1007/978-3-030-30493-5_67

suspension of human liver microsomes is used to assess intrinsic clearance. An important measure to optimize for a bioactive molecule is its plasma exposure after oral administration that is often expressed as "area under the curve" (AUC), i.e., the concentration of the active molecule in blood plasma integrated over time. Bioavailability depends on multiple properties of the molecule including cell layer permeability and clearance in the liver.

Medicinal chemists need quantitative models that would allow to prioritize the most promising molecules for biological testing. Quantitative structure-activity relationships (QSAR) is a central technology in drug discovery and have been investigated in multiple publications [1, 2]. But the chronological occurrence of information in the project workflows of drug discovery has rarely been investigated. A chemical series in drug discovery starts usually with one or a few molecules, often, with modest activity on the target protein. The starting compounds are modified by medicinal chemists to improve their properties. In a maturing drug discovery project, compounds become "smarter", because they contain more information from previous measurements. Here, we show how to set up quantitative machine learning models in an industrial drug discovery context in order to predict biological properties.

2 Methods

2.1 Data Sets (Table 1)

Each data set contained molecules from the same chemical series that shared an identical 'backbone' chemical substructure. Our first dataset contained measurements of half-maximal effective concentration (EC50) on the target protein for 1400 molecules. Much fewer data points were available for the other four datasets originating from active drug discovery programs. PPB is an important measure to assess the concentration of the free (unbound) molecule in the blood.

Table 1. Data sets used for machine learning models.

Data set ID	Dataset name	Data response (min, median, max)	Unit	Number of molecules
1	Activity on target protein	45, 1100, 20000	nmol/l	1400
2	Plasma protein binding	57, 94, 100	%	129
3	Permeability in MDCK I	0.06, 2.7, 39.4	10–6 cm/s	89
4	Intrinsic clearance HLM	65, 313, 1250	l/(min mg)	179
5	Bioavailability AUC	0, 540, 31600	ng h/ml	182

Dataset 2 contained 129 molecules. The permeability MDCK I (dataset 3) contained data from 89 molecules that were tested in a cell permeability assay. High permeability values are desired when the compound should pass the intestinal membranes in humans. The intrinsic clearance (dataset 4) contained data for 179 molecules. To determine the stability of the molecules they were measured in a human liver microsomes (HLM) assay. Dataset 5 contained area under the concentration vs time curve (AUC) values from rats for 182 compounds representing the overall bioavailability of the compound. After the compound was administered to rats, blood samples were taken and the concentration of the test compound in the blood was measured.

To create quantitative computer models for molecules it is necessary to encode the molecules as vectors $\mathbf{x}$. We decided to use the Skeleton Spheres descriptor [3], where one row in the matrix $\mathbf{X}$ represents one molecule. For each molecule in a data set, there is a single response value y_i. In every data set, the molecules were highly similar in descriptor space as they were in chemical space.

2.2 Machine Learning Techniques

Five modeling techniques were applied to construct regression models for the data sets: KNN regression, PLSR, PLSR with power transformation, random forest regression, and support vector (SVM) regression. All parameters for these machine learning models were optimized by an exhaustive search. The median model was used as a baseline model. Any successful machine learning model should be significantly better than the baseline model. Almost as simple was the k next neighbor model for regression. Partial least square regression (PLSR) is a multivariate linear regression technique [4] that requires the number of factors as the only input parameter. PLSR with power transformation includes a Box Cox transformation. It is often used to model biological data, which are notorious to be not normally distributed [5]. For random forests, we used the implementation from Li [6]. The Java program library libsvm was used for the support vector machine regression [7].

2.3 Successive Regression

To assess the predictive power of a machine learning tool in a drug discovery project it is necessary to consider the point in time a compound was made. Therefore, we ordered all molecules in a dataset according the point in time it was synthesized. A two-step process was implemented to ensure an unbiased estimation for the predictive power of a model. The first step was the selection of one meta-parameter set for every machine learning technique. The algorithm started with the first 20% of the molecule descriptors $\mathbf{X}_{0,0.2}$, $\mathbf{y}_{0,0.2}$ together with the measured response values to determine the meta parameters of the machine learning models via an exhaustive search. An eleven-fold Monte Carlo cross validation was employed to split all data into the training and validation datasets [8]. A left out of 10% was chosen as the size of the validation dataset. With this setup, the average error for all meta-parameter sets was calculated. For each machine learning technique t, the meta-parameter set $\mathbf{M}_{min,t}$ was chosen that showed the minimum average error. This meta-parameter set was used to construct a model from all data in $\mathbf{X}_{0,0.2}$, $\mathbf{y}_{0,0.2}$. In the second step, an independent test set was compiled from the next 10% of data, $\mathbf{X}_{0.3}$, $\mathbf{y}_{0.3}$. The average prediction error of $\widehat{\mathbf{y}_{0.3}}$ gave an unbiased estimator for the model, because the machine learning algorithm $\mathbf{M}_{min,t,0.2}$ had not seen these data before prediction. Subsequently, step one was repeated, this time with the data set $\mathbf{X}_{0.3}$, $\mathbf{y}_{0.3}$. So, the former test data were added to $\mathbf{X}_{0,0.2}$, $\mathbf{y}_{0,0.2}$. The meta parameter for the machine learning algorithms $\mathbf{M}_{min,t,0.3}$ were now determined with $X_{0,0.3}$, $y_{0,0.3}$. So, the prediction was done for $y_{0.4}$. This process was repeated eight times, up to a model size with $X_{0,0.9}$, $y_{0,0.9}$ and a prediction for $y_{1.0}$. With this method, we assessed how the predictive power and depends on the time point when the data were obtained in a drug discovery project. The 10% test set, next in time, was an unbiased estimator of the model's quality.

3 Results and Conclusions

For all five data sets, increasing portions of the chronologically sorted biological data were used as training data to build models that predicted the next 10% of the data. For the largest data set (Fig. 1) already all 'first-step models' had more predictive power than the median model. As the project developed in time, it could be observed that the variance in the EC50 values declined. This was indicated by the smaller median error.

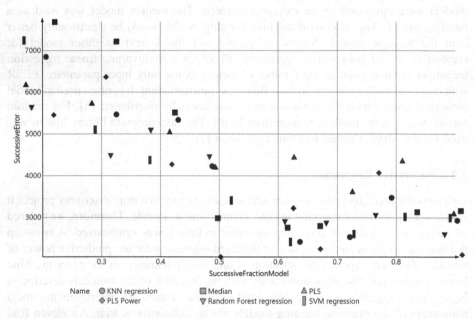

Fig. 1. Results for the successive prediction of EC50 values for dataset 1. The x-axis specifies the fraction of the full dataset used as training data. The y-axis indicates the average error in nmol/l for the prediction of EC50.

For the PPB dataset, the first training data set with $X_{0,0.2}$, and $y_{0,0.2}$ contained only 26 molecules. All 'first-step models' predicted the test data better than the median model. From $X_{0,0.5}$, and $y_{0,0.5}$ all predictions were superior compared to the median model.

The successive prediction of intrinsic clearance was more successful than the prediction of the permeability MDCK I dataset. The machine learning models outperformed the median model (data not shown). On the AUC (bioavailability) dataset 5, no machine learning technique performed better that the median model (data not shown).

Summary. All vv inner similarity in the SkeletonSpheres descriptor domain. No time-dependent learning curve was observed for the five biological datasets. The introduction of newly designed compounds increased the model error in several cases, even though the added compounds shared larger parts of their molecular structure with the training dataset compounds.

Conclusions. The uncertainty in prediction is correlated with the underlying biological complexity of the modeled parameter. Meaningful models for most of the PPB test datasets were created by all techniques. The permeability MDCK I (dataset 3) can be partially explained by diffusion that is relatively easy to model. However, MDCK I cell membranes contain active transporter proteins. Whether a molecule will be substrate for a transporter is hard to predict. The intrinsic clearance (data set 4) depends on the activity of approximately 20 enzymes, which makes the modeling more challenging than activity prediction for a single target enzyme. The bioavailability measured as AUC (dataset 5), is the result of multiple processes in animals, including cell layer permeability and intrinsic clearance. Consequently, the bioavailability AUC model had the highest uncertainty, followed by the target protein bioactivity data, and the intrinsic clearance model, while the most reliable models were created for the PPB and the permeability dataset. Naturally, all predictions include the uncertainty of the measurements of the response data. We presented time-series prediction results for four important measurements in drug discovery: PPB, permeability MDCK I, intrinsic clearance for human liver microsomes, and oral bioavailability as AUC in rats. The chronology-based prediction gives a good estimation of how well results of biological tests can be modeled in drug discovery projects.

References

1. Cherkasov, A., et al.: Qsar modeling: where have you been? where are you going to? J. Med. Chem. **57**, 4977–5010 (2014). https://doi.org/10.1021/jm4004285
2. Gramatica, P.: Principles of QSAR models validation: internal and external. QSAR Comb. Sci. **26**, 694–701 (2007). https://doi.org/10.1002/qsar.200610151
3. Boss, C., et al.: The screening compound collection: a key asset for drug discovery. Chimia (Aarau) **71**, 667–677 (2017). https://doi.org/10.2533/chimia.2017.667
4. de Jong, S.: SIMPLS: an alternative approach to partial least squares regression. Chemometr. Intell. Lab. Syst. **18**, 251–263 (1993). https://doi.org/10.1016/0169-7439(93)85002-X
5. Sakia, R.: The box-cox transformation technique: a review. J. Roy. Stat. Soc.: Ser. D (Stat.) **41**, 169–178 (1992). https://doi.org/10.2307/2348250
6. https://haifengl.github.io/smile/. Accessed 02 July 2019

7. Chang, C.-C., Lin, C.-J.: LIBSVM: a library for support vector machines. ACM Trans. Intell. Syst. Technol. (TIST) **2**, 27 (2011). https://doi.org/10.1145/1961189.1961199
8. Xu, Q.-S., Liang, Y.-Z.: Monte carlo cross validation. Chemometr. Intell. Lab. Syst. **56**, 1–11 (2001). https://doi.org/10.1016/S0169-7439(00)00122-2

Open Access This chapter is licensed under the terms of the Creative Commons Attribution 4.0 International License (http://creativecommons.org/licenses/by/4.0/), which permits use, sharing, adaptation, distribution and reproduction in any medium or format, as long as you give appropriate credit to the original author(s) and the source, provide a link to the Creative Commons license and indicate if changes were made.

The images or other third party material in this chapter are included in the chapter's Creative Commons license, unless indicated otherwise in a credit line to the material. If material is not included in the chapter's Creative Commons license and your intended use is not permitted by statutory regulation or exceeds the permitted use, you will need to obtain permission directly from the copyright holder.

Improving Deep Generative Models with Randomized SMILES

Josep Arús-Pous[1,3]([⊠]) [iD], Simon Johansson[1] [iD], Oleksii Prykhodko[1] [iD],
Esben Jannik Bjerrum[1] [iD], Christian Tyrchan[2] [iD], Jean-Louis Reymond[3] [iD],
Hongming Chen[1] [iD], and Ola Engkvist[1] [iD]

[1] Hit Discovery, Discovery Sciences, IMED Biotech Unit, AstraZeneca,
Gothenburg, Sweden
[2] Medicinal Chemistry, Cardiovascular, Renal and Metabolism, IMED Biotech Unit,
AstraZeneca, Gothenburg, Sweden
[3] Department of Chemistry and Biochemistry, University of Bern,
Bern, Switzerland
josep.arus@dcb.unibe.ch

Abstract. A Recurrent Neural Network (RNN) trained with a set of
molecules represented as SMILES strings can generate millions of different valid and meaningful chemical structures. In most of the reported
architectures the models have been trained using a canonical (unique for
each molecule) representation of SMILES. Instead, this research shows
that when using randomized SMILES as a data amplification technique,
a model can generate more molecules and those are going to accurately
represent the training set properties. To show that, an extensive benchmark study has been conducted using research from a recently published
article which shows that models trained with molecules from the GDB-
13 database (975 million molecules) achieve better overall chemical space
coverage when the posterior probability distribution is as uniform as possible. Specifically, we created models that generate nearly all the GDB-13
chemical space using only 1 million molecules as training set. Lastly, models were also trained with smaller training set sizes and show substantial
improvement when using randomized SMILES compared to canonical.

Keywords: Cheminformatics · Molecular generative models ·
Randomized SMILES · Molecular databases ·
Recurrent Neural Networks · Benchmarking

1 Introduction

Molecular deep generative models have emerged as a powerful tool to generate
chemical space [6] and obtain optimised compounds [2,5]. Models trained with a
set of drug-like molecules can generate molecules that are similar but not equal
to those in the training set, thus spanning a bigger chemical space than that of
training data. The most popular architecture uses Recurrent Neural Networks
(RNNs) and the SMILES syntax [7] to represent molecules. Nevertheless, a recent

© The Author(s) 2019
I. V. Tetko et al. (Eds.): ICANN 2019, LNCS 11731, pp. 747–751, 2019.
https://doi.org/10.1007/978-3-030-30493-5_68

publication [1] shows that this architecture introduces bias to the generated chemical space. To be able to prove that, models were created with a subset of GDB-13 [4], a database that holds most drug-like molecules up to 13 heavy atoms, and sampled with replacement 2 billion times. At most, only 68% of GDB-13 could be obtained from a theoretical maximum of 87%, which would be from a sample of the same size from an ideal model that has a uniform probability of obtaining each molecule from GDB-13.

This study uses the previous research as a starting point and focuses on benchmarking RNN with SMILES trained with subsets of GDB-13 of different sizes (1 million and 1000 molecules) and with different variants of the SMILES notation. One of those variants, randomized SMILES, can be used as a data amplification technique and is shown to generate more diversity [3]. When the right data representations and hyperparameter combinations are chosen, models are able to generate more diversity and learn to better generalise the training set information.

2   Methods

The model architecture used is similar to the one used in [1,5]. The training set sequences are pre-processed, and for each training epoch the entire training set is shuffled and subdivided in batches. The encoded SMILES strings of each batch are input token by token to an embedding layer, followed by several layers of RNN cells. Between the inner RNN layers there can be dropout layers. Then, the output from the cells is squeezed to the vocabulary size by a linear layer and a *softmax* is performed to obtain the probabilities of sampling each token in the next position. This is repeated for each token in the entire sequence.

Table 1. Hyperparameter combinations for both the 1M model and the 1 K model. Notice that the 1 K model also optimises the network topology, this was possible due to shorter training times.

Model	Cells	Num. layers	Layer size	Dropout	Batch
1M	GRU, LSTM	3	512	0, 0.25, 0.5	64, 128, 256, 512
1K	LSTM	2, 3, 4	128, 192, 256	0, 0.25, 0.5	4, 8, 16

The models were optimised for the hyperparameter combinations shown in Table 1. Also, training sets were set up with canonical SMILES and randomized SMILES. In the case of the randomized SMILES, each training epoch had a different permutation. For each combination of hyperparameters a model was trained and a sample with replacement of 2 billion SMILES strings was performed (Fig. 1). Then, three ratios were calculated from the percentages obtained that characterise the three main properties that the output domain should have: uniformity (even posterior probability for each molecule), completeness (all molecules from GDB-13) and closeness (no molecules outside of GDB-13 should be generated). Lastly, the UCC, a ratio obtained from the other three was used as a sorting criteria for all the models.

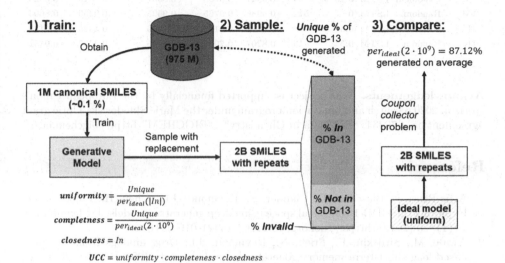

Fig. 1. Training and sampling process used for each model in the benchmark and the formulas for the ratios calculated from the sample.

3 Results

Table 2 shows the results for the models with highest UCC score of each training set size with each SMILES variant. 1M models trained with randomized SMILES are overall better than those trained with canonical SMILES. This might be due to the additional information the model has from molecules in the training set when they are input as different randomized SMILES each epoch. Notice especially that the completeness is at 0.95, which indicates that the model is theoretically able to reproduce mostly all of GDB-13 given enough sampling. On the other hand, models trained with 1000 SMILES have much lower performance, as there is not enough information in the training sets to be able to generalise the entire database. Nevertheless, the randomized SMILES model has an even better performance compared to the canonical SMILES one. Namely, a model trained with canonical SMILES can only reach 52% valid molecules, whereas the randomized SMILES model learns much better (82%). This shows that randomized SMILES add more information to the model and effectively increase its learning capability without having to add additional data to the training set.

Table 2. Results for the best canonical and randomized SMILES models for both the 1M and 1 K training set benchmarks.

Model	SMILES	Cell	Dropout	Batch	Validity	Uniformity	Completeness	Closeness	UCC
1M	Canonical	LSTM	0.25	64	0.9941	0.8788	0.8361	0.8613	0.6328
1M	Random	LSTM	0	512	0.9986	0.9765	0.9525	0.9250	0.8604
1K	Canonical	LSTM	0.5	4	0.5236	0.6114	0.1669	0.1325	0.0135
1K	Random	LSTM	0.5	16	0.8207	0.7902	0.3915	0.2757	0.0852

Acknowledgements. This project is supported financially by the European Union's Horizon 2020 research and innovation program under the Marie Skłodowska-Curie grant agreement no. 676434, "Big Data in Chemistry" ("BIGCHEM" http://bigchem.eu).

References

1. Arús-Pous, J., Blaschke, T., Ulander, S., Reymond, J.L., Chen, H., Engkvist, O.: Exploring the GDB-13 chemical space using deep generative models. J. Cheminform. **11**(1), 20 (2019). https://doi.org/10.1186/s13321-019-0341-z
2. Awale, M., Sirockin, F., Stiefl, N., Reymond, J.l.: Drug analogs from fragment based long short-term memory generative neural networks (2018). https://doi.org/10.26434/chemrxiv.7277354.v1, https://chemrxiv.org/articles/Drug_Analogs_from_Fragment_Based_Long_Short-Term_Memory_Generative_Neural_Networks/7277354
3. Bjerrum, E.J.: SMILES Enumeration as Data Augmentation for Neural Network Modeling of Molecules. arXiv March 2017. http://arxiv.org/abs/1703.07076
4. Blum, L.C., Reymond, J.L.: 970 million druglike small molecules for virtual screening in the chemical universe database GDB-13. J. Am. Chem. Soc. **131**(25), 8732–8733 (2009). https://doi.org/10.1021/ja902302h
5. Olivecrona, M., Blaschke, T., Engkvist, O., Chen, H.: Molecular de novo design through deep reinforcement learning. J. Cheminform. **9**(1) (2017). https://doi.org/10.1186/s13321-017-0235-x, http://arxiv.org/abs/1704.07555
6. Segler, M.H.S., Kogej, T., Tyrchan, C., Waller, M.P.: Generating focused molecule libraries for drug discovery with recurrent neural networks. ACS Cent. Sci. **4**(1), 1–17 (2018). https://doi.org/10.1021/acscentsci.7b00512, http://arxiv.org/abs/1701.01329
7. Weininger, D.: SMILES, a chemical language and information system: 1: introduction to methodology and encoding rules. J. Chem. Inf. Comput. Sci. **28**(1), 31–36 (1988). https://doi.org/10.1021/ci00057a005

Open Access This chapter is licensed under the terms of the Creative Commons Attribution 4.0 International License (http://creativecommons.org/licenses/by/4.0/), which permits use, sharing, adaptation, distribution and reproduction in any medium or format, as long as you give appropriate credit to the original author(s) and the source, provide a link to the Creative Commons license and indicate if changes were made.

The images or other third party material in this chapter are included in the chapter's Creative Commons license, unless indicated otherwise in a credit line to the material. If material is not included in the chapter's Creative Commons license and your intended use is not permitted by statutory regulation or exceeds the permitted use, you will need to obtain permission directly from the copyright holder.

Attention and Edge Memory Convolution
for Bioactivity Prediction

Michael Withnall[(✉)] (iD), Edvard Lindelöf, Ola Engkvist, and Hongming Chen

Hit Discovery, Discovery Sciences, IMED Biotech Unit, AstraZeneca,
Gothenburg, Sweden
followup@withnall.org.uk

Abstract. We present some augmentations to literature Message Passing Neural Network (MPNN) architectures and benchmark their performances against a wide range of chemically and pharmaceutically relevant datasets. We analyse the effects of activation function for regularisation, we propose a new graph attention mechanism, and we implement a new edge-based memory system that should maximise the effectiveness of hidden state usage by directing and isolating information flow around the graph. We compare our results to the MolNet [14] benchmarking paper results on graph-based techniques, and also investigate the effect of method performance as a function of dataset preprocessing.

Keywords: Graph convolution · Cheminformatics · Deep learning

1 Introduction

Many fields and research areas over the past decade have benefit greatly from the rise of deep learning [3]. AI has risen in popularity notably in the pharmaceutical industry, for activities such as bioactivity and physical-chemical property prediction, *de novo* design, synthesis prediction and image analysis, to name a few. The rapid growth of accessible computing power thanks to graphically-accelerated computing, and the ever increasing quantity of available chemical and biochemical data, have lead to a natural desire for data-hungry machine learning techniques such as deep learning to attempt to exploit this information to the greatest possible extent.

1.1 Graph Convolution

In Graph Convolutional Networks (GCNs), information propagates through a given graph much like how convolutional neural networks (CNNs) treat grid data (e.g. image data, text strings etc.). In contrast to image-data, however, graphs have irregular local connectivity, are not necessarily shift-invariant, and are not from a Euclidean domain (Fig. 1). CNNs exploit these properties for

Funding from the EU H2020 MSC Grant 676434 "BigChem".

© The Author(s) 2019

I. V. Tetko et al. (Eds.): ICANN 2019, LNCS 11731, pp. 752–757, 2019.
https://doi.org/10.1007/978-3-030-30493-5_69

their powerful performance, and to match this on graphs these problems must be surmounted, in a manner that is invariant to the ordered representation of the graph.

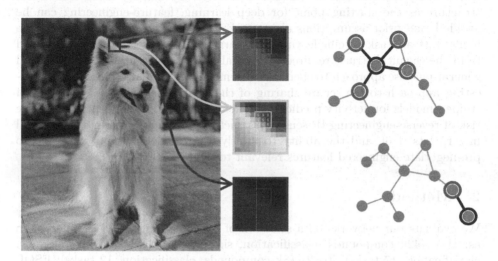

Fig. 1. Convolutional Neural Networks operating on e.g. image data (left) have a regular Euclidean representation, with a fixed dimensionality of neighbours to each data point (vertical, horizontal, and channel-depth). With graph based data, however, each node can have a variable number of neighbours (irregular), and the graph can be traversed in any order (isomorphic representation) without an easily-describable canonical representation

Message-Passing Neural Networks. Graph Neural Networks began in 2005 by Gori *et al.* [9], and in 2013 the first Graph Convolutional Network schema based on spectral graph theory was published by Bruna *et al.* [2]. Our work is focussed on the framework presented by Google – the Message Passing Neural Network [7], which was developed to generalize and be able to represent a selection of previously-published graph-based techniques [2,4,5,10–13]. We analyse the performance effects of activation function (SELU)-based normalisation, and chemically-based dataset preprocessing, and propose two new novel architectures as extensions to the MPNN framework:

Attention MPNN (AMPNN) in which attention is performed over hidden state vector elements, dependent on edge type, allowing weighted summation in the message-passing function.

An Edge-Memory network, in which hidden states belong to directed edges and can only propagate in a single direction, designed to naturally allow for asymmetric bias and to maximise useful hidden memory information when propagating a node's neighbourhood.

Low-Level Features from Graph Structure. Unlike traditional cheminformatic approaches to Machine Learning tasks, which use feature engineering, graphs are one of the lowest-level representations of chemical structures from which many features can be calculated directly. By directly using a chemical structure as the starting point for deep learning, feature-engineering can be avoided, and prior assumptions about task-specific knowledge don't need to be made. Instead, task-specific features are learned within the network, and derived from the chemical structure directly. This allows for a potentially very powerful general-purpose approach to chemical task modelling, and also presents an interesting approach to the secure sharing of chemical data – the dissemination of trained models for activity prediction in lieu of chemical data itself, without the risk of reverse-engineering IP-sensitive structural information from e.g. chemical fingerprints [1,6], and the ability to jointly-train models without the need to pre-negotiate engineered features relevant to the task.

2   Method

We evaluate our networks on a selection of benchmarking datasets, referred to as: HIV (42k compounds, classification, single-task); MUV (93k compounds, classification, 17 tasks); Tox21 (8k compounds, classification, 12 tasks); ESOL (1k compounds, regression, single-task); QM8 (22k compounds, regression, 12 tasks); SIDER (1.4k compounds, classification, 27-task), LIPO (4k compounds, regression, single-task) and BBBP (2k compounds, classification, single-task). Datasets were split and tested according to previous MolNet benchmarking [14] and hyperparameter optimisation was performed using Bayesian Optimisation in parallel with Local Penalisation [8].

3   Results

We present results for models trained on benchmarking datasets both as presented verbatim, referred to as Original Dataset, and with custom preprocessing (Charge-Parent Missing-Data – CPMD). Results are in general on-par with state-of-the-art, beating classification performance on the MUV dataset, and obtaining lower error on the smaller LIPO regression set. The charge-parent aspect of database preprocessing was found to be negligible (no performance difference between e.g. SIDER models or single-task models, with no missing data values), but the introduction of missing data values and a suitable masking loss function was found to have a strong positive effect on performance on highly sparse sets (MUV), over tripling performance relative to MolNet for the Attention, Edge and SELU networks, bringing them on-par with SVM and beating SVM with the Edge-based approach. The charge-parent aspect of the preprocessing was done to investigate how robust the model is to ionic complexes, such as those shown in Fig. 4. As the network does not model ionic bonds, it was unknown whether disjoint graphs would interfere with the message propagation, and ions such as sodium would act as noise in the training. However, due to the lack of performance difference between the two sets when all data is present, it can be assumed that the readout function safely bridges these gaps and does not interfere with models' performance (Figs. 2 and 3).

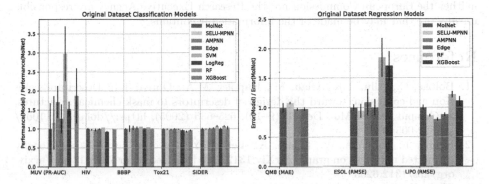

Fig. 2. Relative Performance of Classification (left) and Relative Error of Regression (right) models against the best presented MolNet model, on the original datasets. Unless otherwise stated, classification sets were evaluated using the ROC-AUC metric.

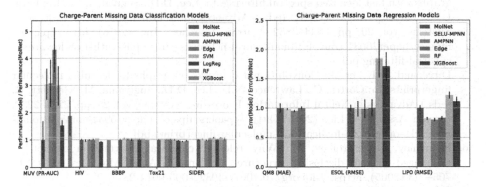

Fig. 3. Relative Performance of Classification (left) and Relative Error of Regression (right) models against the best presented MolNet model, on the CPMD datasets. Unless otherwise stated, classification sets were evaluated using the ROC-AUC metric.

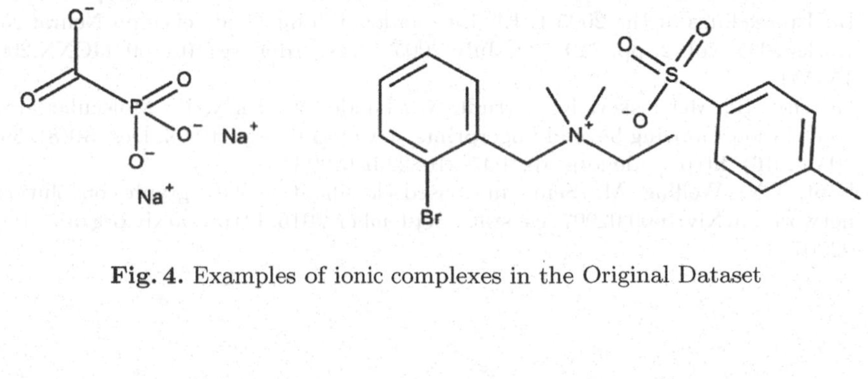

Fig. 4. Examples of ionic complexes in the Original Dataset

Acknowledgements. The project leading to this article received funding from the European Union's Horizon 2020 research and innovation program under the Marie Skłodowska-Curie grant agreement No. 676434, "Big Data in Chemistry" ("BIGCHEM", http://bigchem.eu). The article reflects only the authors' view, and neither the European Commission nor the Research Executive Agency are responsible for any use that may be made of the information it contains.

References

1. Bologa, C., Allu, T.K., Olah, M., Kappler, M.A., Oprea, T.I.: Descriptor collision and confusion: toward the design of descriptors to mask chemical structures. J. Comput. Aided Mol. Des. **19**(9–10), 625–635 (2005). https://doi.org/10.1007/s10822-005-9020-4
2. Bruna, J., Zaremba, W., Szlam, A., LeCun, Y.: Spectral networks and locally connected networks on graphs. arXiv:1312.6203 [cs], December 2013. http://arxiv.org/abs/1312.6203
3. Chen, H., Engkvist, O., Wang, Y., Olivecrona, M., Blaschke, T.: The rise of deep learning in drug discovery. Drug Discov. Today **23**(6), 1241–1250 (2018). https://doi.org/10.1016/j.drudis.2018.01.039, http://www.sciencedirect.com/science/article/pii/S1359644617303598
4. Defferrard, M., Bresson, X., Vandergheynst, P.: Convolutional neural networks on graphs with fast localized spectral filtering. In: Lee, D.D., Sugiyama, M., Luxburg, U.V., Guyon, I., Garnett, R. (eds.) Advances in Neural Information Processing Systems, vol. 29, pp. 3844–3852. Curran Associates Inc. (2016). http://papers.nips.cc/paper/6081-convolutional-neural-networks-on-graphs-with-fast-localized-spectral-filtering.pdf
5. Duvenaud, D.K., et al.: Convolutional networks on graphs for learning molecular fingerprints. In: Cortes, C., Lawrence, N.D., Lee, D.D., Sugiyama, M., Garnett, R. (eds.) Advances in Neural Information Processing Systems, vol. 28, pp. 2224–2232. Curran Associates Inc. (2015). http://papers.nips.cc/paper/5954-convolutional-networks-on-graphs-for-learning-molecular-fingerprints.pdf
6. Filimonov, D., Poroikov, V.: Why relevant chemical information cannot be exchanged without disclosing structures. J. Comput. Aided Mol. Des. **19**(9–10), 705–713 (2005). https://doi.org/10.1007/s10822-005-9014-2
7. Gilmer, J., Schoenholz, S.S., Riley, P.F., Vinyals, O., Dahl, G.E.: Neural message passing for quantum chemistry. arXiv:1704.01212 [cs], April 2017. http://arxiv.org/abs/1704.01212
8. González, J., Dai, Z., Hennig, P., Lawrence, N.D.: Batch Bayesian Optimization via Local Penalization. arXiv:1505.08052 [stat], May 2015. http://arxiv.org/abs/1505.08052
9. Gori, M., Monfardini, G., Scarselli, F.: A new model for learning in graph domains. In: Proceedings of the 2005 IEEE International Joint Conference on Neural Networks 2005, vol. 2, pp. 729–734, July 2005. https://doi.org/10.1109/IJCNN.2005.1555942
10. Kearnes, S., McCloskey, K., Berndl, M., Pande, V., Riley, P.: Molecular graph convolutions: moving beyond fingerprints. J. Comput. Aided Mol. Des. **30**(8), 595–608 (2016). https://doi.org/10.1007/s10822-016-9938-8
11. Kipf, T.N., Welling, M.: Semi-supervised classification with graph convolutional networks. arXiv:1609.02907 [cs, stat], September 2016. http://arxiv.org/abs/1609.02907

12. Li, Y., Tarlow, D., Brockschmidt, M., Zemel, R.: Gated Graph Sequence Neural Networks. arXiv:1511.05493 [cs, stat], November 2015. http://arxiv.org/abs/1511.05493
13. Schütt, K.T., Arbabzadah, F., Chmiela, S., Müller, K.R., Tkatchenko, A.: Quantum-chemical insights from deep tensor neural networks. Nat. Commun. **8**, 13890 (2017). https://doi.org/10.1038/ncomms13890
14. Wu, Z., et al.: MoleculeNet: a benchmark for molecular machine learning. Chem. Sci. **9**(2), 513–530 (2018). https://doi.org/10.1039/C7SC02664A, https://pubs.rsc.org/en/content/articlelanding/2018/sc/c7sc02664a

Open Access This chapter is licensed under the terms of the Creative Commons Attribution 4.0 International License (http://creativecommons.org/licenses/by/4.0/), which permits use, sharing, adaptation, distribution and reproduction in any medium or format, as long as you give appropriate credit to the original author(s) and the source, provide a link to the Creative Commons license and indicate if changes were made.

The images or other third party material in this chapter are included in the chapter's Creative Commons license, unless indicated otherwise in a credit line to the material. If material is not included in the chapter's Creative Commons license and your intended use is not permitted by statutory regulation or exceeds the permitted use, you will need to obtain permission directly from the copyright holder.

Application of Materials Informatics Tools to the Analysis of Combinatorial Libraries of All Metal-Oxides Photovoltaic Cells

Hanoch Senderowitz[1](✉) ⓘ, Abraham Yosipof[2], and Omer Kaspi[1]

[1] Department of Chemistry, Bar Ilan University, 5290002 Ramat Gan, Israel
omerkaspi@gmail.com, hsenderowitz@gmail.com,
hanoch.senderowitz@biu.ac.il
[2] Department of Information Systems, College of Law and Business,
Ramat-Gan, Israel
avi.yosipof@gmail.com

Abstract. Material informatics is engaged with the application of informatics tools, frequently in the form of machine learning algorithms, to gain insight into structure properties relationships of materials and to design new materials with desired properties. Here we describe the application of such algorithms to the analysis of solar cell (i.e., photovoltaic; PV) libraries made entirely from metal oxides (MOs). MOs-based solar cells hold the potential to provide clean and affordable energy if their power conversion efficiencies are improved. We demonstrate the power of dimensionality reduction methods to visualize the MOs-based solar cell space and the power of several algorithms to develop predictive models for key PV properties. We stress the importance of conducting such studies in collaboration with experimentalists.

Keywords: Materials informatics · QSAR · Solar cells ·
Dimensionality reduction · RANSAC · Genetic programming · kNN

1 Introduction

Materials informatics is an emerging field of research primarily engaged with the application of informatic principles to materials science for the purpose of discovering and developing new materials [1, 2]. Several factors contribute to the continuous development of materials informatics including the rapid growth of available databases that contain experimental and computational information on structures and properties of materials and the multiple similarities between materials informatics and e.g., chemo- and bio-informatics. Thus, this younger field could draw from the experience and large arsenal of tools developed in the more established fields [2].

Central to materials informatics is the application of data mining techniques and in particular machine learning approaches, often referred to as Quantitative Structure Activity Relationship (QSAR) modeling, to derive predictive models for a variety of materials-related "activities". Such models could accelerate the development of new materials with favorable properties and provide insight into the factors governing these

© The Author(s) 2019
I. V. Tetko et al. (Eds.): ICANN 2019, LNCS 11731, pp. 758–763, 2019.
https://doi.org/10.1007/978-3-030-30493-5_70

properties. In this work we present the application of several machine learning tools to the analysis of solar cells.

Growth in energy demands, coupled with the movement towards clean energy, are likely to make solar cells an important part of future energy resources. At present, most solar cells are based on silicon yet new alternatives are constantly emerging with examples including organic photovoltaic (OPV) cells, [3] dye sensitized solar cells (DSSC), [4] and pervoskites [5]. Another appealing alternative is presented by solar cells entirely made of metal oxides (MOs). Such cells are reasonably cheap to manufacture due to the abundance of their constituting elements, are environmentally friendly and are stable over time. Yet, their performances in terms of their ability to convert sunlight into electricity need to be improved [6]. Such improvements require the development of new MOs which could benefit from combining combinatorial materials science for producing solar cells libraries with machine learning approaches to analyze the resulting libraries and direct synthesis efforts.

MO-based solar cell libraries are produced through a combinatorial materials science approach involving the non-uniform deposition of two or more layers of different metal oxides on top of a glass support, followed by the introduction of the appropriate metal contacts. A typical library synthesized in this form consists of multiple cells (169 in our case) each of which is typically characterized by multiple composition and photovoltaic (PV) parameters including, for example, the thickness of the different layers that make up the cell and the thickness ratios between them, the band gap of the absorber layer, the short circuit photocurrent density (Jsc), the open circuit photovoltage (Voc) and the internal quantum efficiency (IQE) [6, 7].

2 Results and Discussion

In this work we present the application of several machine learning tools to the analysis of single and multiple MO-based solar cell libraries. Some of these tools were recently incorporated into a newly developed MATLAB-based decision support system (DSS) called "PV Analyzer" [8]. PV Analyzer integrates tools common to chemo- and materials-informatics such as simple bi-parametric correlations, heat-maps and principle component analysis (PCA) with a workflow for the development of predictive quantitative structure activity relationship (QSAR) models based on the RANSAC algorithm which was originally developed for computer vision (see below). The PV Analyzer workflow is composed of three blocks: (1) Input data; (2) Data pre-processing and visualization; (3) RANSAC modeling (Fig. 1). Importantly, PV Analyzer is a modular system which allows for the facile incorporation of additional tools.

The PV libraries studied in this work were gathered over time from multiple literature reports [6, 7, 9, 10 and references cited therein]. Importantly, all were manufactured by the same lab using largely similar methods. First, looking at individual libraries, we demonstrated the ability of standard PCA to identify outliers in the PV space and together with self organizing maps (SOM) to unveil local and global effects of a Molybdenum Oxide layer on PV properties [9]. Next, five libraries were compiled into a unified dataset for a total of 1165 solar cells where each solar cell was characterized by seven experimentally measured PV properties including short circuit

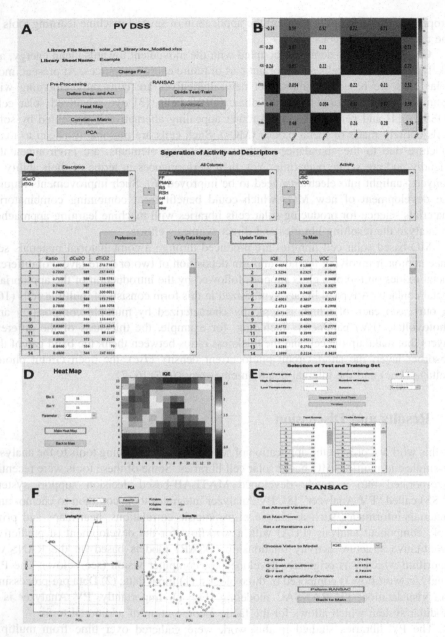

Fig. 1. Screenshots describing the application of the DSS to a specific solar cell library. (A) PV Analyzer's main screen. From this screen the user can control the tasks available in the DSS. (B) Correlation matrix screen showing the correlations between all descriptors and all activities (C) Activities/descriptors assignment and data curation screen. (D) Heat map screen showing the heat-map obtained for *IQE*. (E) Test set/Training set selection screen. (F) PCA screen showing the loading plot (left) and scoring plot (right) resulting from the PCA performed on the descriptors. (G) RANSAC control screen. Taken from reference [8].

photocurrent density (*Jsc*), open circuit photovoltage (*Voc*), internal quantum efficiency (*IQE*), maximum photovoltaic power (*Pmax*), fill factor (*FF*), series resistance (*Rs*) and Shunt resistance (*Rsh*). This database was subjected to several linear and non linear dimensionality reduction methods (PCA, kernel-PCA, Isomap and Diffusion map) to allow for a facile visualization of the resulting "PV space". Furthermore, we studied the relative performances of the different methods in terms of their ability to separate, in the reduced space, the five original libraries which were used to make up the space and maintain their original neighborhood [10]. We found that all methods were able to segregate the different libraries into unique regions of the reduced space (Fig. 2) but that PCA performed best in terms of the ability to correctly maintain the local environment of samples whereas Isomap did the best job in assigning class membership based on the identity of nearest neighbors, i.e., it was the best classifier. In addition, diffusion map identified the smallest number of outliers. We also found that many of the outliers identified by all methods could be rationalized.

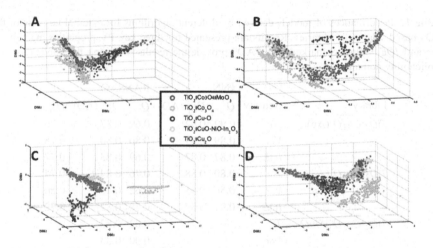

Fig. 2. 3D representation of the integrated database after different dimensionality reduction methods. Subfigures A-D represent the different libraries plotted in the reduced space following PCA, Kernel PCA, Isomap and Diffusion map, respectively. Taken from reference [10].

Next we used several machine learning algorithms such as k nearest neighbors (*k*NN), genetic programming (GP) [11] and RANSAC to develop predictive QSAR models for key PV properties including *Jsc*, *Voc* and *IQE* for several libraries [7, 12]. In particular, the Random Sample Consensus (RANSAC) algorithm is a predictive modeling tool widely used in the image processing field for cleaning datasets from noise. In this work we demonstrated that RANSAC could be used as a "one stop shop" algorithm for the development and validation of QSAR models, performing multiple tasks including outlier removal, descriptors selection, model development and predictions of test set samples while using the concept of applicability domain [13]. Contrary to most conventional QSAR models, the input descriptors used in these studies were experimentally measured rather than theoretically computed and included the thicknesses of the different MO layers used to construct the cells and the thickness ratios between them. The need to rely on measured rather than on calculated descriptors is because when the (well defined) different MOs are combined (see above for a short description of the manufacturing process), the exact composition/structure of the resulting solar cell is not well defined and consequently is not amenable to the calculation of structure-based descriptors. Our results (Table 1) demonstrate that QSAR models with good prediction statistics could be developed by any of these methods for most PV properties (except for several poor models developed for Voc) and that these models highlight important factors affecting these properties in accord with experimental findings. Specifically, the thicknesses of the different MO layers play a pivotal role in determining the PV properties. The resulting models are therefore suitable for designing better solar cells. Interestingly, the success of developing kNN-based models demonstrates that the similar property principle which is well established in the small molecules/pharmaceuticals hyperspace also holds in the PV space.

Table 1. Performances of models derived by different algorithms for external test sets of three MOs-based solar cells libraries. Results are presented for three key PV properties, namely, Jsc, Voc and IQE. Columns marked with "AD" provide the results following the application of an applicability domain.

Library	Activity	kNN Q^2_{ext}	Q^2_{ext} (AD)	RANSAC Q^2_{ext}	Q^2_{ext} (AD)	GP Q^2_{ext}
$TiO_2|Cu_2O$ (Ag)	Jsc	0.92	0.92	0.69	0.87	0.76
	Voc	0.89	0.89	0.80	0.80	0.78
	IQE	0.87	0.87	0.69	0.83	0.72
$TiO_2|Cu_2O$ (Ag|Cu)	Jsc	0.89	0.88	0.76	0.84	0.74
	Voc	0.56	0.55	0.62	0.73	0.50
	IQE	0.91	0.89	0.78	0.82	0.72
$TiO_2|Co_3O_4|MoO_3$	Jsc			0.82	0.82	
	Voc			0.00	0.33	
	IQE			0.79	0.79	

3 Conclusions

In this work we demonstrated the power of materials informatics methods to visualize the PV space of MO-based solar cell libraries and to develop predictive QSAR models for key PV properties. Importantly, we wish to emphasize the need to perform studies of the type described in this work in close collaboration with experimentalists in order to both provide physics/chemistry based explanation to the observed trends/effects/correlations and to capitalize on the results.

References

1. Rajan, K.: Mater. Today **8**, 38–45 (2005). https://doi.org/10.1016/S1369-7021(05)71123-8
2. Senderowitz, H., Tropsha, A.: J. Chem. Inf. Model. **58**, 1313–1314 (2018). https://doi.org/10.1021/acs.jcim.8b00016
3. Scharber, M.C., Sariciftci, N.S.: Prog. Polym. Sci. **38**, 1929–1940 (2013). https://doi.org/10.1016/j.progpolymsci.2013.05.001
4. Faccio, R., Fernandez-Werner, L., Pardo, H., Mombre, A.W.: Recent Pat. Nanotechnol. **5**, 46–61 (2011). https://doi.org/10.2174/187221011794474930
5. Rong, Y., et al.: Science **361**(6408) (2018). https://doi.org/10.1126/science.aat8235
6. Rühle, S., et al.: Phys. Chem. Chem. Phy. **16**, 7066–7073 (2014). https://doi.org/10.1039/C4CP00532E
7. Yosipof, A., Nahum, O.E., Anderson, A.Y., Barad, H., Zaban, A., Senderowitz, H.: Mol. Inf. **34**, 367–379 (2015). https://doi.org/10.1002/minf.201400174
8. Kaspi, O., Yosipof, A., Senderowitz, H.: Mol. Inf. **37**(9–10), e1800067 (2018). https://doi.org/10.1002/minf.201800067
9. Yosipof, A., Kaspi, O., Majhi, K., Senderowitz, H.: Mol. Inf. **35**, 622–628 (2016). https://doi.org/10.1002/minf.201600050
10. Kaspi, O., Yosipof, A., Senderowitz, H.: J. Chem. Inf. Model. **58**, 2428–2439 (2018). https://doi.org/10.1021/acs.jcim.8b00552
11. Mitchell, M.: An Introduction to Genetic Algorithms. MIT Press, London (1998). ISBN 0-262-13316-4
12. Kaspi, O., Yosipof, A., Senderowitz, H.: J. Chemoinf. **9**, 34 (2017). https://doi.org/10.1186/s13321-017-0224-0
13. Fischler, M.A., Bolles, R.C.: Commun. ACM **24**, 381–395 (1981). https://doi.org/10.1145/358669.358692

Open Access This chapter is licensed under the terms of the Creative Commons Attribution 4.0 International License (http://creativecommons.org/licenses/by/4.0/), which permits use, sharing, adaptation, distribution and reproduction in any medium or format, as long as you give appropriate credit to the original author(s) and the source, provide a link to the Creative Commons license and indicate if changes were made.

The images or other third party material in this chapter are included in the chapter's Creative Commons license, unless indicated otherwise in a credit line to the material. If material is not included in the chapter's Creative Commons license and your intended use is not permitted by statutory regulation or exceeds the permitted use, you will need to obtain permission directly from the copyright holder.

Analysis and Modelling of False Positives in GPCR Assays

Dipan Ghosh[1](✉), Igor Tetko[2] ⓘ, Bert Klebl[1], Peter Nussbaumer[1], and Uwe Koch[1]

[1] Lead Discovery Center GmbH, Otto-Hahn-Straße 15,
44227 Dortmund, Germany
{ghosh,klebl,nussbaumer,koch}@lead-discovery.de
[2] Institute of Structural Biology, Helmholtz Zentrum München – German
Research Center for Environmental Health (GmbH), Ingolstaedter Landstrasse 1,
85764 Neuherberg, Germany
i.tetko@helmholtz-muenchen.de

Abstract. G-Protein Coupled Receptors (GPCR) are involved in all the major signaling pathways. As a result, they often serve as potential target for therapeutic drugs. In this study we analyze publicly available assays involving different classes of GPCR to identify false positives. Using the latest developments in Machine Learning, we then build models that can predict such compounds with high confidence. Given the ubiquity of GPCR assays, we believe such models will be very helpful in flagging potential false positives for further testing.

Keywords: Neural Networks · Least squares SVM · Frequent hitters

1 Introduction

G-Protein Coupled Receptors (GPCR) are the largest family of cell surface receptors [1]. These plasma membrane bound receptors have evolved to recognize a variety of extracellular physical and chemical signals and, upon recognition, act as the proximal stimulus in cell signaling pathways. With over ~ 800 members [2], GPCRs are involved in almost every physiological function, from sensation to growth to hormone responses. Due to their widespread physiological relevance and presence of druggable sites, GPCRs are one of the major targets of therapeutic drugs. A 2017 study notes that 475 drugs act at 108 unique GPCRs. Approximately 321 agents are currently in clinical trials, of which $\sim 20\%$ target 66 potentially novel GPCR targets. GPCRs also account for $\sim 27\%$ of the global market share of therapeutic drugs, with aggregated sales for 2011–2015 of $\sim$ US\$890 billion.

As promising drug targets, assays involving a member of the GPCR family are commonly employed in high throughput screening (HTS) campaigns. There are a plethora of different techniques and a wide range of commercial kits available, many of which are suitable for High Throughput Screening (HTS) [3]. In such HTS, identifying false positives is a challenge. False positives may be compounds that interfere with the assay detection technology in some way, such as inhibiting luciferase in luciferase-

© The Author(s) 2019

I. V. Tetko et al. (Eds.): ICANN 2019, LNCS 11731, pp. 764–769, 2019.
https://doi.org/10.1007/978-3-030-30493-5_71

based system [4], or quenching fluorescence where it is the final readout [5]. There may also be compounds that are not specific to the target protein, but are promiscuous, either to a narrow or broad class of proteins [6].

In the previous study we developed a machine learning method to flag potential frequent hitters for luciferase assays [4]. In this study we investigated whether the developed methodology can be extended to identify false positives for GPCR assays.

2 Data

2.1 Data Description

Our initial goal was to explore the available data and find suitable assays that we can then use for further analysis. On PUBCHEM, we identified 92 assays with more than 500 compounds for GPCR agonists and antagonists. We separated the two and decided to focus on the agonists. This was just to narrow down the scope of the study. From the list of available agonist screenings we selected the 20 assays with the highest number of active compounds. This is because as we are looking for false positives. Assays that have little to no positives are less relevant for us. For further selection particular assays, we focused on the GPCR subtypes as described below.

2.2 Data Collection

The GPCR family is commonly classified into five different families based on their structural and sequence similarity. The families are then further classified into a family tree [7, 8]. Of these five major families, the Rhodopsin class in the largest. For selecting assays for our analysis, we mapped the target proteins onto this family tree (Fig. 1), and selected assays with set of representative proteins distant from each other in the family tree. This ensures that compounds that are frequently active, are not preferential agonists of a subtype of GPCR, but are more likely a result of assay artifact.

Using these criteria we chose a set of 12 assays and looked for compounds that are frequently active in these assays (see Methods section), i.e. actives across all of the various different subtypes and assay technologies and thus frequent hitters of the Rhodopsin class of GPCR. However, only 59 out of 373,131 compounds matched our definition of being frequently active. Upon closer examination we found that these compounds were tested only thrice, and therefore are more likely to be an artifact of selection criteria rather than a GPCR frequent hitter or assay artifact.

To further refine our search, we next focused on different detection technologies that were used in the assays. We found that half of assays (six) used fluorescence while other six assays used bioluminescence. Only 71 compounds were frequently active in the bioluminescence group. In the fluorescence group, although the number of datapoints and active compounds was very similar, 502 compounds were frequently active (Table 1). This indicates that fluorescence technology contributes many more artifacts and these 502 compounds were used for further analysis.

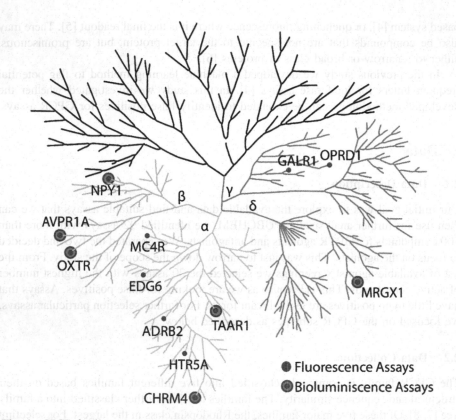

Fig. 1. GPCR family tree represented as a tree and dots mapping the protein targets in identified assays. The colored part of the tree represents the Rhodopsin class of the GPCR family and various subfamilies of the Rhodopsin class are marked with different colors. (Color figure online)

All data were harvested from PUBCHEM [9], manually or by using the PUBCHEM REST API with Python. All data were obtained and stored locally in the CSV format to be analyzed later with various python scripts.

Table 1. Statistics of compounds for the datasets used in the study.

	Inactive	Active	Frequently active
All assays	352685	20446	59
Fluorescence assays	363459	9605	502
Bioluminiscence assays	358770	10841	71

2.3   Frequent Hitter Flagging

We defined frequent hitters as compounds that were active according to our criteria in more than half of the assays they were tested in. Additionally each compound had to be tested at least in three different assays. Compounds satisfying both criteria were identified using a Python script and flagged as frequent hitters.

3   Methods

Using the freely accessible platform On-line Chemical and Modeling Environment (OCHEM) [10], was used to build models for our data. Different descriptors available in OCHEM include CDK, Dragon 6 and 7, ISHIDA fragmentor, among others. Their detailed description can be found elsewhere [4]. Associative Neural Networks (ASNN) [11], Deep Neural Network (DNN) [12], Extreme Gradient Boost (XGBOOST) [13], and Least Squares Support Vector Machine (LSSVM) [14] algorithms were analyzed for training the models. The methods were used with default parameters as specified on the OCHEM web site.

4   Results and Discussion

4.1   Machine Learning

The analyzed methods were used in combination with different descriptors sets. LSSVM provided on average the highest accuracy amid the chosen algorithms (Table 2). We selected LSSVM models with the highest accuracy based on their ROC-AUC score for building a consensus model. The consensus model had ROC-AUC score of 0.93 with balanced accuracy of 86%.

To test our model, we constructed an independent dataset by looking up GPCR agonist assays in PUBCHEM that we did not use for the training set. We found five relevant assays with 4323 active compounds. Our frequent hitter analysis identified 157 compounds from these 5 assays. Our consensus model predicted the molecules from this set with a balanced accuracy of 76% and an AUC score of 0.85. The consensus model which was based only subset of 2D descriptors provided a very similar accuracy of 75% and AUC score of 0.85 thus indicating the importance of only 2D information for this analysis.

Table 2. The performance of models built using the GPCR dataset. The ROC-AUC scores are calculated using 5-fold stratified cross-validation. Models marked with asterisk were used to build the consensus model.

Descriptors/methods	DNN	ASNN	XGBOOST	LSSVM
ALogPS, OEstate (2D)	0.84	0.84	0.87	0.89*
CDK2 (3D)	0.79	0.85	0.86	0.87*
ChemaxonDescriptors (3D)	0.82	0.82	0.84	0.88*
Dragon6 (2D blocks)	0.83	0.87	0.88	0.91
Dragon6 (3D, all blocks)	0.87	0.85	0.89	0.91*
Fragmentor (2D)	0.85	0.83	0.88	0.89*
GSFrag (2D)	0.81	0.8	0.86	0.85
InductiveDescriptors (3D)	0.79	0.78	0.79	0.83
JPlogP (2D)	0.82	0.79	0.85	0.84
Mera, Mersy (3D)	0.69	0.76	0.8	0.81
PyDescriptor (3D)	0.89	0.86	0.85	0.89*
QNPR (2D)	0.81	0.82	0.87	0.86
RDKIT (2D, all blocks)	0.88	0.88	0.87	0.91*
RDKIT (3D, all blocks)	0.88	0.88	0.87	0.91
SIRMS (2D)	0.86	0.83	0.86	0.87*
Spectrophores (3D)	0.63	0.69	0.72	0.68
StructuralAlerts (2D)	0.79	0.79	0.77	0.78
alvaDesc (2D blocks)	0.86	0.85	0.87	0.91*
alvaDesc (3D, all blocks)	0.88	0.86	0.88	0.91

5   Conclusion

In this study, we analyzed GPCR assays from PUBCHEM with the aim to identify frequent hitters. We found that fluorescence-based assays are more susceptible to false positives than bioluminescence. Compounds that were frequent hitters at fluorescence-based assays did not appear as frequent hitters in bioluminescence assays. A predictive machine-learning model to identify such compounds for GPCR assays was developed. The provided analysis can help to interpret HTS screening using GPCR assays.

Acknowledgement. The project leading to this report has received funding from the European Union's Horizon 2020 research and innovation program under the Marie Skłodowska-Curie grant agreement No 676434, "Big Data in Chemistry". The article reflects only the author's view and neither the European Commission nor the Research Executive Agency (REA) are responsible for any use that may be made of the information it contains. We thank Michael Withnall for English correction. The authors thank ChemAxon (http://www.chemaxon.com) for Academic license of software tools (Standartizer, ChemAxon plugins) as well as AlvaScience (http://alvascience.com), Molecular Networks GmbH (http://mn-am.com) and Chemosophia (http://chemosophia.com) for providing descriptors and Corina 2D to 3D conversion program used in this study.

References

1. Hauser, A.S., Attwood, M.M., Rask-Andersen, M., Schioth, H.B., Gloriam, D.E.: Trends in GPCR drug discovery: new agents, targets and indications. Nat. Rev. Drug Discov. **16**, 829–842 (2017). https://doi.org/10.1038/nrd.2017.178
2. Fredriksson, R., Lagerström, M.C., Lundin, L.-G., Schiöth, H.B.: The G-protein-coupled receptors in the human genome form five main families. phylogenetic analysis, paralogon groups, and fingerprints. Mol. Pharmacol. **63**, 1256–1272 (2003). https://doi.org/10.1124/mol.63.6.1256
3. Zhang, R., Xie, X.: Tools for GPCR drug discovery. Acta Pharmacol. Sin. **33**, 372–384 (2012). https://doi.org/10.1038/aps.2011.173
4. Ghosh, D., Koch, U., Hadian, K., Sattler, M., Tetko, I.V.: Luciferase advisor: high-accuracy model to flag false positive hits in luciferase HTS assays. J. Chem. Inf. Model. **58**, 933–942 (2018). https://doi.org/10.1021/acs.jcim.7b00574
5. Schorpp, K., Rothenaigner, I., Salmina, E., Reinshagen, J., Low, T., et al.: Identification of small-molecule frequent hitters from AlphaScreen high-throughput screens. J. Biomol. Screen. **19**, 715–726 (2014). https://doi.org/10.1177/1087057113516861
6. Roche, O., Schneider, P., Zuegge, J., Guba, W., Kansy, M., et al.: Development of a virtual screening method for identification of "frequent hitters" in compound libraries. J. Med. Chem. **45**, 137–142 (2002). https://doi.org/10.1021/jm010934d
7. Hu, G.M., Mai, T.L., Chen, C.M.: Visualizing the GPCR network: classification and evolution. Sci. Rep. **7**, 15495 (2017). https://doi.org/10.1038/s41598-017-15707-9
8. Stevens, R.C., Cherezov, V., Katritch, V., Abagyan, R., Kuhn, P., et al.: The GPCR network: a large-scale collaboration to determine human GPCR structure and function. Nat. Rev. Drug Discov. **12**, 25 (2012). https://doi.org/10.1038/nrd3859
9. Kim, S., Chen, J., Cheng, T., Gindulyte, A., He, J., et al.: PubChem 2019 update: improved access to chemical data. Nucleic Acids Res. **47**, D1102–D1109 (2018). https://doi.org/10.1093/nar/gky1033
10. Sushko, I., Novotarskyi, S., Körner, R., Pandey, A.K., Rupp, M., et al.: Online chemical modeling environment (OCHEM): web platform for data storage, model development and publishing of chemical information. J. Comput. Aided Mol. Des. **25**, 533–554 (2011). https://doi.org/10.1007/s10822-011-9440-2
11. Tetko, I.V.: Associative neural network. Neural Process. Lett. **16**, 187–199 (2002). https://doi.org/10.1023/a:1019903710291
12. Sosnin, S., Karlov, D., Tetko, I.V., Fedorov, M.V.: Comparative study of multitask toxicity modeling on a broad chemical space. J. Chem. Inf. Model. **59**, 1062–1072 (2019). https://doi.org/10.1021/acs.jcim.8b00685
13. Chen, T., Guestrin, C.: XGBoost. In: Proceedings of the 22nd ACM SIGKDD International Conference on Knowledge Discovery and Data Mining - KDD 2016, San Francisco, California, USA, pp. 785–94. ACM (2016). https://doi.org/10.1145/2939672.2939785
14. Suykens, J.A.K., Vandewalle, J.: Least squares support vector machine classifiers. Neural Process. Lett. **9**, 293–300 (1999). https://doi.org/10.1023/a:1018628609742

770   D. Ghosh et al.

Open Access This chapter is licensed under the terms of the Creative Commons Attribution 4.0 International License (http://creativecommons.org/licenses/by/4.0/), which permits use, sharing, adaptation, distribution and reproduction in any medium or format, as long as you give appropriate credit to the original author(s) and the source, provide a link to the Creative Commons license and indicate if changes were made.

The images or other third party material in this chapter are included in the chapter's Creative Commons license, unless indicated otherwise in a credit line to the material. If material is not included in the chapter's Creative Commons license and your intended use is not permitted by statutory regulation or exceeds the permitted use, you will need to obtain permission directly from the copyright holder.

Characterization of Quantum Derived Electronic Properties of Molecules: A Computational Intelligence Approach

Julio J. Valdés[✉]🆔 and Alain B. Tchagang🆔

Digital Technologies Research Centre, National Research Council Canada,
M50, 1200 Montreal Rd., Ottawa K1A0R6, Canada
{julio.valdes,alain.tchagang}@nrc-cnrc.gc.ca

Abstract. The availability of BIG molecular databases derived from quantum mechanics computations represent an opportunity for computational intelligence practitioners to develop new tools with same accuracy but much lower computational complexity compared to the costly Schrödinger equation. In this study, unsupervised and supervised learning methods are applied to investigate the internal structure of the data and to learn the mapping between the atomic coordinates of molecules and their properties. Low dimensional spaces revealed a well defined clustering structure as defined by the measures used for comparing molecules based their atom distributions and chemical composition. Supervised learning techniques were applied on the original predictor variables, as well as on a subset of selected variables found using evolutionary algorithms guided by residual variance analysis (Gamma Test). Black and white box modeling approaches were used (random forests, neural networks and model trees and adaptive regression respectively). All of them delivered good performance, error and correlation-wise, with neural networks producing the best results. In particular white box techniques obtained explicit functional dependencies, some of them achieving considerably reduction of the feature set and expressed as simple models.

Keywords: Computational intelligence · Quantum mechanics · Molecules · 3D visualization · Random forests · Neural networks · Model trees · Multivariate adaptive regression · Black box models · White box models

1 Introduction

Computational quantum mechanics derived from first principle has traditionally been used for the discovery and design of de-novo molecules and for the study of their structures and electronic properties [1]. More recently, the availability of huge molecular databases derived from quantum mechanics computations has given rise to new methods based machine learning [2–4]. These quantum mechanics machine learning models have shown great promises, approaching the same

© Crown 2019
I. V. Tetko et al. (Eds.): ICANN 2019, LNCS 11731, pp. 771–782, 2019.
https://doi.org/10.1007/978-3-030-30493-5_72

accuracy as first principle quantum mechanics computations at a much lower computational cost.

There are two main models in machine learning: discriminative (forward) and generative (inverse). In the context of quantum mechanics and discriminative learning which is the aim of this study, the goal is to learn a mapping from molecule x to a property y. In general, given a dataset $[x_i \rightarrow y_i$, with i = 1 to N] that consists of N molecules (x_i) with their associated properties (y_i), the discriminative model will learn a mapping from $x_i \rightarrow y_i$ and use that mapping to predict new molecules. Several approaches have been explored in the literature to tackle these problems [4].

In these approaches, observations are described in terms of collections of variables/attributes, having several kinds of mutual dependencies, redundancies and noise. However, such a description may affect performance statistical and machine learning procedures because of the curse of dimensionality. Often the data concentrate in low dimension nonlinear manifolds, embedded within the high dimensional space in which the data is represented, either using an instance-attribute (the present case) or a dissimilarity representation. The implication is that in fact the data is not often really high dimensional. The dimension of those manifolds is considered to be the intrinsic dimension and usually it is much smaller than that of the original data representation space. Learning and uncovering these manifolds is important and useful for understanding the internal structure of the data, as well as for improving the performance of data analytic methods like clustering, classification and regression. In this study, we explored unsupervised and supervised learning approaches to map the atomic coordinates of the molecules to their electronics properties as follows. (a) From the atomic coordinate of each molecule, its Coulomb matrix is computed [2]. (b) Rows and columns of these matrices are sorted in decreasing order according to their column norm. This gives rise to an N × M feature matrix, where N is the total number of molecules and M the number of atoms that make the largest molecule in the set. (c) The Gamma test is performed to estimate the level of noise in the data. (d) M5 Model trees, Random Forests, adaptive regression and Neural networks are used to learn a mapping from the feature matrix to the electronics properties of the molecules with correlation coefficient close to 0.996.

The rest of this paper is organized as follows. In Sect. 2 we described the molecular dataset used in this study. Section 3 presents the machine learning techniques. Section 4 presents the experimental settings and the results obtained while we conclude in Sect. 5.

2   Molecules

The QM7 dataset used in this study is a subset of the GDB-13 dataset and was downloaded from [2]. This set consists of 7102 small organic molecules and their associated atomization energy. Initial Cartesian coordinates were generated and subsequently relaxed using the Universal Force Field106 as implemented in OpenBabel107 (version 2.3.2). Structures were further relaxed and self-consistent field energies calculated at the density functional level of theory (DFT) using the Perdew-Burke-Ernzerhof (PBE0) functional with def2-TZVP basis set as implemented in Gaussian (version 09 rev. D.01). Atomization energies were then obtained by subtracting free atom energies computed in the same fashion. More information relative to this dataset can be obtained at [2].

2.1 Coulomb Matrix

The inputs to our machine learning models are the same descriptors that also enter the Schrodinger equations i.e. the nuclear charges Z_i and the atomic positions R_i. Our machine learning model, instead of finding the wavefunction which maps the system's Hamiltonian to its energy, it directly learns a mapping from the system to energy based on examples given for training. The Coulomb matrix used in this case is directly obtained from Z_i and R_i.

$$Z_{ij} = \begin{cases} 0.5Z_i^{2.4} & i = j \\ \frac{Z_i Z_j}{|R_i - R_j|} & i \neq j \end{cases} \tag{1}$$

Z_i is the atomic number or nuclear charge of atom i, and R_i is its position in atomic units. The Coulomb matrix M is symmetric and has as many rows and columns as the number of atoms in the molecule.

While the Coulomb matrix is invariant to translation and rotation of the molecule, it is not invariant to re-indexing of its atoms. One remedy is to sort the columns and rows of the Coulomb matrices by descending order relative to their norm 2 [2]. That is, for each molecule in the dataset, compute its Coulomb matrix. Pad each matrix to the right and bottom with zeros so they all have the same size that is 23×23, which is the maximum number of atoms per molecule in the QM7 dataset. Compute the norm-2 of each molecules and sort rows and columns in descending order. Given that the Coulomb matrix is symmetrical, only the lower triangular part is kept. Finally they are unfolded into 1D vector representation of the molecule. For the 7102 QM7 molecules the representation has a matrix of 7102×276 feature, where each row represents the signature of a molecule. This matrix was extended by including five extra features given by the chemical composition of the molecule with respect to the number of atoms of Carbon, Hydrogen, Nitrogen, and Sulfur. Both matrices were converted to z-scores (column-wise) by subtracting the mean and dividing by the standard deviation. The final data matrix was composed of 7102×282 features (281 predictors and the target property: Atomization energy).

3   Machine Learning Techniques

3.1   Unsupervised Analysis and Data Exploration

Low Dimensional Spaces for Data Exploration. It is possible to create spaces for data exploration and visualization by computing low dimensional spaces that preserve chosen properties of the original dissimilarity matrix describing the objects [17]. Many techniques have been developed based on different principles. Among them are the Sammon nonlinear mapping and the t-distributed Stochastic Neighbor Embedding (t-SNE).

Sammon's nonlinear mapping [15] transforms vectors of two spaces of different dimensions $(D > m)$ by means of a transformation like $\varphi : \mathbb{R}^D \rightarrow \mathbb{R}^m$ which maps vectors $\boldsymbol{x} \in \mathbb{R}^D$ to vectors $\boldsymbol{y} \in \mathbb{R}^m$, $\boldsymbol{y} = \varphi(\boldsymbol{x})$. Sammon error $= \frac{1}{\sum_{i<j} \delta_{ij}} \sum_{i<j} \frac{(\delta_{ij} - d(\boldsymbol{y}_i, \boldsymbol{y}_j))^2}{\delta_{ij}}$, where typically d is an Euclidean distance in $\mathbb{R}^m$. The weight term δ_{ij}^{-1} gives more importance to the preservation of smaller distances rather than larger ones and is determined by the dissimilarity distribution in the data space. Moreover, they are fixed, which is referred to as lack of plasticity.

t-SNE is an enhancement of SNE [10], where the mapping from higher dimensional space to lower dimensional space is based on the consideration of the similarity of conditional probabilities between datapoints. A conditional probability $p_j|i$ is the probability of datapoint x_i to have x_j as a neighbor based on a Gaussian distribution $p_j|i = \frac{exp(-\|x_i - x_j\|^2 / 2\sigma_i^2)}{\sum_{k \neq i} exp(-\|x_i - x_k\|^2 / 2\sigma_i^2)}$, where σ_i^2 is the variance of datapoint x_i and k is a perplexity parameter related to selected local neighbors size. For the lower dimensional space, SNE utilizes conditional probabilities $q_{j|i}$ of datapoints x_i based on another Gaussian distribution. The goal is to minimize the difference between the probability distributions of the two spaces, expressed as the sum of Kullback-Leibler divergences: $C = \sum_i \sum_j p_{j|i} log \frac{p_{j|i}}{q_{j|i}}$. One drawback of SNE, is the low cost when representing widely separated points. t-SNE applies a symmetric cost function and uses the Student's t-distribution in the target space, which has a heavier tail [18]. These modifications represent a notable improvement.

3.2 Supervised Analysis

Gamma Test. The Gamma test is a nonparametric technique aimed at estimating the variance of the noise present in a dataset [6,12,16], very useful in the construction of data-driven models. Noise is any source of variation in the target variable that cannot be explained by a smooth function (model) relating the target with the predictor variables. The gamma estimate indicates whether it is possible to explain the target variable by a smooth deterministic model based on the predictor variables. From this, an assessments can be made on *(i)* whether it is hopeful or hopeless to find a smooth model to the data, *(ii)* whether more explanatory variables should be incorporated to the data, *(iii)* how many observations are minimally required in order to build a model, *(iv)* appropriate thresholds in order to avoid overfitting during training and *(v)* what is the overall quality of the data. The most important assumptions of the procedure are *(i)* the model function f is continuous within the input space, *(ii)* the noise is independent of the input vector $\overleftarrow{x}$ and *(iii)* the function f has bounded first and second partial derivatives.

Let $\mathcal{S}$ be a system described in terms of a set of variables and with $y \in \mathbb{R}$ being a variable of interest, potentially related to a set of m variables $\overleftarrow{x} \in \mathbb{R}^m$

expressed as $y = f(\overleftarrow{x}) + r$, where f is a smooth unknown function representing the system, $\overleftarrow{x}$ is a set of predictor variables and r is a random variable representing noise or unexplained variation. Let M be the number of observations and p is the number of nearest neighbors considered. If $\overleftarrow{x}_{N\lfloor i,k\rfloor}$ is the k-th nearest neighbor of object $\overleftarrow{x}_i$, for every $k \in [1,p]$, a sequence of estimates of $\mathbf{E}\left(\frac{1}{2}(y' - y)^2\right)$ based on sample means is computed as

$$\gamma_M(k) = \frac{1}{2M} \sum_{i=1}^{M} |y_{N\lfloor i,k\rfloor} - y_i|^2 \tag{2}$$

$$\delta_M(k) = \frac{1}{M} \sum_{i=1}^{M} |\overleftarrow{x}_{N\lfloor i,k\rfloor} - \overleftarrow{x}_i|^2$$

where $\mathbf{E}$ denotes the mathematical expectation and $|.|$ Euclidean distance. The relationship between $\gamma_M(k)$ and $\delta_M(k)$ is assumed linear as $\delta_M(k) \to 0$ and an estimate for the variance of the noise $\Gamma = Var(r)$ is obtained by linear regression of $\delta_M(k)$ vs. $\gamma_M(k)$

$$\gamma_M(k) = \Gamma + G\,\delta_M(k) \tag{3}$$

From Eq. 3 the vRatio (V_r) is defined as a normalized Γ value with respect to the variance of the target variable. Since $V_r \in [0,1]$, it allows comparisons across different datasets:

$$V_r = \frac{\Gamma}{Var(y)} \tag{4}$$

Assessing the relevance of the predictor variables is approached by searching for subsets with good Γ-statistics. In real-world cases the search space is determined by the power set of the predictor variables and evolutionary computation methods provide an alternative to the prohibitive brute force. A genetic algorithms explores subsets of predictors represented as binary vectors $\overleftarrow{\vartheta} = \{0,1\}^m \in \mathbb{R}^m$ (masks). Each represents a subset determined by the predictors present in the vector and the target y. The potential of each subset of variables is given by the Γ-statistics, which could be specified in different ways. A single-objective cost function can be formulated as a linear combination of partial fitness coming from (i) the MSE as associated to V_r (the I_f term), (ii) 'model smoothness' as associated to G (the G_f term) and (iii) 'model complexity' given by the relative number of predictors (the L_f term).

$$F(\overleftarrow{\vartheta}) = W_i * I_f(\overleftarrow{\vartheta}) + W_g * G_f(\overleftarrow{\vartheta}) + W_l * L_f(\overleftarrow{\vartheta}) \tag{5}$$

where $W_i = 0.8$, $W_g = 0.1$, $W_l = 0.1$ are the weights of the contributing fitness terms, the largest of which is given to I_f, directly related to the estimated MSE.

$$I_f(\overleftarrow{\vartheta}) = \begin{cases} 1 - (1 - 10 * V_r(\overleftarrow{\vartheta}))^{-1} \text{ if } V_r(\overleftarrow{\vartheta}) < 0 \\ 2 - 2(1 + V_r(\overleftarrow{\vartheta}))^{-1} \quad \text{otherwise} \end{cases}$$

$$G_f(\overleftarrow{\vartheta}) = 1 - (1 + |G(\overleftarrow{\vartheta})|/range(y))^{-1}$$

$$L_f(\overleftarrow{\vartheta}) = \sum \overleftarrow{\vartheta}/m \tag{6}$$

The choice of the weights $\{W_i, W_g, W_l\}$ is a compromise between the importance given to the partial fitness components coming from the subset's V_r, the model complexity G and the model's cardinality (the smaller, the simpler, since it contains fewer predictors). The practical form of $\{I_f, G_f, L_f\}$ in (6) is a heuristic emerging from many different applications. This use of *GammaTest* statistics has been very successful elsewhere [21–23].

3.3 Modeling Techniques

Several black and white box approaches have been used for learning predictive models for the Atomization Energy property. Namely, Neural networks (fully connected multilayer perceptrons) and Random Forests [5,13] as black box representatives, with M5 model trees [11,14,19,20] and Multivariate adaptive regression splines (MARS)[1] [7,8] as white box instances.

4   Results

4.1   Unsupervised Analysis and Data Exploration

Since it is not possible to properly display 3D content on hard media, snapshots from fixed perspectives are presented. In order to simplify the representation, the original 7102 objects were pre-clustered using the leader algorithm [9] with an Euclidean distance threshold of 12.33 , which produced 1003 clusters (leaders). They are shown as semi-transparent spheres with sizes proportional to the cluster sizes. For both methods, the Sammon and the t-SNE mappings, the 3D transformations clearly exhibit the presence of well defined structures composed of different clusters. In the case of Sammon mapping (Fig. 1, Top), there is a lower density structure (upper right, mostly composed of outlying elements), well differentiated from a left area of much higher density composed of a sequence of clusters which progressively become more sparse. Under t-SNE, the mapping exhibits outlying elements at the top and the right respectively (Fig. 1, Bottom). Several clusters are also well defined and they correspond to the major structures of the Sammon mapping.

This initial exploration of the data using unsupervised visualization techniques reveals the existence of well differentiated classes of molecules, determined by their Coulomb matrices and atomic composition. These structures would be exploited by supervised techniques aiming at predicting molecular properties.

---

[1] MARS is trademarked and licensed to Salford Systems.

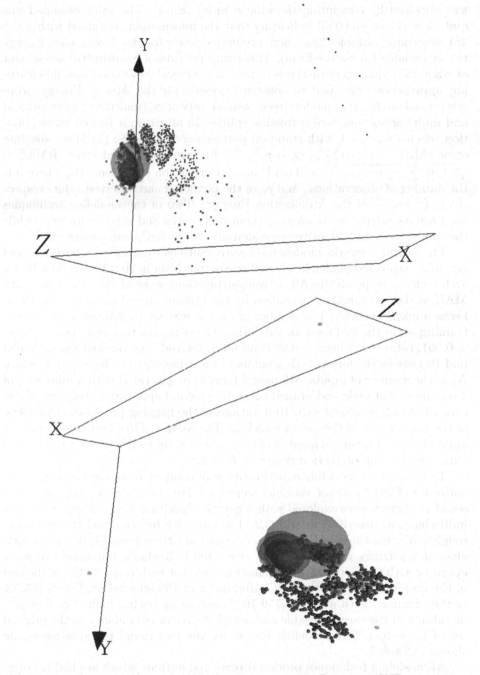

Fig. 1. Low dimensional spaces. Top: Sammon mapping. Bottom: t-SNE mappings.

4.2   Supervised Analysis

An orientative estimation of the predictive information contained in the data was obtained by computing the vRatio index (Eq. 4). The value obtained was quite low ($V_r = -0.0056$) indicating that the information contained within the 281 descriptor variables has a high predictive power for the Atomization Energy target variable (in z-score form). This result provides a quantitative assessment of what the exploratory methods of Sect. 3.1. Several supervised machine learning approaches were used to construct models for the Atomic Energy property: random forests, model trees, neural networks (multilayer perceptrons) and multivariate adaptive regression splines. In all cases, a 10-fold crossvalidation scheme was used, with standard performance measures: (1) Mean absolute error (MAE $= (1/n) \sum_{i=1}^{n} |o_i - p_i|$), (2) Root mean squared error (RMSE $= \sqrt{(1/n) \sum_{i=1}^{n} (o_i - p_i)^2})$) and (3) Pearson Correlation coefficient (R), where n is the number of observations , and p_i, o_i the predicted and observed values respectively (z-scores) of the Atomization Energy . Two of the modeling techniques used are considered as 'black box' (neural networks and random forests), while the model trees and adaptive regression are 'white box', transparent ones.

The neural networks models used were multilayer perceptrons with one and two fully connected hidden layers and one output layer in $20 \times 15 \times 1$, $30 \times 15 \times 1$ architectures respectively. All activation functions were of the relu type, with MAE as the loss function, optimized by the 'Adam' procedure (using a python-keras implementation). The number of epochs was set to 200, with 10% of the training set in the fold used for validation. Other parameters were: learning rate $= 0.001$, beta$_1 = 0.9$, beta$_2 = 0.999$ and no AMSGrad. The random forests model had 10 trees in the forest, with a number of features equal to $log(N_i) + 1$, where N_i is the number of inputs. M5 model trees were generated with a minimum of 4 instances/leaf node and pruned rules. In a second application, bagging of this type of models was used with 10 iterations of the bagging procedure, and 100% of the training set as the size of each bag. The Adaptive Regression models were applied with a maximum number of terms allowed set to 100 in two variants and with a maximum interaction degree of $K = 5$.

Two rounds of modeling experiments were conducted. In the first round the entire set of 281 predictor variables were used. For the second round, the power set of predictors were explored with a genetic algorithm using binary vectors as individuals, as described in Sect. 3.2. The objective function and the balancing weights W were those of Eq. 5, using the partial fitness from Eq. 6. The genetic algorithm settings were: population size $= 100$ individuals, one-point crossover operator with rate $= 0.5$ and bit mutation operator with rate $= 0.05$. At the end of the evolution process, the best individual had 156 selected predictors (55.5% of the originals) with a $V_r = 5.759 \ 10^{-8}$, indicating both a high degree of predictability of the target variable and a high degree of redundancy in the original set of predictors. The modeling results for the two round of experiments are shown in Table 1.

All modeling techniques produced predicted outputs which are highly correlated with the observed Atomization Energies. The minimum correlation coeffi-

Table 1. Modeling results with different machine learning techniques (10-fold Cross-validation)

ROUND 1 (281 original predictors)	MAE	RMSE	R
NN (20 × 15 × 1)	0.0220	0.0340	0.9995
NN (30 × 15 × 1)	0.0230	0.0330	0.9995
Random forest	0.0639	0.0857	0.9964
M5 model trees	0.0612	0.0915	0.9958
M5 model trees (bagging)	0.0515	0.0742	0.9973
Adaptive regression	0.0658	0.0876	0.9961
ROUND 2 (156 selected predictors)	MAE	RMSE	R
NN (20 × 15 × 1)	0.034	0.048	0.9990
NN (30 × 15 × 1)	0.035	0.051	0.9985
Random forest	0.0659	0.0874	0.9962
M5 model trees	0.0631	0.0982	0.9952
M5 model trees (bagging)	0.0538	0.0771	0.9971
Adaptive regression	0.0689	0.0926	0.9957

cient was 0.9952, corresponding to M5 model trees using the reduced set of variables. The highest correlation was obtained with the neural network (0.9995), closely followed by the bagged M5 model trees (0.9973) Overall, using only 55.5% of the predictors (Round 2) barely affected the correlation results. From the point of view of the error measures, the best models were the neural networks, in particular the 20 × 15 × 1 architecture, with a cross-validation MAE of 0.0220 on the z-scores of Atomization Energy. More complex layouts (30 × 15 × 1) did not differ significantly in performance. This kind of behavior has been observed elsewhere, when working with neural networks. It is noteworthy that the bagged M5 model trees performed consistently better than Random Forest for all of the measured considered. Moreover, the later was matched by the Adaptive Regression model, which is an explicit, deterministic representation of the functional dependencies. While random forests are notoriously opaque, M5 and Adaptive Regression models are totally transparent. The single M5 model tree is composed of 100 rules when using 281 predictors and 92 when using the 156 selected by the evolutionary algorithm. Altogether, the model composition indicates a high level of irrelevancies in the set of descriptor variables, which could be removed without losing predictive performance.

In the case of Adaptive Regression models, they did not achieve a competitive error-wise performance with respect to neural networks. However, they ranked similarly with Random Forest with respect to both error and correlation measures. The later is specially important when looking at the model structure. Both Adaptive Regression models for data with 281 and 156 predictors respectively, used only 5 variables (1.78% and 3.2% respectively). Considering the original high dimensionality of the descriptor space, this represents a considerable reduction from a feature selection point of view. An important aspect is that this reduction is performed simultaneously with learning the underlying functional dependencies. The explicit models when using data with 281 and 156 predictors respectively are

$$
\begin{aligned}
zscores(Atom.Energy) = &-k_1 * x_{276} - k_2 * x_{277} + k_3 * Max(0, -x_1 - k_4) + \\
&k_5 * Max(0, x_1 + k_6) + k7 * Max(0, -x_{19} - k_8) * Max(0, -x_{26} - k_9) - \\
&k_{10} * Max(0, x_{19} + k_{11}) * Max(0, -x_{26} - k_{12}) + k_{13} * Max(0, -x_{26} - k_{14}) * \\
&Max(0, -x_{26} - k_{16}) + k_{17} * Max(0, -x_{26} - k_{18}) * Max(0, x_{26} + k_{19}) + k_{20} * \\
&Max(0, -x_{26} - k_{21}) - k_{22} * Max(0, x_{26} + k_{23}) - k_{25}
\end{aligned}
$$

and

$$
\begin{aligned}
zscores(Atom.Energy) = &-K_1 * x_{152} - K_2 * x_{153} + K_3 * Max(0, -x_1 - K_4) + \\
&K_5 * Max(0, x_1 + K_6) + K_7 * Max(0, -x_{11} - K_8) - K_9 * Max(0, x_{11} + K_{10}) + \\
&K_{11} * Max(0, -x_{24} - K_{12}) - K_{13} * Max(0, x_{24} + K_{14}) - K_{15},
\end{aligned}
$$

where x $k_i, K_i \in \mathbb{R}$, $i \in \mathbb{N}^+$ are constants found during the learning process and $Max(p, q)$ is the maximum between p and q. In models of this kind, it is possible not only to explicit the important predictor variables, but also the way in which they interact, which is transparently shown by the multiplicative terms involving the basis functions (e.g. $max(0, -x_1 + k_{11})max(0, x_1 - k_{12})$). As discussed above, this model has a performance that matches the one from a random forest, which is a widely used, well established machine learning technique.

From the point of view of performance, neural networks outperformed all other models. However, explicit, transparent models were capable of providing good results, at the level of other well established (black box) techniques, while working with significantly fewer number of predictors, with the advantage of exposing the nature of their interrelations and producing simple models.

5   Conclusions

The visualization of low-dimensional mappings from Coulomb matrices and atomic composition provided understanding of the structure of the data. They revealed the existence of well defined clusters from the point of view of both local distance preservation and consistency of conditional probability distributions between the original and the target spaces. The results obtained with different machine learning techniques aiming at modeling Atomization Energy (random forests, neural networks, model trees and adaptive regression), proved effective at capturing the functional dependencies between molecular structure and composition, and molecular properties, like Atomization Energy. Black and white models were produced that combine performance with transparency and explanation, identifying small subsets of relevant variables. Performance-wise, neural network models were superior, but adaptive regression in particular, produced relatively compact and transparent models, with accuracies comparable to those provided by well established techniques like random forests. Future work will expand the studies to other molecular properties, as well as to mechanisms for deriving molecular structure from desired properties.

References

1. Baldi, P., Müller, K.R., Schneider, G.: Charting chemical space: challenges and opportunities for artificial intelligence and machine learning. Mol. Inf. **30**(9), 751 (2011)
2. Rupp, M.: Machine learning for quantum mechanics in a nutshell. Int. J. Quantum Chem. **115**, 1058–1073 (2015)
3. Montavon, G., et al.: Machine learning of molecular electronic properties in chemical compound space. New J. Phys. **15**(9), 095003 (2013)
4. Belisle, E., Huang, Z., Le Digabel, S., Gheribi, A.E.: Evaluation of machine learning interpolation techniques for prediction of physical properties. Comput. Mater. Sci. **98**, 170–177 (2015)
5. Breiman, L.: Random forests. Mach. Learn. **45**(1), 5–32 (2001)
6. Evans, D., Jones, A.J.: A proof of the gamma test. Proc. Roy. Soc. Lond. A **458**, 1–41 (2002)
7. Friedman, J.: Multivariate adaptive regression splines. Ann. Stat. **19**(1), 1–67 (1991)
8. Friedman, J.: Fast mars. Technical Report 110, Stanford University, Department of Statistics (1993)
9. Hartigan, J.A.: Clustering Algorithms. Wiley, New York (1975)
10. Hinton, G.E., Roweis, S.T.: Stochastic neighbor embedding. In: Advances in Neural Information Processing Systems, vol. 15, pp. 833–840 (2002)
11. Holmes, G., Hall, M., Frank, E.: Generating rule sets from model trees. In: Foo, N. (ed.) AI 1999. LNCS (LNAI), vol. 1747, pp. 1–12. Springer, Heidelberg (1999). https://doi.org/10.1007/3-540-46695-9_1
12. Jones, A.J., Evans, D., Margetts, S., Durrant, P.: The Gamma Test. In: Sarker, R., Abbass, H., Newton, S. (eds.) Heuristic and Optimization for Knowledge Discovery. Idea group Publishing (2002)
13. Kuncheva, L.I.: Combining Pattern Classifiers, Methods and Algorithms. Wiley, New York (2005)
14. Quinlan, J.R.: Learning with continuous classes. In: 5th Australian Joint Conference on Artificial Intelligence, Singapore, pp. 343–348 (1992)
15. Sammon, J.W.: A nonlinear mapping for data structure analysis. IEEE Trans. Comput. **C−18**(5), 401–409 (1969)
16. Stefánsson, A., Končar, N., Jones, A.J.: A note on the gamma test. Neural Comput. Appl. **5**, 131–133 (1997)
17. Valdés, J.J.: Virtual reality representation of information systems and decision rules: an exploratory technique for understanding data and knowledge structure. In: Wang, G., Liu, Q., Yao, Y., Skowron, A. (eds.) RSFDGrC 2003. LNCS (LNAI), vol. 2639, pp. 615–618. Springer, Heidelberg (2003). https://doi.org/10.1007/3-540-39205-X_101
18. Maaten, L.V.D., Hinton, G.: Visualizing high-dimensional data using t-SNE. J. Mach. Learn. Res. **9**, 2579–2605 (2008)
19. Wang, Y., Witten, I.H.: Induction of model trees for predicting continuous classes. In: Proceedings European Conferenve on Machine Learning, Prague, pp. 128–137 (1997)

20. Witten, I.H., Frank, E.: Data Mining: Practical Machine Learning Tools and Techniques. Morgan Kaufmann, San Francisco (2005)
21. Valdés, J.J., Cheung, C., Wang, W.: Evolutionary computation methods for helicopter loads estimation. In: Proceedings of the 2011 IEEE Congress on Evolutionary Computation, New Orleans, USA, 5–11 June 2011 (2011)
22. Valdés, J.J., Cheung, C., Li, M.: Towards conservative helicopter loads prediction using computational intelligence techniques. In: Proceedings of the 2012 IEEE World Congress on Computational Intelligence, International Convention Centre, Brisbane, Australia, 10–15 June 2012, pp. 1853–1860 (2012)
23. Valdés, J.J., Cheung, C., Li, M.: Sensor dynamics in high dimensional phase spaces via nonlinear transformations: application to helicopter loads monitoring. In: 2014 IEEE Symposium Series on Computational Intelligence (IEEE SSCI 2014), Orlando, 9–12 December 2014 (2014)

Open Access This chapter is licensed under the terms of the Creative Commons Attribution 4.0 International License (http://creativecommons.org/licenses/by/4.0/), which permits use, sharing, adaptation, distribution and reproduction in any medium or format, as long as you give appropriate credit to the original author(s) and the source, provide a link to the Creative Commons license and indicate if changes were made.

The images or other third party material in this chapter are included in the chapter's Creative Commons license, unless indicated otherwise in a credit line to the material. If material is not included in the chapter's Creative Commons license and your intended use is not permitted by statutory regulation or exceeds the permitted use, you will need to obtain permission directly from the copyright holder.

Using an Autoencoder for Dimensionality Reduction in Quantum Dynamics

Sebastian Reiter[ID], Thomas Schnappinger[ID], and Regina de Vivie-Riedle[(✉)][ID]

Department of Chemistry, Ludwig-Maximilians-Universität München,
Butenandtstr. 5–13, Munich, Germany
{sebastian.reiter,thomas.schnappinger,regina.de_vivie}@cup.uni-muenchen.de
https://www.cup.uni-muenchen.de/pc/devivie/index.html

Abstract. A key step in performing quantum dynamics for a chemical system is the reduction of dimensionality to allow a numerical treatment. Here, we introduce a machine learning approach for the (semi)automatic construction of reactive coordinates. After generating a meaningful data set from trajectory calculations, we train an autoencoder to find a low-dimensional set of non-linear coordinates for use in molecular quantum dynamics. We compare the wave packet dynamics of proton transfer reactions in both linear and non-linear coordinate spaces and find significant improvement for physical properties like reaction timescales.

Keywords: Quantum dynamics · Machine learning · Autoencoder · Dimensionality reduction

1 Introduction

Quantum dynamics is a powerful computational tool to study (photo)chemical processes happening on a femto- to picosecond timescale. The time-dependent Schrödinger equation (TDSE)

$$i\hbar\frac{\partial}{\partial t}\Psi = \hat{\mathcal{H}}\Psi \tag{1}$$

describes the quantum mechanical motion of the atomic nuclei, where the Hamiltonian $\hat{\mathcal{H}}$ acting on the wave function Ψ contains the kinetic and the potential energy of the system. The TDSE is commonly solved on a discrete spatial grid which is constructed by displacing a molecular geometry along predefined reactive coordinates. Including more and more degrees of freedom, the number of grid points scales exponentially, a challenge known as the *curse of dimensionality*. Without serious approximations, this means for example that a full-dimensional quantum mechanical representation of the internal motion of a small organic molecule is already out of reach in terms of computational effort.

This issue can be addressed by transitioning from full dimensionality to a low-dimensional subspace of internal coordinates that cover a large proportion

© The Author(s) 2019
I. V. Tetko et al. (Eds.): ICANN 2019, LNCS 11731, pp. 783–787, 2019.
https://doi.org/10.1007/978-3-030-30493-5_73

of the process in question. The prevalent way to construct such a subspace is the manual selection of either pure normal modes or linear combinations thereof. Although tried and tested [2,4], this procedure relies heavily on the chemical intuition of the researcher and requires a high amount of *a priori* knowledge of the molecular system. The first step to automate this process is the generation of a data set based on semiclassical trajectories. We introduce two coordinate reduction schemes [5,6] to evaluate these data sets and provide an automatic alternative to manual coordinate construction.

2   Data Set Generation

The preparation of a meaningful data set is always a challenge in a machine learning project. There are many different approaches to produce the training data set for chemical problems. Since we are interested in the temporal evolution of the system, one possible way is to use semiclassical trajectories to sample the accessible space for the process in question. The trajectories are started from the intrinsic reaction coordinate (IRC) or in general from any minimum energy path (MEP). Using the IRC/MEP has the advantage, that it is one of the most easily available sources of information about a chemical reaction. To generate data points, we cannot simply let trajectories run along the IRC/MEP, because they will follow its descent and not explore the space orthogonal to it. Using a constraint algorithm, the motion along a certain direction can be restricted. These constraints can introduce spurious rotations and translations, which must be removed. For an efficient sampling the trajectories are initialized from points along the IRC/MEP with random momentum orthogonal to this path. In order to avoid redundancy in the data set, a minimum full-dimensional Euclidean distance between the points is enforced. With this approach [5,6] we ensure that the sampling covers both the desired reaction path itself as well as its vicinity in all accessible dimensions.

Figure 1 shows a projection of trajectory data points (corresponding to different geometries) produced by the described procedure together with the IRC of an example reaction. The orthogonality of the trajectories to the IRC is clearly visible.

3   Coordinate Construction

The obtained data set can be used directly by dimensionality reduction algorithms. In this work, we present both the construction of linear coordinates [5] by means of a principal component analysis (PCA) and the generation of nonlinear coordinates [6] with an autoencoder.

For the former, the basic idea is to apply a PCA to the data set and extract the degrees of freedom with the largest molecular motion. PCA is already being employed in molecular dynamics to extract so called essential dynamics of larger systems, in particular proteins. Here, the dimensionality reduction is a helpful tool for interpretation and allows a reduction of computational cost [1,3].

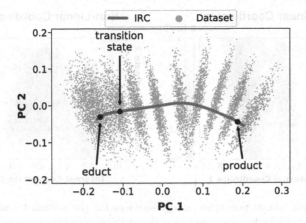

Fig. 1. IRC (blue) and training data points (orange) projected onto the first two principal components of the data set. (Color figure online)

One of the drawbacks of linear coordinates is that chemical processes often involve complex deformations which would require a large number of purely linear coordinates to resolve. Non-linear coordinate spaces typically allow the use of fewer dimensions to investigate the same reactions. In this context, an autoencoder (Fig. 2) provides a convenient way to reduce the dimensionality of a chemical problem to a small number of the most relevant non-linear coordinates.

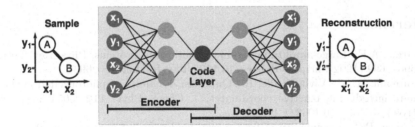

Fig. 2. Topology of an autoencoder to reduce the dimensionality of chemical reaction dynamics. Cartesian molecular coordinates are used as input to the neural network.

The autoencoder is trained to reproduce the Cartesian coordinates of the trajectory data set. After training is completed, an equidistant grid is constructed in the low-dimensional space of the code layer. Using the decoder part of the neural network, this latent grid is projected onto the Cartesian space to generate a grid of molecular coordinates. The amount of non-linearity contained in the resulting grid can be tuned by employing different activation functions in the neural network. Both the potential and the kinetic energy are evaluated on this resulting grid, enabling the numerical solution of the TDSE (1).

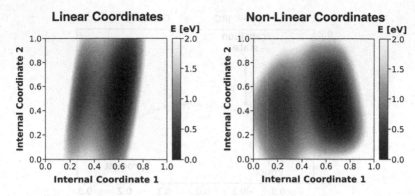

Fig. 3. Two-dimensional potential energy surfaces for the proton transfer reaction in (Z)-Chloromalonaldehyde, generated in a linear PCA coordinate space (left) and in a non-linear coordinate space (right).

To compare both approaches for automatic coordinate construction, we focus on proton transfer reactions in small organic molecules. Figure 3 illustrates two-dimensional potential energy surfaces for the proton transfer reaction in (Z)-Chloromalonaldehyde, constructed in a linear and a non-linear coordinate space. It is discernible from the compressed shape of the left potential that two linear coordinates cover less of the reactive space than two non-linear coordinates. In this case, a third linear coordinate would be necessary to yield a potential comparable to that in a two-dimensional non-linear subspace.

References

1. García, A.E.: Large-amplitude nonlinear motions in proteins. Phys. Rev. Lett. **68**, 2696–2699 (1992). https://doi.org/10.1103/PhysRevLett.68.2696
2. Hofmann, A., de Vivie-Riedle, R.: Quantum dynamics of photoexcited cyclohexadiene introducing reactive coordinates. J. Chem. Phys. **112**, 5054–5059 (2000). https://doi.org/10.1063/1.481059
3. Nguyen, P.H.: Complexity of free energy landscapes of peptides revealed by non-linear principal component analysis. Proteins Struct., Funct., Bioinf. **65**, 898–913 (2006). https://doi.org/10.1002/prot.21185
4. Thallmair, S., Roos, M.K., de Vivie-Riedle, R.: Design of specially adapted reactive coordinates to economically compute potential and kinetic energy operators including geometry relaxation. J. Chem. Phys. **144**, 234104 (2016). https://doi.org/10.1063/1.4953667
5. Zauleck, J.P.P., Thallmair, S., Loipersberger, M., de Vivie-Riedle, R.: Two new methods to generate internal coordinates for molecular wave packet dynamics in reduced dimensions. J. Chem. Theory Comput. **12**, 5698–5708 (2016). https://doi.org/10.1021/acs.jctc.6b00800
6. Zauleck, J.P.P., de Vivie-Riedle, R.: Constructing grids for molecular quantum dynamics using an autoencoder. J. Chem. Theory Comput. **14**, 55–62 (2017). https://doi.org/10.1021/acs.jctc.7b01045

Open Access This chapter is licensed under the terms of the Creative Commons Attribution 4.0 International License (http://creativecommons.org/licenses/by/4.0/), which permits use, sharing, adaptation, distribution and reproduction in any medium or format, as long as you give appropriate credit to the original author(s) and the source, provide a link to the Creative Commons license and indicate if changes were made.

The images or other third party material in this chapter are included in the chapter's Creative Commons license, unless indicated otherwise in a credit line to the material. If material is not included in the chapter's Creative Commons license and your intended use is not permitted by statutory regulation or exceeds the permitted use, you will need to obtain permission directly from the copyright holder.

Conformational Oversampling as Data Augmentation for Molecules

Jennifer Hemmerich⬚, Ece Asilar⬚, and Gerhard F. Ecker(✉)⬚

University of Vienna, 1090 Vienna, Austria
{jennifer.hemmerich,ece.asilar,gerhard.f.ecker}@univie.ac.at

Abstract. Toxicological datasets tend to be small and imbalanced. This quickly causes models to overfit and disregard the minority class. To solve this issue we generate conformations of molecules. Thereby, we can balance datasets as well as increase their size. Using this approach on the Tox21 Challenge data we observed conformational oversampling to be a viable approach to train datasets, increasing the balanced accuracy of trained models.

Keywords: Deep learning · Toxicity · Chemoinformatics · Imbalanced learning

1 Introduction

Safety evaluation during drug development would greatly benefit from models reliably predicting specific toxicities. Such models could reduce time and cost as well as animal testing by indicating hazards in the first stage of drug design. The Merck Kaggle competition and the Tox21 Challenge demonstrated the superiority of neural networks over traditional machine learning approaches for biological activity predictions [7,8,11,12]. Hence, neural networks are a viable tool to generate predictive models for toxicities. However, toxicological datasets often are small and imbalanced. This oftentimes results in overfitting of models or neglection of the minority class, predicting almost all compounds as the majority class. Especially for toxicity predictions, it is important that models exhibit high sensitivity and specificity, that is, identify hazards but not flagging every single compound.

With the aim to solve both, imbalance and overfitting, we developed a method named COVER (Conformational OVERsampling). This method was derived from augmentation as used in image recognition, where images are transformed to increase generalization and dataset size (e.g. [3,5,10,13]). Instead of transforming an image we "transform" molecule representations by generating multiple conformations along with a 3D based description of molecules. This enables us to balance datasets for neural network training, as well as increase the dataset size.

© The Author(s) 2019
I. V. Tetko et al. (Eds.): ICANN 2019, LNCS 11731, pp. 788–792, 2019.
https://doi.org/10.1007/978-3-030-30493-5_74

Table 1. Datasets used for training with the number of molecules per class and the overall dataset size, each conformation is counted as separate molecule

| | No. of conformations per | | No. of molecules | | |
Dataset	Inactive	Active	Inactive	Active	Overall
1–1	1	1	5502	341	5843
1–16	1	16	5502	5428	10930
2–2	2	2	11001	680	11681
2–32	2	32	11001	10865	21866
5–5	5	5	27504	1698	29202
5–80	5	80	27504	27145	54649

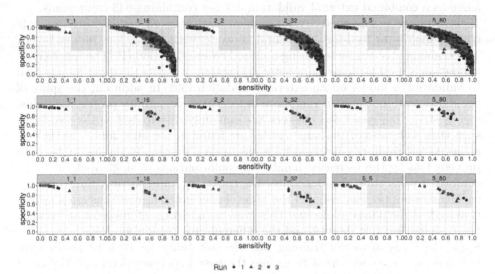

Fig. 1. Sensitivity versus specificity of the trained models. Each datapoint presents the evaluation of one model. The captions of the subplots denote the amount of oversampling done, e.g. 1_16 denotes 1 conformation for the negative class and 16 conformations for the positive class. Each shape with the respective color denotes an independent run of the cross-validation scheme. The area in the upper right corner shows the region of predictive models (model with a sensitivity and specificity higher than 0.5). (A) Plot of the models obtained during hyperparameter selection, (B) performance of the final models on the external fold, (C) performance of the final models on the external test set. (Color figure online)

2   Method

For model training we assembled a dataset using the the endpoint p53 activation ("SR-p53") from the Tox21 Challenge data which contains 5843 compounds with binary annotation. The endpoint has an imbalance ratio of 1:16, denoting that the dataset contains 16 times more inactive than active molecules. In the next step we generated conformations of the molecules using the ETKDG Algorithm with energy minimisation as implemented in RDKit [6,9]. In total, we generated 6 datasets (see Table 1). For each dataset m conformations were generated for the inactives and n conformations for the actives, resulting in the label "m-n dataset". Therefore, to generate m conformations the conformation generation was run m times, without any further processing. The datasets were prepared with different goals: Firstly to evaluate the training on the original dataset ("1–1 dataset"), secondly to evaluate training on a balanced dataset ("1–16 dataset"), thirdly to evaluate whether oversampling alone is beneficial ("2–2" and "5–5 dataset") or if balancing is mandatory ("2–32" and "5–80 dataset"). After conformer generation 1145 3D descriptors were calculated, which are used as input for the models. In addition, we used the two test datasets from the Tox21 Challenge as a combined external validation dataset containing 643 compounds.

To erase increased predictivity by sophisticated network architectures we only trained multilayer feed-forward neural networks with two to four hidden layers. The training was conducted using a nested cross-validation [1]. The inner cross-validation is used for hyperparameter grid search, whereas the outer loop is used to validate the model on an external validation fold. In addition, the splits of the dataset are not chosen randomly but decided via affinity propagation clustering [4,7]. The generated clusters are randomly distributed to the five folds, such that molecules of the same cluster are found in the same fold. The thereby achieved low similarity between folds ensures a reduction in the model bias. To prevent bias by the increased number of conformations of single molecules, the conformer generation was done after the clustering and splitting. Subsequently, each conformation was assigned the same cluster number, and thus cross validation fold, as the parent molecule. This prevents a leakage of information between the training and test dataset and the different cross validation folds.

Since the focus of model training was the suitability for toxicity predictions, balanced accuracy was used to choose the best hyperparameter set. Balanced accuracy is calculated as the harmonic mean between sensitivity and specificity.

3   Results and Discussion

For all datasets we could generate the necessary number of conformations. For rigid molecules, which constituted about 10% of the dataset, we obtained very similar or the same conformation. Nevertheless, the conformers of one molecule had an average root mean square deviation (RMSD) of about 1.7–1.9 Å.

Using the dataset with one conformation per molecule (1–1 dataset) we observed that training yielded good results with respect to the area under the receiver operating curve (AUC) but, in most cases, balanced accuracy was lower than 0.6. In these cases, the models have a high specificity but lack sensitivity. Evaluating the predictions showed that mostly all molecules were classified as the majority class. Applying COVER, we observed considerable changes. Specifically, trained models gained sensitivity with only slight loss of specificity and no loss in the AUC. This indicates the models are able to predict both classes.

Figure 1A illustrates oversampling by itself does not increase predictivity of models trained with the 2–2 or 5–5 dataset. Contrarily, models trained with the 1–16, 2–32 and 5–80 datasets show an increase in the sensitivity, and thus predictivity. The results for the external fold of the cross-validation can be seen in Fig. 1B. It shows models trained with balanced datasets do not suffer from low sensitivity. To ascertain the models also work for an external dataset we used the test set from the Tox21 challenge. No model showed a decrease in predictivity, which is depicted in Fig. 1C.

Our approach demonstrates creation of multiple conformations of a molecule facilitates training of neural networks. This is achieved using only information inherent in the dataset without having to create artificial samples as is often done in traditional machine learning (e.g. SMOTE [2]). Since our idea originated from image augmentation, we do not view the conformations as biological relevant, therefore all conformation have the same label as the parent molecule. Rather we hypothesize that through increase in training space the model is able to generalize better. This is similar to augmented images helping a network in the generalization. Therefore, rather than regarding the images as meaningful representations, they present the variety of the real world to a model. Thus no conformation selection process is implemented. In the future it would be very interesting to investigate the influence of a more sophisticated conformer selection on model training. However, this makes balancing harder as there might not be enough distinct conformations per molecule.

In general, we observed COVER is only beneficial when also balancing the dataset. By oversampling, the training space of the network is increased, facilitating training. Subsequently, models cannot disregard the minority class. Our final validation showed that models trained on datasets balanced with COVER have a much higher sensitivity than those trained on unbalanced datasets.

Acknowledgements. This project has received funding from the Innovative Medicines Initiative 2 Joint Undertaking under grant agreement No 777365 ("eTRANSAFE"). This Joint Undertaking receives support from the European Union's Horizon 2020 research and innovation programme and EFPIA.

References

1. Baumann, D., Baumann, K.: Reliable estimation of prediction errors for QSAR models under model uncertainty using double cross-validation. J. cheminf. **6**(1), 47 (2014). https://doi.org/10.1186/s13321-014-0047-1
2. Chawla, N.V., Bowyer, K.W., Hall, L.O., Kegelmeyer, W.P.: SMOTE: synthetic minority over-sampling technique. J. Artif. Intell. Res. **16**, 321–357 (2002). https://doi.org/10.1613/jair.953
3. Ciresan, D.C., Meier, U., Gambardella, L.M., Schmidhuber, J.: Deep big simple neural nets excel on handwritten digit recognition. Neural Comput. **22**(12), 3207–3220 (2010). https://doi.org/10.1162/NECO_a_00052
4. Frey, B.J., Dueck, D.: Clustering by passing messages between data points. Science **315**(5814), 972–976 (2007). https://doi.org/10.1126/science.1136800

5. Krizhevsky, A., Sutskever, I., Hinton, G.E.: ImageNet classification with deep convolutional neural networks. In: Pereira, F., Burges, C.J.C., Bottou, L., Weinberger, K.Q. (eds.) Advances in Neural Information Processing Systems 25, pp. 1097–1105. Curran Associates, Inc., New York (2012). http://papers.nips.cc/paper/4824-imagenet-classification-with-deep-convolutional-neural-networks.pdf
6. Landrum, G.: RDKit: open-source cheminformatics (2006). http://www.rdkit.org/
7. Mayr, A., Klambauer, G., Unterthiner, T., Hochreiter, S.: DeepTox: toxicity prediction using deep learning. Front. Environ. Sci. **3**, 80 (2016). https://doi.org/10.3389/fenvs.2015.00080
8. MerckKaggle: Merck Molecular Activity Challenge (2012). https://www.kaggle.com/c/MerckActivity
9. Riniker, S., Landrum, G.A.: Better informed distance geometry: using what we know to improve conformation generation. J. Chem. Inf. Model. **55**(12), 2562–2574 (2015). https://doi.org/10.1021/acs.jcim.5b00654
10. Simard, P.Y., Steinkraus, D., Platt, J.C.: Best practices for convolutional neural networks applied to visual document analysis. In: Proceedings of the Seventh International Conference on Document Analysis and Recognition 2003, pp. 958–963, August 2003. https://doi.org/10.1109/ICDAR.2003.1227801
11. Team, K.: Deep Learning How I Did It: Merck 1st place interview, November 2012. http://blog.kaggle.com/2012/11/01/deep-learning-how-i-did-it-merck-1st-place-interview/
12. Tox21: Tox21 Data Challenge 2014 (2014). https://tripod.nih.gov/tox21/challenge/
13. Wong, S.C., Gatt, A., Stamatescu, V., McDonnell, M.D.: Understanding data augmentation for classification: when to warp? In: 2016 International Conference on Digital Image Computing: Techniques and Applications (DICTA), pp. 1–6, November 2016. https://doi.org/10.1109/DICTA.2016.7797091

Open Access This chapter is licensed under the terms of the Creative Commons Attribution 4.0 International License (http://creativecommons.org/licenses/by/4.0/), which permits use, sharing, adaptation, distribution and reproduction in any medium or format, as long as you give appropriate credit to the original author(s) and the source, provide a link to the Creative Commons license and indicate if changes were made.

The images or other third party material in this chapter are included in the chapter's Creative Commons license, unless indicated otherwise in a credit line to the material. If material is not included in the chapter's Creative Commons license and your intended use is not permitted by statutory regulation or exceeds the permitted use, you will need to obtain permission directly from the copyright holder.

Prediction of the Atomization Energy of Molecules Using Coulomb Matrix and Atomic Composition in a Bayesian Regularized Neural Networks

Alain B. Tchagang[✉] ⓘ and Julio J. Valdés ⓘ

Digital Technologies Research Centre, National Research Council Canada,
M-50, 1200 Montréal Road, Ottawa, ON K4A 0S2, Canada
{alain.tchagang, julio.valdes}@nrc-cnrc.gc.ca

Abstract. Exact calculation of electronic properties of molecules is a fundamental step for intelligent and rational compounds and materials design. The
intrinsically graph-like and non-vectorial nature of molecular data generates a
unique and challenging machine learning problem. In this paper we embrace a
learning from scratch approach where the quantum mechanical electronic
properties of molecules are predicted directly from the raw molecular geometry,
similar to some recent works. But, unlike these previous endeavors, our study
suggests a benefit from combining molecular geometry embedded in the Coulomb matrix with the atomic composition of molecules. Using the new combined features in a Bayesian regularized neural networks, our results improve
well-known results from the literature on the QM7 dataset from a mean absolute
error of 3.51 kcal/mol down to 3.0 kcal/mol.

Keywords: Atomization energy · Atomic composition ·
Bayesian regularization · Coulomb matrix · Electronic properties · Molecules ·
Neural networks

1 Introduction

Finding new molecules, compounds or materials with desired properties is strategic to
the innovation and progress of many chemical, agrochemical and pharmaceutical
industries. One of the major challenges consists of making quantitative estimates in the
chemical compound space at moderate computational cost (milliseconds per compound
or faster). Currently only high level quantum-chemistry calculations, which can take
days per molecule depending on property and system, yield the desired chemical
accuracy of 1 kcal/mol required for computational molecular and material design [1].

Recent technological advances have shown that data-to-knowledge approaches are
beginning to show enormous promise within materials science. Intelligent exploration
and exploitation of the vast materials property space has the potential to alleviate the
cost, risks, and time involved in trial-by-error approach experiment cycles used by
current techniques to identify useful compounds [2]. For example, obtaining
atomization energies from the Schrödinger equation solver is computationally

© Crown 2019
I. V. Tetko et al. (Eds.): ICANN 2019, LNCS 11731, pp. 793–803, 2019.
https://doi.org/10.1007/978-3-030-30493-5_75

expensive and, as a consequence, only a fraction of the molecules in the chemical compound space can be labeled. By training a machine learning algorithm on the few label ones, the trained quantum mechanics machine learning (QM/ML) model can be used to generalize from these few data points to unseen molecules [3]. One of the central questions in QM/ML is how to represent molecules in a way that makes prediction of molecular properties feasible and accurate [4]. This question has already been extensively discussed in the cheminformatics literature, and many so-called molecular descriptors exist [5]. Unfortunately, they often require a substantial amount of domain knowledge and engineering. Furthermore, they are not necessarily transferable across the whole chemical compound space [1].

In this paper, we follow a more direct approach introduced in [6], and adopted by several other authors. We learn the mapping between the molecule and its atomization energy from scratch using the Coulomb matrix as a low-level molecular descriptor [6, 7]. Coulomb matrix is invariant to translation and rotation but not to permutations or re-indexing of the atoms. Methods to tackle this issue have been proposed. Examples include Coulomb sorted eigenspectrum [1], Coulomb sorted L2 norm of the matrix's columns [7], Coulomb bag of bonds [8], and random Coulomb matrices [1, 3]. Our study extends the work of [1, 3, 6, 7]. Unlike these previous authors, we show that by combining the molecular geometry embedded in the Coulomb matrix with atomicity or atomic composition of molecules (i.e. atom counts of each type in a molecule), the outcome of the QM/ML models can be significantly improved.

To test this new hypothesis, five representations are constructed: (1) sorted Coulomb matrix, (2) atomic composition of molecules, (3) Coulomb eigenspectrum, (4) the combination of the Coulomb eigenspectrum and the atomic composition of molecules, and (5) the combination of the sorted Coulomb matrix and the atomic composition of molecules. Each one is used as input to a well-defined multilayer Bayesian regularized neural networks [9–13]. Results obtained using the combination of either the sorted Coulomb matrix or the Coulomb eigenspectrum with the atomic composition showed better predictions by a difference of more than 1.5 kcal/mol compared to when the sorted Coulomb matrix or the Coulomb eigenspectrum is used solely. More interestingly, the mean absolute error (MAE) = 3.0 kcal/mol obtained in this study is lower than the 3.51 kcal/mol well-known results obtained in [1, 3]. These results confirm the efficacy of using the atomic composition of molecules in a QM/ML model for their electronic properties predictions. Furthermore, the Bayesian regularized neural network is shown to be a suitable candidate for the modeling of molecular data.

The rest of this paper is organized as follows. In Sect. 2, the dataset used in this study is described. Section 3 provides a detailed description of the proposed method. Section 4 presents the results and Sect. 5 the conclusions.

2 Materials

The QM7 dataset used in this study is a subset of the GDB-13 dataset [14]. The version used here is the one published in [7] consisting of 7102 small organic molecules and their associated atomization energy. These molecules are composed of a maximum of 23 atoms. Molecules are converted to a suitable Cartesian coordinates representation using universal forcefield method [15] as implemented in the software OpenBabel [16]. Atomization energies are calculated for each molecule and ranging from -800 to -2000 kcal/mol. Note that all the 7102 molecules are unique and there are no isomers in the set.

3 Methods

Sorted Coulomb matrix, Coulomb eigenspectrum and atomic composition of each molecule are computed using the atomic coordinates and the chemical formulae of each molecule respectively as described in the QM7 dataset. Next, atomic composition, Coulomb eigenspectrum and sorted Coulomb matrix are either combined or used separately as input to a regularized Bayesian neural network for the prediction of the atomization energy.

3.1 Atomicity, Atom Counts or Atomic Composition

Let's define $\Omega = \{\Omega_1, \Omega_2, \ldots, \Omega_m, \ldots, \Omega_M\}$, the set of possible molecules in the chemical compound space (CCS). By construction, this space is very large. In this study, we will assume that it is bounded by M. Let's define A the set of unique atoms that make Ω. A is bounded by K and it is defined as: $A = \{A^1, A^2, \ldots, A^k, \ldots, A^K\}$. Let's define a chemical operator "." that combines atoms among them in a specific numbers α_k^m and according to the laws of chemistry to form a stable molecule Ω_m. The chemical formulae of Ω_m can be written as: $\Omega_m = \alpha_1^m A^1 . \alpha_2^m A^2 \ldots \alpha_k^m A^k \ldots \alpha_K^m A^K$, or as in chemical textbook.

$$\Omega_m \equiv A_{\alpha_1^m}^1 A_{\alpha_2^m}^2 \ldots A_{\alpha_k^m}^k \ldots A_{\alpha_K^m}^K \tag{1}$$

The atomic composition (AC) of molecule Ω_m in the atomic space $[A^1 \, A^2 \ldots A^k \ldots A^K]$ is defined as $[\alpha_1^m \alpha_2^m \ldots \alpha_k^m \ldots \alpha_K^m]$, where α_k^m is a positive integer that represents the number of atom A^k in molecule Ω_m. The AC of the M molecules in the atomic space $[A^1 \, A^2 \ldots A^k \ldots A^K]$ can be viewed as an M $\times$ K matrix α, Eq. (2).

$$\alpha = \begin{bmatrix} \Omega_1 \\ \Omega_2 \\ \vdots \\ \Omega_m \\ \vdots \\ \Omega_M \end{bmatrix} = \begin{bmatrix} \alpha_1^1 & \alpha_2^1 & \cdots & \alpha_k^1 & \cdots & \alpha_K^1 \\ \alpha_1^2 & \alpha_2^2 & \cdots & \alpha_k^2 & \cdots & \alpha_K^2 \\ \vdots & \vdots & \cdots & \vdots & \cdots & \vdots \\ \alpha_1^m & \alpha_2^m & \cdots & \alpha_k^m & \cdots & \alpha_K^m \\ \vdots & & \cdots & & \cdots & \vdots \\ \alpha_1^M & \alpha_2^M & \cdots & \alpha_k^M & \cdots & \alpha_K^M \end{bmatrix} \tag{2}$$

Row $\alpha(m,:)$ of α corresponds to the AC of the mth molecule (Ω_m). Column $\alpha(:,k)$ corresponds to the number of atom A^k in each molecule of Ω. K is an integer and correspond to the number of unique atoms that makes Ω. For example, given a set of seven molecules: $\Omega = \{CH_4, C_2H_2, C_3H_6, C_2NH_3, OC_2H_2, ONC_3H_3, SC_3NH_3\}$. The set of unique atoms that makes Ω is $A = \{C, H, N, O, S\}$. The matrix α is then:

$$\alpha = \begin{array}{c} CH_4 \\ C_2H_2 \\ C_3H_6 \\ C_2NH_3 \\ OC_2H_2 \\ ONC_3H_3 \\ SC_3NH_3 \end{array} = \begin{bmatrix} 1 & 4 & 0 & 0 & 0 \\ 2 & 2 & 0 & 0 & 0 \\ 3 & 6 & 0 & 0 & 0 \\ 2 & 3 & 1 & 0 & 0 \\ 2 & 2 & 0 & 1 & 0 \\ 3 & 3 & 1 & 1 & 0 \\ 3 & 3 & 1 & 0 & 1 \end{bmatrix} \tag{3}$$

It is obvious that this representation is not unique. That is two molecules with identical atomic composition may have different electronic properties. Isomers are great examples in this case. They are compound with the same molecular formulas but that are structurally different in some way, and they can have different chemical, physical and biological properties [17]. It is also worth to note that such molecular representation had been explored in the past in quantitative structure activity relationship and correspond to a different form of the Atomistic index developed by Burden [13].

3.2   Coulomb Matrix

The Coulomb matrix (CM) has recently been widely used as molecular descriptors in the QM/ML models [1, 3, 6, 7]. Given a molecule its Coulomb matrix $CM = [c_{ij}]$ is defined by Eq. (4).

$$c_{ij} = \begin{cases} 0.5Z_i^{2.4} & \text{for } i = j \\ \frac{Z_i Z_j}{\|R_i - R_j\|} & \text{for } i \neq j \end{cases} \tag{4}$$

Z_i is the atomic number of atom i, and R_i is its position in atomic units [7]. CM is of size $I \times I$, where I corresponds to the number of atoms in the molecule. It is symmetric and has as many rows and columns as there are atoms in the molecule. As we mentioned earlier, the Coulomb matrix is invariant to rotation, translation but not to permutation of its atoms. Several techniques to tackle this issue have been explored in the literature. Examples include sorted Coulomb matrix and Coulomb eigenspectrum.

Sorted Coulomb Matrix (SCM). This approach sorts the CMs by descending order with respect to the norm-2 of their columns and simultaneously permuting their rows and columns accordingly. After the ordering step and given the symmetry of these matrices, it is customary to only consider their lower triangular part [6, 7], and to unfold them row-wise in a 1-dimensional (1D) vector of length $L = \sum_{i=0}^{I} (I - i)$, where I here corresponds to the number of atoms of the largest molecule. In this study, the 1D vector is called the SCM signal $x(m,:) = x_m[l]$, with $l = 1$ to L and m corresponds to molecule Ω_m. For a set of M molecules, their 1D SCM signals can be organized in an M × L matrix x:

$$x = \begin{bmatrix} x_{11} & x_{12} & \cdots & x_{1l} & \cdots & x_{1L} \\ x_{21} & x_{22} & \cdots & x_{2l} & \cdots & x_{2L} \\ \vdots & \vdots & \cdots & \vdots & \cdots & \vdots \\ x_{m1} & x_{m2} & \cdots & x_{ml} & \cdots & x_{mL} \\ \vdots & \cdots & \cdots & \vdots & \cdots & \vdots \\ x_{M1} & x_{M2} & \cdots & x_{Ml} & \cdots & x_{ML} \end{bmatrix} \tag{5}$$

The mth row of x represents the 1D SCM signal of the mth molecule. Given that molecules have different number of atoms, the short ones are padded with zeros so that all the 1D SCM signals have the same length L.

Coulomb Eigenspectrum (CES). The Coulomb eigenspectrum [6, 7] is obtained by solving the eigen value problem $Cv = \lambda v$, under constraint $\lambda_i \geq \lambda_{i+1}$ where $\lambda_i > 0$. The spectrum $(\lambda_1, ..., \lambda_I)$ is used as the representation and it corresponds to a 1D signal: $z(m,:) = z_m[n]$, with $n = 1$ to N and m corresponds to molecule Ω_m. For a set of M molecules, their 1D CES signals can be organized in an M × N matrix z:

$$z = \begin{bmatrix} z_{11} & z_{12} & \cdots & z_{1n} & \cdots & z_{1N} \\ z_{21} & z_{22} & \cdots & z_{2n} & \cdots & z_{2N} \\ \vdots & \vdots & \cdots & \vdots & \cdots & \vdots \\ z_{m1} & z_{m2} & \cdots & z_{mn} & \cdots & z_{mN} \\ \vdots & \cdots & \cdots & \vdots & \cdots & \vdots \\ z_{M1} & z_{M2} & \cdots & z_{Mn} & \cdots & z_{MN} \end{bmatrix} \tag{6}$$

The mth row of z represents the 1D CES signal of the mth molecule. Given that molecules have different number of atoms, the short ones are padded with zeros so that all the 1D CES signals have the same length N.

3.3   Input of the QM/ML Model

Let's define X as the input to the neural network defined below. In order to test the usefulness of the AC in the prediction of the electronic properties of molecules, we have considered five different inputs and compared them against each other. The five inputs are: X = α (only the AC is used), X = z (only the CES is used), X = x (only the SCM is used), X = [α z] (AC and CES are combined and used as inputs), and finally X = [α x] (AC and SCM are combined and used as inputs). By combining AC, CES and SCM, taking the Z-scores of X prior to its utilization as input to the ML model becomes an obvious choice. The Z-score of X will return a matrix of same size X', where each column of X' has mean 0 and a standard deviation of 1 [18].

3.4   Output – Atomization Energy of Molecules

The output to the QM/ML is the atomization energy E. It quantifies the potential energy stored in all chemical bonds. As such, it is defined as the difference between the potential energy of a molecule and the sum of potential energies of its composing isolated atoms. The potential energy of a molecule is the solution to the electronic Schrödinger equation $H\Phi = E\Phi$, where H is the Hamiltonian of the molecule and Φ is the state of the system. The atomization energy of molecules are organized in an $M \times 1$ column vector $y = [y_1\ y_2\ ...\ y_m\ ...\ y_M]^T$. The superscript T indicates the transpose operator. The entry y_m is a real number that corresponds to the atomization energy of the mth molecule.

3.5   Bayesian Regularized Neural Networks

Neural networks (NN) are universal function approximators that can be applied to a wide range of problems such as classification and model building. It is already a mature field within machine learning and there are many different NN paradigms. Multilayer feed-forward networks are the most popular and a large number of training algorithms have been proposed. Compared to other non-linear techniques, in multilayer NNs, the measure of similarity is learned essentially from data and implicitly given by the mapping onto increasingly many layers. In general, NNs are more flexible and make fewer assumptions about the data. However, it comes at the cost of being more difficult to train and regularize [3]. In this paper, we used the Bayesian regularization method to train our NNs [9–12].

Bayesian methods are optimal methods for solving learning problems. Any other method not approximating them should not perform as well on average. They are very useful for comparison of data models as they automatically and quantitatively embody "Occam's Razor" [19]. Complex models are automatically self-penalizing under Bayes' Rule. Bayesian methods are complementary to NNs as they overcome the tendency of an over flexible network to discover nonexistent, or overly complex, data models.

Unlike a standard back-propagation NN training method where a single set of parameters (weights, biases, etc.) are used, the Bayesian approach to NN modeling considers all possible values of network parameters weighted by the probability of each set of weights. Bayesian inference is used to determine the posterior probability distribution of weights and related properties from a prior probability distribution according to updates provided by the training set D using the Bayesian regularized NN model, H_i. Where orthodox statistics provide several models with several different criteria for deciding which model is best, Bayesian statistics only offer one answer to a well-posed problem.

$$P(w|D, H_i) = \frac{P(D|w, H_i)P(w|H_i)}{P(D|H_i)} \tag{7}$$

Bayesian methods can simultaneously optimize the regularization constants in NNs, a process which is very laborious using cross-validation [9].

4 Results and Discussions

As we mentioned earlier, the version of the QM7 dataset used in this study is the one published in [7] and it is composed of M = 7102 molecules and contains up to five types of atoms: Carbon (C), Hydrogen (H), Oxygen (O), Nitrogen (N), and Sulfur (S). Therefore the set of unique atoms is A = {C, H, N, O, S}. The matrix α is of size M × K = 7102 × 5. The largest molecule is made of I = 23 atoms. Thus the CES matrix z is of size M × N = 7102 × 23, and the SCM matrix x is of size M × L = 7102 × 276, because L = $\sum_{i=0}^{23}(23 - i) = 276$. The column vector of atomization energy y is of size 7102 × 1. The QM7 dataset is randomly divided into 80% training and 20% testing sets. Performance is measured using the root mean square error (RMSE), Eq. (8), the mean absolute error (MAE), Eq. (9), and the Pearson correlation coefficient r_{ppe}, Eq. (10).

$$RMSE = \sqrt{\frac{1}{M}\sum_{m=1}^{M}(y_m - y_m^e)^2} \tag{8}$$

$$MAE = \frac{1}{M}\sum_{m=1}^{M}|y_m - y_m^e| \tag{9}$$

$$r_{PP^e} = \frac{\sum_{m=1}^{M}(y_m - \bar{y})(y_m^e - \bar{y}^e)}{\sqrt{\sum_{m=1}^{M}(y_m - \bar{y})^2}\sqrt{\sum_{m=1}^{M}(y_m^e - \bar{y}^e)^2}} \tag{10}$$

4.1 Results

We used the Matlab implementation of the regularized Bayesian network to model the relationship between the inputs (X) and the output (y). Figure 1 for example shows the Matlab architecture of one of the networks used.

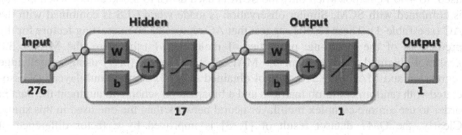

Fig. 1. Matlab representation of one hidden layer neural networks with 17 neurons when the SCM is used as input (276 inputs and 1 output).

In the literature, there is no clear and rational approach on how to select the number of neurons and the number of hidden layers of a NN. The middle ground is usually to select an architecture that will neither under-fit nor over-fit the model. In this study we tested several architecture based on some empirical observations also coming from the literature with the goal for avoiding under-fitting and overfitting of the model. Table 1 shows the results obtained using different NN architectures with the data partitioned into 90% training and 10% for validation and testing.

Table 1. Statistics of the results using three different network architectures, each trained using Bayesian regularization. MAE and RMSE are in kcal/mol.

Network architecture	Input*	Statistics		
		MAE	RMSE	r_{ppe}
[17]	α	13.82	18.05	0.9967
	z	9.40	12.29	0.9985
	x	5.02	6.72	0.9995
	[α z]	8.38	11.09	0.9988
	[α x]	3.70	5.0	0.9997
[16 × 8 × 4]	α	13.82	18.05	0.9967
	z	8.70	11.35	0.9987
	x	4.83	6.58	0.9996
	[α z]	7.59	10.06	0.9990
	[α x]	3.42	4.73	0.9998
[18 × 9 × 3]	α	13.80	18.04	0.9967
	z	8.57	11.16	0.9987
	x	4.40	5.95	0.9996
	[α z]	7.38	9.80	0.9990
	[α x]	**3.0**	**4.22**	**0.9998**

*α = Atomic Composition (AC), z = Coulomb Eigen Spectrum (CES), x = Sorted Coulomb Matrix (SCM)

The results obtained show that the association of the AC with either CES or SCM significantly improved the prediction accuracy. For example, with the three hidden layer network [18 × 9 × 3], the MAE goes from 13.80 kcal/mol when only the AC is used, to 4.40 kcal/mol when only the SCM is used down to 3.0 kcal/mol when the AC is combined with SCM. Similar observation is made when CES is combined with the AC (see Table 1). These results suggest that AC represents an interesting feature for the predictions of the electronic properties of molecules. Furthermore, the MAE = 3.0 kcal/mol obtained is lower than the MAE of 9.9 kcal/mol [6, 7] using kernel ridge regression and MAE of 3.51 kcal/mol obtained in [1, 3] using a multilayer NN associated with random coulomb matrices and a binarization scheme to augment the data in order to use a more complex multilayer neural network than the one used in this study. Clearly, the QM7 dataset result of [1, 3] is improved by a factor difference of 0.5 kcal/mol in this study. Our result is close to the acceptable 1 kcal/mol chemical accuracy.

4.2 Discussions

Predicting molecular energies quickly and accurately across the CCS is an important problem as the QM calculations take days and do not scale well to more complex molecules. ML is a good candidate for solving this problem as it encourages the framework to focus on solving the problem of interest rather than solving the more general Schrödinger equations. In this paper, we have developed further the learning-from-scratch approach initiated in [6] and provided a new ingredient for learning a successful mapping between raw molecular geometries and atomization energies. Our results suggest important discoveries and open new venues for future research.

Atomicity, Atom Counts or Atomic Composition Represents an Interesting Feature for QM/ML Models. Atomic composition (AC), i.e. atom counts of each type in a molecule is a representation that does not contain any molecular structural information. But our analysis suggests a correlation between the AC representation and the atomization energy. The combination of AC with CES or SCM yields a new molecular representation which inherits all the properties of either CES or SCM representation respectively. Even though the AC representation is not unique (case of isomers as we mentioned earlier), by combining it with the SCM for example, the pair [AC SCM] inherit all the properties of SCM and becomes a representation that is uniquely defined, invariant to rotation, translation and re-indexing of the atoms, given that the SCM had already been sorted in decreasing order to tackle the non-invariance to atom re-indexing. Similar observation can be made with the CES.

Bayesian Regularized Neural Networks are Suited for Molecular Data. The Bayesian regularization approach used in this study seems to fit molecular data very well. Similar observation was made in [12] when developing quantitative structure activity relationship (QSAR) model of compounds active at the benzodiazepine and muscarinic receptors. The results obtained here further prove the point that Bayesian regularized neural networks possess several properties useful for the analysis of molecular data. One advantage of the Bayesian regularized neural networks is that the number of effective parameters used in the model is less than the number of weights, as some weights do not contribute to the models. This minimizes the likelihood of overfitting. The concerns about overfitting and overtraining are also removed by this method so that the production of a definitive and reproducible model is attained [9–12].

5 Conclusions

In this study, we show that by combining the atomic composition of molecules with their Coulomb matrix representation, the output of the quantum mechanics machine learning model can be significantly improved. Using the QM7 dataset as a test case, our results show a decrease by a difference of 1.5 kcal/mol when the sorted Coulomb matrix representation is combined with the atomic composition compared to when the sorted Coulomb matrix is used alone. Furthermore, our results improve well-known results from the literature on the QM7 dataset from a mean absolute error of 3.51 kcal/mol down to 3.0 kcal/mol. These results suggest that the atomic composition of molecules contain interesting information useful for quantum mechanics machine learning model and should not be neglected.

References

1. Montavon, G., et al.: Learning invariant representations of molecules for atomization energy prediction. In: Pereira, F., Burges, C.J.C., Bottou, L., Weinberger, K.Q. (eds.) Proceedings of the 25th International Conference on Neural Information Processing Systems - (NIPS 2012), Curran Associates Inc., USA, vol. 1, pp. 440–448 (2012). doi:2999134.2999184
2. Xue, D., Balachandran, P.V., Hogden, J., Theiler, J., Xue, D., Lookman, T.: Accelerated search for materials with targeted properties by adaptive design. Nat. Commun. **15**(7), 11241 (2016). https://doi.org/10.1038/ncomms11241
3. Montavon, G., et al.: Machine learning of molecular electronic properties in chemical coumpound space. New J. Phys. **15**(9), 095003 (2013). https://doi.org/10.1088/1367-2630/15/9/095003
4. Von Lilienfeld, O.A., Tuckerman, M.E.: Molecular grand-canonical ensemble density functional theory and exploration of chemical space. J. Chem. Phys. **125**(15), 154104 (2006). https://doi.org/10.1063/1.2338537
5. Mauri, A., Consonni, V., Todeschini, R.: Molecular descriptors. In: Leszczynski, J. (ed.) Handbook of Computational Chemistry. Springer, Dordrecht (2016). https://doi.org/10.1007/978-94-007-6169-8_51-1
6. Rupp, M., Tkatchenko, A., Müller, K.-R., von Lilienfeld, O.A.: Fast and accurate modeling of molecular atomization energies with machine learning. Phys. Rev. Lett. **108**, 058301 (2012). https://doi.org/10.1103/physrevlett.108.058301
7. Rupp, M.: Machine learning for quantum mechanics in a nutshell. Int. J. Quan. Chem. **115**, 1058–1073 (2015). https://doi.org/10.1002/qua.24954
8. Hansen, K., et al.: Machine learning predictions of molecular properties: accurate many-body potentials and nonlocality in chemical space. J. Phys. Chem. Lett. **6**, 2326–2331 (2015). https://doi.org/10.1021/acs.jpclett.5b00831
9. MacKay, D.J.C.: A practical bayesian framework for backprop networks. Neural Comput. **4**, 415–447 (1992). https://doi.org/10.1162/neco.1992.4.3.448
10. Mackay, D.J.C.: Probable networks and plausible predictions - a review of practical bayesian methods for supervised neural networks. Comput. Neural Sys. **6**, 469–505 (1995). https://doi.org/10.1088/0954-898x_6_3_011
11. Mackay, D.J.C.: Bayesian interpolation. Neural Comput. **4**, 415–447 (1992). https://doi.org/10.1162/neco.1992.4.3.415
12. Buntine, W.L., Weigend, A.S.: Bayesian back-propagation. Complex Sys. 5, 603–643, (1991). https://doi.org/10.1007/s00138-012-0450-4
13. Burden, F.R.: Robust QSAR models using bayesian regularized neural networks. J. Med. Chem. **42**, 3183–3187 (1999). https://doi.org/10.1021/jm980697n
14. Blum, L.C., Reymond, J.-L.: 970 million druglike small molecules for virtual screening in the chemical universe database GDB-13. J. Am. Chem. Soc. **131**, 8732–8733 (2009). https://doi.org/10.1021/ja902302h
15. Rappé, A.K., Casewit, C.J., Colwell, K.S., Goddard, W.A., Skiff, W.M.: UFF, a full periodic table force field for molecular mechanics and molecular dynamics simulations. J. Am. Chem. Soc. **114**(25), 10024–10035 (1992). https://doi.org/10.1021/ja00051a040
16. Guha, R., et al.: The blue obelisk, interoperability in chemical informatics. J. Chem. Inf. Model. **46**(3), 991–998 (2006). https://doi.org/10.1021/ci050400b

17. Gorzynski, S.J.: General Organic and Biological Chemistry, 2nd edn, p. 450. McGraw-Hill, New York (2010)
18. Aho, K.A.: Foundational and Applied Statistics for Biologists, 1st edn. CRC Press, Boca Raton (2014)
19. Hugh, G., Gauch, H.G.: Scientific Method in Practice. Cambridge University Press, Cambridge (2003)

Open Access This chapter is licensed under the terms of the Creative Commons Attribution 4.0 International License (http://creativecommons.org/licenses/by/4.0/), which permits use, sharing, adaptation, distribution and reproduction in any medium or format, as long as you give appropriate credit to the original author(s) and the source, provide a link to the Creative Commons license and indicate if changes were made.

The images or other third party material in this chapter are included in the chapter's Creative Commons license, unless indicated otherwise in a credit line to the material. If material is not included in the chapter's Creative Commons license and your intended use is not permitted by statutory regulation or exceeds the permitted use, you will need to obtain permission directly from the copyright holder.

Deep Neural Network Architecture
for Drug-Target Interaction Prediction

Nelson R. C. Monteiro$^{(\boxtimes)}$, Bernardete Ribeiro , and Joel P. Arrais

Department of Informatics Engineering (DEI),
Center for Informatics and Systems of the University of Coimbra (CISUC),
University of Coimbra, Coimbra, Portugal
nelsonrcm117@gmail.com

Abstract. The discovery of potential Drug-Target Interactions (DTIs) is a determining step in the drug discovery and repositioning process, as the effectiveness of the currently available antibiotic treatment is declining. Successful approaches have been presented to solve this problem but seldom protein sequences and structured data are used together. We present a deep learning architecture model, which exploits the particular ability of Convolutional Neural Networks (CNNs) to obtain 1D representations from protein amino acid sequences and SMILES (Simplified Molecular Input Line Entry System) strings. The results achieved demonstrate that using CNNs to obtain representations of the data, instead of the traditional descriptors, lead to improved performance.

Keywords: Drug repositioning · Drug-Target Interaction ·
Deep learning · Convolutional Neural Network ·
Fully Connected Neural Network

1 Introduction

The discovery of new and potential drugs is declining, as there is an increase of the misuse of the available medicine, causing a resistance effect to these kinds of agents [1]. Therefore, establishing effective computational methods is decisive to find new leads. Computational methods for DTI prediction are divided into 3 main approaches [4], namely ligand based, docking simulation and chemogenomic. Ligand based approaches are built upon the concept that similar molecules have similar properties and therefore should bind to the same group of proteins [6]. Docking Simulation is used for structure based drug design, where the interaction is simulated and scored using 3D structures [5]. Chemogenomic approaches are based on the chemical, genomic and/or the pharmacological space [8]. Due to the amount of available data and computational power, machine learning [3] and deep learning [9] are pursued over the traditional methods. We propose a deep learning approach to predict DTIs using 1D raw data, amino acids sequences and SMILES. We exploit the particular ability of CNNs

© The Author(s) 2019
I. V. Tetko et al. (Eds.): ICANN 2019, LNCS 11731, pp. 804–809, 2019.
https://doi.org/10.1007/978-3-030-30493-5_76

to obtain 1D representations, which are features that express local dependencies or patterns, that can then be used in a Fully Connected Neural Network (FCNN), acting as a binary classifier. Coelho et al. (2016) [7] dataset was used to evaluate and validate the model. Additionally, we compared our model with different approaches, specifically random forest (RF), a FCNN architecture and support vector machine (SVM).

2 Methods

2.1 Data

The protein sequences were extracted from UniProt and the SMILES strings were collected from PubChem exclusively, in their canonical format. Since we are using protein sequences and SMILES strings directly, each amino acid and character, respectively, is considered as a feature. Therefore, it was necessary to define a threshold based on their length. An information threshold of 95% was used, resulting in a maximum length of 1205 for the protein sequences and 90 for the SMILES. All entries duplicated or containing missing characters in one of the datasets were also removed, resulting in 16011 (5839 positive and 10712 negative) samples for training and 7926 (3012 positive and 4914 negative) for testing. Table 1 summarizes the amount of unique drugs, targets and drug-target interactions extracted from the databases used to create the datasets and Table 2 the amount of unique proteins, drugs and number of targets for the training and testing datasets. Plus, only Yamanishi et al. (2008) [8] and DrugBank positive entries were used for training and testing, respectively.

Table 1. Unique drugs, targets and DTIs used to create the datasets.

	Positive		Negative	
	DrugBank	Yamanishi et al. (2008)	BioLip	BindingDB
Drugs	1328	790	894	12454
Targets	706	1371	636	404
DTI	3530	7206	1223	14985

Table 2. Unique drugs, targets and number of targets for the training and testing datasets.

	Unique		Number of targets	
	Targets	Drugs	1	>1
Training	1790	9583	8026	1557
Testing	1068	5718	4884	834

2.2 Data Representation

We used Yu et al. (2010) [10] protein substitution table, which organizes amino acids into 7 groups according to their physicochemical properties. Each amino acid was encoded into an integer based on the corresponding group. In the case of SMILES strings, a simple integer encoding was used to transform each character of the strings into an integer.

2.3   Model

The proposed approach is based on a deep learning architecture (Fig. 1) to predict DTIs using directly protein sequences and SMILES (1D raw data). One-Hot Layer was used to assign a binary variable for each unique integer value, converting every integer into a binary vector. Two series of 1D convolutional layers were used, one for the protein sequences and another for the SMILES. A global max pooling layer was applied after each series of convolutional layers to reduce the spatial size of each feature map to its maximum representative feature. The obtained deep representations were concatenated into a single feature vector, characterizing a DTI pair. The resulting feature vectors were then used as the input of a FCNN architecture. Dropout was applied between each fully connected layer to reduce the overfitting by deactivating a percentage of neurons. This architecture was followed by an output layer.

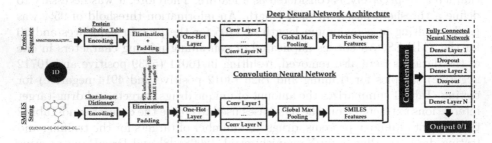

Fig. 1. Drug-Target Interaction model architecture.

2.4   Hyperparameter Optimization Approach

Two simultaneous methods, combined with grid search, were used to determine the best model, early stopping and model checkpoint. Considering the fact that dividing the training set into training and validation and applying cross validation led to high scores for every model architecture and set of parameters in both training and validation, it was not possible to select the best model using this approach, as every model was supposedly good in the validation set but the results were inconsistent when applied to the testing set. Therefore, we decided to use all the training set for training and the testing set to evaluate the model performance in each epoch. Since the testing set is highly imbalanced, F1-score was used for this evaluation (Fig. 2). Table 3 summarizes the hyper-parameters obtained from grid search.

3   Results

We applied grid search for all the models in order to accurately compare and evaluate the performance. The descriptors used were the same as the original

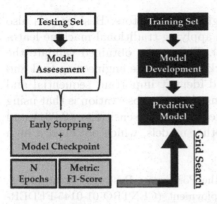

Fig. 2. Hyperparameter optimization model based on grid search.

Table 3. Parameter settings for the proposed model.

Parameters	Value
Number of convolutional layers	3
Number of dense layers (FC)	3
Number of filters	[128, 256, 384]
Filter length (proteins)	[3,4,5]
Filter length (compounds)	[3,4,5]
Epochs*	500
Hidden neurons	[128,128,128]
Batch size	256
Dropout rate	0.5
Optimizer	Adam
Learning rate	0.0001
Loss function	Binary cross entropy
Activation function (CNN)	ReLU
Activation function (FC)	ReLU
Activation function (output)	Sigmoid
Class weights (imbalanced classes)	{0: 0.36, 1: 0.64}

Table 4. Prediction results of testing set.

		Model					
		CNN + FCNN	FCNN	Random forest		SVM RBF	
		CNN representations	Descriptors	Descriptors	CNN representations	Descriptors	CNN representations
Metric	Sensitivity	0.861	0.827	0.809	0.821	0.739	0.769
	Specificity	0.961	0.963	0.989	0.992	0.989	0.993
	F1-Score	0.895	0.876	0.886	0.896	0.842	0.864
	Accuracy	0.923	0.911	0.921	0.927	0.894	0.908

work [7], which contains a total of 432 protein descriptors and 323 drug descriptors, and collected using PyDPI package [2]. The protein descriptors are divided into amino acid composition, Moran autocorrelation and CTD (Composition, Transition, Distribution) descriptors. On the other hand, drug descriptors are divided into molecular constitutional, molecular connectivity, molecular property, kappa shape and charge descriptors, molecular access system (MACCS) keys and E-state fingerprints.

Due to the fact that the traditional split of the training set into training and validation led to inconclusive results, as mentioned in Sect. 2.4, the results obtained are related to "internal validation". The testing set was used to discover the best set of parameters, thus there is not an external validation set. Nonetheless, given the disparity of the training and testing set and the low similarity of the drug pairs that constitute them, the results are considered as valid and relevant.

The differences in performance between all models can be interpreted as a result of the difference between using deep representations, obtained from protein sequences and SMILES strings, and global descriptors. Besides, it's also possible to highlight the difference between applying traditional machine learning and deep learning approaches (Table 4). The results obtained validate the effectiveness of convolutional neural networks as a feature engineering tool and their capacity to automatically surmise and identify important sequential and structural regions for drug-target interactions. Another observation is that using an end-to-end deep learning method resulted in a high sensitivity (0.861) and specificity (0.961) when compared to the other models, which obtained a high specificity and a low sensitivity.

Funding. This research has been funded by the Portuguese Research Agency FCT, through D4 - Deep Drug Discovery and Deployment (CENTRO-01-0145-FEDER-029266).

References

1. Aslam, B., et al.: Antibiotic resistance: a rundown of a global crisis. Infect. Drug Resist. **11**, 1645–1658 (2018). https://doi.org/10.2147/IDR.S173867. 30349322[pmid]
2. Cao, D.S., Liang, Y.Z., Yan, J., Tan, G.S., Xu, Q.S., Liu, S.: PyDPI: freely available python package for chemoinformatics, bioinformatics, and chemogenomics studies. J. Chem. Inf. Model. **53**(11), 3086–3096 (2013). https://doi.org/10.1021/ci400127q
3. Cao, D.S., et al.: Computational prediction of drug target interactions using chemical, biological, and network features. Mol. Inform. **33**(10), 669–681 (2014). https://doi.org/10.1002/minf.201400009
4. Chen, R., Liu, X., Jin, S., Lin, J., Liu, J.: Machine learning for drug-target interaction prediction. Molecules **23**(9), 2208 (2018). https://doi.org/10.3390/molecules23092208
5. Cheng, A.C., et al.: Structure-based maximal affinity model predicts small-molecule druggability. Nat. Biotechnol. **25**(1), 71 (2007). https://doi.org/10.1038/nbt1273
6. Cheng, F., Zhou, Y., Li, J., Li, W., Liu, G., Tang, Y.: Prediction of chemical-protein interactions: multitarget-QSAR versus computational chemogenomic methods. Mol. BioSyst. **8**(9), 2373–2384 (2012). https://doi.org/10.1039/C2MB25110H
7. Coelho, E.D., Arrais, J.P., Oliveira, J.L.: Computational discovery of putative leads for drug repositioning through drug-target interaction prediction. PLOS Comput. Biol. **12**(11), 1–17 (2016). https://doi.org/10.1371/journal.pcbi.1005219. 11
8. Gutteridge, A., Araki, M., Kanehisa, M., Honda, W., Yamanishi, Y.: Prediction of drug-target interaction networks from the integration of chemical and genomic spaces. Bioinformatics **24**(13), i232–i240 (2008). https://doi.org/10.1093/bioinformatics/btn162
9. Tian, K., Shao, M., Wang, Y., Guan, J., Zhou, S.: Boosting compound-protein interaction prediction by deep learning. Methods **110**, 64–72 (2016). https://doi.org/10.1016/j.ymeth.2016.06.024
10. Yu, C.Y., Chou, L.C., Chang, D.T.H.: Predicting protein-protein interactions in unbalanced data using the primary structure of proteins. BMC Bioinform. **11**(1), 167 (2010). https://doi.org/10.1186/1471-2105-11-167

Open Access This chapter is licensed under the terms of the Creative Commons Attribution 4.0 International License (http://creativecommons.org/licenses/by/4.0/), which permits use, sharing, adaptation, distribution and reproduction in any medium or format, as long as you give appropriate credit to the original author(s) and the source, provide a link to the Creative Commons license and indicate if changes were made.

The images or other third party material in this chapter are included in the chapter's Creative Commons license, unless indicated otherwise in a credit line to the material. If material is not included in the chapter's Creative Commons license and your intended use is not permitted by statutory regulation or exceeds the permitted use, you will need to obtain permission directly from the copyright holder.

Mol-CycleGAN - A Generative Model for Molecular Optimization

Łukasz Maziarka[1,2(✉)], Agnieszka Pocha[2], Jan Kaczmarczyk[1], Krzysztof Rataj[1], and Michał Warchoł[1]

[1] Ardigen, Kraków, Poland
lukasz.maziarka@ardigen.com
[2] Jagiellonian University, Kraków, Poland

Abstract. During the drug design process, one must develop a molecule, which structure satisfies a number of physicochemical properties. To improve this process, we introduce Mol-CycleGAN – a CycleGAN-based model that generates compounds optimized for a selected property, while aiming to retain the already optimized ones. In the task of constrained optimization of penalized logP of drug-like molecules our model significantly outperforms previous results.

Keywords: Drug design · Generative models · Molecular optimization

1 Introduction

The principal goal of the drug design process is to find new chemical compounds that are able to modulate the activity of a given target in a desired way [13]. However, finding such molecules in the high-dimensional chemical space of all molecules without any prior knowledge is nearly impossible. *In silico* methods have been introduced to leverage the existing knowledge, thus forming a new branch of science - computer-aided drug design (CADD) [1,12].

The recent advancements in deep learning have encouraged its application in CADD [4]. One of the main approaches is *de novo design*, that is using generative models to propose new molecules that are likely to possess the desired properties [3,5,15,17].

In the center of our interest are the hit-to-lead and lead optimization phases of the compound design process. Their goal is to optimize the drug candidates identified in the previous steps in terms of the desired activity profile and their physicochemical and pharmacokinetic properties.

To address this problem, we introduce **Mol-CycleGAN** – a generative model based on CycleGAN [19]. Given a starting molecule, it generates a structurally similar one but with a desired characteristic. We show that our model generates molecules that possess desired properties while retaining their structural similarity to the starting compound. Moreover, thanks to employing graph-based representation, our algorithm always returns valid compounds.

© The Author(s) 2019
I. V. Tetko et al. (Eds.): ICANN 2019, LNCS 11731, pp. 810–816, 2019.
https://doi.org/10.1007/978-3-030-30493-5_77

To assess the model's utility for compound design we evaluate its ability to maximize penalized logP property. Penalized logP is chosen because it is often selected as a testing ground for molecule optimization models [7,18], due to its relevance in the drug design process. In the optimization of penalized logP for drug-like molecules our model significantly outperforms previous results. To the best of our knowledge, Mol-CycleGAN is the first approach to molecule generation that uses the CycleGAN architecture.

2   Mol-CycleGAN

Mol-CycleGAN is a novel method of performing compound optimization by learning from the *sets* of molecules with and without the desired molecular property (denoted by the sets X and Y). Our approach is to train a model to perform the transformation $G : X \to Y$ (and $F : Y \to X$) which returns the optimized molecules. In the context of compound design X (Y) can be, e.g., the set of inactive (active) molecules.

To represent the sets X and Y our approach requires an embedding of molecules which is reversible, i.e. enables both encoding and decoding of molecules. For this purpose we use the latent space of Junction Tree Variational Autoencoder (JT-VAE) [7] – we represent each molecule as a point in the latent space, given by the mean of the variational encoding distribution [9]. This approach has the advantage that the distance between molecules (required to calculate the loss function) can be defined directly in the latent space.

Our model works as follows: (i) we define the sets X and Y (e.g., inactive/active molecules); (ii) we introduce the mapping functions $G : X \to Y$ and $F : Y \to X$; (iii) we introduce discriminator D_X (and D_Y) which forces the generator F (and G) to generate samples from a distribution close to the distribution of X (or Y). The components F, G, D_X, and D_Y are modeled by neural networks (see subsect. 2.1 for technical details).

The main idea is to: (i) take the prior molecule x without a specified feature (e.g. activity) from set X, and compute its latent space embedding; (ii) use the generative neural network G to obtain the embedding of molecule $G(x)$, that has this feature (as if the $G(x)$ molecule came from set Y) but is also similar to the original molecule x; (iii) decode the latent space coordinates given by $G(x)$ to obtain the optimized molecule. Thereby, the method is applicable in *lead optimization* processes, as the generated compound $G(x)$ remains structurally similar to the input molecule.

To train the Mol-CycleGAN we use the following loss function:

$$L(G, F, D_X, D_Y) = L_{\text{GAN}}(G, D_Y, X, Y) + L_{\text{GAN}}(F, D_X, Y, X) \\ + \lambda_1 L_{\text{cyc}}(G, F) + \lambda_2 L_{\text{identity}}(G, F),$$ (1)

and aim to solve

$$G^*, F^* = \arg \min_{G,F} \max_{D_X, D_Y} L(G, F, D_X, D_Y).$$ (2)

We use the adversarial loss introduced in LS-GAN [11]:

$$L_{\text{GAN}}(G, D_Y, X, Y) = \frac{1}{2}\,\mathbb{E}_{y \sim p_{\text{data}}(y)}[(D_Y(y) - 1)^2] + \frac{1}{2}\,\mathbb{E}_{x \sim p_{\text{data}}(x)}[(D_Y(G(x)))^2],\tag{3}$$

which ensures that the generator G (and F) generates samples from a distribution close to the distribution of Y (or X).

The cycle consistency loss:

$$L_{\text{cyc}}(G, F) = \mathbb{E}_{y \sim p_{\text{data}}(y)}[\|G(F(y)) - y\|_1] + \mathbb{E}_{x \sim p_{\text{data}}(x)}[\|F(G(x)) - x\|_1],\tag{4}$$

reduces the space of possible mapping functions, such that for a molecule x from set X, the GAN cycle brings it back to a molecule similar to x, i.e. $F(G(x))$ is close to x (and analogously $G(F(y))$ is close to y).

Finally, to ensure that the generated (optimized) molecule is close to the starting one, we use the identity mapping loss [19]:

$$L_{\text{identity}}(G, F) = \mathbb{E}_{y \sim p_{\text{data}}(y)}[\|F(y) - y\|_1] + \mathbb{E}_{x \sim p_{\text{data}}(x)}[\|G(x) - x\|_1],\tag{5}$$

which further reduces the space of possible mapping functions and prevents the model from generating molecules that lay far away from the starting molecule in the latent space of JT-VAE.

In our experiments, we use the hyperparameters $\lambda_1 = 0.3$ and $\lambda_2 = 0.1$. Note that these parameters control the balance between *improvement* in the optimized property and *similarity* between the generated and the starting molecule.

2.1 Workflow

We conduct experiments to test if the proposed model is able to generate molecules that are close to the starting ones and possess increased octanol-water partition coefficient (logP) penalized by the synthetic accessibility (SA) score. We optimize penalized logP, while constraining the degree of deviation from the starting molecule. The similarity between molecules is measured with Tanimoto similarity on Morgan Fingerprints [14].

We use the ZINC-250K dataset used in similar studies [7,10] which contains 250000 drug-like molecules extracted from the ZINC database [16]. The sets X_{train} and Y_{train} are random samples of size 80000 from ZINC-250K, where the compounds' penalized logP values are below and above the median, respectively. X_{test} is a separate, non-overlapping dataset, consisting of 800 molecules with the lowest values of penalized logP in ZINC-250K.

All networks are trained using the Adam optimizer [8] with learning rate 0.0001, batch normalization [6] and leaky-ReLU with $\alpha = 0.1$. The models are trained for 300 epochs. Generators are built of four fully connected residual layers, with 56 units. Discriminators are built of 7 dense layers of the following sizes: 48, 36, 28, 18, 12, 7, 1 units.

Table 1. Results of the constrained optimization for JT-VAE [7], Graph Convolutional Policy Network (GCPN) [18] and Mol-CycleGAN.

δ	JT-VAE		GCPN		Mol-CycleGAN	
	Improvement	Similarity	Improvement	Similarity	Improvement	Similarity
0	1.91 ± 2.04	0.28 ± 0.15	4.20 ± 1.28	0.32 ± 0.12	**8.30 ± 1.98**	0.16 ± 0.09
0.2	1.68 ± 1.85	0.33 ± 0.13	4.12 ± 1.19	0.34 ± 0.11	**5.79 ± 2.35**	0.30 ± 0.11
0.4	0.84 ± 1.45	0.51 ± 0.10	2.49 ± 1.30	0.47 ± 0.08	**2.89 ± 2.08**	0.52 ± 0.10
0.6	0.21 ± 0.75	0.69 ± 0.06	0.79 ± 0.63	0.68 ± 0.08	**1.22 ± 1.48**	0.69 ± 0.07

3 Results

We optimize the penalized logP under the constraint that the similarity between the original and the generated molecule is higher than a fixed threshold (denoted as δ). This is a realistic scenario in drug discovery, where the development of new drugs usually starts with known molecules such as existing drugs [2].

We maximize the penalized logP coefficient and use the Tanimoto similarity with the Morgan fingerprint to define the threshold of similarity. We compare our results with previous similar studies [7,18].

In our optimization procedure, each molecule is fed into the generator to obtain the 'optimized' molecule $G(x)$. The pair $(x, G(x))$ defines an 'optimization path' in the latent space of JT-VAE. To be able to make a comparison with the previous research [7] we start the procedure from the 800 molecules with the lowest values of penalized logP in ZINC-250K and then we decode molecules from 80 points along the path from x to $G(x)$ in equal steps. From the resulting set of molecules we report the molecule with the highest penalized logP score

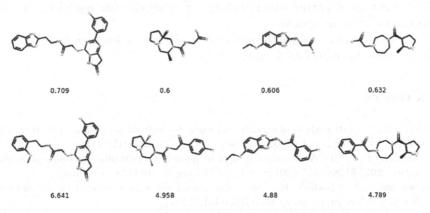

Fig. 1. Molecules with the highest improvement of the penalized logP for $\delta \geq 0.6$. In the top row we show the starting and in the bottom row the optimized molecules. Upper row numbers indicate Tanimoto similarities between the starting and the final molecule. The improvement in the score is given below the generated molecules.

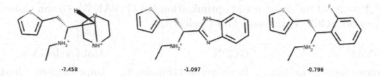

Fig. 2. Evolution of a selected exemplary molecule during constrained optimization. We only include the steps along the path where a change in the molecule is introduced. We show values of penalized logP below the molecules.

that satisfies the similarity constraint. In the task of optimizing penalized logP of *drug-like* molecules, our method significantly outperforms the previous results in the mean improvement of the property (Table 1) and achieves a comparable mean similarity in the constrained scenario (for $\delta > 0$).

Molecules with highest improvement of logP are presented in Fig. 1 with the improvement given below the generated molecules.

Figure 2 shows starting and final molecules, together with all molecules generated along the optimization path and their values of penalized logP.

4   Conclusions

In this work, we introduce Mol-CycleGAN – a new model based on CycleGAN which can be used for the *de novo* generation of molecules. The advantage of the proposed model is the ability to learn transformation rules from the *sets* of compounds with desired and undesired values of the considered property. The model can generate molecules with desired properties, as shown on the example of penalized logP. The generated molecules are close to the starting ones and the degree of similarity can be controlled via a hyperparameter. In the task of constrained optimization of drug-like molecules our model significantly outperforms previous results.

The code used to produce the reported results can be found online at https://github.com/ardigen/mol-cycle-gan.

References

1. Bajorath, J.: Integration of virtual and high-throughput screening. Nat. Rev. Drug Discov. **1**(11), 882–894 (2002). https://doi.org/10.1038/nrd941
2. Besnard, J., et al.: Automated design of ligands to polypharmacological profiles. Nature **492**(7428), 215 (2012). https://doi.org/10.1038/nature11691
3. Bjerrum, E.J., Threlfall, R.: Molecular generation with recurrent neural networks (RNNs). arXiv preprint. arXiv:1705.04612 (2017)
4. Chen, H., Engkvist, O., Wang, Y., Olivecrona, M., Blaschke, T.: The rise of deep learning in drug discovery. Drug Discov. Today **23**(6), 1241–1250 (2018). https://doi.org/10.1016/j.drudis.2018.01.039

5. Gupta, A., Müller, A.T., Huisman, B.J., Fuchs, J.A., Schneider, P., Schneider, G.: Generative recurrent networks for de novo drug design. Mol. Inform. **37**(1–2), 1700111 (2018). https://doi.org/10.1002/minf.201700111
6. Ioffe, S., Szegedy, C.: Batch normalization: accelerating deep network training by reducing internal covariate shift. In: Proceedings of the 32nd International Conference on Machine Learning, vol. 37, pp. 448–456. ICML 2015, JMLR.org (2015). http://dl.acm.org/citation.cfm?id=3045118.3045167
7. Jin, W., Barzilay, R., Jaakkola, T.: Junction tree variational autoencoder for molecular graph generation. In: Dy, J., Krause, A. (eds.) Proceedings of the 35th International Conference on Machine Learning. Proceedings of Machine Learning Research, vol. 80, pp. 2323–2332. PMLR, Stockholmsmässan, Stockholm (10–15 July 2018)
8. Kingma, D.P., Ba, J.: Adam: a method for stochastic optimization. arXiv preprint. arXiv:1412.6980 (2014)
9. Kingma, D.P., Welling, M.: Auto-encoding variational bayes. arXiv preprint. arXiv:1312.6114 (2013)
10. Kusner, M.J., Paige, B., Hernández-Lobato, J.M.: Grammar variational autoencoder. In: Proceedings of the 34th International Conference on Machine Learning, vol. 70, pp. 1945–1954. JMLR.org (2017)
11. Mao, X., Li, Q., Xie, H., Lau, R.Y., Wang, Z., Paul Smolley, S.: Least squares generative adversarial networks. In: 2017 IEEE International Conference on Computer Vision (ICCV), pp. 2794–2802 (2017). https://doi.org/10.1109/ICCV.2017.304
12. Rao, V.S., Srinivas, K.: Modern drug discovery process: an in silico approach. J. Bioinform. Seq. Anal. **2**(5), 89–94 (2011)
13. Ratti, E., Trist, D.: The continuing evolution of the drug discovery process in the pharmaceutical industry. Farmaco **56**(1–2), 13–19 (2001). https://doi.org/10.1016/S0014-827X(01)01019-9
14. Rogers, D., Hahn, M.: Extended-connectivity fingerprints. J. Chem. Inf. Model. **50**(5), 742–754 (2010). https://doi.org/10.1021/ci100050t
15. Segler, M.H., Kogej, T., Tyrchan, C., Waller, M.P.: Generating focused molecule libraries for drug discovery with recurrent neural networks. ACS Cent. Sci. **4**(1), 120–131 (2017). https://doi.org/10.1021/acscentsci.7b00512
16. Sterling, T., Irwin, J.J.: Zinc 15-ligand discovery for everyone. J. Chem. Inf. Model. **55**(11), 2324–2337 (2015). https://doi.org/10.1021/acs.jcim.5b00559
17. Winter, R., Montanari, F., Noé, F., Clevert, D.A.: Learning continuous and data-driven molecular descriptors by translating equivalent chemical representations. Chem. Sci. **10**(6), 1692–1701 (2019). https://doi.org/10.1039/C8SC04175J
18. You, J., Liu, B., Ying, Z., Pande, V., Leskovec, J.: Graph convolutional policy network for goal-directed molecular graph generation. In: Advances in Neural Information Processing Systems, pp. 6410–6421 (2018)
19. Zhu, J.Y., Park, T., Isola, P., Efros, A.A.: Unpaired image-to-image translation using cycle-consistent adversarial networks. In: Proceedings of the IEEE International Conference on Computer Vision, pp. 2223–2232 (2017)

816   Ł. Maziarka et al.

Open Access This chapter is licensed under the terms of the Creative Commons
Attribution 4.0 International License (http://creativecommons.org/licenses/by/4.0/),
which permits use, sharing, adaptation, distribution and reproduction in any medium
or format, as long as you give appropriate credit to the original author(s) and the
source, provide a link to the Creative Commons license and indicate if changes were
made.

The images or other third party material in this chapter are included in the chapter's
Creative Commons license, unless indicated otherwise in a credit line to the material. If
material is not included in the chapter's Creative Commons license and your intended
use is not permitted by statutory regulation or exceeds the permitted use, you will
need to obtain permission directly from the copyright holder.

A Transformer Model for Retrosynthesis

Pavel Karpov[1,3](✉) ⓘ, Guillaume Godin[2] ⓘ, and Igor V. Tetko[1,3] ⓘ

[1] Helmholtz Zentrum München – Research Center
for Environmental Health (GmbH), Institute of Structural Biology,
Ingolstädter Landstraße 1, 85764 Neuherberg, Germany
pavel.karpov@helmholtz-muenchen.de
[2] Research and Development Division, Firmenich International SA,
Route des Jeunes 1, 1227 Les Acacias, Switzerland
guillaume.godin@firmenich.com
[3] BigChem GmbH, Ingolstädter Landstraße 1, b. 60w, 85764 Neuherberg, Germany
itetko@bigchem.de

Abstract. We describe a Transformer model for a retrosynthetic reaction prediction task. The model is trained on 45 033 experimental reaction examples extracted from USA patents. It can successfully predict the reactants set for 42.7% of cases on the external test set. During the training procedure, we applied different learning rate schedules and snapshot learning. These techniques can prevent overfitting and thus can be a reason to get rid of internal validation dataset that is advantageous for deep models with millions of parameters. We thoroughly investigated different approaches to train Transformer models and found that snapshot learning with averaging weights on learning rates minima works best. While decoding the model output probabilities there is a strong influence of the temperature that improves at $T = 1.3$ the accuracy of models up to 1–2%.

Keywords: Retrosynthesis prediction ·
Computer aided synthesis planning · Character-based models ·
Transformer

1 Introduction

New chemical compounds drive technological advances in material, agricultural, environmental, and medical sciences, thus, embracing all fields of scientific activities which have been bringing social and economic benefits throughout human history. Design of chemicals with predefined properties is an arena of QSAR/QSPR (Quantitative Structure Activity/Property Relationships) approaches aimed at finding correlations between molecular structures and their desired outcomes and then applying these models to optimise activity/property of compounds.

The advent of deep learning [3,5] gave a new impulse for virtual modeling and also opened a venue for a promising set of generative methods based on Recurrent

ⓒ The Author(s) 2019
I. V. Tetko et al. (Eds.): ICANN 2019, LNCS 11731, pp. 817–830, 2019.
https://doi.org/10.1007/978-3-030-30493-5_78

Neural Networks [10], Variational Autoencoders [13], and Generative Adversarial Networks trained with reinforcement learning [14,23]. These techniques are changing the course of QSAR studies from the observation to the invention: from a virtual screening of available compounds to direct synthesis of new candidates. Generative models can produce big sets of promising molecules and impaired with SMILES-based QSAR methods [18] provide a strong foundation for creating highly optimized focussed libraries, but estimation of synthetic availability of these compounds is an open question though several approaches based on fragmentation [11] and machine learning [7] approaches have been developed. To synthesize a molecule, one should have a plan of a multi-step synthesis and also a set of available reactants. Finding an optimal combination of reactants, reactions, and conditions to obtain the compound with good yield, sufficient quality, and quantity is not a trivial task even for experts in organic chemistry. Recent advances in the computer-aided synthesis planning are reviewed in [2,6,9].

The retrosynthetic analysis worked out by Corey [8] tries to account for all factors while deriving the synthetic route. It iteratively decomposes the molecule on simpler blocks till all of them become available either by purchase or by synthesis described in the literature. At each step, Fig. 1, all possible disconnections (rules) with known reactions simplify the target molecule bringing to the scene less complex compounds. Some of them may be already available, while the others undergo the next step of retrosynthesis decomposition. Due to the recursive nature of the procedure, it can deal with thousands of putative compounds so computational retrosynthetic approaches can greatly help chemists in finding the best routes. Managing of the database of such rules is complicated and more critical the models based on it are not ready to accommodate new reactions and will always be outdated. Unfortunately, almost more than 60 years of developing rule-based systems ended with no remarkable success in synthesis planning programs [28]. Another approach to tackle the problem is to use so-called template-free methods inspired by the success of machine-translation. They don't require the database of templates and rules due to an inherent possibility to derive this information during training directly from a database of organic reactions with clearly designated roles of reactants, products, reagents, and conditions.

The analogy between machine translation and retrosynthesis is evident: each target molecule has its predecessors from which it can be synthesized as every meaningful sentence one can translate from source language to target one. If all parts of a reaction are written in SMILES notation, then our source and target sentence are composed of valid SMILES tokens as words. The main goal of the work is to build a model which could for a given target molecule for our example[1] COC(=O)c1cccc(−c2nc3cccnc3[nH]2)c1 in Fig. 1 correctly predict the set of reactants. Namely, it should predict Nc1cccnc1N.COC(=O)c1cccc(C(=O)O)c1 in this case.

Neural sequence-to-sequence (seq2seq) approach has been recently applied for a direct reaction prediction task [26,27] with outstanding statistical parameters of final models – 90.4% of accuracy on test set. Seq2seq modeling has been also

---

[1] This reaction is in the test set and it was correctly predicted by our model.

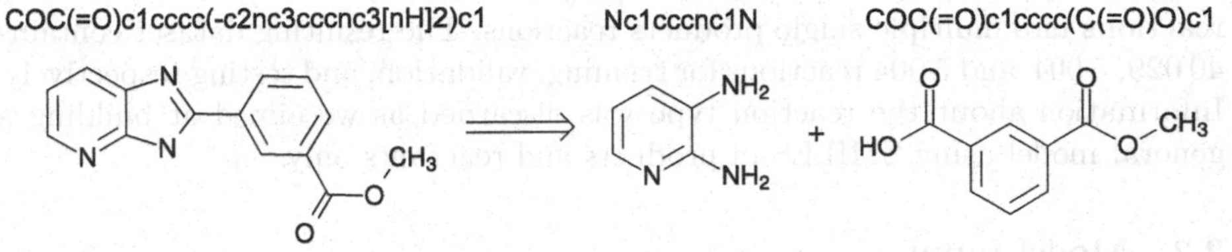

COC(=O)c1cccc(-c2nc3cccnc3[nH]2)c1   Nc1cccnc1N   COC(=O)c1cccc(C(=O)O)c1

Fig. 1. An example of a retrosynthetic reaction: on the left side of the arrow the target molecule is depicted, and on the right side the one possible set of reactants that can lead to the target is shown in common chemistry-like scheme and using SMILES notation. Here two successive amidation reactions result in cyclisation and aromatization.

tested on retrosynthesis task [21], but due to the complex nature of retrosynthesis itself and difficulty in estimating the correct predictions of reactants[2], accuracy on the test set was moderate 37.4% but still comparable to rule-based systems 35.4%. We questioned about the possibility of improvement models for one-step retrosynthesis utilizing modern neural network architectures and training techniques. Applying the Transformer Model [29], together with cyclical learning rate schedule [24], resulted in a model with accuracy 42.7%, that is >5% higher compare to the baseline model [21].

Our main contributions are:

- We show that Transformer can be efficiently used for a retrosynthesis prediction task.
- We show that for this particular task there is no advantage to use a validation dataset for early-stopping or other parameters optimization. We trained all the parameters directly from the training dataset.
- Applying weights averaging and snapshot learning helped to train the most precise model for one-step retrosynthesis prediction. We averaged weights on 5 successive cycles of learning rate schedule.
- Increasing the temperature while performing a beam-search procedure improves the accuracy up to 2%.

2 Approach

2.1 Dataset

In this study we used the same dataset of reactions as in [21]. This dataset was filtered from the USPTO database [22] originally derived from the USA patents and contains 50 000 reactions classified into 10 reaction types [25]. The authors [21] further preprocessed the database by splitting multiple products

---

[2] A target molecule usually can be synthesized with different reactions starting from different sets of reactants. The predictions of the model may be correct from organic chemist point of view but differ from the reactant set in ground truth. This may lead to underestimation of effectiveness of models.

reactions into multiple single products reactions. The resulting dataset contains 40 029, 5 004, and 5 004 reactions for training, validation, and testing respectively. Information about the reaction type was discarded as we aimed at building a general model using SMILES of products and reactants only.

2.2   Model Input

The seq2seq models were developed to support machine translation where the input is a sentence in one language, and the output is a sentence with approximately the same meaning but in another language. String nature of data implies some tokenization procedures similar to word2vec to be used for preprocessing the input. Most of works in cheminformatics dealing with SMILES tokenize the input with a regexp equal or similar to [26].

```
token_regex= "(\[[^\]]+]|Br?|Cl?|N|O|S|P|F|I|b|c|n|o|s|p|\(|\)|
\.|=|#|-|\+|\\\\|\/|:|~|@|\?|>|\*|\$|\%[0-9]{2}|[0-9])".
```

Though such tokenization is more similar to way chemists think, it also has some drawbacks that confuse network by putting forward low represented molecular parts. For example, after applying this regexp to the database one can see some not frequent moieties such as [C@@], [C@@H], [S@@], [C@], [C@H], [N@@+], [se], [C−], [Cl+3]. The thing in brackets according to SMILES specification can be quite a complex gathering not only the element's name itself, but also its isotopic value, stereochemistry configuration, the formal charge, and the number of hydrogens[3]. Strictly speaking, to do tokenization right one should also parse the content of brackets just increasing the number of possible words in the vocabulary what eventually leads to the simplest tokenization only with letters. We tried different schemes of tokenization in this work but did not see any improvements in using them over simple character-based method.

Our final vocabulary has length of 66 symbols[4]:

```
chars = "  ^#%()+-.0123456789=@ABCDEFGHIKLMNOPRSTVXYZ[\\]
         abcdefgilmnoprstuy$"
```

To convert a token to a dense vector we used a trainable embedding[5] of size 64. It is well known that training neural networks in batches is more stable, faster, and leads to more accurate models. To facilitate batch training we also used masks of input strings of shape (batch_size, max_length) with elements equal to 1 for those positions where are valid SMILES symbols and 0 everywhere else.

---

[3] We do not want to exclude stereochemistry information from our model as well as charges and explicit hydrogens that will lead to reducing of the dataset. Moreover, work in generative models showed excellent abilities of models to close cycles, for example, c1cc(COC)cccc1. If the model can capture such a long distance relation why should it be cracked on much simplier substrings enclosed by brackets?

[4] This vocabulary derived from the complete USPTO set and is a little bit wider than needed for this study. But for future extending of the models it is better to fix the input shape to the biggest possible value.

[5] The encoder and the decoder share embeddings in this study.

2.3   Transformer Model

We used a promising Transformer [29] model for this study which is a new generation of encoder-decoder neural networks family. The architecture is suited for exploration of the internal representation of data by deriving questions (Q) the data could be asked for, keys for its indexed knowledge (K), and answers written as values (V) corresponding to queries and keys. Technically these three entities are simply matrixes learned during the network training. Multiplying them with the input (X) gives keys (k), questions (q), and values (v) relevant to a current batch. Equipped with these calculated parameters of the input the self-attention layers transforms it pointing out to some encoding (decoding) parts based on the attention vector.

The Transformer has wholly got rid of any recurrences or convolutional operations. To tackle distances between elements of a string a positional encoding matrix was proposed with elements equal to the values of trigonometric functions depending on the position in a string and also the position in the embedding direction. Summed with learned embeddings positional encodings do their job linking far located parts of the input together. The output of self-attention layers is then mixed with original data, layer-wise normalized, and passed position-wise through a couple of ordinary dense layers to go further either in next level of self-attention layers or to a decoder as an information-rich vector representing the input. The decoder part of Transformer resembles the encoder but has an additional self-attention layer which corresponds to encoder's output.

Transformer model shows the state-of-the-art results in machine translation and reaction prediction outcomes [27]. The latter work showed that training the Transformer on large and noisy datasets results in a model that can outperform not only other machine models but also well qualified and experienced organic chemists.

2.4   Model Inference

The model estimates the probability of the next symbol over the model's vocabulary given all previous symbols in the string. Technically, the Transformer model first calculates logits, z_i, and then transforms them to probabilities.

$$z_i = Transformer(\{x_1, x_2, x_3, ..., x_L\}, \{y_1, y_2, y_3, ..., y_{i-1}\}) \tag{1}$$

Here x_i is the input of the models at i position; L – the length of the input string; y_i is the decoded output of the model up to position $(i-1)$; and z_i – logits that are to be converted to probabilities:

$$q_i = \frac{exp(z_i/T)}{\sum_{j=0}^{V} exp(z_j/T)} \tag{2}$$

where V is the size of the vocabulary (66 in this work) and T stands for the temperature[6] usually assigned to 1.0 in standard softmax layers. With higher T

---

[6] Similar to formula of Boltzmann (Gibbs) distribution used in statistical mechanics.

the landscape of the probability distribution becomes more smooth. During the training the model adapts its weights to better predict q_i, so $y_i = q_i$.

During the inference however we have several possibilities how to convert q_i into y_i, namely greedy and beam search. The first one picks up a symbol with maximum probability whereas the second one at each step holds $top - K$ (K = beam's size) suggestions of the model and summarises the overall likelihood for each of K final decodings. The beam search allows better inference and the probability landscape exploration compared to the greedy search because at a particular step of decoding it may choose a symbol with less than maximum probability, but the total likelihood of the result can be higher due to more significant probabilities on the next steps.

2.5   Training Heuristics

Training a Transformer model is a challenge, and several heuristics have been proposed [24], some of them were used in this study:

Using as Bigger Batch Size as Possible. Due to our hardware limitations we could not set the batch size more then 64^7;

Increasing the learning rate at the beginning of training up to warmup steps[8]. The authors of the original Transformer paper [29] used 4 000 steps for warming. The Transformer model for reaction prediction task from [27] used 8 000 steps. We analysed different values for warmup and eventually found that 16 000 works well with our model.

Applying Cyclic Learning Rate Schedules. This tips can generally improve any model [17] through better loss landscape exploration with bigger learning rates after the optimiser fell down to some local minima. For this study we used the following scheme for learning rate calculation depending on the step:

$$u(step) = \begin{cases} warmup + (step \bmod cycle), & \text{if } step \geq cycle \\ step, & \text{otherwise} \end{cases}$$

where *cycle* stands for the number of steps while the learning rate is decreasing before raising to the maximum again.

$$\lambda(step) = factor * \frac{min(1.0, u(step)/warmup)}{max(u(step), warmup)} \tag{3}$$

where *factor* is just a constant. Big values of *factor* introduce numerical instability during training, so after several trials we set *factor* = 20.0. The curve for learning rate in this study is shown in Fig. 2, plot (4, f).

---

[7] Our first implementation of the model required a lot of memory to deal with masks of reactants and products. Though later we improved the code we still remained this size for consistency of the results.

[8] In our implementation 1 step is equivalent to 1 batch. The number of reactions for training is 40 029 + 5 004, so one epoch is equal to 704 batches.

Averaging weights during last steps (usually 10–20) of training or at minima of learning rates in case of snapshop learning [16]. Also with cyclic learning rate schedules it is possible to average weights of those models that have minimum in losses just before increasing of the rate. Such approach leads to more stable and plain region in loss landscapes [17].

3 Results

3.1 Learning Details and Evaluation of Models for Retrosynthesis Prediction

For this study, we implemented The Transformer model in Tensorflow [1] library to support its integration in our in-house programs set (https://github.com/bigchem/retrosynthesis). All values reported are averages for three repetitive runs. Preliminary modeling showed that the architecture with 3 layers and 8 attention heads works well for the datasets, though we tried combinations of 2, 3, 4, 5 layers with 6, 8, 10, 12 heads. So all calculations were performed with these values fixed. The number of learnable parameters of the model is 1 882 176, embedding layer common for product and reactants has size 64.

Following the standard machine learning protocol, we trained our first models (T1) using three datasets for training, validation, and external testing (8:1:1) as was done in [21]. Learning curves for T1 are depicted in Fig. 2, (c) and (d) for training and validation loss, respectively, (a) shows the original learning rate schedule developed by the authors of the Transformer but with 16 000 warmup steps. On reaching cross-entropy loss about 0.1 on the validation dataset, it stagnates without noticeable fluctuations as training loss steadily decreases. After warming up phase the learning rate begins fading and eventually after 1 000 epochs its value reaches $2.8 * 10^{-5}$ inevitable causing to stop training because of too small updates.

During the decoding procedure, we explored the influence of the temperature parameter on the final quality of prediction and found that inferring at higher temperatures gives better result then at $T = 1$. This observation similarly repeated for all our models. Figure 3 shows the influence of this parameter on the reactants prediction of the part of the training set. Clearly, at $T = 1.3$ the model reaches the maximum of chemically-based accuracy. This fact one can explain that at higher temperatures the landscape of output probabilities of the model is softer letting the beam-search procedure to find more suitable ways during decoding. Of course, the temperature influences only relative distances between peaks, so it does not affect the greedy search method.

If we applied the early stopping technique, the training of a model is stopped around 200 epoch[9]. Effectiveness of such a model marked $T1_1$ in Table 1 resulted in TOP-1 37.9% on the test set. If we chose the last one model obtained at 1 000 epoch, then the model $T1_2$ gave us better value – 39.8%. In this case, we did not see any need of the validation dataset and keeping in mind that our model has

---

[9] Though we trained our models for 1 000 epochs we also saved their weights after each epoch and for imitating early stopping technique selected those weights that correspond to minimum in validation loss function.

almost 2 millions of parameters we decided to combine training and validation sets and train our next models on both data, e.g., without validation. The model T2 was trained on all data and with the same learning rate schedule as T1. The results obtained when applying T2 to the test set are better than for T1 model namely 41.8% vs. 39.8%, respectively.

Then we trained our model with cyclic learning rate schedule, Eq. 3, Fig. 2 (b) for better exploration of loss landscape. During training, we also saved the character-based accuracy of the model, Fig. 2, (f). This snapshot training regime [16] produces a set of different weights at each minimum of learning rate. Averaging them is to some extent equivalent to a consensus of models but within one model [17]. We tried different averaging regimes for T3 and found that averaging five last cycles gives better results.

Our final T3 model outperforms [21] by 5.3% with beam search and more critical it is also effective with greedy search 40.6%. The latter one is much faster and consequently more suitable for virtual screening campaigns.

It worth to notice that TOP-5 accuracy reaches almost 70%. That means the model can correctly predict reactants but sometimes scoring is wrong and

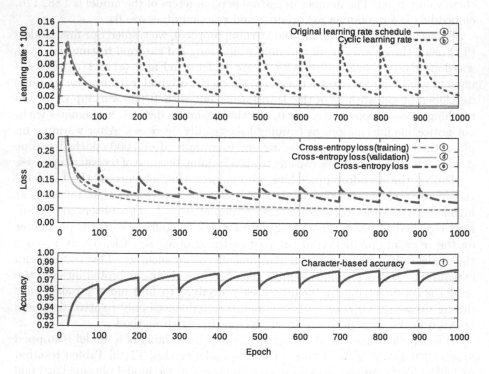

Fig. 2. Summary of learning curves for the Transformer model: (a) original learning rate schedule with warmup; (b) cyclic learning rate with warmup; (c) cross-entropy loss for training and (d) validation; (e) cross-entropy loss and (f) character-based accuracy for training a model wit cyclic learning schedule.

TOP-1 is much less. We tried to improve TOP-1 scoring with internal confidence estimation.

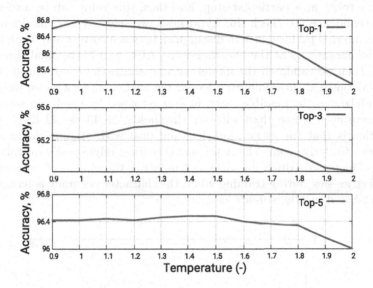

Fig. 3. Dependence of the beam search on temperature. For better exploration, higher temperatures are more useful. In this study we explored T = 1.3. Bigger values significantly worse for Top-3, and approximately the same for Top-1 and Top-5. This curve was derived from the training dataset.

Table 1. Accuracy (%) of the models on test set when all reactants were correctly predicted.

Model	Greedy	Top-1	Top-3	Top-5	Description
Seq2Seq		37.4	52.4	57.0	Literature result from [21] based on Seq2Seq architecture
T1	34.4	37.9	57.3	62.7	Transformer Model trained with validation control set (early stopping, ˜200 epochs)
T1$_1$	37.3	39.8	59.1	63.9	The same as T1, but without early stopping (1000 epochs)
T2	39.3	41.8	61.3	67.2	Transformer Model trained on both training and validation sets for 1000 epochs
T3	**40.6**	**42.7**	**63.9**	**69.8**	Transformer Model trained with cyclic learning rate schedule for 1000 epochs. Averaging cycles 6, 7, 8, 9, and 10

3.2 Internal Scoring

The beam search calculates the sum of negative logarithms of probabilities of selecting a token at a particular step, and thus, this value can be a measure of internal confidence. To check this hypothesis, we selected T3-2 model and estimated its internal performance to distinguish between correct and invalid predictions. The parameters of the classifier were: AUC = 0.77, optimal threshold = 0.00678. Then we validated the model with an additional condition: if the score is less than optimal threshold we selected the answer, otherwise we went to the next candidate in the possible reactant sets returned by the beam search. The results were even worse than without thresholds, 28.45 vs. 42.42. A possible explanation is that the estimation does not deal with organic chemistry. The model tries to derive some character-based scoring relying only on tokens in a string and increasing this value does not influence the quality of prognosis. The same effect we saw during training when the character accuracy is 98% whereas chemistry-based metric is much lower.

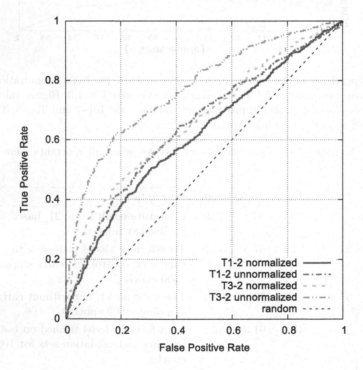

Fig. 4. Internal classification performance.

Estimation of optimal thresholds on training sets almost always a bad idea due to the biasing of a model to its source data. The correct way is to use validation dataset instead. We built the classifier for the T1-2 with characteristics: AUC = 0.65, optimal threshold 0.00396, and applied it for testing the model.

The results were again worse, 14.1% vs. 40.85%. There are no significant differencies of accuracies when using unnormalized or normalized on the length of the reactants string scores. Figure 4 shows ROC curves for T1-2 and T3-2 models derived at T = 1.3. Evidently one cannot use this estimation to improve TOP-1 scoring.

4 Discussion

Much attention paid in the scientific literature for rule-based approaches [4, 28]. Since the authors of [20] have described the algorithm of automatic rule extraction from mapped reaction database several implementations of the procedure appeared, and then widely accepted by researchers. However, it should be noticed that, first, there is no algorithm to make atom-mapping [2] if it is absent (the typical situation with laboratory notebooks (ELN) for example). Second, all available information on synthesis usually contains only positive reactions, so all binary classification accuracies are inevitable overestimated because of artificial negative sets exploited in studies. Finally, the absence of commonly accepted dataset for testing makes the results of different groups practically disparate and biased to those problems the authors tried to solve. The authors of [4] selected 40 molecules from DrugBank database to test their multiscale models, whereas [21] used database specially prepared for classification [25].

Our model can correctly predict reactant set in TOP-5 with accuracy 69.8%. Internal confidence estimation cannot guarantee a correct ordering of reactants sets, so different scoring methods should be developed. One of the promising ways is to use a forward reaction prediction model to estimate whether it is possible to assemble a target molecule from reactants proposed. The scoring model should have excellent characteristics and probably it is possible to apply the same cycling learning rate and snapshot averaging to build it.

First work on applying reinforcement learning for the whole retrosynthetic path [28] showed superior performance compared to the rule-based methods developed before. More important if can deal with several steps of synthesis. But the policy learned during the training again used extracted rules limiting the method. Thus, the development of models for direct estimation of reactants is still of prime importance. During the encoding process, the Transformer finds an internal representation of a reaction which can be useful for multicomponent QSAR [19] for predicting rate constants [12] and yields of reactions. Embedding such systems in policy networks within reinforcement learning paradigm can bring forward an entirely data-driven approach to solve challenging organic synthesis problems.

5 Conclusions

We have described a Transformer model for retrosynthesis one-step prediction task. Our final model trained with cyclic learning rate schedule and its weights were averaged during last five loss minimum. The model outperforms the previous published retrosynthetic character-based model by 5.3%. It also does not require the extraction of specific rules, atom mappings, and reaction types in reaction dataset. We believe it is possible to improve the model further applying knowledge distillation method [15] for example. The current model can be used as a building block for reinforcement learning aimed at solving complex organic problems.

All source code and also models built are available online via github

https://github.com/bigchem/retrosynthesis

Acknowledgments. This study has been partially supported by ERA-CVD (https://era-cvd.eu) "Cardio-Oncology" project, BMBF 01KL1710 and by European Union's Horizon 2020 research and innovation program under the Marie Skłodowska-Curie grant agreement No. 676434, "Big Data in Chemistry" ("BIGCHEM", http://bigchem.eu). The authors thank NVIDIA Corporation for donating Quadro P6000 and Titan Xp and V graphics cards for this research.

References

1. Abadi, M., et al.: TensorFlow: large-scale machine learning on heterogeneous systems (2015). https://www.tensorflow.org/
2. Baskin, I.I., Madzhidov, T.I., Antipin, I.S., Varnek, A.A.: Artificial intelligence in synthetic chemistry: achievements and prospects. Russ. Chem. Rev. **86**(11), 1127–1156 (2017). https://doi.org/10.1070/RCR4746
3. Baskin, I.I., Winkler, D., Tetko, I.V.: A renaissance of neural networks in drug discovery. Expert Opin. Drug Discov. **11**(8), 785–795 (2016). https://doi.org/10.1080/17460441.2016.1201262
4. Baylon, J.L., Cilfone, N.A., Gulcher, J.R., Chittenden, T.W.: Enhancing retrosynthetic reaction prediction with deep learning using multiscale reaction classification. J. Chem. Inf. Model. **59**(2), 673–688 (2019). https://doi.org/10.1021/acs.jcim.8b00801
5. Chen, H., Engkvist, O., Wang, Y., Olivecrona, M., Blaschke, T.: The rise of deep learning in drug discovery. Drug Discov. Today **23**(6), 1241–1250 (2018). https://doi.org/10.1016/j.drudis.2018.01.039
6. Coley, C.W., Green, W.H., Jensen, K.F.: Machine learning in computer-aided synthesis planning. Acc. Chem. Res. **51**(5), 1281–1289 (2018). https://doi.org/10.1021/acs.accounts.8b00087
7. Coley, C.W., Rogers, L., Green, W.H., Jensen, K.F.: SCScore: synthetic complexity learned from a reaction corpus. J. Chem. Inf. Model. **58**(2), 252–261 (2018). https://doi.org/10.1021/acs.jcim.7b00622
8. Corey, E.J., Cheng, X.M.: The Logic of Chemical Synthesis. Wiley, Hoboken (1995)
9. Engkvist, O., et al.: Computational prediction of chemical reactions: current status and outlook. Drug Discov. Today **23**(6), 1203–1218 (2018). https://doi.org/10.1016/j.drudis.2018.02.014
10. Ertl, P., Lewis, R., Martin, E., Polyakov, V.: In silico generation of novel, drug-like chemical matter using the LSTM neural network. arXiv (2017). arXiv:1712.07449
11. Ertl, P., Schuffenhauer, A.: Estimation of synthetic accessibility score of drug-like molecules based on molecular complexity and fragment contributions. J. Cheminform. **1**(1), 8 (2009). https://doi.org/10.1186/1758-2946-1-8

12. Gimadiev, T., et al.: Bimolecular nucleophilic substitution reactions: predictive models for rate constants and molecular reaction pairs analysis. Mol. Inform. **37**, 1800104 (2018). https://doi.org/10.1002/minf.201800104

13. Gómez-Bombarelli, R., et al.: Automatic chemical design using a data-driven continuous representation of molecules. ACS Cent. Sci. **4**(2), 268–276 (2018). https://doi.org/10.1021/acscentsci.7b00572

14. Guimaraes, G.L., Sanchez-Lengeling, B., Outeiral, C., Farias, P.L.C., Aspuru-Guzik, A.: Objective-reinforced generative adversarial networks (ORGAN) for sequence generation models. arXiv (2017). arXiv:1705.10843

15. Hinton, G., Vinyals, O., Dean, J.: Distilling the knowledge in a neural network. arXiv (2015). arXiv:1503.02531

16. Huang, G., Li, Y., Pleiss, G., Liu, Z., Hopcroft, J.E., Weinberger, K.Q.: Snapshot ensembles: train 1, get M for free. arXiv (2017). arXiv:1704.00109

17. Izmailov, P., Podoprikhin, D., Garipov, T., Vetrov, D., Wilson, A.G.: Averaging weights leads to wider optima and better generalization. arXiv (2018). arXiv:1803.05407

18. Kimber, T.B., Engelke, S., Tetko, I.V., Bruno, E., Godin, G.: Synergy effect between convolutional neural networks and the multiplicity of SMILES for improvement of molecular prediction. arXiv (2018). arXiv:1812.04439

19. Kravtsov, A.A., Karpov, P.V., Baskin, I.I., Palyulin, V.A., Zefirov, N.S.: Prediction of rate constants of SN2 reactions by the multicomponent QSPR method. Dokl. Chem. **440**(2), 299–301 (2011). https://doi.org/10.1134/S0012500811100107

20. Law, J., et al.: Route designer: a retrosynthetic analysis tool utilizing automated retrosynthetic rule generation. J. Chem. Inf. Model. **49**(3), 593–602 (2009). https://doi.org/10.1021/ci800228y

21. Liu, B., et al.: Retrosynthetic reaction prediction using neural sequence-to-sequence models. ACS Cent. Sci. **3**(10), 1103–1113 (2017). https://doi.org/10.1021/acscentsci.7b00303

22. Lowe, D.M.: Extraction of chemical structures and reactions from the literature. Ph.D. thesis, Pembroke College (2012). https://www.repository.cam.ac.uk/handle/1810/244727

23. Olivecrona, M., Blaschke, T., hongming Chen, O.E.: Molecular de-novo design through deep reinforcement learning. J Cheminform. **9**(48), 1758–2946 (2017). https://doi.org/10.1186/s13321-017-0235-x

24. Popel, M., Bojar, O.: Training tips for the transformer model. arXiv (2018). https://doi.org/10.2478/pralin-2018-0002

25. Schneider, N., Stiefl, N., Landrum, G.A.: What's what: the (nearly) definitive guide to reaction role assignment. J. Chem. Inf. Model. **56**(12), 2336–2346 (2016). https://doi.org/10.1021/acs.jcim.6b00564

26. Schwaller, P., Gaudin, T., Lanyi, D., Bekas, C., Laino, T.: Found in translation: predicting outcomes of complex organic chemistry reactions using neural sequence-to-sequence models. arXiv (2018). arXiv:1711.04810

27. Schwaller, P., Laino, T., Gaudin, T., Bolgar, P., Bekas, C., Lee, A.A.: Molecular transformer for chemical reaction prediction and uncertainty estimation. arXiv (2018). arXiv:1811.02633

28. Segler, M.H., Preuss, M., Waller, M.P.: Planning chemical synthesis with deep neural networks and symbolic AI. Nature **555**, 604–610 (2018). https://doi.org/10.1038/nature25978

29. Vaswani, A., et al.: Attention is all you need. arXiv (2017). arXiv:1706.03762

830   P. Karpov et al.

Open Access This chapter is licensed under the terms of the Creative Commons Attribution 4.0 International License (http://creativecommons.org/licenses/by/4.0/), which permits use, sharing, adaptation, distribution and reproduction in any medium or format, as long as you give appropriate credit to the original author(s) and the source, provide a link to the Creative Commons license and indicate if changes were made.

The images or other third party material in this chapter are included in the chapter's Creative Commons license, unless indicated otherwise in a credit line to the material. If material is not included in the chapter's Creative Commons license and your intended use is not permitted by statutory regulation or exceeds the permitted use, you will need to obtain permission directly from the copyright holder.

Augmentation Is What You Need!

Igor V. Tetko[1(✉)] [iD], Pavel Karpov[1] [iD], Eric Bruno[2] [iD],
Talia B. Kimber[3] [iD], and Guillaume Godin[3] [iD]

[1] Institute of Structural Biology, Helmholtz Zentrum Muenchen -
German Research Center for Environmental Health (GmbH) and BIGCHEM
GmbH, Neuherberg, Germany
{i.tetko,pavel.karpov}@helmholtz-muenchen.de
[2] Expedia, Geneva, Switzerland
ebruno@expedia.com
[3] Research and Development Division, Firmenich International SA,
Geneva, Switzerland
talia.kimber@gmail.com, guillaume.godin@firmenich.com

Abstract. We investigate the effect of augmentation of SMILES to increase the performance of convolutional neural network models by extending the results of our previous study [1] to new methods and augmentation scenarios. We demonstrate that augmentation significantly increases performance and this effect is consistent across investigated methods. The convolutional neural network models developed with augmented data on average provided better performances compared to those developed using calculated molecular descriptors for both regression and classification tasks.

Keywords: Augmentation · Convolutional neural networks · Descriptor representation · QSAR · Chemoinformatics · Regression · Classification

1 Introduction

The renaissance of neural networks methods [2, 3] has brought a wide variety of neural network types including methods that are capable to analyze chemical structure represented as graphs or text (e.g., SMILES) in chemistry. These methods are particularly interesting since their internal representations, latent variables, in principle do not lose information about the chemical structures as compared to methods using chemical descriptors. These latent variables can be used to decode the chemical structures and address the problem of inverse Quantitative Structure Activity Relationship (QSAR) [4], which has been a challenge for chemoinformatics since first QSAR models. However, the practical question arises whether the prediction power of such methods is similar to those based on descriptors. In our previous study [1] we have demonstrated that the Convolutional Neural Fingerprint (CNF) which is based on ideas of text processing originally proposed by [5], provided similar performance to models developed using three descriptors sets. The high prediction accuracy of the CNF was achieved thanks to the augmentation technique, which was originally proposed in

© The Author(s) 2019
I. V. Tetko et al. (Eds.): ICANN 2019, LNCS 11731, pp. 831–835, 2019.
https://doi.org/10.1007/978-3-030-30493-5_79

computer vision and was recently introduced to QSAR studies [6]. It is worth mentioning that a related technique to enhance accuracy of QSAR models by considering symmetry of molecules was proposed more than 20 years ago [7]. Typically so called canonical SMILES produced according to some rule-defined enumeration of atoms, are used to train the model. The augmentation procedure generates a number of unique SMILES for the same molecule, e.g., by starting enumeration of atoms from a random atom and/or traversing molecular graph path in random order. Augmentation is employed during both training and inference steps: during training, augmentation increases the dataset size by providing for each structure a number n of distinct SMILES; during inference, the prediction for a given structure is averaged over predictions of m distinct SMILES of that structure.

The goal of this study was to clarify whether the performance of augmented models is similar to those developed using a large set of descriptors typically used in QSAR studies. We also include a result for TextCNN [8] which is a DeepChem implementation of the Char-CNN [5], but using a different architecture.

2 Methods

The CNF method from our previous study [1] as well as the TextCNN method as implemented in DeepChem [8] were used as convolutional methods. Associative Neural Networks (ASNN) [9], which is a shallow neural network was used as a traditional method to develop models using descriptors. Early stopping was used to prevent overfitting of neural networks [10] for all three analyzed methods.

Augmentation. The augmentation used for convolutional methods was generated with RDKit and included: (1) **no-augmentation** – canonical SMILES was used, (2) **off-line augmentation** with n = 10 SMILES generated before the neural network training and (3) **on-line augmentation** in which new SMILES were generated for each training epoch (only for CNF). The augmentation with n = 10 SMILES (which was selected based on results in [1]) was also applied during the prediction step and the average value was used as the final model prediction. It was shown that model performance decreased when model developed with canonical SMILES was used to predict augmented data [1]. Therefore for models developed with the first protocol only canonical SMILES were used during the prediction step.

Hyperparameter Optimization. The methods were used with their default parameters as available on the On-line Chemical Database and Modeling Environment (http://ochem.eu). For CNF we used the same parameters as in [1] with an exception of the convolutional filter, which was increased to 5. We tried to optimize neural network parameters, such as the number of neurons, architecture, activation function, etc. but did not see improvement as compared to the defaults. The full run of optimization required more than 12 h on six GPU cards (GeForce RTX 2070 and 1070, Quadro P6000, Titan Xp and V) and thus exhaustive investigation of all options was impossible. Since training with augmented data was about 10 times longer for each epoch, the number of epochs for on-line training was respectively increased 10 times to allow networks to use about the same number of training steps.

Descriptors. In total 16 sets of descriptors, namely ALogPS + OEstate, CDK2, ChemaxonDescriptors, Dragon7, Fragmentor, GSFrag, InductiveDescriptors, JPlogP, Mera + Mersy, PyDescriptor, QNPR, RDKIT, SIRMS, Spectrophores, StructuralAlerts and alvaDesc were used with their default settings. The descriptors are described on the OCHEM web site and were used in multiple previous studies. Many of the descriptors have their own hyper parameters, e.g., size of fragments for fragmental descriptors, which can be also optimized. We did not perform such optimization for this study but instead used default hyperparameters that were found to be optimal ones in the previous studies.

Model Validation. Five-fold cross-validation was used to test performance of all models.

Datasets. The same 9 regression and 9 classification sets from our previous study [1] were used.

Statistical Parameters. The regression models were compared using the coefficient of determination

$$r^2 = 1 - \sum (f_i - y_i)^2 \Big/ \sum (\bar{y} - y_i)^2$$

where $\bar{y}$ is the average value across all samples, while f_i and y_i are predicted and target values for sample i, respectively. The classification results were compared using Area Under the Curve (AUC).

3 Results

The augmentation dramatically improved results for CNF method but also contributed better models for TextCNN for both regression and classification datasets as shown on Figs. 1 and 2.

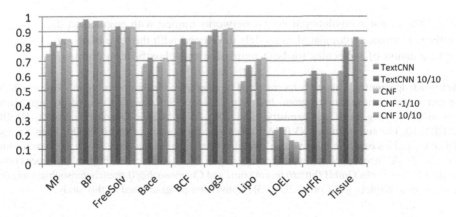

Fig. 1. Coefficient of determination, r^2, for regression tasks. With an exception of LOEL dataset, which had the lowest r^2, the training with augmentation improved the accuracy of models for all datasets. "10/10" and "−1/10" indicate off-line and on-line augmentations, respectively.

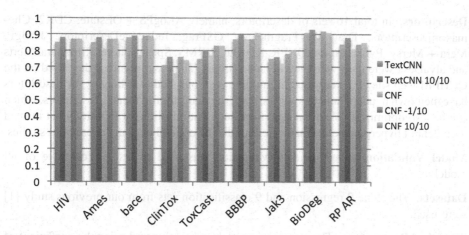

Fig. 2. For classification tasks the augmentation consistently improved AUC values for both convolutional neural networks.

For comparison of performances of models developed using descriptors and convolutional neural networks, we counted the number of models for which ASNN (using any descriptor set) or one of augmented convolutional models provided better results. Such comparison was biased towards ASNN since the best result for this method was selected from 16 models corresponding to the used descriptor sets versus only three (two off-line and one on-line) models for SMILES-based approaches. For three datasets the best models for both approaches had the same performance. For remaining data, the SMILES-based approaches contributed better models in 11 cases while descriptor-based approaches did it for 4 models.

4 Conclusions

We showed that convolutional neural networks trained with augmented data provide better performances compared to models developed with the state-of-the-art descriptor representation of molecules for both regression and classification problems.

Acknowledgements. This study has received funding from the European Union's Horizon 2020 research and innovation program under the Marie Skłodowska-Curie grant agreement No. 676434, "Big Data in Chemistry" and ERA-CVD "Cardio-Oncology" project, BMBF 01KL1710. The authors thank NVIDIA Corporation for donating Quadro P6000, Titan Xp and Titan V graphics cards for this research. We also thank ChemAxon (http://www.chemaxon.com) for Academic license of software tools as well as AlvaScience (http://alvascience.com), Molecular Networks GmbH (http://mn-am.com) and Chemosophia (http://chemosophia.com) for providing descriptors and Corina 2D to 3D conversion program used in this study.

References

1. Kimber, T.B., Engelke, S., Tetko, I.V., Bruno, E., Godin, G.: Synergy effect between convolutional neural networks and the multiplicity of SMILES for improvement of molecular prediction. eprint. arXiv:1812.04439 (2018)
2. Baskin, I.I., Winkler, D., Tetko, I.V.: A renaissance of neural networks in drug discovery. Expert Opin. Drug Discov. **11**(8), 785–795 (2016). https://doi.org/10.1080/17460441.2016.1201262
3. Chen, H., Engkvist, O., Wang, Y., Olivecrona, M., Blaschke, T.: The rise of deep learning in drug discovery. Drug Discov. Today **23**(6), 1241–1250 (2018). https://doi.org/10.1016/j.drudis.2018.01.039
4. Blaschke, T., Olivecrona, M., Engkvist, O., Bajorath, J., Chen, H.: Application of generative autoencoder in De Novo molecular design. Mol. Inform. **37**(1–2), 1700123 (2018). https://doi.org/10.1002/minf.201700123
5. Zhang, X., LeCun, Y.: Text understanding from scratch. eprint. arXiv:1502.01710 (2015)
6. Bjerrum, J.E.: SMILES enumeration as data augmentation for neural network modeling of molecules. eprint. arXiv:1703.07076 (2017)
7. Baskin, I.I., Halberstam, N.M., Mukhina, T.V., Palyulin, V.A., Zefirov, N.S.: The learned symmetry concept in revealing quantitative structure-activity relationships with artificial neural networks. SAR QSAR Environ. Res. **12**(4), 401–416 (2001). https://doi.org/10.1080/10629360108033247
8. Ramsundar, B., Eastman, P., Walters, P., Pande, V.: Deep Learning for the Life Sciences: Applying Deep Learning to Genomics, Microscopy, Drug Discovery, and More. O'Reilly Media, Newton (2019)
9. Tetko, I.V.: Associative neural network. Methods Mol. Biol. **458**, 185–202 (2008). https://doi.org/10.1007/978 1 60327 101-1_10
10. Tetko, I.V., Livingstone, D.J., Luik, A.I.: Neural network studies. 1. Comparison of overfitting and overtraining. J. Chem. Inf. Comput. Sci. **35**(5), 826–833 (1995). https://doi.org/10.1021/ci00027a006

Open Access This chapter is licensed under the terms of the Creative Commons Attribution 4.0 International License (http://creativecommons.org/licenses/by/4.0/), which permits use, sharing, adaptation, distribution and reproduction in any medium or format, as long as you give appropriate credit to the original author(s) and the source, provide a link to the Creative Commons license and indicate if changes were made.

The images or other third party material in this chapter are included in the chapter's Creative Commons license, unless indicated otherwise in a credit line to the material. If material is not included in the chapter's Creative Commons license and your intended use is not permitted by statutory regulation or exceeds the permitted use, you will need to obtain permission directly from the copyright holder.

Abstracts of the BIGCHEM Session

Diversify Libraries Using Generative Topographic Mapping

Arkadii Lin[1,2] (iD), Bernd Beck[2], Dragos Horvath[1] (iD),
and Alexandre Varnek[1(✉)] (iD)

[1] Laboratory of Chemoinformatics, Faculty of Chemistry,
University of Strasbourg, 4, Blaise Pascal str., 67008 Strasbourg, France
arkadiyl18@gmail.com
[2] Department of Medicinal Chemistry, Boehringer Ingelheim Pharma
GmbH & Co. KG, Birkendorferstrasse 65, 88397 Biberach an der Riss, Germany

Abstract. Generative Topographic Mapping (GTM) approach was used to investigate the possibility to structurally enrich the in-house collection of Boehringer Ingelheim (BI) with structures from the Aldrich-Market Select (AMS) database of purchasable compounds. For this purpose, a tool named "AutoZoom" was implemented which investigates the zones of a map and extracts the corresponding maximum common substructures. As a result, AutoZoom detected 45.5 K new interesting and diverse substructures within the AMS database compared to the BI collection. The corresponding compounds were retrieved and checked for drug-likeness and biological activity.

Keywords: Generative topographic mapping · Library enrichment · Big data

Structural library enrichment (diversity) is an important task for pharmaceutical industry. The number of hits found in screening or in virtual screening campaigns, like molecular docking and/or QSAR studies, depends on the quality and the diversity of the underlying screening set. To be efficient in drug-discovery, the existing screening pool has to be updated on a regular basis.

Recently, Generative Topographic Mapping (GTM) [1] method was shown as an effective tool for large chemical libraries comparison [2]. A new technique called *"Zooming"* (training a new GTM using the compounds extracted from the chosen area on the main map) was applied in order to increase the map resolution, and, in turn, to improve the ability of GTM to distinguish different chemotypes and to extract specific Maximum Common Substructures (MCS).

In this study, the *Zooming* technique was automatized and a new MCSs extraction protocol was implemented. The developed tool was applied to enrich the in-house collection of Boehringer Ingelheim (BI Pool) containing more than 1.7 M structures using publicly available Aldrich-Market Select (AMS) collection of purchasable compounds. AMS collection containing more than 8.2 M items was taken here as a source of potentially new chemotypes (http://www.aldrichmarketselect.com).

The data was standardized by ChemAxon's Standardizer tool, and ISIDA fragment descriptors were computed. The manifold has been constructed on a frame set containing 25.000 molecules selected from AMS. Then, the entire AMS and BI Pool

© Springer Nature Switzerland AG 2019
I. V. Tetko et al. (Eds.): ICANN 2019, LNCS 11731, pp. 839–841, 2019.
https://doi.org/10.1007/978-3-030-30493-5

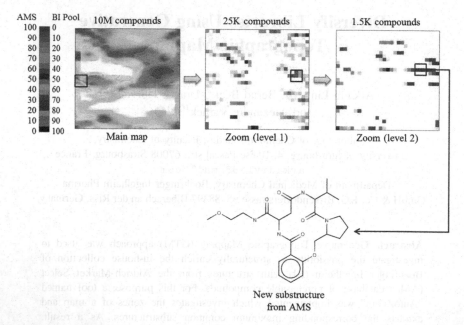

Fig. 1. An example of zooming analysis. Here, a new substructure from AMS collection was discovered using 2-levels zooming. The white space means non-populated areas, and the transparency corresponds to the density of population.

collections were projected onto the manifold. The zooming algorithm revealed 187 clusters, each representing a square agglomeration of 9 pixels. Out of these, 151 clusters were zoomed up to the 3^d level (see example of 2-levels zoom in Fig. 1).

Detailed analysis of the maps helped us to discover more than 45.5 K AMS substructures absent in BI Pool. They correspond to 402 K structures which were extracted from the AMS collection. These structures were checked for druglikeness using Lipinski rules and profiled against 749 ChEMBL targets using a virtual screening service on a web-server of the Laboratory of Chemoinformatics in the University of Strasbourg (http://infochim.u-strasbg.fr/webserv/VSEngine.html). MCS extracted for selected zones on the maps helped us to generalize the list of chemotypes required to the enrichment of the in-house collection.

Funding Sources

The project leading to this article has received funding from the European Union's Horizon 2020 research and innovation program under the Marie Skłodowska-Curie grant agreement No 676434, "Big Data in Chemistry" ("BIGCHEM", http://bigchem.eu).

Acknowledgment. The authors thank Boehringer Ingelheim Pharma GmbH & Co KG for the provided data.

References

1. Bishop, C.M., Svensén, M., Williams, C.K.: GTM: The generative topographic mapping. Neural Comput. **10**(1), 215–234 (1998)
2. Lin, A., Horvath, D., Afonina, V., Marcou, G., Reymond, J.L., Varnek, A.: Mapping of the available chemical space versus the chemical universe of lead-like compounds. ChemMedChem **13**(6), 540–554 (2018)

Detection of Frequent-Hitters Across Various HTS Technologies

Laurianne David[1,2](✉) ⓘ, Jarrod Walsh[3], Jürgen Bajorath[2] ⓘ,
and Ola Engkvist[1] ⓘ

[1] Hit Discovery, Discovery Sciences, R&D BioPharmaceuticals, AstraZeneca,
Gothenburg, Sweden
laurianne.david1@gmail.com

[2] Department of Life Science Informatics, B-IT, LIMES Program
Unit Chemical Biology and Medicinal Chemistry,
Rheinische Friedrich-Wilhelms-Universität Bonn, Bonn, Germany

[3] Hit Discovery, Discovery Sciences, R&D BioPharmaceuticals, AstraZeneca,
Cambridge, UK

Abstract. A major conundrum in High-Throughput Screening studies is the presence of frequent-hitters, which include non-selective compounds and molecules that are false positives in many screens. In our study, we introduce a method to detect frequent-hitters specific to an assay technology using historical compounds' structural information. Results from historic HTS campaigns, including artefact assays, are curated on a cooperate database. Structural fingerprints are generated for each compound and used to train a machine-learning model able to predict the behavior of novel compounds.

Keywords: Frequent-hitter · High-throughput screening ·
Machine-learning

1 Introduction

The emergence of the 'big data' concept is having an impact in the field of drug discovery. Computational methodologies are evolving to handle and process the volume of data generated by modern drug discovery endeavors. Advances in the fields of biology and chemistry have been aided by the development of High-Throughput Screening (HTS) [3] techniques. HTS is the process of testing a large number of chemical compounds against a biological target to identify 'hits' that will be investigated and optimized to become drugs. This process implies that activity data related to thousands of compounds can be generated in very little time, leading to two problems: (1) it is crucial to be able to investigate computationally the data generated, which requires an increase in computational power and an improvement in general analysis methods, (2) it is quite beneficial to select more efficiently promising compounds prior to their testing in HTS as this would save research and computational time and resources.

© Springer Nature Switzerland AG 2019
I. V. Tetko et al. (Eds.): ICANN 2019, LNCS 11731, pp. 842–844, 2019.
https://doi.org/10.1007/978-3-030-30493-5

A major challenge in HTS is to identify compounds referred to as 'Frequent-Hitters' [6] which include problematic (e.g. reactive, impure or aggregating) compounds, promiscuous compounds that exhibit the desired mode of action on the target but are non-selective across targets, and compounds that are false actives across many assays, as a result of an interference with the assay technology. Such interference may arise through e.g., inhibition of a coupled enzyme in the assay, or fluorescence quenching properties [2]. A wide variety of technologies are available in HTS and their choices depend on the type of biochemical pathway under investigation [4].

An existing tool for the identification of frequent-hitter across HTS technologies is BSF [5], a score based on historical HTS data that can be calculated for every compound tested in a screen and estimates the probability of the compound to be a frequent-hitter. Various substructure filters are available to identify frequent-hitters, the most popular one being the Pan-Assay Interference Compounds filter (PAINS) [1] which contains 480 suspicious substructures that were derived from a dataset based on one specific assay technology.

2 Method

In our study, we introduce a method to detect frequent-hitters by structural analysis of known frequent-hitters and non frequent-hitters of specific assay technologies. For three assay technologies (AlphaScreen, FRET and TRF), two types of assays were extracted from the AstraZeneca screen collection: primary assays, which aim to test the activity of compounds on a target at a single concentration, and artefact assays in which the same compounds are usually tested without any target. The aim of this last test is to identify compounds interfering with the assay technology. Comparing the activity of the compounds in both assays allows for the classification of compounds as frequent-hitters or non frequent-hitters. For each compound of both classes, structural fingerprints are generated and used to train a random forest classifier (RFC) model able to predict the behavior of new compounds. RFC is a classification algorithm constituting a consensus of decision trees, each of them giving one prediction per compound and the most popular prediction being the final one. The performance of the RFC model is compared to the BSF score as well as to the PAINS filters.

3 Result

FHs identified by RFC and BSF are complementary and the models can therefore be used in tandem. Furthermore, RFC can infer predictions for compounds that were never tested before, which is a non-supported task for BSF. RFC provides high-quality predictions and relies on high-quality data, while BSF requires huge volume of data to outperform RFC. Thus RFC is especially useful for novel HTS technologies with very little primary assays data.

Acknowledgements. Laurianne David has received funding from the European Union's Horizon 2020 research and innovation program under the Marie Sklodowska Curie grant agreement No 676434, "Big Data in Chemistry" ("BIGCHEM", http://bigchem.eu). The article reflects only the authors view and neither the European Commission nor the Research Executive Agency (REA) are responsible for any use that may be made of the information it contains.

References

1. Baell, J.B., Holloway, G.A.: New substructure filters for removal of pan assay interference compounds (PAINS) from screening libraries and for their exclusion in bioassays. J. Med. Chem. (2010). https://doi.org/10.1021/jm901137j
2. Dahlin, J.L., Nissink, J.W.M., Strasser, J.M., Francis, S., Higgins, L., Zhou, H., Zhang, Z., Walters, M.A.: PAINS in the assay: chemical mechanisms of assay interference and promiscuous enzymatic inhibition observed during a sulfhydryl-scavenging HTS. J. Med. Chem. (2015). https://doi.org/10.1021/jm5019093
3. Hertzberg, R.P., Pope, A.J.: High-throughput screening: new technology for the 21st century. Curr. Opin. Chem. Biol. **4**(4), 445–451 (2000). https://doi.org/10.1016/S1367-5931(00)00110-1
4. Janzen, W.: Screening technologies for small molecule discovery: the state of the art. Chem. Biol. **21**(9), 1162–1170 (2014). https://doi.org/10.1016/j.chembiol.2014.07.015
5. M Nissink, J.W., Blackburn, S.: Quantification of frequent-hitter behavior based on historical high-throughput screening data. Future Med. Chem. (2014). https://doi.org/10.4155/fmc.14.72
6. Roche, O., et al.: Development of a virtual screening method for identification of "frequent hitters" in compound libraries. J. Med. Chem. **45**(1), 137–142 (2002). https://doi.org/10.1021/jm010934d

Message Passing Neural Networks Scoring Functions for Structure-Based Drug Discovery

Dmitry S. Karlov[1]([⊠]) [iD], Petr Popov[1] [iD], Sergey Sosnin[1,2] [iD],
and Maxim V. Fedorov[1,2] [iD]

[1] Skolkovo Institute of Science and Technology, Skolkovo Innovation Center,
Moscow 143026, Russia
d.karlov@skoltech.ru
[2] Skolkovo Innovation Center, Syntelly LLC, 42 Bolshoy Boulevard,
Moscow 143026, Russia

Abstract. The scoring function for ranking protein-ligand complexes by their estimated binding affinity was developed based on the Message Passing Neural Network (MPNN). Behler-Parrinello Symmetric functions were utilized as descriptors of the atomic environment. The performance on the CASF 2016 benchmark reveals better results, as compared to the other methods.

Keywords: Neural networks · Graph convolutions ·
Scoring functions · Affinity prediction

A lot of efforts had been made to facilitate the drug discovery process by the aid of computer methods. At the same time, one of the central problems of computer-aided molecular design (the problem of prediction of ligand binding affinity) is still waiting for satisfactory solution [2]. So-called "scoring functions" (SF) based on different machine learning methods and interaction descriptors were developed [7, 8]. They were found useful for correct pose prediction in a docking task or to select a subset from the database of diverse chemicals enriched in active compounds in virtual screening computational experiment (classification task: active/inactive). It should be noted that more sophisticated and computationally intensive methods (FEP, MM-PBSA) considered more accurate than scoring functions often give similar results to SF. The Free Energy Perturbation (FEP) technique [1] is based on the alchemical transformations and allows to achieve in some cases very good results for the affinity ranking in a series of closely related compounds. MM-PB(GB)SA techniques [4] being not so demanding for computational resources can be considered as a cheaper but a "dirtier" alternative to FEP. The application of the two above mentioned methods requires a lot of computational time and may be useful for the optimization of the active molecule.

There is still a need in more rigorous scoring functions with an ability to rank the active molecules by their affinity. Deep learning approaches were applied

© Springer Nature Switzerland AG 2019
I. V. Tetko et al. (Eds.): ICANN 2019, LNCS 11731, pp. 845–847, 2019.
https://doi.org/10.1007/978-3-030-30493-5

in this area. Recently, two Deep Neural Network based scoring functions were constructed using the Convolutional Neural Network architecture [10]. Kdeep [6] trained using the PDBBind data set and aimed to predict absolute binding affinities takes a set of 3D grids representing a map of certain structural features (hydrophobic, aromatic, H-bond acceptors, etc.) as a result of the protein and ligand coordinate transformation. And the model developed by Ragoza et al. [9] with the goal to improve virtual screening results act as a classification model and operates with 3D maps defined by a set of predefined atom types.

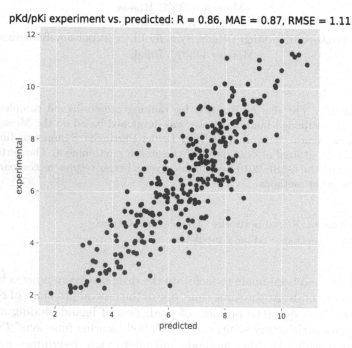

Fig. 1. The performance of the MPNN scoring function on CASF 2016 test set

Here we present a scoring function based on MPNN paradigm [5]. The descriptor sets for each atom in ligand were defined as Behler-Parrinello Symmetric functions (BPS) [3]. The input data is defined by the 3D structure of a protein in PDB format and the SDF structure of ligand binding pose (molecular graph connectivity information is necessary for MPNN processing). The training was performed on the PDBBind 2018 core dataset with the addition of the IC_{50} data. Five-fold cross-validation resulted in five trained models used for the test set assessment by comparison with the average value of five predictions. The test set is predefined in PDBbind 2018 and is used for Comparative Assessment of Scoring Functions (CASF) [11]. The results of the test set performance are presented in Fig. 1 and shows that it performs better than Kdeep (Pearson R = 0.82, RMSE = 1.27). The test set contain 58 different protein targets with 4–6 ligands for each one. We compared distributions of Pearson correlation coefficients

MPNNScore and Kdeep and noticed the valuable improvement: the number of Pearson's $R < 0.5$ is 7 for MPNNScore and is 20 for Kdeep and the number of Pearson's $R < 0.0$ is one for MPNNScore and is 6 for Kdeep.

Usually, the prediction time is about 5–10 seconds, and the 95 % of this time is spent on the descriptors calculation. That is why we recommend the usage of this function for the post-processing of the virtual screening results rather than for docking.

References

1. Abel, R., Wang, L., Harder, E.D., Berne, B.J., Friesner, R.A.: Advancing drug discovery through enhanced free energy calculations. Acc. Chem. Res. **50**(7), 1625–1632 (2017)
2. Ain, Q.U., Aleksandrova, A., Roessler, F.D., Ballester, P.J.: Machine-learning scoring functions to improve structure-based binding affinity prediction and virtual screening. Wiley Interdiscip. Rev. Comput. Mol. Sci. **5**(6), 405–424 (2015)
3. Behler, J., Parrinello, M.: Generalized neural-network representation of high-dimensional potential-energy surfaces. Phys. Rev. Lett. **98**, 146401 (2007)
4. Genheden, S., Ryde, U.: The MM/PBSA and MM/GBSA methods to estimate ligand-binding affinities. Expert Opin. Drug Discov. **10**(5), 449–461 (2015)
5. Gilmer, J., Schoenholz, S.S., Riley, P.F., Vinyals, O., Dahl, G.E.: Neural message passing for quantum chemistry. arXiv: 1704.01212 (2017)
6. Jiménez, J., Škalič, M., Martínez-Rosell, G., De Fabritiis, G.: Kdeep: Protein-ligand absolute binding affinity prediction via 3d-convolutional neural networks. J. Chem. Inf. Model. **58**(2), 287–296 (2018)
7. Kadukova, M., Grudinin, S.: Convex-pl: a novel knowledge-based potential for protein-ligand interactions deduced from structural databases using convex optimization. J. Comput. Aided Mol. Des. **31**(10), 943–958 (2017)
8. Mitchell, J.B.O., Ballester, P.J.: A machine learning approach to predicting protein-ligand binding affinity with applications to molecular docking. Bioinformatics **26**(9), 1169–1175 (2010)
9. Ragoza, M., Hochuli, J., Idrobo, E., Sunseri, J., Koes, D.R.: Protein-ligand scoring with convolutional neural networks. J. Chem. Inf. Model. **57**(4), 942–957 (2017)
10. Rawat, W., Wang, Z.: Deep convolutional neural networks for image classification: a comprehensive review. Neural Comput. **29**(9), 2352–2449 (2017)
11. Su, M., Yang, Q., Du, Y., Feng, G., Liu, Z., Li, Y., Wang, R.: Comparative assessment of scoring functions: the CASF-2016 update. J. Chem. Inf. Model. **59**(2), 895–913 (2019)

Author Index

Printed in the United States
By Bookmasters

Printed in the United States
By Bookmasters